ANALOG INTEGRATED CIRCUIT DESIGN

David Johns
Ken Martin
University of Toronto

John Wiley & Sons, Inc.

Acquisitions Editor Charity Robey
Marketing Manager Jay Kirsch
Production Manager Lucille Buonocore
Senior Production Editor Tracey Kuehn
Designer Kevin Murphy
Manufacturing Manager Mark Cirillo
Illustration Editor Sigmund Malinowski

This book was set in Times Roman by Publication Services and printed and bound by R.R. Donnelley/Crawfordsville. The cover was printed by Lehigh Press.

Recognizing the importance of preserving what has been written, it is a policy of John Wiley & Sons, Inc., to have books of enduring value published in the United States printed on acid-free paper, and we exert our best efforts to that end.

The paper in this book was manufactured by a mill whose forest management programs include sustained yield harvesting of its timberlands. Sustained yield harvesting principles ensure that the number of trees cut each year does not exceed the amount of new growth.

Library of Congress Cataloging-in-Publication Data:

Johns, David, 1958–
Analog integrated circuit design / David Johns, Ken Martin.
p. cm.
Includes bibliographical references.
ISBN 0-471-14448-7 (cloth : alk. paper)
1. Linear integrated circuits—Design and construction.
I. Martin, Kenneth W. (Kenneth W.) 1952– . II. Title.
TK7874.J65 1996
621.3815—dc20 96-34365
CIP

ISBN 0-471-14448-7

Printed in the United States of America

11 12 13 14 15

To Cecilia, Christopher, Timothy and Victoria
To Elisabeth and Jeremy

Preface

For the past twenty years, numerous people have predicted there soon would be little need for analog circuitry because the world would rely on digital circuits. However, although many applications have indeed replaced much analog circuitry with their digital counterparts (such as digital audio), the need for good analog circuit design remains strong. For example, when digitizing physical signals, analog-to-digital and digital-to-analog converters are always needed, together with their associated anti-aliasing and reconstruction filters. In addition, new applications continue to appear in which speed and power-consumption requirements often demand the use of high-speed analog front ends, such as digital communications over copper wires or wireless communication channels. Also as integrated circuits become larger due to system integration, it is much more likely that at least some portion of a modern integrated circuit will include analog circuitry required to interface to the real world. Often this analog circuitry, although it constitutes only a small portion of the total chip area, may be the limiting factor on overall system performance and the most difficult part of the IC to design. As a result, a strong industrial need for analog circuit designers continues. The purpose of this book is to help develop excellent analog circuit designers by presenting a concise treatment of the wide array of knowledge required by an integrated circuit designer.

Many refer to the designing and testing of high-performance analog circuits as a "mystical art". In other words, whereas digital design is relatively systematic, analog design appears to be much more confusing and based on gut feelings. In addition, analog testing may sometimes seem to depend more on the time of day and phase of the moon rather than on concrete electrical properties. These thoughts about analog circuits usually occur when one is not familiar with the many fundamentals required to create high-performance analog circuits. A major goal of this book is to help take the mystery out of analog integrated circuit design. The authors believe that most experienced electrical engineers are capable of good design if they are familiar with the most important design principles. We have attempted to highlight these principles throughout this text. Although many circuits and techniques are described, we have emphasized the most important and fundamental principles involved in realizing state-of-the-art analog circuits. Throughout this book, we give physical and intuitive explanations, and, although mathematical quantitative analyses of many circuits have necessarily been presented, we have attempted not to miss seeing the forest because of the trees. In other words, this book attempts to present the critical underlying concepts without becoming entangled in tedious and overcomplicated circuit analyses.

INTENDED AUDIENCE

This book is primarily intended for use as a graduate-level textbook and as a reference for practicing engineers, although portions of this text are also useful for senior-level undergraduate courses. To appreciate the material in this book, it is expected that the reader has had at least one basic introductory course in analog circuits. Specifically, the reader should be familiar with the concept of small-signal analysis and have been exposed to basic transistor amplifier circuits. In addition, the reader should be comfortable working in the frequency domain (i.e., should be familiar with the Laplace transform) with possibly some knowledge of discrete-time signals.

In a graduate course, this book might be used in many ways. The chapters of this book have intentionally been made mostly independent so that some chapters can be covered while others are skipped. Also, it has been found to be very easy to change the order of presentation. For example, if readers have a good modelling background, they might skip Chapter 1, and if their discrete-time knowledge is good, Chapter 9 might be assigned only as review. We believe that such flexibility is essential in presenting textbooks for the later years of study.

Here at the University of Toronto, we present the material in this book in a few courses. A senior-level undergraduate course in analog circuits assigns Chapters 1, 3, 4, and 5, with the fundamentals and selected topics briefly covered from Chapters 6 and 7, as well. Another senior-level undergraduate course on analog filters uses Chapter 9 and the filter portion of Chapter 10 after presenting filter basics. In a first-level graduate course, we presently assign Chapters 1–3 and 5–7, while a second-level graduate course (often taken concurrently with the first course) assigns Chapters 4, 8–10, and 16. Chapters 9, 10, and 15 are presently covered in a separate graduate course with other material on advanced analog filters, whereas Chapters 11–14 (together with journal publications) are assigned in a data converters course. Furthermore, we have often modified this selection and order from year-to-year with little difficulty. Note that some chapters are used in both undergraduate and graduate courses since many of our graduate students do not come from our undergraduate program.

A secondary audience for this book includes recently graduated electrical engineers who wish to rapidly increase their knowledge of modern analog circuit-design techniques. In fact, much of the material covered in this text was originally taught and refined, over many years, in popular short courses offered to working engineers who realized the importance to their careers of upgrading their knowledge in analog circuit design. For this audience, we have put much effort into highlighting the most important considerations when designing the various circuits, and we have also tried to include modern, well-designed examples and references to sources for further study.

TEXT OUTLINE

In Chapter 1, the basic physical behavior and modelling for three integrated devices are presented—diodes, MOSFET transistors, and bipolar transistors. Here, many of the modelling equations are derived to give the reader some appreciation of model parameters and how they are affected as processes evolve. Models for the three devices are summarized in a table format for quick reference.

In Chapter 2, issues associated with the manufacturing of a microcircuit are discussed. Emphasis is placed on CMOS fabrication, where details are given for a generic process. Layout and design rules are presented with respect to analog circuits where matching and noise considerations are important issues. This chapter concludes with a description of a destructive phenomenon known as latch-up.

Fundamental building blocks of analog integrated circuits are discussed in Chapter 3. Specifically, MOS current mirrors are presented as well as single-stage amplifiers. A point to note here is that only active-load amplifiers are considered since these types of amplifiers occur more often than other types in integrated circuits. The last section in Chapter 3 presents SPICE simulation results for selected examples within the chapter.

Noise analysis and modelling are discussed in Chapter 4. Here, we assume the reader has not previously been exposed to random-signal analysis, and thus basic concepts in analyzing random signals are first presented. Noise models are then presented for basic circuit elements. Finally, a variety of circuits are analyzed from a noise perspective to give the reader some experience in noise analysis.

In Chapter 5, the fundamental principles of basic opamp design are presented. To illustrate many of these principles, the design of a classic two-stage CMOS opamp is thoroughly discussed. A systematic compensation procedure, which also applies to many other opamp designs, is described. Also discussed here are methods to make the compensation and transconductance values quite stable. Finally, the chapter concludes with SPICE simulation results for selected examples.

Chapter 6 begins with a look at two advanced current-mirror approaches useful in low-power supply operation and in short-channel processes. Next, two modern high-speed opamps are discussed followed by a presentation of fully differential design techniques and common-mode feedback circuits. These fully differential opamps are used in many modern industrial applications where high speed and noise are important considerations. Next, current-feedback opamps are discussed. Their main advantage is that they maintain relatively constant bandwidth while the closed-loop gain is changed. Once again, SPICE simulation examples are found at the end of this chapter.

In Chapter 7, comparator design is discussed. Comparators are perhaps the second most common analog building block after opamps. Here, the practical limitations of comparators are described as well as circuit techniques to improve performance. In addition, examples of modern high-speed comparators are presented.

In Chapter 8, some additional analog building blocks are covered. Specifically, sample-and-hold circuits are first covered, followed by bandgap voltage references, and the chapter concludes with bipolar translinear-gain and multiplier circuits. At the end of this chapter, all the main analog building blocks have been covered and the remaining material in the text deals with more system-level analog blocks.

The basics of discrete-time signals and filters are presented in Chapter 9. This material is essential for understanding the operation of many modern analog circuits such as switched-capacitor filters and oversampling converters. The approach taken here is to show the close relationship between the Z-transform and the Laplace transform, thereby building on the reader's experience in the continuous-time domain.

In Chapter 10, the basics of switched-capacitor circuits are described. Switched-capacitor techniques are the most common approach for realizing integrated filters due to their high degree of accuracy and linearity. The chapter concludes with a description of other switched-capacitor circuits, such as gain stages, modulators, and voltage-controlled oscillators.

In Chapter 11, the fundamentals of data converters are presented. Ideal converters are discussed first, and it is shown that quantization noise does not occur in D/A converters but is a fundamental limit in A/D converters. Signed codes are then presented, and the chapter concludes with a discussion of performance limitations.

Popular Nyquist-rate D/A architectures are discussed in Chapter 12, with various approaches for realizing Nyquist-rate A/D converters described in Chapter 13. The importance of data conversion cannot be overemphasized in today's largely digital world, and these two chapters discuss the main advantages and design issues of many modern approaches.

Oversampling converters are presented separately in Chapter 14 due to the large amount of signal-processing concepts needed to properly describe these converters. Here, digital issues (such as decimation filters) are also presented since good overall system knowledge is needed to properly design these types of converters. In addition, practical issues and recent new approaches (such as the use of multibit conversion) are also discussed. This chapter concludes with a third-order A/D converter example.

Continuous-time filters are presently being applied to many new products, particularly in high-speed data-communication applications, and they are the focus of Chapter 15. Bipolar, CMOS, and BiCMOS approaches are described, followed by some tuning approaches. Finally, dynamic range measurement aspects of continuous-time filters are discussed, such as THD, spurious-free dynamic range, and the third-order intercept point.

Finally, the text concludes with phase-locked loops in Chapter 16. Phase-locked loops are essential components for clock recovery in many digital and data communication circuits. Here, basic concepts and a simulation technique are presented for such systems.

ACKNOWLEDGMENTS

The authors would like to acknowledge many of their colleagues who participated in short courses in which much of the material for this text was originally taught and refined. In particular, Gabor C. Temes is acknowledged as well as instructors Jim McCreary and Bill Black. In addition, the authors acknowledge that much of the material and concepts originated with many practicing engineers they have worked with over the years as well as in the publications cited in the references section at the end of each chapter. As much as possible, appropriate references for original concepts are cited in the text, but the authors have been working in the area of analog circuits for so many years that often the original sources of popular and important concepts have been forgotten. For any reference omissions, they sincerely apologize.

Contents

CHAPTER 1 INTEGRATED-CIRCUIT DEVICES AND MODELLING 1

1.1 Semiconductors and pn Junctions 1
1.2 MOS Transistors 16
1.3 Advanced MOS Modelling 39
1.4 Bipolar-Junction Transistors 42
1.5 Device Model Summary 56
1.6 SPICE-Modelling Parameters 61
1.7 Appendix 65
1.8 References 78
1.9 Problems 78

CHAPTER 2 PROCESSING AND LAYOUT 82

2.1 CMOS Processing 82
2.2 Bipolar Processing 95
2.3 CMOS Layout and Design Rules 96
2.4 Analog Layout Considerations 105
2.5 Latch-Up 118
2.6 References 121
2.7 Problems 121

CHAPTER 3 BASIC CURRENT MIRRORS AND SINGLE-STAGE AMPLIFIERS 125

3.1 Simple CMOS Current Mirror 125
3.2 Common-Source Amplifier 128
3.3 Source-Follower or Common-Drain Amplifier 129
3.4 Common-Gate Amplifier 132
3.5 Source-Degenerated Current Mirrors 135
3.6 High-Output-Impedance Current Mirrors 137
3.7 Cascode Gain Stage 140
3.8 MOS Differential Pair and Gain Stage 142
3.9 Bipolar Current Mirrors 146
3.10 Bipolar Gain Stages 149

3.11 Frequency Response 154
3.12 SPICE Simulation Examples 169
3.13 References 176
3.14 Problems 176

CHAPTER 4 NOISE ANALYSIS AND MODELLING 181

4.1 Time-Domain Analysis 181
4.2 Frequency-Domain Analysis 186
4.3 Noise Models for Circuit Elements 196
4.4 Noise Analysis Examples 204
4.5 References 216
4.6 Problems 217

CHAPTER 5 BASIC OPAMP DESIGN AND COMPENSATION 221

5.1 Two-Stage CMOS Opamp 221
5.2 Feedback and Opamp Compensation 232
5.3 SPICE Simulation Examples 251
5.4 References 252
5.5 Problems 253

CHAPTER 6 ADVANCED CURRENT MIRRORS AND OPAMPS 256

6.1 Advanced Current Mirrors 256
6.2 Folded-Cascode Opamp 266
6.3 Current-Mirror Opamp 273
6.4 Linear Settling Time Revisited 278
6.5 Fully Differential Opamps 280
6.6 Common-Mode Feedback Circuits 287
6.7 Current-Feedback Opamps 291
6.8 SPICE Simulation Examples 295
6.9 References 299
6.10 Problems 300

CHAPTER 7 COMPARATORS 304

7.1 Using an Opamp for a Comparator 304
7.2 Charge-Injection Errors 308
7.3 Latched Comparators 317
7.4 Examples of CMOS and BiCMOS Comparators 321
7.5 Examples of Bipolar Comparators 328
7.6 References 330
7.7 Problems 331

CHAPTER 8 SAMPLE AND HOLDS, VOLTAGE REFERENCES, AND TRANSLINEAR CIRCUITS 334

8.1 Performance of Sample-and-Hold Circuits 334
8.2 MOS Sample-and-Hold Basics 336
8.3 Examples of CMOS S/H Circuits 343
8.4 Bipolar and BiCMOS Sample and Holds 349
8.5 Bandgap Voltage Reference Basics 353
8.6 Circuits for Bandgap References 357
8.7 Translinear Gain Cell 364
8.8 Translinear Multiplier 366
8.9 References 368
8.10 Problems 370

CHAPTER 9 DISCRETE-TIME SIGNALS 373

9.1 Overview of Some Signal Spectra 373
9.2 Laplace Transforms of Discrete-Time Signals 374
9.3 z-Transform 377
9.4 Downsampling and Upsampling 379
9.5 Discrete-Time Filters 382
9.6 Sample-and-Hold Response 389
9.7 References 391
9.8 Problems 391

CHAPTER 10 SWITCHED-CAPACITOR CIRCUITS 394

10.1 Basic Building Blocks 394
10.2 Basic Operation and Analysis 398
10.3 First-Order Filters 409
10.4 Biquad Filters 415
10.5 Charge Injection 423
10.6 Switched-Capacitor Gain Circuits 427
10.7 Correlated Double-Sampling Techniques 433
10.8 Other Switched-Capacitor Circuits 434
10.9 References 441
10.10 Problems 443

CHAPTER 11 DATA CONVERTER FUNDAMENTALS 445

11.1 Ideal D/A Converter 445
11.2 Ideal A/D Converter 447
11.3 Quantization Noise 448
11.4 Signed Codes 452

11.5 Performance Limitations 454
11.6 References 461
11.7 Problems 461

CHAPTER 12 NYQUIST-RATE D/A CONVERTERS 463

12.1 Decoder-Based Converters 463
12.2 Binary-Scaled Converters 469
12.3 Thermometer-Code Converters 475
12.4 Hybrid Converters 481
12.5 References 484
12.6 Problems 484

CHAPTER 13 NYQUIST-RATE A/D CONVERTERS 487

13.1 Integrating Converters 487
13.2 Successive-Approximation Converters 492
13.3 Algorithmic (or Cyclic) A/D Converter 504
13.4 Flash (or Parallel) Converters 507
13.5 Two-Step A/D Converters 513
13.6 Interpolating A/D Converters 516
13.7 Folding A/D Converters 519
13.8 Pipelined A/D Converters 523
13.9 Time-Interleaved A/D Converters 526
13.10 References 527
13.11 Problems 528

CHAPTER 14 OVERSAMPLING CONVERTERS 531

14.1 Oversampling without Noise Shaping 531
14.2 Oversampling with Noise Shaping 538
14.3 System Architectures 547
14.4 Digital Decimation Filters 551
14.5 Higher-Order Modulators 555
14.6 Bandpass Oversampling Converters 557
14.7 Practical Considerations 559
14.8 Multi-Bit Oversampling Converters 565
14.9 Third-Order A/D Design Example 568
14.10 References 571
14.11 Problems 572

CHAPTER 15 CONTINUOUS-TIME FILTERS 574

15.1 Introduction to G_m-C Filters 575
15.2 Bipolar Transconductors 584

15.3 CMOS Transconductors Using Triode Transistors 597
15.4 CMOS Transconductors Using Active Transistors 607
15.5 BiCMOS Transconductors 616
15.6 MOSFET-C Filters 620
15.7 Tuning Circuitry 626
15.8 Dynamic Range Performance 635
15.9 References 643
15.10 Problems 645

CHAPTER 16 PHASE-LOCKED LOOPS 648

16.1 Basic Loop Architecture 648
16.2 PLLs with Charge-Pump Phase Comparators 663
16.3 Voltage-Controlled Oscillators 670
16.4 Computer Simulation of PLLs 680
16.5 Appendix 689
16.6 References 692
16.7 Problems 693

INDEX 696

CHAPTER

1 Integrated-Circuit Devices and Modelling

In this chapter, the operation and modelling of semiconductor devices are described. Although it is possible to do simple integrated-circuit design with a basic knowledge of semiconductor device modelling, for high-speed state-of-the-art design, an in-depth understanding of the second-order effects of device operation and their modelling is considered critical.

It is assumed that most readers have been introduced to transistors and their basic modelling in a previous course. Thus, fundamental semiconductor concepts are only briefly reviewed. Section 1.1 describes pn junctions (or diodes). This section is important in understanding the parasitic capacitances in many device models, such as junction capacitances. Section 1.2 covers MOS transistors and modelling. It should be noted that this section relies to some degree on the material previously presented in Section 1.1, in which depletion capacitance is covered. Section 1.4 covers bipolar-junction transistors and modelling. A summary of device models and important equations is presented in Section 1.5. This summary is particularly useful for a reader who already has a good background in transistor modelling, in which case the summary can be used to follow the notation used throughout the remainder of this book. In addition, a brief description is given of the most important process-related parameters used in SPICE modelling. Finally, this chapter concludes with an Appendix containing derivations of the more physically based device equations.

1.1 SEMICONDUCTORS AND pn JUNCTIONS

A semiconductor is a crystal lattice structure that can have free electrons (which are negative carriers) and/or free holes (which are an absence of electrons and are equivalent to positive carriers). The type of semiconductor typically used is silicon (commonly called *sand*). This material has a valence of four, implying that each atom has four free electrons to share with neighboring atoms when forming the covalent bonds of the crystal lattice. *Intrinsic* silicon (i.e., undoped silicon) is a very pure crystal structure having equal numbers of free electrons and holes. These free carriers are those electrons or holes that have gained enough energy due to thermal agitation to escape their bonds. At room temperature, there are approximately 1.5×10^{10} carriers of each type per cm^3, or equivalently 1.5×10^{16} carriers/m^3. The number of carriers approximately doubles for every 11 °C increase in temperature.

If one dopes silicon with a pentavalent impurity (i.e., atoms of an element having a valence of five, or equivalently five electrons in the outer shell, available when bonding with neighboring atoms), there will be almost one extra free electron for every impurity atom.[1] These free electrons can be used to conduct current. A pentavalent impurity is said to *donate* free electrons to the silicon crystal, and thus the impurity is known as a *donor*. Examples of donor elements are phosphorus, P, and arsenic, As. These impurities are also called n-type dopants since the free carriers resulting from their use have negative charge. When an n-type impurity is used, the total number of negative carriers or electrons is almost the same as the doping concentration, and is much greater than the number of free electrons in intrinsic silicon. In other words,

$$n_n = N_D \tag{1.1}$$

where n_n denotes the free-electron concentration in n-type material and N_D is the doping concentration (with the subscript D denoting donor). On the other hand, the number of free holes in n-doped material will be much less than the number of holes in intrinsic silicon and can be shown [Sze, 1981] to be given by

$$p_n = \frac{n_i^2}{N_D} \tag{1.2}$$

Here, n_i is the carrier concentration in intrinsic silicon.

Similarly, if one dopes silicon with atoms having a valence of three, for example, boron (B), the concentration of positive carriers or holes will be approximately equal to the *acceptor* concentration, N_A,

$$p_p = N_A \tag{1.3}$$

and the number of negative carriers in the p-type silicon, n_p, is given by

$$n_p = \frac{n_i^2}{N_A} \tag{1.4}$$

EXAMPLE 1.1

Intrinsic silicon is doped with boron at a concentration of 10^{26} atoms/m^3. At room temperature, what are the concentrations of holes and electrons in the resulting doped silicon? Assume that $n_i = 1.5 \times 10^{16}$ carriers/m^3.

Solution

The hole concentration, p_p, will approximately equal the doping concentration ($p_p = N_A = 10^{26}$ holes/m^3). The electron concentration is found from (1.4) to be

1. In fact, there will be slightly fewer mobile carriers than the number of impurity atoms since some of the free electrons from the dopants have recombined with holes. However, since the number of holes of intrinsic silicon is much less than typical doping concentrations, this inaccuracy is small.

$$n_p = \frac{(1.5 \times 10^{16})^2}{10^{26}} = 2.3 \times 10^6 \text{ electrons/m}^3 \quad (1.5)$$

Such doped silicon is referred to as p type since it has many more free holes than free electrons.

Diodes

To realize a diode, or, equivalently, a pn junction, one part of a semiconductor is doped n type, and a closely adjacent part is doped p type, as shown in Fig. 1.1. Here the diode, or junction, is formed between the p^+ region and the n region. It should be noted that the superscripts indicate the relative doping levels. For example, the p^- bulk region might have an impurity concentration of 5×10^{21} carriers/m^3, whereas the p^+ and n^+ regions would be doped more heavily to a value around 10^{25} to 10^{27} carriers/m^3. Also, note that the metal contacts to the diode (in this case, aluminum) are connected to a heavily doped region as opposed to a lightly doped region; otherwise a *Schottky diode* would occur. (Schottky diodes are discussed on page 15.) Thus, in order not to make a Schottky diode, the connection to the n region is actually made via the n^+ region.

In the p^+ side, a large number of free positive carriers are available, whereas in the n side, many free negative carriers are available. The holes in the p^+ side will tend to disperse or diffuse into the n side, whereas the free electrons in the n side will tend to diffuse to the p^+ side. This process is very similar to two gases randomly diffusing together. This diffusion lowers the concentration of free carriers in the region between the two sides. As the two types of carriers diffuse together, they recombine. Every electron that diffuses from the n side to the p side leaves behind a *bound* positive charge close to the transition region. Similarly, every hole that diffuses from the p side leaves behind a bound electron near the transition region. The end result is shown in Fig. 1.2. This diffusion of free carriers creates a *depletion region* at the junction of the two sides where no free carriers exist, and which has a net negative charge on the p^+ side and a net positive charge on the n side. The total amount of exposed or bound

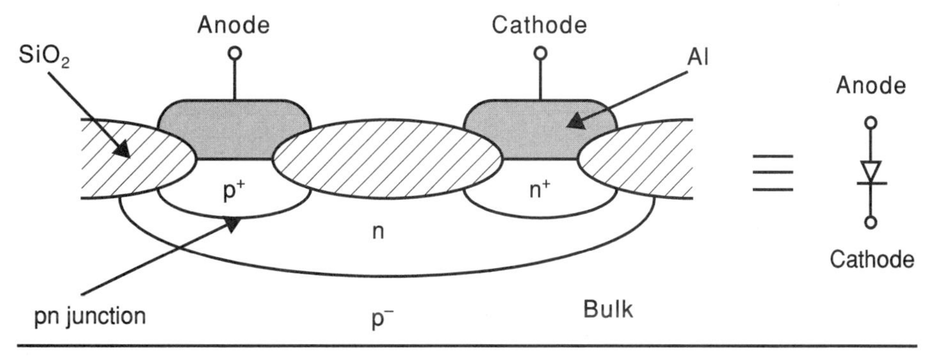

Fig. 1.1 A cross section of a pn diode.

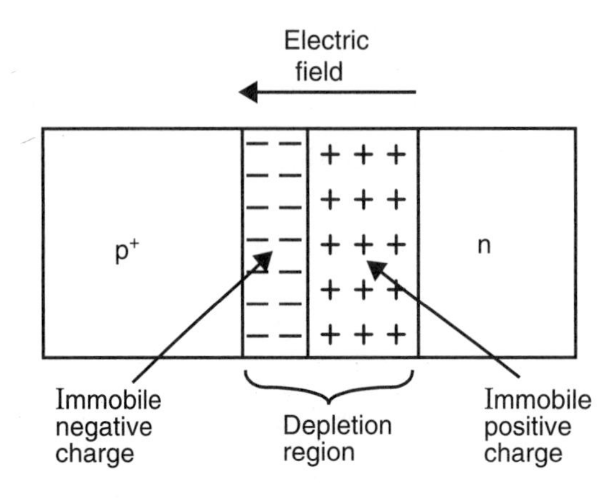

Fig. 1.2 A simplified model of a diode. Note that a depletion region exists at the junction due to diffusion and extends farther into the more lightly doped side.

charge on the two sides of the junction must be equal for charge neutrality. This requirement causes the depletion region to extend farther into the more lightly doped n side than into the p^+ side.

As these bound charges are exposed, an electric field develops going from the n side to the p side. This electric field is often called the built-in potential of the junction. It opposes the diffusion of free carriers until there is no net movement of charge under open-circuit and steady-state conditions. The built-in voltage of an open-circuit pn junction is given by Sze [1981] as

$$\Phi_0 = V_T \ln\left(\frac{N_A N_D}{n_i^2}\right) \tag{1.6}$$

where

$$V_T = \frac{kT}{q} \tag{1.7}$$

with T being the temperature in degrees Kelvin ($\cong$ 300 °K at room temperature), k being Boltzmann's constant (1.38×10^{-23} JK^{-1}), and q being the charge of an electron (1.602×10^{-19} C). At room temperature, V_T is approximately 26 mV.

EXAMPLE 1.2

A pn junction has $N_A = 10^{25}$ holes/m^3 and $N_D = 10^{22}$ electrons/m^3. What is the built-in junction potential? Assume that $n_i = 1.5 \times 10^{16}$ carriers/m^3.

Solution

Using (1.6), we obtain

$$\Phi_0 = 0.026 \times \ln\left(\frac{10^{25} \times 10^{22}}{(1.5 \times 10^{16})^2}\right) = 0.88 \text{ V} \tag{1.8}$$

This is a typical value for the built-in potential of a junction with one side heavily doped. As an approximation, we will normally use $\Phi_0 \cong 0.9\ \text{V}$ for the built-in potential of a junction having one side heavily doped.

Reverse-Biased Diodes

A silicon diode having an anode-to-cathode (i.e., p side to n side) voltage of 0.4 V or less will not be conducting appreciable current. In this case, it is said to be *reverse biased.* If a diode is reverse biased, current flow is primarily due to thermally generated carriers in the depletion region, and it is extremely small. Although this reverse-biased current is only weakly dependent on the applied voltage, *the reverse-biased current is directly proportional to the area of the diode junction.* However, an effect that should not be ignored, particularly at high frequencies, is the junction capacitance of a diode. In reverse-biased diodes, this junction capacitance is due to varying charge storage in the depletion regions and is modelled as a *depletion capacitance.*

To determine the depletion capacitance, we first state the relationship between the depletion widths and the applied reverse voltage, V_R [Sze, 1981].

$$x_n = \left[\frac{2K_s\varepsilon_0(\Phi_0 + V_R)}{q}\frac{N_A}{N_D(N_A + N_D)}\right]^{1/2} \tag{1.9}$$

$$x_p = \left[\frac{2K_s\varepsilon_0(\Phi_0 + V_R)}{q}\frac{N_D}{N_A(N_A + N_D)}\right]^{1/2} \tag{1.10}$$

Here, ε_0 is the permittivity of free space (equal to $8.854 \times 10^{-12}\ \text{F/m}$), V_R is the reverse-bias voltage of the diode, and K_s is the relative permittivity of silicon (equal to 11.8). It should be noted that these equations assume that the doping changes abruptly from the n to the p side.

From the above equations, we see that if one side of the junction is more heavily doped than the other, the depletion region will extend mostly on the lightly doped side. For example, if $N_A >> N_D$ (i.e., if the p region is more heavily doped), we can approximate (1.9) and (1.10) as

$$x_n \cong \left[\frac{2K_s\varepsilon_0(\Phi_0 + V_R)}{qN_D}\right]^{1/2} \qquad x_p \cong \left[\frac{2K_s\varepsilon_0(\Phi_0 + V_R)N_D}{qN_A^2}\right]^{1/2} \tag{1.11}$$

Indeed, for this case

$$\frac{x_n}{x_p} \cong \frac{N_A}{N_D} \tag{1.12}$$

This special case is called a *single-sided diode.*

EXAMPLE 1.3

For a pn junction having $N_A = 10^{25}$ holes/m^3 and $N_D = 10^{22}$ electrons/m^3, what are the depletion-layer depths for a 5-V reverse-bias voltage?

Solution

Since $N_A >> N_D$ and we already have found in Example 1.2 that $\Phi_0 = 0.9$ V, we can use (1.11) to find

$$x_n = \left[\frac{2 \times 11.8 \times 8.854 \times 10^{-12} \times 5.9}{1.6 \times 10^{-19} \times 10^{22}}\right]^{1/2} = 0.88\ \mu\text{m} \tag{1.13}$$

$$x_p = \frac{x_n}{(N_A/N_D)} = 0.88\ \text{nm} \tag{1.14}$$

Note that the depletion width in the lightly doped n region is 1,000 times greater than that in the more heavily doped p region.

The charge stored in the depletion region, per unit cross-sectional area, is found by multiplying the depletion-region width by the concentration of the immobile charge (which is approximately equal to q times the impurity doping density). For example, on the n side, we find the charge in the depletion region to be given by multiplying (1.9) by qN_D, resulting in

$$Q^+ = \left[2qK_s\varepsilon_0(\Phi_0 + V_R)\frac{N_A N_D}{N_A + N_D}\right]^{1/2} \tag{1.15}$$

This amount of charge must also equal Q^- on the p side since there is charge equality. In the case of a single-sided diode when $N_A >> N_D$, we have

$$Q^- = Q^+ \cong [2qK_s\varepsilon_0(\Phi_0 + V_R)N_D]^{1/2} \tag{1.16}$$

Note that this result is independent of the impurity concentration on the heavily doped side. Thus, we see from the above relation that the charge stored in the depletion region is dependent on the applied reverse-bias voltage. *It is this charge-voltage relationship that is modelled by a nonlinear depletion capacitance.*

For small changes in the reverse-biased junction voltage, about a bias voltage, we can find an equivalent *small-signal* capacitance, C_j, by differentiating (1.15) with respect to V_R. Such a differentiation results in

$$C_j = \frac{dQ^+}{dV_R} = \left[\frac{qK_s\varepsilon_0}{2(\Phi_0 + V_R)}\frac{N_A N_D}{N_A + N_D}\right]^{1/2} = \frac{C_{j0}}{\sqrt{1 + \frac{V_R}{\Phi_0}}} \tag{1.17}$$

where C_{j0} is the depletion capacitance per unit area at $V_R = 0$ and is given by

$$C_{j0} = \sqrt{\frac{qK_s\varepsilon_0}{2\Phi_0}\frac{N_A N_D}{N_A + N_D}} \tag{1.18}$$

In the case of a one-sided diode with $N_A >> N_D$, we have

$$C_j = \left[\frac{qK_s\varepsilon_0 N_D}{2(\Phi_0 + V_R)}\right]^{1/2} = \frac{C_{j0}}{\sqrt{1 + \frac{V_R}{\Phi_0}}} \tag{1.19}$$

where now

$$C_{j0} = \sqrt{\frac{qK_s\varepsilon_0 N_D}{2\Phi_0}} \tag{1.20}$$

It should be noted that many of the junctions encountered in integrated circuits are one-sided junctions with the lightly doped side being the substrate or sometimes what is called the *well.* The more heavily doped side is often used to form a contact to interconnecting metal. From (1.20), we see that, for these one-sided junctions, the depletion capacitance is approximately independent of the doping concentration on the heavily doped side, and is proportional to the square root of the doping concentration of the more lightly doped side. Thus, smaller depletion capacitances are obtained for more lightly doped substrates—a strong incentive to strive for lightly doped substrates.

Finally, note that by combining (1.15) and (1.18), we can express the equation for the immobile charge on either side of a reverse-biased junction as

$$Q = 2C_{j0}\Phi_0\sqrt{1 + \frac{V_R}{\Phi_0}} \tag{1.21}$$

As seen in Example 1.6, this equation is useful when one is approximating the large-signal charging (or discharging) time for a reverse-biased diode.

EXAMPLE 1.4

For a pn junction having $N_A = 10^{25}$ holes/m^3 and $N_D = 10^{22}$ electrons/m^3, what is the total zero-bias depletion capacitance for a diode of area $10\ \mu m \times 10\ \mu m$? What is its depletion capacitance for a 3-V reverse-bias voltage?

Solution

Making use of (1.20), we have

$$C_{j0} = \sqrt{\frac{1.6 \times 10^{-19} \times 11.8 \times 8.854 \times 10^{-12} \times 10^{22}}{2 \times 0.9}} = 304.7\ \mu F/m^2 \tag{1.22}$$

Since the diode area is $100 \times 10^{-12}\ \mathrm{m}^2$, the total zero-bias depletion capacitance is

$$C_{T\text{-}j0} = 100 \times 10^{-12} \times 304.7 \times 10^{-6} = 30.5\ \mathrm{fF} \tag{1.23}$$

At a 3-V reverse-bias voltage, we have from (1.19)

$$C_{T\text{-}j} = \frac{30.5\ \mathrm{fF}}{\sqrt{1 + \left(\frac{3}{0.9}\right)}} = 14.7\ \mathrm{fF} \tag{1.24}$$

As expected, we see a decrease in junction capacitance as the width of the depletion region is increased.

Graded Junctions

All of the above equations assumed an abrupt junction where the doping concentration changes quickly from p to n over a small distance. Although this is a good approximation for many integrated circuits, it is not always true. For example, the collector-to-base junction of a bipolar transistor is most commonly realized as a *graded* junction. In the case of graded junctions, the exponent 1/2 in Eq. (1.15) is inaccurate, and a better value to use is an exponent closer to unity, perhaps 0.6 to 0.7. Thus, for graded junctions, (1.15) is typically written as

$$Q = \left[2qK_s\varepsilon_0(\Phi_0 + V_R)\frac{N_A N_D}{N_A + N_D}\right]^{1-m} \tag{1.25}$$

where m is a constant typically around 1/3.

Differentiating (1.25) to find the depletion capacitance, we have

$$C_j = (1-m)\left[2qK_s\varepsilon_0\frac{N_A N_D}{N_A + N_D}\right]^{1-m}\frac{1}{(\Phi_0 + V_R)^m} \tag{1.26}$$

This depletion capacitance can also be written as

$$C_j = \frac{C_{j0}}{\left(1 + \frac{V_R}{\Phi_0}\right)^m} \tag{1.27}$$

where

$$C_{j0} = (1-m)\left[2qK_s\varepsilon_0\frac{N_A N_D}{N_A + N_D}\right]^{1-m}\frac{1}{\Phi_0^m} \tag{1.28}$$

From (1.27), we see that a graded junction results in a depletion capacitance that is less dependent on V_R than the equivalent capacitance in an abrupt junction. In other words, since m is less than 0.5, the depletion capacitance for a graded junction is more linear than that for an abrupt junction. Correspondingly, increasing the reverse-bias voltage for a graded junction is not as effective in reducing the depletion capacitance as it is for an abrupt junction.

Finally, as in the case of an abrupt junction, the depletion charge on either side of the junction can also be written as

$$Q = \frac{C_{j0}}{1-m}\Phi_0\left(1+\frac{V_R}{\Phi_0}\right)^{1-m} \quad (1.29)$$

EXAMPLE 1.5

Repeat Example 1.4 for a graded junction with $m = 0.4$.

Solution

Noting once again that $N_A >> N_D$, we approximate (1.28) as

$$C_{j0} = (1-m)[2qK_s\varepsilon_0N_D]^{1-m}\frac{1}{\Phi_0^m} \quad (1.30)$$

resulting in

$$C_{j0} = 81.5\ \mu F/m^2 \quad (1.31)$$

which, when multiplied by the diode's area of $10\ \mu m \times 10\ \mu m$, results in

$$C_{T\text{-}j0} = 8.1\ fF \quad (1.32)$$

For a 3-V reverse-bias voltage, we have

$$C_{T\text{-}j} = \frac{8.1 fF}{(1+3/0.9)^{0.4}} = 4.5\ fF \quad (1.33)$$

Large-Signal Junction Capacitance

The equations for the junction capacitance given above are only valid for small changes in the reverse-bias voltage. This limitation is due to the fact that C_j depends on the size of the reverse-bias voltage instead of being a constant. As a result, it is extremely difficult and time consuming to accurately take this nonlinear capacitance into account when calculating the time to charge or discharge a junction over a large voltage change. A commonly used approximation when analyzing the transient response for large voltage changes is to use an *average size* for the junction capacitance by calculating the junction capacitance at the two extremes of the reverse-bias voltage. Unfortunately, a problem with this approach is that when the diode is forward biased with $V_R \cong -\Phi_0$, Eq. (1.17) "blows up" (i.e., is equal to infinity). To circumvent this problem, one can instead calculate the charge stored in the junction for the two extreme values of applied voltage (through the use of (1.21)), and then through the use of $Q = CV$, calculate the average capacitance according to

$$C_{j\text{-av}} = \frac{Q(V_2) - Q(V_1)}{V_2 - V_1} \tag{1.34}$$

where V_1 and V_2 are the two voltage extremes [Hodges, 1988].

From (1.21), for an abrupt junction with reverse-bias voltage V_i, we have

$$Q(V_i) = 2C_{j0}\Phi_0\sqrt{1 + \frac{V_i}{\Phi_0}} \tag{1.35}$$

Therefore,

$$C_{j\text{-av}} = 2C_{j0}\Phi_0\frac{\left(\sqrt{1 + \frac{V_2}{\Phi_0}} - \sqrt{1 + \frac{V_1}{\Phi_0}}\right)}{V_2 - V_1} \tag{1.36}$$

One special case often encountered is charging a junction from 0 V to 5 V. For this special case, and using $\Phi_0 = 0.9$ V, we find that

$$C_{j\text{-av}} = 0.56C_{j0} \tag{1.37}$$

Thus, as a rough approximation to quickly estimate the charging time of a junction capacitance from 0 V to 5 V (or vice versa), one can use

$$C_{j\text{-av}} = \frac{C_{j0}}{2} \tag{1.38}$$

It will be seen in the following example that (1.37) compares well with a SPICE simulation.

EXAMPLE 1.6

For the circuit shown in Fig. 1.3, where a reverse-biased diode is being charged from 0 V to 5 V, through a 10-kΩ resistor, calculate the time required to charge the diode from 0 V to 3.5 V. Assume that $C_{j0} = 0.2\ \text{fF}/(\mu\text{m})^2$ and that the diode has an area of $20\ \mu\text{m} \times 5\ \mu\text{m}$. Compare your answer to that obtained using SPICE. Repeat the question for the case of the diode being discharged from 5 V to 1.5 V.

Solution

The total small-signal capacitance of the junction at 0-V bias voltage is obtained by multiplying $0.2\ \text{fF}/(\mu\text{m})^2$ by the junction area to obtain

$$C_{j0} = 0.2 \times 10^{-15} \times 20 \times 5 = 0.02\ \text{pF} \tag{1.39}$$

Using (1.37), we have

$$C_{j\text{-av}} = 0.56 \times 0.02 = 0.011\ \text{pF} \tag{1.40}$$

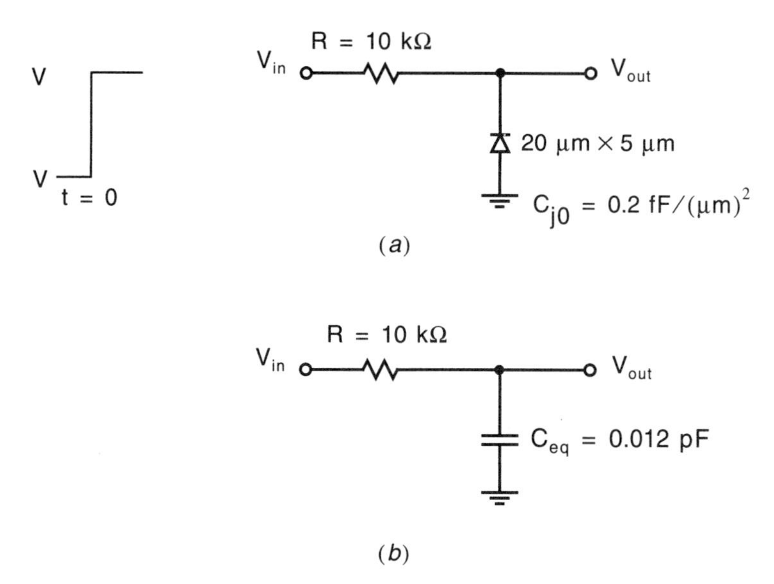

Fig. 1.3 (*a*) The circuit used in Example 1.6; (*b*) its RC approximate equivalent.

resulting in a time constant of

$$\tau = RC_{j\text{-}av} = 0.11 \text{ ns} \tag{1.41}$$

It is not difficult to show that the time it takes for a first-order circuit to rise (or fall) 70 percent of its final value is equal to 1.2τ. Thus, in this case,

$$t_{70\%} = 1.2\tau = 0.13 \text{ ns} \tag{1.42}$$

As a check, the circuit of Fig. 1.3(*a*) was analyzed using SPICE. The input data file was as follows:

```
R 1 2 10k
D 0 2 DMOD
*
VIN 1 0 dc 2.5 PULSE (0 5 0 10p 10p 0.49n 1.0n)
*
.MODEL DMOD D(CJO=0.02E-12)
*
.OPTIONS NUMDGT=5 ITL1=500
.WIDTH OUT=80
.TRAN 0.01n 1.0n
.PRINT TRAN V(2)
.END
```

The SPICE simulation gave a 0-V to 3.5-V rise time of 0.14 ns and a 5-V to 1.5-V fall time of 0.12 ns. These times compare favorably with the 0.13 ns predicted. The reason for the different values of the rise and fall times is the nonlinearity of the junction capacitance. For smaller bias voltages it is larger than that predicted by (1.37), whereas for larger bias voltages it is smaller. If we use the more accurate approximation of (1.36) for the rise time with $V_2 = 3.5$ and $V_1 = 0$ V, we find

$$C_{j\text{-}av} = 2 \times 0.02 \times \frac{0.9}{3.5}\left(\sqrt{1 + \frac{3.5}{0.9}} - 1\right) = 0.012 \text{ pF} \tag{1.43}$$

Also, for the fall time, we find that

$$C_{j\text{-}av} = 2 \times 0.02 \times \frac{0.9}{1.5 - 5}\left(\sqrt{1 + \frac{1.5}{0.9}} - \sqrt{1 + \frac{5}{0.9}}\right) = 0.010 \text{ pF} \tag{1.44}$$

These more accurate approximations result in

$$t_{+70\%} = 0.144 \text{ ns} \tag{1.45}$$

and

$$t_{-70\%} = 0.114 \text{ ns} \tag{1.46}$$

in closer agreement with SPICE. Normally, the extra accuracy that results from using (1.36) instead of (1.37) is not worth the extra complication because one seldom knows the area of C_{j0} to better than 20 percent accuracy.

Forward-Biased Junctions

A positive voltage applied from the p side to the n side of a diode reduces the electric field opposing the diffusion of the free carriers across the depletion region. It also reduces the width of the depletion region. If this forward-bias voltage is large enough, the carriers will start to diffuse across the junction, resulting in a current flow from the anode to the cathode. For silicon, appreciable diode current starts to occur for a forward-bias voltage around 0.5 V. For germanium and gallium arsenide semiconductor materials, current conduction starts to occur around 0.3 V and 0.9 V, respectively.

When the junction potential is sufficiently lowered for conduction to occur, the carriers diffuse across the junction due to the large gradient in the mobile carrier concentrations. Note that there are more carriers diffusing from the heavily doped side to the lightly doped side than from the lightly doped side to the heavily doped side.

After the carriers cross the depletion region, they greatly increase the *minority charge* at the edge of the depletion region. These minority carriers will diffuse away from the junction toward the bulk. As they diffuse, they recombine with the majority carriers, thereby decreasing their concentration. This concentration gradient of the minority charge (which decreases the farther one gets from the junction) is responsible for the current flow near the junction.

The majority carriers that recombine with the diffusing minority carriers come from the metal contacts at the junctions because of the forward-bias voltage. These majority carriers flow across the bulk, from the contacts to the junction, due to an electric field applied across the bulk. This current flow is called *drift*. It results in small potential drops across the bulk, especially in the lightly doped side. Typical values of this voltage drop might be 50 mV to 0.1 V, depending primarily on the doping concentration of the lightly doped side, the distance from the contacts to the junction, and the cross-sectional area of the junction.

In the forward-bias region, the current-voltage relationship is exponential and can be shown (see Appendix) to be

$$I_D = I_S e^{V_D/V_T} \tag{1.47}$$

where V_D is the voltage applied across the diode and

$$I_S \propto A_D\left(\frac{1}{N_A} + \frac{1}{N_D}\right) \tag{1.48}$$

I_S is known as the *scale current* and is seen to be proportional to the area of the diode junction, A_D, and inversely proportional to the doping concentrations.

Junction Capacitance of Forward-Biased Diode

When a junction changes from reverse biased (with little current through it) to forward biased (with significant current flow across it), the charge being stored near and across the junction changes. Part of the change in charge is due to the change in the width of the depletion region and therefore the amount of immobile charge stored in it. This change in charge is modelled by the depletion capacitance, C_j, similar to when the junction is reverse biased. An additional change in charge storage is necessary to account for the change of the minority carrier concentration close to the junction required for the diffusion current to exist. For example, if a forward-biased diode current is to double, then the slopes of the minority charge storage at the diode junction edges must double, and this, in turn, implies that the minority charge storage must double. This component is modelled by another capacitance, called the *diffusion capacitance,* and denoted C_d.

The diffusion capacitance can be shown (see Appendix) to be

$$C_d = \tau_T \frac{I_D}{V_T} \tag{1.49}$$

where τ_T is the transit time of the diode. Normally τ_T is specified for a given technology, so that one can calculate the diffusion capacitance. *Note that the diffusion capacitance of a forward-biased junction is proportional to the diode current.*

The total capacitance of the forward-biased junction is the sum of the diffusion capacitance, C_d, and the depletion capacitance, C_j. Thus, the total junction capacitance is given by

$$C_T = C_d + C_j \tag{1.50}$$

For a forward-biased junction, the depletion capacitance, C_j, can be roughly approximated by $2C_{j0}$. The accuracy of this approximation is not critical since the diffusion capacitance is typically much larger than the depletion capacitance.

Finally, it should be mentioned that as a diode is turned off for a short period of time a current will flow in the negative direction until the minority charge is removed. This behavior does not occur in Schottky diodes since they do not have minority charge storage.

Small-Signal Model of a Forward-Biased Diode

A small-signal equivalent model for a forward-biased diode is shown in Fig. 1.4. A resistor, r_d, models the change in the diode voltage, V_D, that occurs when I_D changes. Using (1.47), we have

$$\frac{1}{r_d} = \frac{dI_D}{dV_D} = I_S\frac{e^{V_D/V_T}}{V_T} = \frac{I_D}{V_T} \tag{1.51}$$

This resistance is called the incremental resistance of the diode. For very accurate modelling, it is sometimes necessary to add the series resistance due to the bulk and also the resistance associated with the contacts. Typical values for the contact resistance (caused by the work-function[2] difference between metal and silicon) might be $20\ \Omega$ to $40\ \Omega$.

By combining (1.49) and (1.51), we see that an alternative equation for the diffusion capacitance, C_d, is

$$C_d = \frac{\tau_T}{r_d} \tag{1.52}$$

Since for moderate forward-bias currents, $C_d >> C_j$, the total small-signal capacitance is $C_T \cong C_d$, and

$$r_d C_T \cong \tau_T \tag{1.53}$$

Thus, for charging or discharging a forward-biased junction with a current source having an impedance much larger than r_d, the time constant of the charging is approximately equal to the transit time of the diode and is independent of the diode current. For smaller diode currents, where C_j becomes important, the charging or discharging time constant of the circuit becomes larger than τ_T.

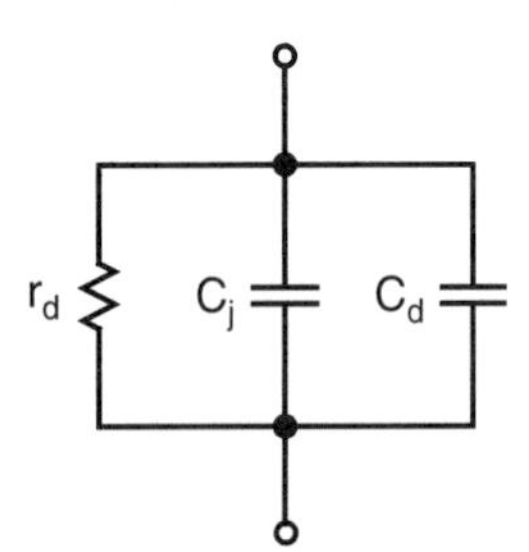

Fig. 1.4 The small-signal model for a forward-biased junction.

2. The work-function of a material is defined as the minimum energy required to remove an electron at the Fermi level to the outside vacuum region.

EXAMPLE 1.7

A given diode has a transit time of 100 ps and is biased at 1 mA. What are the values of its small-signal resistance and diffusion capacitance? Assume room temperature, so that $V_T = kT/q = 26\ \text{mV}$.

Solution

We have

$$r_d = \frac{V_T}{I_D} = \frac{26\ \text{mV}}{1\ \text{mA}} = 26\ \Omega$$

and

$$C_d = \frac{\tau_T}{r_d} = \frac{100\ \text{ps}}{26\ \Omega} = 3.8\ \text{pF}$$

Note that this diffusion capacitance is over 100 times larger than the total depletion capacitance found in Examples 1.4 and 1.5.

Schottky Diodes

A different type of diode, one often used in microcircuit design, is realized by contacting metal to a lightly doped semiconductor region (rather than a heavily doped region) as shown in Fig. 1.5. Notice that the aluminum anode is in direct contact with a relatively lightly doped n^- region. Because the n^- region is relatively lightly doped, the work-function difference between the aluminum contact and the n^- silicon is larger than would be the case for aluminum contacting to an n^+ region, as occurs at the cathode. This causes a depletion region and, correspondingly, a diode to occur at the interface between the aluminum anode and the n^- silicon region. This diode has different characteristics than a normal pn junction diode. First, its voltage drop when forward

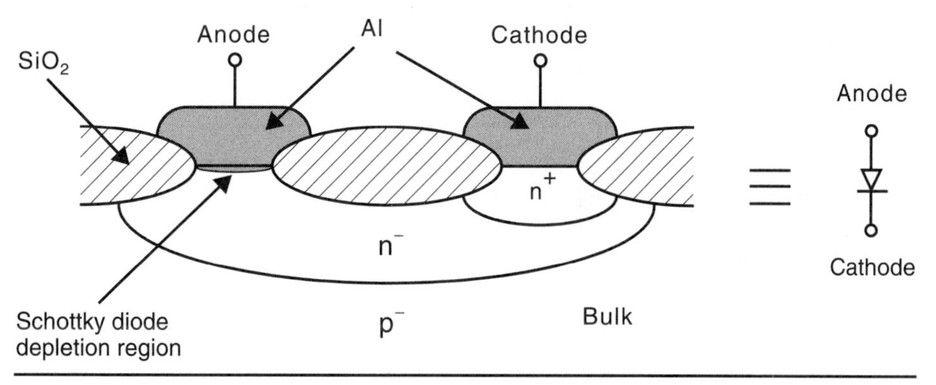

Fig. 1.5 A cross section of a Schottky diode.

biased is smaller. This voltage drop is dependent on the metal used; for aluminum it might be around 0.5 V. More importantly, when the diode is forward biased, there is no minority-charge storage in the lightly doped n^- region. Thus, the small-signal model of a forward-biased Schottky diode has $C_d = 0$ (with reference to Fig. 1.4). The absence of this diffusion capacitance makes the diode much faster. It is particularly faster when turning off, because it is not necessary to remove the minority charge first. Rather, it is only necessary to discharge the depletion capacitance through about 0.2 V.

Schottky diodes have been used extensively in bipolar logic circuits. They are also used in a number of high-speed analog circuits, particularly those realized in gallium arsenide (GaAs) technologies, rather than silicon technologies.

1.2 MOS TRANSISTORS

Presently, the most popular technology for realizing microcircuits makes use of MOS transistors. Unlike most bipolar junction transistor (BJT) technologies, which make dominant use of only one type of transistor (npn transistors in the case of BJT processes[3]), MOS circuits normally use two complementary types of transistors—n-channel and p-channel. While n-channel devices conduct with a positive gate voltage, p-channel devices conduct with a negative gate voltage. Moreover, electrons are used to conduct current in n-channel transistors, while holes are used in p-channel transistors. Microcircuits containing both n-channel and p-channel transistors are called CMOS circuits, for *complementary MOS*. The acronym MOS stands for *metal-oxide semiconductor,* which historically denoted the gate, insulator, and channel region materials, respectively. However, most present CMOS technologies utilize polysilicon gates rather than metal gates.

Before CMOS technology became widely available, most MOS processes made use of only n-channel transistors (NMOS). However, often two different types of n-channel transistors could be realized. One type, enhancement n-channel transistors, is similar to the n-channel transistors realized in CMOS technologies. Enhancement transistors require a positive gate-to-source voltage to conduct current. The other type, depletion transistors, conduct current with a gate-source voltage of 0 V. Depletion transistors were used to create high-impedance loads in NMOS logic gates.

A typical cross section of an n-channel enhancement-type MOS transistor is shown in Fig. 1.6. With no voltage applied to the gate, the n^+ source and drain regions are separated by the p^- substrate. The separation between the drain and the source is called the channel length, L. In present MOS technologies, the minimum channel length is typically between $0.3\ \mu m$ and $1.0\ \mu m$. It should be noted that there is no physical difference between the drain and the source.[4] *The source terminal of an*

---

3. Most BJT technologies can also realize low-speed lateral pnp transistors. Normally these would only be used to realize current sources as they have low gains and poor frequency responses. Recently, bipolar technologies utilizing high-speed vertical pnp transistors, as well as high-speed npn transistors, have become available and are growing in popularity. These technologies are called complementary bipolar technologies.

4. Large MOS transistors used for power applications might not be realized with symmetric drain and source junctions.

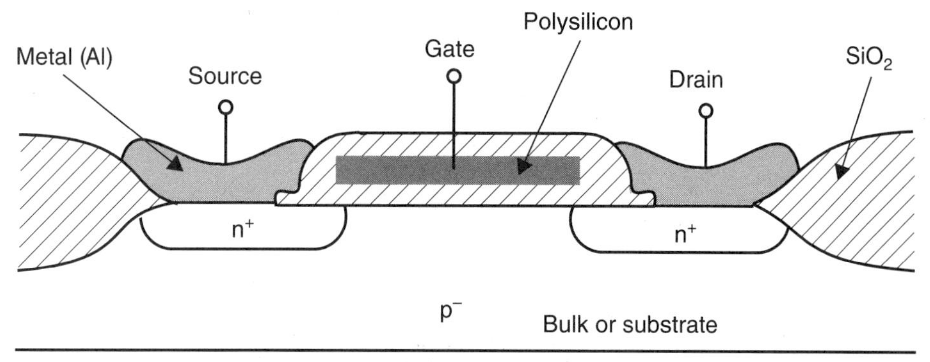

Fig. 1.6 A cross section of a typical n-channel transistor.

n-channel transistor is defined as whichever of the two terminals has a lower voltage. For a p-channel transistor, the source would be the terminal with the higher voltage. When a transistor is turned on, current flows from the drain to the source in an n-channel transistor and from the source to the drain in a p-channel transistor. In both cases, the true carriers travel from the source to drain, but the current directions are different because n-channel carriers (electrons) are negative, whereas p-channel carriers (holes) are positive.

The gate is normally realized using polysilicon, which is heavily doped noncrystalline (or amorphous) silicon. Polysilicon gates are used nowadays (instead of metal) because polysilicon allows the dimensions of the transistor to be realized much more accurately during the patterning of the transistor, which involves what is called a self-aligned process. This higher geometric accuracy results in smaller, faster transistors.

The gate is physically separated from the surface of the silicon by a thin insulator made of silicon dioxide (SiO_2). Thus, the gate is electrically isolated from the channel and affects the channel (and hence, the transistor current) only through electrostatic coupling, similar to capacitive coupling. A typical thickness for the SiO_2 insulator between the gate and the channel is presently from 0.01 μm to 0.03 μm. Since the gate is electrically isolated from the channel, it never conducts dc current. Indeed, the excellent isolation results in leakage currents being almost undetectable. However, because of the inherent capacitances in MOS transistors, transient gate currents do exist when gate voltages are quickly changing.

Normally the p^- substrate (or bulk) is connected to the most negative voltage in a microcircuit. In analog circuits, this might be the negative power supply, but in digital circuits it is normally ground or 0 V. This connection results in all transistors placed in the substrate being surrounded by reverse-biased junctions, which electrically isolate the transistors and thereby prevent conduction through the substrate between transistors (unless, of course, they are connected together through some other means).

Symbols for MOS Transistors

Many symbols have been used to represent MOS transistors. Figure 1.7 shows some of the symbols that have been used to represent n-channel MOS transistors. The symbol in Fig. 1.7(*a*) is often used; note that there is nothing in the symbol to specify

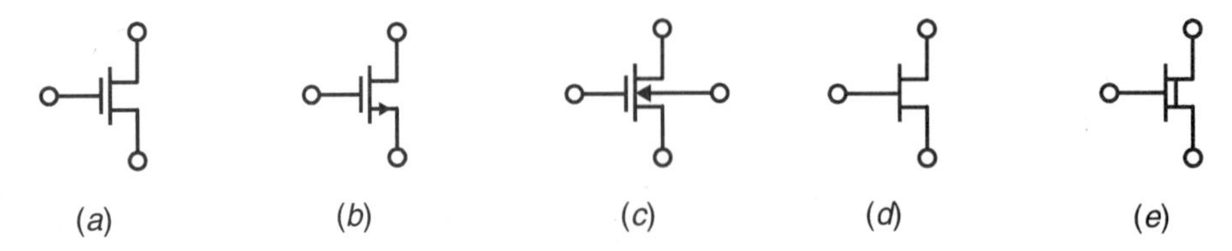

Fig. 1.7 Commonly used symbols for n-channel transistors.

whether the transistor is n-channel or p-channel. A common rule is to assume, when in doubt, that the transistor is an n-channel enhancement transistor. Figure 1.7(*b*) is the most commonly used symbol for an n-channel enhancement transistor and is used throughout this text. The arrow pointing outward on the source indicates that the transistor is n-channel, similar to the convention used for npn transistors, and indicates the direction of hole current.

MOS transistors are actually four-terminal devices, with the substrate being the fourth terminal. In n-channel devices, the p^- substrate is normally connected to the most negative voltage in the microcircuit, whereas for p-channel devices, the n^- substrate is normally connected to the most positive voltage. In these cases the substrate connection is normally not shown in the symbol. However, for CMOS technologies, at least one of the two types of transistors will be formed in a *well* substrate that need not be connected to one of the power supply nodes. For example, an n-well process would form n-channel transistors in a p^- substrate encompassing the entire microcircuit, while the p-channel transistors would be formed in many n-well substrates. In this case, most of the n-well substrates would be connected to the most positive power supply, while some might be connected to other nodes in the circuit (often the well is connected to the source of a transistor that is not connected to the power supply). In these cases, the symbol shown in Fig. 1.7(*c*) can be used to show the substrate connection explicitly. It should be noted that this case is not encountered often in digital circuits and is more common in analog circuits. Sometimes, in the interest of simplicity, the isolation of the gate is not explicitly shown, as is the case of the symbol of Fig. 1.7(*d*). This simple notation is more common for digital circuits in which a large number of transistors are present. Since this symbol is also used for JFET transistors, it will never be used to represent MOS transistors in this text. The last symbol, shown in Fig. 1.7(*e*), denotes an n-channel depletion transistor. The extra line is used to indicate that a physical channel exists for a 0-V gate-source voltage. Depletion transistors were used in older NMOS technologies but are not typically available in CMOS processes.

Figure 1.8 shows some commonly used symbols for p-channel transistors. In this text, the symbol of Fig. 1.8(*a*) will be most often used. The symbol in Fig. 1.8(*c*) is

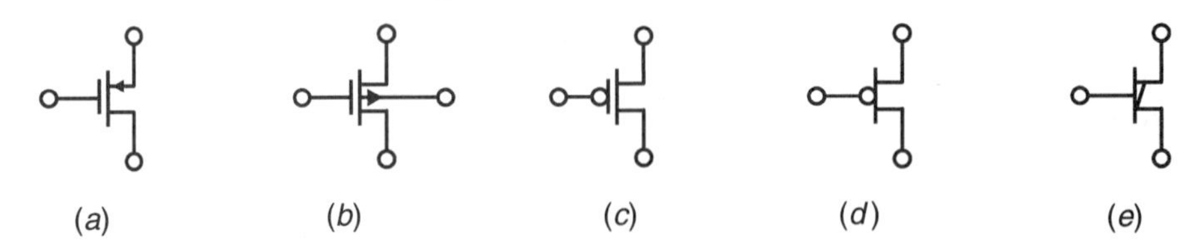

Fig. 1.8 Commonly used symbols for p-channel transistors.

sometimes used in digital circuits, where the circle indicates that a low voltage on the gate turns the transistor on, as opposed to a high voltage for an n-channel transistor (Fig. 1.7(*a*)). The symbols of Fig. 1.8(*d*) or Fig. 1.8(*e*) might be used in larger circuits where many transistors are present, to simplify the drawing somewhat. They will not be used in this text.

Basic Operation

The basic operation of MOS transistors will be described with respect to an n-channel transistor. First, consider the simplified cross sections shown in Fig. 1.9, where the source, drain, and substrate are all connected to ground. In this case, the MOS transistor operates similarly to a capacitor. The gate acts as one plate of the capacitor, and the surface of the silicon, just under the thin insulating SiO_2, acts as the other plate.

If the gate voltage is very negative, as shown in Fig. 1.9(*a*), positive charge will be attracted to the channel region. Since the substrate was originally doped p^-, this negative gate voltage has the effect of simply increasing the channel doping to p^+,

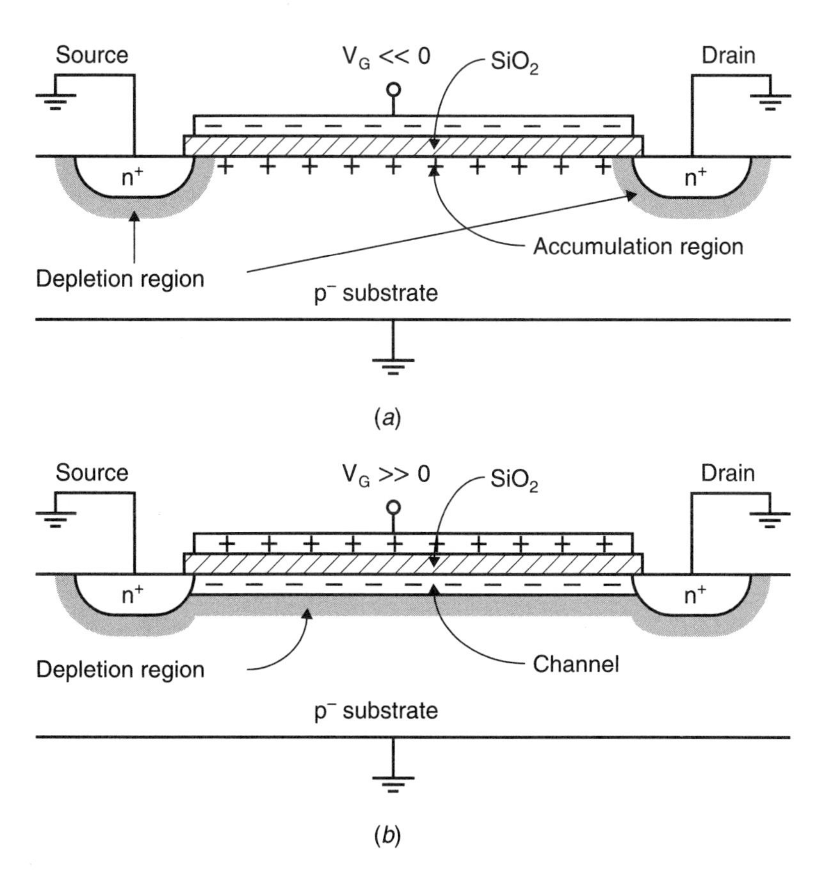

Fig. 1.9 An n-channel MOS transistor. (*a*) $V_G << 0$, resulting in an accumulated channel (no current flow); (*b*) $V_G >> 0$, and the channel is present (current flow possible from drain to source).

resulting in what is called an *accumulated channel.* The n^+ source and drain regions are separated from the p^+-channel region by depletion regions, resulting in the equivalent circuit of two back-to-back diodes. Thus, only leakage current will flow even if one of the source or drain voltages becomes large (unless the drain voltage becomes so large as to cause the transistor to break down).

In the case of a positive voltage being applied to the gate, the opposite situation occurs, as shown in Fig. 1.9(*b*). For small positive gate voltages, the positive carriers in the channel under the gate are initially repulsed and the channel changes from a p^- doping level to a depletion region. As a more positive gate voltage is applied, the gate attracts negative charge from the source and drain regions, and the channel becomes an n region with mobile electrons connecting the drain and source regions.[5] In short, a sufficiently large positive gate-source voltage changes the channel beneath the gate to an n region, and the channel is said to be *inverted.*

The gate-source voltage, for which the concentration of electrons under the gate is equal to the concentration of holes in the p^- substrate far from the gate, is commonly referred to as the *transistor threshold voltage* and denoted V_{tn} (for n-channel transistors). For gate-source voltages larger than V_{tn}, there is an n-type channel present, and conduction between the drain and the source can occur. For gate-source voltages less than V_{tn}, it is normally assumed that the transistor is off and no current flows between the drain and the source. However, it should be noted that this assumption of zero drain-source current for a transistor that is off is only an approximation. In fact, for gate voltages around V_{tn}, there is no abrupt current change, and for gate-source voltages slightly less than V_{tn}, small amounts of *subthreshold current* can flow, as discussed in Section 1.3.

When the gate-source voltage, V_{GS}, is larger than V_{tn}, the channel is present. As V_{GS} is increased, the density of electrons in the channel increases. Indeed, the carrier density, and therefore the charge density, is proportional to $V_{GS} - V_{tn}$, which is often called the *effective gate-source voltage* and denoted V_{eff}. Specifically, define

$$V_{eff} \equiv V_{GS} - V_{tn} \tag{1.54}$$

The charge density of electrons is then given by

$$Q_n = C_{ox}(V_{GS} - V_{tn}) = C_{ox}V_{eff} \tag{1.55}$$

Here, C_{ox} is the gate capacitance per unit area and is given by

$$C_{ox} = \frac{K_{ox}\varepsilon_0}{t_{ox}} \tag{1.56}$$

where K_{ox} is the relative permittivity of SiO_2 (approximately 3.9) and t_{ox} is the thickness of the thin oxide under the gate. A point to note here is that (1.55) is only accurate when both the drain and the source voltages are zero.

5. The drain and source regions are sometimes called diffusion regions or junctions for historical reasons. This use of the word *junction* is not synonymous with our previous use, in which it designated a pn interface of a diode.

To obtain the total gate capacitance, (1.56) should be multiplied by the effective gate area, WL, where W is the gate width and L is the effective gate length. These dimensions are shown in Fig. 1.10. Thus the total gate capacitance, C_{gs}, is given by

$$C_{gs} = WLC_{ox} \tag{1.57}$$

and the total charge of the channel, Q_{T-n}, is given by

$$Q_{T-n} = WLC_{ox}(V_{GS} - V_{tn}) = WLC_{ox}V_{eff} \tag{1.58}$$

The gate capacitance, C_{gs}, is one of the major load capacitances that circuits must be capable of driving. Gate capacitances are also important when one is calculating *charge injection,* which occurs when a MOS transistor is being turned off because the channel charge, Q_{T-n}, must flow from under the gate out through the terminals to other places in the circuit.

Next, if the drain voltage is increased above 0 V, a drain-source potential difference exists. This difference results in current flowing from the drain to the source.[6] The relationship between V_{DS} and the drain-source current, I_D, is the same as for a resistor, assuming V_{DS} is small. This relationship is given [Sze, 1981] by

$$I_D = \mu_n Q_n \frac{W}{L} V_{DS} \tag{1.59}$$

where $\mu_n \cong 0.06\ m^2/Vs$ is the mobility of electrons near the silicon surface, and Q_n is the charge concentration of the channel per unit area (looking from the top down). Note that as the channel length increases, the drain-source current decreases, whereas this current increases as either the charge density or the transistor width increases. Using (1.58) and (1.59) results in

$$I_D = \mu_n C_{ox} \frac{W}{L}(V_{GS} - V_{tn})V_{DS} = \mu_n C_{ox} \frac{W}{L} V_{eff} V_{DS} \tag{1.60}$$

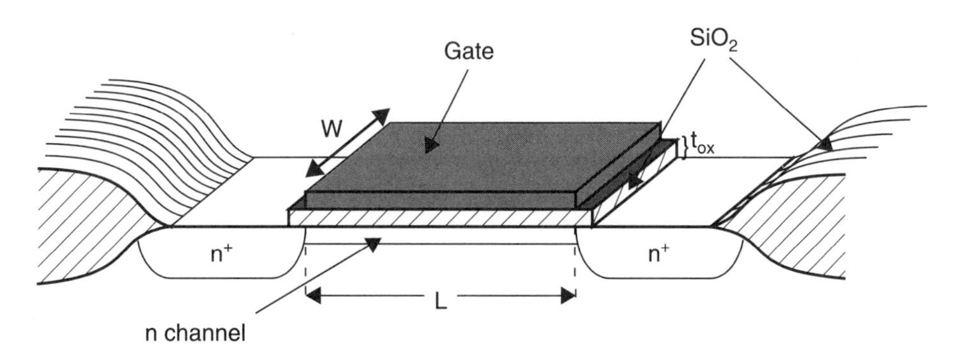

Fig. 1.10 The important dimensions of a MOS transistor.

6. The current is actually conducted by negative carriers (electrons) flowing from the source to the drain. Negative carriers flowing from source to drain results in a positive current from drain to source, I_{DS}.

where it should be emphasized that this relationship is only valid for drain-source voltages near zero (i.e., V_{DS} much smaller than V_{eff}).

As the drain-source voltage increases, the channel charge concentration decreases at the drain end. This decrease is due to the smaller gate-to-channel voltage difference across the thin gate oxide as one moves closer to the drain. In other words, since the drain voltage is assumed to be at a higher voltage than the source, there is an increasing voltage gradient from the source to the drain, resulting in a smaller gate-to-channel voltage near the drain. Since the charge density at a distance x from the source end of the channel is proportional to $V_G - V_{ch}(x) - V_{tn}$, as $V_G - V_{ch}(x)$ decreases, the charge density also decreases.[7] This effect is illustrated in Fig. 1.11.

Note that at the drain end of the channel, we have

$$V_G - V_{ch}(L) = V_{GD} \tag{1.61}$$

For small V_{DS}, we saw from (1.60) that I_D was linearly related to V_{DS}. However, as V_{DS} increases, and the charge density decreases near the drain, the relationship becomes nonlinear. In fact, the linear relationship for I_D versus V_{DS} flattens for larger V_{DS}, as shown in Fig. 1.12.

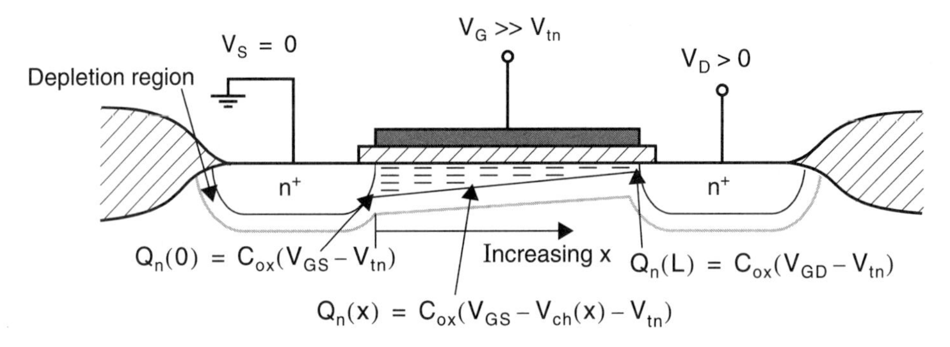

Fig. 1.11 The channel charge density for $V_{DS} > 0$.

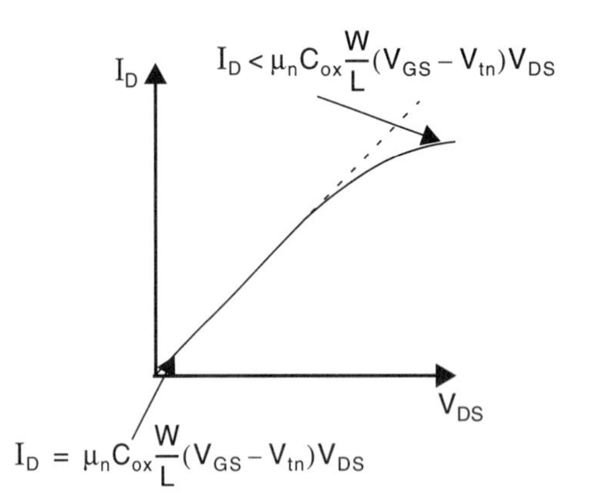

Fig. 1.12 For V_{DS} not close to zero, the I_D versus V_{DS} relationship is no longer linear.

7. $V_G - V_{CH}(x)$ is the gate-to-channel voltage drop at distance x from the source end, with V_G being the same everywhere in the gate, since the gate material is highly conductive.

As the drain voltage is increased, at some point the gate-to-channel voltage at the drain end will decrease to the threshold value V_{tn} — the minimum gate-to-channel voltage needed for n carriers in the channel to exist. Thus, at the drain end, the channel becomes *pinched off*, as shown in Fig. 1.13. This pinch-off occurs at $V_{GD} = V_{tn}$, since the channel voltage at the drain end is simply equal to V_D. Thus, pinch-off occurs for

$$V_{DG} > -V_{tn} \tag{1.62}$$

Denoting $V_{DS\text{-sat}}$ as the drain-source voltage when the channel becomes pinched off, we can substitute $V_{DG} = V_{DS} - V_{GS}$ into (1.62) and find an equivalent pinch-off expression

$$V_{DS} > V_{DS\text{-sat}} \tag{1.63}$$

where $V_{DS\text{-sat}}$ is given[8] by

$$V_{DS\text{-sat}} = V_{GS} - V_{tn} = V_{eff} \tag{1.64}$$

The electron carriers travelling through the pinched-off drain region are velocity saturated, similar to a gas under pressure travelling through a very small tube. If the drain-gate voltage rises above this critical pinch-off voltage of $-V_{tn}$, the charge concentration in the channel remains constant (to a first-order approximation) and the drain current no longer increases with increasing V_{DS}. The result is the current-voltage relationship shown in Fig. 1.14 for a given gate-source voltage. In the region of operation where $V_{DS} > V_{DS\text{-sat}}$, the drain current is independent of V_{DS} and is called the *active region*.[9] The region where I_D changes with V_{DS} is called the *triode region*. When MOS transistors are used in analog amplifiers, they almost always are biased in the active region. When they are used in digital logic gates, they often operate in both regions.

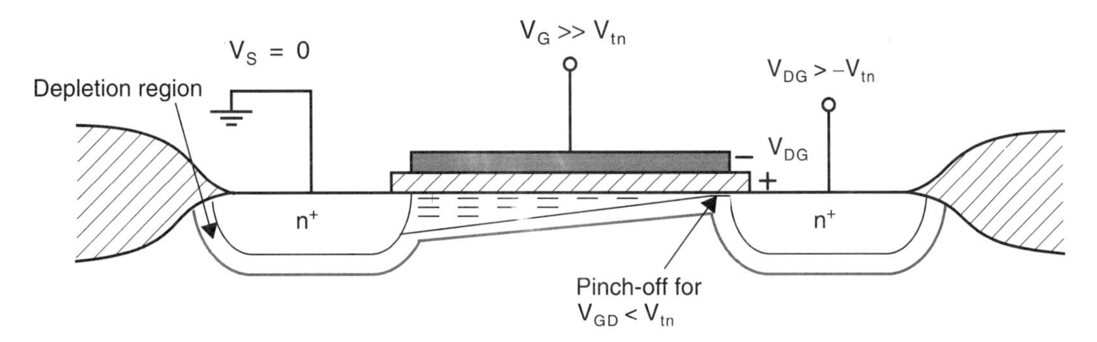

Fig. 1.13 When V_{DS} is increased so that $V_{GD} < V_{tn}$, the channel becomes pinched off at the drain end.

8. Because of the body effect, the threshold voltage at the drain end of the transistor is increased, resulting in the true value of $V_{DS\text{-sat}}$ being slightly lower than V_{eff}.

9. Historically, the active region was called the saturation region, but this led to confusion because in the case of bipolar transistors, the saturation region occurs for small V_{CE}, whereas for MOS transistors it occurs for large V_{DS}. The renaming of the saturation region to the active region is becoming widely accepted.

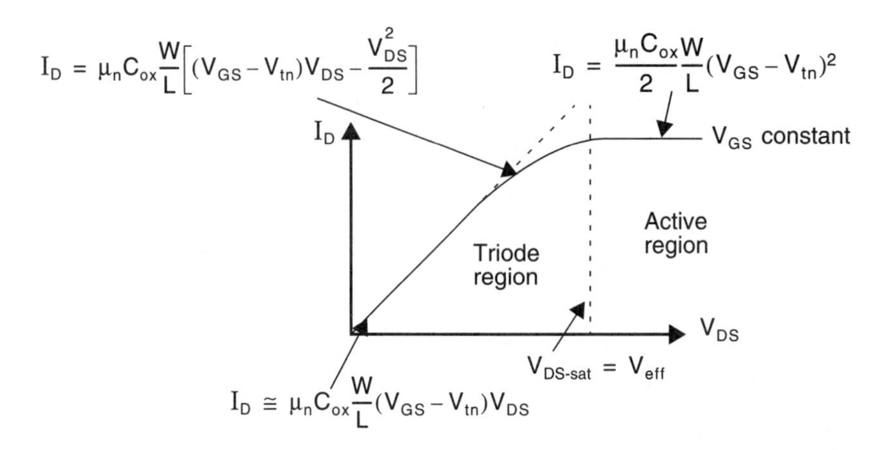

Fig. 1.14 The I_D versus V_{DS} curve for an ideal MOS transistor. For $V_{DS} > V_{DS\text{-sat}}$, I_D is approximately constant.

Before proceeding, it is worth discussing the terms *weak, moderate,* and *strong inversion.* As just discussed, a gate-source voltage greater than V_{tn} results in an inverted channel, and drain-source current can flow. However, as the gate-source voltage is increased, the channel does not become inverted (i.e., n-region) suddenly, but rather gradually. Thus, it is useful to define three regions of channel inversion with respect to the gate-source voltage. In most circuit applications, noncutoff MOSFET transistors are operated in strong inversion, with $V_{eff} > 100\ \text{mV}$ (many prudent circuit designers use a minimum value of 200 mV). As the name suggests, strong inversion occurs when the channel is strongly inverted. It should be noted that all the equation models in this section assume strong inversion operation. Weak inversion occurs when V_{GS} is approximately 100 mV or more below V_{tn} and is discussed as subthreshold operation in Section 1.3. Finally, moderate inversion is the region between weak and strong inversion.

Large-Signal Modelling

The *triode region equation* for a MOS transistor relates the drain current to the gate-source and drain-source voltages. It can be shown (see Appendix) that this relationship is given by

$$I_D = \mu_n C_{ox}\left(\frac{W}{L}\right)\left[(V_{GS} - V_{tn})V_{DS} - \frac{V_{DS}^2}{2}\right] \tag{1.65}$$

As V_{DS} increases, I_D increases until the drain end of the channel becomes pinched off, and then I_D no longer increases. This pinch-off occurs for $V_{DG} = -V_{tn}$, or approximately,

$$V_{DS} = V_{GS} - V_{tn} = V_{eff} \tag{1.66}$$

Right at the edge of pinch-off, the drain current resulting from (1.65) and the drain current in the active region (which, to a first-order approximation, is constant with respect to V_{DS}) must have the same value. Therefore, the *active region equation* can be found by substituting (1.66) into (1.65), resulting in

$$I_D = \frac{\mu_n C_{ox}}{2}\left(\frac{W}{L}\right)(V_{GS} - V_{tn})^2 \tag{1.67}$$

For $V_{DS} > V_{eff}$, the current stays constant at the value given by (1.67), ignoring second-order effects such as the finite output impedance of the transistor. This equation is perhaps the most important one that describes the large-signal operation of a MOS transistor. It should be noted here that (1.67) represents a squared current-voltage relationship for a MOS transistor in the active region. In the case of a BJT transistor, an exponential current-voltage relationship exists in the active region.

As just mentioned, (1.67) implies that the drain current, I_D, is independent of the drain-source voltage. This independence is only true to a first-order approximation. The major source of error is due to the channel length shrinking as V_{DS} increases. To see this effect, consider Fig. 1.15, which shows a cross section of a transistor in the active region. A pinched-off region with very little charge exists between the drain and the channel. The voltage at the end of the channel closest to the drain is fixed at $V_{GS} - V_{tn} = V_{eff}$. The voltage difference between the drain and the near end of the channel lies across a short depletion region often called the *pinch-off region.* As V_{DS} becomes larger than V_{eff}, this depletion region surrounding the drain junction increases its width in a square-root relationship with respect to V_{DS}. This increase in the width of the depletion region surrounding the drain junction decreases the effective channel length. In turn, this decrease in effective channel length increases the drain current, resulting in what is commonly referred to as *channel-length modulation.*

To derive an equation to account for channel-length modulation, we first make use of (1.11) and denote the width of the depletion region by x_d, resulting in

$$\begin{aligned} x_d &\cong k_{ds}\sqrt{V_{D\text{-}ch} + \Phi_0} \\ &= k_{ds}\sqrt{V_{DG} + V_{tn} + \Phi_0} \end{aligned} \tag{1.68}$$

where

$$k_{ds} = \sqrt{\frac{2K_s\varepsilon_0}{qN_A}} \tag{1.69}$$

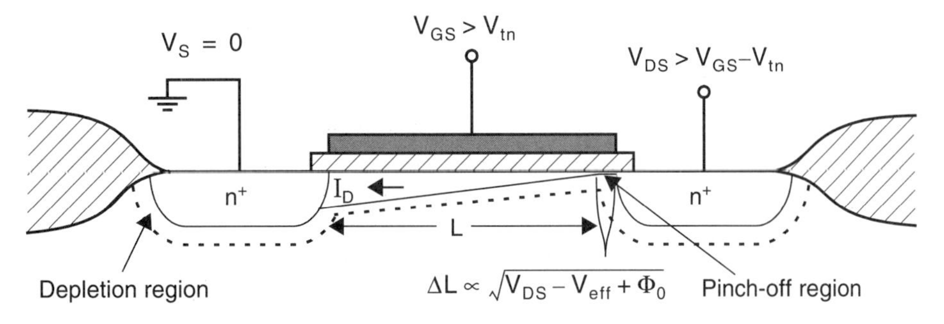

Fig. 1.15 Channel length shortening for $V_{DS} > V_{eff}$.

and has units of $m/\sqrt{V}$. Note that N_A is used here since the n-type drain region is more heavily doped than the p-type channel (i.e., $N_D >> N_A$). By writing a Taylor approximation for I_D around its operating value of $V_{DS} = V_{GS} - V_{tn} = V_{eff}$, we find I_D to be given by

$$I_D = I_{D\text{-sat}} + \left(\frac{\partial I_D}{\partial L}\right)\left(\frac{\partial L}{\partial V_{DS}}\right)\Delta V_{DS} \cong I_{D\text{-sat}}\left(1 + \frac{k_{ds}(V_{DS} - V_{eff})}{2L\sqrt{V_{DG} + V_{tn} + \Phi_0}}\right) \tag{1.70}$$

where $I_{D\text{-sat}}$ is the drain current when $V_{DS} = V_{eff}$, or equivalently, the drain current when the channel-length modulation is ignored. Note that in deriving the final equation of (1.70), we have used the relationship $\partial L/\partial V_{DS} = -\partial x_d/\partial V_{DS}$. Usually, (1.70) is written as

$$I_D = \frac{\mu_n C_{ox}}{2}\left(\frac{W}{L}\right)(V_{GS} - V_{tn})^2[1 + \lambda(V_{DS} - V_{eff})] \tag{1.71}$$

where λ is the output impedance constant (in units of V^{-1}) given by

$$\lambda = \frac{k_{ds}}{2L\sqrt{V_{DG} + V_{tn} + \Phi_0}} = \frac{k_{ds}}{2L\sqrt{V_{DS} - V_{eff} + \Phi_0}} \tag{1.72}$$

Equation (1.71) is accurate until V_{DS} is large enough to cause second-order effects, often called *short-channel effects.* For example, (1.71) assumes that current flow down the channel is not *velocity-saturated* (i.e., increasing the electric field no longer increases the carrier speed). Velocity saturation commonly occurs in new technologies that have very short channel lengths and therefore large electric fields. If V_{DS} becomes large enough so short-channel effects occur, I_D increases more than is predicted by (1.71). Of course, for quite large values of V_{DS}, the transistor will eventually break down.

A plot of I_D versus V_{DS} for different values of V_{GS} is shown in Fig. 1.16. Note that in the active region, the small (but nonzero) slope indicates the small dependence of I_D on V_{DS}.

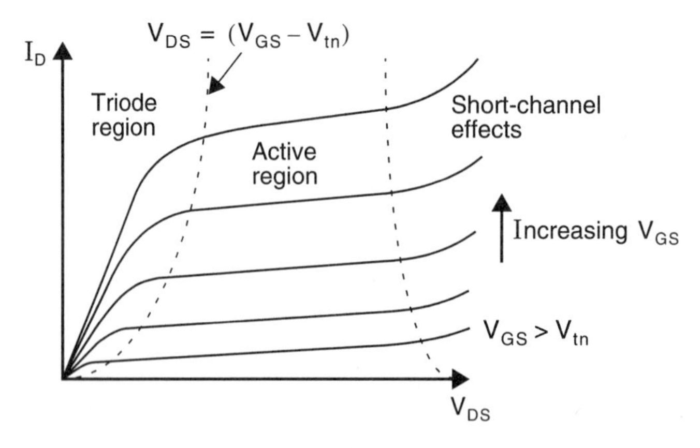

Fig. 1.16 I_D versus V_{DS} for different values of V_{GS}.

EXAMPLE 1.8

Find I_D for an n-channel transistor that has doping concentrations of $N_D = 10^{25}$, $N_A = 10^{22}$, $\mu_n C_{ox} = 92\ \mu A/V^2$, $W/L = 20\ \mu m/2\ \mu m$, $V_{GS} = 1.2\ V$, $V_{tn} = 0.8\ V$, and $V_{DS} = V_{eff}$. Assuming λ remains constant, estimate the new value of I_D if V_{DS} is increased by 0.5 V.

Solution

From (1.69), we have

$$k_{ds} = \sqrt{\frac{2 \times 11.8 \times 8.854 \times 10^{-12}}{1.6 \times 10^{-19} \times 10^{22}}} = 362 \times 10^{-9}\ m/\sqrt{V}$$

which is used in (1.72) to find λ as

$$\lambda = \frac{362 \times 10^{-9}}{2 \times 2 \times 10^{-6} \times \sqrt{0.9}} = 95.3 \times 10^{-3}\ V^{-1}$$

Using (1.71), we find for $V_{DS} = V_{eff} = 0.4\ V$,

$$I_{D1} = \left(\frac{92 \times 10^{-6}}{2}\right)\left(\frac{20}{2}\right)(0.4)^2(1) = 73.6\ \mu A$$

In the case where $V_{DS} = V_{eff} + 0.5\ V = 0.9\ V$, we have

$$I_{D2} = 73.6\ \mu A \times (1 + \lambda \times 0.5) = 77.1\ \mu A$$

Note that this example shows almost a 5 percent increase in drain current for a 0.5 V increase in drain-source voltage.

Body Effect

The large-signal equations in the preceding section were based on the assumption that the source voltage was the same as the substrate (i.e., bulk) voltage. However, often the source and substrate can be at different voltage potentials. In these situations, a second-order effect exists that is modelled as an increase in the threshold voltage, V_{tn}, as the source-to-substrate reverse-bias voltage increases. This effect, typically called the *body effect,* is more important for transistors in a well of a CMOS process where the substrate doping is higher. It should be noted that the body effect is often important in analog circuit designs and should not be ignored without consideration.

To account for the body effect, it can be shown (see Appendix at the end of this chapter) that the threshold voltage of an n-channel transistor is now given by

$$V_{tn} = V_{tn0} + \gamma(\sqrt{V_{SB} + |2\phi_F|} - \sqrt{|2\phi_F|}) \tag{1.73}$$

where V_{tn0} is the threshold voltage with zero V_{SB} (i.e., source-to-substrate voltage),

and

$$\gamma = \frac{\sqrt{2qN_A K_s \varepsilon_0}}{C_{ox}} \tag{1.74}$$

The factor γ is often called the *body-effect constant* and has units of $\sqrt{V}$. Notice that γ is proportional to $\sqrt{N_A}$,[10] so the body effect is larger for transistors in a well where typically the doping is higher than the substrate of the microcircuit.

p-Channel Transistors

All of the preceding equations have been presented for n-channel enhancement transistors. *In the case of p-channel transistors, these equations can also be used if a negative sign is placed in front of every voltage variable.* Thus, V_{GS} becomes V_{SG}, V_{DS} becomes V_{SD}, V_{tn} becomes $-V_{tp}$, and so on. The condition required for conduction is now $V_{SG} > V_{tp}$, where V_{tp} is now a negative quantity for an enhancement p-channel transistor.[11] The requirement on the source-drain voltage for a p-channel transistor to be in the active region is $V_{SD} > V_{SG} + V_{tp}$. The equations for I_D, in both regions, remain unchanged, because all voltage variables are squared, resulting in positive hole current flow from the source to the drain in p-channel transistors. For n-channel depletion transistors, the only difference is that $V_{td} < 0$ V. A typical value might be $V_{td} = -2$ V.

Small-Signal Modelling in the Active Region

The most commonly used small-signal model for a MOS transistor operating in the active region is shown in Fig. 1.17. We first consider the dc parameters in which all the capacitors are ignored (i.e., replaced by open circuits). This leads to the low-frequency, small-signal model shown in Fig. 1.18. The voltage-controlled current source, $g_m v_{gs}$, is the most important component of the model, with the transistor transconductance g_m defined as

$$g_m = \frac{\partial I_D}{\partial V_{GS}} \tag{1.75}$$

In the active region, we use (1.67), which is repeated here for convenience,

$$I_D = \frac{\mu_n C_{ox}}{2}\left(\frac{W}{L}\right)(V_{GS} - V_{tn})^2 \tag{1.76}$$

10. For an n-channel transistor. For a p-channel transistor, γ is proportional to N_D.

11. It is possible to realize depletion p-channel transistors, but these are of little value and seldom worth the extra processing involved. Depletion n-channel transistors are also seldom encountered in CMOS microcircuits, although they might be worth the extra processing involved in some applications, especially if they were in a well.

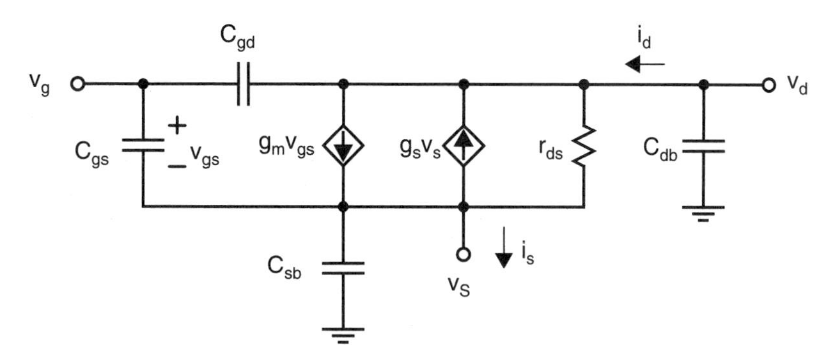

Fig. 1.17 The small-signal model for a MOS transistor in the active region.

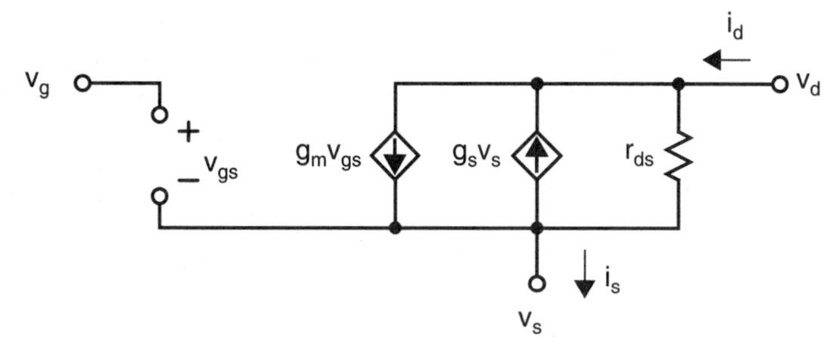

Fig. 1.18 The low-frequency, small-signal model for an active MOS transistor.

and we apply the derivative shown in (1.75) to obtain

$$g_m = \frac{\partial I_D}{\partial V_{GS}} = \mu_n C_{ox} \frac{W}{L}(V_{GS} - V_{tn}) = \mu_n C_{ox} \frac{W}{L} V_{eff} \tag{1.77}$$

or equivalently,

$$g_m = \mu_n C_{ox} \frac{W}{L} V_{eff} \tag{1.78}$$

where the effective gate-source voltage, V_{eff}, is defined as $V_{eff} \equiv V_{GS} - V_{tn}$. Thus, we see that the transconductance of a MOS transistor is directly proportional to V_{eff}.

Sometimes it is desirable to express g_m in terms of I_D rather than V_{GS}. From (1.76), we have

$$V_{GS} = V_{tn} + \sqrt{\frac{2I_D}{\mu_n C_{ox}(W/L)}} \tag{1.79}$$

The second term in (1.79) is the effective gate-source voltage, V_{eff}, where

$$V_{eff} = V_{GS} - V_{tn} = \sqrt{\frac{2I_D}{\mu_n C_{ox}(W/L)}} \tag{1.80}$$

Substituting (1.80) in (1.78) results in an alternate expression for g_m.

$$g_m = \sqrt{2\mu_n C_{ox} \frac{W}{L} I_D} \tag{1.81}$$

Thus, the transistor transconductance is proportional to $\sqrt{I_D}$ for a MOS transistor, whereas it is proportional to I_C for a BJT.

A third expression for g_m is found by rearranging (1.81) and then using (1.80) to obtain

$$g_m = \frac{2I_D}{V_{eff}} \tag{1.82}$$

Note that this expression is independent of $\mu_n C_{ox}$ and W/L, and it relates the transconductance to the ratio of drain current to effective gate-source voltage. This simple relationship can be quite useful during an initial circuit design.

The second voltage-controlled current-source in Fig. 1.18, shown as $g_s v_s$, models the body effect on the small-signal drain current, i_d. When the source is connected to small-signal ground, or when its voltage does not change appreciably, then this current source can be ignored. When the body effect cannot be ignored, we have

$$g_s = \frac{\partial I_D}{\partial V_{SB}} = \frac{\partial I_D}{\partial V_{tn}} \frac{\partial V_{tn}}{\partial V_{SB}} \tag{1.83}$$

From (1.76) we have

$$\frac{\partial I_D}{\partial V_{tn}} = -\mu_n C_{ox}\left(\frac{W}{L}\right)(V_{GS} - V_{tn}) = -g_m \tag{1.84}$$

Using (1.73), which gives V_{tn} as

$$V_{tn} = V_{tn0} + \gamma(\sqrt{V_{SB} + |2\phi_F|} - \sqrt{|2\phi_F|}) \tag{1.85}$$

we have

$$\frac{\partial V_{tn}}{\partial V_{SB}} = \frac{\gamma}{2\sqrt{V_{SB} + |2\phi_F|}} \tag{1.86}$$

The negative sign of (1.84) is eliminated by subtracting the current $g_s v_s$ from the major component of the drain current, $g_m v_{gs}$, as shown in Fig. 1.18. Thus, using (1.84) and (1.86), we have

$$g_s = \frac{\gamma g_m}{2\sqrt{V_{SB} + |2\phi_F|}} \tag{1.87}$$

Note that although g_s is nonzero for $V_{SB} = 0$, if the source is connected to the bulk, ΔV_{SB} is zero, and so the effect of g_s does not need to be taken into account. However, if the source happens to be biased at the same potential as the bulk but is not directly connected to it, then the effect of g_s should be taken into account since ΔV_{SB} is not necessarily zero.

The resistor, r_{ds}, shown in Fig. 1.18, accounts for the finite output impedance (i.e., it models the channel-length modulation and its effect on the drain current due to changes in V_{DS}). Using (1.71), repeated here for convenience,

$$I_D = \frac{\mu_n C_{ox}}{2}\left(\frac{W}{L}\right)(V_{GS} - V_{tn})^2[1 + \lambda(V_{DS} - V_{eff})] \tag{1.88}$$

we have

$$\frac{1}{r_{ds}} = g_{ds} = \frac{\partial I_D}{\partial V_{DS}} = \lambda\left(\frac{\mu_n C_{ox}}{2}\right)\left(\frac{W}{L}\right)(V_{GS} - V_{tn})^2 = \lambda I_{D\text{-sat}} \cong \lambda I_D \tag{1.89}$$

where the approximation assumes λ is small, such that we can approximate the drain bias current as being the same as $I_{D\text{-sat}}$. Thus,

$$r_{ds} \cong \frac{1}{\lambda I_D} \tag{1.90}$$

where

$$\lambda = \frac{k_{ds}}{2L\sqrt{V_{DS} + (-V_{eff}) + \Phi_0}} \tag{1.91}$$

and

$$k_{ds} = \sqrt{\frac{2K_s\varepsilon_0}{qN_A}} \tag{1.92}$$

It should be noted here that (1.90) is often empirically adjusted to take into account second-order effects.

EXAMPLE 1.9

Derive the low-frequency model parameters for an n-channel transistor that has doping concentrations of $N_D = 10^{25}$, $N_A = 10^{22}$, $\mu_n C_{ox} = 92\ \mu A/V^2$, $W/L = 20\ \mu m/2\ \mu m$, $V_{GS} = 1.2$ V, $V_{tn} = 0.8$ V, and $V_{DS} = V_{eff}$. Assume $\gamma = 0.5\ \sqrt{V}$ and $V_{SB} = 0.5$ V. What is the new value of r_{ds} if the drain-source voltage is increased by 0.5 V?

Solution

Since these parameters are the same as in Example 1.8, we have

$$g_m = \frac{2I_D}{V_{eff}} = \frac{2 \times 73.6\ \mu A}{0.4\ V} = 0.368\ mA/V$$

and from (1.87), we have

$$g_s = \frac{0.5 \times 0.368 \times 10^{-3}}{2\sqrt{0.5 + 1.8}} = 0.061 \text{ mA/V}$$

Note that this source-bulk transconductance value is about 1/6 that of the gate-source transconductance.

For r_{ds}, we use (1.90) to find

$$r_{ds} = \frac{1}{95.3 \times 10^{-3} \times 73.6 \times 10^{-6}} = 143 \text{ k}\Omega$$

At this point, it is interesting to calculate the gain $g_m r_{ds} = 52.6$, which is the largest voltage gain this single transistor can achieve for these operating bias conditions. As we will see, this gain of 52.6 is much smaller than the corresponding single-transistor gain in a bipolar transistor.

Recalling that $V_{eff} = 0.4$ V, if V_{DS} is increased to 0.9 V, the new value for λ is

$$\lambda = \frac{362 \times 10^{-9}}{2(2 \times 10^{-6})\sqrt{1.4}} = 76.4 \times 10^{-3} \text{ V}^{-1}$$

resulting in a new value of r_{ds} given by

$$r_{ds} = \frac{1}{\lambda I_{D2}} = \frac{1}{76.4 \times 10^{-3} \times 77.1 \ \mu\text{A}} = 170 \text{ k}\Omega$$

An alternate low-frequency model, known as a T model, is shown in Fig. 1.19. This *T model* can often result in simpler equations and is most often used by experienced designers for a quick analysis. At first glance, it might appear that this model allows for nonzero gate current, but a quick check confirms that the drain current must always equal the source current, and, therefore, the gate current must always be zero. For this reason, when using the T model, one assumes from the beginning that the gate current is zero.

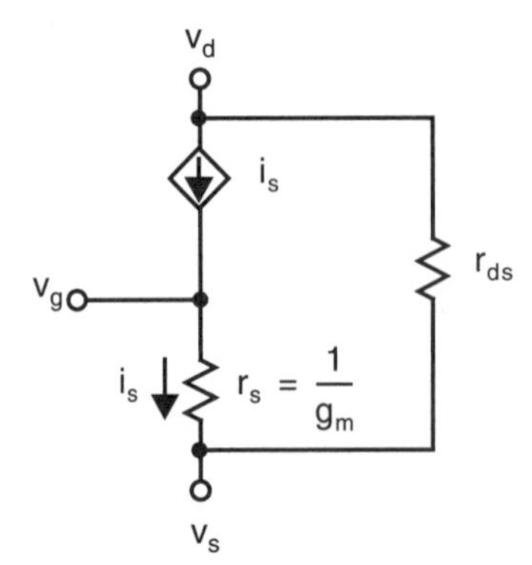

Fig. 1.19 The small-signal, low-frequency T model for an active MOS transistor (the body effect is not modelled).

EXAMPLE 1.10

Find the T model parameter, r_s, for the transistor in Example 1.9.

Solution

The value of r_s is simply the inverse of g_m, resulting in

$$r_s = \frac{1}{g_m} = \frac{1}{0.368 \times 10^{-3}} = 2.72\ \text{k}\Omega$$

The value of r_{ds} remains the same, either 143 kΩ or 170 kΩ, depending on the drain-source voltage.

Most of the capacitors in the small-signal model are related to the physical transistor. Shown in Fig. 1.20 is a cross section of a MOS transistor, where the parasitic capacitances are shown at the appropriate locations. The largest capacitor in Fig. 1.20 is C_{gs}. This capacitance is primarily due to the change in channel charge as a result of a change in V_{GS}. It can be shown [Tsividis, 1987] that C_{gs} is approximately given by

$$C_{gs} \cong \frac{2}{3} W L C_{ox} \tag{1.93}$$

When accuracy is important, an additional term should be added to (1.93) to take into account the overlap between the gate and source junction, which should include the *fringing capacitance* (fringing capacitance is due to boundary effects). This additional component is given by

$$C_{ov} = W L_{ov} C_{ox} \tag{1.94}$$

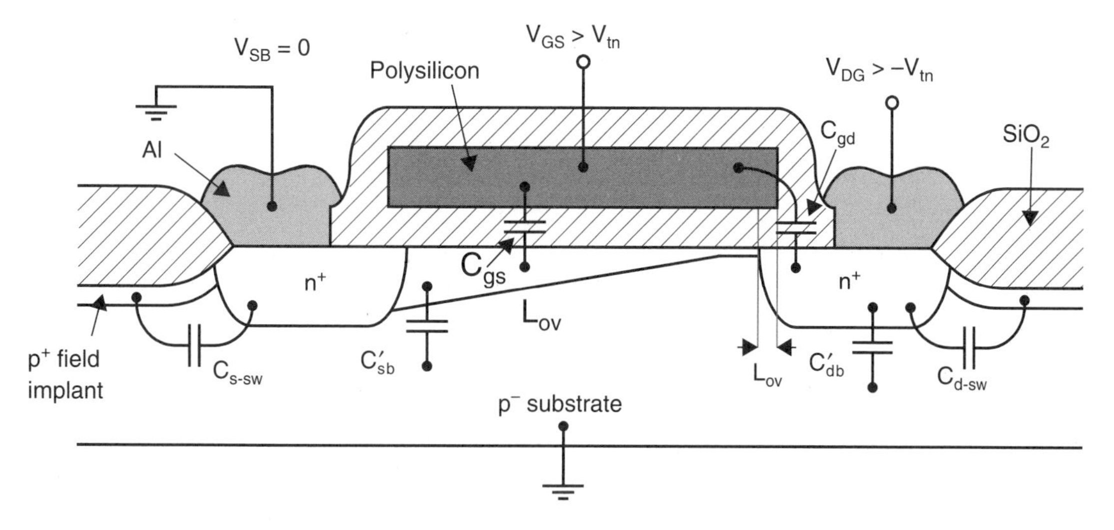

Fig. 1.20 A cross section of an n-channel MOS transistor showing the small-signal capacitances.

where L_{ov} is the overlap distance and is usually empirically derived. Thus,

$$C_{gs} = WC_{ox}\left(\frac{2}{3}L + L_{ov}\right) \tag{1.95}$$

when higher accuracy is needed.

The next largest capacitor in Fig. 1.20 is C'_{sb}, the capacitor between the source and the substrate. This capacitor is due to the depletion capacitance of the reverse-biased source junction, and it includes the channel-to-bulk capacitance (assuming the transistor is on). Its size is given by

$$C'_{sb} = (A_s + A_{ch})C_{js} \tag{1.96}$$

where A_s is the area of the source junction, A_{ch} is the area of the channel (i.e., WL) and C_{js} is the depletion capacitance of the source junction, given by

$$C_{js} = \frac{C_{j0}}{\sqrt{1 + \frac{V_{SB}}{\Phi_0}}} \tag{1.97}$$

Note that the total area of the effective source includes the original area of the junction (when no channel is present) plus the effective area of the channel.

The depletion capacitance of the drain is smaller because it does not include the channel area. Here, we have

$$C'_{db} = A_d C_{jd} \tag{1.98}$$

where

$$C_{jd} = \frac{C_{j0}}{\sqrt{1 + \frac{V_{DB}}{\Phi_0}}} \tag{1.99}$$

and A_d is the area of the drain junction.

The capacitance C_{gd}, sometimes called the *Miller-capacitor,* is important when the transistor is being used in circuits with large voltage gain. C_{gd} is primarily due to the overlap between the gate and the drain and fringing capacitance. Its value is given by

$$C_{gd} = C_{ox}WL_{ov} \tag{1.100}$$

where, once again, L_{ov} is usually empirically derived.

Two other capacitors are often important in integrated circuits. These are the source and drain *sidewall capacitances,* $C_{s\text{-}sw}$ and $C_{d\text{-}sw}$. These capacitances can be large because of some highly doped p^+ regions under the thick field oxide called *field implants.* The major reason these regions exist is to ensure there is no leakage current between transistors. Because they are highly doped and they lie beside the highly doped source and drain junctions, the sidewall capacitances can result in large additional capacitances that must be taken into account in determining C_{sb} and C_{db}. The sidewall capacitances are especially important in modern technologies as dimensions shrink. For the source, the sidewall capacitance is given by

$$C_{s\text{-}sw} = P_s C_{j\text{-}sw} \tag{1.101}$$

where P_s is the length of the perimeter of the source junction, excluding the side adjacent to the channel, and

$$C_{j\text{-}sw} = \frac{C_{j\text{-}sw0}}{\sqrt{1 + \frac{V_{SB}}{\Phi_0}}} \tag{1.102}$$

It should be noted that $C_{j\text{-}sw0}$, the sidewall capacitance per unit length at 0-V bias voltage, can be quite large because the field implants are heavily doped.

The situation is similar for the drain sidewall capacitance, $C_{d\text{-}sw}$,

$$C_{d\text{-}sw} = P_d C_{j\text{-}sw} \tag{1.103}$$

where P_d is the drain perimeter excluding the portion adjacent to the gate.

Finally, the source-bulk capacitance, C_{sb}, is given by

$$C_{sb} = C'_{sb} + C_{s\text{-}sw} \tag{1.104}$$

with the drain-bulk capacitance, C_{db}, given by

$$C_{db} = C'_{db} + C_{d\text{-}sw} \tag{1.105}$$

EXAMPLE 1.11

An n-channel transistor is modelled as having the following capacitance parameters: $C_j = 2.4 \times 10^{-4}\ \text{pF}/(\mu\text{m})^2$, $C_{j\text{-}sw} = 2.0 \times 10^{-4}\ \text{pF}/\mu\text{m}$, $C_{ox} = 1.9 \times 10^{-3}\text{pF}/(\mu\text{m})^2$, $C_{gs\text{-}ov} = C_{gd\text{-}ov} = 2.0 \times 10^{-4}\ \text{pF}/\mu\text{m}$. Find the capacitances C_{gs}, C_{gd}, C_{db}, and C_{sb} for a transistor having $W = 100\ \mu\text{m}$ and $L = 2\ \mu\text{m}$. Assume the source and drain junctions extend $4\ \mu\text{m}$ beyond the gate, so that the source and drain areas are $A_s = A_d = 400\ (\mu\text{m})^2$ and the perimeter of each is $P_s = P_d = 108\ \mu\text{m}$.

Solution

We calculate the various capacitances as follows:

$$C_{gs} = \left(\frac{2}{3}\right) WLC_{ox} + C_{gs\text{-}ov} \times W = 0.27\ \text{pF}$$

$$C_{gd} = C_{gd\text{-}ov} \times W = 0.02\ \text{pF}$$

$$C_{sb} = C_j(A_s + WL) + (C_{j\text{-}sw} \times P_s) = 0.17\ \text{pF}$$

$$C_{db} = (C_j \times A_d) + (C_{j\text{-}sw} \times P_d) = 0.12\ \text{pF}$$

Note that the source-bulk and drain-bulk capacitances are significant compared to the gate-source capacitance. Thus, for high-speed circuits, it is important to keep the areas and perimeters of drain and source junctions as small as possible (possibly by sharing junctions between transistors, as seen in the next chapter).

Small-Signal Modelling in the Triode and Cutoff Regions

The low-frequency, small-signal model of a MOS transistor in the triode region (which is sometimes referred to as the linear region) is a resistor. Using (1.65), the large-signal equation for I_D in the triode region,

$$I_D = \mu_n C_{ox}\left(\frac{W}{L}\right)\left[(V_{GS} - V_{tn})V_{DS} - \frac{V_{DS}^2}{2}\right] \quad (1.106)$$

results in

$$\frac{1}{r_{ds}} = g_{ds} = \frac{dI_D}{dV_{DS}} = \mu_n C_{ox}\left(\frac{W}{L}\right)(V_{GS} - V_{tn} - V_{DS}) \quad (1.107)$$

where r_{ds} is the small-signal drain-source resistance (and g_{ds} is the conductance). For the common case of V_{DS} near zero, we have

$$g_{ds} = \frac{1}{r_{ds}} = \mu_n C_{ox}\left(\frac{W}{L}\right)(V_{GS} - V_{tn}) = \mu_n C_{ox}\left(\frac{W}{L}\right)V_{eff} \quad (1.108)$$

which is similar to the I_D-versus-V_{DS} relationship given earlier in (1.60).

EXAMPLE 1.12

For the transistor of Example 1.9, find the triode model parameters when V_{DS} is near zero.

Solution

From (1.108), we have

$$g_{ds} = 92 \times 10^{-6} \times \left(\frac{20}{2}\right) \times 0.4 = 0.368 \text{ mA/V}$$

Note that this conductance value is the same as the transconductance of the transistor, g_m, in the active region. The resistance, r_{ds}, is simply $1/g_{ds}$, resulting in $r_{ds} = 2.72 \text{ k}\Omega$.

The accurate modelling of the high-frequency operation of a transistor in the triode region is nontrivial (even with the use of a computer simulation). A moderately accurate model is shown in Fig. 1.21, where the gate-to-channel capacitance

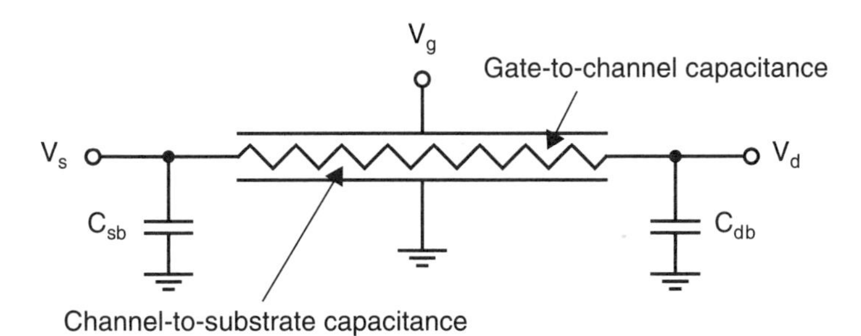

Fig. 1.21 A distributed RC model for a transistor in the active region.

and the channel-to-substrate capacitance are modelled as distributed elements. However, the I-V relationships of the distributed RC elements are highly nonlinear because the junction capacitances of the source and drain are nonlinear depletion capacitances, as is the channel-to-substrate capacitance. Also, if V_{DS} is not small, then the channel resistance per unit length should increase as one moves closer to the drain. This model is much too complicated for use in hand analysis.

A simplified model often used for small V_{DS} is shown in Fig. 1.22, where the resistance, r_{ds}, is given by (1.108). Here, the gate-to-channel capacitance has been evenly divided between the source and drain nodes,

$$C_{gs} = C_{gd} = \frac{A_{ch}C_{ox}}{2} = \frac{WLC_{ox}}{2} \tag{1.109}$$

Note that this equation ignores the gate-to-junction overlap capacitances, as given by (1.94), which should be taken into account when accuracy is very important. The channel-to-substrate capacitance has also been divided in half and shared between the source and drain junctions. Each of these capacitors should be added to the junction-to-substrate capacitance and the junction-sidewall capacitance at the appropriate

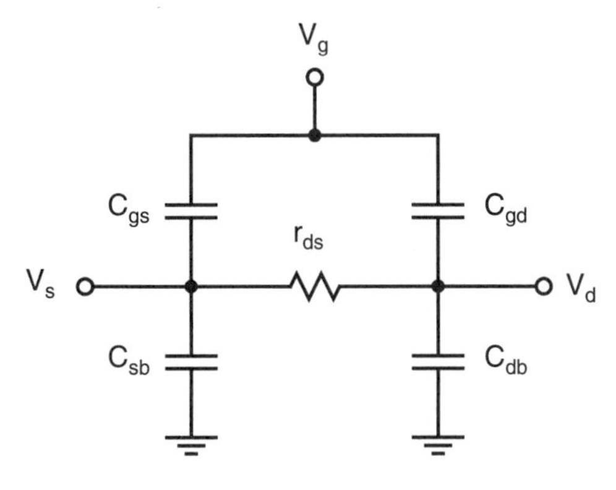

Fig. 1.22 A simplified triode-region model valid for small V_{DS}.

node. Thus, we have

$$C_{sb\text{-}0} = C_{j0}\left(A_s + \frac{A_{ch}}{2}\right) + C_{j\text{-}sw0}P_s \tag{1.110}$$

and

$$C_{db\text{-}0} = C_{j0}\left(A_d + \frac{A_{ch}}{2}\right) + C_{j\text{-}sw0}P_d \tag{1.111}$$

Also,

$$C_{sb} = \frac{C_{sb\text{-}0}}{\sqrt{1 + \frac{V_{sb}}{\Phi_0}}} \tag{1.112}$$

and

$$C_{db} = \frac{C_{db\text{-}0}}{\sqrt{1 + \frac{V_{sb}}{\Phi_0}}} \tag{1.113}$$

It might be noted that C_{sb} is often comparable in size to C_{gs} due to its larger area and the sidewall capacitance.

When the transistor turns off, the model changes considerably. A reasonable model is shown in Fig. 1.23. Perhaps the biggest difference is that r_{ds} is now infinite. Another major difference is that C_{gs} and C_{gd} are now much smaller. Since the channel has disappeared, these capacitors are now due to only overlap and fringing capacitance. Thus, we have

$$C_{gs} = C_{gd} = WL_{ov}C_{ox} \tag{1.114}$$

However, the reduction of C_{gs} and C_{gd} does not mean that the total gate capacitance is necessarily smaller. We now have a "new" capacitor, C_{gb}, which is the gate-

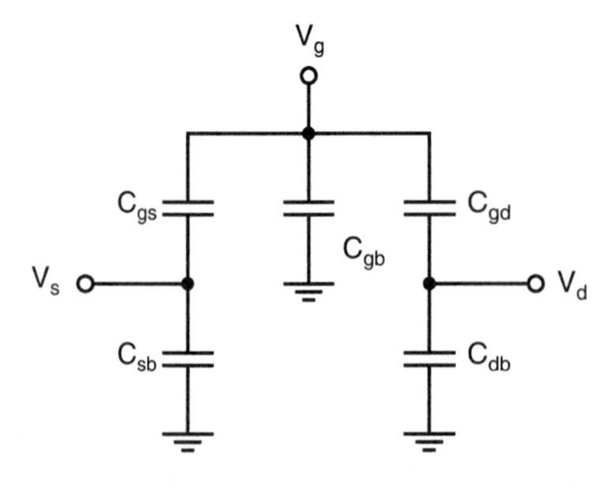

Fig. 1.23 A small-signal model for a MOSFET that is turned off.

to-substrate capacitance. This capacitor is highly nonlinear and dependent on the gate voltage. If the gate voltage has been very negative for some time and the gate is accumulated, then we have

$$C_{gb} = A_{ch}C_{ox} = WLC_{ox} \tag{1.115}$$

If the gate-to-source voltage is around 0 V, then C_{gb} is equal to C_{ox} in series with the channel-to-bulk depletion capacitance and is considerably smaller, especially when the substrate is lightly doped. Another case where C_{gb} is small is just after a transistor has been turned off, before the channel has had time to accumulate. Because of the complicated nature of correctly modelling C_{gb} when the transistor is turned off, equation (1.115) is usually used for hand analysis as a worst-case estimate.

The capacitors C_{sb} and C_{db} are also smaller when the channel is not present. We now have

$$C_{sb\text{-}0} = A_sC_{j0} \tag{1.116}$$

and

$$C_{db\text{-}0} = A_dC_{j0} \tag{1.117}$$

1.3 ADVANCED MOS MODELLING

In this section, we look at three advanced modelling concepts that a microcircuit designer is likely to encounter—short-channel effects, subthreshold operation, and leakage currents.

Short-Channel Effects

A number of short-channel effects degrade the operation of MOS transistors as device dimensions are scaled down. These effects include mobility degradation, reduced output impedance, and hot-carrier effects (such as oxide trapping and substrate currents). These short-channel effects will be briefly described here. For more detailed modelling of short-channel effects, see [Wolf, 1995].

Transistors that have short channel lengths and large electric fields experience a degradation in the effective mobility of their carriers due to several factors. One of these factors is the large lateral electric field (which has a vector in a direction perpendicular from the gate into the silicon) caused by large gate voltages and short channel lengths. This large lateral field causes the effective channel depth to change and also causes more electron collisions, thereby lowering the effective mobility. Another factor causing this degradation is that, due to large electric fields, carrier velocity begins to saturate. A first-order approximation that models this carrier-velocity saturation for electrons is given by

$$\upsilon_d \cong \frac{\mu_n E}{1 + E/E_c} \tag{1.118}$$

where E is the electric field and E_c is the critical electrical field, which might be on the order of 1.5×10^6 V/m. Using this equation in the derivation of the I_D-V_{eff}

characteristics of a MOS transistor, it can be shown [Gray, 1993] that the drain current is now given by

$$I_D = \frac{\mu_n C_{ox}}{2[1 + \theta V_{eff}]} \frac{W}{L} V_{eff}^2 \tag{1.119}$$

where $\theta = 1/(LE_c)$ and, for a 0.8-μm technology, might have a typical value of 0.6 V^{-1}. It can be shown that this mobility degradation is equivalent to a finite series source resistance given by

$$R_{SX} = \frac{1}{E_C} \frac{1}{\mu_n C_{ox}} \frac{1}{W} \tag{1.120}$$

For $\mu_n C_{ox} = 90\ \mu A/V^2$, this resistance might be on the order of 6 kΩ per μm of width (again, for a 0.8-μm-long transistor). This equivalent series source resistance is typically larger than the physical source resistance. This saturation causes the square-law characteristic of the current-voltage relationship to be inaccurate, and the true relationship will be somewhere between linear and square. In many voltage-to-current conversion circuits that rely on the square-law characteristic, this inaccuracy can be a major source of error. Taking channel lengths larger than the minimum allowed helps to minimize this degradation.

Transistors with short channel lengths also experience a reduced output impedance because depletion region variations at the drain end (which affect the effective channel length) have an increased proportional effect on the drain current. In addition, a phenomenon known as drain-induced barrier lowering (DIBL) effectively lowers V_t as V_{DS} is increased, thereby further lowering the output impedance of a short-channel device. This lower output impedance is the main reason that cascode current mirrors are becoming increasingly popular.

Another important short-channel effect is due to *hot carriers*. These high-velocity carriers can cause harmful effects, such as the generation of electron-hole pairs by impact ionization and avalanching. These extra electron-hole pairs can cause currents to flow from the drain to the substrate, as shown in Fig. 1.24. This effect can be mod-

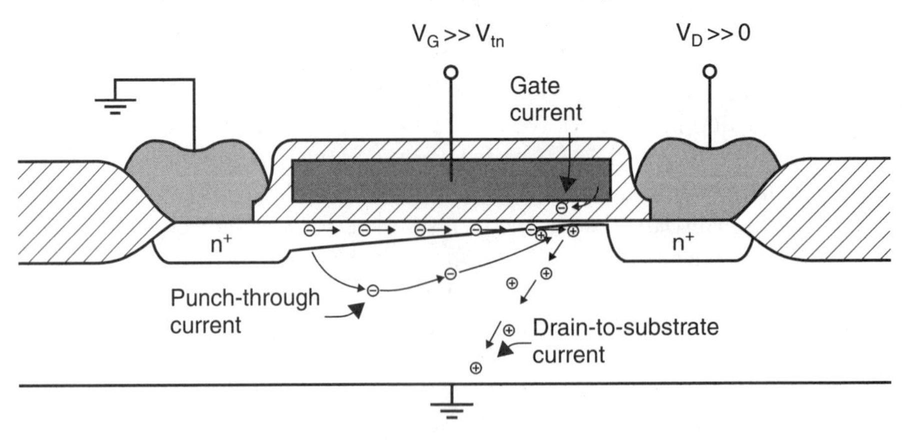

Fig. 1.24 Drain-to-substrate current caused by electron-hole pairs generated by impact ionization at drain end of channel.

elled by a finite drain-to-ground impedance. As a result, this effect is one of the major limitations on achieving very high output impedances of cascode current sources. In addition, this current flow can cause voltage drops across the substrate and possibly cause latch-up, as the next section describes.

Another hot-carrier effect occurs when electrons gain energies high enough so they can tunnel into and possibly through the thin gate oxide. Thus, this effect can cause dc gate currents. However, often more harmful is the fact that any charge trapped in the oxide will cause a shift in transistor threshold voltage. As a result, hot carriers are one of the major factors limiting the long-term reliability of MOS transistors.

A third hot-carrier effect occurs when electrons with enough energy *punch through* from the source to the drain. As a result, these high-energy electrons are no longer limited by the drift equations governing normal conduction along the channel. This mechanism is somewhat similar to punch-through in a bipolar transistor, where the collector depletion region extends right through the base region to the emitter. In a MOS transistor, the channel length becomes effectively zero, resulting in unlimited current flow (except for the series source and drain impedances, as well as external circuitry). This effect is an additional cause of lower output impedance and possibly transistor breakdown.

It should be noted that all of the hot-carrier effects just described are more pronounced for n-channel transistors than for their p-channel counterparts because electrons have larger velocities than holes.

Finally, it should be noted that short-channel transistors have much larger subthreshold currents than long-channel devices.

Subthreshold Operation

The device equations presented for MOS transistors in the preceding sections are all based on the assumption that V_{eff} (i.e., $V_{GS} - V_t$) is greater than about 100 mV and the device is in strong inversion. When this is not the case, the accuracy of the square-law equations is poor. If $V_{eff} < -100$ mV, the transistor is in weak inversion and is said to be operating in the *subthreshold region*. In this region, the transistor is more accurately modelled by an exponential relationship between its control voltage and current, somewhat similar to a bipolar transistor. In the subthreshold region, the drain current is approximately given by the exponential relationship [Geiger, 1990]

$$I_D \cong I_{D0}\left(\frac{W}{L}\right)e^{(qV_{GS}/nkT)} \tag{1.121}$$

where

$$n = \frac{C_{ox} + C_{depl}}{C_{ox}} \cong 1.5 \tag{1.122}$$

and it has been assumed that $V_S = 0$ and $V_{DS} > 75$ mV. The constant I_{D0} might be around 20 nA.

Although the transistors have an exponential relationship in this region, the transconductances are still small because of the small bias currents, and the transistors are slow because of small currents for charging and discharging capacitors. In addition, matching between transistors suffers because it now strongly depends on transistor-threshold-voltage matching. Normally, transistors are not operated in the subthreshold region, except in very low-frequency and low-power applications.

Leakage Currents

An important second-order device limitation in some applications is the leakage current of the junctions. For example, this leakage can be important in estimating the maximum time a sample-and-hold circuit or a dynamic memory cell can be left in hold mode. The leakage current of a reverse-biased junction (not close to breakdown) is approximately given by

$$I_{lk} \cong \frac{qA_j n_i}{2\tau_0} x_d \tag{1.123}$$

where A_j is the junction area, n_i is the intrinsic concentration of carriers in undoped silicon, τ_0 is the effective minority carrier lifetime, and x_d is the thickness of the depletion region. τ_0 is given by

$$\tau_0 \cong \frac{1}{2}(\tau_n + \tau_p) \tag{1.124}$$

where τ_n and τ_p are the electron and hole lifetimes. Also, x_d is given by

$$x_d = \sqrt{\frac{2K_s\varepsilon_0}{qN_A}(\Phi_0 + V_r)} \tag{1.125}$$

and n_i is given by

$$n_i \cong \sqrt{N_C N_V}\ e^{(-E_g)/(kT)} \tag{1.126}$$

where N_C and N_V are the densities of states in the conduction and valence bands and E_g is the difference in energy between the two bands.

Since the intrinsic concentration, n_i, is a strong function of temperature (it approximately doubles for every temperature increase of 11 °C for silicon), the leakage current is also a strong function of temperature. Roughly speaking, the leakage current also doubles for every 11 °C rise in temperature. Thus, the leakage current at higher temperatures is much larger than at room temperature. This leakage current imposes a maximum time on how long a dynamically charged signal can be maintained in a high impedance state.

1.4 BIPOLAR-JUNCTION TRANSISTORS

In the early electronic years, the majority of microcircuits were realized using bipolar-junction transistors (BJTs). However, in the late 1970s, microcircuits that used MOS transistors began to dominate the industry, with BJT microcircuits remaining popular for high-speed applications. More recently, bipolar CMOS (BiCMOS) technologies, where both bipolar and MOS transistors are realized in the same microcircuit, have grown in popularity. BiCMOS technologies are particularly attractive for mixed analog-digital applications, and thus it is important for an analog designer to become familiar with bipolar devices.

Modern bipolar transistors can have unity-gain frequencies as high as 15 to 45 GHz or more, compared to unity-gain frequencies of only 1 to 4 GHz for MOS transistors that use a technology with similar lithography resolution. Unfortunately, in bipolar transistors, the *base* control terminal has a nonzero input current when the transistor is conducting current (from the collector to the emitter for an npn transistor; from the emitter to the collector for a pnp transistor). Fortunately, at low frequencies, the base current is much smaller than the collector-to-emitter current—it may be only 1/100 of the collector current for an npn transistor. For lateral pnp transistors, the base current may be as large as 1/20 of the emitter-to-collector current.

A typical cross section of an npn bipolar-junction transistor is shown in Fig. 1.25. Although this structure looks quite complicated, it corresponds approximately to the equivalent structure shown in Fig. 1.26. In a good BJT transistor, the width of the base

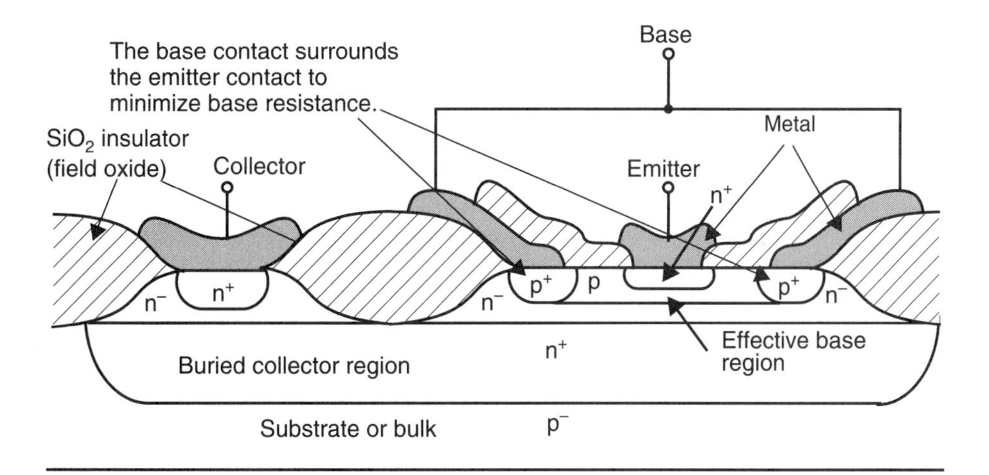

Fig. 1.25 A cross section of an npn bipolar-junction transistor.

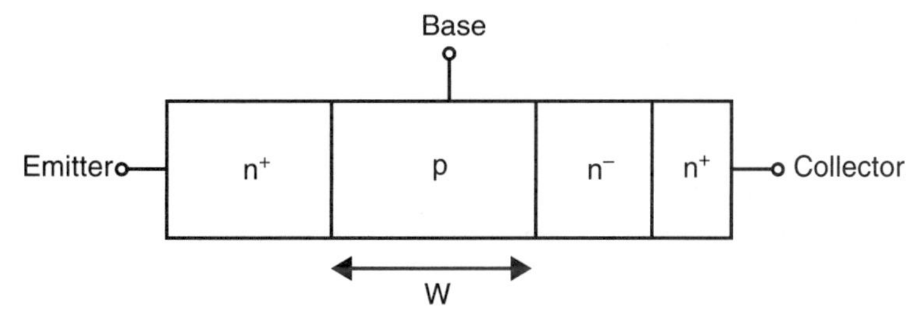

Fig. 1.26 A simplified structure of an npn transistor.

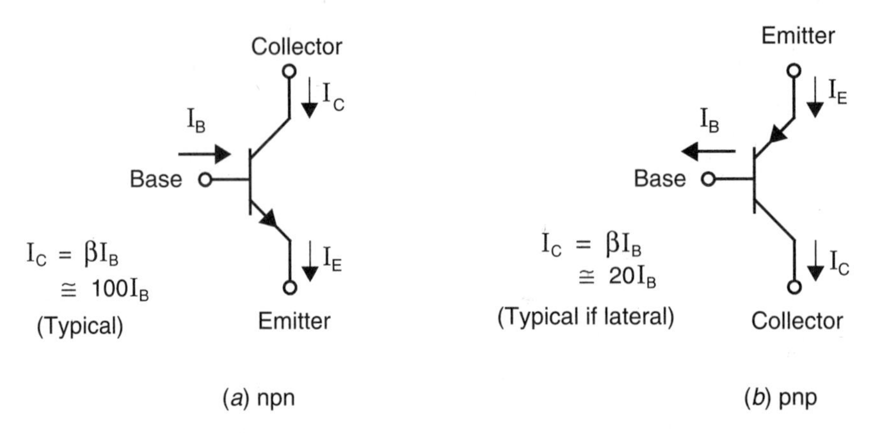

Fig. 1.27 The symbols representing (*a*) an npn bipolar-junction transistor and (*b*) a pnp bipolar-junction transistor.

region, W, is small (typically, less than 1 μm). Also, as we will see, the base must be more lightly doped than the emitter.

The circuit symbols used to represent npn and pnp transistors are shown in Fig. 1.27.

Basic Operation

To understand the operation of bipolar transistors, we consider here an npn transistor with the emitter connected to ground, as shown in Fig. 1.28. If the base voltage, V_B, is less than about 0.5 V, the transistor will be cut off, and no current will flow. We will see that when the base-emitter pn junction becomes forward biased, current will start to flow from the base to the emitter, but, partly because the base width is small, a much larger proportional current will flow from the collector to the emitter. Thus, the npn transistor can be considered a current amplifier at low frequencies. In other

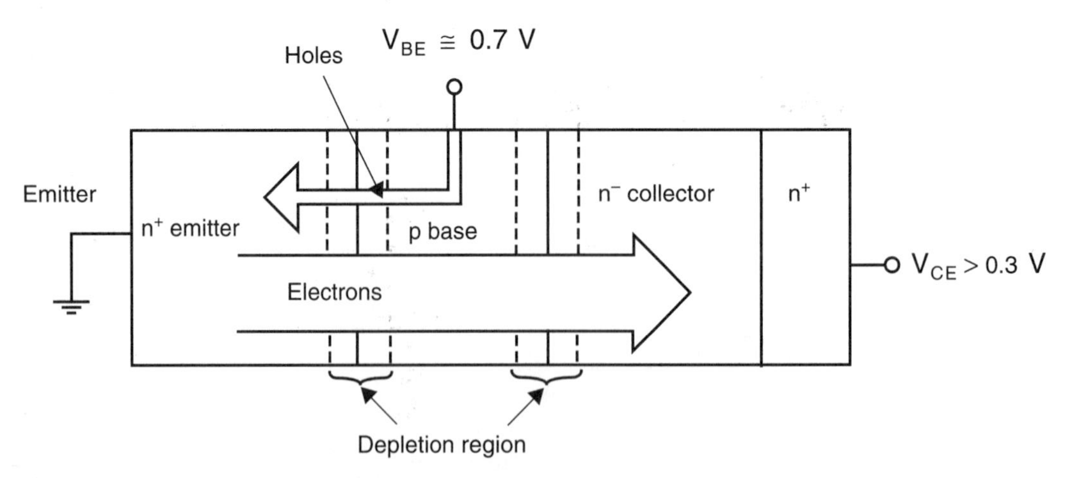

Fig. 1.28 Various components of the currents of an npn transistor.

words, if the transistor is not cut off and the collector-base junction is reverse biased, a small base current controls a much larger collector-emitter current.

A simplified overview of how an npn transistor operates follows: When the base-emitter junction becomes forward biased, it starts to conduct, similar to any forward-biased junction. The current consists of majority carriers from the base (in this case, holes) and majority carriers from the emitter (in this case, electrons) diffusing across the junction. Because the emitter is more heavily doped than the base, there are many more electrons injected from the emitter than there are holes injected from the base. Assuming the collector voltage is large enough so that the collector-base junction is reverse biased, no holes from the base will go to the collector. However, the electrons that travel from the emitter to the base, where they are now minority carriers, diffuse away from the base-emitter junction because of the minority-carrier concentration gradient in the base region. Any of these minority electrons that get close to the collector-base junction will immediately be "whisked" across the junction due to the large positive voltage on the collector, which attracts the negatively charged electrons. In a properly designed bipolar transistor, such as that shown in Fig. 1.25, the vertical base width, W, is small, and almost all of the electrons that diffuse from the emitter to the base reach the collector-base junction and are swept across the junction, thus contributing to current flow in the collector. The result is that the collector current very closely equals the electron current flowing from the emitter to the base. The much smaller base current very closely equals the current due to the holes that flow from the base to the emitter. The total emitter current is the sum of the electron collector current and the hole base current, but since the hole current is much smaller than the electron current, the emitter current is approximately equal to the collector current.

Since the collector current is approximately equal to the electron current flowing from the emitter to the base, and the amount of this electron current is determined by the base-emitter voltage, it can be shown (see Appendix at the end of this chapter) that the collector current is exponentially related to the base-emitter voltage by the relationship

$$I_C \cong I_{CS} e^{V_{BE}/V_T} \tag{1.127}$$

where I_{CS} is the *scale current.* This scale current is proportional to the area of the base-emitter junction. The base current, determined by the hole current flowing from the base to the emitter, is also exponentially related to the base-emitter voltage, resulting in the ratio of the collector current to the base current being a constant that, to a first-order approximation, is independent of voltage and current. This ratio, typically denoted as β, is defined to be

$$\beta \equiv \frac{I_C}{I_B} \tag{1.128}$$

where I_C and I_B are the collector and base currents. Typical values of β are between 50 and 200.

Note that (1.127) implies that the collector current is independent of the collector voltage. This independence ignores second-order effects such as the decrease in effective base width, W, due to the increase in the width of the collector-base depletion

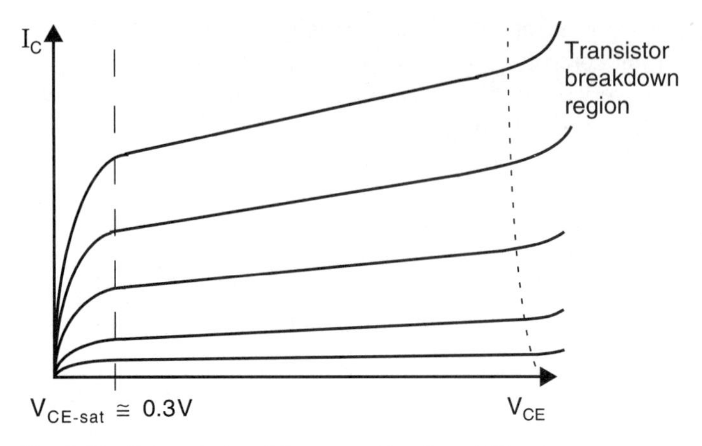

Fig. 1.29 Typical plot of I_C versus V_{CE} for a BJT.

region when the collector bias voltage is increased. To illustrate this point, a typical plot of the collector current, I_C, as a function of collector-to-emitter voltage, V_{CE}, for different values of I_B is shown in Fig. 1.29 for a practical transistor. The fact that the curves are not flat for $V_{CE} > V_{CE\text{-sat}}$ indicates the dependence of I_C on V_{CE}. Indeed, to a good approximation, the dependence is linear with a slope that intercepts the V_{CE} axis at $V_{CE} = -V_A$ for all values of I_B. The intercept voltage value, V_A, is called the *Early voltage* for bipolar transistors, with a typical value being from 50 V to 100 V. This dependency results in a finite output impedance (as in a MOS transistor) and can be modelled by modifying equation (1.127) [Sze, 1981] to be

$$I_C \cong I_{CS} e^{V_{BE}/V_T}\left(1 + \frac{V_{CE}}{V_A}\right) \tag{1.129}$$

Large-Signal Modelling

A conducting BJT that has a V_{CE} greater than $V_{CE\text{-sat}}$ (which is approximately 0.3 V) is said to be operating in the active region. Such a collector-emitter voltage is required to ensure that none of the holes from the base go to the collector. A large-signal model of a BJT operating in the active region is shown in Fig. 1.30.

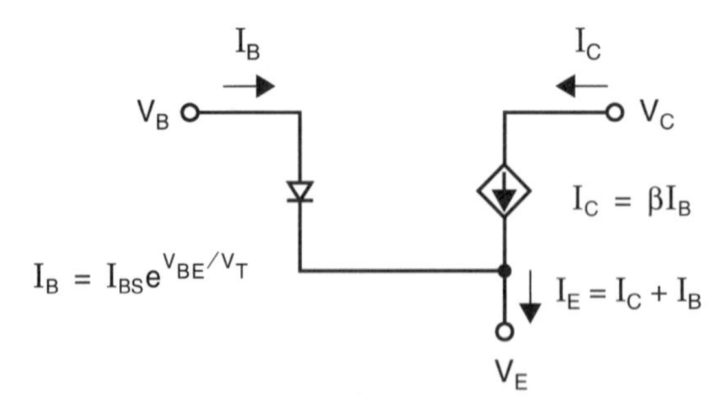

Fig. 1.30 A large-signal model for a BJT in the active region.

Since $I_B = I_C/\beta$, we have

$$I_B = \frac{I_{CS}}{\beta} e^{V_{BE}/V_T} = I_{BS} e^{V_{BE}/V_T} \tag{1.130}$$

which is similar to a diode equation, but with a multiplying constant of $I_{CS}/\beta = I_{BS}$. Since $I_E = I_B + I_C$, we have

$$I_E = I_{CS}\left(\frac{\beta + 1}{\beta}\right) e^{V_{BE}/V_T} = I_{ES} e^{V_{BE}/V_T} \tag{1.131}$$

or equivalently

$$I_C = \alpha I_E \tag{1.132}$$

where α has been defined as

$$\alpha = \frac{\beta}{\beta + 1} \tag{1.133}$$

and for large values of β, can be approximated as

$$\alpha \cong 1 - \frac{1}{\beta} \cong 1 \tag{1.134}$$

If the effect of V_{CE} on I_C is included in the model, the current-controlled source, βI_B, should be replaced by a current source given by

$$I_C = \beta I_B\left(1 + \frac{V_{CE}}{V_A}\right) \tag{1.135}$$

where V_A is the Early-voltage constant. This additional modelling of the finite output impedance is normally not done in large-signal analysis without the use of a computer due to its complexity.

As the collector-emitter voltage approaches $V_{CE\text{-sat}}$ (typically around 0.2 to 0.3 V), the base-collector junction becomes forward biased, and holes from the base will begin to diffuse to the collector. A common model for this case, when the transistor is *saturated* or in the *saturation region,* is shown in Fig. 1.31. It should be noted that the value of $V_{CE\text{-sat}}$ decreases for smaller values of collector current.

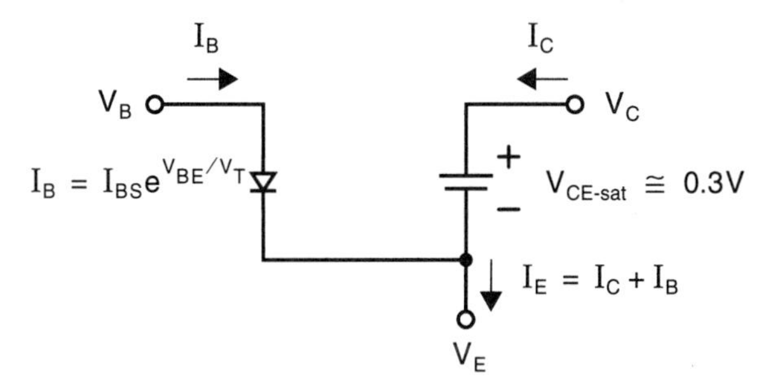

Fig. 1.31 A large-signal model for a BJT in the saturation region.

Base-Charge Storage in the Active Region

When a transistor is in the active region, many minority carriers are stored in the base region (electrons are stored in an npn transistor). Recall that this minority charge is responsible for I_C, so this charge must be removed (through the base contact) before a transistor can turn off. As in a forward-bias diode, this charge can be modelled as a diffusion capacitance, C_d, between the base and emitter given by (see Appendix at the end of this chapter)

$$C_d = \tau_b \frac{I_C}{V_T} \tag{1.136}$$

where τ_b is the base-transit-time constant. Thus, we see that the diffusion capacitance is proportional to I_C. The total base-emitter capacitance, C_{be}, will include the base-emitter depletion capacitance, C_j, in parallel with C_d. Normally, however, C_j is much less than C_d, unless the transistor current is small, and can often be ignored.

Base-Charge Storage of a Saturated Transistor

When a transistor becomes saturated, the minority-charge storage in the base and, even more so, in the lightly doped region of the collector, increases drastically. The major component of this charge storage is due to holes diffusing from the base, through the collector junction, and continuing on through the lightly doped n^- *epitaxial region* of the collector to the n^+ collector region. The n^- epitaxial region is so named because it is epitaxially grown on a p region. Most of the charge storage occurs in this region. Also, additional charge storage occurs because electrons that diffused from the collector are stored in the base, but this charge is normally smaller. The magnitude of the additional charge stored by a transistor that is saturated is given by

$$Q_s = \tau_s\left(I_B - \frac{I_C}{\beta}\right) \tag{1.137}$$

where the base overdrive current, defined to be $I_B - I_C/\beta$, is approximately equal to the hole current from the base to the collector. Normally, in saturation, $I_B >> I_C/\beta$, and (1.137) can be approximated by

$$Q_s \cong \tau_s I_B \tag{1.138}$$

The constant τ_s is approximately equal to the epitaxial-region transit time, τ_E (ignoring the storage of electrons in the base that diffused from the collector). Since the epitaxial region is much wider than the base, the constant τ_s is normally much larger than the base transit time, the constant τ_b, often by up to two orders of magnitude. The specific value of τ_s is usually found empirically for a given technology.

When a saturated transistor is being turned off, first the base current will reverse. *However, before the collector current will change, the saturation charge, Q_s, must be removed.* After Q_s is removed, the base minority charge, Q_b, will be removed. During this time, the collector current will decrease until the transistor shuts off. Typically, the time to remove Q_S greatly dominates the overall charge removal.

If the time required to remove the base saturation charge, t_S, is much shorter than the epitaxial-region transit time, τ_E, then one can derive a simple expression for the time required to remove the saturation charge. If the reverse base current (when the saturation charge is being removed), denoted by I_{BR}, remains constant while Q_s is being removed, then we have [Hodges, 1988]

$$t_s \cong \frac{Q_s}{I_{BR}} \cong \frac{\tau_s\left(I_B - \frac{I_C}{\beta}\right)}{I_{BR}} \cong \tau_s \frac{I_B}{I_{BR}} \tag{1.139}$$

where $\tau_s \cong \tau_E$.

Normally, the forward base current during saturation, I_B, will be much smaller than the reverse base current during saturation-charge removal, I_{BR}. If this were not the case, then our original assumption that $t_s \ll \tau_E \cong \tau_s$ would not be true. In this case, the turn-off time of the BJT would be so slow as to make the circuit unusable in most applications. Nevertheless, the turn-off time for this case, when t_s is not much less than τ_E, is given by [Hodges, 1988]

$$t_s = \tau_s \ln\left[\frac{I_{BR} + I_B}{I_{BR} + \frac{I_C}{\beta}}\right] \tag{1.140}$$

The reader should verify that for $I_{BR} \gg I_B$ and $I_{BR} \gg I_C/\beta$, the expression in (1.140) is approximately equivalent to the much simpler one in (1.139).

In both of the cases just described, the time required to remove the storage charge of a saturated transistor is much larger than the time required to turn off a transistor in the active region. *In high-speed microcircuit designs, one never allows bipolar transistors to saturate, to avoid the long turn-off time that would result.*

EXAMPLE 1.13

For $\tau_b = 0.2$ ns, $\tau_s = 100$ ns (a small value for τ_s), $I_B = 0.2$ mA, $I_C = 1$ mA, $\beta = 100$, and $I_{BR} = 1$ mA, calculate the time required to remove the base saturation charge using (1.139), and compare it to the time obtained using the more accurate expression of (1.140). Compare these results to the time required to remove the base minority charge for the same I_{BR}.

Solution

Using (1.139), we have

$$t_s = \frac{10^{-7}(2 \times 10^{-4})}{10^{-3}} = 20 \text{ ns} \tag{1.141}$$

Using (1.140), we have

$$t_s = 10^{-7}\ln\left[\frac{10^{-3} + 2\times 10^{-4}}{10^{-3} + \frac{10^{-3}}{100}}\right] = 17.2 \text{ ns} \tag{1.142}$$

which is fairly close to the first result.

The time required for an active transistor to remove the base minority charge, Q_b, is given by

$$t_A = \frac{Q_b}{I_{BR}} = \frac{\tau_b I_C}{I_{BR}} = 0.2 \text{ ns} \tag{1.143}$$

This is approximately 100 times shorter than the time for removing the base saturation charge!

Small-Signal Modelling

The most commonly used small-signal model is the hybrid-π model. This model is similar to the small-signal model used for MOS transistors, except it includes a finite base-emitter impedance, r_π, and it has no emitter-to-bulk capacitance. The hybrid-π model is shown in Fig. 1.32. As in the MOS case, we will first discuss the transconductance, g_m, and the small-signal resistances, and then we will discuss the parasitic capacitances.

The transistor transconductance, g_m, is perhaps the most important parameter of the small-signal model. The transconductance is the ratio of the small-signal collector current, i_c, to the small-signal base-emitter voltage, v_{be}. Thus, we have

$$g_m = \frac{i_c}{v_{be}} = \frac{\partial I_C}{\partial V_{BE}} \tag{1.144}$$

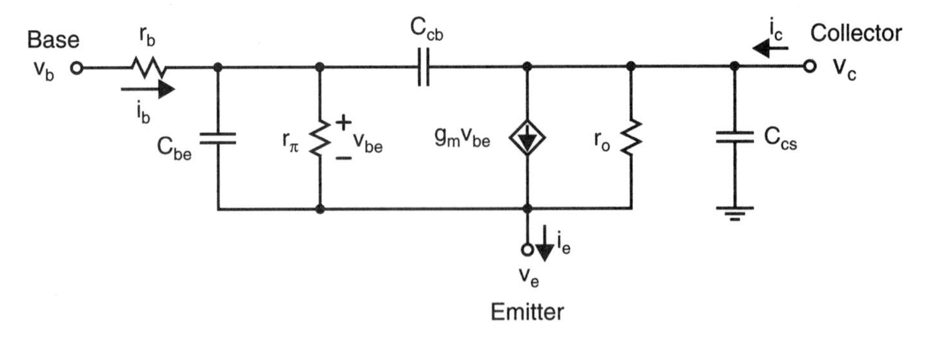

Fig. 1.32 The small-signal model of an active BJT.

Recall that in the active region

$$I_C = I_{CS}e^{V_{BE}/V_T} \tag{1.145}$$

Then

$$g_m = \frac{\partial I_C}{\partial V_{BE}} = \frac{I_{CS}}{V_T}e^{V_{BE}/V_T} \tag{1.146}$$

Using (1.145) again, we obtain

$$g_m = \frac{I_C}{V_T} \tag{1.147}$$

where V_T is given by

$$V_T = \frac{kT}{q} \tag{1.148}$$

and is approximately 26 mV at a room temperature of $T = 300$ °K. Thus, the transconductance is proportional to the bias current of a BJT. In integrated-circuit design, it is important that the transconductance (and hence speed) remain temperature independent, so the bias currents are usually made proportional to absolute temperature (since V_T is proportional to absolute temperature).

The presence of the resistor r_π reflects the fact that the base current is nonzero. We have

$$r_\pi = \frac{\partial V_{BE}}{\partial I_B} \tag{1.149}$$

Because from (1.130) we have

$$I_B = \frac{I_C}{\beta} = \frac{I_{CS}}{\beta}e^{V_{BE}/V_T} \tag{1.150}$$

we therefore have

$$\frac{1}{r_\pi} = \frac{\partial I_B}{\partial V_{BE}} = \frac{I_{CS}}{\beta V_T}e^{V_{BE}/V_T} \tag{1.151}$$

Using (1.150) again, we have

$$r_\pi = \frac{V_T}{I_B} \tag{1.152}$$

or equivalently,

$$r_\pi = \beta\frac{V_T}{I_C} = \frac{\beta}{g_m} \tag{1.153}$$

Since

$$i_e = i_c + i_b \tag{1.154}$$

we also have

$$\frac{\partial I_E}{\partial V_{BE}} = \frac{\partial I_c}{\partial V_{BE}} + \frac{\partial I_B}{\partial V_{BE}} = g_m + \frac{g_m}{\beta} = g_m\left(\frac{1+\beta}{\beta}\right) = \frac{g_m}{\alpha} \tag{1.155}$$

Some alternative models, usually called T models (see page 55), use the emitter resistance, r_e, where

$$r_e = \frac{\partial V_{BE}}{\partial I_E} = \frac{\alpha}{g_m} \tag{1.156}$$

Continuing, we have

$$\frac{1}{r_o} = \frac{\partial I_C}{\partial V_{CE}} \tag{1.157}$$

The small-signal resistance, r_o, models the dependence of the collector current on the collector-emitter voltage. Repeating (1.129) here for convenience,

$$I_C = I_{CS} e^{V_{BE}/V_T}\left(1 + \frac{V_{CE}}{V_A}\right) \tag{1.158}$$

we have

$$\frac{1}{r_o} = \frac{\partial I_C}{\partial V_{CE}} = \frac{I_{CS}}{V_A} e^{V_{BE}/V_T} \tag{1.159}$$

Thus,

$$r_o = \frac{V_A}{I_C} \tag{1.160}$$

which is inversely proportional to the collector current. As an aside, note that $g_m r_o = V_A/V_T$ is a constant value independent of the transistor operating point. This constant is usually between 2,000 and 8,000 for an npn BJT and is an upper limit on the attainable voltage gain for a single-transistor amplifier.

The resistor r_b models the resistance of the semiconductor material between the base contact and the effective base region due to the moderately lightly doped base p material (see Fig. 1.25). This resistor, although small (typically a few hundred ohms), can be important in limiting the speed of very-high-frequency low-gain BJT circuits and is a major source of noise.

EXAMPLE 1.14

For $I_C = 1$ mA, $\beta = 100$, and $V_A = 100$ V, calculate g_m, r_π, r_e, r_o, and $g_m r_o$.

Solution

We have

$$g_m = \frac{I_C}{V_T} = \frac{10 \times 10^{-3}\ \text{A}}{0.026\ \text{V}} = 38.5\ \text{mA/V} \tag{1.161}$$

$$r_\pi = \frac{\beta}{g_m} = 2.6\ \text{k}\Omega \tag{1.162}$$

$$r_e = \frac{\alpha}{g_m} = \left(\frac{100}{101}\right)26 = 25.7\ \Omega \tag{1.163}$$

$$r_o = \frac{V_A}{I_C} = \frac{100}{10^{-3}} = 100\ \text{k}\Omega \tag{1.164}$$

and $g_m r_o$, the maximum possible gain with a single-transistor amplifier, is given by

$$g_m r_o = \frac{V_A}{V_T} = 3{,}846 \tag{1.165}$$

Note that this gain is much higher than the 52.6 that was found for a single MOS transistor in Example 1.9. Also note that this BJT maximum gain is independent of the bias current. For MOS transistors, it can be shown that the maximum gain decreases with larger bias currents in a square-root relationship. This is one of the reasons why it is possible to realize a single-transistor BJT amplifier with a much larger gain than would result if a MOS transistor were used, especially at high current levels (and therefore at high frequencies).

The high-frequency operation of a BJT is limited by the capacitances of the small-signal model. We have already encountered one of these capacitances, C_{be}, in Section 1.1. Recapping, we have

$$C_{be} = C_j + C_d \tag{1.166}$$

where C_j is the depletion capacitance of the base-emitter junction. For a forward-biased junction, a rough approximation for C_j is

$$C_j \cong 2A_E C_{je0} \tag{1.167}$$

The diffusion capacitance, C_d, is given in (1.136) as

$$C_d = \tau_b \frac{I_C}{V_T} = g_m \tau_b \tag{1.168}$$

The capacitor, C_{cb}, models the depletion capacitance of the collector-base junction. Since this is a graded junction, we can approximate C_{cb} by

$$C_{cb} = \frac{A_C C_{jc0}}{\left(1 + \frac{V_{CB}}{\Phi_{c0}}\right)^{1/3}} \tag{1.169}$$

where A_C is the effective area of the collector-base interface.

Due to the lower doping levels in the base and especially in the collector (perhaps 5×10^{22} acceptors/m^3 and 10^{21} donors/m^3, respectively), Φ_{c0}, the built-in potential for the collector-base junction, will be less than that of the base-emitter junction (perhaps 0.75 V as opposed to 0.9 V). It should be noted that the cross-sectional area of the collector-base junction, A_C, is typically much larger than the effective area of the base-emitter junction, A_E, which is shown in Fig. 1.25. This size differential results in $A_C C_{jc0}$ being larger than $A_E C_{je0}$, the base-emitter junction capacitance at 0 V bias, despite the lower doping levels.

Finally, another large capacitor is C_{cs}, the capacitance of the collector-to-substrate junction. Since this area is quite large, C_{cs}, which is the depletion capacitance that results from this area, will be much larger than either C_{cb} or the depletion capacitance component of C_{be}, that is, C_j. The value of C_{cs} can be calculated using

$$C_{cs} = \frac{A_T C_{js0}}{\left(1 + \frac{V_{CS}}{\Phi_{s0}}\right)^{1/2}} \tag{1.170}$$

where A_T is the effective transistor area and C_{js0} is the collector-to-substrate capacitance per unit area at 0-V bias voltage.

A common indicator for the *speed* of a BJT is the frequency at which the transistor's current gain drops to unity, when its collector is connected to a small-signal ground. This frequency is denoted f_t and is called the transistor *unity-gain frequency*. We can see how this frequency is related to the transistor model parameters by analyzing the small-signal circuit of Fig. 1.33. In the simplified model in part *b*, the resistor r_b is ignored because it has no effect on i_b since the circuit is being driven by a perfect current source. We have

$$V_{be} = i_b \left(r_\pi \left\| \frac{1}{sC_{be}} \right\| \frac{1}{sC_{cb}} \right) \tag{1.171}$$

and

$$i_c = g_m V_{be} \tag{1.172}$$

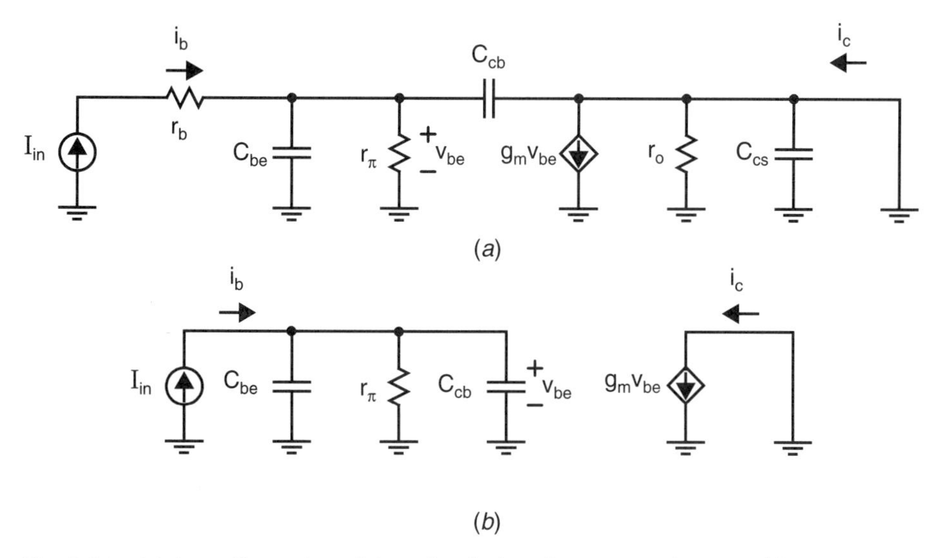

Fig. 1.33 (*a*) A small-signal model used to find f_t; (*b*) an equivalent simplified model.

Solving for i_c / i_b gives

$$\frac{i_c}{i_b} = \frac{g_m r_\pi}{1 + s(C_{be} + C_{cb}) r_\pi} \tag{1.173}$$

At low frequencies, the current gain is $g_m r_\pi$, which equals the expected value of β (using (1.153)). At high frequencies, i_c / i_b is approximately given by

$$\left|\frac{i_c}{i_b}(\omega)\right| \cong \frac{g_m r_\pi}{\omega(C_{be} + C_{cb}) r_\pi} = \frac{g_m}{\omega(C_{be} + C_{cb})} \tag{1.174}$$

To find the unity-gain frequency, we set $|(i_c/i_b)(\omega_t)| = 1$ and solve for ω_t, which results in

$$\omega_t = \frac{g_m}{C_{be} + C_{cb}} \tag{1.175}$$

or

$$f_t = \frac{g_m}{2\pi(C_{be} + C_{cb})} \tag{1.176}$$

Often, either f_t, ω_t, or $\tau_t = 1/\omega_t$ will be specified for a transistor at a particular bias current. These values indicate an upper limit on the maximum frequency at which the transistor can be effectively used.

The hybrid-π model is only one of a number of small-signal models that can be used. One common alternative is the low-frequency T model shown in Fig. 1.34. Use of this T model often results in a much simplified analysis, compared to use of the hybrid-π model, and thus it is useful for hand analysis.

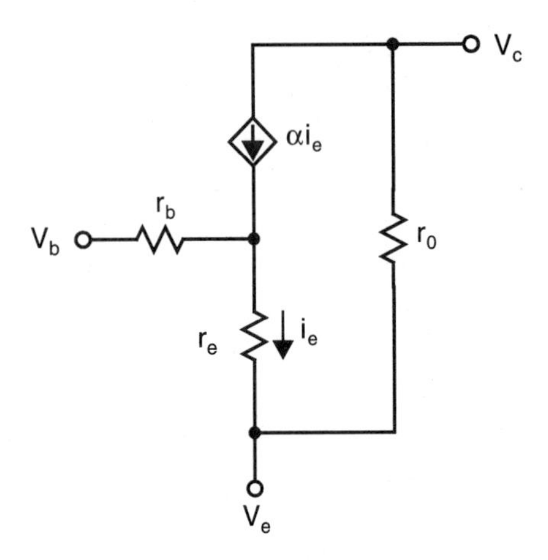

Fig. 1.34 A low-frequency, small-signal T model for an active BJT.

1.5 DEVICE MODEL SUMMARY

As a useful aid, all of the equations for the large-signal and small-signal modelling of diodes, MOS transistors, and bipolar transistors, along with values for the various constants, are listed in the next few pages.

Constants

$q = 1.602 \times 10^{-19}$ C	$k = 1.38 \times 10^{-23}$ JK^{-1}
$n_i = 1.1 \times 10^{16}$ carriers/m^3 at $T = 300$ °K	$\varepsilon_0 = 8.854 \times 10^{-12}$ F/m
$K_{ox} \cong 3.9$	$K_s \cong 11.8$
$\mu_n = 0.05$ m^2/V · s	$\mu_p = 0.02$ m^2/V · s

Diode Equations

Reverse-Biased Diode (Abrupt Junction)

$C_j = \dfrac{C_{j0}}{\sqrt{1 + \dfrac{V_R}{\Phi_0}}}$	$Q = 2C_{j0}\Phi_0\sqrt{1 + \dfrac{V_R}{\Phi_0}}$
$C_{j0} = \sqrt{\dfrac{qK_s\varepsilon_0}{2\Phi_0}\dfrac{N_D N_A}{N_A + N_D}}$	$C_{j0} = \sqrt{\dfrac{qK_s\varepsilon_0 N_D}{2\Phi_0}}$ if $N_A >> N_D$
$\Phi_0 = \dfrac{kT}{q}\ln\left(\dfrac{N_A N_D}{n_i^2}\right)$	

Forward-Biased Diode

$I_D = I_S e^{V_D/V_T}$	$I_S = A_D q n_i^2 \left(\frac{D_n}{L_n N_A} + \frac{D_p}{L_p N_D} \right)$
$V_T = \frac{kT}{q} \cong 26 \text{ mV at } 300\,^{\circ}\text{K}$	

Small-Signal Model of Forward-Biased Diode

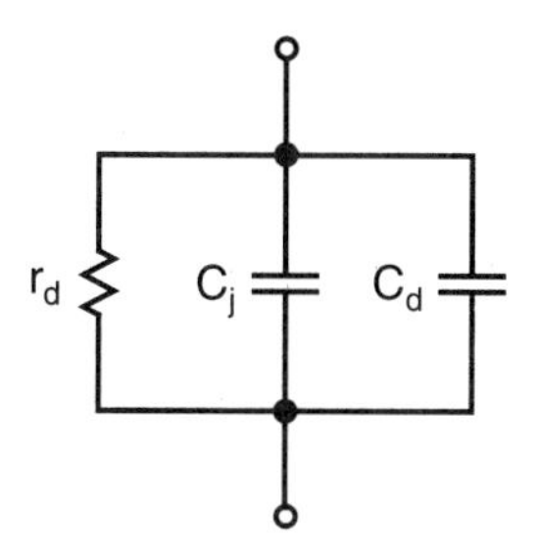

$r_d = \frac{V_T}{I_D}$	$C_T = C_d + C_j$
$C_d = \tau_T \frac{I_D}{V_T}$	$C_j \cong 2C_{j0}$
$\tau_T = \frac{L_n^2}{D_n}$	

MOS Transistor Equations

The following equations are for n-channel devices—for p-channel devices, put negative signs in front of all voltages. These equations do not account for short-channel effects (i.e., $L < 2L_{min}$).

Triode Region ($V_{GS} > V_{tn}$, $V_{DS} \le V_{eff}$)

$I_D = \mu_n C_{ox} \left(\frac{W}{L} \right) \left[(V_{GS} - V_{tn}) V_{DS} - \frac{V_{DS}^2}{2} \right]$	
$V_{eff} = V_{GS} - V_{tn}$	$V_{tn} = V_{tn\text{-}0} + \gamma(\sqrt{V_{SB} + 2\phi_F} - \sqrt{2\phi_F})$

$\phi_F = \frac{kT}{q}\ln\left(\frac{N_A}{n_i}\right)$	$\gamma = \frac{\sqrt{2qK_{si}\varepsilon_0 N_A}}{C_{ox}}$
$C_{ox} = \frac{K_{ox}\varepsilon_0}{t_{ox}}$	

Small-Signal Model in Triode Region (for $V_{DS} << V_{eff}$)

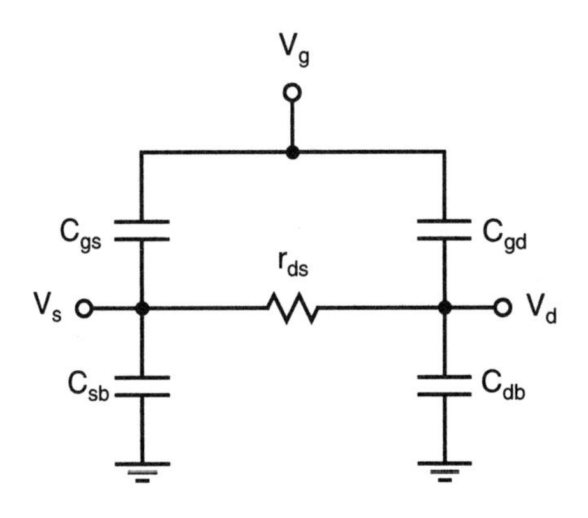

$r_{ds} = \frac{1}{\mu_n C_{ox}\left(\frac{W}{L}\right)V_{eff}}$	
$C_{gd} = C_{gs} \cong \frac{1}{2}WLC_{ox} + WL_{ov}C_{ox}$	$C_{sb} = C_{db} = \frac{C_{j0}(A_s + WL/2)}{\sqrt{1 + \frac{V_{sb}}{\Phi_0}}}$

Active (or Pinch-Off) Region ($V_{GS} > V_{tn}$, $V_{DS} \geq V_{eff}$)

$I_D = \frac{\mu_n C_{ox}}{2}\frac{W}{L}(V_{GS} - V_{tn})^2[1 + \lambda(V_{DS} - V_{eff})]$	
$\lambda \propto \frac{1}{L\sqrt{V_{DS} - V_{eff} + \Phi_0}}$	$V_{tn} = V_{tn\text{-}0} + \gamma(\sqrt{V_{SB} + 2\phi_F} - \sqrt{2\phi_F})$
$V_{eff} = V_{GS} - V_{tn} = \sqrt{\frac{2I_D}{\mu_n C_{ox} W/L}}$	

Small-Signal Model (Active Region)

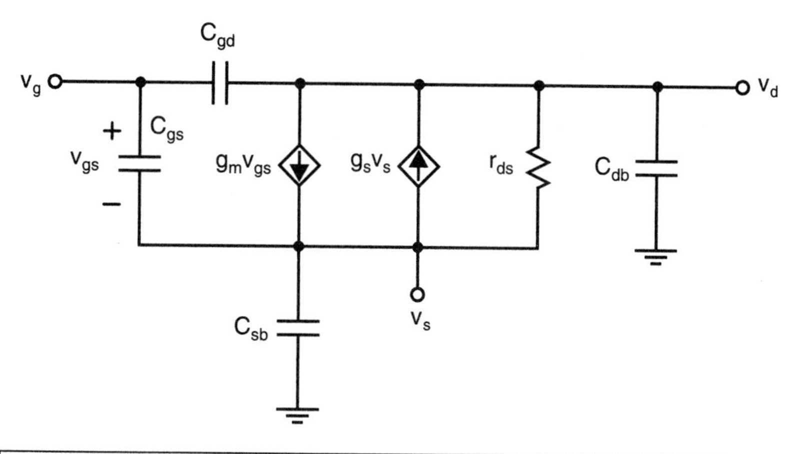

$g_m = \mu_n C_{ox}\left(\frac{W}{L}\right)V_{eff}$	$g_m = \sqrt{2\mu_n C_{ox}(W/L)I_D}$
$g_m = \frac{2I_D}{V_{eff}}$	$g_s = \frac{\gamma g_m}{2\sqrt{V_{SB} + \lvert 2\phi_F \rvert}}$
$r_{ds} = \frac{1}{\lambda I_D}$	$g_s \cong 0.2 g_m$
$\lambda = \frac{k_{rds}}{2L\sqrt{V_{DS} - V_{eff} + \Phi_0}}$	$k_{rds} = \sqrt{\frac{2K_s\varepsilon_0}{qN_A}}$
$C_{gs} = \frac{2}{3}WLC_{ox} + WL_{ov}C_{ox}$	$C_{gd} = WL_{ov}C_{ox}$
$C_{sb} = (A_s + WL)C_{js} + P_sC_{j\text{-}sw}$	$C_{js} = \frac{C_{j0}}{\sqrt{1 + V_{SB}/\Phi_0}}$
$C_{db} = A_dC_{jd} + P_dC_{j\text{-}sw}$	$C_{jd} = \frac{C_{j0}}{\sqrt{1 + V_{DB}/\Phi_0}}$

Typical Values for a 0.8-μm Process

$V_{tn} = 0.8\ \text{V}$	$V_{tp} = -0.9\ \text{V}$
$\mu_n C_{ox} = 90\ \mu\text{A}/\text{V}^2$	$\mu_p C_{ox} = 30\ \mu\text{A}/\text{V}^2$

$C_{ox} = 1.9 \times 10^{-3}\ pF/(\mu m)^2$	$C_j = 2.4 \times 10^{-4}\ pF/(\mu m)^2$
$C_{j\text{-}sw} = 2.0 \times 10^{-4}\ pF/\mu m$	$C_{gs(overlap)} = 2.0 \times 10^{-4}\ pF/\mu m$
$\phi_F = 0.34\ V$	$\Phi_0 = 0.9\ V$
$\gamma = 0.5\ V^{1/2}$	$t_{ox} = 0.02\ \mu m$
$N_B = 6 \times 10^{21}\ impurities/m^3$	

Bipolar-Junction Transistors

Active Transistor

$I_C = I_{CS} e^{V_{BE}/V_T}$	$V_T = \frac{kT}{q} \cong 26\ mV$ at 300 °K
For more accuracy, $I_C = I_{CS} e^{V_{BE}/V_T}\left(1 + \frac{V_{CE}}{V_A}\right)$	
$I_{CS} = \frac{A_E q D_n n_i^2}{W N_A}$	$I_B = \frac{I_C}{\beta}$
$I_E = \left(1 + \frac{1}{\beta}\right) I_C = \frac{1}{\alpha} I_C = (\beta + 1) I_B$	$\beta = \frac{I_C}{I_B} = \frac{D_n N_D L_p}{D_p N_A W} \cong 2.5 \frac{N_D L_p}{N_A W}$
$\alpha = \frac{\beta}{1 + \beta}$	

Small-Signal Model of an Active BJT

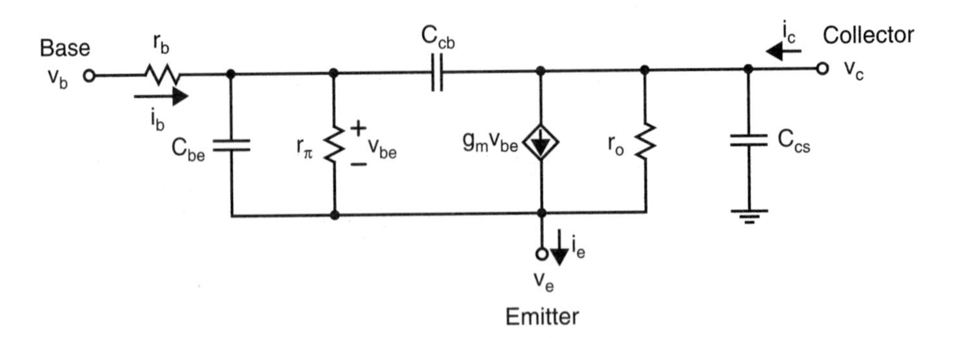

$g_m = \frac{I_C}{V_T}$	$r_\pi = \frac{V_T}{I_B} = \frac{\beta}{g_m}$
$r_e = \frac{\alpha}{g_m}$	$r_o = \frac{V_A}{I_C}$
$g_m r_o = \frac{V_A}{V_T}$	$C_d = \tau_b \frac{I_C}{V_T} = g_m \tau_b$
$C_{be} = C_j + C_d$	$C_{cs} = \frac{A_T C_{js0}}{\left(1 + \frac{V_{CS}}{\Phi_{s0}}\right)^{1/2}}$
$C_{cb} = \frac{A_C C_{jc0}}{\left(1 + \frac{V_{CB}}{\Phi_{c0}}\right)^{1/3}}$	

1.6 SPICE-MODELLING PARAMETERS

This section briefly describes some of the important model parameters for diodes, bipolar transistors, and MOS transistors used during a SPICE simulation. It should be noted here that not all SPICE model parameters are described. However, enough are described to enable the reader to understand the relationship between the relative parameters and the corresponding constants used when doing hand analysis.

Diode Model

There are a number of important dc parameters. The constant I_S is specified using either the parameter IS or JS in SPICE. These two parameters are synonyms, and only one should be specified. A typical value specified for I_S might be between 10^{-18} A and 10^{-15} A for small diodes in a microcircuit. Another important parameter is called the *emission coefficient,* n. This constant multiplies V_T in the exponential diode I-V relationship given by

$$I_D = I_S e^{V_{BE}/(nV_T)} \tag{1.177}$$

The SPICE parameter for n is N and is defaulted to 1 when not specified (1 is a reasonable value for junctions in a microcircuit). A third important dc characteristic is the *series resistance,* which is specified in SPICE using RS. It should be noted here that some SPICE programs allow the user to specify the area of the diode, whereas others expect absolute parameters that already take into account the effective area. The manual for the program being used should be consulted.

The diode transit time is specified using the SPICE parameter TT. The most important capacitance parameter specified is CJ. CJO and CJ are synonyms—one should never specify both. This parameter specifies the capacitance at 0-V bias. Once again, it may be specified as absolute or as relative to the area (i.e., F/m^2), depending on the version of SPICE used. Also, the area junction grading coefficient, MJ, might be specified to determine the exponent used in the capacitance equation. Typical values are 0.5 for abrupt junctions and 0.33 for graded junctions. In some SPICE versions, it might also be possible to specify the sidewall capacitance at 0-V bias as well as its grading junction coefficient. Finally, the built-in potential of the junction, which is also used in calculating the capacitance, can be specified using PB. PHI, VJ, and PHA are all synonyms of PB.

Reasonably accurate diode simulations can usually be obtained by specifying only IS, CJ, MJ, and PB. However, most modern versions of SPICE have many more parameters that can be specified if one wants accurate temperature and noise simulations. Users should consult their manuals for more information.

Table 1.1 summarizes some of the more important diode parameters. This set of parameters constitutes a minimal set for reasonable simulation accuracy under ordinary conditions.

MOS Transistors

Modern MOS models are quite complicated, so only some of the more important MOS parameters used in SPICE simulations are described here. These parameters are used in what are called the Level 2 or Level 3 models. The model level can be chosen by setting the SPICE parameter LEVEL to either 2 or 3. The oxide thickness, t_{ox}, is specified using the SPICE parameter TOX. If it is specified, then it is not necessary to specify the thin gate-oxide capacitance (C_{ox}, specified by parameter COX). The mobility, μ_n, can be specified using UO. If UO is specified, the intrinsic transistor conductance ($\mu_n C_{ox}$) will be calculated automatically, unless this automatic calculation is overridden by specifying either KP (or its synonym, BETA). The transistor threshold voltage at $V_S = 0$ V, V_{tn}, is specified by VTO. The body-effect parameter, γ,

Table 1.1 Important SPICE parameters for modelling diodes

SPICE Parameter	Model Constant	Brief Description	Typical Value
IS	I_S	Transport saturation current	10^{-17} A
RS	R_d	Series resistance	30 Ω
TT	τ_T	Diode transit time	12 ps
CJ	C_{j0}	Capacitance at 0-V bias	0.01 pF
MJ	m_j	Diode grading coefficient exponent	0.5
PB	Φ_0	Built-in diode contact potential	0.9 V

can be specified using GAMMA, or it will be automatically calculated if the substrate doping, N_A, is specified using NSUB. Normally, one would not want SPICE to calculate γ because the effective substrate doping under the channel can differ significantly from the substrate doping in the bulk due to threshold-voltage adjust implants. The output impedance constant, λ, can be specified using LAMBDA. Normally, LAMBDA should not be specified since it takes precedence over internal calculations and does not change the output impedance as a function of different transistor lengths or bias voltages (which should be the case). Indeed, modelling the transistor output impedance is one of weakest points in SPICE. If LAMBDA is not specified, it is calculated automatically. The surface inversion potential, $|2\phi_F|$, can be specified using PHI, or it will be calculated automatically. Another parameter usually specified is the lateral diffusion of the junctions under the gate, L_D, which is specified by LD. For accurate simulations, one might also specify the resistances in series with the source and drain by specifying RS and RD (typically only the source resistance is important). Many other parameters exist to model such things as short-channel effects, subthreshold effects, and channel-width effects, but these parameters are outside the scope of this book.

The modelling of parasitic capacitances in SPICE is quite involved. Originally, this modelling was not very accurate since it did not include charge conservation for the gate charge. However, this modelling has greatly improved in recent commercial versions of SPICE. The capacitances under the junctions per unit area at 0-V bias, (i.e., C_{j0}) can be specified using CJ or can be calculated automatically from the specified substrate doping. The sidewall capacitances at 0 V, $C_{j\text{-}sw0}$, should normally be specified using CJSW because this parameter is used to calculate significant parasitic capacitances. The bulk grading coefficient specified by MJ can usually be defaulted to 0.5. Similarly, the sidewall grading coefficient specified by MJSW can usually be defaulted to 0.33 (SPICE assumes a graded junction). The built-in bulk-to-junction contact potential, Φ_0, can be specified using PB or defaulted to 0.8 V (note that 0.9 V would typically be more accurate, but the resulting simulation differences are small). Sometimes the gate-to-source or drain-overlap capacitances can be specified using CGSO or CGDO, but normally these would be left to be calculated automatically using COX and LD.

Some of the more important parameters that should result in reasonable simulations (except for modelling short-channel effects) are summarized in Table 1.2 for both n- and p-channel transistors. Table 1.2 lists reasonable parameters for a typical 0.8-μm technology.

Bipolar Junction Transistors

For historical reasons, most parameters for modelling bipolar transistors are specified absolutely. Also, rather than specifying the emitter area of a BJT in $(\mu m)^2$ on the line where the individual transistor connections are specified, most SPICE versions have multiplication factors. These multiplication factors can be used to automatically multiply parameters when a transistor is composed of several transistors connected in parallel. This multiplying parameter is normally called M.

Table 1.2 A reasonable set of MOS parameters for a typical 0.8-μm technology

SPICE Parameter	Model Constant	Brief Description	Typical Value
VTO	$V_{tn}:V_{tp}$	Transistor threshold voltage (in V)	0.7:–0.9
UO	$\mu_n:\mu_p$	Carrier mobility in bulk (in $cm^2/V{\cdot}s$)	500:175
TOX	t_{ox}	Thickness of gate oxide (in m)	1.8×10^{-8}
LD	L_D	Lateral diffusion of junction under gate (in m)	6×10^{-8}
GAMMA	γ	Body-effect parameter	0.5: 0.8
NSUB	$N_A:N_D$	The substrate doping (in cm^{-3})	$3 \times 10^{16}:7.5 \times 10^{16}$
PHI	$\lvert 2\phi_F \rvert$	Surface inversion potential (in V)	0.7
PB	Φ_0	Built-in contact potential of junction to bulk (in V)	0.9
CJ	C_{j0}	Junction-depletion capacitance at 0-V bias (in F/m^2)	$2.5 \times 10^{-4}:4.0 \times 10^{-4}$
CJSW	$C_{j\text{-}sw0}$	Sidewall capacitance at 0-V bias (in F/m)	$2.0 \times 10^{-10}:2.8 \times 10^{-10}$
MJ	m_j	Bulk-to-junction exponent (grading coefficient)	0.5
MJSW	$m_{j\text{-}sw}$	Sidewall-to-junction exponent (grading coefficient)	0.3

The most important dc parameters are the transistor current gain, β, specified by the SPICE parameter BF; the transistor-transport saturation current, I_{CS}, specified using the parameter IS; and the Early-voltage constant, specified by the parameter VAF. Typical values for these might be 100, 10^{-17} A, and 50 V, respectively. If one wants to model the transistor in reverse mode (where the emitter voltage is higher than the collector voltage for an npn), then one might specify BR, ISS, and VAR, as well; these are the parameters that correspond to BIF, IS, and VAF in the reverse direction. Typically, this reverse-mode modelling is not important for most circuits. Some other important dc parameters for accurate simulations are the base, emitter, and collector resistances, which are specified by RB, RE, and RC, respectively. It is especially important to specify RB (which might be 200 Ω to 500 Ω).

The important capacitance parameters and their corresponding SPICE parameters include the depletion capacitances at 0-V bias voltage, CJE, CJC, CJS; their grading coefficients, MJE, MJC, MJS; and their built-in voltages, VJE, VJC, VJS, for base-emitter, base-collector, and collector-substrate junctions. Again, the 0-V depletion capacitances should be specified in absolute values for a unit-sized transistor. Normally the base-emitter and base-collector junctions are graded (i.e., MJE, MJC = 0.33), whereas the collector-substrate junction may be either abrupt (MJS = 0.5) or graded (MJS = 0.5), depending on processing details. Typical built-in voltages might be 0.75 V to 0.8 V. In addition, for accurate simulations, one should specify the forward-base transit time, τ_F, specified by TF, and, if the transistor is to be operated in reverse mode or under saturated conditions, the reverse-base transit time, τ_R, specified by TR.

The most important of the model parameters just described are summarized in Table 1.3.

Once again, many other parameters can be specified if accurate simulation is desired. Other parameters might include those to model β degradation under high or

Table 1.3 The most important SPICE parameters for modelling BJTs

SPICE Parameter	Model Constant	Brief Description	Typical Value
BF	β	Transistor current gain in forward direction	100
ISS	I_{CS}	Transport saturation current in forward direction	2×10^{-18} A
VAF	V_A	Early voltage in forward direction	50 V
RB	r_b	Series base resistance	500 Ω
RE	R_E	Series emitter resistance	30 Ω
CJE	C_{je0}	Base-emitter depletion capacitance at 0 V	0.015 pF
CJC	C_{je0}	Base-collector depletion capacitance at 0 V	0.018 pF
CJS	C_{je0}	Collector-substrate depletion capacitance at 0 V	0.040 pF
MJE	m_e	Base-emitter junction exponent (grading factor)	0.30
MJC	m_c	Base-collector junction exponent (grading factor)	0.35
MJS	m_s	Collector-substrate junction exponent (grading factor)	0.29
TF	τ_F	Forward-base transit time	12 ps
TR	τ_R	Reverse-base transit time	4 ns

low current applications and parameters for accurate noise and temperature analysis. Readers should refer to their SPICE manuals for descriptions of these parameters.

1.7 APPENDIX

The purpose of this appendix is to present derivations for device equations that rely heavily on device physics knowledge. Specifically, equations are derived for the exponential relationship and diffusion capacitance of diodes, for the threshold voltage and triode relationship for MOS transistors, and for the exponential relationship and base charge storage for bipolar transistors.

Diode Exponential Relationship

The concentration of minority carriers in the bulk, far from the junction, is given by Eqs. (1.2) and (1.4). Close to the junction, the minority-carrier concentrations are much larger. Indeed, the concentration next to the junction increases exponentially with the external voltage, V_D, that is applied in the forward direction. The concentration of holes in the n side next to the junction, p_n, is given by [Sze, 1981]

$$p_n = p_{n0} e^{V_D/V_T} = \frac{n_i^2}{N_D} e^{V_D/V_T} \tag{1.178}$$

Similarly, the concentration of electrons in the p side next to the junction is given by

$$n_p = n_{p0} e^{V_D/V_T} = \frac{n_i^2}{N_A} e^{V_D/V_T} \tag{1.179}$$

As the carriers diffuse away from the junction, their concentration exponentially decreases. The relationship for holes in the n side is

$$p_n(x) = p_n(0) e^{-x/L_p} \tag{1.180}$$

where x is the distance from the junction and L_p is a constant known as the diffusion length for holes in the n side. Similarly, for electrons in the p side we have

$$n_p(x) = n_p(0) e^{-x/L_n} \tag{1.181}$$

where L_n is a constant known as the diffusion length of electrons in the p side. Note that p_n (0) and n_p (0) are given by (1.178) and (1.179), respectively. Note also that the constants L_n and L_p are dependent on the doping concentrations N_A and N_D, respectively.

The current density of diffusing carriers moving away from the junction is given by the well-known diffusion equations [Sze, 1981]. For example, the current density of diffusing electrons is given by

$$J_{D\text{-}n} = -qD_n \frac{dn_p(x)}{dx} \tag{1.182}$$

where D_n is the diffusion constant of electrons in the p side of the junction. The negative sign is present because electrons have negative charge. Note that $D_n = (kT/q)\mu_n$, where μ_n is the mobility of electrons. Using (1.181), we have

$$\frac{dn_p(x)}{dx} = \frac{n_p(0)}{L_n} e^{-x/L_n} = -\frac{n_p(x)}{L_n} \tag{1.183}$$

Therefore

$$J_{D\text{-}n} = \frac{qD_n}{L_n} n_p(x) \tag{1.184}$$

Thus, the current density due to diffusion is proportional to the minority-carrier concentration. Next to the junction, all the current flow results from the diffusion of minority carriers. Further away from the junction, some of the current flow is due to diffusion and some is due to majority carriers drifting by to replace carriers that recombined with minority carriers or diffused across the junction.

Continuing, we use (1.179) and (1.184) to determine the current density next to the junction of electrons in the p side:

$$\begin{aligned} J_{D\text{-}n} &= \frac{qD_n}{L_n} n_p(0) \\ &= \frac{qD_n}{L_n} \frac{n_i^2}{N_A} e^{V_D/V_T} \end{aligned} \tag{1.185}$$

For the total current of electrons in the p side, we multiply (1.185) by the effective junction area, A_D. The total current remains constant as we move away from the junction since, in the steady state, the minority carrier concentration at any particular location remains constant with time. In other words, if the current changed as we moved away from the junction, the charge concentrations would change with time.

Using a similar derivation, we obtain the total current of holes in the n side, $I_{D\text{-}p}$, as

$$I_{D\text{-}p} = \frac{A_D q D_n n_i^2}{L_p N_D} e^{V_D/V_T} \tag{1.186}$$

where D_n is the diffusion constant of electrons in the p side of the junction, L_p is the diffusion length of holes in the n side, and N_D is the impurity concentration of donors in the n side. This current, consisting of positive carriers, flows in the direction opposite to that of the flow of minority electrons in the p side. However, since electron carriers are negatively charged, the direction of the current flow is the same. Note also that if the p side is more heavily doped than the n side, most of the carriers will be holes, whereas if the n side is more heavily doped than the p side, most of the carriers will be electrons.

The total current is the sum of the minority currents at the junction edges:

$$I_D = A_D q n_i^2 \left(\frac{D_n}{L_n N_A} + \frac{D_p}{L_p N_D} \right) e^{V_D/V_T} \tag{1.187}$$

Equation (1.187) is often expressed as

$$I_D = I_S e^{V_D/V_T} \tag{1.188}$$

where

$$I_S = A_D q n_i^2 \left(\frac{D_n}{L_n N_A} + \frac{D_p}{L_p N_D} \right) \tag{1.189}$$

Equation (1.188) is the well-known exponential current-voltage relationship of forward-biased diodes.

The concentrations of minority carriers near the junction and the direction of current flow are shown in Fig. 1.35.

Diode-Diffusion Capacitance

To find the diffusion capacitance, C_d, we first find the minority charge close to the junction, Q_d, and then differentiate it with respect to V_D. The minority charge close to the junction, Q_d, can be found by integrating either (1.180) or (1.181) over a few diffusion lengths. For example, if we assume n_{p0}, the minority electron concentration in the p side far from the junction is much less than $n_p(0)$, the minority electron con-

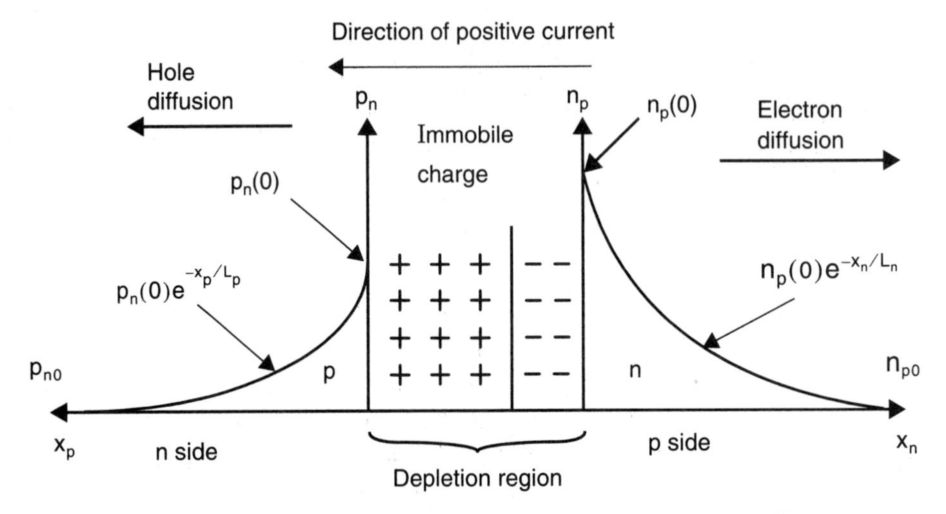

Fig. 1.35 The concentration of minority carriers and the direction of diffusing carriers near a forward-biased junction.

centration at the junction edge, we can use (1.181) to obtain

$$\begin{aligned} Q_n &= qA_D\int_0^{\infty} n_p(x)\,dx \\ &= qA_D\int_0^{\infty} n_p(0)e^{-x/L_n}\,dx \\ &= qA_DL_nn_p(0) \end{aligned} \tag{1.190}$$

Using (1.4) for $n_p(0)$ results in

$$Q_n = \frac{qA_DL_nn_i^2}{N_A}e^{V_D/V_T} \tag{1.191}$$

In a similar manner, we also have

$$Q_p = \frac{qA_DL_nn_i^2}{N_D}e^{V_D/V_T} \tag{1.192}$$

For a typical junction, one side will be much more heavily doped than the other side, and therefore the minority charge storage in the heavily doped side can be ignored since it will be much less than that in the lightly doped side. Assuming the n side is heavily doped, we find the total charge, Q_d, to be approximately given by Q_n, the minority charge in the p side. Thus, the small-signal diffusion capacitance, C_d, is given by

$$C_d = \frac{dQ_d}{dV_D} \cong \frac{dQ_n}{dV_D} = \frac{qA_DL_nn_i^2}{N_AV_T}e^{V_D/V_T} \tag{1.193}$$

Using (1.187) and again noting that $N_D >> N_A$, we have

$$C_d = \frac{L_n^2}{D_n} \frac{I_D}{V_T} \tag{1.194}$$

Equation (1.194) is often expressed as

$$C_d = \tau_T \frac{I_D}{V_T} \tag{1.195}$$

where τ_T is the transit time of the diode given by

$$\tau_T = \frac{L_n^2}{D_n} \tag{1.196}$$

for a single-sided diode in which the n side is more heavily doped.

MOS Threshold Voltage and the Body Effect

Many factors affect the gate-source voltage at which the channel becomes conductive. These factors are as follows:

1. The work-function difference between the gate material and the substrate material
2. The voltage drop between the channel and the substrate required for the channel to exist
3. The voltage drop across the thin oxide required for the depletion region, with its immobile charge, to exist
4. The voltage drop across the thin oxide due to unavoidable charge trapped in the thin oxide
5. The voltage drop across the thin oxide due to implanted charge at the surface of the silicon. The amount of implanted charge is adjusted in order to realize the desired threshold voltage.

The first factor affecting the transistor threshold voltage, V_{th}, is the built-in *Fermi potential* due to the different materials and doping concentrations used for the gate material and the substrate material. If one refers these potentials to that of intrinsic silicon [Tsividis, 1987], we have

$$\phi_{F\text{-}Gate} = \frac{kT}{q} \ln\left(\frac{N_D}{n_i}\right) \tag{1.197}$$

for a polysilicon gate with doping concentration N_D, and

$$\phi_{F\text{-}Sub} = \frac{kT}{q} \ln\left(\frac{n_i}{N_A}\right) \tag{1.198}$$

for a p substrate with doping concentration N_A. The work-function difference is then given by

$$\phi_{MS} = \phi_{F\text{-}Sub} - \phi_{F\text{-}Gate} = \frac{kT}{q}\ln\left(\frac{N_D N_A}{n_i^2}\right) \tag{1.199}$$

The next factor that determines the transistor threshold voltage is the voltage drop from the channel to the substrate, which is required for the channel to exist. The question of exactly when the channel exists does not have a precise answer. Rather, the channel is said to exist when the concentration of electron carriers in the channel is equal to the concentration of holes in the substrate. At this gate voltage, the channel is said to be *inverted.* As the gate voltage changes from a low value to the value at which the channel becomes inverted, the voltage drop in the silicon also changes, as does the voltage drop in the depletion region between the channel and the bulk. After the channel becomes inverted, any additional increase in gate voltage is closely equal to the increase in voltage drop across the thin oxide. In other words, after channel inversion, gate voltage variations have little effect on the voltage drop in the silicon or the depletion region between the channel and the substrate.

The electron concentration in the channel is equal to the hole concentration in the substrate when the voltage drop from the channel to the substrate is equal to two times the difference between the Fermi potential of the substrate and intrinsic silicon, ϕ_F, where

$$\phi_F = -\frac{kT}{q}\ln\left(\frac{N_A}{n_i}\right) \tag{1.200}$$

Equation (1.200) is a factor in several equations used in modelling MOS transistors. For typical processes, ϕ_F can usually be approximated as 0.35 V for typical doping levels at room temperature.

The third factor that affects the threshold voltage is due to the immobile negative charge in the depletion region left behind after the p mobile carriers are repelled. This effect gives rise to a voltage drop across the thin oxide of $-Q_B/C_{ox}$, where

$$Q_B = -qN_A x_d \tag{1.201}$$

and x_d is the width of the depletion region. Since

$$x_d = \sqrt{\frac{2K_s\varepsilon_0|2\phi_F|}{qN_A}} \tag{1.202}$$

we have

$$Q_B = -\sqrt{2qN_A K_s\varepsilon_0|2\phi_F|} \tag{1.203}$$

The fourth factor that determines V_{tn} is due to the unavoidable charge trapped in the thin oxide. Typical values for the effective ion density of this charge, N_{ox}, might be 2×10^{14} to 10^{15} ions/m^3. These ions are almost always positive. This effect gives rise to a voltage drop across the thin oxide, V_{ox}, given by

$$V_{ox} = \frac{-Q_{ox}}{C_{ox}} = \frac{-qN_{ox}}{C_{ox}} \tag{1.204}$$

The *native transistor threshold voltage* is the threshold voltage that would occur naturally if one did not include a special ion implant used to adjust the threshold voltage. This value is given by

$$V_{t\text{-native}} = \phi_{MS} - 2\phi_F - \frac{Q_B}{C_{ox}} - \frac{Q_{ox}}{C_{ox}} \tag{1.205}$$

A typical native threshold value might be around –0.1 V. It should be noted that transistors that have native transistor threshold voltages are becoming more important in circuit design where they might be used in transmission gates or in source-follower buffers.

The fifth factor that affects threshold voltage is a charge implanted in the silicon under the gate to change the threshold voltage from that given by (1.205) to the desired value, which might be 0.7 V for an n-channel transistor.

For the case in which the source-to-substrate voltage is increased, the effective threshold voltage is increased. This is known as the *body effect.* The body effect occurs because, as the source-bulk voltage, V_{SB}, becomes larger, the depletion region between the channel and the substrate becomes wider, and therefore more immobile negative charge becomes uncovered. This increase in charge changes the third factor in determining the transistor threshold voltage. Specifically, instead of using (1.203) to determine Q_B, one should now use

$$Q_B = -\sqrt{2qN_AK_s\varepsilon_0(V_{SB} + |2\phi_F|)} \tag{1.206}$$

If the threshold voltage when $V_{SB} = 0$ is denoted V_{tn0}, then, using (1.205) and (1.206), one can show that

$$\begin{aligned} V_{tn} &= V_{tn0} + \Delta V_{tn} \\ &= V_{tn0} + \frac{\sqrt{2qN_AK_s\varepsilon_0}}{C_{ox}}\left[\sqrt{V_{SB} + |2\phi_F|} - \sqrt{|2\phi_F|}\right] \\ &= V_{tn0} + \gamma\left(\sqrt{V_{SB} + |2\phi_F|} - \sqrt{|2\phi_F|}\right) \end{aligned} \tag{1.207}$$

where

$$\gamma = \frac{\sqrt{2qN_AK_s\varepsilon_0}}{C_{ox}} \tag{1.208}$$

The factor γ is often called the *body-effect constant.*

MOS Triode Relationship

The current flow in a MOS transistor is due to *drift* current rather than diffusion current. This type of current flow is the same mechanism that determines the current in a resistor. The current density, J, is proportional to the electrical field, E, where the constant of proportionality, σ, is called the *electrical permittivity.* Thus,

$$J = \sigma E \tag{1.209}$$

This constant for an n-type material is given by

$$\sigma = qn\mu_n \tag{1.210}$$

where n is the concentration per unit volume of negative carriers and μ_n is the mobility of electrons. Thus, the current density is given by

$$J = qn\mu_n E \tag{1.211}$$

Next, consider the current flow through the volume shown in Fig. 1.36, where the volume has height H and width W. The current is flowing perpendicular to the plane $H \times W$ down the length of the volume, L. The current, I, everywhere along the length of the volume is given by

$$I = JWH \tag{1.212}$$

The voltage drop along the length of the volume in the direction of L for a distance dx is denoted dV and is given by

$$dV = E(x)\,dx \tag{1.213}$$

Combining (1.211), (1.212), and (1.213), we obtain

$$q\mu_n WHn(x)\,dV = I\,dx \tag{1.214}$$

where the carrier density $n(x)$ is now assumed to change along the length L and is therefore a function of x.

As an aside, we examine the case of a resistor where $n(x)$ is usually constant. A resistor of length L would therefore have a current given by

$$I = \frac{q\mu_n WH}{L}\Delta V \tag{1.215}$$

Thus, the resistance is given by

$$R = \frac{L}{q\mu_n WH} \tag{1.216}$$

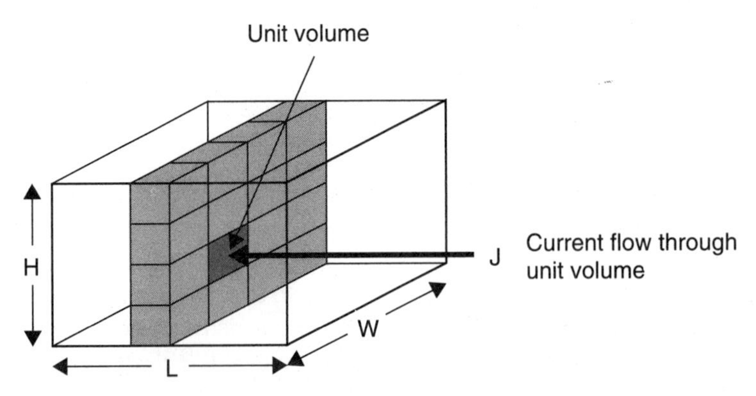

Fig. 1.36 Current flowing through a unit volume.

Often this resistance is presented in a relative manner, in which the length and width are removed (since they can be design parameters) but the height remains included. In this case, the resulting expression is commonly referred to as the *resistance per square* and designated as $R_\square$ where

$$R_\square = q\mu_n H \tag{1.217}$$

The total resistance is then given by

$$R_{total} = R_\square \frac{L}{W} \tag{1.218}$$

This equation is important when calculating the resistance of interconnects used in integrated circuits.

In the case of a MOS transistor, the charge density is not constant down the channel. If, instead of the carrier density per unit volume, one expresses $n(x)$ as a function of charge density per square area from the top looking down, we have

$$Q_n(x) = qHn(x) \tag{1.219}$$

Substituting (1.219) into (1.214) results in

$$\mu_n W Q_n(x)\,dV = I\,dx \tag{1.220}$$

Equation (1.220) applies to drift current through any structure that has varying charge density in the direction of the current flow. It can also be applied to a MOS transistor in the triode region to derive its I-V relationship. It should be noted here that in this derivation, it is assumed the source voltage is the same as the substrate voltage.

Since the transistor is in the triode region, we have $V_{DG} < -V_{tn}$. This requirement is equivalent to $V_{DS} < V_{GS} - V_{tn} = V_{eff}$. It is assumed that the effective channel length is L. Assuming the voltage in the channel at distance x from the source is given by $V_{ch}(x)$, from Fig. 1.37, we have

$$Q_n(x) = C_{ox}[V_{GS} - V_{ch}(x) - V_{tn}] \tag{1.221}$$

Substituting (1.221) into (1.220) results in

$$\mu_n W C_{ox}[V_{GS} - V_{ch}(x) - V_{tn}]dV_{ch} = I_D\,dx \tag{1.222}$$

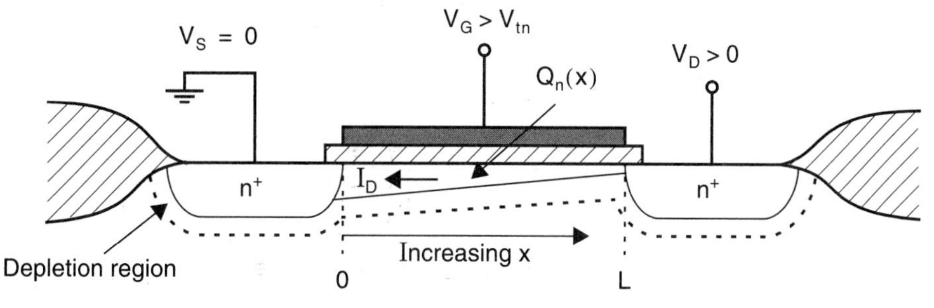

Fig. 1.37 The transistor definitions used in developing the transistor's I-V relationship.

Integrating both sides of (1.222), and noting that the total voltage along the channel of length L is V_{DS}, we obtain

$$\int_0^{V_{DS}} \mu_n W C_{ox}[V_{GS} - V_{ch}(x) - V_{tn}]\, dV_{ch} = \int_0^L I_D\, dx \tag{1.223}$$

which results in

$$\mu_n W C_{ox}\left[(V_{GS} - V_{tn})V_{DS} - \frac{V_{DS}^2}{2}\right] = I_D L \tag{1.224}$$

Thus, solving for I_D results in the well-known triode relationship for a MOS transistor:

$$I_D = \mu_n C_{ox}\left(\frac{W}{L}\right)\left[(V_{GS} - V_{tn})V_{DS} - \frac{V_{DS}^2}{2}\right] \tag{1.225}$$

It should be noted that taking into account the body effect along the channel, the triode model of (1.225) is modified to

$$I_D = \mu C_{ox}\left(\frac{W}{L}\right)\left[(V_{GS} - V_{tn})V_{DS} - \alpha\frac{V_{DS}^2}{2}\right] \tag{1.226}$$

where $\alpha \cong 1.7$ [Tsividis, 1987].

Bipolar Transistor Exponential Relationship

The various components of the base, collector, and emitter currents were shown in Fig. 1.28, on page 44. Figure 1.38 shows plots of the minority-carrier concentrations

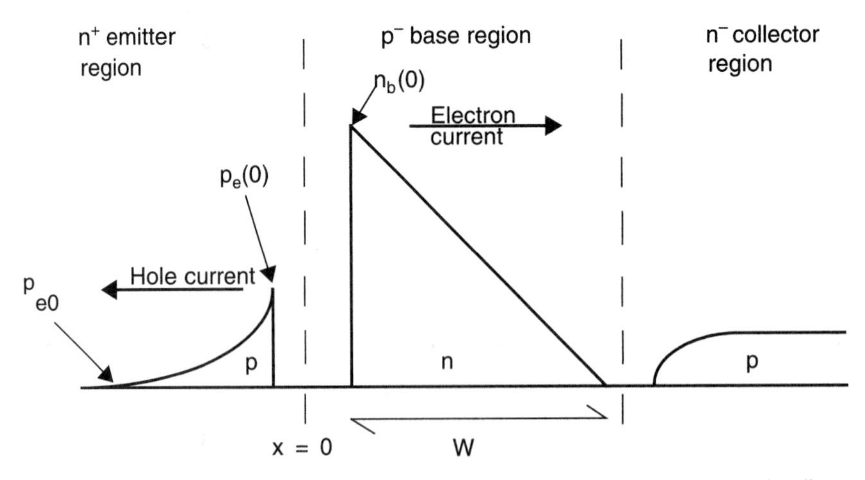

Fig. 1.38 The concentrations of minority carriers in the emitter, base, and collector.

in the emitter, base, and collector regions. The current flow of these minority carriers is due to diffusion. By calculating the gradient of the minority-carrier concentrations near the base-emitter junction in a manner similar to that used for diodes, it is possible to derive a relationship between the electron current and the hole current of Fig. 1.28.

The concentration of holes in the emitter at the edge of the base-emitter depletion region is denoted $p_e(0)$. This concentration decreases exponentially the farther one gets from the junction, in a manner similar to that described for diodes. The concentration far from the junction, p_e0, is given by

$$p_{e0} = \frac{n_i^2}{N_D} \tag{1.227}$$

where N_D is the doping density of the n^+ emitter. At a distance x from the edge of the emitter-base depletion region, we have

$$p_e(x) \cong p_e(0)e^{-x/L_p} \tag{1.228}$$

where

$$\begin{aligned} p_e(0) &= p_{e0}e^{V_{BE}/V_T} \\ &= \frac{n_i^2}{N_D}e^{V_{BE}/V_T} \end{aligned} \tag{1.229}$$

and where V_{BE} is the forward-bias voltage of the base-emitter junction.

At the edge of the base-emitter depletion region, the gradient of the hole concentration in the emitter is found, using (1.228), to be

$$\left.\frac{dp_e(x)}{dx}\right|_{x=0} = \frac{p_e(0)}{L_p} \tag{1.230}$$

Using (1.229), we can rewrite this as

$$\left.\frac{dp_e(0)}{dx}\right|_{x=0} = \frac{n_i^2}{L_pN_D}e^{V_{BE}/V_T} \tag{1.231}$$

The hole current is now found using the diffusion equation

$$I_{pe} = A_EqD_p\left.\frac{dp_e(x)}{dx}\right|_{x=0} \tag{1.232}$$

where A_E is the effective area of the emitter. Recall that the minority-hole current in the emitter, I_{pe}, is closely equal to the base current, I_B. After combining (1.231) and (1.232), we obtain

$$I_B = \frac{A_EqD_pn_i^2}{L_pN_D}e^{V_{BE}/V_T} \tag{1.233}$$

The situation on the base side of the base-emitter junction is somewhat different. The concentration of the minority carriers, in this case electrons that diffused from the emitter, is given by a similar equation,

$$n_b(0) = \frac{n_i^2}{N_A} e^{V_{BE}/V_T} \tag{1.234}$$

However, the gradient of this concentration at the edge of the base-emitter depletion region is calculated differently. This difference in gradient concentration is due to the close proximity of the collector-base junction, where the minority carrier (electron) concentration, $n_b(W)$, must be zero. This zero concentration at the collector-base junction occurs because any electrons diffusing to the edge of the collector-base depletion region immediately drift across the junction to the collector, as stated previously. If the base "width" is much shorter than the diffusion length of electrons in the base, L_n, then almost no electrons will recombine with base majority carriers (holes) before they diffuse to the collector-base junction. Given this fact, the decrease in electron or minority concentration from the base-emitter junction to the collector-base junction is a linear relationship decreasing from $n_b(0)$ at the emitter junction to zero at the collector junction in distance W. This assumption ignores any recombination of electrons in the base as they travel to the collector, which is a reasonable assumption for modern transistors that have very narrow bases. Thus, throughout the base region, the gradient of the minority-carrier concentration is closely given by

$$\begin{aligned} \frac{dn_b(x)}{dx} &= -\frac{n_b(0)}{W} \\ &= -\frac{n_i^2}{WN_A} e^{V_{BE}/V_T} \end{aligned} \tag{1.235}$$

Combining (1.235) with the diffusion equation, we obtain

$$\begin{aligned} I_{nb} &= -A_E q D_n \frac{dn_b(0)}{dx} \\ &= \frac{A_E q D_n n_i^2}{WN_A} e^{V_{BE}/V_T} \end{aligned} \tag{1.236}$$

Remembering that I_{nb} is closely equal to the collector current I_C, we have

$$I_C \cong I_{CS} e^{V_{BE}/V_T} \tag{1.237}$$

where

$$I_{CS} = \frac{A_E q D_n n_i^2}{WN_A} \tag{1.238}$$

The ratio of the collector current to the base current, commonly called the *transistor common-emitter current gain* and denoted β, is found using (1.237), (1.238), and

(1.233). We have

$$\beta = \frac{I_C}{I_B} = \frac{D_n N_D L_p}{D_p N_A W} \cong 2.5 \frac{N_D L_p}{N_A W} \tag{1.239}$$

which is a constant independent of voltage and current. Noting that $N_D >> N_A$, $L_p > W$, and $D_n \cong 2.5\ D_p$, we have $\beta >> 1$. A typical value might be between 50 and 200. The derivation of β just presented ignores many second-order effects that make β somewhat current and voltage dependent and are beyond the scope of this book. Interested readers should see Roulston, 1990, for more details. Regardless of second-order effects, equation (1.239) does reflect the approximate relationships among β, doping levels, and base width. For example, (1.239) explains why heavily doped emitters are important to achieve large current gain.

Base Charge Storage of an Active BJT

Figure 1.38 shows a minority-carrier storage in the base region, Q_b, given by

$$Q_b = A_E q \frac{n_b(0) W}{2} \tag{1.240}$$

Using (1.234) for $n_b(0)$, we have

$$Q_b = \frac{A_E q n_i^2 W}{2 N_A} e^{V_{BE}/V_T} \tag{1.241}$$

This equation can be rewritten using (1.237) and (1.238) to obtain

$$Q_b = \frac{W^2}{2 D_n} I_C = \tau_b I_C \tag{1.242}$$

where τ_b, called the *base-transit time constant,* is given approximately by

$$\tau_b = \frac{W^2}{2 D_n} \tag{1.243}$$

ignoring second-order effects. Normally, the base-transit time constant is specified for a given technology and takes into account other charge-storage effects not considered here, and is therefore often denoted τ_T. However, since the base storage of electrons dominates the other effects, we have $\tau_T \cong \tau_b$.

If the current in a BJT changes, the base charge storage must also change. This change can be modelled by a diffusion capacitance, C_d, between the base and the emitter terminals. Using (1.242), we have

$$C_d \cong \frac{dQ_b}{dV_{BE}} = \frac{d(\tau_b I_C)}{dV_{BE}} \tag{1.244}$$

Using (1.244) and $I_C = I_{CS}e^{V_{BE}/V_T}$ results in

$$C_d = \tau_b \frac{I_C}{V_T} \quad (1.245)$$

This equation is similar to that for a diode.

1.8 REFERENCES

R. Geiger, P. Allen, and N. Strader. *VLSI: Design Techniques for Analog and Digital Circuits*. McGraw-Hill, New York, 1990.

P. Gray and R. G. Meyer. *Analog Integrated Circuits,* 3rd. ed. John Wiley & Sons, New York, 1993.

D. Hodges and H. Jackson. *Analysis and Design of Digital Integrated Circuits,* 2nd ed. McGraw-Hill, New York, 1988.

D. Roulston. *Semiconductor Devices*. McGraw-Hill, New York, 1990.

S. M. Sze. *Physics of Semiconductor Devices*. Wiley Interscience, New York, 1981.

Y. Tsividis. *Operation and Modeling of the MOS Transistor*. McGraw-Hill, New York, 1987.

S. Wolf. *Silicon Processing for the VLSI Era—Volume 3: The Submicron MOSFET*. Lattice Press, Sunset Beach, California, 1995.

1.9 PROBLEMS

Unless otherwise stated, assume the following hold throughout the problems section:

- Room temperature = 300 °K
- npn bipolar transistors:
  $\beta = 100$
  $V_A = 80$ V
  $\tau_b = 13$ ps
  $\tau_s = 4$ ns
  $r_b = 330\ \Omega$
- n-channel MOS transistors:
  $\mu_n C_{ox} = 92\ \mu A/V^2$
  $V_{tn} = 0.8$ V
  $\gamma = 0.5\ V^{1/2}$
  $r_{ds}\ (\Omega) = 8000 L\ (\mu m)/I_D\ (mA)$ in active region
  $C_j = 2.4 \times 10^{-4}\ pF/(\mu m)^2$
  $C_{j\text{-}sw} = 2.0 \times 10^{-4}\ pF/\mu m$
  $C_{ox} = 1.9 \times 10^{-3}\ pF/(\mu m)^2$
  $C_{gs(overlap)} = C_{gd(overlap)} = 2.0 \times 10^{-4}\ pF/\mu m$
- p-channel MOS transistors:
  $\mu_p C_{ox} = 30\ \mu A/V^2$.
  $V_{tp} = -0.9$ V
  $\gamma = 0.8\ V^{1/2}$
  $r_{ds}\ (\Omega) = 12{,}000\ L\ (\mu m)/I_D\ (mA)$ in active region

$C_j = 4.5 \times 10^{-4}\ \text{pF}/(\mu\text{m})^2$
$C_{j\text{-sw}} = 2.5 \times 10^{-4}\ \text{pF}/\mu\text{m}$
$C_{ox} = 1.9 \times 10^{-3}\ \text{pF}/(\mu\text{m})^2$
$C_{gs(overlap)} = C_{gd(overlap)} = 2.0 \times 10^{-4}\ \text{pF}/\mu\text{m}$

1.1 Estimate the hole and electron concentrations in silicon doped with arsenic at a concentration of 10^{25} atoms/m^3 at a temperature 22 °C above room temperature. Is the resulting material n type or p type?

1.2 For the pn junction of Example 1.2, does the built-in potential, Φ_0, increase or decrease when the temperature is increased 11 °C above room temperature?

1.3 Calculate the amount of charge per $(\mu\text{m})^2$ in each of the n and p regions of the pn junction of Example 1.2 for a 5-V reverse-bias voltage. How much charge on each side would be present in a $10\ \mu\text{m} \times 10\ \mu\text{m}$ diode?

1.4 A silicon diode has $\tau_t = 12$ ps and $C_{j0} = 15$ fF. It is biased by a 43-kΩ resistor connected between the cathode of the diode and the input signal, as shown in Fig. P1.4. Initially, the input is 5 V, and then at time 0 it changes to 0 V. Estimate the time it takes for the output voltage to change from 5 V to 1.5 V (i.e., the $\Delta t_{-70\%}$ time). Repeat for an input voltage change from 0 V to 5 V and an output voltage change from 0 V to 3.5 V.

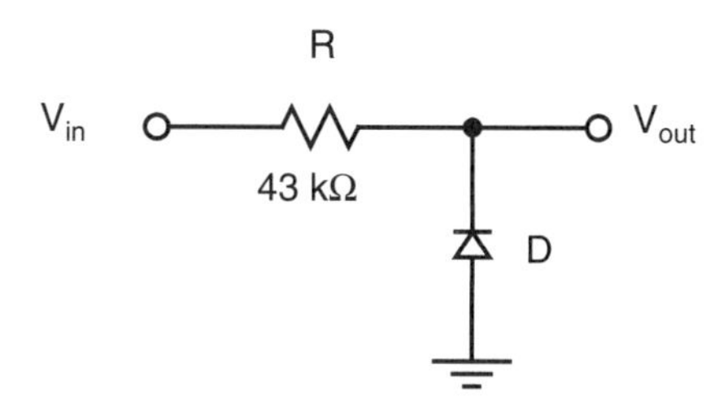

Fig. P1.4

1.5 Compare your answers for Problem 1.4 to those obtained using a SPICE simulation.

1.6 Verify that when $V_{DS} = V_{eff}$ is used in the triode equation for a MOS transistor, the current equals that of the active region equation given in (1.67).

1.7 Find I_D for an n-channel transistor having doping concentrations of $N_D = 10^{25}$ and $N_A = 10^{22}$ with $W = 50\ \mu\text{m}$, $L = 1.5\ \mu\text{m}$, $V_{GS} = 1.1\ \text{V}$, and $V_{DS} = V_{eff}$. Assuming λ remains constant, estimate the new value of I_D if V_{DS} is increased by $0.3\ \text{V}$.

1.8 A MOS transistor in the active region is measured to have a drain current of $20\ \mu\text{A}$ when $V_{DS} = V_{eff}$. When V_{DS} is increased by 0.5 V, I_D increases to $23\ \mu\text{A}$. Estimate the output impedance, r_{ds}, and the output impedance constant, λ.

1.9 Derive the low-frequency model parameters for an n-channel transistor having doping concentrations of $N_D = 10^{25}$ and $N_A = 10^{22}$ with $W = 10\ \mu\text{m}$, $L = 1.2\ \mu\text{m}$, $V_{GS} = 1.1$ V, and $V_{DS} = V_{eff}$. Assume that $V_{SB} = 1.0$ V.

1.10 Find the capacitances C_{gs}, C_{gd}, C_{db}, and C_{sb} for an active transistor having $W = 50\ \mu m$ and $L = 1.2\ \mu m$. Assume that the source and drain junctions extend $4\ \mu m$ beyond the gate, resulting in source and drain areas being $A_s = A_d = 200\ (\mu m)^2$ and the perimeter of each being $P_s = P_d = 58\ \mu m$.

1.11 Consider the circuit shown in Fig. P1.11, where V_{in} is a dc signal of 1 V. Taking into account only the channel charge storage, determine the final value of V_{out} when the transistor is turned off, assuming half the channel charge goes to C_L.

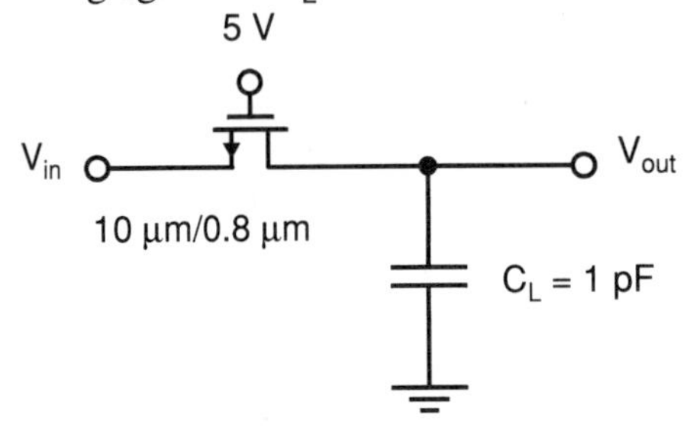

Fig. P1.11

1.12 For the same circuit as in Problem 1.11, the input voltage has a step voltage change at time 0 from 1 V to 1.2 V (the gate voltage remains at 5 V). Find its 99 percent settling time (the time it takes to settle to within 1 percent of the total voltage change). You may ignore the body effect and all capacitances except C_L. Also assume that $V_{tn} = V_{tn0}$. Repeat the question for V_{in} changing from 3 V to 3.1 V.

1.13 Repeat Problem 1.12, but now take into account the body effect on V_{tn}.

1.14 For an npn transistor having $I_C = 0.1$ mA, calculate g_m, r_π, r_e, r_0, and $g_m r_0$.

1.15 A bipolar junction transistor has the following SPICE parameters (the SPICE name for the parameter is included in parentheses):

I_S (IS) $= 2.0 \times 10^{-18}$
B_F (BF) $= 100$
B_R (BR) $= 1$
V_A (VA) $= 50$ V
τ_F (TF) $= 12 \times 10^{-12}$ s
τ_R (TR) $= 4 \times 10^{-9}$ s
C_{je0} (CJE) $= 15 \times 10^{-15}$ F
Φ_e (VJE) $= 0.9$ V
m_e (MJE) $= 0.27$
C_{jc0} (CJC) $= 18 \times 10^{-15}$ F
Φ_c (VJC) $= 0.7$ V
m_c (MJC) $= 0.37$
C_{js0} (CJS) $= 40 \times 10^{-15}$ F
Φ_s (VJS) $= 0.64$ V
m_s (MJS) $= 0.29$
R_e (RE) $= 30$

R_b (RB) = 500
R_c (RC) = 90

Initially, the circuit shown in Fig. P1.15 has a 0-V input. At time 0 its input changes to 5 V. Estimate the time it takes its output voltage to saturate, using the concepts of average capacitance and first-order transient solutions for each node. The time constants of the individual nodes can be added to arrive at an overall time constant for the approximate first-order transient response of the circuit. Next, assume that the input changes from 5 V to 0 V at time 0. How long does it take the output voltage to change to 3.5 V?

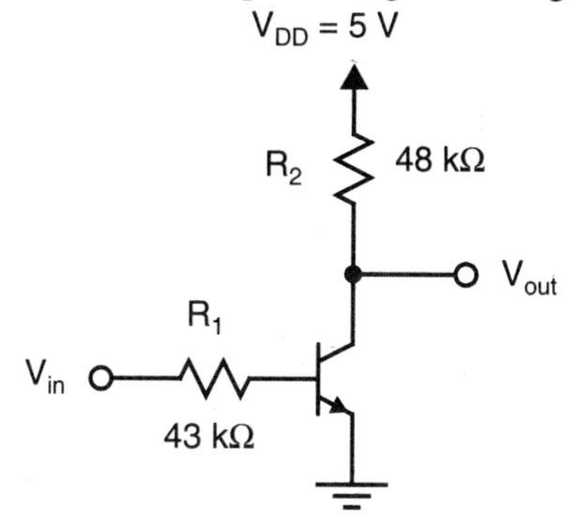

Fig. P1.15

1.16 Compare your answers to Problem 1.15 to those obtained using SPICE.

1.17 Verify that for $I_{BR} >> I_B$ and $I_{BR} >> I_C / \beta$, Eq. (1.140) simplifies to Eq. (1.139).

CHAPTER

2 Processing and Layout

This chapter describes the steps and processes used in realizing modern integrated circuits. While emphasis is placed on CMOS processing, the technology required for BJT circuits is also described. After processing is presented, circuit layout is covered. Layout is the design portion of integrated-circuit manufacturing, in which the geometry of circuit elements and wiring connections is defined. This process leads to the development of photographic masks used in manufacturing a microcircuit. The concepts of design rules and their relationship to microcircuits are emphasized. Next, circuit layout is related to the transistor models. Here, it is shown that once the layout is completed, the values of certain elements in the transistor models can be determined. This knowledge is necessary for accurate computer simulation of integrated circuits. It is also shown that, by using typical design rules, one can make reasonable assumptions to approximate transistor parasitic components before the layout has been done. Analog layout issues are then discussed, including matching and noise considerations. Finally, this chapter concludes with a discussion of a destructive mechanism in CMOS circuits known as latch-up.

2.1 CMOS PROCESSING

In this section, the basic steps involved in processing a CMOS microcircuit are presented. For illustrative purposes, we describe here an example n-well process (with, of course, a p substrate) and two layers of metal. Although the list is not complete, this section also describes many of the possible variations during processing.

The Silicon Wafer

The first step in realizing a microcircuit is to fabricate a defect-free, single-crystalline, lightly doped wafer. To realize such a wafer, one starts by creating metallurgical-grade silicon through the use of a high-temperature chemical process in an electrode-arc furnace. Although metallurgical-grade silicon is about 98 percent pure, it has far too many impurities for use in realizing microcircuits. Next, a silicon-containing gas is formed and then reduced. Pure silicon is precipitated onto thin rods of single-crystalline silicon. This deposited electronic-grade silicon is very pure but, unfortunately, it is also polycrystalline. To obtain single-crystalline silicon, the silicon is melted once again and allowed to cool. As it cools, a single-crystalline *ingot* is slowly pulled and turned from the molten silicon using the *Czochralski method.* The Czochralski method starts with a seed of single crystal silicon, and the pull rate and speed of rotation determine the diameter of the crystalline rod or ingot. Typical diameters might be 10 to 20 cm (i.e., 4 to 8 inches) with lengths usually longer than 1 meter. Producing a silicon ingot can take several days.

Normally, heavily doped silicon is added to the *melt* before the single-crystalline ingot is pulled. After the doped silicon diffuses through the molten silicon, a lightly doped silicon ingot results. In our example process, boron impurities would be added to produce a p^- ingot.

The ingot is cut into wafers using a large diamond saw. A typical wafer might have a thickness of about 1 mm. After the ingot is sawed into wafers, each wafer is polished with Al_2O_3, chemically etched to remove mechanically damaged material, and then fine-polished again with SiO_2 particles in an aqueous solution of NaOH.

Very often, the company that produces the silicon wafers is not the same company that eventually patterns them into monolithic circuits. Sometimes, the surface of the wafer might be doped more heavily, and a single-crystal epitaxial layer of the opposite type might be grown over its surface before the wafer-manufacturing company delivers the wafers to the processing company. This layered approach results in an *epitaxial wafer.*

A starting wafer of p^- might be doped around the level of $N_A \cong 2 \times 10^{21}$ donor/m^3. Such a doping level would give a *resistivity* of 10 to 20 $\Omega \cdot$ cm.

Photolithography and Well Definition

Photolithography is a technique in which selected portions of a silicon wafer can be *masked out* so that some type of processing step can be applied to the remaining areas. Although photolithography is used throughout the manufacturing of an integrated circuit, here we describe this photographic process in the context of preparing the wafer for defining the well regions.[1]

Selective coverage for well definition is performed as follows. First, a glass mask, M_1, is created, which defines where the well regions will be located. The glass mask is created by covering the mask in photographic materials and exposing it to an electron beam, or *e beam,* in the regions corresponding to the well locations. Such exposure results in the well regions on the glass mask turning opaque, or dark. As a result, the glass mask can be thought of as a negative of one layer of the microcircuit. In a typical microcircuit process, ten to twenty different masks might be required. The typical cost for these masks is currently around $50,000. Because a high degree of precision is required in manufacturing these masks, often a company other than the processing

1. Wells are doped regions that will contain one of the two types of transistors realized in a CMOS process. Nowadays, wells are normally n type and contain p-channel transistors.

company makes the masks. The exposure of the opaque regions of the mask, by the electron beam, is controlled by a computer dependent on the contents of a database. The database required for the e beam is derived from the layout database produced by the designer, using a computer and a layout CAD software program.

The first step in masking the surface of the wafer is to thermally grow a thin layer of silicon-dioxide (SiO_2) to protect the surface of the microcircuit. Details of this step are discussed later. On top of the SiO_2, a negative *photoresist*, PR_1, is evenly applied, while spinning the microcircuit to a thickness of around 1 μm. Photoresist is a light-sensitive polymer (similar to latex). In the next step, the mask, M_1, is placed in close proximity to the wafer, and ultraviolet light is projected through the mask onto the photoresist. Wherever the light strikes, the polymers cross-link, or *polymerize*. This change makes these regions insoluble to an organic solvent. This step is shown in Fig. 2.1. The regions where the mask was opaque (i.e., the well regions) are not exposed. The photoresist is removed in these areas using an organic solvent. Next, the remaining photoresist is baked to harden it. After the photoresist in the well regions is removed, the uncovered SiO_2 may also be removed using an acid etch. (However, in some processes where this layer is very thin, it may not be removed.) In the next step, the dopants needed to form the well are introduced into the silicon using either diffusion or ion implantation (directly through the thin oxide, in cases where it has not been removed).

The procedure just described involves a *negative photoresist,* where the exposed photoresist remains after the masking. There are also *positive photoresists,* in which the exposed photoresist is dissolved by the organic solvents. In this case, the photoresist remains where the mask was opaque. By using both positive and negative resists,

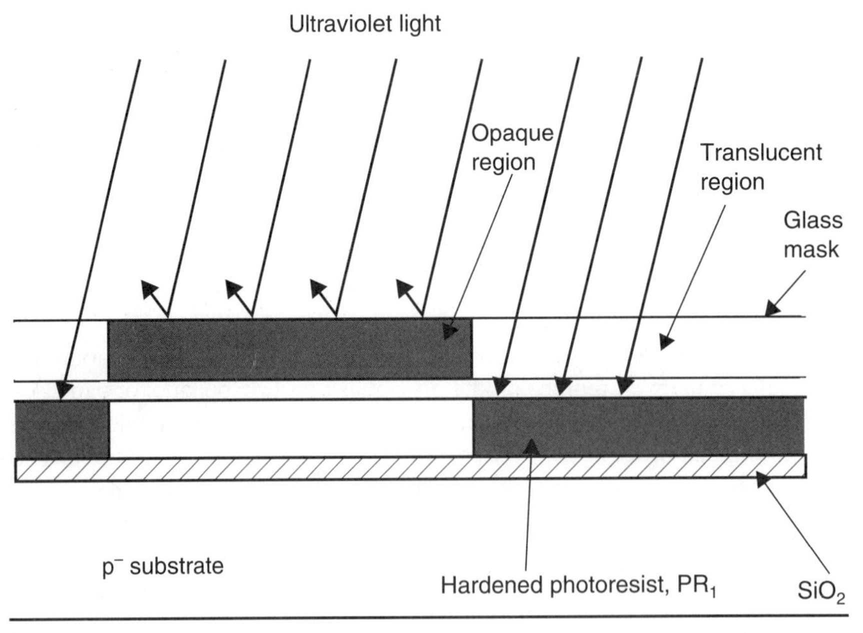

Fig. 2.1 Selectively hardening a region of photoresist using a glass mask.

a single mask can sometimes be used for two steps—first, to protect one region and implant the complementary region and second, to protect the complementary region and implant the original region.

Diffusion and Ion Implantation

After the photoresist over the well region has been removed, the next step is to introduce dopants through the opening where the well region will be located. As just mentioned, there are two approaches for introducing these dopants—diffusion and ion implantation.

In both implantation methods, usually the SiO_2 in the well region will first be removed using an acid etch. Next, the remaining hardened photoresist is stripped using acetone. This leaves SiO_2 that was protected by the hardened photoresist to mask all of the nonwell (i.e., substrate) regions.

In diffusion implantation, the wafers are placed in a quartz tube inside a heated furnace. A gas containing the dopant is introduced into the tube. In the case of forming an n well, the dopant in the gas would probably be phosphorus. Arsenic could also be used, but it takes a much longer time to diffuse. The high temperature of the diffusion furnace, typically 900 to 1,100 °C, causes the dopants to diffuse into the silicon both vertically and horizontally. The dopant concentration will be greatest at the surface and will decrease in a *Gaussian* profile further into the silicon. If a p well had been desired, then boron would have been used as the dopant. The resulting cross section, after diffusing the n well, is shown in Fig. 2.2.

An alternative technique for introducing dopants into the silicon wafer is ion implantation. This technique is now largely replacing diffusion because it allows more independent control over the dopant concentration and the thickness of the doped region. In ion implantation, dopants are introduced as ions into the wafer, as shown in the functional representation of an ion implanter in Fig. 2.3. The ions are generated by bombarding a gas with electrons from an arc-discharge or cold-cathode source. The ions are then focused and sent through a mass separator. This mass sepa-

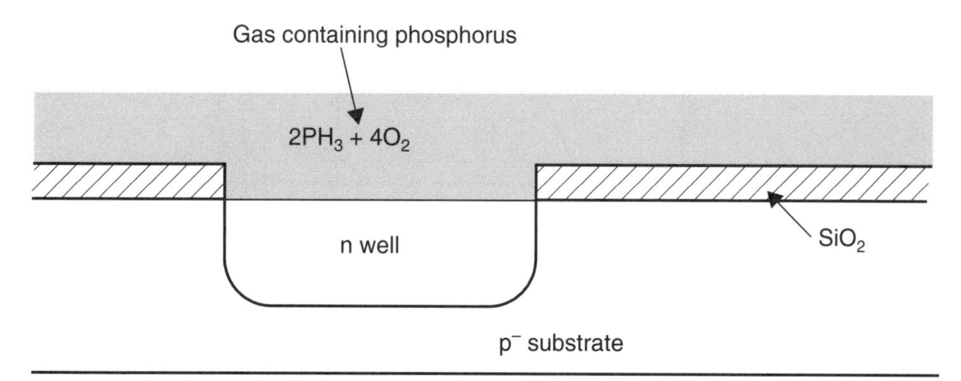

Fig. 2.2 Forming an n well by diffusing phosphorus from a gas into the silicon, through the opening in the SiO_2.

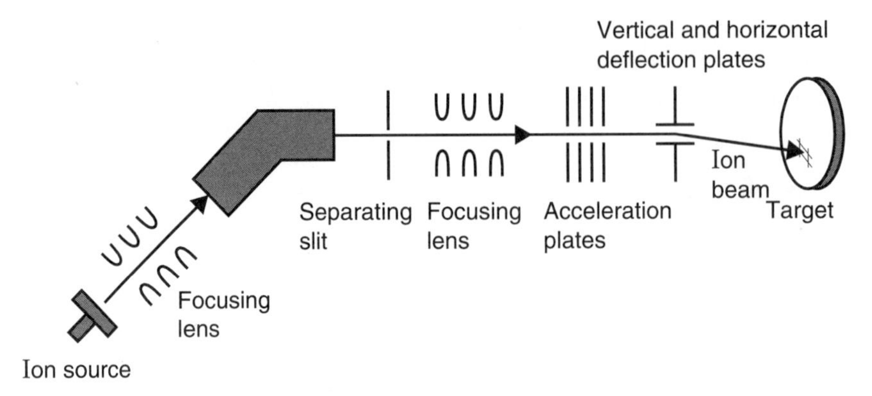

Fig. 2.3 An ion-implantation system.

rator bends the ion beam and sends it through a narrow slit. Since only ions of a specific mass pass through the slit, the beam is purified. Next, the beam is again focused and accelerated to between 10 keV and 1 MeV. The ion current might range from 10 μA to 2 mA. The deflection plates sweep the beam across the wafer (which is often rotated at the same time) and the acceleration potential controls how deeply the ions are implanted. The beam current and time of implantation determine the amount of dosage. Thus, depth and dosage are controlled independently. Two problems that occur with ion implantation are lattice damage and a narrow doping profile. The lattice damage is due to nuclear collisions that result in the displacement of substrate atoms. The narrow profile results in a heavy concentration over a narrow distance, as is shown in Fig. 2.4. For example, arsenic ions with an acceleration voltage of 100 keV might penetrate approximately 0.06 μm into the silicon, with the majority of

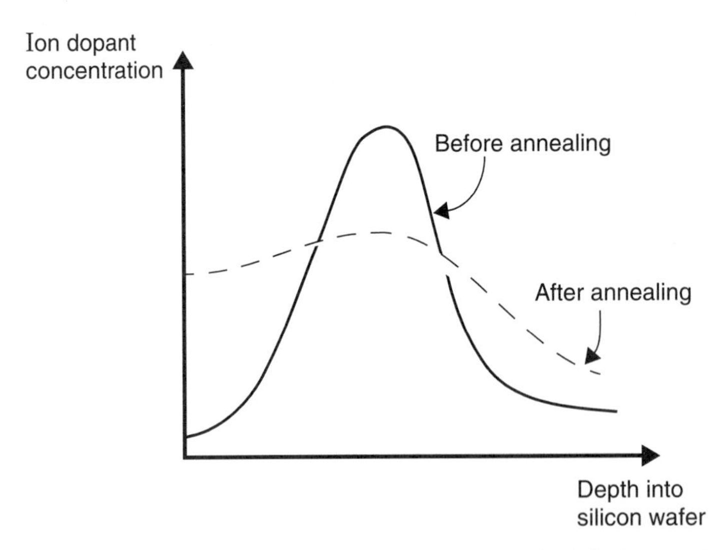

Fig. 2.4 Dopant profiles after ion implantation both before and after annealing.

ions being at $0.06\ \mu m \pm 0.02\ \mu m$. Both of these problems are largely solved by annealing.

Annealing is a step in which the wafer is heated to about 1,000 °C, perhaps for 15 to 30 minutes, and then allowed to cool slowly. This heating stage thermally vibrates the atoms, which allows the bonds to reform. Annealing also broadens the concentration profile, making the doping levels more uniform, as shown in Fig. 2.4. It should be noted that annealing is performed only once during processing, after all the implantation steps have been performed but before any metal layers have been created.[2]

For n-type dopants, arsenic is used for shallow implantations, such as the source or drain junctions. Phosphorus might be used for the well. Boron is always used to form the p regions.

Although more expensive, ion implantation has been largely replacing diffusion for forming n and p regions because of its greater control over doping levels. Another important advantage of ion implantation is the much smaller sideways diffusion, which allows devices to be more closely spaced and, more importantly for MOS transistors, minimizes the overlap between the gate-source or gate-drain regions.

Chemical Vapor Deposition and Defining the Active Regions

The next few steps use the field-oxide mask, M_2, to form the thick field-oxide as well as the field implants (used to isolate devices). These steps result in a thin layer of thermal SiO_2, as well as a layer of silicon nitride (Si_3N_4), everywhere the field-oxide is not desired.

Often, this step will be done using positive photoresist such that, wherever the mask M_2 is not opaque, the photoresist will be softened. In other words, the photoresist is left intact after the organic dissolution step, under the opaque regions of the mask where the field-oxide is not desired. A thin layer of thermal SiO_2 is first grown everywhere to protect the surface of the silicon lattice.

Next, Si_3N_4 is deposited everywhere during a gas-phase reaction in which energy is supplied by heat (at about 850 °C). This process is called *chemical vapor deposition,* or CVD. After this step, the positive photoresist is deposited, exposed through the mask, M_2, dissolved, and hardened. The hardened photoresist is left on top of the Si_3N_4 to protect it where the field-oxide is not desired. Next, the Si_3N_4, wherever it is not protected by the photoresist, 's removed by etching it away with a hot phosphoric acid. The SiO_2 is then removed with a hydrofluoric acid etch. Finally, the remaining photoresist is chemically removed with a process that leaves the remaining Si_3N_4 intact. The remaining Si_3N_4 will act as a mask to protect the active regions when the thick field-oxide is being grown in the field-oxide regions.

Field-Implants and the Field-Oxide

The next step in our example process is to implant the *field-implants* under where the *field-oxide* will be grown. For example, boron will be implanted under the field-oxide

2. If annealing were done after deposition of a metal layer, the metal would melt.

everywhere except in the well regions. This implant guarantees that the silicon under the field-oxide will never invert (or become n) when a conductor over the field-oxide has a large voltage. If this implant were not performed, leakage currents could occur between the junctions of separate n-channel transistors in the substrate region, which are intended to be unconnected.

For the field-oxide in the well regions, where p-channel transistors will eventually reside, an n-type implant such as arsenic (As) could be used. Often, it is not necessary to include field-implants under the field-oxide of the well regions because the heavier doping of the well (compared to that of the substrate) normally guarantees that the silicon will never invert under the field-oxide in these regions.

When implanting the field-implants in the substrate regions, it is necessary to first cover the wells with a protective photoresist, PR_3, so the n-well regions do not receive the p implant. This can be done using the same mask, M_1, that was originally used for implanting the n wells, but now a positive photoresist is used. This positive photoresist remains where the mask is opaque (i.e., dark), which corresponds to the well regions.

After the exposed photoresist has been dissolved, we now have the cross section shown in Fig. 2.5. Notice that at this step, all the active regions, where eventually the transistors will reside, are protected from the field implants by the Si_3N_4, SiO_2, and PR_2. Additionally, the complete well regions are also protected by PR_3. The field-implant will bc a high-cnergy implant with a fairly high doping level.

Growing the Field-Oxide

The next step is to grow the field-oxide, SiO_2. There are two different ways that SiO_2 can be grown. In a wet process, water vapor is introduced over the surface at a moderately high temperature. The water vapor diffuses into the silicon and, after some intermediate steps, reacts according to the formula

$$Si + 2H_2O \rightarrow SiO_2 + 2H_2 \tag{2.1}$$

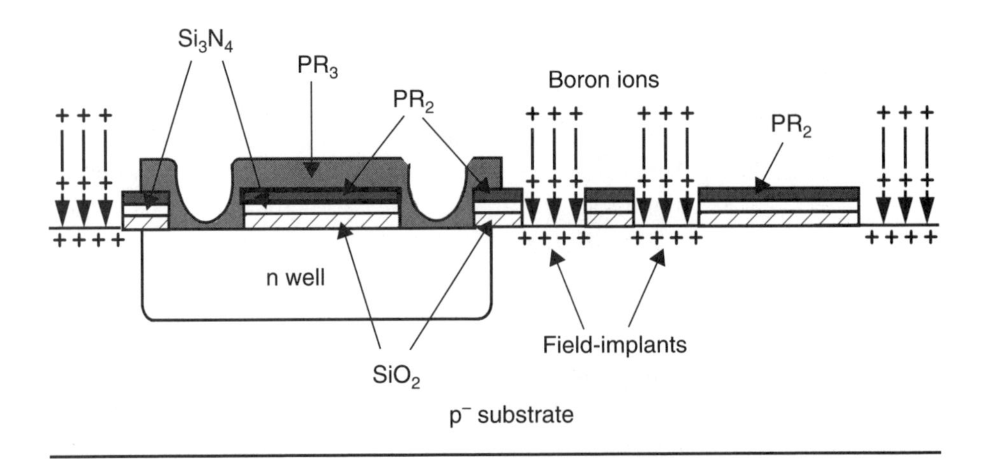

Fig. 2.5 The cross section when the field-implants are being formed.

In a dry process, oxygen is introduced over the wafer, normally at a slightly higher temperature than that used in the wet process, and reacts according to the formula

$$Si + O_2 \rightarrow SiO_2 \tag{2.2}$$

Since both of these processes occur at high temperatures, around 800 to 1200 °C, the oxide that results is sometimes called a *thermal oxide.*

Before the field-oxide is grown, PR_2 and PR_3 are removed, but the silicon-nitride–silicon-dioxide sandwich is left. The reaction does not occur wherever CVD-deposited Si_3N_4 remains, because the Si_3N_4 is relatively inert to both water and oxygen. Wherever the process does occur, the volume increases because oxygen atoms have been added. Specifically, SiO_2 takes up approximately 2.2 times the volume of the original silicon. This increase will cause the SiO_2 to extend approximately 45 percent into, and 55 percent above, what previously was the surface of the silicon. The resulting cross section is shown in Fig. 2.6. Note that in our example process, the field-oxide in the substrate region has field-implants under it, whereas the field-oxide in the wells does not.

When growing thermal SiO_2, the wet process is faster because H_2O diffuses faster in silicon than O_2 does, but the dry process results in denser, higher-quality SiO_2 that is less porous. Sometimes, growing the field-oxide starts with a dry process, changes to a wet process, and finishes with a dry process. When thin layers of SiO_2 are grown, as the next section describes, usually only a dry process is used.

Gate-Oxide and Threshold-Voltage Adjusts

In the next step, the Si_3N_4 is removed using hot phosphoric acid. If a thin layer of SiO_2 is under the Si_3N_4, protecting the surface, as shown in Fig. 2.6, this SiO_2 is also removed, usually with hydrofluoric acid. The high-quality, thin gate-oxide is then grown using a dry process. It is grown everywhere over the wafer to a thickness of about 0.01 to 0.03 μm.

After the gate-oxide has been grown, donors are implanted so that the final threshold voltages of the transistors are correct. Note that this implantation is performed directly through the thin gate-oxide since it now covers the entire surface. It should be mentioned here that many processes differ while realizing the threshold-adjust step.

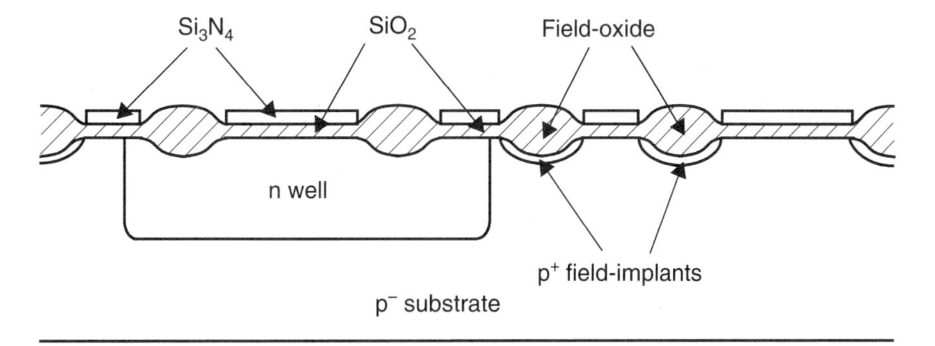

Fig. 2.6 The cross section after the field-oxide has been grown.

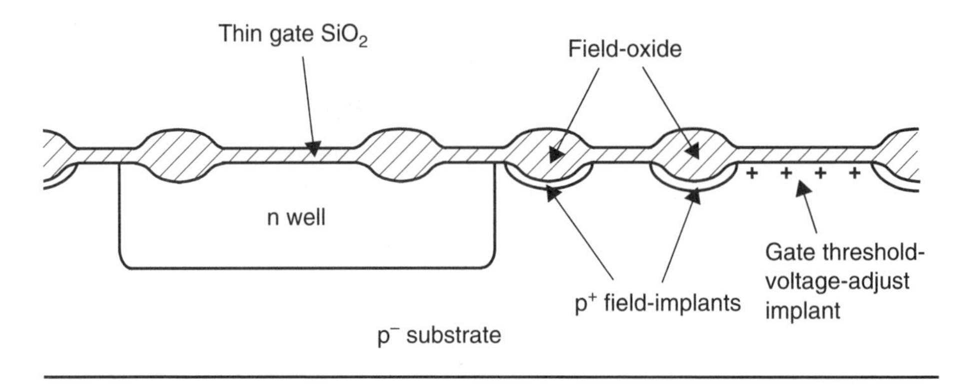

Fig. 2.7 Cross section after the thin gate-oxide growth and threshold-adjust implant.

In a simple process, the threshold voltages of both the p- and n-channel transistors are adjusted at the same time. We saw in the Appendix of Chapter 1 that the n-channel transistors require a boron implant to change V_{tn} from its native value of around −0.1 V to its desired value of 0.7 to 0.8 V. If the n wells are doped a little heavier than ideal, the native threshold voltage of the p-channel transistors in the well will be around −1.6 V. As a result, the same single boron threshold-adjust implant will bring it to around −0.8 to −0.9 V.

By using a single threshold-voltage-adjust implant for both n-channel and p-channel transistors, two photoresist masking steps are eliminated. If the different transistors are individually implanted, then the second of two types of transistors has to be protected by, say, a negative photoresist while the first type is being implanted. Next, a positive photoresist can be used with the same mask to protect the first type of transistor while the second type is being implanted. The mask used is normally the same mask used in forming the n wells, in other words, M_1. Thus, no additional mask is required, but a number of additional processing steps are needed. The major problem with using a single threshold-adjust implant is that the doping level of the n well is higher than optimum. This higher doping level increases the junction capacitances and the body effect of the transistors in the well. A double threshold adjust allows optimum well doping. Both approaches are currently in commercial use, although the double threshold-adjust implant is growing in favor as device dimensions decrease. The cross section at this stage is shown in Fig. 2.7.

Polysilicon Gate Formation

The next step in the process is the chemical deposition of the polysilicon gate material. One method to create polysilicon is to heat a wafer with silane gas flowing over it so the following reaction occurs

$$SiH_4 \rightarrow Si + 2H_2 \tag{2.3}$$

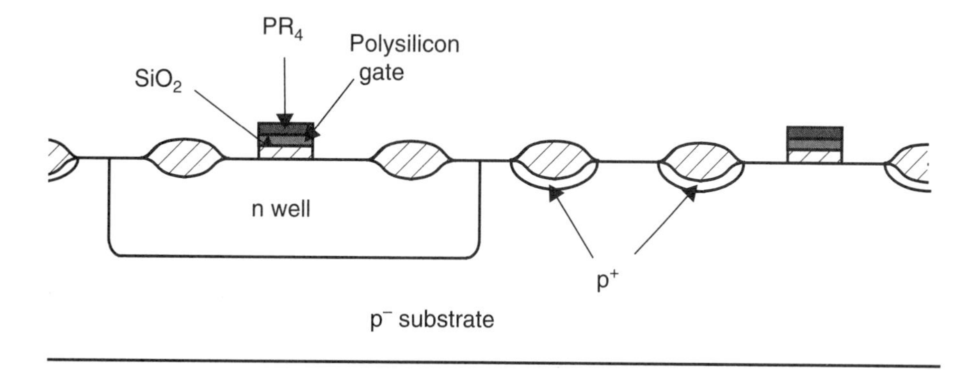

Fig. 2.8 Cross section after depositing and patterning the polysilicon gates.

If this reaction occurs at high temperatures, say, around 1,000 to 1,250 °C, and the original surface of the wafer was single crystal, the deposited silicon will also be single crystal. This approach is used both when epitaxial layers are grown in bipolar processes and in some of the more modern CMOS processes. However, when depositing the polysilicon gates, the original surface is SiO_2 and the wafer is heated only to about 650 °C. As a result, the silicon that is deposited is noncrystalline, or amorphous. Thus, this silicon is often referred to as *polysilicon.* Very often, after the polysilicon is deposited, it will be ion implanted with arsenic to increase its conductivity. A typical final resistivity for polysilicon might be 10 to 30 $\Omega/\square$, and its thickness might be around 0.25 μm.

After the deposition just described, the polysilicon gate material covers the entire wafer. This polysilicon is then patterned using a new mask, M_3, and a positive photoresist, PR_4. The mask is opaque where hardened polysilicon should remain. After the nonhardened photoresist is removed, the polysilicon is etched away using a reactive plasma etch. This etch removes all of the polysilicon not protected by photoresist but removes very little of the underlying SiO_2. This thin gate-oxide layer is used to protect the surface during the next step of junction implantation. The cross section at this phase is shown in Fig. 2.8.

Implanting the Junctions, Depositing SiO_2, and Opening Contact Holes

The next step involves the ion implantation of the junctions. In our example process, the p^+ junctions are formed first by placing positive photoresist, PR_5, everywhere except where the p^+ regions are desired. A new mask, M_4, is used in this step. The p^+ regions are then ion implanted, possibly through a thin oxide in some processes. The cross section at this stage is shown in Fig. 2.9.

Notice that the p^+ junctions of the p-channel transistors are defined on one edge by the field-oxide and, more importantly, next to the active gate area by the edge of the polysilicon gate. During the implantation of the boron, it was the gate polysilicon and the photoresist over it that protected the channel region from the p^+ implant.

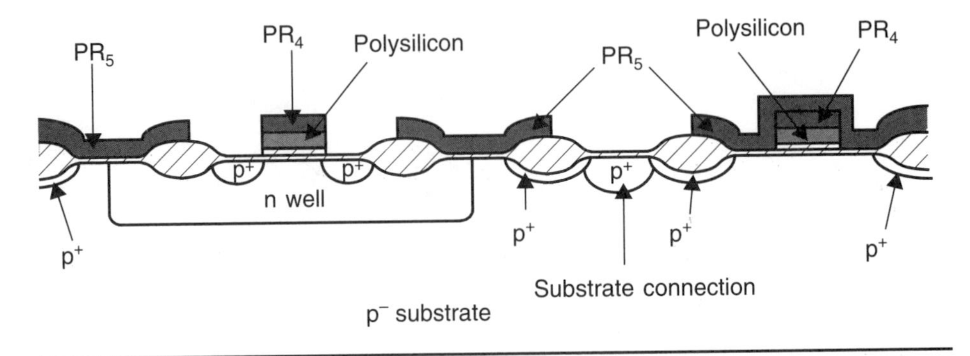

Fig. 2.9 Cross section after ion-implanting the p^+ junctions.

Thus, the p^+ junctions are *self-aligned* to the polysilicon gates, resulting in very little overlap (i.e., a small L_{ov}, as defined in Chapter 1). Also, note that the effective channel areas of the transistors are defined by the intersection of the gate-defining mask, M_3, and the mask used in defining the active regions, M_2 (i.e., the mask used in defining where Si_3N_4 remains). Thus, these are the two most important masks in any MOS process. The development of this self-aligned process has proven to be an important milestone in realizing small high-speed transistors, both in MOS processes and in BJT processes.

Also notice that a p^+ junction has been implanted in the substrate region. This junction, called a *substrate tie,* is used to connect the substrate to ground in microcircuits. These substrate ties are liberally placed throughout the microcircuit to help prevent latch-up, a problem discussed at the end of this chapter. In addition, the underside of the wafer would normally be connected to ground as well, through a package connection.

Next, the photoresists are all removed using acetone. The p^+ active regions are then protected using the same mask, M_4, that was used for the previous step, but now a negative photoresist, PR_6, is used. The n^+ junctions are then implanted using arsenic. The cross section at the end of this stage is shown in Fig. 2.10.

After the junctions have been implanted and PR_6 has been removed, the complete wafer is covered in CVD SiO_2. This protective *glass* layer can be deposited at moder-

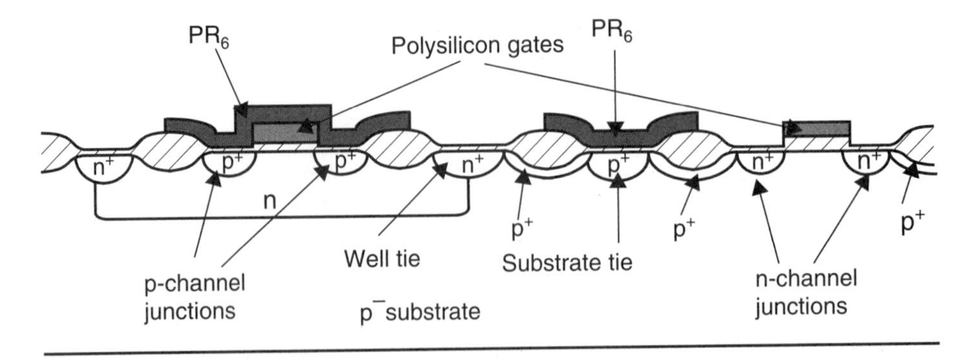

Fig. 2.10 Cross section after ion-implanting the n^+ junctions.

ately low temperatures of 500 °C or lower. The deposited SiO_2 might be 0.25 to 0.5 μm.

The next step is to open contact holes through the deposited SiO_2. The contact holes are defined using mask M_5 and positive resist PR_7.

Annealing, Depositing and Patterning Metal, and Overglass Deposition

After the first layer of CVD SiO_2 has been deposited, the wafer is annealed. As mentioned earlier in this section, annealing entails heating the wafer in an inert gas (such as nitrogen) for some period of time (say, 15 to 30 minutes) at temperatures up to 1,000 °C. The resulting thermal vibrations heal the lattice damage sustained during all the ion implantations, broaden the concentration profiles of the implanted dopants, and increase the density of the deposited SiO_2.

Next, interconnect metal is deposited everywhere. Historically, aluminum (Al) has been used for the interconnect. However, recently other metals have been used that have less of a tendency to diffuse into the silicon during electrical operation of the microcircuit. The metal is deposited using evaporation techniques in a vacuum. The heat required for evaporation is normally produced by using electron-beam bombarding, or possibly ion bombarding in a sputtering system. After the metal is deposited on the entire wafer, it is patterned using mask M_6 and positive photoresist PR_8, and then it is etched.

At this time, a low-temperature annealing might take place to give better bonds between the metal and the silicon. The temperature of this annealing must be less than 550 °C so the aluminum doesn't melt.

Next, an additional layer of CVD SiO_2 is deposited, additional contact holes are formed using mask M_7 and photoresist PR_9, and then a second layer of metal is deposited and etched using mask M_8 and photoresist PR_{10}. Often the primary use of this top layer of metal might be to distribute the power supply voltages. The bottom layer would be used more often for local interconnects in gates. In some modern processes, this process is repeated a third and possibly a fourth time to give up to four levels of metal, which allow for much denser interconnect.

After the last level of metal is deposited, a final *passivation*, or *overglass,* is deposited for protection. This layer would be CVD SiO_2, although often an additional layer of Si_3N_4 might be deposited because it is more impervious to moisture.

The final microcircuit processing step is to etch openings to the pads used for wire bonding. This final step would use mask M_9 and photoresist PR_{11}. A cross section of the final microcircuit for our example process is shown in Fig. 2.11.

This example process is a fairly representative CMOS process. However, many variations, often involving additional masks, are possible. Some of the possible variations are as follows:

1. Two wells may exist—one for p-channel transistors and one for n-channel transistors. This *twin-tub process* allows both wells to be optimally doped.
2. An additional polysilicon layer may be deposited over the first layer. This extra poly layer can be used to realize highly linear poly-to-poly capacitors in which a thin thermal oxide is used to separate the two layers.

3. An additional polysilicon layer might be formed that has an extremely high resistivity (say, $1\ G\Omega/\square$). This high resistivity is used to realize resistor loads in four-transistor, static random-access memory (SRAM) cells.
4. Field-implants may exist under the field-oxide in the well regions as well as under the field-oxide in the substrate regions.
5. Often, the n-channel and the p-channel transistors might have separate threshold-voltage-adjust implants.
6. The microcircuit might have three, four, or five layers of metal.
7. In a multimetal-layer process, it is usually necessary to add several additional steps so that the surface is made smoother, or *planarized,* after each metal-patterning step. This is normally done by a reactive etching process in which the metal is covered with SiO_2 and the hills are etched faster than the valleys.
8. Different metals might be used for the contacts than for the interconnect, to obtain better fill in and less diffusion into the silicon surface.
9. Thin-film nichrome resistors may exist under the top layer of metal.
10. The transistors may be realized in an epitaxial layer. In this case, the substrate would be n^-, and a p^--epitaxial layer would be grown. Before the epitaxial layer is grown, the top of the substrate would be doped p^+. This type of wafer is similar to that used in processing bipolar transistors, but of the opposite type, and is becoming more common in CMOS processing. The advantages of this type of transistor are that it is more immune to a destructive phenomenon called latch-up (described at the end of this chapter), and it is also more immune to gamma radiation in space. Finally, it greatly minimizes substrate noise in microcircuits that have both analog and digital circuits, (i.e., mixed-mode microcircuits).
11. Additional processing steps may be used to ensure that bipolar transistors can be included in the same microcircuit as MOS transistors. This type of process is called a BiCMOS process and is becoming particularly popular for both analog and digital high-speed microcircuits.

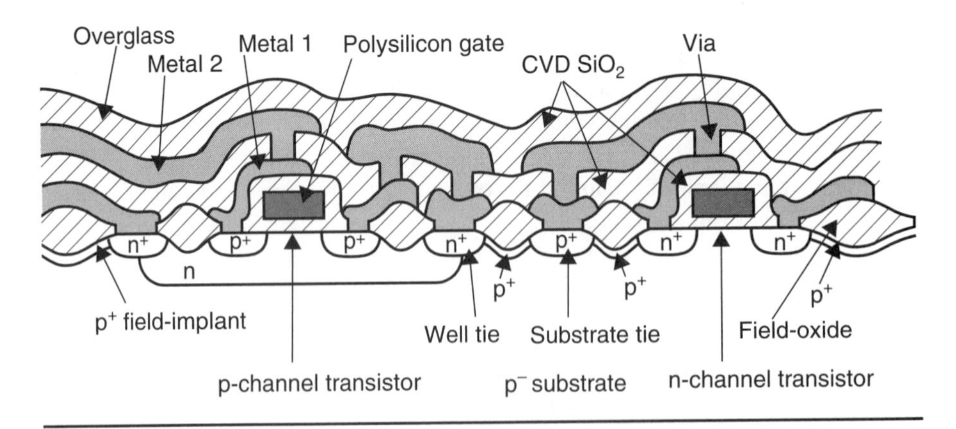

Fig. 2.11 Final cross section of an example CMOS microcircuit.

2.2 BIPOLAR PROCESSING

The processing steps required for realizing bipolar transistors are similar to those used for realizing MOS transistors, with some modifications. Thus, rather than presenting the complete realization of modern bipolar transistors, we briefly discuss some of the modifications needed for realizing them.

A bipolar process normally starts with a p^- substrate. The first masking step involves the diffusion (or ion implantation) of n^+ regions into the substrate wherever transistors are desired. These n^+ regions are used to lower the series collector resistance. Next, an n^- single-crystal epitaxial layer is deposited.

In the next basic step, the field-oxide is formed for isolation. However, before the field-oxide is grown and after the openings in the Si_3N_4-SiO_2-photoresist sandwich have been made, the surface of the silicon is typically etched to form empty cavities. This extra etching step allows the field-oxide to extend further down into the epitaxial region.

After the field oxide is grown, the n^+-collector contact region is implanted. This region extends from the surface down to the n^+-buried region under the transistor.

In a modern process, polysilicon is used to contact the emitter, the base, and possibly the collector, as Fig. 2.12 shows for a typical npn BJT structure. Normally, the base p^+ polysilicon is deposited first. This polysilicon is heavily doped p^+ so that later, during a high-temperature step, the boron dopant from the polysilicon contact diffuses into the silicon underneath the base polysilicon to make the underlying region p^+. The base polysilicon is removed in the active area of the transistor. Next, using one of a variety of possible methods, the base polysilicon is covered with a thin layer of

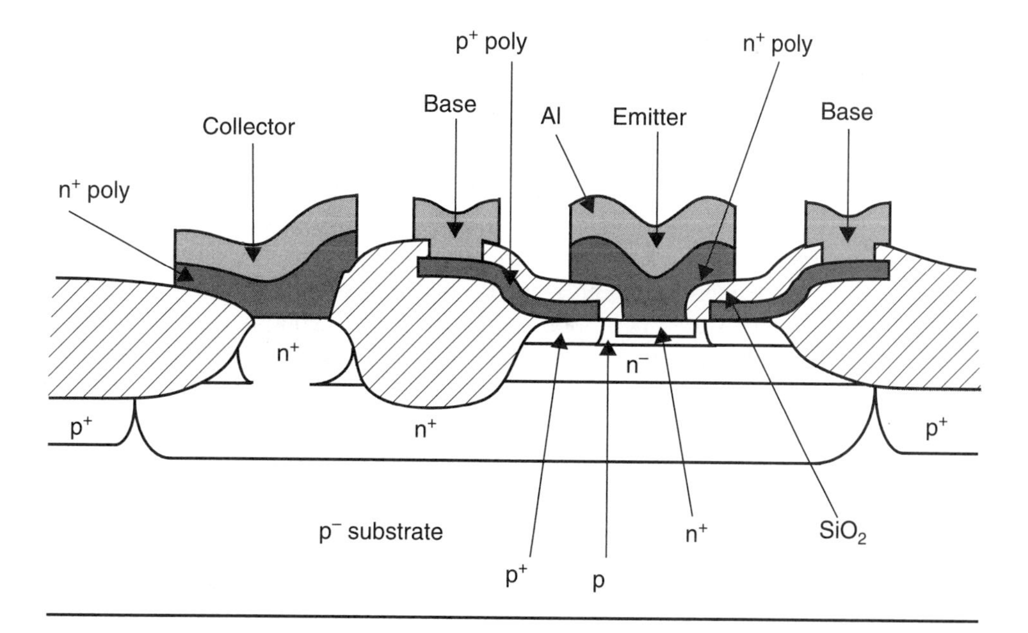

Fig. 2.12 Cross section of a modern, self-aligned bipolar transistor with oxide isolation. The term "poly" refers to polysilicon.

SiO_2, perhaps 0.5 μm in thickness, leaving an opening over the active area of the transistor. This SiO_2 *spacer,* which separates the base polysilicon from the emitter polysilicon, as shown in Fig. 2.12, allows the base polysilicon (or contact) to be very close to the emitter polysilicon (or contact), thereby minimizing the base resistance. Next, the base is ion-implanted to p -type silicon, and then n^+ polysilicon is deposited for the emitter. At this point, the true emitter has not yet been formed—only the emitter n^+ polysilicon has been laid down. However, when the wafer is annealed, the n^+ from the emitter polysilicon diffuses into the base p silicon to form the true emitter region. During annealing, the p^+ dopants from the base polysilicon also diffuse into the extrinsic base region. As a result, this procedure results in a self-aligned process since the use of the SiO_2 spacer allows the base polysilicon to determine where the emitter is finally located. The importance of this process is that, through the use of self-aligned contacts and field-oxide isolation, very small, high-frequency bipolar transistors can be realized using methods similar to those used in realizing modern MOS transistors.

2.3 CMOS LAYOUT AND DESIGN RULES

It is the designer's responsibility to determine the geometry of the various masks required during processing. The process of defining the geometry of these masks is known as *layout* and is done using a computer and a CAD program. Here, we describe some typical layout design rules and the reasons for these rules.

When designing the layout, typically the designer does not need to produce the geometry for all of the masks because some of the masks are automatically produced by the layout program. For example, the p^+ and n^+ masks used for the source and drain regions are usually generated automatically. Also, the program might allow the designer to work in the final desired dimensions. The layout program then automatically sizes the masks to account for any lateral diffusion or etching loss; this sizing produces larger- or smaller-dimension masks. For example, a designer might draw a polysilicon line so that a transistor would have a 1.2-μm length. The program might then produce a mask that had a 1.6-μm line width. This increased mask sizing would account for the junction overlap due to lateral diffusion and the polysilicon loss due to etching.

In a modern layout program, the layout of some circuit *cells* might already be performed and stored in a library. During overall layout, these cells are then parametrically adapted to a required size, and the corresponding geometries for every layer are automatically generated. Often, when the cells are being connected, they might be automatically *placed and routed*, or connected, by the program. The designer might then interactively modify this automatically generated layout. Thus, as time goes on, the layout becomes more automated as more cells become available. However, the designer must still take direct control of the layout of critical cells, especially when the layout must be small or the resulting circuits must be fast. For example, one would rarely allow a computer to automatically generate the layout of a memory cell where space and capacitive loading of the connecting buses are critical. Thus, a digital microcircuit designer must be knowledgeable about the design rules that govern the layout required for the process used.

The two most important masks are those for the active region and for the gate polysilicon. The intersection of these two masks becomes the channel region of MOS transistors. For example, consider Fig. 2.13(*a*), which shows a simplified view of a MOS transistor, and Fig. 2.13(*b*), which shows the corresponding layout of the active mask and the polysilicon, or poly, mask. In Fig. 2.13(*b*), the poly mask runs vertically. The length of the poly that intersects the active-region mask is the transistor width, W, and the width of the poly line is the transistor length, L, as Fig. 2.13 shows.

The design rules for laying out transistors are often expressed in terms of a quantity, λ, where λ is 1/2 the gate length. This generalization allows many of the design rules to be simply expressed, independent of the true value for the minimum channel length (i.e., 2λ). Figure 2.13(*b*) shows the smallest possible transistor that can be realized when a contact must be made to each junction. Also shown are many of the minimum dimensions in terms of λ.

When we express design rules in terms of λ, we assume that each mask has a worst-case alignment of under 0.75λ. Thus, we can guarantee that the relative misalignment between any two masks is under 1.5λ. If an overlap between any two regions of a microcircuit would cause a destructive short circuit, then a separation

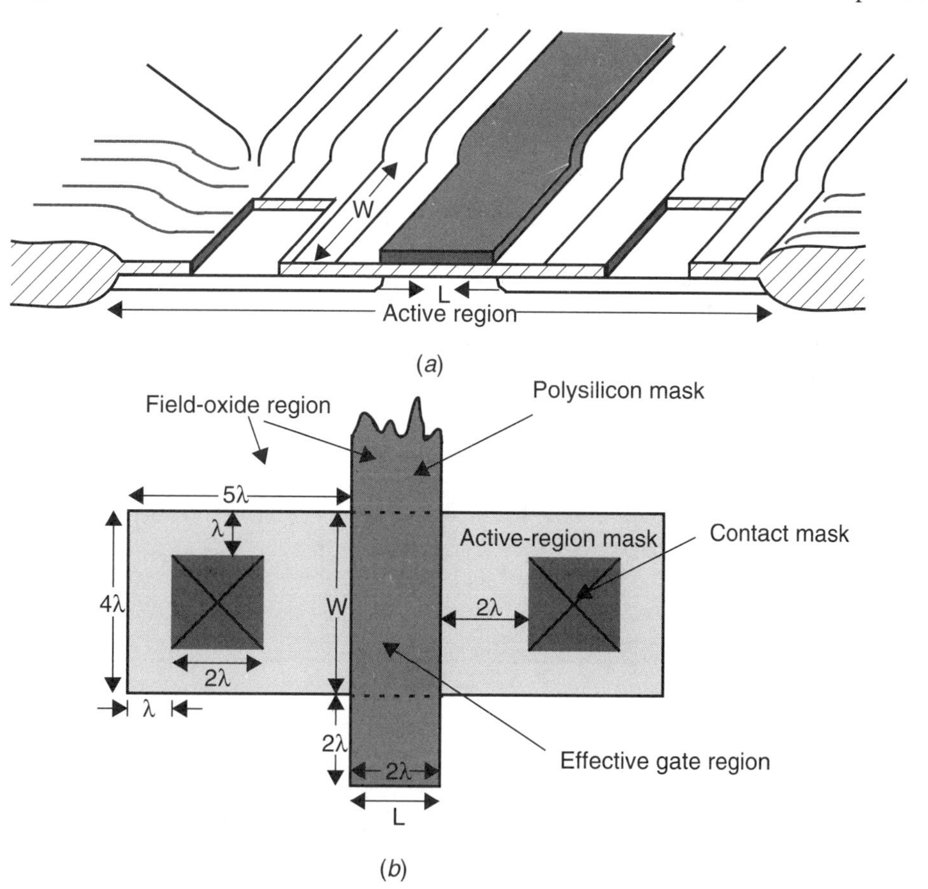

Fig. 2.13 (*a*) A simplified view of a partially finished transistor and (*b*) the corresponding layout of the active, polysilicon, and contact masks.

between the corresponding regions in a layout of 2λ guarantees this will never happen. For example, consider the poly mask and the contact mask in Fig. 2.13(*b*). If these two regions overlap in the microcircuit, then the metal used to contact the source junction is also short-circuited to the gate poly, causing the transistor to be always turned off, as shown in Fig. 2.14. If the source happens to be connected to ground, this error also short-circuits the gate-to-ground. To prevent this type of short, the contact openings must be kept at least 2λ away from the polysilicon gates.

Another example of a catastrophic failure due to misalignment is a gate that does not fully cross the active region (also shown in Fig. 2.14). Since the junctions are implanted everywhere in the active region except under the gate, this misalignment causes a short circuit between the source and the drain—thus the design rule that polysilicon must always extend at least 2λ past the active region.

Another design rule is that active regions should surround contacts by at least 1λ. If, in reality, an overlap exists between the edge of the active-region mask and the contact mask, no disastrous shorts occur. The circuit still works correctly as long as sufficient overlap exists between the contact and the active masks so that a good connection is made between the aluminum interconnect and the junction. Since the maximum relative misalignment is 1.5λ, having the source (or drain) region surround the contact by 1λ guarantees an overlap of at least 1.5λ (the minimum contact width is 2λ).

The few design rules just described are sufficient to allow one to estimate the minimum dimensions of a junction area and perimeter before a transistor has been laid out. For example, assume that in Fig. 2.13 a contact is to be made to a junction; then the active region must extend past the polysilicon region by at least 5λ. Thus, the

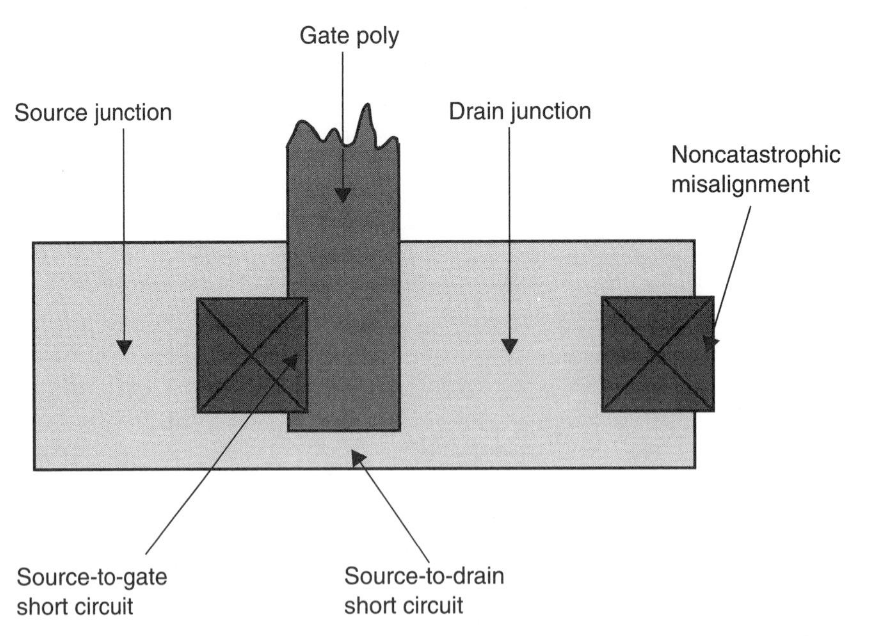

Fig. 2.14 Mask misalignment that results in catastrophic short circuits and an example of a noncatastrophic misalignment.

minimum area of a small junction with a contact to it is

$$A_s = A_d = 5\lambda W \tag{2.4}$$

where W is the transistor width. Similarly, in Fig. 2.13, the perimeter of a junction[3] with a contact is given by

$$P_s = P_d = 10\lambda + W \tag{2.5}$$

These estimates may be used when estimating the parasitic capacitances in the transistor models. They may also be used in SPICE to simulate circuits so the parasitic capacitances are determined more accurately. However, note that they are only estimates; the true layout will differ somewhat from these rough estimates.

Sometimes, when it is important to minimize the capacitance of a junction, a single junction can be shared between two transistors. For example, consider the series connection of two transistors shown in Fig. 2.15(*a*). The active, poly, and contact

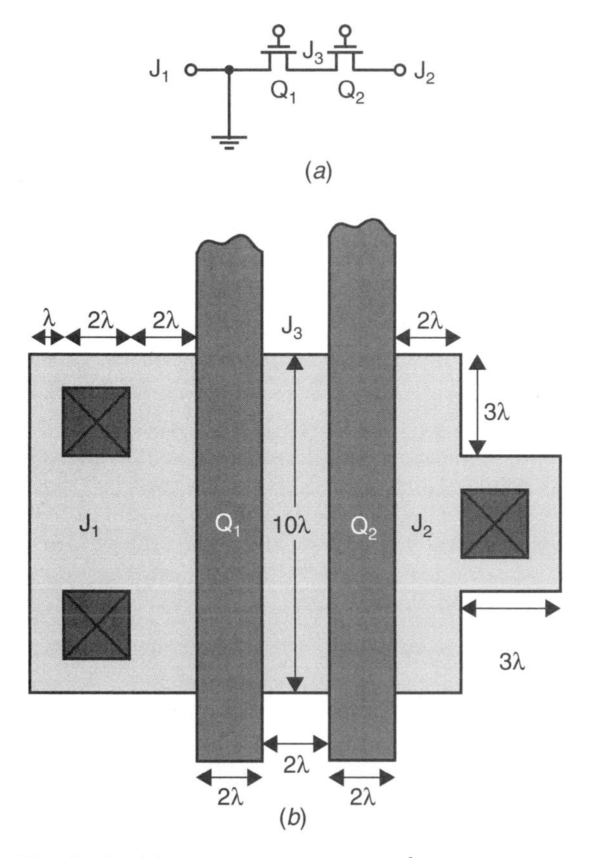

Fig. 2.15 (*a*) A series connection of two transistors and (*b*) a possible layout.

3. Note that the perimeter does not include the edge between the junction and the active channel separating the junction and the gate because there is no field-implant along this edge, and the sidewall capacitance is therefore smaller along that edge.

masks might be laid out as shown in Fig. 2.15(*b*). Notice that a single junction is shared between transistors Q_1 and Q_2. The area, and especially the perimeter of this junction, are much smaller than those given by equations (2.4) and (2.5). Also, in a SPICE simulation, the area and perimeter should be divided by 2 when they are specified in each transistor description, since the junction is shared. Alternatively, all of the area and perimeter could be specified in one transistor description, and the area and perimeter of the other junction could be specified as zero.

Since the junction sidewall capacitance is directly proportional to the junction perimeter, and since this capacitance can be a major part of the total junction capacitance (because of the heavily doped field-implants), minimizing the perimeter is important. It is of interest to note that as transistor dimensions shrink, the ratio of the perimeter to the area increases and the sidewall capacitance becomes more important.

EXAMPLE 2.1

Assuming $\lambda = 0.5\ \mu m$, find the area and perimeters of junctions J_1, J_2, and J_3 for the circuit in Fig. 2.15.

Solution

Since the width and length are shown as 10λ and 2λ, respectively, and $\lambda = 0.5\ \mu m$, the physical sizes are $W = 5\ \mu m$ and $L = 1\ \mu m$.

Thus, for junction J_1, using the formulas of (2.4) and (2.5), we have

$$A_{J1} = 5\lambda W = 5(0.5)5(\mu m)^2 = 12.5(\mu m)^2 \tag{2.6}$$

and

$$P_{J1} = 10\lambda + W = [10(0.5) + 5]\ \mu m = 10\ \mu m \tag{2.7}$$

Since this junction is connected to ground, its parasitic capacitance is unimportant and little has been done to minimize its area. Contrast this case with junction J_2, where we have

$$\begin{aligned} A_{J2} &= 2\lambda W + 12\lambda^2 \\ &= W + 12(0.5)^2 \\ &= 8(\mu m)^2 \end{aligned} \tag{2.8}$$

The perimeter is unchanged, resulting in $P_{J2} = 10\ \mu m$. Thus, we have decreased the junction area by using the fact that the transistor is much wider than the single contact used. However, sometimes wide transistors require additional contacts to minimize the contact impedance. For example, the two contacts used for junction J_1 result in roughly half the contact impedance of junction J_2.

Next, consider the shared junction. Here we have a junction area given by

$$\begin{aligned} A_{J3} &= 2\lambda W \\ &= 5(\mu m)^2 \end{aligned} \tag{2.9}$$

Since this is a shared junction, in a SPICE simulation we would use

$$\begin{aligned} A_s &= A_d = \lambda W \\ &= 2.5(\mu m)^2 \end{aligned} \tag{2.10}$$

for each of the two transistors, which is much less than 12.5 $(\mu m)^2$. The reduction in the perimeter is even more substantial. Here we have

$$\begin{aligned} P_{J3} &= 4\lambda \\ &= 2\ \mu m \end{aligned} \tag{2.11}$$

for the shared junction; so sharing this perimeter value over the two transistors would result in

$$\begin{aligned} P_S &= P_d = 2\lambda \\ &= 1\ \mu m \end{aligned} \tag{2.12}$$

for the appropriate junction of each transistor when simulating it in SPICE. This result is much less than the 10-μm perimeter for node J_1.

Because minimizing the junction capacitance is so important, one of the first steps an experienced designer takes before laying out important high-speed cells is first to identify the most critical nodes and then to investigate possible layouts that minimize the junction capacitance of these nodes.[4]

An additional design rule has been implicitly introduced in the previous example. Notice that for junction J_2 in Fig. 2.15, part of the active region boundary is only 2λ away from the gate. This minimum junction area is the typical design rule for this case.

Several design rules are required in addition to those just mentioned. Some of these are described next, with reference to the layout of a digital inverter, shown in Fig. 2.16. Notice that the n well surrounds the p-channel active region, and therefore the p^+ junctions of the p-channel transistors, by at least 3λ. Notice also that the minimum spacing between the n well and the junctions of n-channel transistors, in the substrate, is 5λ. This large spacing is required because of the large lateral diffusion of the n well and the fact that if the n-channel junction became short-circuited to the n well, which is connected to V_{DD}, the circuit would not work. Conversely, a p^+-substrate tie can be much closer to a well because it is always connected to ground and is separated from the well by a reverse-biased junction. A typical dimension here might be 2λ. Since a p-channel junction must be inside the well by at least 3λ and an n-channel junction must be outside the well by 5λ, the closest an n-channel transistor can be placed to a p-channel transistor is 8λ.

Notice in Fig. 2.16 that metal is used to connect the junctions of the p-channel and n-channel transistors. Normally, the metal must overlap any underlying contacts

4. Note that it is not possible to share junctions between n-channel and p-channel transistors. This limitation is one of the reasons for the larger parasitic capacitances sometimes encountered in CMOS microcircuits as opposed to NMOS microcircuits, where only n-channel transistors are used.

by at least λ. A typical minimum width for first-level metal might be 2λ, the same as the minimum width for polysilicon. However, it can be wider as in Fig. 2.16, where it is 4λ wide.

Notice also in Fig. 2.16 that a single contact opening, known as a *butting contact*, can be used to contact both the p-channel transistor source and an n^+-well tie, because both will be connected to V_{DD}. Although the outlines of the p^+ and n^+ masks are not shown in Fig. 2.16, under the contact, one half will be doped p^+ (the p-channel junction) and one half will be doped n^+ (the well tie). Also, for the n-channel transistor, a

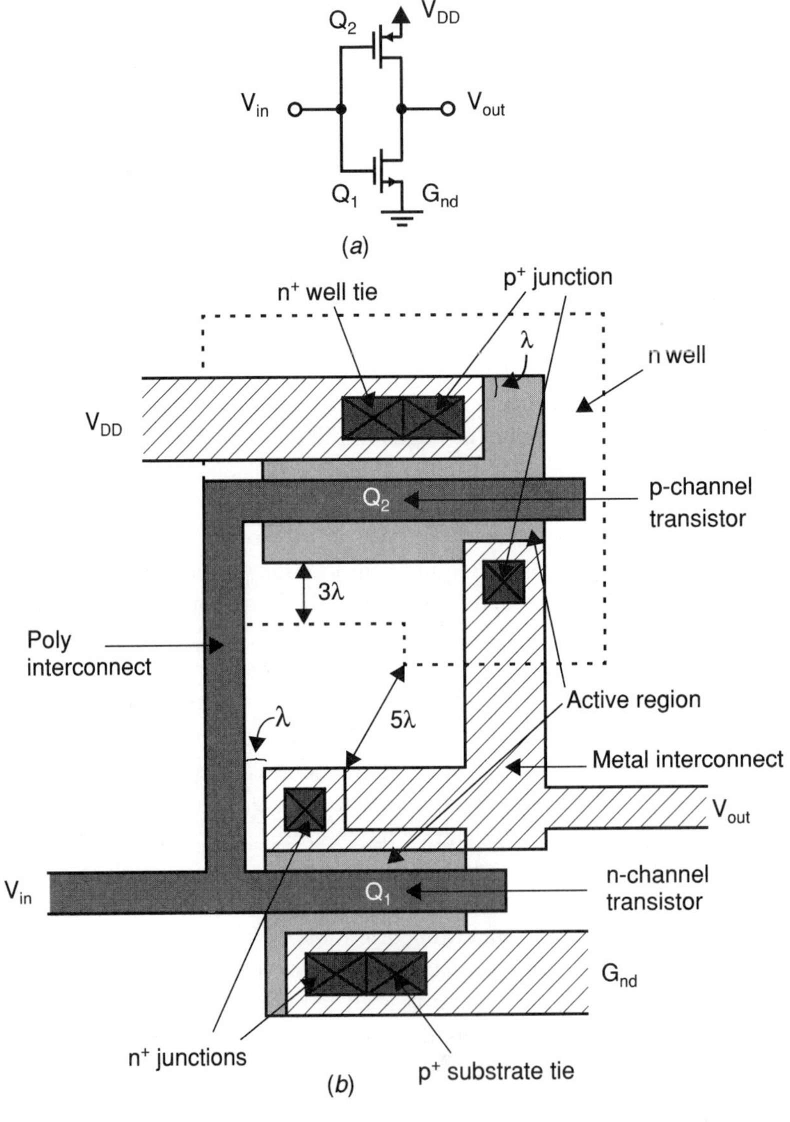

Fig. 2.16 (*a*) A CMOS digital inverter and (*b*) a possible layout with several design rules illustrated.

butting contact was used to connect the n-channel source to a p^+-substrate tie, and both will be connected to ground. In a typical set of design rules, a maximum distance between transistors and well (or substrate) ties is specified, and a maximum distance between substrate ties is also specified. For example, the rules might specify that no transistor can be more than 100λ from a substrate tie. These rules are necessary to prevent latch-up, a phenomenon described at the end of this chapter.

As a final example, we describe the layout of a large transistor. Normally, a wide transistor is composed of smaller transistors connected in parallel. A simplified layout of this approach is shown in Fig. 2.17(*a*), where four transistors that have a common gate are connected in parallel. Figure 2.17(*b*) shows the circuit corresponding to the layout in Fig. 2.17(*a*), where the transistors have been drawn in the same relative positions. Figure 2.17(*c*) shows the same circuit redrawn differently, where it is clear that the circuit consists of four transistors connected in parallel. Notice that the second and fourth junction regions are connected by metal to node 1, whereas the first, third, and fifth junction regions are connected by metal to realize node 2. Because it has a larger total junction area and especially a larger perimeter, node 2 will have a much greater junction capacitance than node 1. Thus, when the equivalent transistor is connected to a circuit, node 1 should be connected to the more critical node. Also notice the large number of contacts used to minimize the contact impedance. Normally, some of these are butting contacts if either node 1 or node 2 is connected to an appropriate power supply. The use of many contacts in wide junction regions greatly minimizes voltage drops that would otherwise occur due to the relatively high resistivity of silicon junctions compared to the resistivity of the metal that overlays the junctions and connects them.[5]

Design rules also specify the minimum pitch between polysilicon interconnects, metal 1 interconnects, and metal 2 interconnects. These might be 2λ, 2λ, and 3λ, respectively. Metal 2 requires a larger minimum pitch because it resides further from the silicon surface where the topography is less even. The minimum widths of poly, metal 1, and metal 2 might also be 2λ, 2λ, and 3λ, respectively.

This concludes our brief introduction to layout and design rules. In a modern process, many more design rules are used than those just described. However, the reasons for using and the methods of applying these rules is similar to that which has been described. Finally, note that when one does modern microcircuit layout, the design rules are usually available to the layout program and are automatically checked as layout progresses.

EXAMPLE 2.2

Consider the transistor shown in Fig. 2.17, where the total width of the five parallel transistors is 80λ, its length is 2λ, and $\lambda = 0.5\ \mu m$. Assuming node 2 is the source, node 1 is the drain, and the device is in the active region, find the

5. The use of metal to overlay a higher-resistivity interconnect, such as polysilicon or heavily-doped silicon, is recommended to lower the resistivity of the interconnect.

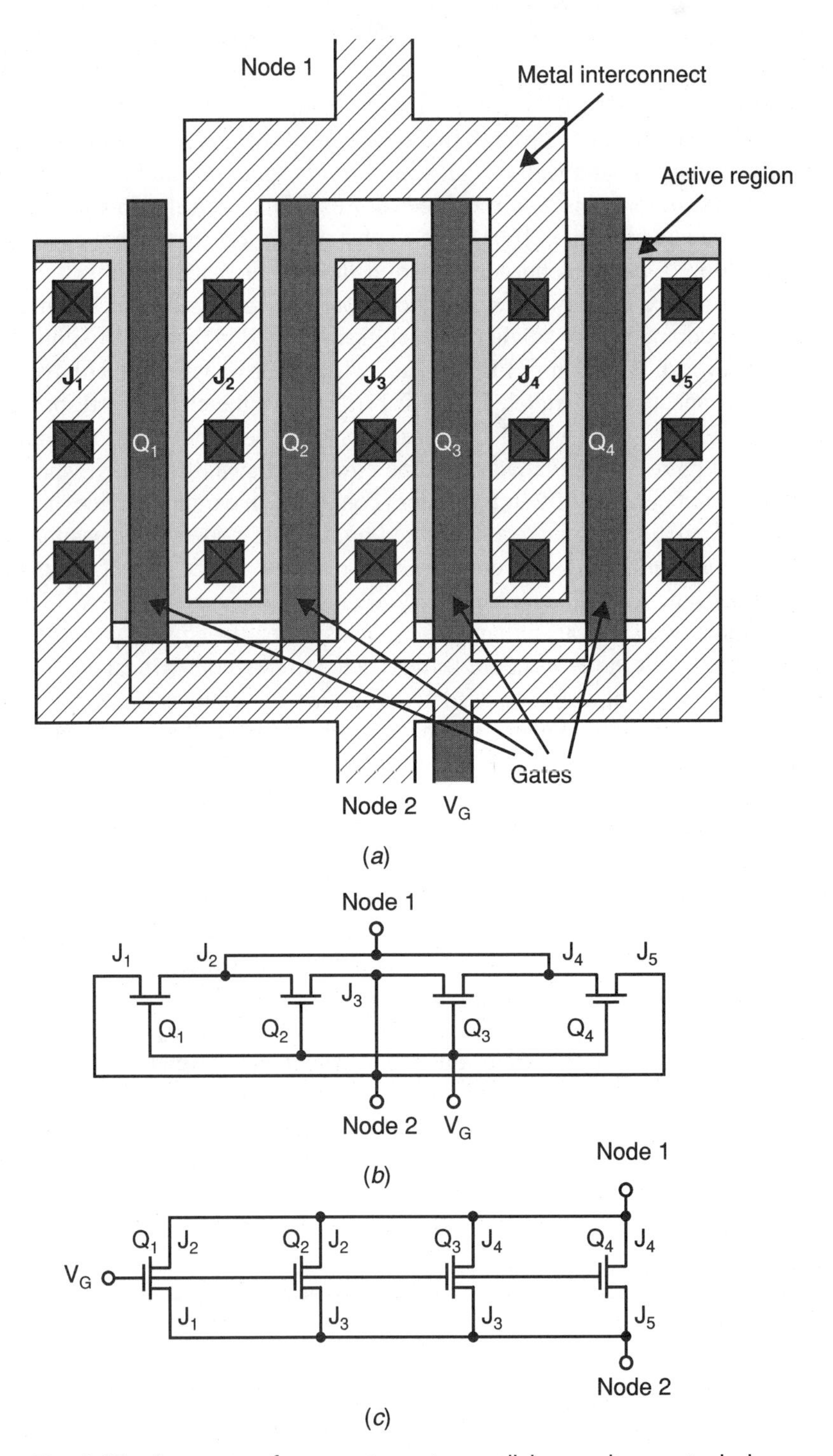

Fig. 2.17 Connecting four transistors in parallel to realize a single large transistor: (*a*) the layout, (*b*) the schematic drawn in the same relative positions as the layout, and (*c*) the circuit redrawn to make the parallel transistors more obvious.

source-bulk and drain-bulk capacitances given the parameters $C_j = 2.4 \times 10^{-4}\ \text{pF}/(\mu\text{m})^2$ and $C_{j\text{-sw}} = 2.0 \times 10^{-4}\ \text{pF}/\mu\text{m}$. Also find the equivalent capacitances if the transistor were realized as a single device with source and drain contacts still evenly placed.

Solution

Starting with node 1, the drain, we find that the areas of the junctions are equal to

$$A_{J2} = A_{J4} = 6\lambda \times 20\lambda = 120\lambda^2 = 30\ (\mu\text{m})^2$$

Ignoring the gate side, the perimeters are given by

$$P_{J2} = P_{J4} = 6\lambda + 6\lambda = 12\lambda = 6\ \mu\text{m}$$

As a result, C_{db} can be estimated to be

$$C_{db} = 2(A_{J2}C_j + P_{J2}C_{j\text{-sw}}) = 0.017\ \text{pF}$$

For node 2, the source, we have

$$A_{J1} = A_{J5} = 5\lambda \times 20\lambda = 100\lambda^2 = 25\ (\mu\text{m})^2$$

and

$$A_{J3} = A_{J2} = 30\ (\mu\text{m})^2$$

The perimeters are found to be

$$P_{J1} = P_{J5} = 5\lambda + 5\lambda + 20\lambda = 30\lambda = 15\ \mu\text{m}$$

and

$$P_{J3} = P_{J2} = 6\ \mu\text{m}$$

resulting in an estimate for C_{sb} of

$$C_{sb} = (A_{J1} + A_{J3} + A_{J5} + WL)C_j + (P_{J1} + P_{J3} + P_{J5})C_{j\text{-sw}} = 0.036\ \text{pF}$$

It should be noted that, even without the additional capacitance due to the WL gate area, node 1 has less capacitance than node 2 since it has less area and perimeter.

In the case where the transistor is a single wide device, rather than four transistors in parallel, we find

$$A_J = 5\lambda \times 80\lambda = 400\lambda^2 = 100\ (\mu\text{m})^2$$

and

$$P_J = 5\lambda + 5\lambda + 80\lambda = 90\lambda = 45\ \mu\text{m}$$

resulting in $C_{db} = 0.033$ pF and $C_{sb} = 0.043$ pF. Note that in this case, C_{db} is nearly twice what it is when four parallel transistors are used.

2.4 ANALOG LAYOUT CONSIDERATIONS

When one designs analog circuits, several important layout issues should be considered to realize high-quality circuits. These issues can be broadly divided into two categories—matching and noise issues.

Matching Issues

When integrated components are realized using lithographic techniques, a variety of two-dimensional effects can cause the effective sizes of the components to differ from the sizes of the glass layout masks. Some examples of these effects are illustrated in Fig. 2.18. For example, Fig. 2.18(*a*) shows how an effective well area will typically be larger than its mask due to the lateral diffusion that occurs not just during ion implantation but also during later high-temperature steps, such as annealing. Another effect, known as *overetching,* occurs when layers such as polysilicon or metal are being etched. Figure 2.18(*b*), for example, shows overetching that occurs under the SiO_2 protective layer at the polysilicon edges and causes the polysilicon layer to be smaller than the corresponding mask layout. A third effect is shown in Fig. 2.18(*c*), where an n-channel transistor is shown as we look along the channel from the drain to the source. The width of the transistor is defined by the width of the active region (as opposed to the width of the polysilicon line), and this width is determined by the field-oxide. The p^+ field implant under the field-oxide causes the effective substrate doping to be greater at the sides of the transistors than elsewhere. This increased doping raises the effective transistor threshold voltage near the sides of the transistors and therefore decreases the channel-charge density at the edges. The result is that the effective width of the transistor is less than the width drawn on thc layout mask. Thc examples shown illustrate typical sizing effects, but many other second-order effects influence the realized compo-

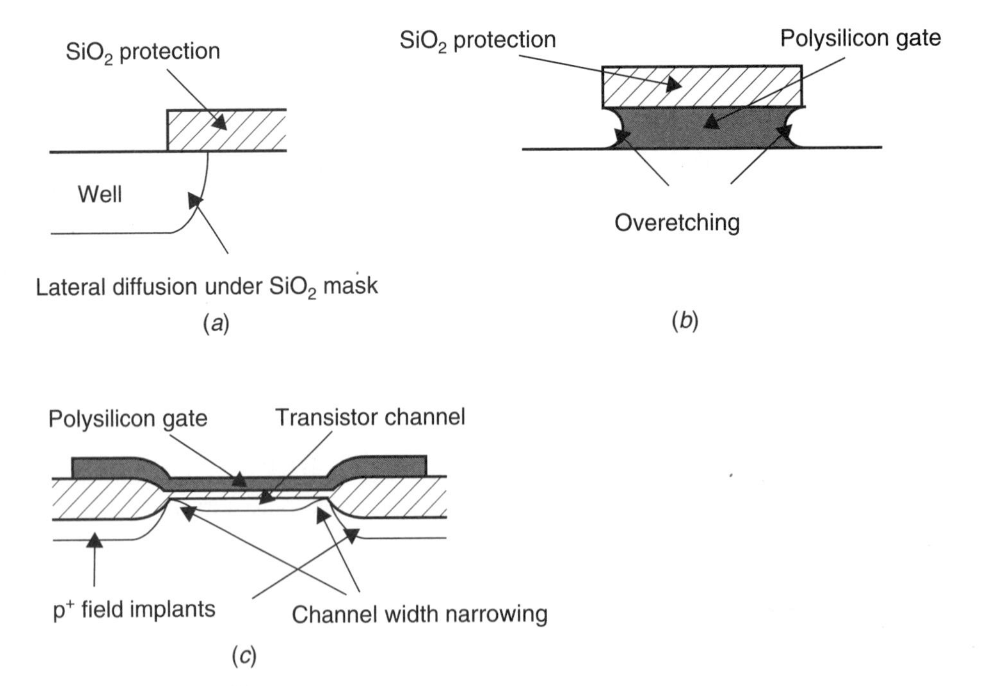

Fig. 2.18 Various two-dimensional effects causing sizes of realized microcircuit components to differ from sizes of layout masks.

nent sizes. These other effects include those caused by boundary conditions of an object, the size of the opening in a protective layout through which etching occurs, and the unevenness of the surface of the microcircuit [Maloberti, 1994]. For these reasons, the absolute sizes of microcircuit components can seldom be accurately determined. These inaccuracies also affect the ratios of sizes, although to a lesser degree, when the ratio is not unity (unless the second-order size error effects are matched). *Matching second-order size error effects is done mainly by making larger objects out of several unit-sized components connected together. Also, for best accuracy, the boundary conditions around all objects should be matched, even when this means adding extra unused components.* Next, we give some examples of how to apply these principles with regard to realizing transistors, capacitors, and resistors.

Transistor Layouts

Transistors in analog circuits are typically much wider than transistors in digital circuits. For this reason, they are commonly laid out using multiple-gate fingers similar to the layout shown in Fig. 2.17. When precision matching between transistors is required, then not only should the individual transistors be realized by combining a single-sized unit transistor, but the fingers for one transistor should be interdigitated with the fingers of the second transistor. This approach, known as *common-centroid layout,* helps match errors caused by gradient effects across a microcircuit, such as the temperature or the gate-oxide thickness changing across the microcircuit. An example of a common-centroid layout of two identical matched transistors whose sources are connected is shown in Fig. 2.19 in simplified form [O'Leary, 1991; Maloberti, 1994].

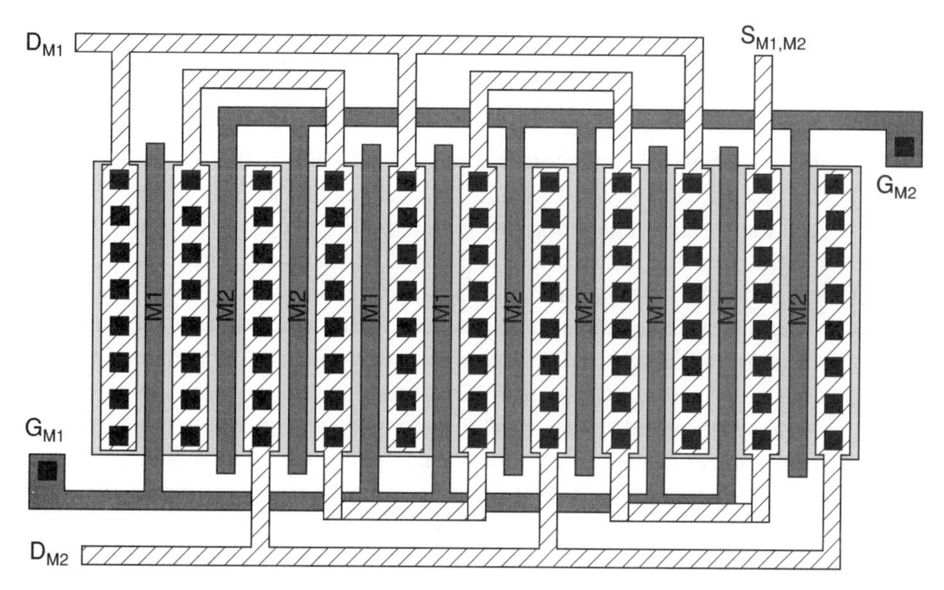

Fig. 2.19 A common-centroid layout for a differential source-coupled pair.

Each of the two transistors is composed of five separate transistor fingers connected in parallel. The outside fingers have separate second-order size effects, and therefore one outside finger is used for M_1 and one is used for M_2. Inside the structure, the fingers occur in doubles—two for M_2, two for M_1, two for M_2, and so on. Thus, the layout is symmetric in both the x and y axes, and any gradients across the microcircuit would affect both M_1 and M_2 in the same way. This layout technique greatly minimizes nonidealities such as opamp input-offset voltage errors when using a differential pair in the input stage of an opamp.

When current mirrors with ratios other than unity are required, again, each of the individual transistors should be realized from a single unit-sized transistor. For example, if a current ratio of 1:2 were desired, then the input transistor might be made from four fingers, whereas the output transistor might be realized using eight identical fingers. In addition, for the greatest accuracy, all fingers should be inside fingers only. Outside, or dummy, fingers would only be included for better matching accuracy and would have no other function. The gates of these dummy fingers are normally connected to the most negative power-supply voltage to ensure they are always turned off (or they are connected to the positive power supply in the case of p-channel transistors).

Capacitor Matching

Very often, analog circuits require precise ratios of capacitors. Ideally, capacitor size is given by

$$C_1 = \frac{\varepsilon_{ox}}{t_{ox}} A_1 = C_{ox} x_1 y_1 \tag{2.13}$$

The major sources of errors in realizing capacitors are due to overetching (which causes the area to be smaller than the area of the layout masks) and an oxide-thickness gradient across the surface of the microcircuit. The former effect is usually dominant and can be minimized by realizing larger capacitors from a parallel combination of smaller, unit-sized capacitors, similar to what is usually done for transistors. For example, to realize two capacitors that have a ratio of 4:6, the first capacitor might be realized from four unit-sized capacitors, whereas the second capacitor might be realized by six unit-sized capacitors. Errors due to the gradient of the oxide thickness can then be minimized by interspersing the unit-sized capacitors in a common-centroid layout so the gradient changes affect both capacitors in the same way. Since oxide-thickness variations are not usually large in a reasonably small area, this common-centroid layout is not typically used except where very accurate capacitors are required.

If only unit-sized capacitors are used, then any overetching will leave the capacitor ratio unaffected. Thus, good designers strive to realize circuits in which only unit-sized capacitors are needed. Unfortunately, this situation is not always possible. When it is not, *overetching error can still be minimized by realizing a nonunit-sized capacitor with a specific perimeter-to-area ratio.* To determine the correct ratio, first note that the error due to overetching is roughly proportional to the perimeter of the capacitor. Specifically, if we assume that a capacitor has an absolute overetching given by

Δe and that its ideal dimensions are given by x_1 and y_1, then its true dimensions are given by $x_{1a} = x_1 - 2\Delta e$ and $y_{1a} = y_1 - 2\Delta e$, and the true capacitor size is given by

$$C_a = C_{ox}x_{1a}y_{1a} = C_{ox}(x_1 - 2\Delta e)(y_1 - 2\Delta e) \tag{2.14}$$

This situation is illustrated in Fig. 2.20. Thus, the error in the true capacitance is given by

$$\Delta C_t = C_{ox}x_{1a}y_{1a} - C_{ox}x_1y_1 = C_{ox}[-2\Delta e(x_1 + y_1) + 4\Delta e^2] \tag{2.15}$$

When this error is small, then the second-order error term can be ignored and (2.15) can be approximated by

$$\Delta C_t \cong -2\Delta e(x_1 + y_1)C_{ox} \tag{2.16}$$

The relative error in the capacitor is therefore given by

$$\varepsilon_r = \frac{\Delta C_t}{C_{ideal}} \cong \frac{-2\Delta e(x_1 + y_1)}{x_1y_1} \tag{2.17}$$

Thus, the relative capacitor error is approximately proportional to the negative of the ratio of the ideal perimeter to the ideal area (assuming only small errors exist, which is reasonable since, if the errors were not small, then that capacitor sizing would probably not be used). When we realize two capacitors that have different sizes, usually the ratio of one capacitor to the other is important, rather than their absolute sizes. This ratio is given by

$$\frac{C_{1a}}{C_{2a}} = \frac{C_1(1 + \varepsilon_{r1})}{C_2(1 + \varepsilon_{r2})} \tag{2.18}$$

If the two capacitors have the same relative errors (i.e., $\varepsilon_{r1} = \varepsilon_{r2}$), then their true ratio is equal to their ideal ratio even when they are not the same sizes. Using (2.17), we see that the relative errors are the same if they both have the same perimeter-to-area ratio. This leads to the following result: *To minimize errors in capacitor ratios due to overetching, their perimeter-to-area ratios should be kept the same, even when the capacitors are different sizes.*

Normally, the unit-sized capacitor will be taken square. *When a non-unit-sized capacitor is required, it is usually set to between one and two times the unit-sized capacitor and is rectangular in shape, so that it has the same number of corners as*

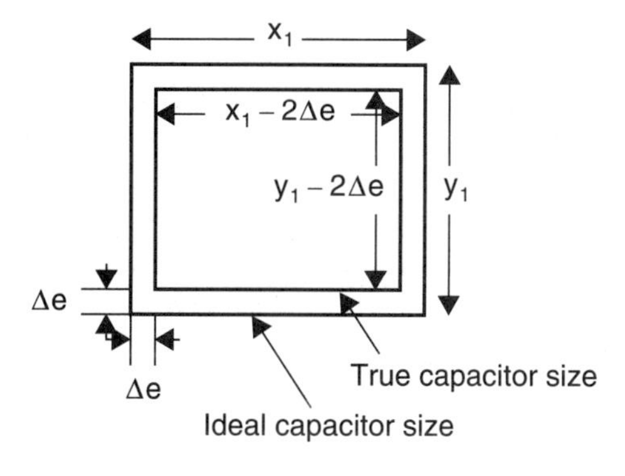

Fig. 2.20 Capacitor errors due to overetching.

the unit-sized capacitor. Defining K to be a desired non-unit-sized capacitor ratio, we have

$$K \equiv \frac{C_2}{C_1} = \frac{A_2}{A_1} = \frac{x_2 y_2}{x_1^2} \quad (2.19)$$

where C_1, A_1, and x_1 represent the capacitance, area, and side-length of a unit-sized capacitor, respectively. Variables C_2, A_2, x_2, and y_2 are similarly defined, except this non-unit-sized capacitor is now rectangular. Equating the ratios of the perimeters-to-areas implies that

$$\frac{P_2}{A_2} = \frac{P_1}{A_1} \quad (2.20)$$

where P_1 and P_2 represent the perimeters of the two capacitors. Rearranging (2.20), we have

$$\frac{P_2}{P_1} = \frac{A_2}{A_1} = K \quad (2.21)$$

which implies that K can also be written as the ratio of perimeters,

$$K = \frac{x_2 + y_2}{2x_1} \quad (2.22)$$

This can be rearranged to become

$$x_2 + y_2 = 2Kx_1 \quad (2.23)$$

Also rearranging (2.19), we have

$$x_2 = \frac{Kx_1^2}{y_2} \quad (2.24)$$

Combining (2.23) and (2.24), we find the quadratic equation

$$y_2^2 - 2Kx_1 y_2 + Kx_1^2 = 0 \quad (2.25)$$

which can be solved to give

$$y_2 = x_1(K \pm \sqrt{K^2 - K}) \quad (2.26)$$

Recall that K is assumed to be greater than one, which ensures that the square root in (2.26) is applied to a positive number. The value for x_2 is then given by (2.24).

EXAMPLE 2.3

Show a layout that might be used to match two capacitors of size 4 and 2.314 units, where a unit-sized capacitor is $10\ \mu m \times 10\ \mu m$.

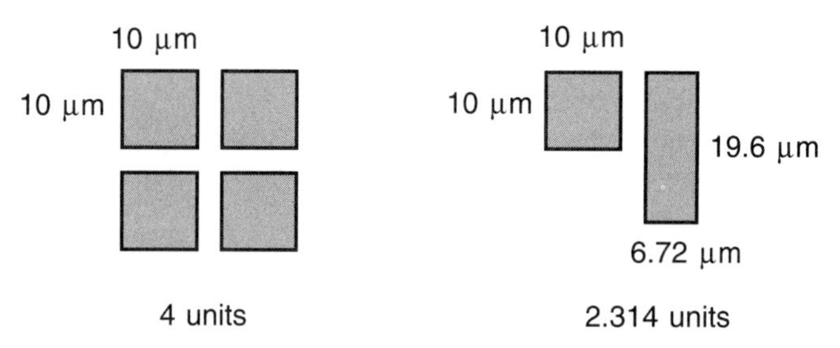

Fig. 2.21 A capacitor layout with equal perimeter-to-area ratios of 4 units and 2.314 units.

Solution

Four units are simply laid out as four unit-sized capacitors in parallel. We break the 2.314-unit capacitor up into one unit-sized capacitor in parallel with another rectangular capacitor of size 1.314 units. The lengths of the sides for this rectangular capacitor are found from (2.26), resulting in

$$y_2 = 10\,\mu\text{m}\left(1.314 \pm \sqrt{1.314^2 - 1.314}\right) = 19.56\,\mu\text{m or } 6.717\,\mu\text{m}$$

Either of these results can be chosen for y_2, and the other result becomes x_2; in other words, the choice of sign affects only the rectangle orientation. Thus, we have the capacitor layout as shown in Fig. 2.21 Note that the ratio of the area of the rectangular capacitor to its perimeter equals 2.5, which is the same as the ratio for the unit-sized capacitor.

Several other considerations should be followed when realizing accurate capacitor ratios. Usually the bottom plate of capacitors will be the first layer of polysilicon or, in some technologies, an ion-implanted region. This region is usually common to many unit-size capacitors.[6] The interconnection of the top plates, which are almost always polysilicon, can often be done in first-level metal with contacts to the polysilicon plates. In some technologies this is not possible, and the polysilicon plates must be interconnected through polysilicon tabs at the sides. These tabs will contact to metal in a region where the bottom-plate polysilicon is not present. The parasitic capacitances of these tabs should be matched as much as possible. This matching often entails adding additional tabs that are not connected anywhere. Another common matching technique is to ensure that the boundary conditions around the unit-sized capacitors match. This boundary-condition matching is accomplished by adding additional top-plate polysilicon around the outside boundaries of unit-sized capacitors at the edge of an array. Many of these principles are illustrated in the simplified layout of two capacitors shown in Fig. 2.22. Each capacitor in the

6. It is usually possible to realize the bottom plate over the top of a well region that has many contacts connected to a low-noise power-supply voltage. This well region acts as a shield to help keep substrate noise out of the bottom plate of the capacitor.

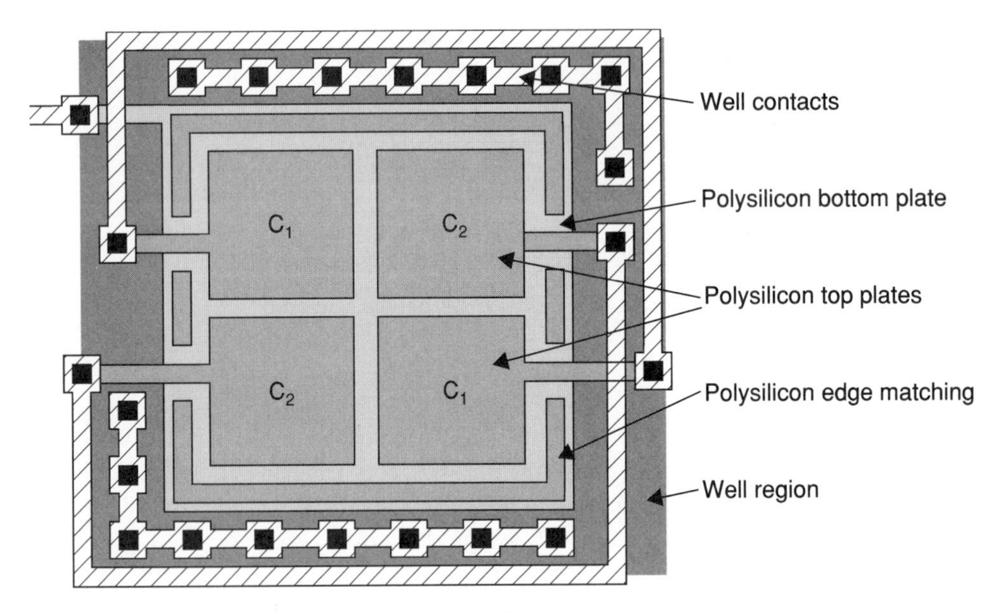

Fig. 2.22 A simplified layout of a capacitor array.

figure consists of two unit-sized capacitors. An additional technique sometimes used (not shown in Fig. 2.22) is to cover the top plates with the first layer of metal, which is then connected to the bottom plate. This sandwich-like structure not only gives additional capacitance per area but more importantly, shields the top level of polysilicon from electromagnetic interference in the air. The capacitor plate formed from the "sandwiched" top level of polysilicon is then connected to critical nodes, such as the virtual inputs of amplifiers.

An important consideration when using capacitors in switched-capacitor circuits is that the bottom plate usually has more noise coupled into it than the top plate because of the large parasitic capacitance between the bottom plate and the silicon region. Therefore the top plate should be connected to critical nodes such as the virtual inputs of opamps, whereas the bottom plate can be connected to less critical nodes such as opamp outputs.

For more details concerning the realization of integrated capacitors, the interested reader can see [Allstot, 1983; O'Leary, 1991; Maloberti, 1994].

Resistor Layout

Integrated resistors can be realized using a wide variety of different conductors. A popular choice is polysilicon, which is a deposited and etched material. Other choices include diffused or ion-implanted regions such as junctions, wells, or base regions. Another possibility is deposited and etched thin-film resistors such as nichrome (consisting of 80 percent nickel and 20 percent chromium) or tantalum. The temperature coefficient of ion-implanted or diffused resistors tends to be positive and large (especially for larger resistivities) with values as large as 1,000 to 3,000 ppm/°C. On the other hand, the temperature coefficient for thin-film resistors can be as small as 100 ppm/°C. Polysilicon resistors usually have large positive temperature coefficients (say, 1,000 ppm/°C) for low-resistivity polysilicon ranging to moderately large, negative temperature coefficients for specially doped, high-resistivity polysilicon. The positive temperature coefficients are primarily due to mobility degradation that results from temperature increases. In implanted and diffused resistors, nonlinear resistance varies greatly with voltage because the depletion-region width is dependent on voltage in the more heavily doped conductive region. This depletion-region width variation is substantially smaller in a polysilicon resistor, which is one of the major reasons polysilicon resistors are preferred over implanted resistors even though they often require more area due to the low resistivity. When thin-film resistors are available in a particular technology, they are almost always the preferred type—unfortunately, they are seldom available.

Regardless of the type of resistor used, the equations governing the resistance (see the Appendix of Chapter 1) are given by

$$R_{\square} = \frac{\rho}{t} \tag{2.27}$$

where $R_{\square}$ is the resistance per square, $\rho = 1/(q\mu_n N_D)$ is the resistivity,[7] t is the thickness of the conductor, and N_D is the concentration of carriers, which we assume are electrons. The total resistance is then given by

$$R = \frac{L}{W} R_{\square} \tag{2.28}$$

where L is the length of the resistor and W is the width of the resistor.

Often, the typical resistivity encountered in integrated circuits is small. For example, the typical resistivity of the polysilicon layer used to form transistor gates is about 20 $\Omega/\square$. Polysilicon is the layer most commonly used to form resistors inside microcircuits.[8] To obtain medium-sized resistors, one must usually use a serpentine layout similar to that shown in Fig. 2.23. When calculating the resistance of such a structure, one must make allowance for the bends and for the end contacts. For example, the contact structure chosen contributes 0.14 squares [Reinhard, 1987; Grebene, 1984], and each bend contributes 2.11 squares[9] [Glasser, 1977]. For the structure shown, each finger contributes 10 squares each for a total of 70 squares, the two contacts contribute 0.28 squares, and the six bends contribute 12.66 squares. Thus, the total is 82.94 squares. If the resistivity of the resistor is 20 $\Omega/\square$, then the total resistance is

$$R = 82.94\ \square \times 20\ \Omega/\square = 1.659\ \text{k}\Omega \tag{2.29}$$

---

7. This equation is valid for resistors with electron carriers. For resistors with holes as carriers, $\rho = 1/(q\mu_p N_A)$.

8. In a modern process, where the gates are formed from a sandwich of a refractory metal over polysilicon, i.e., a polycide, or silicide, then the resistivity is much smaller, perhaps 1–5 $\Omega/\square$, and the gate layer is not useful for realizing moderate-sized resistors.

9. If the bends had been semicircular rather than square, they would have contributed 2.96 $\square$ each.

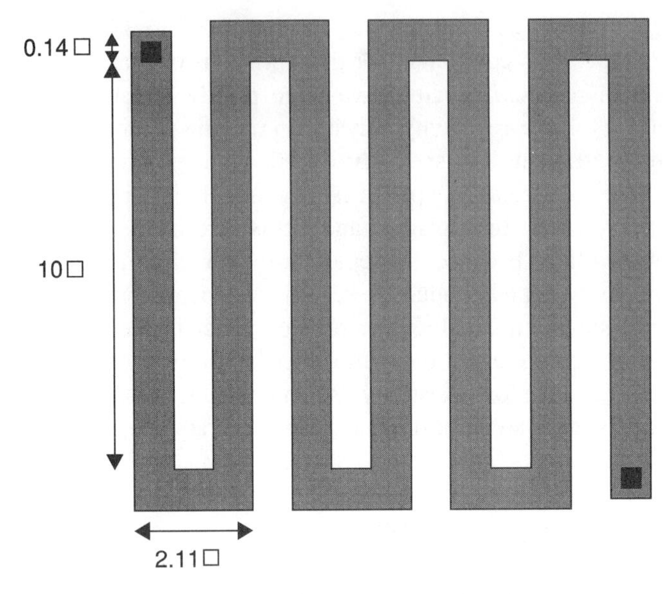

Fig. 2.23 A typical layout for an integrated resistor.

To this value, an additional impedance due to the contacts, about 20 Ω per contact, should be added to give a total resistance of about 1.70 kΩ. Note that quite large structures are required even for relatively small resistances when gate poly is used to realize the resistor.

When very accurate resistor ratios are required and the ratio can be expressed as a ratio of integer values, then an architecture similar to that shown in Fig. 2.24 can be

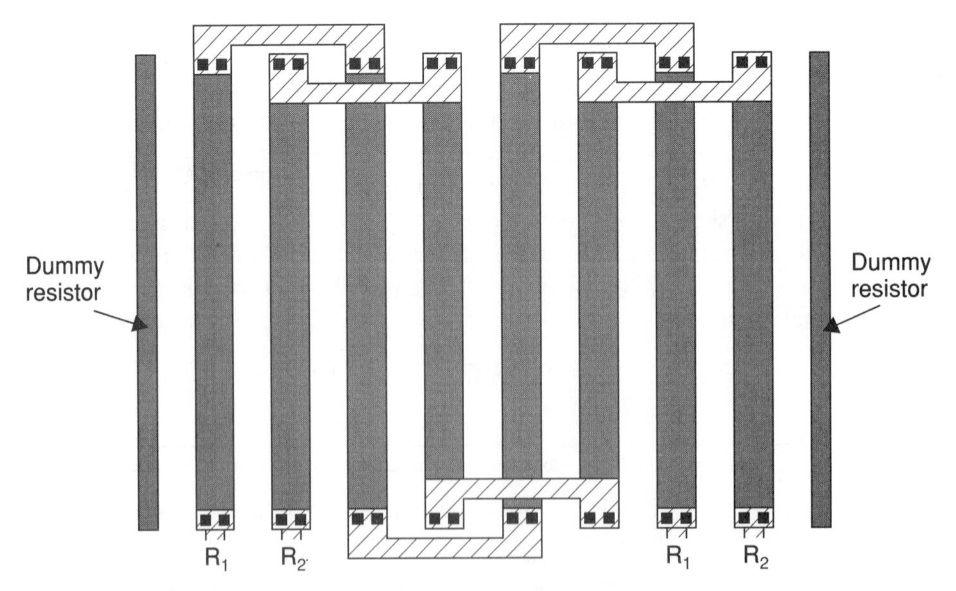

Fig. 2.24 A more accurate, but larger, layout for a resistor.

used. The resistor consists of several fingers connected at their ends using low-resistivity metal. This approach matches errors caused by the contact impedance between R_1 and R_2. Also, two dummy fingers have been included to match boundary conditions. This structure might result in about 0.1 percent matching accuracy of identical resistors if the finger widths are relatively wide (say, 10 μm in a 0.8-μm technology).

As with integrated capacitors, it is a good idea to place a shield under a resistor that is connected to a clean power supply. An appropriate shield might be a well region. This shielding helps keep substrate noise from being injected into the conductive layer. (Noise is due to capacitive coupling between the substrate and a large resistor structure.) Also, the parasitic capacitance between the resistor and the shield should be modelled during simulation. Its second-order effects on circuits such as RC filters can often be eliminated using optimization, which is available in many SPICE-like simulators. For low-noise designs, a metal shield over the top of a resistor may also be necessary, although it will result in a corresponding increase in capacitance.

For more information on realizing accurate resistor ratios, the reader is referred to [O'Leary, 1991; Maloberti, 1994].

Noise Considerations

Some additional layout issues help minimize noise in analog circuits. Most of these issues either attempt to minimize noise from digital circuits coupling into the substrate or analog power supplies, or try to minimize substrate noise that affects analog circuits.

With analog circuits, it is critical that different power-supply connections be used for analog circuits than for digital circuits. Ideally, these duplicate power supplies are connected only off the chip. Where a single I/O pin must be used for the power supply, it is still possible to use two different bonding wires extending from a single-package I/O pin to two separate bonding pads on the integrated circuit. At a very minimum, even if a single bonding pad is used for both analog and digital circuitry, two separated nets from the bonding pad out should be used for the different types of circuitry, as Fig. 2.25 shows. The reason the power-supply interconnects must be separated is that the interconnect does not have zero impedance. Every time a digital gate or buffer changes state, a glitch is injected on the digital power supply and in the surrounding substrate. By having the analog power supplies separate, we prevent this noise from affecting the analog circuitry. In the ideal case, separate pins are used for the positive power supply and for ground in both the digital and analog circuits. In addition, another pair of pins may be used for the supply voltage and ground for digital output buffers, which can inject very large current spikes. Finally, sometimes multiple pins are used for additional supply and grounds for very large microcircuits.

Another common precaution is to lay out the digital and analog circuitry in different sections of the microcircuit. The two sections should be separated by guard rings and wells connected to the power-supply voltages, as Fig. 2.26 shows. The p^+ connections to ground help keep a low-impedance path between the substrate and ground. For modelling purposes, the substrate can be modelled as a number of series-connected resistors with the p^+ ground connections modelled as resistor-dividers

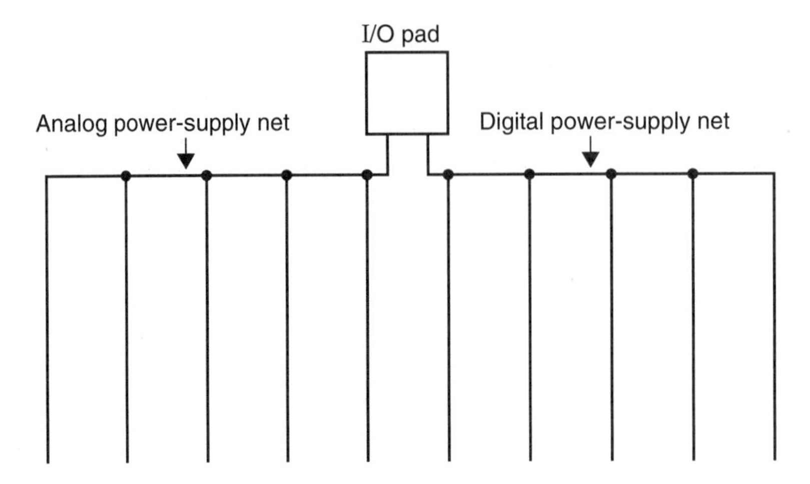

Fig. 2.25 Using separate nets for analog and digital power supplies.

having a small impedance to ground. These low-impedance ground connections help keep substrate noise from propagating through the resistive substrate. The use of the n well between p^+ connections helps to further increase the resistive impedance of the substrate between the analog and digital regions due to graded substrate doping. Specifically, the p substrate often has 10 times higher doping at the surface of the microcircuit compared to the doping level below the n well, which leads to a tenfold increase in substrate resistivity between the two p^+ connections. Finally, the n well also operates as a bypass capacitor to help lower the noise on V_{DD}.

Another important consideration when laying out a circuit that includes both analog and digital circuits is the use of shields connected to either ground or to a separate power-supply voltage. Figure 2.27 shows examples of the use of shields. In this example, an n well is used to shield the substrate from the digital interconnect line. The well is also used to shield an analog interconnect line from any substrate noise. This

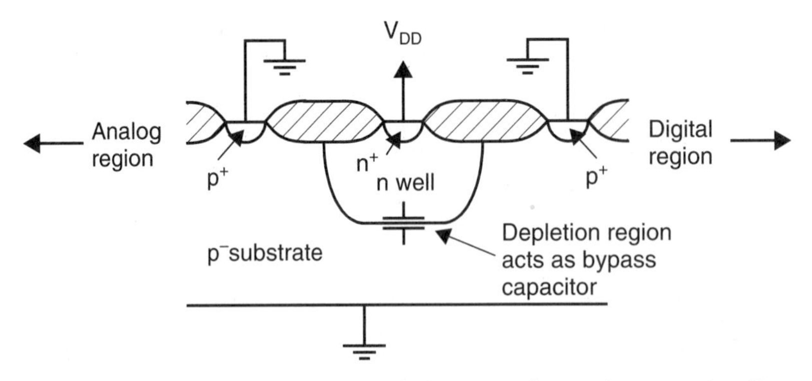

Fig. 2.26 Separating analog and digital areas with guard rings and wells in an attempt to minimize the injection of noise from digital circuits into the substrate under the analog circuit.

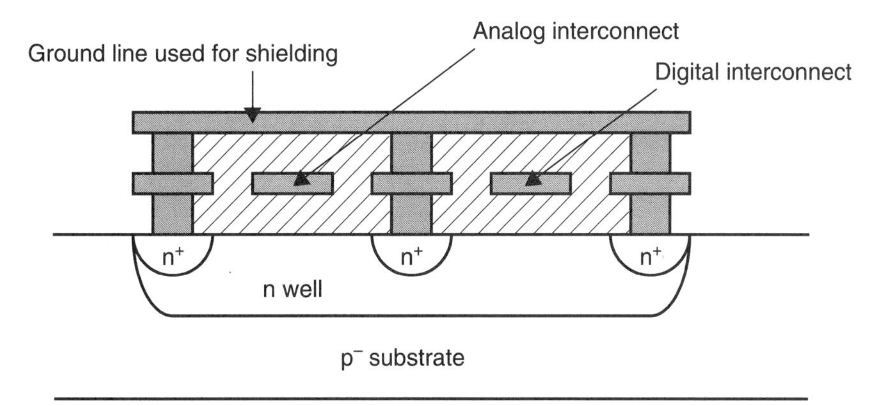

Fig. 2.27 Using shields helps keep noise from being capacitively coupled into and out of the substrate.

shield is ideally connected to a ground net that is used only for shields. If this type of connection is not possible due to layout and space constraints, then the digital ground can be used for the shields, although this is not ideal. In the example, the shield ground is also connected to metal lines that separate the analog and digital lines from each other and from other interconnect lines. Finally, an additional metal shield might be placed above the lines as well. This final shield may be somewhat excessive, but it can often be easily realized in many parts of the microcircuit if ground and power-supply lines are distributed in metal 2, perpendicular to the metal-1 interconnect lines. It should also be mentioned that the n well shield also acts as a bypass capacitor; this helps minimize noise in the substrate, which is connected to V_{DD}. Additional layers that are often used as shields are the polysilicon layers.

Perhaps the most effective technique for minimizing the propagation of substrate noise in a mixed-mode microcircuit containing both analog and digital circuitry is the use of an epitaxial process. An epitaxial process places a conductive layer under all transistors. Any charge flowing through the substrate is attracted to this layer and does not propagate into sensitive analogs regions. Although this process is more expensive, it also helps prevent latch-up. For submicron technologies, this epitaxial process is becoming more common because of its reduced latch-up sensitivity.

Careful thought should go into the overall placement of different blocks in a mixed-mode analog-digital microcircuit. A possible arrangement for an analog section containing switched-capacitor circuits is shown in Fig. 2.28. Notice that an n well shield is being used under the capacitors. Notice also that the clock lines are not only as far from the opamps as possible, but are also separated by two wells and a V_{SS} interconnect that is liberally connected to the substrate, two ground (Gnd) lines, and a V_{DD} line. A well is placed under the clock lines as a shield. This shield is connected to a separate ground line (perhaps digital ground) from the one used in the opamp region because this shield will likely have quite a bit of clock noise coupled into it. Also note that a separate V_{DD} line is used to connect to the n wells under the switches, a region where digital interconnects exist, as is used in the critical opamp section.

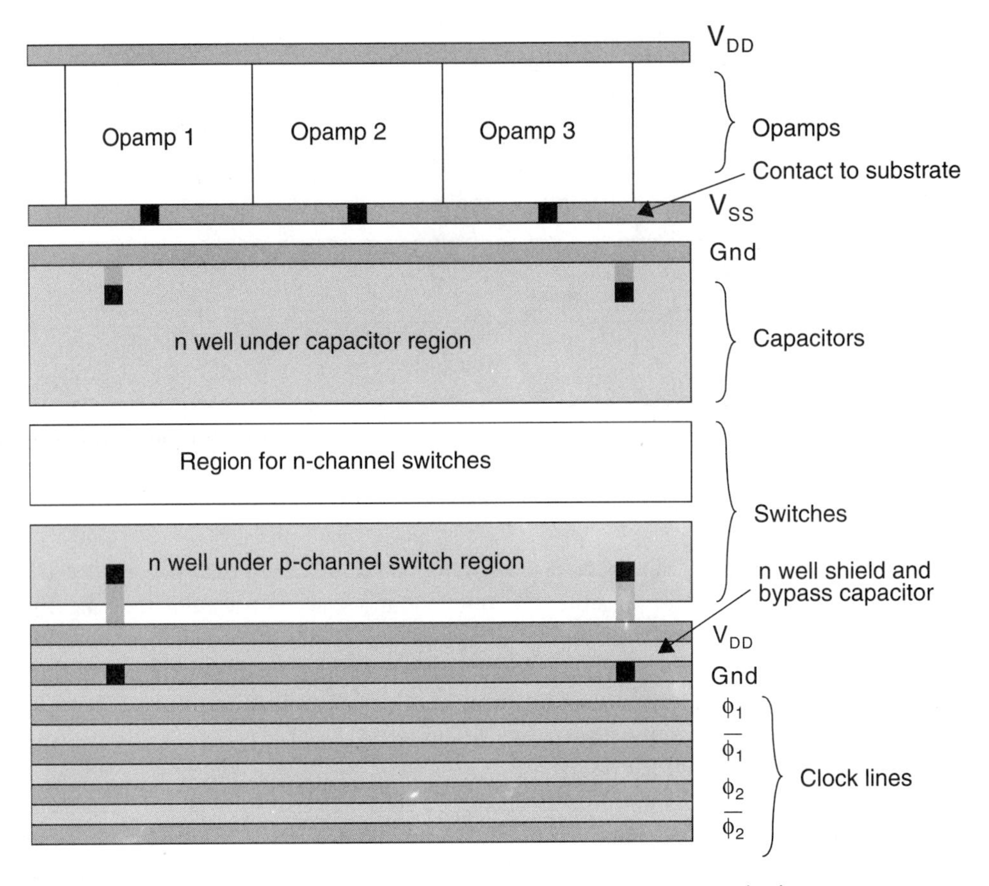

Fig. 2.28 A possible floor plan for an analog section containing switched-capacitor circuits.

One last technique for noise minimization in analog microcircuits should always be used: After layout has been finished, *any unused space should be filled with additional contacts to both the substrate and to the wells, which are used as bypass capacitors.* In a typical microcircuit, this results in a significant increase in bypass capacitance.

Many other techniques have been developed by various companies, but the preceding techniques give the reader a good idea of the types of practical considerations necessary when realizing high-performance analog microcircuits.

2.5 LATCH-UP

One of the effects that CMOS designers must be wary of, especially as dimensions shrink, is latch-up. This effect can occur when there are relatively large substrate or well currents or, equivalently, large substrate or well voltage drops, that might be caused by capacitive coupling. These triggering voltage drops often occur when power is first applied to a CMOS microcircuit.

A latched-up microcircuit is equivalent to a turned-on *silicon-controlled rectifier* (SCR) between the power supply and ground. This SCR effectively short-circuits the power supply on the microcircuit and, unless the supply-current is limited, irreparable damage will probably occur (such as a fused open bonding wire or interconnect line).

To understand latch-up, consider the cross section of the CMOS inverter shown in Fig. 2.29 with the parasitic bipolar transistors Q_1 and Q_2. Q_1 is a lateral npn, with the base being formed by the p^- substrate, whereas Q_2 is a vertical pnp, with the base being formed by the n-well region. The parasitic bipolar circuit has been redrawn in Fig. 2.30 along with some of the parasitic resistances due to the lightly doped substrate and well regions. The circuit realizes two cross-coupled common-emitter amplifiers in a positive feedback loop. This is the equivalent circuit of an SCR, which is sometimes referred to as a crowbar switch.

Normally, the parasitic bipolar transistors are off, and the voltages are as shown in Fig. 2.30(*a*). However, if latch-up is somehow triggered, they turn on when the

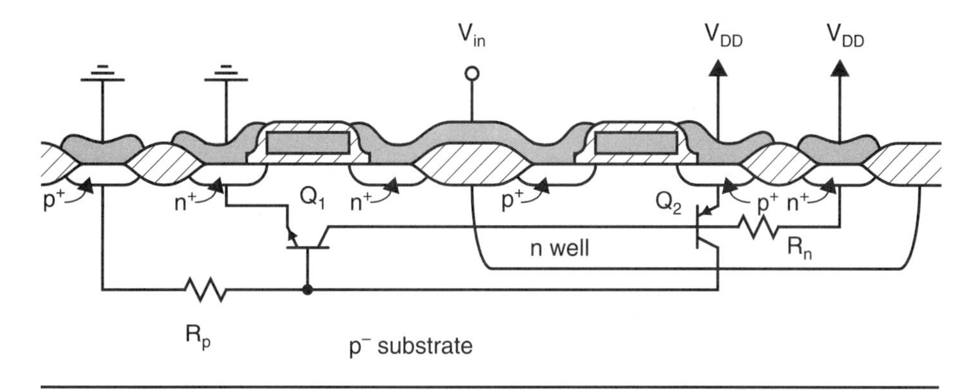

Fig. 2.29 Cross section of a CMOS inverter with superimposed schematic of the parasitic transistors responsible for the latch-up mechanism.

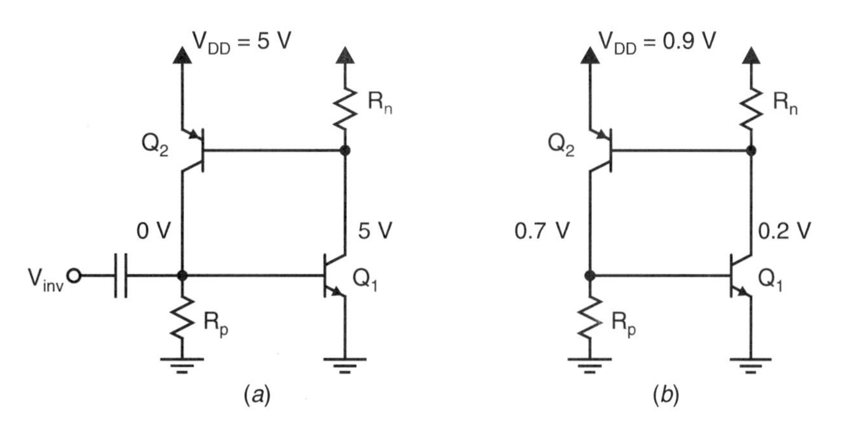

Fig. 2.30 (*a*) The equivalent circuit of the parasitic bipolar transistors, and (*b*) the voltages after latch-up has occurred.

loop gain is larger than unity, and as a result, the voltages are approximately those shown in Fig. 2.30(*b*). This turned-on SCR effectively places a short-circuit across the power-supply voltage and pulls V_{DD} down to approximately 0.9 V. If the power supply does not have a current limit, then excessive current will flow and some portion of the microcircuit may be destroyed.

Latch-up can be triggered in several ways. For example, the output of the CMOS inverter, V_{inv} in Fig. 2.29, is capacitively coupled to the bases of the bipolar transistors by the junction depletion capacitances of the MOS drains. If the inverter is large (as in the case of an output buffer), these capacitances will be large. When the output of the inverter changes, glitches will be capacitively coupled to the base nodes of the parasitic bipolars and may cause latch-up. Alternatively, substrate currents caused by hot-carrier electrons can also result in voltage drops large enough to trigger latch-up.

To prevent latch-up, the loop gain of the cross-coupled bipolar inverters must be kept less than unity. This low loop gain is normally achieved by keeping the current gains of the parasitic bipolars as low as is possible, and most importantly, by keeping shunting resistors R_n and R_p as small as possible. The current gain of the vertical pnp, Q_2, might be 50 to 100 and is difficult to minimize. The current gain of the lateral npn can be decreased by larger spacings between n-channel and p-channel transistors. However, with typically used spacings, the product $\beta_{npn}\beta_{pnp}$ is still normally greater than 1. The loop gain is kept less than unity primarily by decreasing R_n and R_p.

R_n and R_p are decreased mainly by having low-impedance paths between the substrate and well to the power supplies. One way to achieve these low-impedance paths is to have many contacts to the substrate. For example, with an n-well technology, the design rules normally specify a maximum distance between any place in the n-channel region of the microcircuit and the closest p^+ junction, which connects the substrate to ground. Similarly, in the p-channel regions, the maximum distance to the nearest n^+ junction, which connects the n wells to V_{DD}, is specified. Also, after layout is completed, a good designer fills any unused areas with extra connections to the substrate and well regions. In addition, any transistors that conduct large currents are usually surrounded by guard rings. These guard rings are connections to the substrate for n-channel transistors, or to the n well for p-channel transistors, that completely surround the high-current transistors. Also, ensuring that the back of the die is connected to ground through a eutectic gold bond to the package header is helpful.

One of the best ways of preventing latch-up is to use an epitaxial process, especially one with highly doped buried layers. For example, if a p^+ substrate has a p^- epitaxial layer in which the transistors are placed, device performance is only marginally affected, but the highly conductive p^+ substrate has very little impedance to ground contacts and to the package header. Alternatively, one might use a p^- substrate that has n^+- and p-buried regions and an intrinsic epitaxial region that is separately and optimally ion implanted to form the n-channel and p-channel regions. This self-aligned twin-tub technology is very immune to latch-up due to the highly conductive buried layers.

2.6 REFERENCES

D. Allstot and W. Black. "Technological Design Considerations for Monolithic MOS Switched-Capacitor Filtering Systems," *IEEE Proceedings*, Vol. 71, no. 8, pp. 967–986, August 1983.

R. Geiger, P. Allen, and N. Strader. *VLSI: Design Techniques for Analog and Digital Circuits.* McGraw-Hill, New York, 1990.

A. Glasser and G. Subak-Sharpe. *Integrated Circuit Engineering: Design, Fabrication, and Applications.* Addison-Wesley, Reading, Massachusetts, 1977.

A. Grebene, *Bipolar and MOS Analog Integrated Circuits.* Wiley-Interscience, Wiley, New York, 1984.

S. Muraka and M. Peckerar. *Electronic Materials.* Academic Press, New York, 1989, p. 326.

R. Haveman, et al. "A 0.8-μm, 256K BiCMOS SRAM Technology," *Digest of Technical Papers, 1987 Intern. Electron Devices Meeting*, pp. 841–843, December 1987.

R. Haken, R. Haveman, R. Ekund, and L. Hutter. "BiCMOS Process Design," in *BiCMOS Technology and Applications*, ed. A. Alvarez. Kluwer Academic Publishers, Nowell, Massachusetts, 1989.

F. Maloberti. "Layout of Analog and Mixed Analog-Digital Circuits," in *Design of Analog-Digital VLSI Circuits for Telecommunication and Signal Processing*, ed. J. Franca and Y. Tsividis. Prentice Hall, Englewood Cliffs, New Jersey, 1994.

P. O'Leary. "Practical Aspects of Mixed Analogue and Digital Design," in *Analogue-Digital Asics, Circuit Techniques, Design Tools, and Applications*, ed. R. S. Soin, F. Maloberti, and J. Franca. Peter Peregrinus, Stevenage, England, 1991.

D. Reinhard. *Introduction to Integrated Circuit Engineering.* Houghton Mifflin, Boston, 1987.

2.7 PROBLEMS

Unless otherwise stated, assume the following hold throughout the problems section:

- npn bipolar transistors:
  $\beta = 100$
  $V_A = 80\ V$
  $\tau_b = 13\ ps$
  $\tau_s = 4\ ns$
  $r_b = 330\ \Omega$
- n-channel MOS transistors:
  $\mu_n C_{ox} = 92\ \mu A/V^2$
  $V_{tn} = 0.8\ V$
  $\gamma = 0.5\ V^{1/2}$
  $r_{ds}\ (\Omega) = 8{,}000 L\ (\mu m)/I_D\ (mA)$ in active region
  $C_j = 2.4 \times 10^{-4}\ pF/(\mu m)^2$
  $C_{j\text{-}sw} = 2.0 \times 10^{-4}\ pF/\mu m$
  $C_{ox} = 1.9 \times 10^{-3}\ pF/(\mu m)^2$
  $C_{gs(overlap)} = C_{gd(overlap)} = 2.0 \times 10^{-4}\ pF/\mu m$
- p-channel MOS transistors:
  $\mu_p C_{ox} = 30\ \mu A/V^2$
  $V_{tp} = -0.9\ V$
  $\gamma = 0.8\ V^{1/2}$
  $r_{ds}\ (\Omega) = 12{,}000 L\ (\mu m)/I_D\ (mA)$ in active region

$C_j = 4.5 \times 10^{-4}\ \text{pF}/(\mu\text{m})^2$
$C_{j\text{-}sw} = 2.5 \times 10^{-4}\ \text{pF}/\mu\text{m}$
$C_{ox} = 1.9 \times 10^{-3}\ \text{pF}/(\mu\text{m})^2$
$C_{gs(overlap)} = C_{gd(overlap)} = 2.0 \times 10^{-4}\ \text{pF}/\mu\text{m}$

2.1 Discuss briefly the relationships between an ion beam's acceleration potential, the beam current, and the time of implantation on the resulting doping profile.

2.2 Place the following processing steps in their correct order: metal deposition and patterning, field implantation, junction implantation, well implantation, polysilicon deposition and patterning, field-oxide growth.

2.3 What are the major problems associated with a single threshold-voltage-adjust implant?

2.4 What is the reason for using a field implant and why is it often not needed in the well regions?

2.5 What are the major trade-offs in using a wet process or a dry process when growing thermal SiO_2?

2.6 Why is polysilicon rather than metal used to realize gates of MOS transistors?

2.7 Why can't a microcircuit be annealed after metal has been deposited?

2.8 What minimum distance, in terms of λ, would you expect that metal should be separated from polysilicon? Why?

2.9 Find the circuit that the layout shown in Fig. P2.9 realizes. Simplify the circuit, if possible, and give the sizes of all transistors. Assume $L = 2\lambda$, where $\lambda = 1\ \mu\text{m}$.

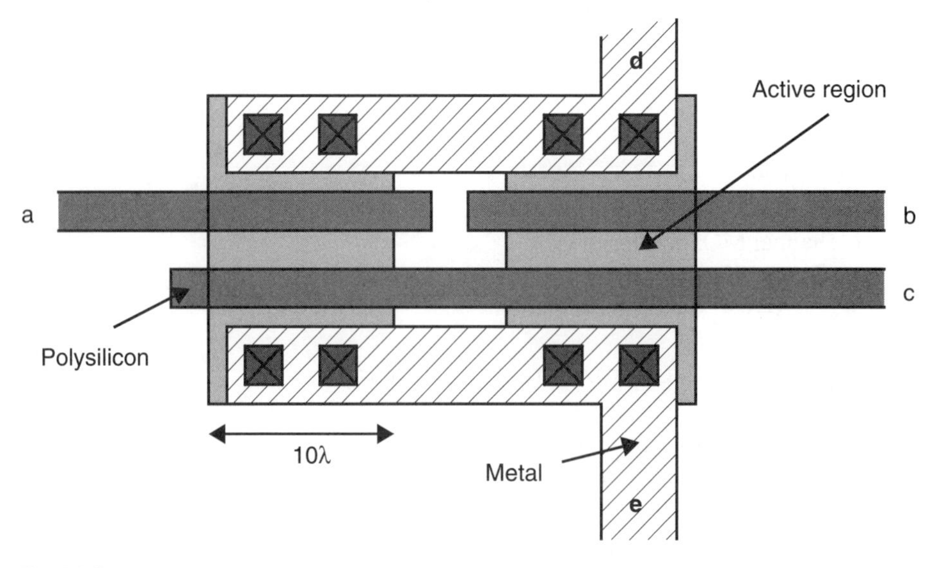

Fig. P2.9

2.10 Find the transistor schematic for the CMOS logic circuit realized by the layout shown in Fig. P2.10. Give the widths of all transistors. Assume $L = 2\lambda$, where $\lambda = 0.4\,\mu\text{m}$. In tabular form, give the area and perimeter of each junction that is not connected to V_{DD} or to ground.

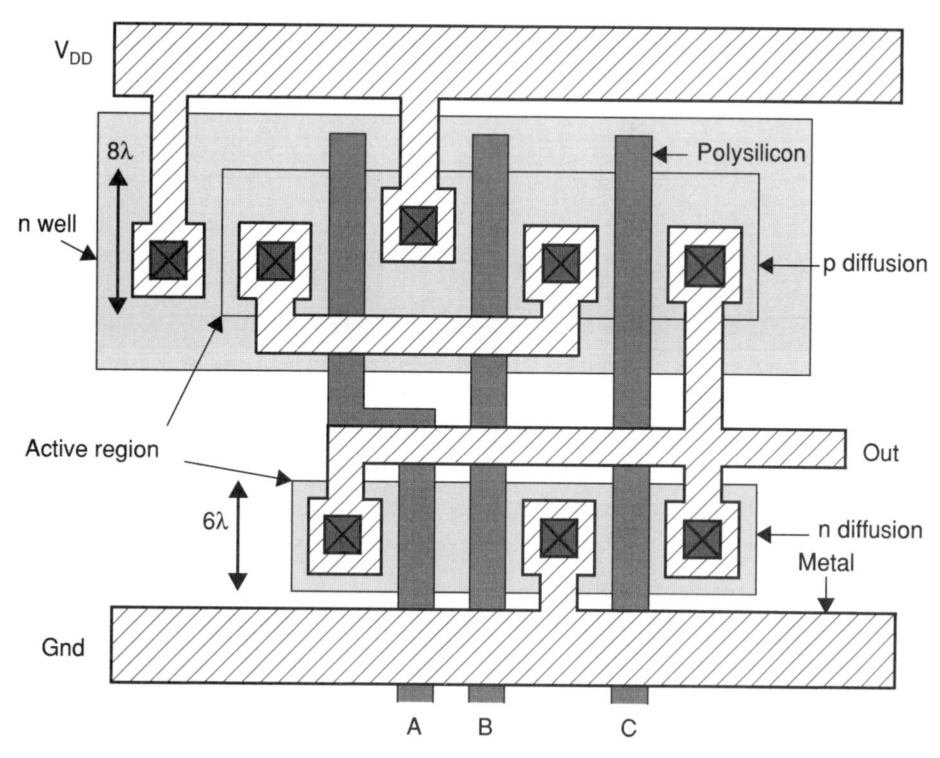

Fig. P2.10

2.11 Repeat Example 2.1 for the case in which the two transistors do not physically share any junction, but each junction is realized in a way similar to junction J_2.

2.12 Repeat Example 2.2 where an overall transistor width of 80λ is still desired, but assume 8 parallel transistors, each of width 10λ, are used.

2.13 Repeat Example 2.2 where an overall transistor width of 80λ is still desired, but assume 2 parallel transistors, each of width 40λ, are used.

2.14 We desire to match two capacitors of relative sizes 9 and 4.523. Sketch a layout for the two capacitors such that their ratio will be maintained during overetching.

2.15 Given that a unit-sized capacitor is $5\ \mu\text{m} \times 5\ \mu\text{m}$ and that the e-beam lithography rounds all sizes to $0.1\ \mu\text{m}$, what is the matching accuracy due to rounding of the capacitors found in Problem 2.14? What is the new matching accuracy if the capacitor sizes are doubled to 18 and 9.046?

2.16 Given that a polysilicon layer has $7\ \Omega/\square$, what is the resistance of a long line that is $2\ \mu\text{m}$ wide and $1{,}000\ \mu\text{m}$ long? (Ignore any contact resistance.)

2.17 Repeat Problem 2.16 for the case in which the resistive line is laid out in a serpentine manner, as shown in Fig. 2.23, where enough bends are used such that a line drawn along the middle of the serpentine resistor has length 1,000 μm (the last finger length might be short). What is the resulting height and width of the overall resistor layout?

CHAPTER

Basic Current Mirrors and Single-Stage Amplifiers

In this chapter, fundamental building blocks are described. These blocks include a variety of current mirrors, single-stage amplifiers with active loads, and differential pairs. A good knowledge of these building blocks is critical to understanding many subjects in the rest of this book and for analog IC design in general. Although some bipolar circuits are described, CMOS mirrors and gain stages are emphasized because modern designs mostly make use of CMOS technology. Fortunately, most of the small-signal analyses presented can be applied to bipolar circuits with little change. In addition, rather than using resistive loads and ac coupling, the gain stages covered are shown with current-mirror active loads since such loads are almost always used in integrated circuits. Finally, single-transistor amplifiers are described here for completeness, although it is assumed that most readers are already somewhat familiar with them.

3.1 SIMPLE CMOS CURRENT MIRROR

A simple CMOS current mirror is shown in Fig. 3.1, in which it is assumed that both transistors are in the active region. If the finite output impedances of the transistors are ignored, and it is assumed that both transistors are the same size, then Q_1 and Q_2 will have the same current since they both have the same gate-source voltage, V_{gs}. However, when finite output impedance is considered, whichever transistor has a larger drain-source voltage will also have a larger current. In addition, the finite output impedance of the transistors will cause the small-signal output impedance of the current mirror, that is, the small-signal impedance looking into the drain of Q_2, to be less than infinite. To find the output impedance of the current mirror, r_{out}, the small-signal circuit is analyzed after placing a signal source, v_x, at the output node. Then by definition, r_{out} is given by the ratio v_x/i_x, where i_x is the current flowing out of the source and into the drain of Q_2.

Before finding r_{out}, consider the small-signal model for Q_1 alone, as shown in Fig. 3.2(*a*). Note that Q_1 is diode connected (i.e., its drain and gate are connected) and that I_{in} does not exist in the small-signal model; I_{in} was replaced with an open

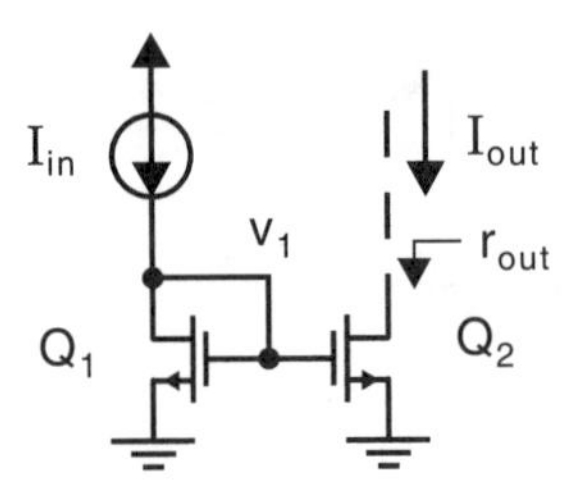

Fig. 3.1 A simple CMOS current mirror.

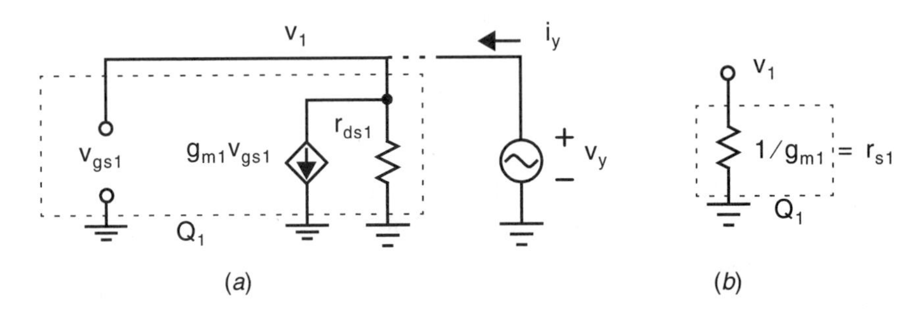

Fig. 3.2 (*a*) The small-signal model for Q_1 and (*b*) the equivalent small-signal model for Q_1.

circuit because it is an *independent* current source. Also note that a low-frequency small-signal model is used for Q_1 (i.e., all the capacitors are ignored in the model). This small-signal model can be further reduced by finding the *Thévenin-equivalent* circuit. The Thévenin-equivalent output voltage is 0 since the circuit is stable and contains no input signal. This circuit's Thévenin-equivalent output impedance is found by applying a test signal voltage, v_y, at v_1 and measuring the signal current, i_y, as shown. Here, the current i_y is given by

$$i_y = \frac{v_y}{r_{ds1}} + g_{m1}v_{gs1} = \frac{v_y}{r_{ds1}} + g_{m1}v_y \tag{3.1}$$

and recalling that the output impedance is given by v_y/i_y, the output impedance equals $1/g_{m1} \| r_{ds1}$. Because typically $r_{ds1} >> 1/g_{m1}$, we approximate the output impedance to be simply $1/g_{m1}$ (which is also defined to be r_{s1}), which results in the equivalent model shown in Fig. 3.2(*b*). This same result holds in the bipolar case and is also equivalent to the small-signal model for a diode—hence the name *diode-connected* transistor.

Using the model just described leads to the simplified small-signal model for the overall current mirror, as shown in Fig. 3.3(*a*), where v_{gs2} has been connected to ground via a resistance of $1/g_{m1}$. Since no current flows through the $1/g_{m1}$ resistor, v_{gs2} equals 0 no matter what voltage level v_x is applied to the current-mirror output. This should come as no surprise, since MOS transistors operate unilaterally at low frequencies. Thus, since $g_{m2}v_{gs2} = 0$, the circuit is simplified to the equivalent small-signal model shown in Fig. 3.3(*b*). The small-signal output impedance, r_{out}, is simply equal to r_{ds2}.

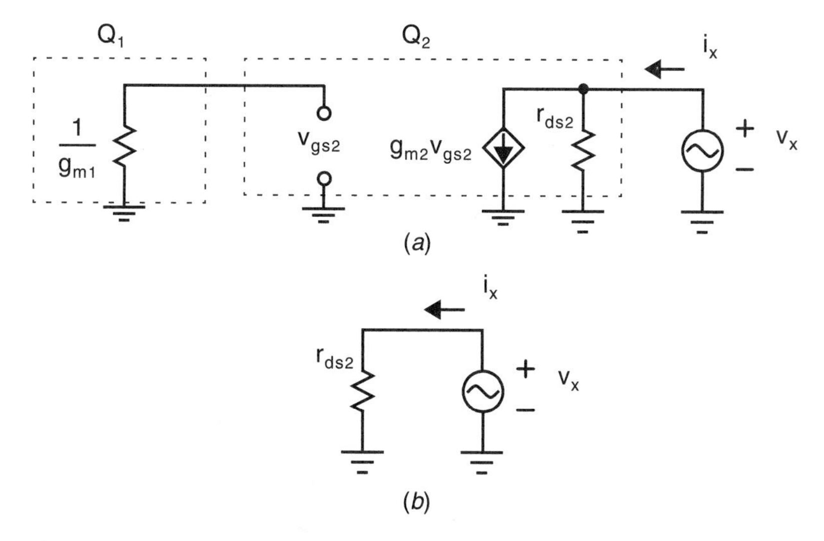

Fig. 3.3 (*a*) A small-signal model for the current mirror of Fig. 3.1 and (*b*) a simplified small-signal model.

EXAMPLE 3.1

Consider the current mirror shown in Fig. 3.1, where $I_{in} = 100\ \mu A$ and each transistor has $W/L = 100\ \mu m/1.6\ \mu m$. Given that $\mu_n C_{ox} = 92\ \mu A/V^2$, $V_{tn} = 0.8\ V$ and $r_{ds} = [8{,}000L\ (\mu m)]/[I_D\ (mA)]$, find r_{out} for the current mirror and the value of g_{m1}. Also, estimate the change in I_{out} for a 0.5 V change in the output voltage.

Solution

Since the W/L ratios of Q_1 and Q_2 are the same, the nominal value of I_{out} equals that of $I_{in} = 100\ \mu A$. Thus, we have

$$r_{out} = r_{ds2} = \frac{8{,}000 \times 1.6}{0.1} = 128\ k\Omega \tag{3.2}$$

The value of g_{m1} is given by

$$g_{m1} = \sqrt{2\mu_n C_{ox}(W/L)I_{D1}} = 1.07\ mA/V \tag{3.3}$$

resulting in $r_{s1} = 1/g_{m1} = 935\ \Omega$. Note that this r_{s1} value is significantly less than r_{ds1}, which equals r_{ds2} in this case.

The change in output current can be estimated, using r_{out}, as

$$\Delta I_{out} = \frac{\Delta V}{r_{out}} = \frac{0.5}{128\ k\Omega} = 3.9\ \mu A \tag{3.4}$$

In other words, if initially I_{out} is measured to be 101 μA (due to mismatch or a larger V_{DS} voltage), then a 0.5 V increase in output voltage would result in a new output current of about 105 μA. Note that this estimate does not account for second-order effects such as the fact that r_{ds} changes as the output current changes.

3.2 COMMON-SOURCE AMPLIFIER

A common use of simple current mirrors is in a single-stage amplifier with an active load, as shown in Fig. 3.4. This common-source topology is the most popular gain stage, especially when high-input impedance is desired.

Here, an n-channel common-source amplifier has a p-channel current mirror used as an active load to supply the bias current for the drive transistor. By using an active load, a high-impedance output load can be realized without using excessively large resistors or a large power-supply voltage. As a result, for a given power-supply voltage, a larger voltage gain can be achieved using an active load than would be possible if a resistor were used for the load. For example, if a 1-MΩ load were required with a 100-μA bias current, a resistive-load approach would require a power-supply voltage of $1\ \text{M}\Omega \times 100\ \mu\text{A} = 100\ \text{V}$. An active load makes use of the nonlinear, large-signal transistor equations to create simultaneous conditions of large bias currents and large small-signal resistances.

A small-signal equivalent circuit for low-frequency analysis of the common-source amplifier of Fig. 3.4 is shown in Fig. 3.5. V_{in} and R_{in} are the Thévenin equivalent of the input source. It is assumed that the bias voltages are such that both transistors are in the active region. The output resistance, R_2, is made up of the parallel combination of the drain-to-source resistance of Q_1, that is, r_{ds1}, and the drain-

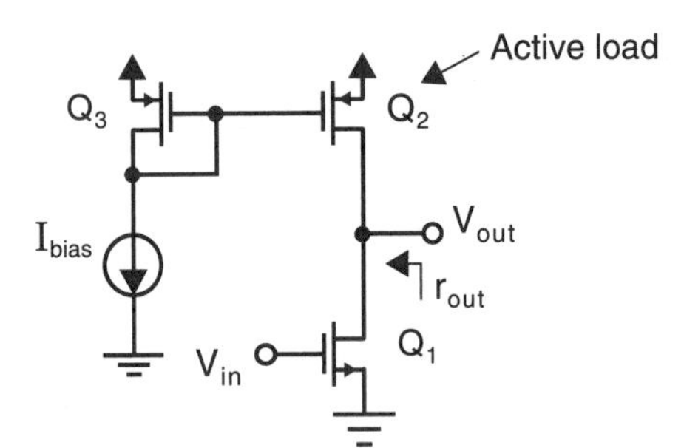

Fig. 3.4 A common-source amplifier with a current-mirror active load.

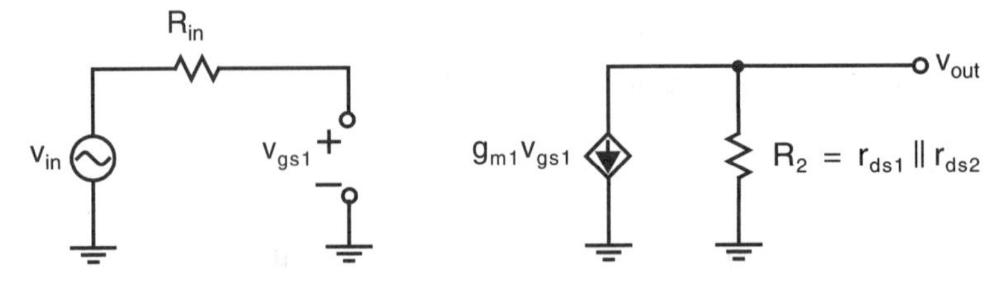

Fig. 3.5 A small-signal equivalent circuit for the common-source amplifier.

to-source resistance of Q_2, that is, r_{ds2}. Notice that the voltage-controlled current source modelling the body effect has not been included since the source is at a small-signal ground, and, therefore, this source always has 0 current.

Using small-signal analysis, we have $v_{gs1} = v_{in}$ and, therefore,

$$A_V = \frac{v_{out}}{v_{in}} = -g_{m1}R_2 = -g_{m1}(r_{ds1} \| r_{ds2}) \tag{3.5}$$

Depending on the device sizes, currents, and the technology used, a typical gain for this circuit is in the range of –10 to –100. To achieve similar gains with resistive loads, much larger power-supply voltages than 5 V must be used. This resistive-load approach also greatly increases the power dissipation. However, it should be mentioned here that for low-gain, high-frequency stages, it may be desirable to use resistor loads (if they do not require much silicon area) because they often have less parasitic capacitances associated with them. They are also typically less noisy than active loads.

EXAMPLE 3.2

Assume all transistors have $W/L = 100\ \mu m/1.6\ \mu m$ in Fig. 3.4, and that $\mu_n C_{ox} = 90\ \mu A/V^2$, $\mu_p C_{ox} = 30\ \mu A/V^2$, $I_{bias} = 100\ \mu A$, $r_{ds\text{-}n}(\Omega) = 8{,}000L\ (\mu m)/I_D\ (mA)$, and $r_{ds\text{-}p}(\Omega) = 12{,}000L\ (\mu m)/I_D\ (mA)$. What is the gain of the stage?

Solution

We have

$$g_{m1} = \sqrt{2\mu_n C_{ox}(W/L)_1 I_{bias}} = 1.06\ mA/V \tag{3.6}$$

Also,

$$r_{ds1} = \frac{8{,}000 \times 1.6\ \mu m}{0.1\ mA} = 128\ k\Omega \tag{3.7}$$

and

$$r_{ds2} = \frac{12{,}000 \times 1.6\ \mu m}{0.1\ mA} = 192\ k\Omega \tag{3.8}$$

Using Eq. (3.5), we have

$$A_V = -g_{m1}(r_{ds1} \| r_{ds2}) = -1.06(128 \| 192) = -81.4 \tag{3.9}$$

3.3 SOURCE-FOLLOWER OR COMMON-DRAIN AMPLIFIER

Another general use of current mirrors is to supply the bias current of source-follower amplifiers, as shown in Fig. 3.6. In this example, Q_1 is the source follower and Q_2 is

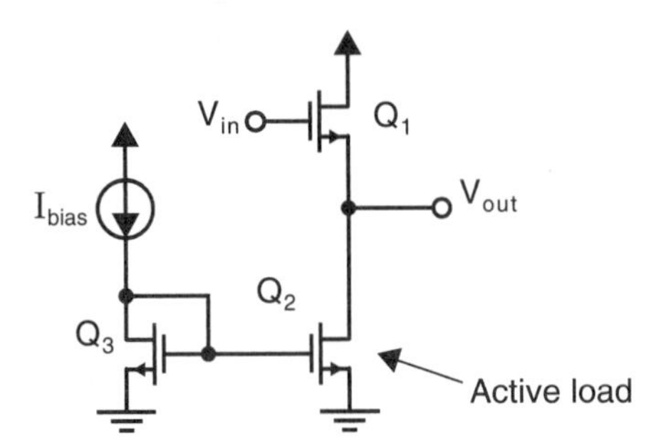

Fig. 3.6 A source-follower stage with a current mirror used to supply the bias current.

an active load that supplies the bias current of Q_1. These amplifiers are commonly used as *voltage buffers* and are therefore commonly called source followers. They are also referred to as common-drain amplifiers, since the input and output nodes are at the gate and source nodes, respectively, with the drain node being at small-signal ground. Although the dc level of the output voltage is not the same as the dc level of the input voltage, ideally the small-signal voltage gain is close to unity. In reality, it is somewhat less than unity. However, although this circuit does not generate voltage gain, it does have the ability to generate current gain.

A small-signal model for low-frequency analysis of this source-follower stage is shown in Fig. 3.7. Note that the voltage-controlled current source that models the body effect of MOS transistors has been included. This body effect is included because the source is not at small-signal ground and the body effect is a major limitation on the small-signal gain. Note that in Fig. 3.7, r_{ds1} is in parallel with r_{ds2}. Notice also that the voltage-controlled current source modelling the body effect produces a current that is proportional to the voltage across it. This relationship makes the body effect equivalent to a resistor of size $1/g_{s1}$, which is also in parallel with r_{ds1} and r_{ds2}. Thus, the small-signal model of Fig. 3.7 is equivalent to the simplified small-signal model of Fig. 3.8, in which $R_{s1} = r_{ds1} \| r_{ds2} \| 1/g_{s1}$. Writing the nodal equation at v_{out}, and noting that $v_{gs1} = v_{in} - v_{out}$, we have

$$v_{out}G_{s1} - g_{m1}(v_{in} - v_{out}) = 0 \tag{3.10}$$

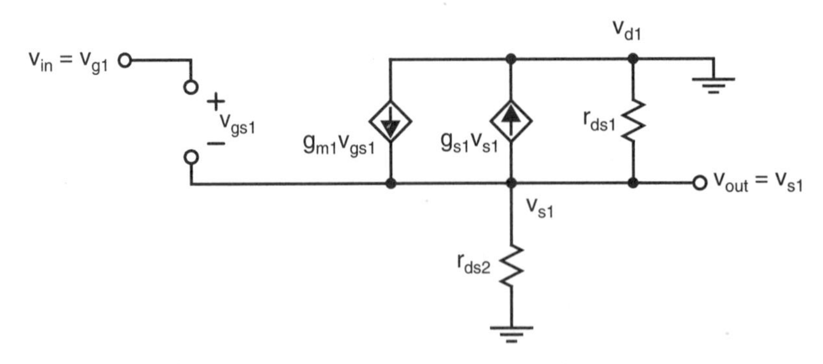

Fig. 3.7 The low-frequency model of the source-follower amplifier.

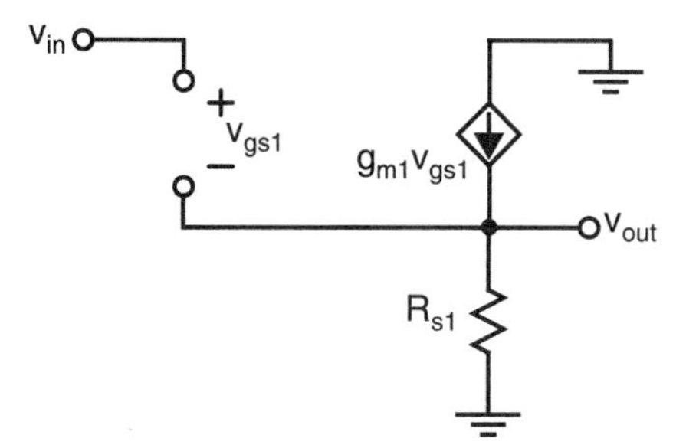

Fig. 3.8 An equivalent small-signal model for the source follower.

where $G_{s1} = 1/R_{s1}$.[1] Before proceeding, it is worth mentioning here that the authors have found that, to minimize circuit equation errors, a consistent methodology should be maintained when writing nodal equations. The methodology employed here is as follows: The first term is always the node at which the currents are being summed. This node voltage is multiplied by the sum of all admittances connected to the node. The next negative terms are the adjacent node voltages, and each is multiplied by the connecting admittance. The last terms are any current sources with a multiplying negative sign used if the current is shown to flow into the node.

Solving for v_{out}/v_{in}, we have

$$A_v = \frac{v_{out}}{v_{in}} = \frac{g_{m1}}{g_{m1} + G_{s1}} = \frac{g_{m1}}{g_{m1} + g_{s1} + g_{ds1} + g_{ds2}} \tag{3.11}$$

Normally, g_{s1} is on the order of one-tenth to one-fifth that of g_{m1}. Also, the transistor output admittances, g_{ds1} and g_{ds2}, might be one-tenth that of the body-effect parameter, g_{s1}. Therefore, it is seen that the body-effect parameter is the major source of error causing the gain to be less than unity. Notice also that at low frequencies the stage is completely unilateral. In other words, there is no signal flow from the output to the input.

EXAMPLE 3.3

Consider the source follower of Fig. 3.6, where all transistors have $W/L = 100\ \mu m/1.6\ \mu m$, $\mu_n C_{ox} = 90\ \mu A/V^2$, $\mu_p C_{ox} = 30\ \mu A/V^2$, $I_{bias} = 100\ \mu A$, $\gamma_n = 0.5\ V^{1/2}$, $r_{ds\text{-}n}(\Omega) = 8{,}000 L\ (\mu m)/I_D\ (mA)$. What is the gain of the stage?

Solution

Notice that many parameters are the same as in Example 3.2. Repeating, we have

$$g_{m1} = \sqrt{2\mu_n C_{ox}(W/L)_1 I_{bias}} = 1.06\ mA/V \tag{3.12}$$

Also,

$$r_{ds1} = r_{ds2} = \frac{8{,}000 \cdot 1.6\ \mu m}{0.1\ mA} = 128\ k\Omega \tag{3.13}$$

1. Whenever a variable is designated G_i, it is assumed that the variable is an admittance and that $G_i = 1/R_i$, where R_i is the resistance of the same component.

The equation for the body-effect parameter, (Eq. 1.87) from Chapter 1, is

$$g_{s1} = \frac{\gamma g_m}{2\sqrt{V_{SB} + |2\phi_F|}} \tag{3.14}$$

To calculate this parameter, we need to know the source-bulk voltage, V_{SB}. Unfortunately, this voltage is dependent on the application and cannot be known accurately beforehand. Here we will assume that 5 V power supplies are being used, and that $V_{SB} \approx 2$ V. This is somewhat arbitrary, but it is the best one can do without more details. We therefore have

$$g_{s1} = \frac{0.5 \cdot g_m}{2\sqrt{2 + 0.7}} = 0.15 g_m = 0.16 \text{ mA/V} \tag{3.15}$$

Using (3.11), we have

$$A_V = \frac{1.06}{1.06 + 0.16 + 1/128 + 1/128} = 0.86 \text{ V/V} \tag{3.16}$$

Note that, as mentioned above, the fact that this result is so far below unity is mainly due to the body-effect parameter, g_s. If the body effect were not present, the gain would be around 0.99 V/V.

3.4 COMMON-GATE AMPLIFIER

A common-gate amplifier with an active load is shown in Fig. 3.9. This stage is commonly used as a gain stage when a relatively small input impedance is desired. For example, it might be designed to have an input impedance of 50 Ω to terminate a 50-Ω transmission line. Another common application for a common-gate amplifier is the first stage of an amplifier designed to amplify current rather than voltage.

If we use straightforward small-signal analysis, when the impedance seen at V_{out} (in this case, the output impedance of the current mirror formed by Q_2) is much less than r_{ds1}, the input impedance, r_{out}, is found to be $1/g_{m1}$ at low frequencies. However, in integrated applications, the impedance seen at V_{out} is often on the same order of magnitude or even much greater than r_{ds1}. In this case, the input impedance at low frequencies can be considerably larger than $1/g_{m1}$. To see this result, consider the small-signal model shown in Fig. 3.10. In this model, the voltage-dependent current source that models the body effect has been included. Notice that $v_{gs1} = -v_{s1}$ and therefore the two current sources can be combined into a single current source, as shown in Fig. 3.11. This simplification is always possible for a transistor that has a grounded gate in a small-signal model, and considerably simplifies taking the body effect into account. Specifically, one can simply ignore the body effect for transistors with grounded gates, and then, after the analysis is complete, simply replace the constants g_{mi} with $g_{mi} + g_{si}$. However, for this example, we include the body-effect parameter throughout the analysis.

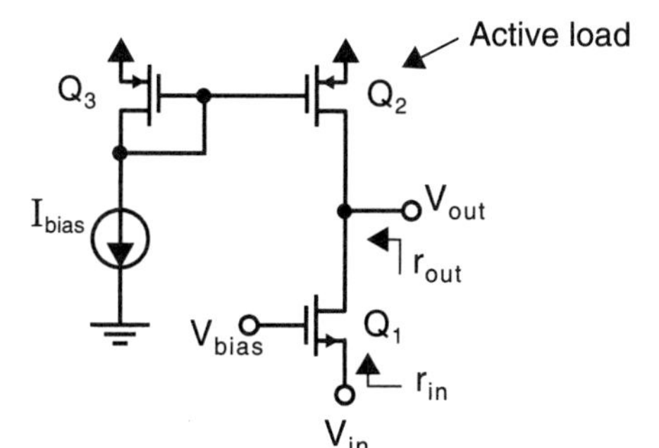

Fig. 3.9 A common-gate amplifier with a current-mirror active load.

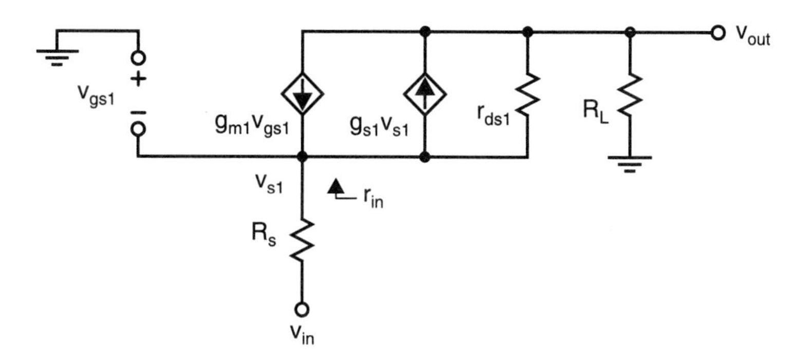

Fig. 3.10 The small-signal model of the common-gate amplifier at low frequencies.

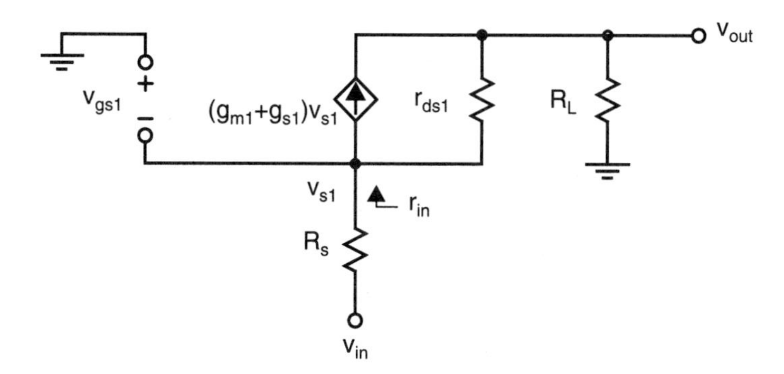

Fig. 3.11 A simplified small-signal model of the common-gate amplifier.

At node v_{out}, we have

$$v_{out}(G_L + g_{ds1}) - v_{s1}g_{ds1} - (g_{m1} + g_{s1})v_{s1} = 0 \tag{3.17}$$

Rearranging slightly, we have

$$\frac{v_{out}}{v_{s1}} = \frac{g_{m1} + g_{s1} + g_{ds1}}{G_L + g_{ds1}} \tag{3.18}$$

where it should be noted that this gain is approximately equal to $g_{m1}/(G_L + g_{ds1})$.

The current going into the source of Q_1 is given by

$$i_s = v_{s1}(g_{m1} + g_{s1} + g_{ds1}) - v_{out}g_{ds1} \tag{3.19}$$

Combining (3.18) and (3.19) to find the input admittance, $y_{in} = 1/r_{in}$, we have

$$y_{in} \equiv \frac{i_s}{v_{s1}} \equiv \frac{g_{m1} + g_{s1} + g_{ds1}}{1 + \frac{g_{ds1}}{G_L}} \cong \frac{g_{m1}}{1 + \frac{g_{ds1}}{G_L}} \tag{3.20}$$

Alternatively, we have

$$r_{in} = \frac{1}{y_{in}} \cong \frac{1}{g_{m1}}\left(1 + \frac{R_L}{r_{ds1}}\right) \tag{3.21}$$

With the p-channel active load shown in Fig. 3.9, $R_L = r_{ds2}$. Since, in this case, R_L is approximately the same magnitude as r_{ds1}, the input impedance, r_{in}, is about $2/g_{m1}$ for low frequencies—twice as large as the expected value of $1/g_{m1}$. This increased input impedance must be taken into account in applications such as transmission-line terminations. In some examples, the current-mirror output impedance realized by Q_2 is much larger than r_{ds1} (i.e., $R_L >> r_{ds1}$), and so *the input impedance for this common-gate amplifier is much larger than* $1/g_{m1}$. This increased input impedance often occurs in integrated circuits and is not commonly known.

The attenuation from the input to the transistor source can be considerable for a common-gate amplifier when R_s is large. This attenuation is given by

$$\frac{v_{s1}}{v_{in}} = \frac{G_s}{G_s + y_{in}} \tag{3.22}$$

using the admittance-divider rule.[2] Using (3.20) to replace y_{in}, we have

$$\frac{v_{s1}}{v_{in}} = \frac{G_s}{G_s + \frac{g_{m1} + g_{s1} + g_{ds1}}{1 + g_{ds1}/G_L}} \tag{3.23}$$

Using (3.18) and (3.23), we find that the overall dc gain is given by

$$A_V = \frac{v_{out}}{v_{in}} = \left[\frac{G_s}{\left(G_s + \frac{g_{m1} + g_{s1} + g_{ds1}}{1 + g_{ds1}/G_L}\right)}\right]\frac{g_{m1} + g_{s1} + g_{ds1}}{G_L + g_{ds1}}$$
$$\cong \left[\frac{G_s}{\left(G_s + \frac{g_{m1}}{1 + g_{ds1}/G_L}\right)}\right]\frac{g_{m1}}{G_L + g_{ds1}} \tag{3.24}$$

2. This rule states that the gain is the ratio of the admittance connected between two nodes, divided by the sum of that admittance and the admittance between the second node and ground.

3.5 SOURCE-DEGENERATED CURRENT MIRRORS

We saw in Section 3.1 that a current mirror can be realized using only two transistors, where the output impedance of this current source was seen to be r_{ds2}. To increase this output impedance, a source-degenerated current mirror can be used, as shown in Fig. 3.12. The small-signal model for this current mirror is shown in Fig. 3.13. Since no current flows into the gate, the gate voltage is 0 V.

Note that the current i_x sourced by the applied voltage source is equal to the current through the degeneration resistor, R_s. Therefore, we have

$$v_s = i_x R_s \tag{3.25}$$

Also, note that

$$v_{gs} = -v_s \tag{3.26}$$

Setting i_x equal to the total current through $g_{m2}v_{gs}$ and r_{ds2} gives

$$i_x = g_{m2}v_{gs} + \frac{v_x - v_s}{r_{ds2}} \tag{3.27}$$

Substituting (3.25) and (3.26) into (3.27) gives

$$i_x = -i_x g_{m2} R_s + \frac{v_x - i_x R_s}{r_{ds2}} \tag{3.28}$$

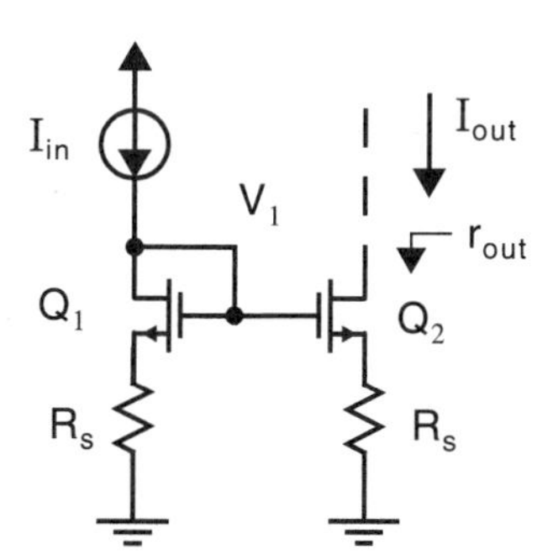

Fig. 3.12 A current mirror with source degeneration.

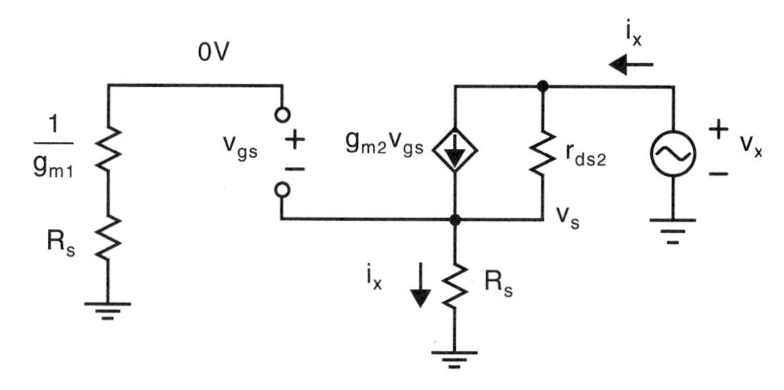

Fig. 3.13 The small-signal model for the source-degenerated current source.

Rearranging, we find the output impedance to be given by

$$r_{out} = \frac{v_x}{i_x} = r_{ds2}[1 + R_s(g_{m2} + g_{ds2})] \cong r_{ds2}(1 + R_s g_{m2}) \tag{3.29}$$

where g_{ds2} is equal to $1/r_{ds2}$, which is much less than g_{m2} (recall that $g_m = 1/r_s$). Thus, the output impedance has been increased by a factor approximately equal to $(1 + R_s g_{m2})$.

This formula can often be applied to moderately complicated circuits to quickly estimate the impedances looking into a node. Such an example follows, in the derivation of the output impedance of cascode current mirrors.

It should be noted that the above derivation ignores the body effect of the transistor, even though the source of the transistor is not connected to a small-signal ground. As discussed earlier, in Section 3.4, since the gate is at a small-signal ground, the body effect can be taken into account by simply replacing g_{m2} in Eq. (3.29) with $g_{m2} + g_{s2}$. This substitution results in

$$r_{out} = \frac{v_x}{i_x} = r_{ds2}[1 + R_s(g_{m2} + g_{s2} + g_{ds2})] \cong r_{ds2}[1 + R_s(g_{m2} + g_{s2})] \tag{3.30}$$

where g_{s2} is the body-effect constant. This result is only slightly different since g_s is roughly one-fifth of g_m.

EXAMPLE 3.4

Consider the current mirror shown in Fig. 3.12, where $I_{in} = 100\ \mu A$, each transistor has $W/L = (100\ \mu m)/(1.6\ \mu m)$, and $R_s = 5\ k\Omega$. Given that $\mu_n C_{ox} = 92\ \mu A/V^2$, $V_{tn} = 0.8\ V$, and $r_{ds} = [8{,}000L\ (\mu m)]/[I_D\ (mA)]$, find r_{out} for the current mirror. Assume the body effect can be approximated by $g_s = 0.2 g_m$.

Solution

Nominally, $I_{out} = I_{in}$, and thus we find the small-signal parameters for this current mirror to be

$$g_{m2} = \sqrt{2\mu_n C_{ox}(W/L)I_{out}} = 1.07\ mA/V \tag{3.31}$$

Also, as in Example 3.1, we have

$$r_{ds2} = \frac{8{,}000 \times 1.6}{0.1} = 128\ k\Omega \tag{3.32}$$

Now, making use of (3.29), the output impedance is given by

$$r_{out} = 128k\left[1 + 5\left(1.07 + 0.2 \times 1.07 + \frac{1}{128}\right)\right] = 955\ k\Omega \tag{3.33}$$

Note that this result is nearly eight times the output impedance for a simple current mirror as found in Example 3.1. Also note that the voltage drop across R_s equals $100\ \mu A \times 5\ k\Omega = 0.5\ V$ due to the dc bias current through it.

3.6 HIGH-OUTPUT-IMPEDANCE CURRENT MIRRORS

In this section two high-output-impedance current mirrors are described—the cascode and the Wilson current mirrors. These current mirrors have output impedances that are larger than that of a simple current mirror by a factor of $g_m r_{ds}$—the maximum gain of a single transistor.

Cascode Current Mirrors

A cascode current mirror is shown in Fig. 3.14. First, note that the output impedance looking into the drain of Q_2 is simply r_{ds2}, which is seen using an analysis very similar to that which was used for the simple current mirror. Thus, the output impedance can be immediately derived by considering Q_4 as a current source with a source-degeneration resistor of value r_{ds2}. Making use of Eq. (3.29), and noting that Q_4 is now the cascode transistor rather than Q_2, we have

$$r_{out} = r_{ds4}[1 + R_s(g_{m4} + g_{s4} + g_{ds4})] \tag{3.34}$$

where now $R_s = r_{ds2}$. Therefore, the output impedance is given by

$$\begin{aligned} r_{out} &= r_{ds4}[1 + r_{ds2}(g_{m4} + g_{s4} + g_{ds4})] \\ &\cong r_{ds4}[1 + r_{ds2}(g_{m4} + g_{s4})] \\ &\cong r_{ds4}(r_{ds2} g_{m4}) \end{aligned} \tag{3.35}$$

Thus, the output impedance has been increased by a factor of $g_{m4} r_{ds2}$, which is an upper limit on the gain of a single-transistor MOS gain-stage, and might be a value between 10 and 100, depending on the transistor sizes and currents and the technology being used. This significant increase in output impedance can be instrumental in realizing single-stage amplifiers with large low-frequency gains.

There is a disadvantage in using a cascode current mirror—it reduces the maximum output-signal swings possible before transistors enter the triode region. To understand this reduction, recall that for an n-channel transistor to be in the active region (also called the saturation or pinch-off region) its drain-source voltage must be greater than V_{eff}; V_{eff} was defined in (1.54) as

$$V_{eff} \equiv V_{GS} - V_{tn} \tag{3.36}$$

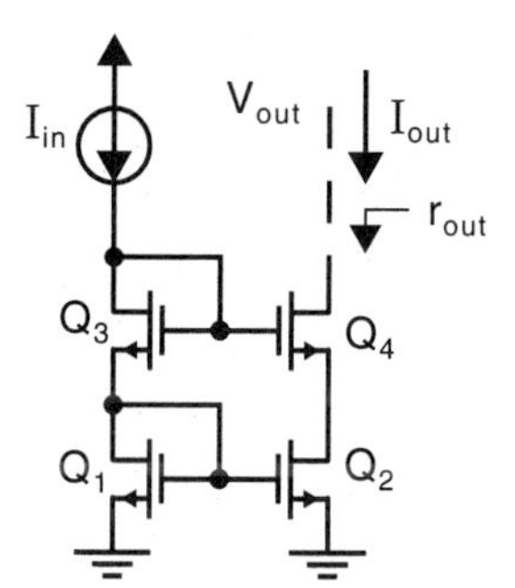

Fig. 3.14 A cascode current mirror.

which was shown in (1.80) to be given by

$$V_{eff} = \sqrt{\frac{2I_D}{\mu_n C_{ox}(W/L)}} \tag{3.37}$$

If we assume all transistors have the same sizes and currents, then they also all have the same V_{eff} and, therefore, the same gate-source voltages, $V_{GSi} = V_{eff} + V_{tn}$. Also, from Fig. 3.14, we see that

$$V_{G3} = V_{GS1} + V_{GS3} = 2V_{eff} + 2V_{tn} \tag{3.38}$$

and

$$V_{DS2} = V_{G3} - V_{GS4} = V_{G3} - (V_{eff} + V_{tn}) = V_{eff} + V_{tn} \tag{3.39}$$

Thus, the drain-source voltage of Q_2 is larger than the minimum needed to place it at the edge of the active region. Specifically, the drain-source voltage of Q_2 is V_{tn} (about 0.8 V) greater than what is required. Since the smallest output voltage, V_{D4}, can be without Q_4 entering the triode region is given by $V_{DS2} + V_{eff}$, the minimum allowed voltage for V_{out} is given by

$$V_{out} > V_{DS2} + V_{eff} = 2V_{eff1} + V_{tn} \tag{3.40}$$

which, again, is V_{tn} greater than the minimum value of $2V_{eff}$. This loss of signal swing is a serious disadvantage when modern technologies are used that might have a maximum allowed power-supply voltage as small as 3 V. In the next chapter, we will see how the cascode current mirror can be modified to maintain large output impedances and yet still allow for near minimum voltages at the output of the mirror.

EXAMPLE 3.5

Consider the cascode current mirror shown in Fig. 3.14, where $I_{in} = 100\ \mu A$ and each transistor has $W/L = (100\ \mu m)/(1.6\ \mu m)$. Given that $\mu_n C_{ox} = 92\ \mu A/V^2$, $V_{tn} = 0.8$ V, and $r_{ds} = [8{,}000L\ (\mu m)]/[I_D\ (mA)]$, find r_{out} for the current mirror (approximate the body effect by $0.2g_m$). Also find the minimum output voltage at V_{out} such that the output transistors remain in the active region.

Solution

Nominally, $I_{out} = I_{in}$, and thus we can find the small-signal parameters for this current mirror to be

$$g_{m4} = \sqrt{2\mu_n C_{ox}(W/L)I_{out}} = 1.07\ mA/V \tag{3.41}$$

We also have

$$r_{ds2} = r_{ds4} = \frac{8{,}000 \times 1.6}{0.1} = 128\ k\Omega \tag{3.42}$$

Now, making use of (3.35), the output impedance is given by

$$r_{out} = 128k[128(1.07 + 0.2 \times 1.07)] = 21\ M\Omega \tag{3.43}$$

To find the minimum output voltage, we first need to determine V_{eff}:

$$V_{eff} = \sqrt{\frac{2I_{out}}{\mu_n C_{ox}(W/L)}} = 0.19\ V \tag{3.44}$$

Thus, the minimum output voltage is determined to be $2 \times 0.19 + 0.8 = 1.18$ V.

Wilson Current Mirror

Another commonly used current mirror is the Wilson current mirror, shown in Fig. 3.15.[3] It is an example of using shunt-series feedback to increase the output impedance [Sedra, 1991]. Basically, Q_2 senses the output current and then mirrors it to I_{D1}, which, in turn, is subtracted from the input current, I_{in}. Note that I_{D1} must precisely equal I_{in}; otherwise the voltage at the gate of Q_3, Q_4 would either increase or decrease, and the negative feedback loop forces this equality. This feedback arrangement increases the output impedance by an amount equal to 1 plus the loop gain. Assuming all devices are matched, the output impedance without the feedback due to Q_1, Q_3 would be $2r_{ds4}$, taking into account that Q_4 has source degeneration equal to $1/g_{m2}$ (i.e., the small-signal impedance of diode-connected Q_2), which is responsible for the 2 factor. The loop gain is approximately given by

$$A_L \cong \frac{g_{m1}(r_{ds1} \parallel r_{in})}{2} \tag{3.45}$$

where r_{in} is the input impedance of the biasing current source I_{in}. The factor of $1/2$ is due to the voltage attenuation from the gate of Q_4 to its source, caused by the source degeneration of the diode-connected Q_2. Assuming r_{in} is approximately equal

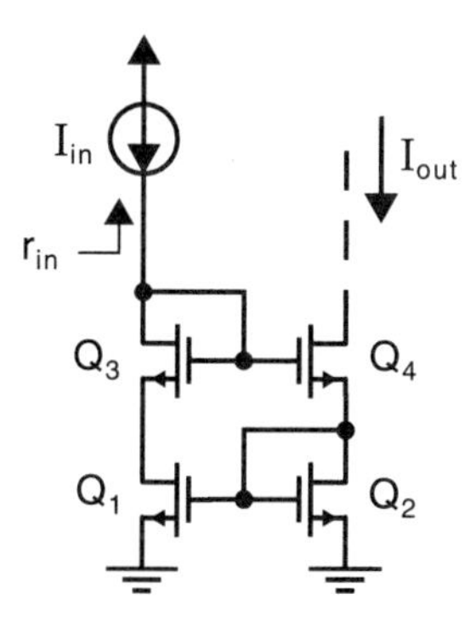

Fig. 3.15 The Wilson current mirror.

3. Actually, this is a modified Wilson current mirror, where Q_3 has been added to make the drain-source voltages of Q_1 and Q_2 approximately equal, which improves the absolute matching of I_{in} and I_{out}.

to r_{ds1}, then the loop gain is given by

$$A_L \cong \frac{g_{m1} r_{ds1}}{4} \tag{3.46}$$

and the output impedance is therefore given by

$$r_{out} \cong 2r_{ds4} \frac{g_{m1}(r_{ds1} \parallel r_{in})}{2} \cong r_{ds4}\left(\frac{g_{m1} r_{ds1}}{2}\right) \tag{3.47}$$

which is roughly one-half the output impedance for that of a cascode current mirror. For this reason, the cascode current mirror is often preferred over the Wilson current mirror. In terms of output voltage swing, the minimum allowed voltage across the current mirror, before Q_4 enters the triode region, is $2V_{eff1} + V_{tn}$, which is similar to that of the cascode current mirror.

Finally, it should be noted that Q_3 is not required in the Wilson current mirror. It has been included to give Q_1 and Q_2 the same drain-source bias voltages, and thus minimizes inaccuracies caused by the large-signal output impedances of the transistors. Without this transistor, the output current would be slightly smaller than the input current because V_{DS1} would be larger than V_{DS2}. However, the small-signal output impedance would remain the same.

3.7 CASCODE GAIN STAGE

In modern IC design, a commonly used configuration for a single-stage amplifier is a cascode configuration. This configuration consists of a common-source-connected transistor feeding into a common-gate-connected transistor. Two examples of cascode amplifiers are shown in Fig. 3.16. The configuration in Fig. 3.16(*a*) has both an n-channel common-source transistor, Q_1, and an n-channel common-gate cascode transistor, Q_2. This configuration is sometimes called a telescopic-cascode amplifier. The configuration shown in Fig. 3.16(*b*) has an n-channel input (or drive) transistor, but a p-channel transistor is used for the cascode (or common-gate) transistor. This configuration is usually called a folded-cascode stage. It allows the dc level of the output signal to be the same as the dc level of the input signal. Unfortunately, it is usually slower than the telescopic-cascode amplifier because although parasitic capacitances at the source of the cascode transistor are similar in both cases, the impedance levels of the folded-cascode stage are roughly three times larger due to the smaller transconductance of p-channel transistors as compared to n-channel transistors.

There are two major reasons for the popularity of cascode stages. The first is that they can have quite large gain for a single stage due to the large impedances at the output. To enable this high gain, the current sources connected to the output nodes are realized using high-quality cascode current mirrors. Normally this high gain is obtained without any degradation in speed, and sometimes with an improvement in speed. The second major reason for the use of cascode stages is that they limit the voltage across the input drive transistor. This minimizes any short-channel effects,

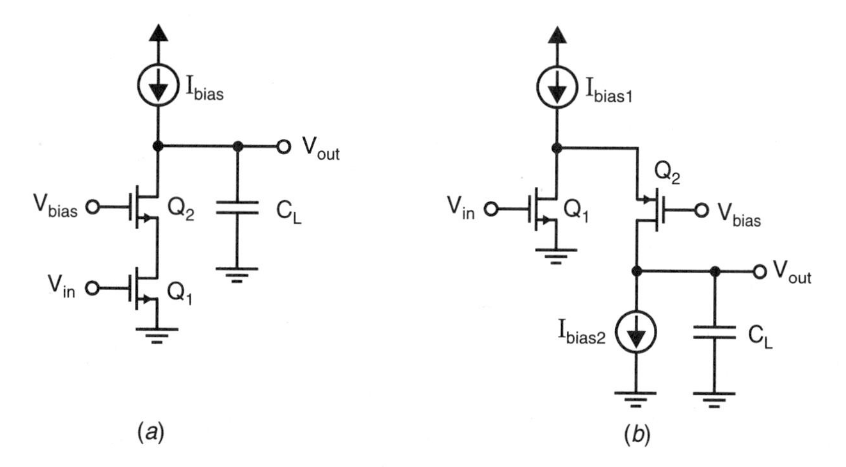

Fig. 3.16 (*a*) A telescopic-cascode amplifier and (*b*) a folded-cascode amplifier.

which becomes more important with modern technologies having very short channel-length transistors.

The analysis of the cascode gain stage is based on the telescopic stage of Fig. 3.16(*a*). The same analysis, with only minor modifications, also applies for the folded-cascode stage of Fig. 3.16(*b*).

From the section on cascode current mirrors, we know that the impedance looking into the drain of cascode transistor Q_2 is approximately given by

$$r_{d2} \cong g_{m2} r_{ds1} r_{ds2} \tag{3.48}$$

The total impedance at the output node is r_{d2} in parallel with R_L, where R_L is the output impedance of the bias current source, I_{bias}. Assuming I_{bias} is a high-quality source with an output impedance on the order of

$$R_L \cong g_{m\text{-}p} r_{ds\text{-}p}^2 \tag{3.49}$$

then the total impedance at the output node is

$$R_{out} \cong \frac{g_m r_{ds}^2}{2} \tag{3.50}$$

We have dropped the indices here under the assumption that the transistors are somewhat matched, and to simplify matters since we are only deriving an approximate solution.

To find the approximate low-frequency gain, we can use part of the analysis done previously for the common-gate stage. Repeating Eq. (3.20) here for convenience, we found that the low-frequency impedance looking into the source of the common-gate, or cascode, transistor, Q_2, was given by

$$Y_{in2} = \frac{g_{m2} + g_{s2} + g_{ds2}}{1 + \dfrac{g_{ds2}}{G_L}} \cong \frac{g_{m2}}{1 + \dfrac{g_{ds2}}{G_L}} \tag{3.51}$$

Note that the indices are changed to reflect the fact that Q_2 is the common-gate transistor rather than Q_1. Substituting (3.49) into (3.51), and again assuming all elements are somewhat matched so that the indices can be dropped, we have

$$Y_{in2} \cong \frac{g_m}{1 + \dfrac{g_{ds}}{g_{ds}^2/g_m}} \cong g_{ds} \tag{3.52}$$

The gain from the input to the source of Q_2 is therefore given by

$$\frac{v_{s2}}{v_{in}} = -\frac{g_{m1}}{g_{ds1} + Y_{in2}} \cong -\frac{g_m}{2g_{ds}} \tag{3.53}$$

The overall gain can then be found using (3.18) and (3.53). We have

$$A_V = \frac{v_{s2}v_{out}}{v_{in}\,v_{s2}} \cong -\frac{g_m}{2g_{ds}} \frac{g_{m2}}{G_L + g_{ds2}} \cong -\frac{g_m}{2g_{ds}} \frac{g_{m2}}{g_{ds2}} \cong -\frac{1}{2}\left(\frac{g_m}{g_{ds}}\right)^2 \tag{3.54}$$

The reader should be cautioned that (3.54) is only approximate, primarily due to the difficulty of accurately determining the output admittance, g_{ds}, for the different transistors. For example, one problem in estimating g_{ds} is that it is voltage dependent. Therefore, prudent designers should never construct a design for which successful operation requires knowing the gain precisely, rather than just knowing that it will be greater than some minimum value.

EXAMPLE 3.6

Assuming g_m is on the order of 0.5 mA/V and r_{ds} is on the order of 100 kΩ, what is the gain of the cascode amplifier?

Solution

Using (3.54), we have $A_V = -1{,}250$. This is a fairly representative number.

3.8 MOS DIFFERENTIAL PAIR AND GAIN STAGE

Most integrated amplifiers have a differential input. To realize this differential input, almost all amplifiers use what is commonly called a differential transistor pair. A differential pair together with a biasing current source is shown in Fig. 3.17. A low-frequency, small-signal model of the differential pair is shown in Fig. 3.18. This small-signal equivalent circuit is based on the T model for a MOS transistor that was described in Chapter 1. To simplify the analysis, we ignore the output impedance of the transistors temporarily. Defining the differential input voltage as $v_{in} \equiv v^+ - v^-$, we have

$$i_{d1} = i_{s1} = \frac{v_{in}}{r_{s1} + r_{s2}} = \frac{v_{in}}{1/g_{m1} + 1/g_{m2}} \tag{3.55}$$

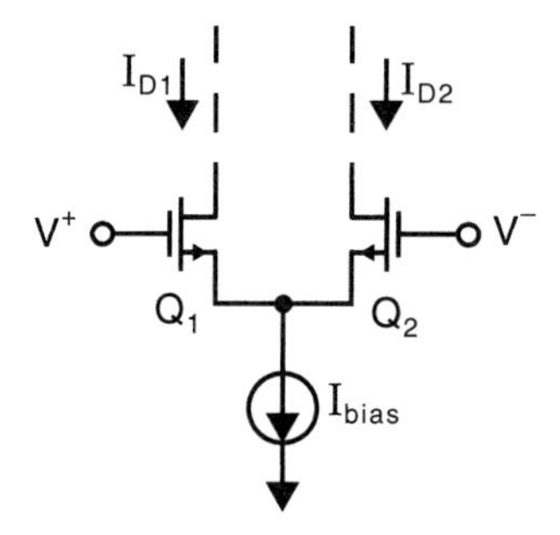

Fig. 3.17 A MOS differential pair.

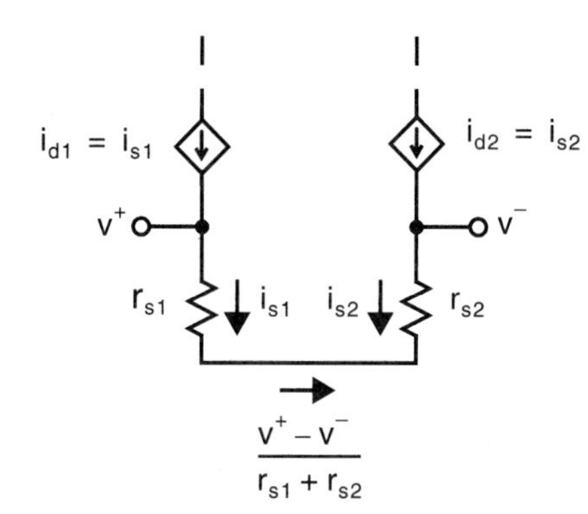

Fig. 3.18 The small-signal model of a MOS differential pair.

Since both Q_1 and Q_2 have the same bias currents, $g_{m1} = g_{m2}$. Therefore, we find

$$i_{d1} = \frac{g_{m1}}{2} v_{in} \tag{3.56}$$

Also, since $i_{d2} = i_{s2} = -i_{d1}$, we find that

$$i_{d2} = -\frac{g_{m1}}{2} v_{in} \tag{3.57}$$

Finally, defining a differential output current, $i_{out} \equiv i_{d1} - i_{d2}$, then the following relationship is obtained:

$$i_{out} = g_{m1} v_{in} \tag{3.58}$$

Thus, if a differential pair has a current mirror as an active load, a complete differential-input, single-ended-output gain stage can be realized, as shown in Fig. 3.19. This circuit is the typical first gain stage of a classical two-stage integrated opamp in which the input differential pair is realized using n-channel transistors and the active current-mirror load is realized using p-channel transistors. From the small-signal analysis of the differential pair, we have

$$i_{d1} = i_{s1} = \frac{g_{m1}}{2} v_{in} \tag{3.59}$$

Also, ignoring transistor output impedances, we have

$$i_{d4} = i_{d3} = -i_{s1} \tag{3.60}$$

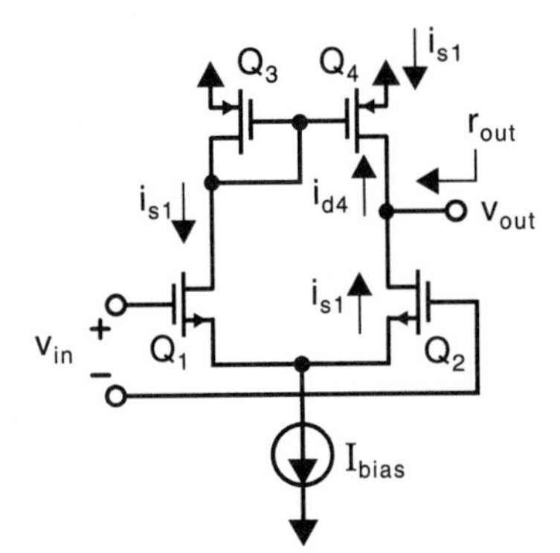

Fig. 3.19 A differential-input, single-ended-output MOS gain stage.

Note that a positive small-signal current is defined as the current going into the drain of a transistor. Using (3.60) and the fact that $i_{d2} = -i_{s1}$, we have

$$v_{out} = (-i_{d2} - i_{d4})r_{out} = 2i_{s1}r_{out} = g_{m1}r_{out}v_{in} \tag{3.61}$$

This result assumes that the output impedance is purely resistive. If there is also a capacitive load, C_L, then the gain is given by

$$A_v \equiv \frac{v_{out}}{v_{in}} = g_{m1}z_{out} \tag{3.62}$$

where $z_{out} = r_{out} \parallel 1/(sC_L)$. Thus, for this differential stage, the very simple model shown in Fig. 3.20 is commonly used. This model implicitly assumes that the time constant at the output node is much larger than the time constant due to the parasitic capacitance at the node at the sources of Q_1 and Q_2. This assumption is usually justified, because the impedance at the output node, r_{out}, is much larger than the impedance at the Q_1, Q_2 source node (i.e., $1/g_{m1} \parallel 1/g_{m2}$). Also, the capacitance at the output node, C_L, is usually larger than the parasitic capacitance at the Q_1, Q_2 source node. However, when high-frequency effects are important (which may be the case when compensating an opamp to guarantee stability), then this assumption may not be justified.

The evaluation of the output resistance, r_{out}, is determined by using the small-signal equivalent circuit and applying a voltage to the output node, as seen in Fig. 3.21. Note that the T model was used for both Q_1 and Q_2, whereas Q_3 was replaced by an equivalent resistance (since it is diode-connected), and the hybrid-π model was used for Q_4.

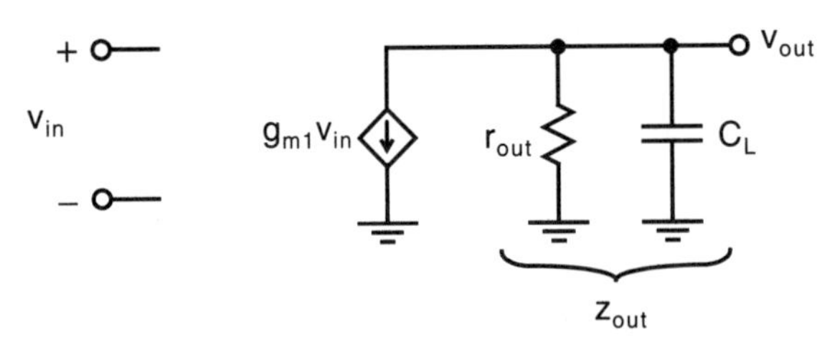

Fig. 3.20 A small-signal model for the differential-input amplifier.

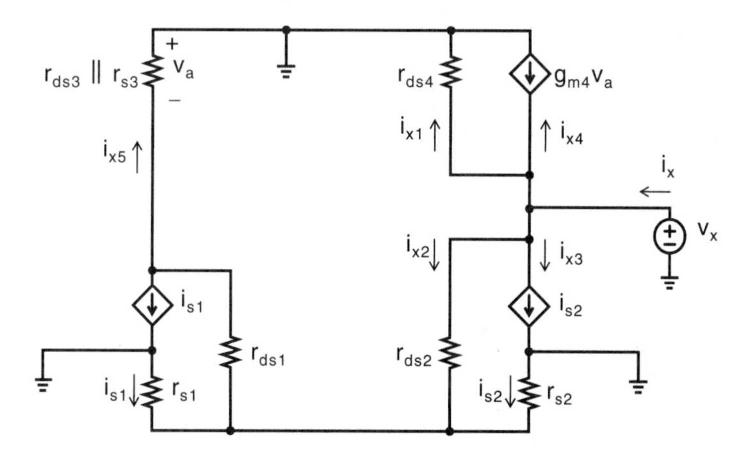

Fig. 3.21 The small-signal model for the calculation of the output impedance of the differential-input, single-ended-output MOS gain stage.

As usual, r_{out} is defined as the ratio v_x/i_x, where i_x is given by the sum $i_x = i_{x1} + i_{x2} + i_{x3} + i_{x4}$. Clearly,

$$i_{x1} = \frac{v_x}{r_{ds4}} \tag{3.63}$$

implying that the resistance seen in the path taken by i_{x1} is equal to r_{ds4}. Now, assuming that the effect of r_{ds1} can be ignored (since it is much larger than r_{s1}), we see that the current i_{x2} is given by

$$i_{x2} \cong \frac{v_x}{r_{ds2} + (r_{s1} \parallel r_{s2})} \cong \frac{v_x}{r_{ds2}} \tag{3.64}$$

where the second approximation is valid, since r_{ds2} is typically much greater than $r_{s1} \parallel r_{s2}$. This i_{x2} current splits equally between i_{s1} and i_{s2} (assuming $r_{s1} = r_{s2}$ and once again ignoring r_{ds1}), resulting in

$$i_{s1} = i_{s2} = \frac{-v_x}{2r_{ds2}} \tag{3.65}$$

However, since the current mirror realized by Q_3 and Q_4 results in $i_{x4} = i_{x5}$ (assuming $g_{m4} = 1/r_{s4} = 1/r_{s3}$ and r_{ds3} is much larger than r_{s3}), the current i_{x4} is given by

$$i_{x4} = -i_{s1} = -i_{s2} = -i_{x3} \tag{3.66}$$

In other words, when the current splits equally between r_{s1} and r_{s2}, the current mirror of Q_3 and Q_4 causes the two currents i_{x3} and i_{x4} to cancel each other. Finally, the output resistance, r_{out}, is given by

$$r_{out} = \frac{v_x}{i_{x1} + i_{x2} + i_{x3} + i_{x4}} = \frac{v_x}{(v_x/r_{ds4}) + (v_x/r_{ds2})} \tag{3.67}$$

which results in the simple relationship

$$r_{out} = r_{ds2} \| r_{ds4} \tag{3.68}$$

Therefore, at low frequencies the gain, A_v, is given by

$$A_v = g_{m1}(r_{ds2} \| r_{ds4}) \tag{3.69}$$

EXAMPLE 3.7

Consider the gain stage shown in Fig. 3.19, where $I_{bias} = 200\ \mu A$ and all transistors have $W/L = (100\ \mu m)/(1.6\ \mu m)$. Given that $\mu_n C_{ox} = 92\ \mu A/V^2$, $V_{tn} = 0.8$ V, and $r_{ds} = [8{,}000L\ (\mu m)]/[I_D\ (mA)]$, find the output impedance, r_{out}, and the gain from the differential input to the output, V_{out}.

Solution

To find the bias currents, we assume that I_{bias} splits evenly between the two sides of the differential circuit, resulting in

$$I_{D1} = I_{D2} = I_{D3} = I_{D4} = 100\ \mu A \tag{3.70}$$

Therefore, the transconductance of the input transistors is equal to

$$g_{m1} = g_{m2} = \sqrt{2\mu_n C_{ox}(W/L)(I_{bias}/2)} = 1.07\ mA/V \tag{3.71}$$

The output impedance of Q_2 and Q_4 is given by

$$r_{ds2} = r_{ds4} = \frac{8{,}000 \times 1.6}{0.1} = 128\ k\Omega \tag{3.72}$$

Thus, the gain for this stage is

$$A_v \equiv \frac{V_{out}}{V_{in}} = g_{m1}(r_{ds2} \| r_{ds4}) = 68.5\ V/V \tag{3.73}$$

3.9 BIPOLAR CURRENT MIRRORS

The most popular bipolar current mirrors are very similar to the MOS current mirrors. A simple bipolar current mirror is shown in Fig. 3.22. This current mirror has an out-

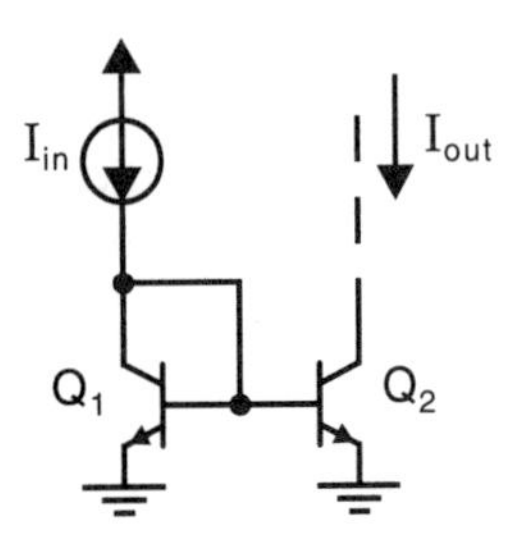

Fig. 3.22 A simple bipolar current mirror.

put current almost equal to its input current. In fact, its output current is slightly smaller than the input current, due to the finite base currents of Q_1 and Q_2. Taking these two base currents into account results in

$$I_{out} = \frac{1}{(1 + 2/\beta)} I_{in} \tag{3.74}$$

where β is the transistor current gain. For large β, this relation is approximately given by $I_{out} \cong (1 - 2/\beta)I_{in}$.

In one commonly used modification of the simple bipolar current mirror, an emitter-follower buffer, Q_3, is added to supply the base currents, as shown in Fig. 3.23. This additional transistor minimizes the errors due to finite base currents, resulting in $I_{out} \cong I_{in}(1 - 2/\beta^2)$. Such an arrangement is almost always used for current mirrors when lateral transistors[4] are used because of their low current gains (i.e., β's on the order of only 10–20). The output impedance of both of the previous current sources is equal to the output impedance of Q_2, which is r_{o2}.

In another often-used variation of the simple current mirror, emitter degeneration is added. This approach results in larger output impedances and also minimizes errors caused by mismatches between Q_1 and Q_2. An example of this current mirror is shown in Fig. 3.24. In an analysis similar to that given for the MOS current mirror with source degeneration, the output impedance is now found to be given by

$$r_{out} \cong r_{o2}(1 + g_{m2}R_e) \tag{3.75}$$

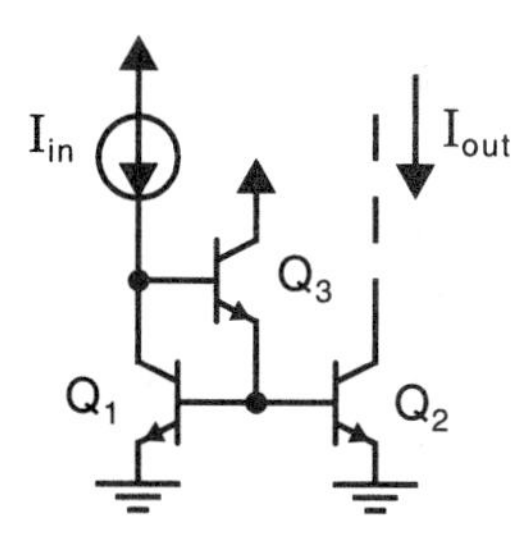

Fig. 3.23 A current mirror with fewer inaccuracies caused by finite base currents.

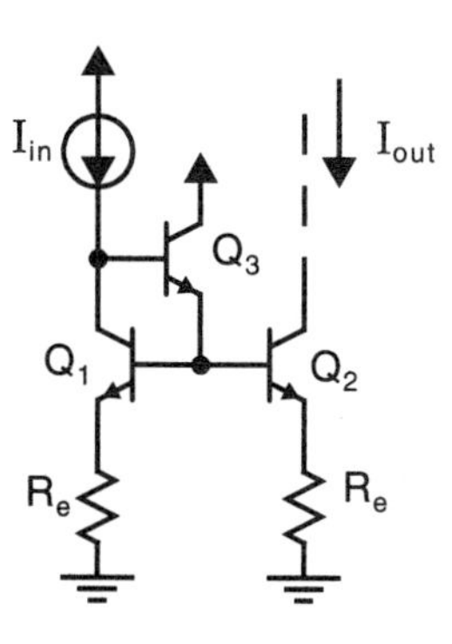

Fig. 3.24 A current mirror with emitter degeneration.

4. In many bipolar processes, pnp transistors are only available as lateral devices.

for $R_e << r_\pi$. Normally, this current mirror is designed so that the bias voltage across R_e, defined to be V_{Re}, is about 0.25 V. This implies that R_e is given by

$$R_e = \frac{V_{Re}}{I_{e2}} \cong \frac{V_{Re}}{I_{c2}} \tag{3.76}$$

Using the relationship $g_{m2} = I_{c2}/V_T$, where $V_T = kT/q \cong 26$ mV at 300 °K, gives

$$r_{out} \cong r_{o2}(1 + g_{m2}R_e) = r_{o2}\left(1 + \frac{V_{Re}}{V_T}\right) \cong 11r_{o2} \tag{3.77}$$

assuming $V_{Re} = 0.25$ V. It should be mentioned here that the addition of these R_e resistors also minimizes the noise output current of this current mirror, generated by the base resistance thermal noise, which is often the major source of noise in bipolar wideband circuits.

To achieve still higher output impedances, one can use either a cascode or a Wilson current mirror, as shown in Fig. 3.25. The Wilson current mirror is preferred in bipolar realizations because the cascode mirror exhibits large errors, due to the fact that the base currents of all of the transistors are supplied by I_{in} only. As a result, the output current is smaller than the input current by a factor roughly equal to $1 - 4/\beta$. For the Wilson current mirror, the base currents of Q_3 and Q_4 are supplied by I_{in}, whereas the base currents of Q_1 and Q_2 come from I_{out}. It can be shown that the errors due to finite base currents are on the order of $2/\beta^2$ [Gray, 1993]. It can also be shown that both of these current mirrors have an output impedance on the order of [Gray, 1993]

$$r_{out} \cong \frac{\beta r_o}{2} \tag{3.78}$$

This concludes the review of commonly used current mirrors. Chapter 6 describes some improved MOS current mirrors that have become popular recently.

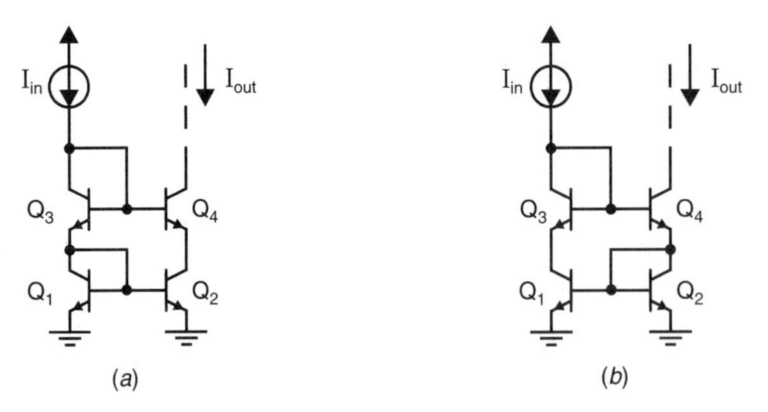

Fig. 3.25 High-output impedance (*a*) cascode and (*b*) Wilson current mirrors.

3.10 BIPOLAR GAIN STAGES

In this section, we look at two bipolar circuits—emitter followers and differential pairs. Although much of the small-signal analyses follow those of their MOS counterparts very closely, one difference is the finite base impedance of bipolar transistors. Here we shall see some simple rules for dealing with this finite base impedance during small-signal analysis. Because of its importance in translinear circuits, we also look at the *large-signal* behavior of the differential pair.

Emitter Follower

A bipolar emitter-follower is very similar to a MOS source follower, except that its input impedance is not infinite and its gain is normally much closer to unity. The analysis of its low-frequency response provides a good illustration of the use of the bipolar T model, and also of some relatively general principles for coming up with quick estimates of input and output impedances of bipolar circuits.

A bipolar emitter follower with a resistive load is shown in Fig. 3.26. Its small-signal model at low frequencies, where the bipolar T model has been used, is shown in Fig. 3.27(*a*). In this small-signal model, the device base resistance r_b has been ignored, since it can be easily taken into account after the fact by simply making R_S slightly larger (i.e., equal to $R_S + r_b$). Notice that R_E and r_o are in parallel. This allows a slightly simplified small-signal model to be analyzed, as shown in Fig. 3.27(*b*), where $R'_E = R_E \parallel r_o$. The analysis of the gain of this circuit is done in two steps. First, the input impedance looking into the base of the transistor, R_b, is found, allowing us to calculate the gain from the input to the base. Second, the gain from the base to the output is found, which allows us to derive the overall gain. The current in the emitter, i_e, is given by

$$i_e = i_b(\beta + 1) \tag{3.79}$$

Therefore,

$$v_b = i_e(r_e + R'_E) = i_b(\beta + 1)(r_e + R'_E) \tag{3.80}$$

This gives

$$R_b = \frac{v_b}{i_b} = (\beta + 1)(r_e + R'_E) \tag{3.81}$$

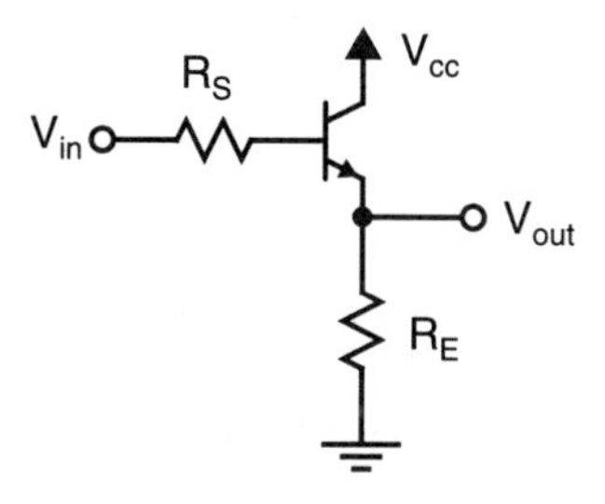

Fig. 3.26 A bipolar emitter follower.

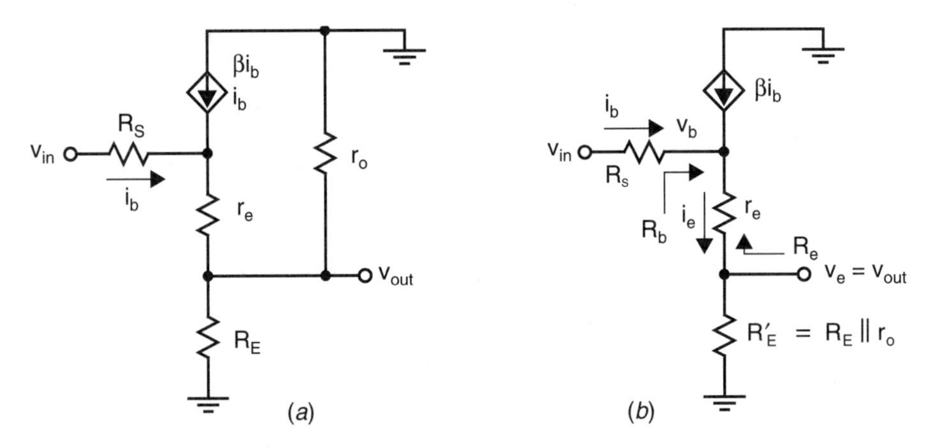

Fig. 3.27 (*a*) The small-signal model for the bipolar emitter follower and (*b*) a simplified model.

Alternatively, noting that

$$(\beta+1)r_e = (\beta+1)\frac{\alpha}{g_m} = (\beta+1)\frac{\beta/(\beta+1)}{g_m} = \frac{\beta}{g_m} = r_\pi \tag{3.82}$$

gives

$$R_b = r_\pi + (\beta+1)\,R'_E \tag{3.83}$$

Equations (3.81) and (3.83) illustrate a principle that is generally applicable for bipolar circuits: *At low frequencies, resistances in series with the emitter appear* $\beta + 1$ *times larger when seen looking into the base or, equivalently, when they are reflected into the base.* Continuing, using the resistor-divider formula, we now have

$$\frac{v_b}{v_{in}} = \frac{R_b}{R_b + R_S} = \frac{(\beta+1)(r_e + R'_E)}{(\beta+1)(r_e + R'_E) + R_S} \tag{3.84}$$

The gain from the base to the emitter, which is the output, is easily found from Fig. 3.26(*b*), again using the resistor-divider formula, to be given by

$$\frac{v_{out}}{v_b} = \frac{R'_E}{R'_E + r_e} \cong \frac{R'_E}{R'_E + 1/g_m} \tag{3.85}$$

Using (3.84) and (3.85), the overall gain is now given by

$$\frac{v_{out}}{v_{in}} = \frac{v_b}{v_{in}}\frac{v_{out}}{v_b} = \left(\frac{(\beta+1)(r_e + R'_E)}{(\beta+1)(r_e + R'_E) + R_S}\right)\left(\frac{R'_E}{R'_E + 1/g_m}\right) \tag{3.86}$$

With a little practice, transfer functions such as that given by (3.86) can be written by simply inspecting the actual circuit, without actually analyzing the small-signal model.

It is also interesting to find the output impedance of the emitter follower at low frequencies excluding R'_E, which is equal to the impedance seen looking into the emitter. The small-signal model for this analysis is shown in Fig. 3.28, where the

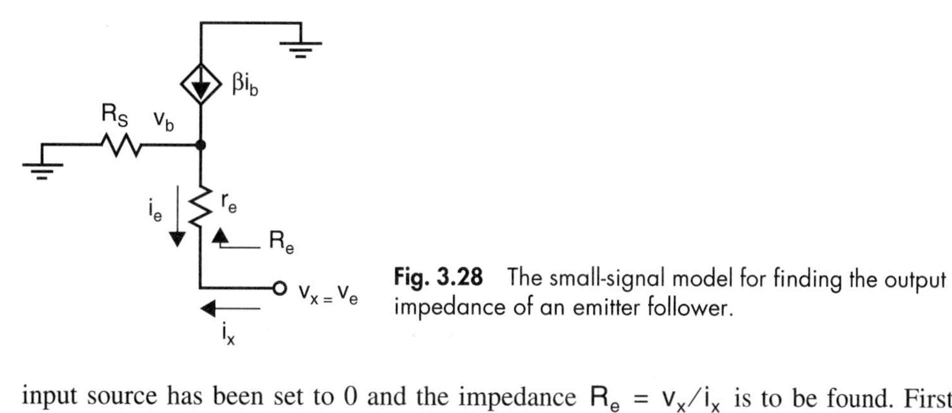

Fig. 3.28 The small-signal model for finding the output impedance of an emitter follower.

input source has been set to 0 and the impedance $R_e = v_x/i_x$ is to be found. First note that

$$i_b = \frac{i_e}{\beta + 1} = \frac{-i_x}{\beta + 1} \tag{3.87}$$

since $i_x = -i_e$. Therefore, we have

$$\begin{aligned} v_x &= v_b + i_x r_e \\ &= -i_b R_S + i_x r_e \\ &= i_x \frac{R_S}{\beta + 1} + i_x r_e \end{aligned} \tag{3.88}$$

This gives the impedance seen looking into the emitter as

$$R_e = \frac{v_x}{i_x} = \frac{R_S}{\beta + 1} + r_e = \frac{R_S}{\beta + 1} + \frac{r_\pi}{\beta + 1} \tag{3.89}$$

This is an example of a general principle for bipolar circuits: *Resistances in series with the base are divided by* $\beta + 1$ *when they are seen looking into the emitter, or, equivalently, are reflected to the emitter.* The total output impedance of the emitter-follower is now simply R_e in parallel with R'_E.

Bipolar Differential Pair—Large-Signal

A bipolar differential pair is shown in Fig. 3.29. This circuit's large-signal behavior can be analyzed by first recalling the exponential relationship for a bipolar transistor,

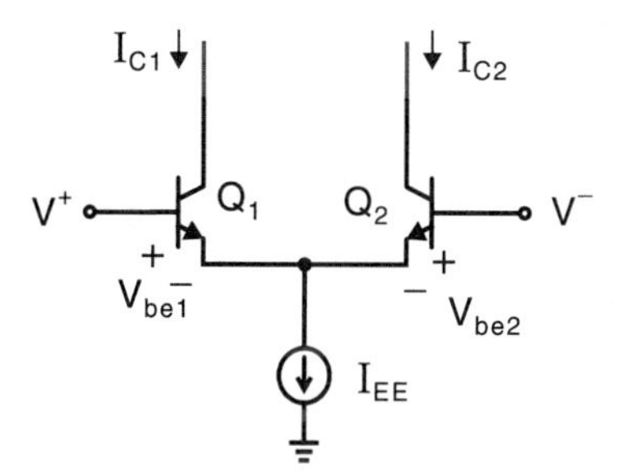

Fig. 3.29 A bipolar differential pair.

$$I_C = I_S e^{(V_{BE}/V_T)} \tag{3.90}$$

which can be used to find the base-emitter voltages

$$V_{BE1} = V_T \ln\left(\frac{I_{C1}}{I_{S1}}\right) \tag{3.91}$$

$$V_{BE2} = V_T \ln\left(\frac{I_{C2}}{I_{S2}}\right) \tag{3.92}$$

Now, writing an equation for the sum of voltages around the loop of input and base-emitter voltages, we have

$$V^+ - V_{BE1} + V_{BE2} - V^- = 0 \tag{3.93}$$

Combining (3.91), (3.92), and (3.93), and assuming $I_{S1} = I_{S2}$, we can write

$$\frac{I_{C1}}{I_{C2}} = e^{(V^+ - V^-)/V_T} = e^{V_{id}/V_T} \tag{3.94}$$

where V_{id} is defined as the difference between V^+ and V^-. In addition, we can write

$$\alpha I_{EE} = I_{C1} + I_{C2} \tag{3.95}$$

where α is defined to be $\beta/(\beta + 1)$ and is due to some currents flowing through the base terminals. Finally, combining (3.94) and (3.95), we have

$$I_{C1} = \frac{\alpha I_{EE}}{1 + e^{-V_{id}/V_T}} \tag{3.96}$$

$$I_{C2} = \frac{\alpha I_{EE}}{1 + e^{V_{id}/V_T}} \tag{3.97}$$

A plot of these two currents with respect to V_{id} is shown in Fig. 3.30, where we note that the currents are split equally at $V_{id} = 0$ and saturate near I_{EE} or 0 for differential input voltages approaching $4\ V_T$ (around 100 mV). Note that the relations shown

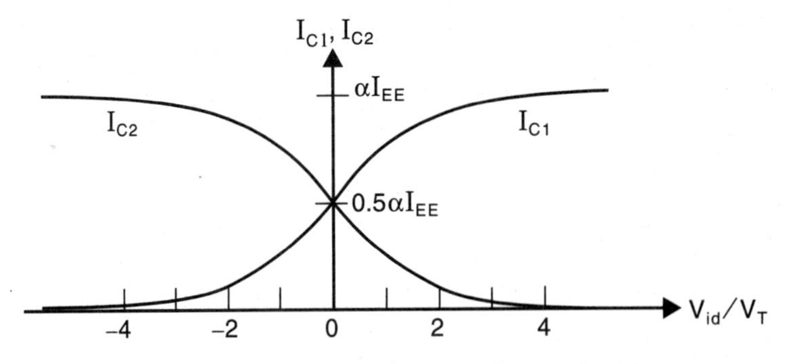

Fig. 3.30 Collector currents for a bipolar differential pair.

in (3.96) and (3.97) results in a hyperbolic tangent function when their difference is taken. Thus this current–voltage relationship for a bipolar differential is commonly referred to as the *tanh relationship.*

Bipolar Differential Pair—Small-Signal

The small-signal model for the bipolar differential pair of Fig. 3.29 is shown in Fig. 3.31. Once again, defining $v_{id} = v^+ - v^-$, we have

$$i_{c1} = \alpha i_{e1} = \frac{\alpha v_{id}}{r_{e1} + r_{e2}} = \frac{\alpha v_{id}}{(\alpha/g_{m1}) + (\alpha/g_{m2})} \tag{3.98}$$

In the case where both transistors have the same bias currents through them (i.e., the nominal differential voltage is 0, resulting in $g_{m1} = g_{m2}$), we find the same result as for a MOS differential pair,

$$i_{c1} = \frac{g_{m1}}{2} v_{id} \tag{3.99}$$

A similar result holds for i_{c2}.

Finally, the input impedance of this differential pair can be found using the impedance-scaling rule just discussed. Specifically, consider the small-signal model shown in Fig. 3.32, where v^- is grounded and we wish to find the impedance r_{id} looking into the base of Q_1. The impedance, r_2, seen looking into the emitter of Q_2 is simply r_{e2},

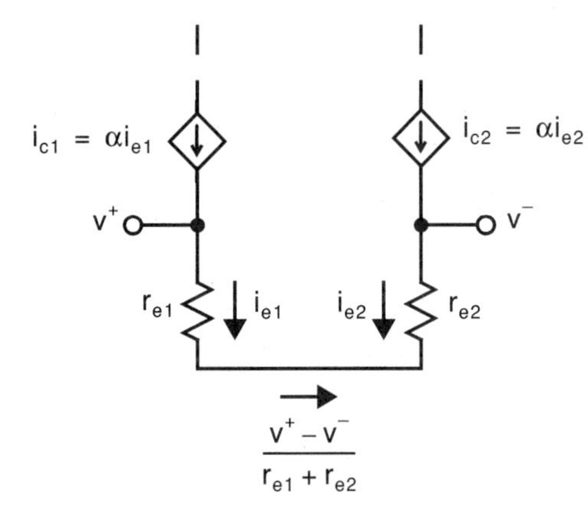

Fig. 3.31 The small-signal model of a bipolar differential pair.

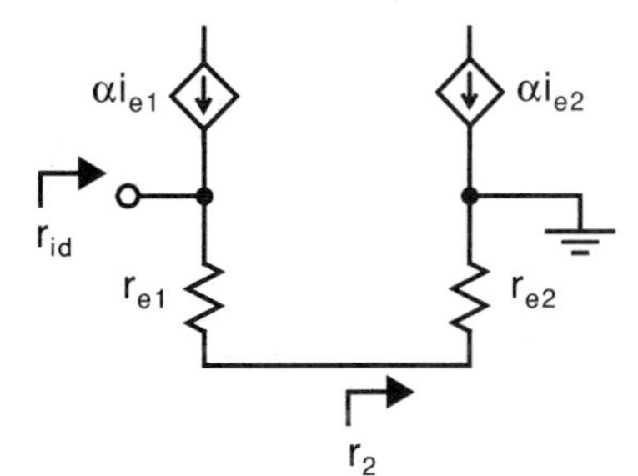

Fig. 3.32 Finding the input impedance of a bipolar differential pair.

since the base is grounded. Thus, the total emitter resistance to the ground of Q_1 is $r_{e1} + r_{e2}$, and using the impedance reflection rule, the impedance, r_{id}, seen looking into the base of Q_1 is equal to $\beta + 1$ times the emitter resistance, or, equivalently,

$$r_{id} = (\beta + 1)(r_{e1} + r_{e2}) \tag{3.100}$$

3.11 FREQUENCY RESPONSE

In this section, we look at the frequency response of many of the previous circuits. Although the precise calculation of frequency responses is most often left to computer simulations, there is much insight that can be obtained by finding the dominant frequency effects in integrated circuits.

Common-Source Amplifier

A small-signal equivalent circuit for high-frequency analysis of the common-source amplifier of Fig. 3.4 is shown in Fig. 3.33. Here, C_{gs1} is the gate-to-source capacitance of Q_1, whereas C_{gd1} is the gate-to-drain capacitance of Q_1. Note that we have assumed that the output capacitance of the input source can be ignored. The capacitance C_2 is made up of the parallel connection of the drain-to-bulk capacitances of Q_1 and Q_2 together with the load capacitance, C_L. Usually, C_L dominates.

To analyze the circuit at high frequencies, one can use nodal analysis. At node v_1, we add all of the currents leaving the node and set the sum equal to zero, to obtain

$$v_1(G_{in} + sC_{gs1} + sC_{gd1}) - v_{in}G_{in} - v_{out}sC_{gd1} = 0 \tag{3.101}$$

where $G_{in} = 1/R_{in}$. Also, at the output node, we have

$$v_{out}(G_2 + sC_{gd1} + sC_2) - v_1 sC_{gd1} + g_{m1}v_1 = 0 \tag{3.102}$$

where $v_1 = v_{gs1}$.

Solving (3.101) and (3.102) (somewhat tediously), we have

$$\frac{v_{out}}{v_{in}} = \frac{-g_{m1}R_2\left(1 - s\dfrac{C_{gd1}}{g_{m1}}\right)}{1 + sa + s^2b} \tag{3.103}$$

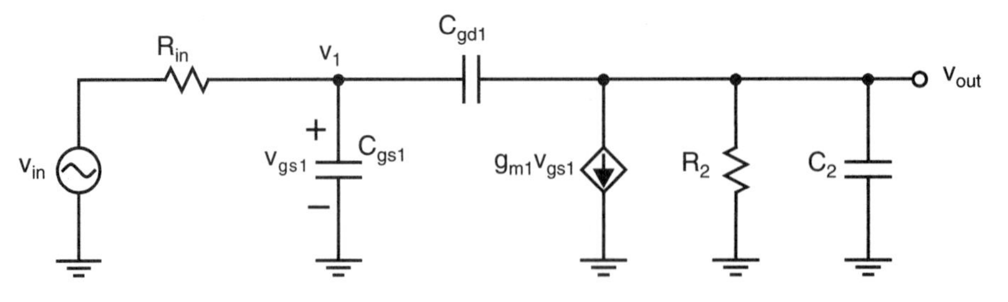

Fig. 3.33 A small-signal model for high-frequency analysis of the common-source amplifier.

where

$$a = R_{in}[C_{gs1} + C_{gd1}(1 + g_{m1}R_2)] + R_2(C_{gd1} + C_2) \tag{3.104}$$

and

$$b = R_{in}R_2(C_{gd1}C_{gs1} + C_{gs1}C_2 + C_{gd1}C_2) \tag{3.105}$$

At frequencies where the gain has started to decrease but is still much greater than unity, the first-order term in the numerator, $-s(C_{gd1}/g_{m1})$, and the second-order term in the denominator, s^2b, can be ignored. For this case, we have

$$A(s) = \frac{V_{out}}{V_{in}} \cong \frac{-g_{m1}R_2}{1 + s\{R_{in}[C_{gs1} + C_{gd1}(1 + g_{m1}R_2)] + R_2(C_{gd1} + C_2)\}} \tag{3.106}$$

The low-frequency gain is as expected, $-g_{m1}R_2$. Also, setting $s = j\omega_{-3\text{ dB}}$ and solving for

$$|A(j\omega_{-3\text{ dB}})| = \frac{1}{\sqrt{2}} \tag{3.107}$$

gives

$$\omega_{-3\text{ dB}} \cong \frac{1}{R_{in}[C_{gs1} + C_{gd1}(1 + g_{m1}R_2)] + R_2(C_{gd1} + C_2)} \tag{3.108}$$

As an aside, it is of interest to note that the result for the –3-dB frequency is the same result that one would obtain if the zero-value time-constant analysis technique were used [Gray, 1993]. In this technique, one calculates a time constant for each capacitor by assuming all other capacitors are zero, replacing the capacitor in question with a voltage source, and then calculating the resistance seen by that capacitor by taking the ratio of the voltage source to the current flowing from the voltage source. The time constant seen by the capacitor is then simply the capacitor multiplied by the resistance seen by that capacitor. The –3-dB frequency for the complete circuit is then 1 divided by the sum of the individual capacitor time constants. For the common-source amplifier, the resistance seen by C_{gs1} is the input source impedance R_{in}, the resistance seen by C_{gd1} is $R_{in}(1 + g_{m1}R_2) + R_2$, and the resistance seen by C_2 is R_2.

Often, unless $R_{in} << R_2$, the first term in the denominator of (3.108) dominates, and we have

$$\omega_{-3\text{ dB}} \cong \frac{1}{R_{in}[C_{gs1} + C_{gd1}(1 + A)]} \tag{3.109}$$

where $A = g_{m1}R_2$ is the magnitude of the low-frequency gain. The term $C_{gd1}(1 + A)$ is often called the Miller capacitance because it is the equivalent capacitance obtained when one uses the Miller approximation [Sedra, 1991]. Because the size of C_{gd1} is effectively multiplied by one plus the gain of the amplifier, C_{gd1} must be small.

At higher frequencies, when the gain is not much greater than unity, the second pole and the zero should be considered. The frequency for the second pole can be found by assuming that the poles are real and widely separated and that the denominator can therefore be expressed as

$$D(s) = \left(1+\frac{s}{\omega_{p1}}\right)\left(1+\frac{s}{\omega_{p2}}\right) \cong 1+\frac{s}{\omega_{p1}}+\frac{s^2}{\omega_{p1}\omega_{p2}} \tag{3.110}$$

The coefficients of (3.110) can then be equated to the coefficients of the denominator of (3.103). This analysis is almost identical to that used later in Section 5.2 for analyzing the two-stage amplifier for its frequency response, so it will not be given here. However, the equation for the approximate frequency of the second pole of the denominator is simply given as

$$\omega_{p2} \cong \frac{g_{m1}C_{gd1}}{C_{gs1}C_{gd1}+C_{gs1}C_2+C_{gd1}C_2} \tag{3.111}$$

It should be mentioned that the Miller approximation results in a very different and incorrect approximation for the frequency of the second pole.

EXAMPLE 3.8

Use the same parameters as in Example 3.2 along with the following: R_{in} = 180 kΩ, C_L = 0.3 pF, C_{gs1} = 0.2 pF, C_{gd1} = 0.015 pF, C_{db1} = 20 fF, and C_{db2} = 36 fF. Estimate the –3-dB frequency of the common-source amplifier in Fig. 3.4.

Solution

We have

$$R_2 = r_{ds1} \| r_{ds2} = 77\ \text{k}\Omega \tag{3.112}$$

and

$$C_2 = C_L + C_{db1} + C_{db2} = 0.36\ \text{pF} \tag{3.113}$$

The time constant due to R_{in}, namely, $R_{in}[C_{gs1} + C_{gd1}(1 + A)]$, is now equal to 0.26 μs. The time constant due to R_2, namely, $R_2(C_{gd1} + C_2)$, is equal to 0.03 μs. The –3-dB frequency (in hertz) is equal to

$$f_{-3\ \text{dB}} \cong \left[\frac{1}{2\pi}\right][R_{in}[C_{gs1}+C_{gd1}(1+g_{m1}R_2)]+R_2(C_{gd1}+C_2)]^{-1} \tag{3.114}$$
$$= 550\ \text{kHz}$$

Source-Follower Amplifier

Before proceeding, it should be stated that the high-frequency analysis of source-follower amplifiers is somewhat involved. We will show that these types of amplifiers can have complex poles, and thus a designer should be careful that the circuit does not exhibit too much overshoot and ringing. We also show a compensation circuit that results in only real axis poles and therefore no overshoot and ringing. However, because the material is complex, we suggest that only advanced readers cover this subsection in detail.

The frequency response of the source follower can be found by modelling the source as a Norton equivalent circuit and adding a load capacitance, as Fig. 3.34 shows. The small-signal model used for this circuit, which includes the parasitic capacitances, is shown in Fig. 3.35. Capacitor C_s includes both the load capacitor, C_L, and the parasitic capacitor, C_{sb1}. Similar to what was done at low frequencies, r_{ds1}, r_{ds2}, and the voltage-controlled current source modelling the body-effect current source can be modelled by a single resistor. This model allows us to analyze the simplified small-signal model shown in Fig. 3.36, where again $R_{s1} = r_{ds1} \| r_{ds2} \| (1)/g_{s1}$ and the input capacitance is given by $C'_{in} = C_{in} + C_{gd1}$.

Nodal analysis is possible, but it is very complicated for this example. The analysis proceeds in four steps. First, the gain from v_{g1} to v_{out} is found. Second, the admittance, Y_g, looking into the gate of Q_1, but not taking into account C_{gd1}, is found. Third, the gain from i_{in} to v_{g1} is found. Finally, the overall gain from v_{in} to v_{out} is found and the results are interpreted.

At node v_{out}, we set the sum of the currents that leave the node to zero, so we have

$$v_{out}(sC_s + sC_{gs1} + G_{s1}) - v_{g1}sC_{gs1} - g_{m1}(v_{g1} - v_{out}) = 0 \tag{3.115}$$

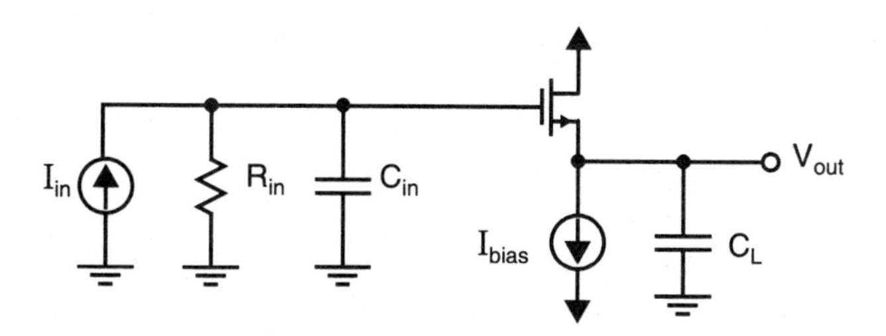

Fig. 3.34 The configuration used to analyze the frequency response of the source follower.

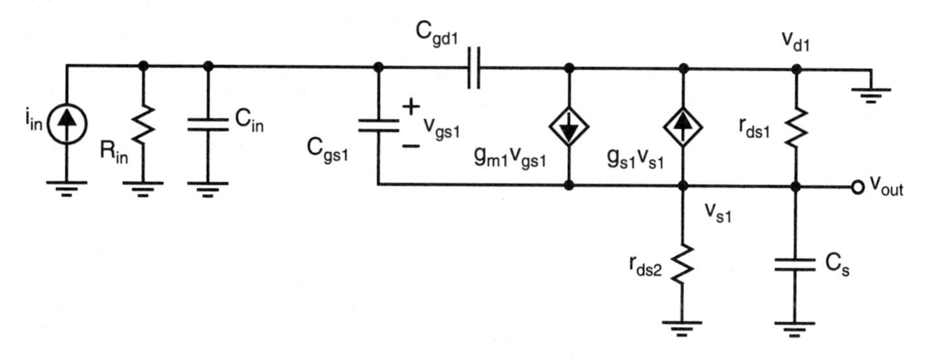

Fig. 3.35 An equivalent small-signal model for the source follower.

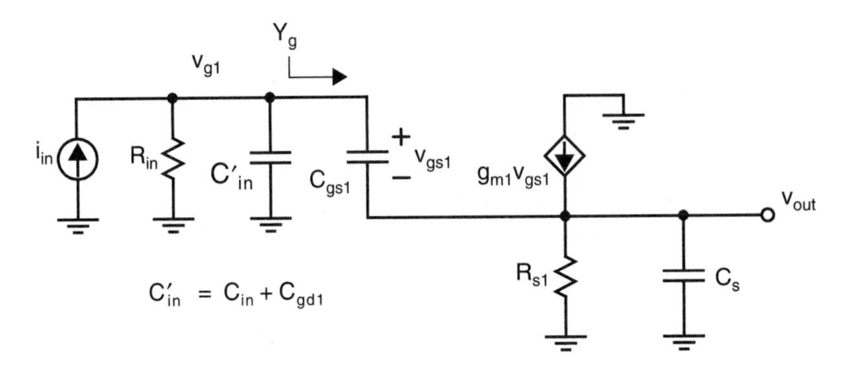

Fig. 3.36 A simplified equivalent small-signal model for the source follower.

Solving for v_{out}/v_{g1}, we have

$$\frac{v_{out}}{v_{g1}} = \frac{sC_{gs1} + g_{m1}}{s(C_{gs1} + C_s) + g_{m1} + G_{s1}} \tag{3.116}$$

The next step is to calculate the admittance, Y_g, looking into the gate of Q_1, but not taking into account the current entering C_{gd1}. The input current is given by

$$i_{g1} = (v_{g1} - v_{out})sC_{gs1} \tag{3.117}$$

Using (3.116) to eliminate v_{out} in (3.117), and solving for $Y_g = i_{g1}/v_{g1}$, we have

$$Y_g = \frac{i_{g1}}{v_{g1}} = \frac{sC_{gs1}(sC_s + G_{s1})}{s(C_{gs1} + C_s) + g_{m1} + G_{s1}} \tag{3.118}$$

We can write an equation relating the input current, i_{in}, to the gate voltage, v_{g1}, as

$$i_{in} = v_{g1}(sC'_{in} + G_{in} + Y_g) \tag{3.119}$$

Substituting (3.118) into (3.119) and rearranging gives

$$\frac{v_{g1}}{i_{in}} = \frac{s(C_{gs1} + C_s) + g_{m1} + G_{s1}}{a + sb + s^2c} \tag{3.120}$$

where

$$\begin{aligned} a &= G_{in}(g_{m1} + G_{s1}) \\ b &= G_{in}(C_{gs1} + C_s) + C'_{in}(g_{m1} + G_{s1}) + C_{gs1}G_{s1} \\ c &= C_{gs1}C_s + C'_{in}(C_{gs1} + C_s) \end{aligned} \tag{3.121}$$

Using (3.116) and (3.120), we then have

$$A(s) = \frac{v_{out}}{i_{in}} = \frac{sC_{gs1} + g_{m1}}{a + sb + s^2c} \tag{3.122}$$

Thus, we see that the transfer function is second order. Specifically, it has two poles (roots of the denominator) that may be either real or complex conjugate. If they are complex conjugate, then the step response of the circuit will exhibit overshoot and possibly ringing. This potential problem is a disadvantage when using source followers.

To determine if the transfer function will exhibit ringing, (3.122) can be written in the form

$$A(s) = A(0)\frac{N(s)}{1 + \frac{s}{\omega_0 Q} + \frac{s^2}{\omega_0^2}} \tag{3.123}$$

where ω_0 and Q can be found by equating the coefficients of (3.123) to the coefficients of (3.122). Here, parameter ω_0 is called the *pole frequency* and parameter Q is called the *Q factor*[5] [Sedra, 1991]. It is well known that if $Q < \sqrt{1/2} \cong 0.707$, then the magnitude of the transfer function has its maximum at dc and no peaking will occur (assuming the zero is at a very high frequency and therefore has negligible effect). Furthermore, for $Q = \sqrt{1/2}$, the –3-dB frequency is equal to ω_0. When the time-domain response is investigated, restrictions on the Q factor can also be found to guarantee no peaking for a step input. Specifically, to have no peaking in the step response, both poles must be real; this is equivalent to the requirement that $Q \le 0.5$. In the case where $Q > 0.5$, the percentage overshoot of the output voltage can be shown to be given by

$$\%\ \text{overshoot} = 100e^{-\pi/\sqrt{4Q^2 - 1}} \tag{3.124}$$

For the source follower, equating the coefficients of (3.123) to the coefficients of (3.122) and solving for ω_0 and Q results in

$$\omega_0 = \sqrt{\frac{G_{in}(g_{m1} + G_{s1})}{C_{gs1}C_s + C'_{in}(C_{gs1} + C_s)}} \tag{3.125}$$

$$Q = \frac{\sqrt{G_{in}(g_{m1} + G_{s1})[C_{gs1}C_s + C'_{in}(C_{gs1} + C_s)]}}{G_{in}C_s + C'_{in}(g_{m1} + G_{s1}) + C_{gs1}G_{s1}} \tag{3.126}$$

If Q is greater than 0.5, the poles will be complex conjugate, and the circuit will exhibit overshoot. Although this Q equation is rather complex, note that if C_s, C'_{in}, or both become large (i.e., if the load, the input capacitor, or both become large), then Q becomes small, and no overshoot will occur (although the circuit will be slow). And when C'_{in} and G_{s1} become small (G_{s1} becomes small when the transistor's source is connected to the substrate, which eliminates the body effect), then the circuit will have a large Q (i.e., large ringing) when G_{in} becomes small and $C_s \cong C_{gs1}$. *In summary, source follower (and emitter follower) circuits can exhibit large amounts of overshoot and ringing under certain conditions.* Fortunately, the parasitic capacitances and output impedances in practical microcircuits typically result in only moderate overshoot for worst-case conditions.

---

5. The Q factor is 1/2 times the inverse of the damping factor. The damping factor is an alternative method of indicating the pole locations in second-order transfer functions.

Finally, note also that the numerator zero of the transfer function lies on the negative real axis at a frequency given by

$$\omega_z = \frac{-g_{m1}}{C_{gs1}} \tag{3.127}$$

and is typically at a much higher frequency than ω_0.

EXAMPLE 3.9

Use the parameters in Example 3.3 and assume that $R_{in} = 180\ k\Omega$, $C_L = 10\ pF$, $C_{gs1} = 0.2\ pF$, $C_{gd1} = 15\ fF$, $C_{sb1} = 40\ fF$, and $C_{in} = 30\ fF$. Find ω_0, Q, and the frequency of the zero for the source follower in Fig. 3.34.

Solution

From Example 3.3, we have $g_{m1} = 1.06\ mA/V$, $r_{ds1} = 128\ k\Omega$, $r_{ds2} = 128\ k\Omega$, and $g_{s1} = 0.16\ mA/V$. Thus, we have

$$C'_{in} = C_{in} + C_{gd1} = 45\ fF \tag{3.128}$$

$$G_{s1} = g_{s1} + g_{ds1} + g_{ds2} = 0.176\ mA/V \tag{3.129}$$

$$C_s = C_L + C_{sb1} = 10.04\ pF \tag{3.130}$$

and so we can find ω_0 as

$$\omega_0 = \sqrt{\frac{G_{in}(g_{m1} + G_{s1})}{C_{gs1}C_s + C'_{in}(C_{gs1} + C_s)}}$$

$$= 5.24 \times 10^7\ rad/s = 2\pi \times 8.34\ MHz \tag{3.131}$$

$$Q = \frac{\sqrt{G_{in}(g_{m1} + G_{s1})[C_{gs1}C_s + C'_{in}(C_{gs1} + C_s)]}}{G_{in}C_s + C'_{in}(g_{m1} + G_{s1}) + C_{gs1}G_{s1}} \tag{3.132}$$

$$= 0.8$$

This results in an overshoot for a step input given by

$$\%\ overshoot = 100e^{-\pi/\sqrt{4Q^2 - 1}} = 8\% \tag{3.133}$$

The zero frequency is found using (3.127) to be 844 MHz, and thus it can almost certainly be ignored.

A bipolar-transistor emitter follower is very similar to a CMOS source follower, and therefore its high-frequency analysis is not included here. However, note that the problem of complex-conjugate poles can be more severe for bipolar emitter followers.

For both CMOS source followers and bipolar emitter followers, when complex-conjugate poles occur, they can be eliminated by adding a compensation network. To see this, note that (3.118) can be rewritten as

$$Y_g = sC_2 + \frac{1}{-R_1 - \frac{1}{sC_1}} \tag{3.134}$$

where

$$\begin{aligned} C_1 &= \frac{C_{gs1}(C_s g_{m1} - C_{gs1}G_{s1})}{(g_{m1}+G_{s1})(C_{gs1}+C_s)} \cong \frac{g_{m1}C_{gs1}C_s}{(g_{m1}+G_{s1})(C_{gs1}+C_s)} \\ R_1 &= \frac{(C_{gs1}+C_s)^2}{C_{gs1}(C_s g_{m1} - C_{gs1}G_{s1})} \cong \frac{(C_{gs1}+C_s)^2}{C_{gs1}C_s g_{m1}} \\ C_2 &= \frac{C_{gs1}C_s}{C_{gs1}+C_s} \end{aligned} \tag{3.135}$$

and the approximations result from the fact that typically $C_s > C_{gs1}$ and $g_{m1} > G_{s1}$. This is the same admittance as the circuit shown in Fig. 3.37. Thus, the input admittance is the same as a capacitor in parallel with a series combination of a negative capacitor and a negative resistor. If a third network consisting of a capacitor of size C_1 and a resistor of size R_1, in series, is connected to the gate of the source follower, as shown in Fig. 3.38, then the negative elements are cancelled. The resulting input admittance is then simply C_2, as given in (3.135). In this case, (3.120) becomes

$$\frac{v_{g1}}{i_{in}} = \frac{1}{G_{in} + s\left(C'_{in} + \frac{C_{gs1}C_s}{C_{gs1}+C_s}\right)} \tag{3.136}$$

and (3.122) becomes

$$A(s) = \frac{v_{out}}{i_{in}} = R_{in}\left(\frac{g_{m1}}{g_{m1}+G_{s1}}\right)\frac{\left(1+s\frac{C_{gs1}}{g_{m1}}\right)}{\left(1+\frac{s}{p_1}\right)\left(1+\frac{s}{p_2}\right)} \tag{3.137}$$

where

$$p_1 = \frac{G_{in}}{C'_{in} + \frac{C_{gs1}C_L}{C_{gs1}+C_L}} \cong \frac{G_{in}}{C'_{in}+C_{gs1}} \tag{3.138}$$

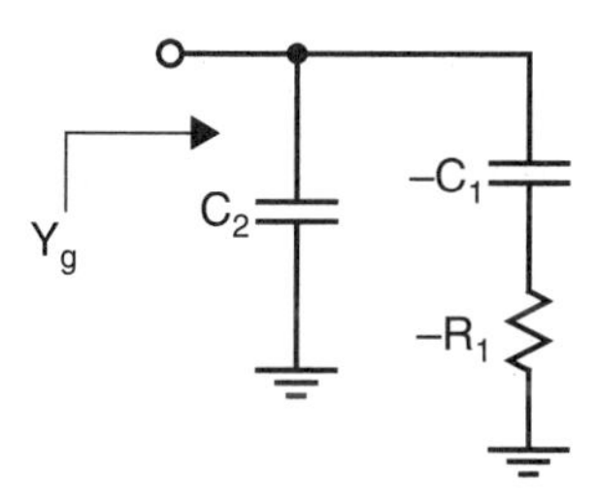

Fig. 3.37 A circuit that has the same admittance as the input impedance looking into the gate of a source follower (and ignoring C_{gd}).

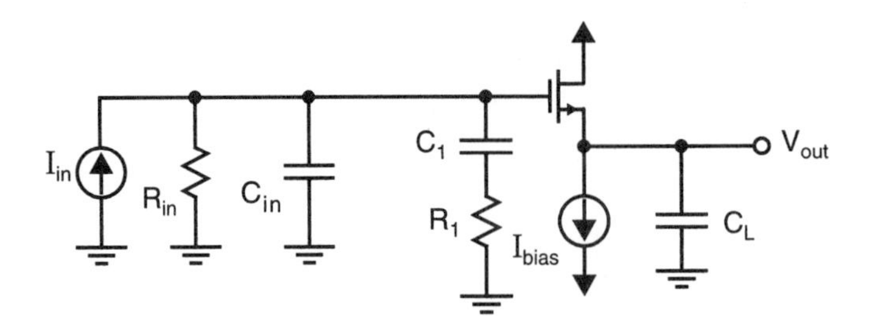

Fig. 3.38 Addition of a compensation network (C_1 and R_1) to compensate for the negative components of the admittance looking into the gate of the source follower.

$$p_2 = \frac{g_{m1} + G_{s1}}{C_{gs} + C_L} \cong \frac{g_{m1} + G_{s1}}{C_L} \tag{3.139}$$

The approximation is accurate when $C_s >> C_{gs1}$. Regardless of the approximation, the poles are now guaranteed real, and no overshoot will occur.

Therefore, when designing source followers (or emitter followers), the recommended procedure is to check to see if the poles are complex by using either (3.126) or a SPICE transient analysis to look for overshoot. When the poles are complex, then increase either C_{in}, C_s, or both, or, alternatively, add the compensation network, as shown Fig. 3.38.

EXAMPLE 3.10

Using the same parameters as in Example 3.9, find the compensation network and the resulting first and second poles of the source follower in Fig. 3.34.

Solution

Using (3.135), we have

$$C_1 \cong \frac{g_{m1}C_{gs1}C_s}{(g_{m1} + G_{s1})(C_{gs1} + C_s)} = 0.170\ \text{pF} \tag{3.140}$$

and

$$R_1 \cong \frac{(C_{gs1} + C_s)^2}{C_{gs1}C_s g_{m1}} \cong 49.3\ \text{k}\Omega \tag{3.141}$$

The capacitor is a reasonable value to be realized on chip. The resistor can be realized by a MOS transistor biased in the triode (i.e., linear) region. Assuming the compensation network is used, the poles of the transfer function then become

$$p_1 \cong \frac{G_{in}}{C_{gs1} + C'_{in}} = 2\pi \times 3.61\ \text{MHz} \tag{3.142}$$

and

$$p_2 = \frac{g_{m1} + G_{s1}}{C_{gs} + C_L} = 2\pi \times 19.3 \text{ MHz} \tag{3.143}$$

The speed penalty paid for using the compensation network is quite high, because the pole frequency without compensation was around 8 MHz whereas here the dominant pole is at 3.6 MHz.

Finally, it should be mentioned that if the source-follower buffer is intended to be used in an opamp (and thus feedback will be placed around the buffer), and if the resonant frequency of the source follower is substantially greater than the unity-gain frequency of the amplifier, then the overshoot can be tolerated, and no compensation network is necessary.

Common-Gate Amplifier

The frequency response of the common-gate stage is usually superior to that of the common-source stage due to the low impedance, r_{in}, at the source node, assuming G_L is not considerably smaller than g_{ds1}. Analysis of the frequency response of the common-gate stage is left as an exercise for the reader.

High-Output Impedance Mirrors

Both the Wilson and the cascode current mirrors introduce high-frequency poles into the signal transfer function. The approximately equivalent time constant of these poles is C_{gs}/g_m. The proof of this statement can been found by doing a high-frequency, small-signal analysis (where the capacitances of the small-signal models are included in the analysis). This analysis is left as an exercise for the reader.

Cascode Gain Stage

The exact high-frequency analysis of a cascode gain stage (Fig. 3.16) is usually left to simulation on a computer; however, an approximate analysis is not too complicated. At high frequencies, the time constant due to the output node almost always dominates since the impedance is so large at that node. The total capacitance at the output node, C_{out}, is the parallel combination of $C_{gd2} + C_{db2}$, the load capacitance C_L, and the output capacitance of the bias current source, C_{bias} (normally, C_L is the major contributor). Assuming that the time constant at the output stage dominates, the –3-dB frequency is approximately equal to the inverse of the time constant. In other words, we have

$$\omega_{-3\text{ dB}} \cong \frac{1}{R_{out}C_L} \cong \frac{2g_{ds}^2}{g_m C_L} \tag{3.144}$$

A more accurate (though still not exact) estimate may be found using the zero-value time-constant analysis method of [Gray, 1993]. The advantage of this estimation technique is that it gives some insight into the relative importance of each capacitor in determining the overall –3-dB frequency. The small-signal model being analyzed is shown in Fig. 3.39, where

$$\begin{aligned} C_{s2} &= C_{db1} + C_{sb2} + C_{gs2} \\ C_{d2} &= C_{gd2} + C_{db2} + C_L + C_{bias} \end{aligned} \tag{3.145}$$

In the zero-value time-constant analysis, all independent sources are set to zero (i.e., here, V_{in} is set to 0 V) and each capacitor is considered in turn with all other capacitors set to zero. The corresponding time constant is found and labeled τ_{Ci}. Then the –3-dB frequency, $\omega_{-3\text{ dB}}$, is estimated to be 1 divided by the sum of all the time constants.

In this circuit, the first time constant found is the one that corresponds to C_{gs1}, and it is labelled τ_{Cgs1}. The resistance seen by C_{gs1} is R_{in}, and therefore,

$$\tau_{Cgs1} = C_{gs1} R_{in} \tag{3.146}$$

The calculation of the time constant corresponding to C_{gd1} is more involved, so formal methods are used in its calculation. C_{gd1} is replaced by a voltage source, V_x. Next, the resistance seen by C_{gd1} is found by calculating the ratio of V_x to i_x (the current leaving V_x). The final time constant is then given by this resistance multiplied by C_{gd1}. The small-signal model for this analysis is shown in Fig. 3.40(*a*), where the resistance R_{d1} is the parallel combination of r_{ds1} and the impedance seen looking into the source of Q_2 (the cascode transistor) at low frequencies. The circuit of Fig. 3.40(*a*) can be redrawn as the equivalent circuit shown in Fig. 3.40(*b*). In this transformation, we changed the ground node (which node is called ground is arbitrary in an analysis) and the direction of the voltage-controlled current source. Notice that the circuit of Fig. 3.40(*b*) is essentially the same as the circuit of Fig. 3.13, which was used for finding the output impedance of a source-degenerated current source. The analysis here is essentially the same as that used in Fig. 3.13, but will be included here for clarity. We have

$$v_y = i_x R_{in} \tag{3.147}$$

Also,

$$i_x = (v_x - v_y) G_{d1} - g_{m1} v_y \tag{3.148}$$

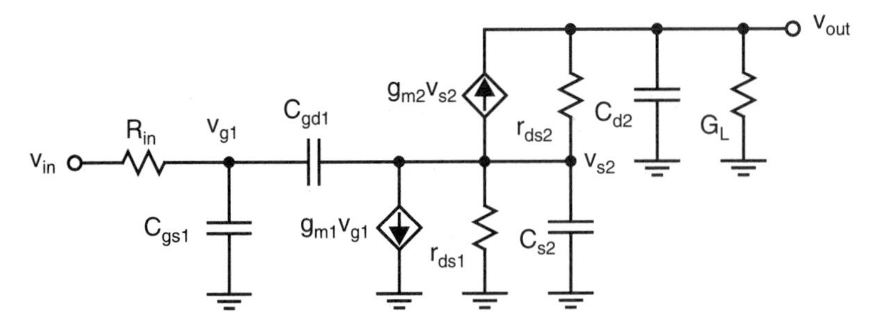

Fig. 3.39 The small-signal model of the cascode gain stage.

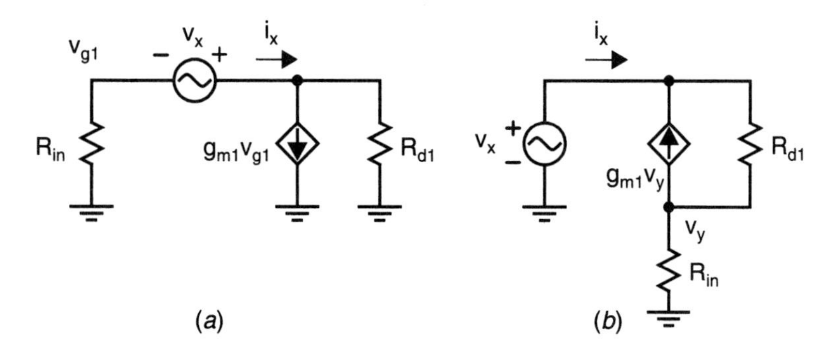

Fig. 3.40 Two equivalent small-signal models for calculating the resistance seen by C_{gd1}.

Substituting (3.147) into (3.148) and solving for v_x/i_x gives

$$r_{Cgd1} = \frac{v_x}{i_x} = R_{d1}[1 + R_{in}(G_{d1} + g_{m1})] \tag{3.149}$$

The admittance looking into the source of the cascode transistor, Q_2, was found previously in (3.52). Namely,

$$Y_{s2} \cong g_{ds} \tag{3.150}$$

The impedance, R_{d1}, is the parallel combination of this admittance and r_{ds1}. Therefore, we have

$$R_{d1} \cong \frac{r_{ds}}{2} \tag{3.151}$$

Substituting this result into (3.149), we have

$$\begin{aligned} r_{Cgd1} &\cong \frac{r_{ds}}{2}[1 + R_{in}(2g_{ds} + g_{m1})] \\ &\cong \frac{r_{ds}}{2}(1 + g_{m1}R_{in}) \end{aligned} \tag{3.152}$$

and therefore

$$\tau_{Cgd1} \cong C_{gd1}\frac{r_{ds}}{2}(1 + g_{m1}R_{in}) \tag{3.153}$$

If R_{in} is large, say, on the order of a transistor output impedance, r_{ds}, then this time constant is approximately given by

$$\tau_{Cgd1} \cong C_{gd1}\frac{g_m r_{ds}^2}{2} \tag{3.154}$$

This time constant can be almost as large as the corresponding time constant for a common-source amplifier—a fact not well known.

The resistance seen by capacitor C_{s2} is r_{ds1} in parallel with the impedance seen looking into the source of Q_2, which, from (3.52), is approximately r_{ds}. Thus, we have

$$\tau_{Cs2} \cong C_{s2}\frac{r_{ds}}{2} \tag{3.155}$$

The resistance seen by C_{d2} is the output impedance of the cascode amplifier, which is approximately given by $(g_m r^2_{ds})/2$ from (3.50). Thus, the time constant due to C_{d2} is given by

$$\tau_{Cd2} \cong C_{d2}\frac{g_m r_{ds}^2}{2} \tag{3.156}$$

Note that this time constant has the same form as (3.154), but normally C_{d2} will be much larger than C_{gd1} (because C_L is often large), making τ_{Cd2} dominate.

The sum of the time constants is then given by

$$\begin{aligned}\tau_{total} &\cong \tau_{Cgs1} + \tau_{Cgd1} + \tau_{Cs2} + \tau_{Cd2} \\ &\cong C_{gs1}R_{in} + C_{gd1}\frac{g_m r_{ds}^2}{2} + C_{s2}\frac{r_{ds}}{2} + C_{d2}\frac{g_m r_{ds}^2}{2}\end{aligned} \tag{3.157}$$

The –3-dB frequency, $\omega_{-3\ dB}$, is estimated to be $1/\tau_{total}$.

EXAMPLE 3.11

Assume that for both the input transistor and the cascode transistor, $g_m = 1\ mA/V$, $r_{ds} = 100\ k\Omega$, $R_{in} = 180\ k\Omega$, $C_L = 5\ pF$, $C_{gs} = 0.2\ pF$, $C_{gd} = 15\ fF$, $C_{sb} = 40\ fF$, $C_{db} = 20\ fF$, and $C_{bias} = 20\ fF$. Estimate the –3-dB frequency of the cascode amplifier of Fig. 3.16(*a*).

Solution

The time constants associated with each capacitor are readily evaluated using (3.157). First, note that

$$\begin{aligned}C_{s2} &= C_{db1} + C_{sb2} + C_{gs2} = 0.26\ pF \\ C_{d2} &= C_{gd2} + C_{db2} + C_L + C_{bias} = 5.055\ pF\end{aligned} \tag{3.158}$$

We have

$$\begin{aligned}\tau_{Cgs1} &= C_{gs1}R_{in} = 36\ ns \\ \tau_{Cgd1} &= C_{gd1}\frac{g_m r_{ds}^2}{2} = 75\ ns \\ \tau_{Cs2} &= C_{s2}\frac{r_{ds}}{2} = 13\ ns \\ \tau_{Cd2} &= C_{d2}\frac{g_m r_{ds}^2}{2} = 25.3\ \mu s\end{aligned} \tag{3.159}$$

As expected, the time constant at the output node dominates, and the second most important time constant is that due to C_{gd1}, although its effect on the –3-dB frequency is negligible. Therefore, the –3-dB frequency is accurately given by $\omega_{-3\text{ dB}} \cong 1/\tau_{Cd2} = 2\pi \times 6.3\text{ kHz}$.

Before leaving this section, some comments should be made about the high-frequency operation of cascode-gain stages. As we just saw, typically one pole dominates, and thus we can reasonably model the amplifier gain as

$$A(s) = \frac{A_v}{1 + s/\omega_{-3\text{ dB}}} \tag{3.160}$$

Thus, at frequencies substantially larger than $\omega_{-3\text{ dB}}$, which is usually the frequency band of operation, the gain is approximately given by

$$A(s) \cong \frac{A_v}{s/\omega_{-3\text{ dB}}} \cong -\frac{g_{m1}}{sC_L} \tag{3.161}$$

using (3.54) and (3.144). It should also be noted that the approximations of (3.144) and (3.161) are quite good, unless either the source impedance or source capacitance is very large. In addition, at frequencies much larger than the –3-dB frequency, the admittance at the source of Q_2 can be found by using (3.51), where G_L is replaced by $G_L + sC_L$. Such a substitution results in

$$\begin{aligned} Y_{in2} &= \frac{g_{m2}}{1 + \dfrac{g_{ds2}}{G_L + sC_L}} \\ &= g_{m2}\left(\frac{G_L + sC_L}{g_{ds2} + G_L + sC_L}\right) \\ &\cong g_{m2}\left(\frac{G_L + sC_L}{g_{ds2} + sC_L}\right) \end{aligned} \tag{3.162}$$

At frequencies where $\omega >> 1/(r_{ds}C_L)$, the terms in s dominate, and $Y_{in2} \cong g_{m2}$. The approximate time constant due to the node at the source of Q_2 is then simply given by the total capacitance at that node divided by g_{m2}. The total capacitance at the source of Q_2 is C_{gs2} in parallel with C_{db1} in parallel with C_{gd1}. Because this capacitance is not excessively large and the impedance at the node, $1/g_{m2}$, is small, the time constant at that node can usually be neglected. However, in amplifiers with a small source impedance, this node would still be the primary factor determining the second pole of a cascode amplifier. It is easy to derive an upper bound on the time constant at the source of Q_2. Almost always, C_{db1} in parallel with C_{gd1} is less than C_{gs2}. Therefore, the total capacitance at the source of Q_2 is equal to KC_{gs2}, where K is between 1 and 2 (usually closer to 1). Using

$$g_{m2} = \mu_p C_{ox}\left(\frac{W}{L}\right)_2 V_{eff2} \tag{3.163}$$

for the folded-cascode amplifier (for the telescopic-cascode amplifier, substitute μ_n for μ_p), and using

$$C_{S2} = KC_{gs2} = K\frac{2}{3}(WL)_2 C_{ox} \tag{3.164}$$

gives the approximate frequency of the second pole (ignoring the time constants at other nondominant nodes):

$$\omega_{p2} \cong \frac{1}{\tau_{S2}} = \frac{g_{m2}}{C_{S2}} = \frac{3\mu_p V_{eff2}}{2KL_2^2} > \frac{3\mu_p V_{eff2}}{4L_2^2} \tag{3.165}$$

This equation is an upper limit on the unity-gain frequency of any amplifier that uses a cascode gain stage. Note that (3.165) is relatively independent of that actual design once V_{eff2} is chosen, and V_{eff2} is usually determined by maximum signal-handling requirements. Also note that ω_{p2} has a very strong dependence on the channel length.

EXAMPLE 3.12

Estimate the lower bound on the frequency of the second pole of a folded-cascode amplifier for a 0.8-μm technology, where a typical value of 0.25 V is chosen for V_{eff2}.

Solution

Normally, the minimum length of a cascode transistor in an analog circuit might be 25 to 50 percent of the minimum length of transistors used in digital circuits. Therefore, assuming $L_2 = 1.5 \cdot 0.8\ \mu m = 1.2\ \mu m$, and using $\mu_p = 0.02\ m^2/V \cdot s$, and $V_{eff2} = 0.25$ V, we have $\omega_{p2} > 1.7 \times 10^9$ rad $= 2\pi \cdot 276$ MHz. For a telescopic-cascode amplifier, the upper bound would be 690 MHz. In most practical opamp designs, the unity-gain frequency of a typical design might be limited to around one-half the frequency of the lower bound second pole. In this example, the typical unity-gain frequencies would be 138 MHz and 345 MHz for the folded-cascode and telescopic-cascode amplifiers, respectively.

Finally, the time constant at the gate of the input transistor can be important when the source resistance is large, though not as important as in a common-source stage, since a cascode gain-stage does not suffer as much from a *Miller effect*. In other words, at high frequencies, the effective impedance at the source of Q_2 has decreased to $1/g_{m2}$, and there is not much gain from the gate of Q_1 to the source of Q_2. Recall that in a common-source amplifier, the effective size of C_{gd1} was magnified by the gain of the common-source amplifier.

3.12 SPICE SIMULATION EXAMPLES

In this section, circuit diagrams, netlists, and simulation results for selected examples in this chapter are presented. Each of the netlists contains the circuit elements and commands used in the simulations, but does not contain transistor model parameters.

Simulation of Example 3.2

The circuit for Example 3.2 is shown in Fig. 3.41 where a 5-V power supply is used. The bias voltage, V_{in}, has to be set such that both Q_1 and Q_2 are in the active region, and approximately 2.5 V appears across the drain-source nodes for each of them.

```
NETLIST:
vdd     1   0               dc 5
ibias   2   0               dc 100u
m3      2   2   1   1       pmos w=100u l=1.6u
m2      3   2   1   1       pmos w=100u l=1.6u
m1      3   4   0   0       nmos w=100u l=1.6u
vin     4   0               dc 0.849 ac 1

.op
.ac dec 10 1k 10000Meg
.print vdb(3)
.option post
.lib "../cmos_models" library
.end
```

The frequency plot for this simulation is shown in Fig. 3.42. Note that the gain is 36 dB, which corresponds to approximately 63 V/V.

Simulation of Example 3.3

The circuit for Example 3.3 is shown in Fig. 3.43, where it should be noted that the body of Q_1 is tied to ground, rather than to its source.

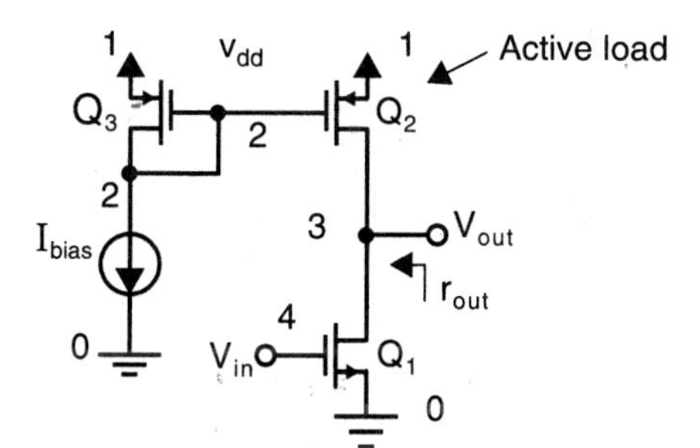

Fig. 3.41 Circuit for Example 3.2.

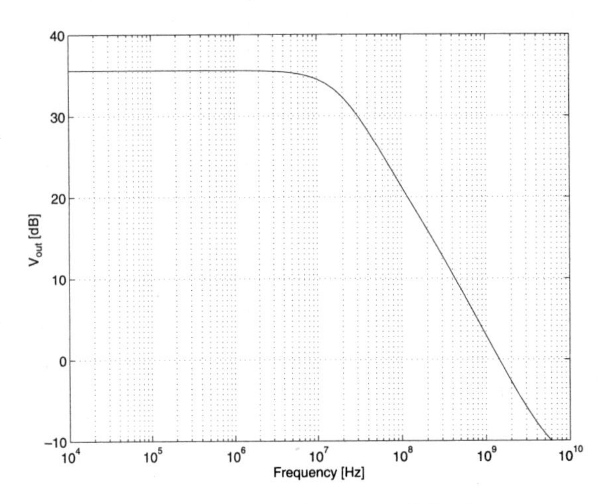

Fig. 3.42 Frequency simulation result for the common-source amplifier.

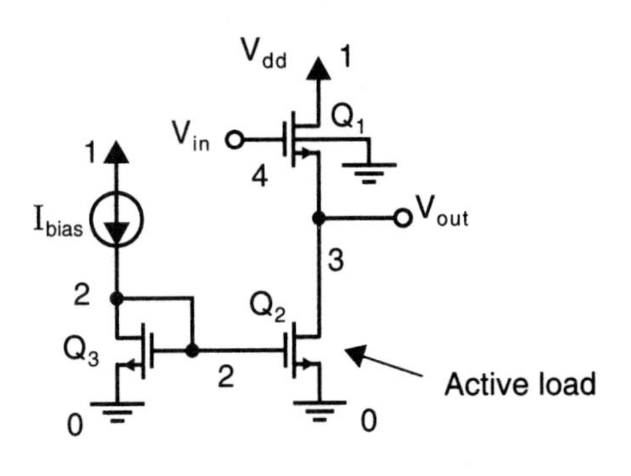

Fig. 3.43 Circuit for Example 3.3.

```
NETLIST:
vdd     1   0               dc 5
ibias   1   2               dc 100u
m3      2   2   0   0       nmos w=100u l=1.6u
m2      3   2   0   0       nmos w=100u l=1.6u
m1      1   4   3   0       nmos w=100u l=1.6u
vin     4   0               dc 2 ac 1

.op
.ac dec 10 1k 1000Meg
.print vdb(3)
.option post
.lib "../cmos_models" library
.end
```

The frequency plot for this simulation is shown in Fig. 3.44. The gain is –1.3 dB, which corresponds to 0.86 V/V.

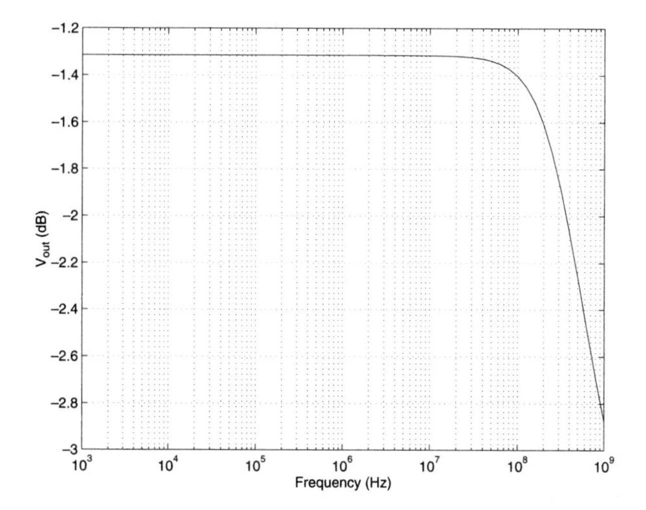

Fig. 3.44 Frequency plot for the source-follower circuit.

Simulation of Example 3.8

The circuit for Example 3.8 is shown in Fig. 3.45.

```
NETLIST:
vdd     1   0               dc 5
ibias   2   0               dc 100u
m3      2   2   1   1       pmos W=100u l=1.6u
m2      3   2   1   1       pmos W=100u l=1.6u
m1      3   4   0   0       nmos W=100u l=1.6u
rin     5   4               180k
vin     5   0               dc 0.849 ac 1
cl      3   0               0.3p

.op
.ac dec 20 1k 100Meg
.print vdb(3)
```

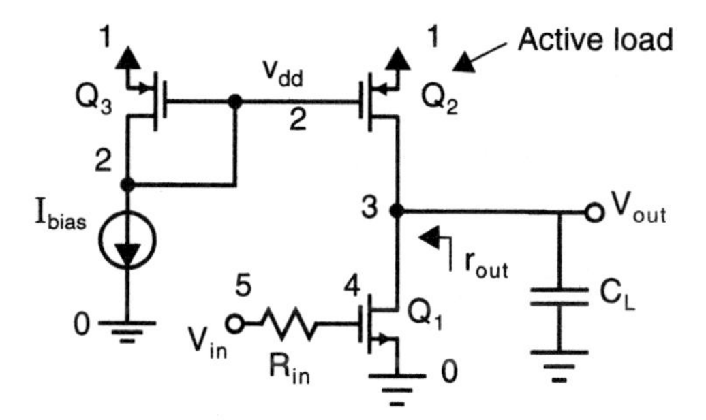

Fig. 3.45 Circuit for Example 3.8.

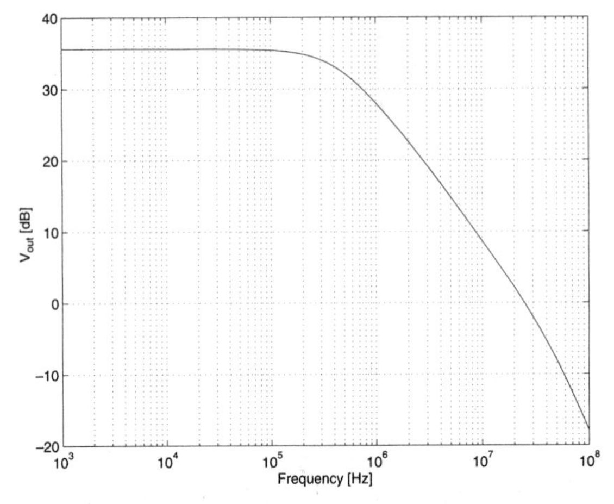

Fig. 3.46 Estimate of the –3-dB frequency for a common-source amplifier.

```
.option post
.lib "../cmos_models" library
.end
```

The frequency plot for this simulation is shown in Fig. 3.46, where we see that the –3-dB frequency occurs around 460 kHz.

Simulation of Example 3.9

The circuit for Example 3.9 is shown in Fig. 3.47.

```
NETLIST:
vdd     1   0        dc 5
vss     2   0        dc -5
ibias   3   2        dc 100u
rin     4   0        180k
```

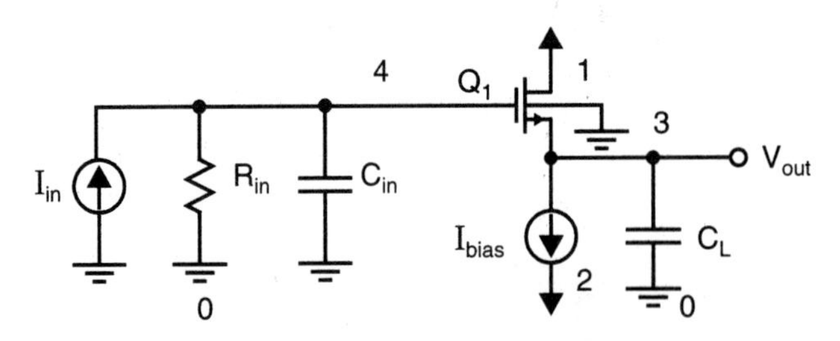

Fig. 3.47 Circuit for Example 3.9.

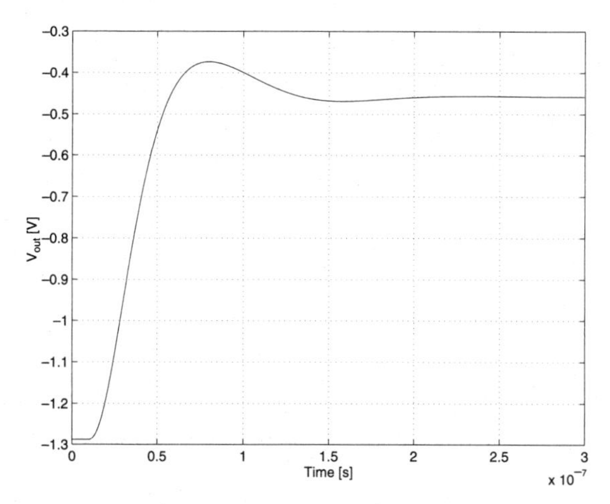

Fig. 3.48 The step response of a source follower, showing a 10-percent overshoot.

```
cin     4   0           30f
cl      3   0           10p
m1      1   4   3   2   nmos w=100u l=1.6u
iin     4   0           pulse (0 –5u 10n 0 0)

.op
.option post
.tran 0.5n 300n
.print v(3)
.lib "../cmos_models" library
.end
```

The step response of the source follower is shown in Fig. 3.48. The overshoot here is about 10 percent.

Simulation of Example 3.10

The circuit for Example 3.10 is shown in Fig. 3.49. The compensation network that eliminates the complex conjugate poles is included. The simulation is used to obtain the frequency plot from which the first and second poles of the resulting source follower are determined.

```
NETLIST:
vdd     1   0       dc 5
vss     2   0       dc –5
ibias   3   2       dc 100u
```

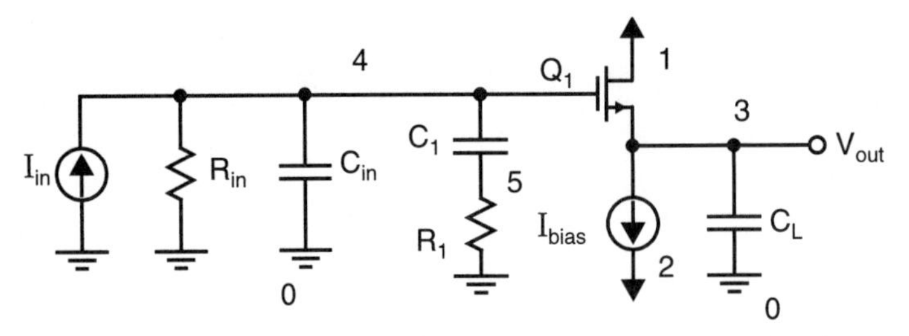

Fig. 3.49 Circuit for Example 3.10.

```
rin   4  0          180k
cin   4  0          30f
cl    3  0          10p
m1    1  4   3  2   nmos W=100u l=1.6u
iin   4  0          dc 0 ac 1
c1    4  5          0.17p
r1    5  0          49.3k

.op
.print vdb(3)
.lib "../cmos_models" library
.end
```

The frequency plot of this source follower is shown in Fig. 3.50. The first pole occurs around 3.6 MHz, whereas the second pole occurs around 16 MHz.

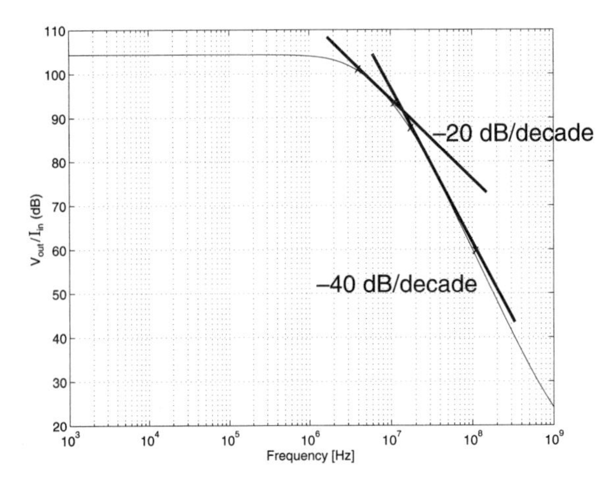

Fig. 3.50 The bode plot of a source follower with a compensation network.

Simulation of Example 3.11

The circuit for Example 3.11 is shown in Fig. 3.51. Notice that Q_3 through Q_6 form a cascode current mirror to produce I_{bias}. The width and length of the p-channel transistors are chosen such that their g_m and r_{ds} are matched to those of the n-channel transistors.

```
NETLIST:
vdd     1  0           dc 5
ibias   6  0           dc 100u
m4      6  6  7  1     pmos w=390u l=2u
m5      7  7  1  1     pmos w=390u l=2u
m6      8  7  1  1     pmos w=390u l=2u
m3      2  6  8  1     pmos w=390u l=2u
m2      2  3  4  0     nmos w=100u l=1.6u
m1      4  5  0  0     nmos w=100u l=1.6u
cl      2  0           5p
vbias   3  0           dc 2.5
vin     5  0           dc 0.8425 ac 1

.op
.ac dec 10 0.1 1000Meg
.print vdb(2)
.option post
.lib "../cmos_models" library
.end
```

The frequency plot of this cascode amplifier is shown in Fig. 3.52. The gain at dc is 80 dB (i.e., 10,000 V/V) and has a –3-dB frequency around 2 kHz.

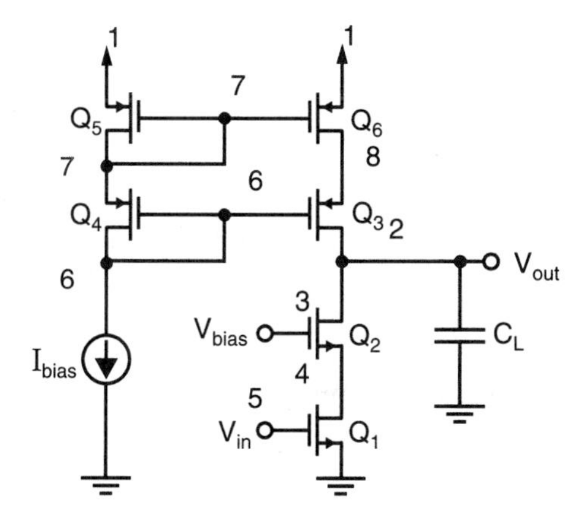

Fig. 3.51 Circuit for Example 3.11.

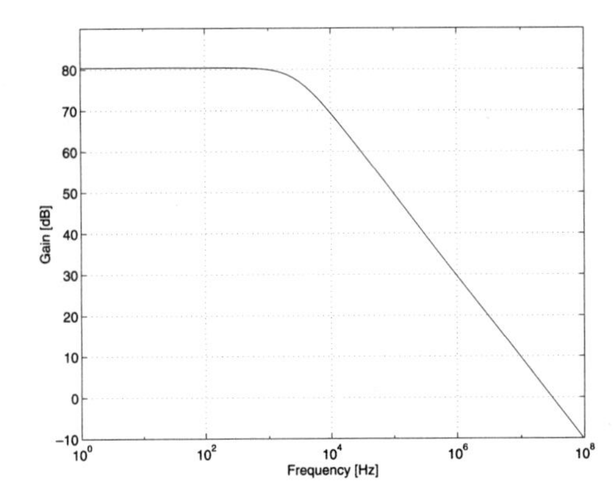

Fig. 3.52 The frequency plot of the cascode amplifier.

3.13 REFERENCES

P. R. Gray and R. G. Meyer. *Analysis and Design of Analog Integrated Circuits,* 3rd ed. John Wiley & Sons, New York, 1993.

A. S. Sedra and K. C. Smith. *Microelectronic Circuits,* 3rd ed. Holt, Rinehart & Winston, New York, 1991.

3.14 PROBLEMS

Unless otherwise stated, assume the following:

- npn bipolar transistors:
  $\beta = 100$
  $V_A = 80$ V
  $\tau_b = 13$ ps
  $\tau_s = 4$ ns
  $r_b = 330\ \Omega$
- n-channel MOS transistors:
  $\mu_n C_{ox} = 92\ \mu A/V^2$
  $V_{tn} = 0.8$ V
  $\gamma = 0.5\ V^{1/2}$
  $r_{ds}\ (\Omega) = 8{,}000\, L\ (\mu m)/I_D\ (mA)$ in active region
  $C_j = 2.4 \times 10^{-4}\ pF/(\mu m)^2$
  $C_{j\text{-}sw} = 2.0 \times 10^{-4}\ pF/\mu m$
  $C_{ox} = 1.9 \times 10^{-3}\ pF/(\mu m)^2$
  $C_{gs(overlap)} = C_{gd(overlap)} = 2.0 \times 10^{-4}\ pF/\mu m$
- p-channel MOS transistors:
  $\mu_p C_{ox} = 30\ \mu A/V^2$
  $V_{tp} = -0.9$ V
  $\gamma = 0.8\ V^{1/2}$

$r_{ds}\ (\Omega) = 12{,}000L\ (\mu m)/I_D\ (mA)$ in active region
$C_j = 4.5 \times 10^{-4}\ pF/(\mu m)^2$
$C_{j\text{-}sw} = 2.5 \times 10^{-4}\ pF/\mu m$
$C_{ox} = 1.9 \times 10^{-3}\ pF/(\mu m)^2$
$C_{gs(overlap)} = C_{gd(overlap)} = 2.0 \times 10^{-4}\ pF/\mu m$

3.1 Consider the current mirror shown in Fig. 3.1, where $I_{in} = 80\ \mu A$, transistor Q_1 has $W/L = 100\ \mu m/1.6\ \mu m$, and transistor Q_2 has $W/L = 25\ \mu m/1.6\ \mu m$. Find the nominal output current as well as the output impedance, r_{out}. Also, find the minimum output voltage such that both transistors remain in the active region.

3.2 For the common-source amplifier in Fig. 3.4, derive the relationship between the bias current, I_{bias}, and the dc gain. Assume all transistor sizes are the same.

3.3 For the common-source amplifier in Fig. 3.4, assume there is a load capacitor, C_L, that dominates the frequency response. Derive the –3-dB frequency of the amplifier. Find the relationship between the –3-dB frequency and the bias current, I_{bias}. Assume all transistor sizes are the same.

3.4 In the high-frequency analysis of a common-source amplifier, show that the transfer function given in equation (3.103) can be derived from nodal equations (3.101) and (3.102).

3.5 Find the –3-dB frequency of the common-source amplifier in Fig. 3.4, assuming that all transistors have $W/L = 75\ \mu m/1.6\ \mu m$ and that $I_{bias} = 75\ \mu A$. The first step is to estimate the areas and peripheries of the junctions based on simple layout rules. In this step, assume each transistor is composed of three gate stripes. Based on this geometry, the next step is to estimate the parasitic capacitances of the small-signal model. In this step, assume the dc bias voltage of the output node is halfway between ground and the 5-V power-supply voltage.

3.6 Derive the low-frequency output impedance of the source follower shown in Fig. 3.6.

3.7 Derive an equation for the impedance looking into the source of a source follower. Use a small-signal model similar to that in Fig. 3.8, but do not take into account C_s and R_{s1}.

3.8 Repeat Example 3.9, but assume $C_L = 0.5\ pF$ when (*a*) the source is not connected to the substrate, and (*b*) the source is connected to the substrate.

3.9 Repeat Example 3.10, but assume $C_L = 0.5\ pF$ when (*a*) the source is not connected to the substrate, and (*b*) the source is connected to the substrate.

3.10 Assume that the common-gate amplifier of Fig. 3.9 has a bias current of $0.1\ mA$ and that all transistors have a W/L of $100\ \mu m/1.6\ \mu m$. Also assume the following:

$r_{ds\text{-}n}\ (\Omega) = 8{,}000L\ (\mu m)/I_D\ (mA)$
$r_{ds\text{-}p}\ (\Omega) = 12{,}000L\ (\mu m)/I_D\ (mA)$

$R_{in} = 180\ \text{k}\Omega$
$C_L = 1\ \text{pF}$
$C_{gs} = 0.2\ \text{pF}$
$C_{gd} = 15\ \text{fF}$
$C_{sb} = 40\ \text{fF}$
$C_{db} = 20\ \text{fF}$

And assume the input source has a 30-fF output capacitance. Find the –3-dB frequency of the amplifier.

3.11 Derive the output impedance of the current mirror shown in Fig. P3.11, where a diode-connected transistor has been included in series with the source of the output transistor. Ignore the body effect.

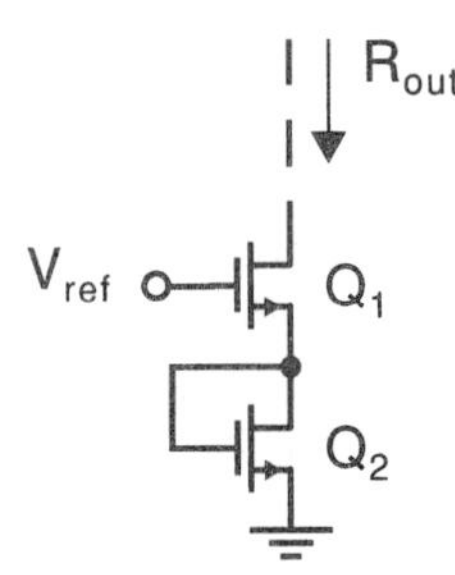

Fig. P3.11

3.12 A MOS n-channel cascode current mirror has a bias current of 0.1 mA and transistor sizes given by $W/L = 50\ \mu\text{m}/1.6\ \mu\text{m}$. What is the minimum output voltage allowable across the current mirror without any transistors entering the triode region?

3.13 Using small-signal analysis, find the output impedance of a MOS cascode current mirror. Include in your analysis the voltage-dependent current source that models the body effect.

3.14 Using small-signal analysis, find the equation for the output impedance, r_{out}, of a MOS Wilson current mirror, and show that it is approximately given by

$$r_{out} \cong r_{ds4}\left(\frac{g_{m1}r_{ds1}}{2}\right)$$

Ignore the body effect. Remember to include the effect of the output impedance of the input current source.

3.15 Repeat Problem 3.14, but include the voltage-dependent current source that models the body effect.

3.16 Repeat Example 3.11, in which a cascode amplifier is analyzed to find its –3-dB frequency, but assume $C_L = 0.5\ \text{pF}$.

3.17 Derive an expression for the frequency at which the magnitude of the gain of a cascode amplifier has decreased to unity. What is this value when the parameters of Example 3.11 are used, except that $C_L = 2\ \text{pF}$?

3.18 Derive the current gain, I_{out}/I_{in}, for the bipolar current mirror shown in Fig. 3.22, and show that, for $\beta >> 1$, this gain is $1 - 2/\beta$. (Neglect finite output impedances.)

3.19 Derive the current gain, I_{out}/I_{in}, for the bipolar current mirror shown in Fig. 3.23, and show that, for $\beta >> 1$, this gain is $1 - 2/\beta^2$. (Neglect finite output impedances.)

3.20 For the bipolar cascode current mirror shown in Fig. 3.25(*a*), derive the current gain, I_{out}/I_{in}, and show that, for large β, this gain is given by

$$\frac{I_{out}}{I_{in}} \approx 1 - \frac{4}{\beta}$$

3.21 For the bipolar Wilson current mirror shown in Fig. 3.25(*b*), show that the current gain, I_{out}/I_{in}, is given by

$$\frac{I_{out}}{I_{in}} = 1 - \frac{2}{\beta^2 + 2\beta + 2}$$

3.22 Derive the output impedance of the simple bipolar current mirror shown in Fig. P3.22. Take the finite output impedance, r_o, into account for both transistors. What is the required value for R, and what is the output impedance if we want the transistors to be biased at 0.2 mA ?

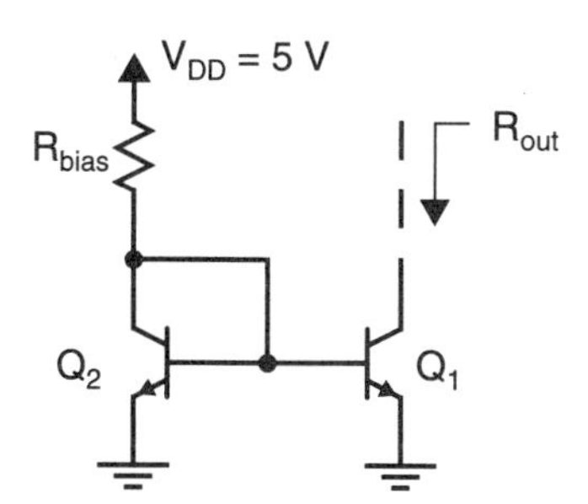

Fig. P3.22

3.23 For a common-source amplifier similar to that in Fig. 3.4 but realized using bipolar transistors, derive the relationship between the bias current, I_{bias}, and the dc gain.

3.24 Derive the –3-dB frequency of a bipolar common-emitter amplifier.

3.25 Derive the low-frequency gain and output impedance of a bipolar emitter follower that is biased with a current mirror. What are these values for a 0.5-mA bias current?

3.26 Derive the low-frequency input impedance and the gain of a bipolar common-base amplifier.

3.27 For the circuit shown in Fig. P3.27, show that the output impedance, r_{out}, is approximately given by

$$r_{out} \cong r_{o2}(1 + g_{m2}(R_e \,\|\, r_\pi))$$

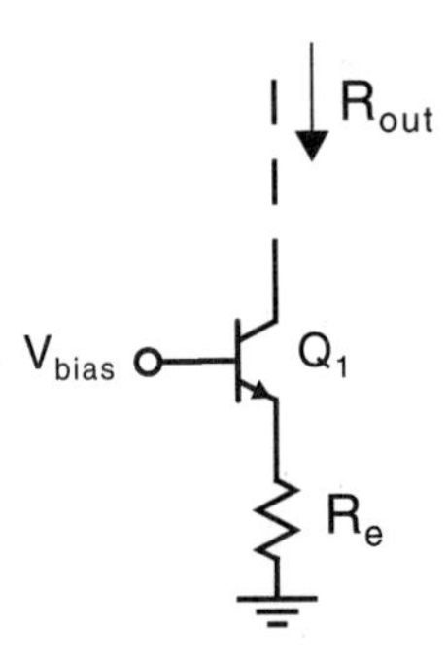

Fig. P3.27

3.28 Taking into account the finite current gain of bipolar transistors, but ignoring the finite output impedance, find the large-signal current gains of cascode and Wilson current mirrors.

3.29 Assuming $\beta >> 1$ and $g_m r_o >> \beta$, show that the output impedance of a bipolar Wilson current mirror is approximately given by

$$r_{out} \cong \frac{\beta r_o}{2}$$

3.30 Repeat Problem 3.29 for a bipolar cascode current mirror.

3.31 For the differential-input stage in Fig. 3.19, assume that $I_{bias} = 0.1\,\text{mA}$, all transistors have $W/L = 100\ \mu\text{m}/1.6\ \mu\text{m}$, and the load capacitance is 100 pF. Find the dc gain and the –3-dB frequency.

CHAPTER

4 Noise Analysis and Modelling

To develop good analog circuit design techniques, a basic understanding of noise sources and analysis is required. Another motivation to study noise analysis is to learn basic concepts of random signals for a proper understanding of oversampling converters. The purpose of this chapter is to present some fundamentals of noise analysis followed by an introduction to electronic noise sources and circuit analysis.

It should be mentioned here that this chapter deals with *inherent noise* as opposed to *interference noise*. Interference noise is a result of unwanted interaction between the circuit and the outside world, or between different parts of the circuit itself. This type of noise may or may not appear as random signals. Examples are power supply noise on ground wires (such as a 60-Hz hum) or electromagnetic interference between wires. Interference noise can be significantly reduced by careful circuit wiring or layout. In contrast, inherent noise refers to random noise signals that can be reduced but never eliminated since this noise is due to fundamental properties of the circuits. Some examples of inherent noise are thermal, shot, and flicker noise. Inherent noise is only moderately affected by circuit wiring or layout, such as using multiple base contacts to change the resistance value of the base of a transistor. However, inherent noise can be significantly reduced through proper circuit design, such as changing the circuit structure or increasing the power consumption.

The outline of this chapter is as follows: First, a time-domain analysis of noise signals (or other random signals) is presented. Here, basic concepts such as rms value, signal-to-noise ratio, and noise summation are presented. Next, a frequency-domain analysis of noise is presented. As with deterministic signals, analysis in the frequency domain results in more powerful analysis tools than does analysis that remains strictly in the time domain. Noise models for circuit elements are then presented, and finally, two circuit noise analyses are performed to give the reader some experience in such analysis.

4.1 TIME-DOMAIN ANALYSIS

Since inherent noise signals are random in nature, we define here some basic tools to effectively deal with random signals. Specifically, in this section, we define the following terms in the time domain: rms value, SNR, dBm, and noise summation.

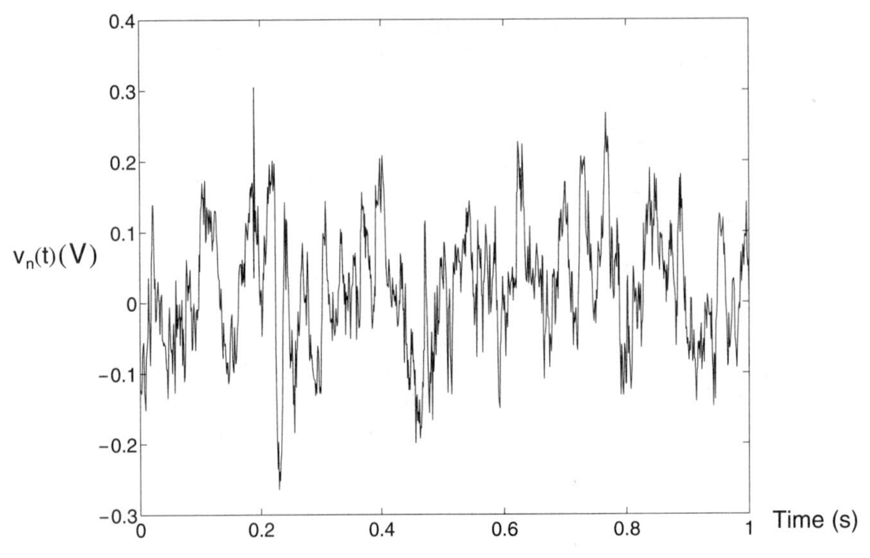

Fig. 4.1 Example of a voltage noise signal (time domain).

An example of a random noise signal in the time domain is shown in Fig. 4.1. Although this signal here is a voltage signal, it could just as easily be current noise or some other quantity. It should be noted that this noise signal appears to have an average value of zero. In fact, *throughout this chapter we will assume all noise signals have a mean value of zero*, which simplifies many of the definitions and is also valid in most physical systems.

Rms Value

Consider a noise voltage, $v_n(t)$, such as that shown in Fig. 4.1, or a noise current, $i_n(t)$. The *rms*, or *root mean square*, voltage value is defined[1] as

$$V_{n(rms)} \equiv \left[\frac{1}{T}\int_0^T v_n^2(t)\, dt\right]^{1/2} \tag{4.1}$$

where T is a suitable averaging time interval. Typically, a longer T gives a more accurate rms measurement. Similarly, the rms current value is defined as

$$I_{n(rms)} \equiv \left[\frac{1}{T}\int_0^T i_n^2(t)\, dt\right]^{1/2} \tag{4.2}$$

The benefit in knowing the rms value of a signal is that it indicates the *normalized noise power* of the signal. Specifically, if the random signal $v_n(t)$ is applied to a 1-Ω resistor, the average power dissipated, P_{diss}, in watts, equals the normalized noise

1. For those more rigorously inclined, we assume throughout this chapter that random signals are ergodic, implying their ensemble averages can be approximated by their time averages.

power and is given by

$$P_{diss} = \frac{V^2_{n(rms)}}{1\ \Omega} = V^2_{n(rms)} \tag{4.3}$$

This relationship implies that the power dissipated by a resistor is the same whether a random signal or a dc level of k volts (rms) is applied across it. For example, a noise signal with an rms value of 1 mV (rms) dissipates the same power across a resistor as a dc voltage of 1 mV.

Similarly, for a noise current, $i_n(t)$, applied to a 1-Ω resistor,

$$P_{diss} = I^2_{n(rms)} \times 1\ \Omega = I^2_{n(rms)} \tag{4.4}$$

As a result, the square of the rms values, $V^2_{n(rms)}$ and $I^2_{n(rms)}$, are sometimes referred to as the normalized noise powers of these two signals.

SNR

The *signal-to-noise ratio (SNR)* value (in dB) of a signal node in a system is defined as

$$\text{SNR} \equiv 10\log\left[\frac{\text{signal power}}{\text{noise power}}\right] \tag{4.5}$$

Thus, assuming a node in a circuit consists of a signal, $v_x(t)$, that has a normalized signal power of $V^2_{x(rms)}$ and a normalized noise power of $V^2_{n(rms)}$, the SNR is given by

$$\text{SNR} = 10\log\left[\frac{V^2_{x(rms)}}{V^2_{n(rms)}}\right] = 20\log\left[\frac{V_{x(rms)}}{V_{n(rms)}}\right] \tag{4.6}$$

Clearly, when the mean-squared values of the noise and signal are the same, then SNR = 0 dB.

Units of dBm

Although dB units relate the relative ratio of two power levels, it is often useful to know a signal's power in dB on an absolute scale. One common measure is that of dBm, where all power levels are referenced by 1 mW. In other words, a 1-mW signal corresponds to 0 dBm, whereas a 1-μW signal corresponds to –30 dBm. When voltage levels are measured, it is also common to reference the voltage level to the equivalent power dissipated if the voltage is applied across either a 50-Ω or a 75-Ω resistor.

EXAMPLE 4.1

Find the rms voltage of a 0-dBm signal referenced to a 50-Ω resistor. What is the level in dBm of a 2-volt rms signal?

Solution

A 0-dBm signal referenced to a 50-Ω resistor implies that the rms voltage level equals

$$V_{(rms)} = \sqrt{(50\ \Omega) \times 1\ \text{mW}} = 0.2236 \tag{4.7}$$

Thus, a 2-volt rms signal corresponds to

$$20 \log\left(\frac{2.0}{0.2236}\right) = 19\ \text{dBm} \tag{4.8}$$

and would dissipate 80 mW across a 50-Ω resistor.

Note that the measured voltage may never be physically applied across any 50-Ω resistor. The measured voltage is referenced only to power levels that would occur if the voltage were applied. Similar results are obtained if the power is referenced to a 75-Ω resistor.

Noise Summation

Consider the case of two noise sources added together, as shown in Fig. 4.2. If the rms values of each individual noise source are known, what can be said about the rms value of the combined signal? We answer this question as follows. Define $v_{no}(t)$ as

$$v_{no}(t) = v_{n1}(t) + v_{n2}(t) \tag{4.9}$$

where $v_{n1}(t)$ and $v_{n2}(t)$ are two noise sources with known rms values $V_{n1(rms)}$ and $V_{n2(rms)}$, respectively. Then we can write

$$V^2_{no(rms)} = \frac{1}{T}\int_0^T [v_{n1}(t) + v_{n2}(t)]^2\, dt \tag{4.10}$$

which, when expanded, gives,

$$V^2_{no(rms)} = V^2_{n1(rms)} + V^2_{n2(rms)} + \frac{2}{T}\int_0^T v_{n1}(t)v_{n2}(t)\, dt \tag{4.11}$$

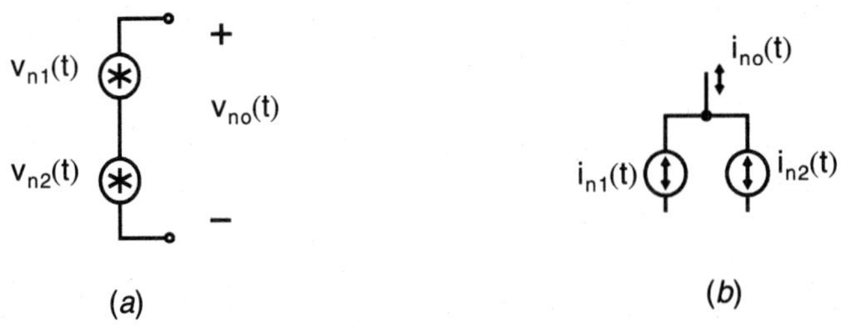

Fig. 4.2 Combining two noise sources, (a) voltage, and (b) current.

Note that the first two terms in the right-hand side of (4.11) are the individual mean-squared values of the noise sources. The last term shows the correlation between the two signal sources, $v_{n1}(t)$ and $v_{n2}(t)$. An alternate way to write (4.11) that better indicates the effects of signal correlation, is to define a correlation coefficient, C, as

$$C \equiv \frac{\frac{1}{T}\int_0^T v_{n1}(t)v_{n2}(t)\,dt}{V_{n1(rms)}V_{n2(rms)}} \tag{4.12}$$

With this definition, (4.11) can also be written as

$$V_{no(rms)}^2 = V_{n1(rms)}^2 + V_{n2(rms)}^2 + 2CV_{n1(rms)}V_{n2(rms)} \tag{4.13}$$

It can be shown that the correlation coefficient always satisfies the condition $-1 \le C \le 1$. Also, a value of $C = \pm 1$ implies the two signals are fully correlated, whereas $C = 0$ indicates the signals are uncorrelated. Values in between imply the signals are partially correlated. Fortunately, we have little reason to analyze partially correlated signals since different inherent noise sources are typically uncorrelated.

In the case of two uncorrelated signals, the mean-squared value of their sum is given by

$$V_{no(rms)}^2 = V_{n1(rms)}^2 + V_{n2(rms)}^2 \tag{4.14}$$

This relationship indicates that two rms values add as though they were vectors at right angles to each other (i.e., orthogonal) when signals are uncorrelated.

It is of interest to contrast this uncorrelated case with that for fully correlated signals. An example of two fully correlated (though deterministic) sources are two sinusoidal signals that have the same frequency and a phase of 0 or 180 degrees with each other. In this fully correlated case, the mean-squared value of their sum is given by

$$V_{no(rms)}^2 = [V_{n1(rms)} \pm V_{n2(rms)}]^2 \tag{4.15}$$

where the sign is determined by whether the signals are in or out of phase with each other. Note that, in this case where the signals are fully correlated, the rms values add linearly (similar to aligned vectors).

EXAMPLE 4.2

Given two uncorrelated noise sources that have $V_{n1(rms)} = 10\ \mu V$ and $V_{n2(rms)} = 5\ \mu V$, find their total output rms value when combined. If we are required to maintain the total rms value at $10\ \mu V$, how much should $V_{n1(rms)}$ be reduced while $V_{n2(rms)}$ remains unchanged?

Solution

Using (4.14) results in

$$V_{no(rms)}^2 = (10^2 + 5^2) = 125(\mu V)^2 \tag{4.16}$$

which results in $V_{no(rms)} = 11.2\ \mu V$.

To maintain $V_{no(rms)} = 10\ \mu V$ and $V_{n2(rms)} = 5\ \mu V$, we have

$$10^2 = V_{n1(rms)}^2 + 5^2 \tag{4.17}$$

which results in $V_{n1(rms)} = 8.7\ \mu V$. Therefore, reducing $V_{n1(rms)}$ by 13 percent is equivalent to eliminating $V_{n2(rms)}$ altogether!

The above example has an important moral. *To reduce overall noise, concentrate on large noise signals.*

4.2 FREQUENCY-DOMAIN ANALYSIS

As with deterministic signals, the frequency-domain techniques are useful for dealing with random signals such as noise. This section presents frequency-domain techniques for dealing with noise signals and other random signals. It should be noted that units of hertz (Hz) (rather than radians/second) are used throughout this chapter since, historically, such units have been commonly used in the measurement of continuous-time spectral densities.

Noise Spectral Density

Although periodic signals (such as a sinusoid) have power at distinct frequency locations, random signals have their power spread out over the frequency spectrum. For example, if the time-domain signal shown in Fig. 4.1 is applied to a spectrum analyzer, the resulting spectrum might look like that shown in Fig. 4.3(*a*). Note here that although the horizontal scale is the usual frequency axis, the vertical scale is in units of microvolts-squared/hertz. In other words, the vertical axis is a measure of the normalized noise power (mean-squared value) over a 1-Hz bandwidth at each frequency point. For example, the measurement at 100 Hz in Fig. 4.3(*a*) indicates that the normalized power between 99.5 Hz and 100.5 Hz is $10\ (\mu V)^2$.

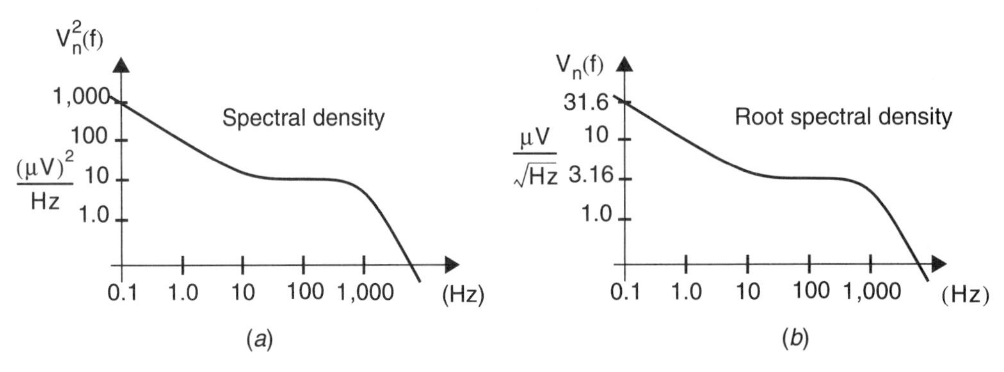

Fig. 4.3 Example of voltage spectral density (frequency domain), for (*a*) spectral density, and (*b*) root spectral density.

Thus, we define the noise spectral density, $V_n^2(f)$ (or, in the case of current, $I_n^2(f)$), as the average normalized noise power over a 1-Hz bandwidth. The units of $V_n^2(f)$ are volts-squared/hertz, whereas those of $I_n^2(f)$ are amps-squared/hertz. Also, $V_n^2(f)$ is a positive real-valued function.

It should be emphasized here that the *mean-squared value of a random noise signal at a single precise frequency is zero*. In other words, the mean-squared value of the signal shown in Fig. 4.3(*a*) measured at 100 Hz is directly proportional to the bandwidth of the bandpass filter used for the measurement. In a laboratory spectrum analyzer, the bandwidth of the bandpass filter is determined by the resolution-bandwidth control. Thus, as the resolution bandwidth goes to zero, the mean-squared value also becomes zero.[2] Conversely, as the resolution bandwidth increases, so does the measured mean-squared value. In either case, the measured mean-squared value should be normalized to the value that would be obtained for a bandwidth of 1 Hz when the noise spectral density is stated in units of $V^2/(Hz)$. For example, if the signal corresponding to Fig. 4.3(*a*) were measured at around 0.1 Hz using a resolution bandwidth of 1 mHz, the mean-squared value measured would be $1\ (\mu V)^2$, which, when scaled to a 1-Hz bandwidth, equals $1{,}000\ (\mu V^2)/Hz$.

An intuitive explanation of how a random noise signal is measured using a spectrum analyzer is as follows. A random noise signal has a frequency that is continually changing through a broad continuum of frequencies. A spectrum analyzer is sensitive to only a narrow frequency range that is the passband of its filter, and it measures the mean-squared value in that range. The filter of the spectrum analyzer reacts with a time constant approximately given by

$$\tau \approx \frac{1}{\pi W} \tag{4.18}$$

where W is the bandwidth of the filter. For some of the time, the random signal has the same frequency as the spectrum analyzer's filter. Thus, the spectrum analyzer effectively measures what percentage of time the random signal is in the frequency range of its filter. The narrower the bandwidth of the filter, the less percentage of time the signal is within its frequency range, and therefore the smaller is the spectrum analyzer's output. However, if the signal is not a random signal wandering in and out of the frequency range of the spectrum analyzer's filter, but is a deterministic signal at the center frequency of the filter, then the spectrum analyzer's reading is independent of the filter's bandwidth.

It is often convenient to plot the square root of the noise spectral density when we deal with filtered noise. Taking a square root results in $V_n(f)$, as shown in Fig. 4.3(*b*). We will refer to $V_n(f)$ as the *root spectral density which is expressed in units of volts/root-hertz* (i.e., $V/\sqrt{Hz}$). In the case of current noise, the resulting units are *amps/root-hertz*. Note that the horizontal axis remains unchanged although there is a root-hertz factor in the vertical axis.

Since the spectral density measures the mean-squared value over a 1-Hz bandwidth, one can obtain the total mean-squared value by integrating the spectral density

2. Such a result would not occur when a 100-Hz sinusoidal waveform is measured.

over the entire frequency spectrum. Thus, the rms value of a noise signal can also be obtained in the frequency domain using the following relationship:

$$V_{n(rms)}^2 = \int_0^\infty V_n^2(f)\, df \tag{4.19}$$

and similarly for current noise,

$$I_{n(rms)}^2 = \int_0^\infty I_n^2(f)\, df \tag{4.20}$$

Finally, it should be mentioned here that the spectral density function, $V_n^2(f)$, is the Fourier transform of the autocorrelation function of the time-domain signal, $v_n(t)$. This relationship is known as the *Wiener–Khinchin theorem*. It should also be noted that the relationship just shown defines a one-sided spectral density function since the noise is integrated only over positive frequencies as opposed to both negative and positive frequencies, again primarily for historical reasons. A two-sided definition results in the spectral density being divided by two since, for real-valued signals, the spectral density is the same for positive and negative frequencies, and the total mean-squared value remains unchanged.

EXAMPLE 4.3

What mean-squared value would be measured on the signal shown in Fig. 4.3 at 100 Hz when a resolution bandwidth of 30 Hz is used on a spectrum analyzer? Answer the same question for a 0.1-Hz resolution bandwidth.

Solution

Since the portion of spectral density function is flat at about 100 Hz, the measured value should simply be proportional to the bandwidth. Since the noise spectral density is $10\ (\mu V)^2/Hz$ at $100\ Hz$, the output of a 30-Hz filter is $30\ Hz \times 10\ (\mu V)^2/Hz = 300\ (\mu V)^2$ (or an rms value of $\sqrt{300}\ \mu V$).

For a 0.1-Hz bandwidth, the measured value would be 10 times smaller than for a 1-Hz bandwidth, resulting in a value of $1\ (\mu V)^2$ (or an rms value of $1\ \mu V$).

White Noise

One common type of noise is *white noise*. A noise signal is said to be white if its spectral density is constant over a given frequency. In other words, a white noise signal would have a flat spectral density, as shown in Fig. 4.4, where $V_n(f)$ is given by

$$V_n(f) = V_{nw} \tag{4.21}$$

and V_{nw} is a constant value.

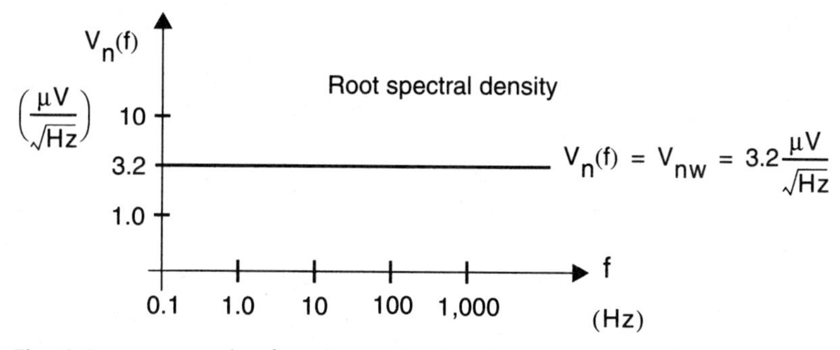

Fig. 4.4 An example of a white noise signal.

1/f, or Flicker, Noise

Another common noise shape is that of *1/f, or flicker,* noise.[3] The spectral density, $V_n^2(f)$, of 1/f noise is approximated by

$$V_n^2(f) = \frac{k_v^2}{f} \tag{4.22}$$

where k_v is a constant. Thus, the *spectral density is inversely proportional to frequency,* and hence the term "1/f noise." In terms of root spectral density, 1/f noise is given by

$$V_n(f) = \frac{k_v}{\sqrt{f}} \tag{4.23}$$

Note that it is inversely proportional to $\sqrt{f}$ (rather than f). An example of a signal having both 1/f and white noise is shown in Fig. 4.5. Note that the 1/f noise falls off at a rate of −10 dB/decade since it is inversely proportional to $\sqrt{f}$. The intersection of the 1/f and white noise curves is often referred to as the *1/f noise corner* (it occurs at 10 Hz in Fig. 4.5).

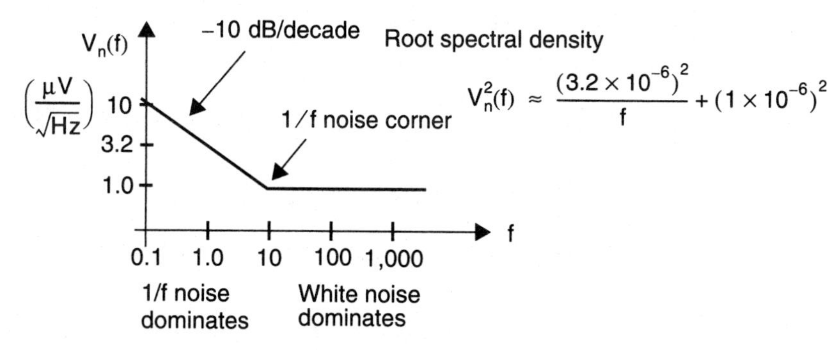

Fig. 4.5 A noise signal that has both 1/f and white noise.

3. 1/f, or flicker, noise is also referred to as *pink noise* since it has a large low-frequency content.

Filtered Noise

Consider the case of a noise signal, $V_{ni}(f)$, being filtered by the transfer function $A(s)$, as shown in Fig. 4.6. Here, $A(s)$ represents a linear transfer function as a result of some circuit amplification, filtering, or both. The following relationship between the input and output signals can be derived using the definition of the spectral density.

$$V_{no}^2(f) = |A(j2\pi f)|^2 V_{ni}^2(f) \tag{4.24}$$

The term $2\pi f$ arises here since, for physical frequencies, we replace s with $j\omega = j2\pi f$. Note that the output spectral density is a function only of the magnitude of the transfer function, and not its phase. As a result of (4.24), the total output mean-squared value is given by

$$V_{no(rms)}^2 = \int_0^\infty |A(j2\pi f)|^2 V_{ni}^2(f)\, df \tag{4.25}$$

If we wish to work with root spectral densities, we can take the square root of both sides of (4.24) resulting in

$$V_{no}(f) = |A(j2\pi f)| V_{ni}(f) \tag{4.26}$$

The relationship in (4.26) should make intuitive sense since it indicates that the transfer function simply shapes the root spectral density of the input noise signal. It is important to note here that the root spectral density is simply shaped by $|A(j2\pi f)|$, whereas the spectral density is shaped by $|A(j2\pi f)|^2$ (as seen by (4.24)). Hence, we see the benefit in dealing with the root spectral density rather than the spectral density. Specifically, straightforward transfer function analysis is applied when using root spectral densities, whereas squared terms are required to deal with spectral densities.

It is also of interest to consider the case of multiple uncorrelated noise sources that are each filtered and summed together. For example, consider the system shown in Fig. 4.7 in which three filtered, uncorrelated noise sources combine to form the total output noise, $V_{no}(f)$. In this case, one can show that if the input random signals are uncorrelated, the filter outputs are also uncorrelated. As a result, the output spectral density is given by

$$V_{no}^2(f) = \sum_{i=1,2,3} |A_i(j2\pi f)|^2 V_{ni}^2(f) \tag{4.27}$$

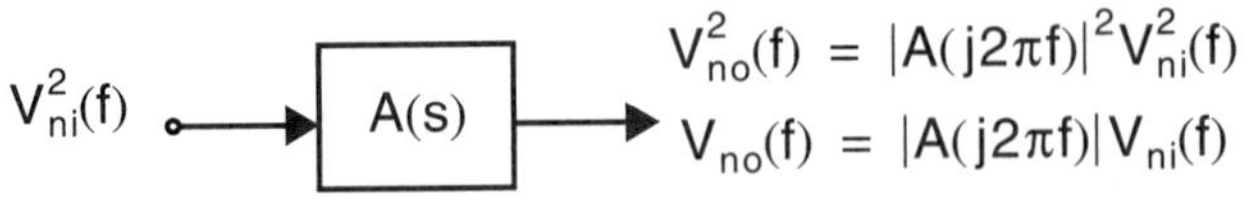

Fig. 4.6 Applying a transfer function (i.e., filter) to a noise signal.

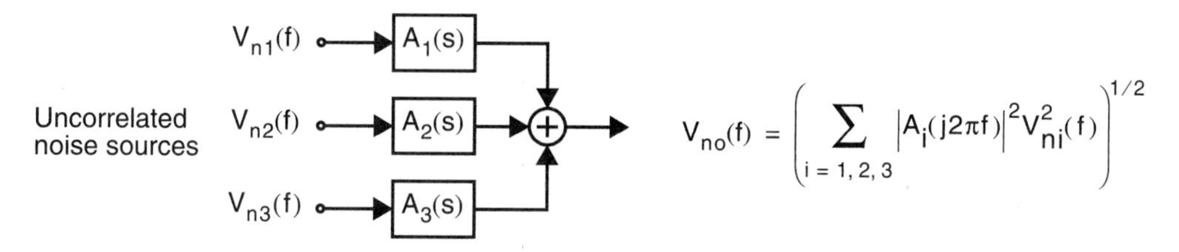

Fig. 4.7 Filtered uncorrelated noise sources contributing to total output noise.

EXAMPLE 4.4

Consider a noise signal, $V_{ni}(f)$, that has a white root spectral density of $20\ nV/\sqrt{Hz}$, as shown in Fig. 4.8(*a*). Find the total noise rms value between dc and 100 kHz. What is the total noise rms value if it is filtered by the RC filter shown in Fig. 4.8(*b*), where it is assumed the RC filter is noise free?

Solution

For the noise mean-square value from dc to 100 kHz of $V_{ni}(f)$, we have

$$V_{ni(rms)}^2 = \int_0^{10^5} 20^2\,df = 4 \times 10^7 (nV)^2 \tag{4.28}$$

resulting in an rms value of $V_{ni(rms)} = 6.3\ \mu V$ rms. Note that, for this simple case, one could also obtain the rms value by multiplying $20\ nV/\sqrt{Hz}$ by the square root of the frequency span, or $\sqrt{100\ kHz}$, resulting in $6.3\ \mu V$ rms.

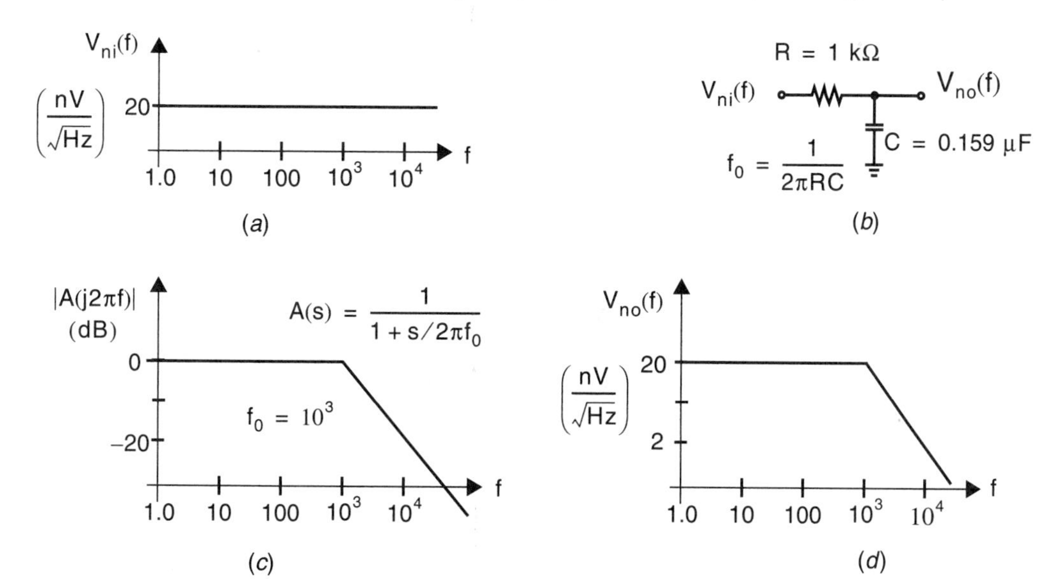

Fig. 4.8 (*a*) Spectral density for $V_{ni}(f)$. (*b*) RC filter to shape noise. (*c*) RC filter frequency response. (*d*) Spectral density for $V_{no}(f)$.

For the filtered signal, $V_{no}(f)$, we find that the RC filter has the frequency response shown in Fig. 4.8(*c*). Therefore, we can multiply the root spectral density of $V_{ni}(f)$ with the frequency response, $|A(j2\pi f)|$, to obtain the root spectral density of $V_{no}(f)$, as shown in Fig. 4.8(*d*). Mathematically, the output root spectral density is given by

$$V_{no}(f) = \frac{20 \times 10^{-9}}{\sqrt{1 + \left(\frac{f}{f_0}\right)^2}} \tag{4.29}$$

where $f_0 = 10^3$. Thus, the noise mean-squared value of $V_{no}(f)$ between dc and 100 kHz is given by

$$V_{no(rms)}^2 = \int_0^{10^5} \frac{20^2}{1 + \left(\frac{f}{f_0}\right)^2} df = 20^2 f_0 \arctan(f/f_0)\Big|_0^{10^5} \tag{4.30}$$

$$= 6.24 \times 10^5 (\text{nV})^2$$

which results in an rms value of $V_{no(rms)} = 0.79\ \mu\text{V rms}$. Note that the noise rms value of $V_{no}(f)$ is almost $1/10$ that of $V_{ni}(f)$ since high-frequency noise above 1 kHz was filtered. The lesson here is that you should not design circuits for larger bandwidths than your signal requires, otherwise noise performance suffers.

Noise Bandwidth

We just saw that the spectral density function is determined by the noise power within each 1-Hz bandwidth. Thus, in theory, one could measure the spectral density function by filtering a noise signal with a brick-wall bandpass filter having a 1-Hz bandwidth. The term brick wall implies here that the 1-Hz bandwidth of the filter is passed with a gain of one, whereas all other frequencies are entirely eliminated. However, practical filters can only approach a brick-wall response as their complexity (i.e., filter order) is increased. Thus, for lower-order filters with a 1-Hz passband, more noise power is passed than what is simply in a 1-Hz bandwidth. To account for the fact that practical filters have more gradual stopband characteristics, the term *noise bandwidth* is defined. *The noise bandwidth of a given filter is equal to the frequency span of a brick-wall filter that has the same output noise rms value that the given filter has when white noise is applied to both filters (peak gains are the same for the given and brick-wall filters).* In other words, given a filter response with peak gain A_0, the noise bandwidth is the width of a rectangular filter that has the same area and peak gain, A_0, as the original filter.

For example, consider a first-order, low-pass response with a 3-dB bandwidth of f_0, as shown in Fig. 4.9(*a*). Such a response would occur from the RC filter shown in Fig. 4.8(*b*) with $f_0 = (1/2\pi RC)$. The transfer function of $A(s)$ is given by

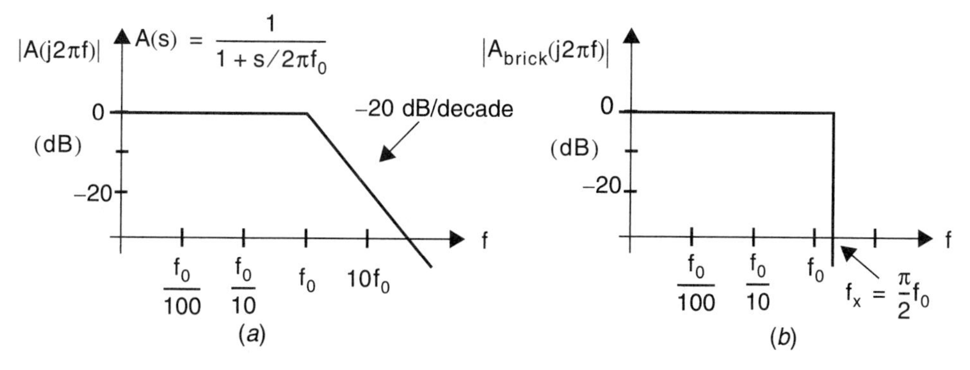

Fig. 4.9 (*a*) A first-order, low-pass response, and (*b*) a brick-wall filter that has the same peak gain and area as the first-order response.

$$A(s) = \frac{1}{1 + \dfrac{s}{2\pi f_0}} \tag{4.31}$$

This results in the magnitude response of $A(s)$ being equal to

$$|A(jf)| = \left(\frac{1}{1 + \left(\dfrac{f}{f_0}\right)^2}\right)^{1/2} \tag{4.32}$$

An input signal, $V_{ni}(f)$, is a white noise source given by

$$V_{ni}(f) = V_{nw} \tag{4.33}$$

where V_{nw} is a constant. The total output noise rms value of $V_{no}(f)$ is equal to

$$V^2_{no(rms)} = \int_0^{\infty} \frac{V^2_{nw}}{1 + \left(\dfrac{f}{f_0}\right)^2} df = V^2_{nw} f_0 \arctan\left(\frac{f}{f_0}\right)\Bigg|_0^{\infty} = \frac{V^2_{nw}\pi f_0}{2} \tag{4.34}$$

If this same input signal, $V_{ni}(f)$, is applied to the filter shown in Fig. 4.9(*b*), then the total output noise rms value equals,

$$V^2_{brick(rms)} = \int_0^{f_x} V^2_{nw}\, df = V^2_{nw} f_x \tag{4.35}$$

Finally, equating the two output noise rms values, $V_{no(rms)} = V_{brick(rms)}$, results in

$$f_x = \frac{\pi f_0}{2} \tag{4.36}$$

Thus, the noise bandwidth of a first-order, low-pass filter with a 3-dB bandwidth of f_0 equals $\pi(f_0/2)$. Note that, for the common case in which a first-order circuit is realized by a capacitor, C, and the resistance seen by that capacitor, R_{eq}, then

$$f_0 = \frac{1}{2\pi R_{eq} C} \tag{4.37}$$

and the noise bandwidth is given by

$$f_x = \frac{1}{4R_{eq}C} \tag{4.38}$$

The advantage of knowing the noise bandwidth of a filter is that, when white noise is applied to the filter input, the total output noise mean-squared value is easily calculated by multiplying the spectral density by the noise bandwidth. Specifically, in the first-order case just described, the total output noise mean-squared value, $V^2_{no(rms)}$, is equal to

$$V^2_{no(rms)} = V^2_{nw} f_x = V^2_{nw}\left(\frac{\pi}{2}\right) f_0 \tag{4.39}$$

Similar results for noise-bandwidth relationships can be obtained for higher-order and bandpass filters.

Piecewise Integration of Noise

Although simulation and computer techniques can analyze circuits or systems quite precisely (depending on their modelling accuracy), it is often convenient to make approximations that are useful in the early design stages. Such approximations are particularly useful in noise analysis, where large inaccuracies occur naturally due to such things as unknown parasitic effects and incomplete noise models. One such approximation is the estimation of total noise by integrating over frequency with the assumption that piecewise-linear Bode diagrams are exact. With such an approximation, integration formulas become much simpler when one needs only to integrate under linear functions and add together the resulting portions. The following example demonstrates this approach.

EXAMPLE 4.5

Consider an input noise signal, $V_{ni}(f)$, being applied to the amplifier $A(s)$, as shown in Fig. 4.10. Find the output noise rms value of $V_{no}(f)$ above 1 Hz.

Solution

As shown, the output root spectral density, $V_{no}(f)$, is determined by the addition of the Bode diagrams for $V_{ni}(f)$ and $A(s)$. To perform piecewise integration of $V_{no}(f)$, the frequency range is broken into four regions, N_1 through N_4, as shown.

For the region N_1, the total mean-square noise is given by

$$N_1^2 = \int_1^{100} \frac{200^2}{f} df = 200^2 \ln(f)\Big|_1^{100} = 1.84 \times 10^5 (\text{nV})^2 \tag{4.40}$$

In the region N_2, we have

$$N_2^2 = \int_{100}^{10^3} 20^2\, df = 20^2 f\Big|_{100}^{10^3} = 3.6 \times 10^5 (\text{nV})^2 \tag{4.41}$$

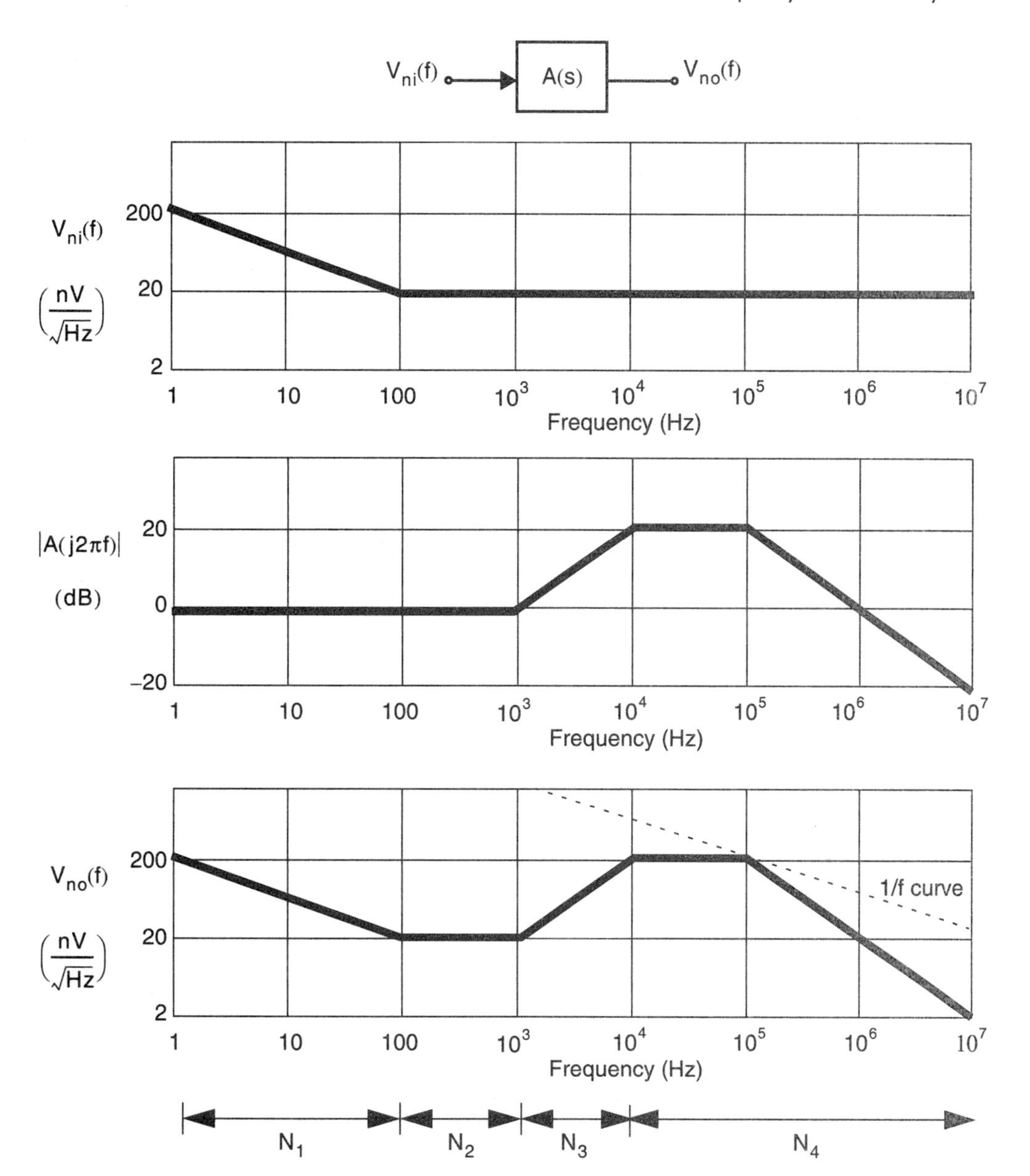

Fig. 4.10 Root spectral densities and amplifier curve example. $V_{no}(f)$ is the output noise that results from applying an input signal with noise $V_{ni}(f)$ to the amplifier $A(s)$.

Region N_3 ramps up rather than down, resulting in

$$N_3^2 = \int_{10^3}^{10^4} \frac{20^2 f^2}{(10^3)^2}\, df = \left(\frac{20}{10^3}\right)^2 \left[\frac{1}{3} f^3 \Big|_{10^3}^{10^4}\right] = 1.33 \times 10^8 (\text{nV})^2 \tag{4.42}$$

Finally, for region N_4, we can use the noise bandwidth result of a first-order, low-pass response and simply remove the noise portion resulting from under

10^4 Hz. Specifically, we have,

$$N_4^2 = \int_{10^4}^{\infty} \frac{200^2}{1+\left(\frac{f}{10^5}\right)^2} df = \int_0^{\infty} \frac{200^2}{1+\left(\frac{f}{10^5}\right)^2} df - \int_0^{10^4} 200^2 df \tag{4.43}$$

$$= 200^2\left(\frac{\pi}{2}\right)10^5 - (200^2)(10^4) = 5.88 \times 10^9 (\text{nV})^2$$

Thus, the total output noise can be estimated to be

$$V_{no(rms)} = (N_1^2 + N_2^2 + N_3^2 + N_4^2)^{1/2} = 77.5\ \mu\text{V rms} \tag{4.44}$$

An interesting point to note here is that in the preceding example, $N_4 = 76.7\ \mu\text{V rms}$ is quite close to the total noise value of $77.5\ \mu\text{V rms}$. Thus, in practice, there is little need to find the noise contributions in the regions N_1, N_2, and N_3. Such an observation leads us to the 1/f noise tangent principle.

1/f Noise Tangent Principle

The *1/f noise tangent principle* is as follows: To determine the frequency region or regions that contribute the dominant noise, *lower a 1/f noise line until it touches the spectral density curve—the total noise can be approximated by the noise in the vicinity of the 1/f line* [Kennedy, 1988]. For example, lowering a 1/f line toward the root spectral density of $V_{no}(f)$ in Fig. 4.10 indicates that the noise around 10^5 dominates. The reason this simple rule works is that a curve proportional to $1/x$ results in equal power over each decade of frequency. Therefore, by lowering this constant power/frequency curve, the largest power contribution will touch it first. However, because the integration of $1/x$ approaches infinity if either the upper bound is infinity or the lower bound is zero, one must be careful in cases where the spectral density curve runs parallel to a 1/f tangent line for an appreciable frequency span. For example, consider once again Fig. 4.10, where the region N_1 runs parallel to the 1/f tangent line. However, in this example, region N_1 does not contribute much noise since the noise was only integrated above 1 Hz. If a much lower frequency bound is used, this region can also contribute appreciable noise power.

4.3 NOISE MODELS FOR CIRCUIT ELEMENTS

There are three main fundamental noise mechanisms—thermal, shot, and flicker. In this section, we discuss noise models for popular circuit elements where all three mechanisms occur. However, first we briefly describe these noise phenomena.

Thermal noise is due to the thermal excitation of charge carriers in a conductor. This noise has a white spectral density and is proportional to absolute temperature. It is not dependent on bias conditions (dc bias current) and it occurs in all resistors (including conductors) above absolute zero. Thus, thermal noise places fundamental limits on the dynamic range achievable in electronic circuits. It should be mentioned here that thermal noise is also referred to as Johnson or Nyquist noise since it was first observed by J. B. Johnson [Johnson, 1928] and analyzed using the second law of thermodynamics by H. Nyquist [Nyquist, 1928].

Shot noise was first studied by W. Schottky using vacuum-tube diodes [Schottky, 1918], but shot noise also occurs in pn junctions. This noise occurs because the dc bias current is not continuous and smooth but instead is a result of pulses of current caused by the individual flow of carriers. As such, shot noise is dependent on the dc bias current. It can also be modelled as a white noise source. Shot noise is also typically larger than thermal noise and is sometimes used to create white noise generators.

Flicker noise is the least understood of the three noise phenomena. It is found in all active devices as well as in carbon resistors,[4] but it occurs only when a dc current is flowing. Flicker noise usually arises due to traps in the semiconductor, where carriers that would normally constitute dc current flow are held for some time period and then released. Flicker noise is also commonly referred to as 1/f noise since it is well modelled as having a $1/f^{\alpha}$ spectral density, where α is between 0.8 and 1.3. Although both bipolar and MOSFET transistors have flicker noise, it is a significant noise source in MOS transistors, whereas it can often be ignored in bipolar transistors.

Resistors

The major source of noise in resistors is thermal noise. As just discussed, it appears as white noise and can be modelled as a voltage source, $V_R(f)$, in series with a noiseless resistor. With such an approach, the spectral density function, $V_R^2(f)$, is found to be given by

$$V_R^2(f) = 4kTR \tag{4.45}$$

where k is Boltzmann's constant ($1.38 \times 10^{-23}\ \mathrm{JK}^{-1}$), T is the temperature in Kelvins, and R is the resistance size.

An alternate way to write (4.45) is to note that a $1\text{-k}\Omega$ resistor exhibits a root spectral density of $4.06\ \mathrm{nV}/\sqrt{\mathrm{Hz}}$ in thermal noise at room temperature (300 °K). Since the root spectral density is proportional to the square root of the resistance, we can also write

$$V_R(f) = \sqrt{\frac{R}{1k}} \times 4.06\ \mathrm{nV}/\sqrt{\mathrm{Hz}} \quad \text{for } 27\ ^\circ\mathrm{C} \tag{4.46}$$

Note that, to reduce the thermal noise due to resistors, one must either lower the temperature or use lower resistance values. The fact that lower resistance values cause less thermal noise becomes much more apparent when we look at kT/C noise in capacitors later in this section.

4. Carbon resistors are not used in integrated-circuit design but are available as discrete elements.

An alternate model can be derived by finding the Norton equivalent circuit. Specifically, the series voltage noise source, $V_R(f)$, can be replaced with a parallel current noise source, $I_R(f)$, given by

$$I_R^2(f) = \frac{V_R^2(f)}{R^2} = \frac{4kT}{R} \tag{4.47}$$

Both resistor models are summarized in Fig. 4.11.

Diodes

Shot noise is typically the dominant noise in diodes and can be modelled with a current source in parallel with the small-signal resistance of the diode, as Fig. 4.11 shows. The spectral density function of the current source is found to be given by

$$I_d^2(f) = 2qI_D \tag{4.48}$$

where q is one electronic charge $(1.6 \times 10^{-19}\ C)$ and I_D is the dc bias current flowing through the diode. The small-signal resistance of the diode, r_d, is given by the usual relationship,

$$r_d = \frac{kT}{qI_D} \tag{4.49}$$

Note that the Thévenin equivalent circuit can also be used, as shown in Fig. 4.11. It should be noted here that the small-signal resistance, r_d, is used for modelling and is not a physical resistor; hence, r_d does not contribute any thermal noise.

Bipolar Transistors

The noise in bipolar transistors is due to the shot noise of both the collector and base currents, the flicker noise of the base current, and the thermal noise of the base resistance. A common practice is to combine all these noise sources into two equivalent noise sources at the base of the transistor, as shown in Fig. 4.11. Here, the equivalent input voltage noise, $V_i(f)$, is given by

$$V_i^2(f) = 4kT\left(r_b + \frac{1}{2g_m}\right) \tag{4.50}$$

where the r_b term is due to the thermal noise of the base resistance and the g_m term is due to collector-current shot noise referred back to the input. The equivalent input current noise, $I_i(f)$, equals

$$I_i^2(f) = 2q\left(I_B + \frac{KI_B}{f} + \frac{I_C}{|\beta(f)|^2}\right) \tag{4.51}$$

where the $2qI_B$ term is a result of base-current shot noise, the KI_B/f term models 1/f noise (K is a constant dependent on device properties), and the I_C term is the input-referred collector-current shot noise (it is often ignored).

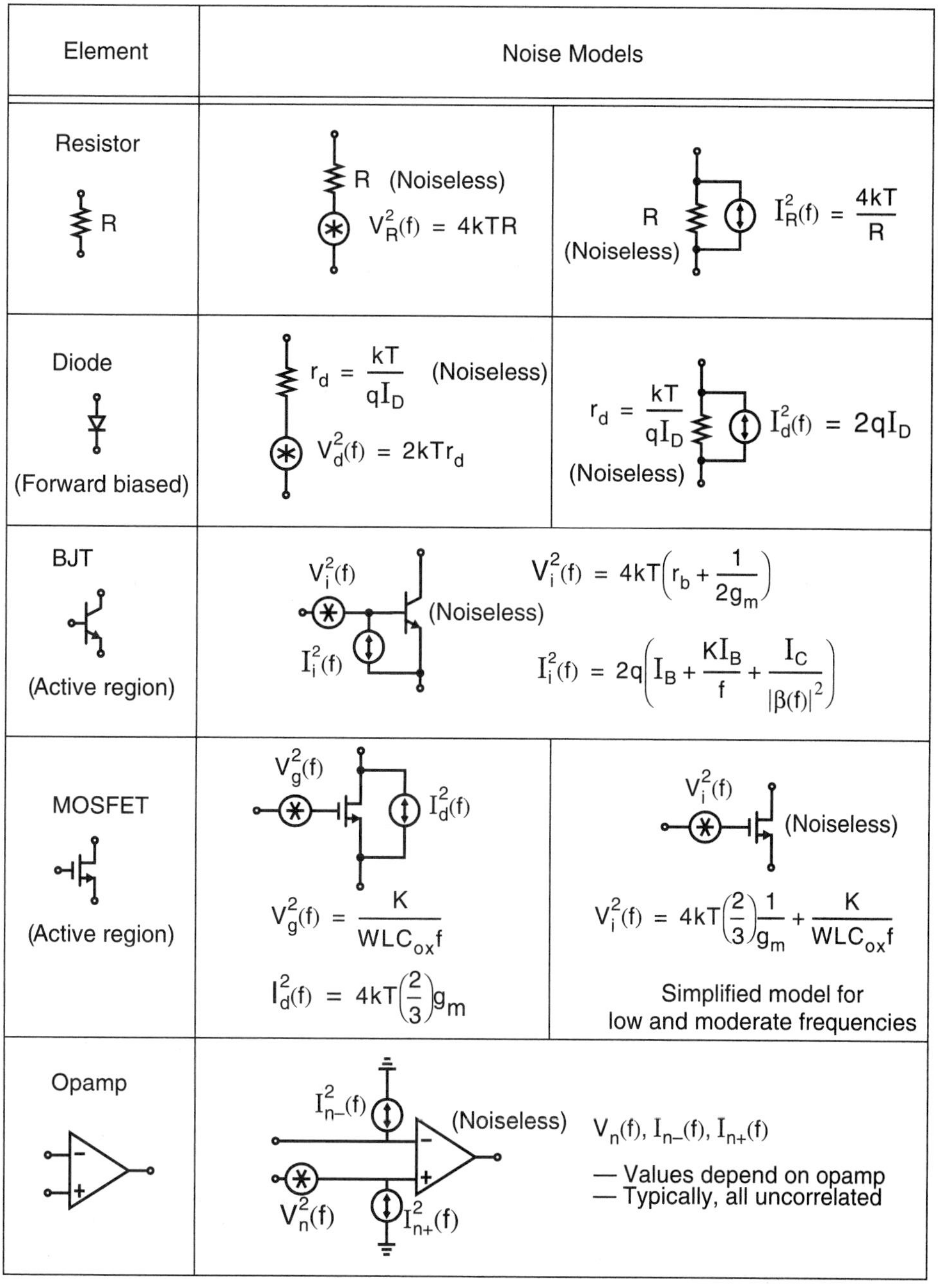

Element	Noise Models	
Resistor R	R (Noiseless) $V_R^2(f) = 4kTR$	R (Noiseless) $I_R^2(f) = \frac{4kT}{R}$
Diode (Forward biased)	$r_d = \frac{kT}{qI_D}$ (Noiseless) $V_d^2(f) = 2kTr_d$	$r_d = \frac{kT}{qI_D}$ (Noiseless) $I_d^2(f) = 2qI_D$
BJT (Active region)	$V_i^2(f)$ $I_i^2(f)$ (Noiseless)	$V_i^2(f) = 4kT\left(r_b + \frac{1}{2g_m}\right)$ $I_i^2(f) = 2q\left(I_B + \frac{KI_B}{f} + \frac{I_C}{\lvert\beta(f)\rvert^2}\right)$
MOSFET (Active region)	$V_g^2(f)$ $I_d^2(f)$ $V_g^2(f) = \frac{K}{WLC_{ox}f}$ $I_d^2(f) = 4kT\left(\frac{2}{3}\right)g_m$	$V_i^2(f)$ (Noiseless) $V_i^2(f) = 4kT\left(\frac{2}{3}\right)\frac{1}{g_m} + \frac{K}{WLC_{ox}f}$ Simplified model for low and moderate frequencies
Opamp	$I_{n-}^2(f)$ (Noiseless) $V_n^2(f)$ $I_{n+}^2(f)$	$V_n(f), I_{n-}(f), I_{n+}(f)$ — Values depend on opamp — Typically, all uncorrelated

Fig. 4.11 Circuit elements and their noise models. Note that capacitors and inductors do not generate noise.

A couple of comments here are that the noise of r_b typically dominates in $V_i(f)$, and the base-current shot noise often dominates in the input-current noise, $I_i^2(f)$. Thus, the equivalent voltage and current noise are not derived from the same noise sources. As a result, it is common practice to assume that the input voltage and current noise sources in a bipolar transistor are uncorrelated.[5]

MOSFETS

The dominant noise sources for active MOSFET transistors are flicker and thermal noise, as shown in Fig. 4.11. The flicker noise is modelled as a voltage source in series with the gate of value

$$V_g^2(f) = \frac{K}{WLC_{ox}f} \tag{4.52}$$

where the constant K is dependent on device characteristics and can vary widely for different devices in the same process. The variables W, L, and C_{ox} represent the transistor's width, length, and gate capacitance per unit area, respectively. *An important point to note here is that the 1/f noise is inversely proportional to the transistor area,* WL. In other words, larger devices have less 1/f noise. 1/f noise is extremely important in MOSFET circuits, because it typically dominates at low frequencies unless switching-circuit techniques are used to reduce its effect. Also, typically p-channel transistors have less noise than their n-channel counterparts since their majority carriers (holes) are less likely to be trapped.

The derivation of the thermal noise term is straightforward and is due to the resistive channel of a MOS transistor in the active region. If the transistor was in triode, the thermal noise current in the drain due to the channel resistance would simply be given by $I_d^2(f) = (4kT)/r_{ds}$, where r_{ds} is the channel resistance. However, when the transistor is in the active region, the channel cannot be considered homogeneous, and thus, the total noise is found by integrating over small portions of the channel. Such an integration results in the noise current in the drain being given by

$$I_d^2(f) = 4kT\left(\frac{2}{3}\right)g_m \tag{4.53}$$

for the case $V_{DS} = V_{GS} - V_T$.

Often, noise analyses are done just by including this noise source between the transistor drain and source. Sometimes, however, analysis may be simplified if it is replaced by an equivalent input noise source. To find the equivalent noise voltage that would cause this drain current, we note that the drain current is equal to the gate voltage times the transconductance of the device, or, mathematically, $I_d(f) = g_m V_i(f)$.

5. An exception to this is high-frequency designs in which the collector shot noise becomes more important because, at high frequencies, β becomes small. In this case, the input current noise source is partially correlated with the input voltage noise source. If neither noise source dominates, then the correct analysis is more difficult and beyond the scope of this text. Fortunately, this case is not often encountered in practice.

Thus, dividing (4.53) by g_m^2 results in the simplified MOSFET model, also shown in Fig. 4.11, where there is now only one voltage noise source. However, one should be aware that this simplified model assumes the gate current is zero. Although this assumption is valid at low and moderate frequencies, an appreciable amount of current would flow through the gate-source capacitance, C_{gs}, at higher frequencies. In summary, although most noise analysis can be performed using the simplified model, if in doubt, one should use the model with the thermal noise placed as a current source in parallel with the drain-source channel.

Finally, it should be noted that no gate leakage noise terms have been included in this noise model since, in modern process, the gate leakage is so small that its noise contribution is rarely significant.

EXAMPLE 4.6

A large MOS transistor consists of ten individual transistors connected in parallel. Considering 1/f noise only, what is the equivalent input voltage noise spectral density for the ten transistors compared to that of a single transistor?

Solution

From (4.52), the 1/f noise can be modelled as a current source going from the drain to the source with a noise spectral density of

$$I_d^2(f) = \frac{Kg_m^2}{WLC_{ox}f} \tag{4.54}$$

where g_m is the transconductance of a single transistor. When ten transistors are connected in parallel, the drain-current noise spectral density is ten times larger, so we have

$$I_{d=10}^2(f) = \frac{10Kg_m^2}{WLC_{ox}f} \tag{4.55}$$

When this noise is referred back to the input of the equivalent transistor, we have

$$V_{i=10}^2(f) = \frac{10Kg_m^2}{WLC_{ox}fg_{m=10}^2} = \frac{K}{10WLC_{ox}f} \tag{4.56}$$

since the transconductance of ten transistors in parallel, $g_{m=10}$, is equal to $10\ g_m$. Thus, the drain current noise spectral density of the equivalent transistor is ten times larger, but the input voltage noise spectral density is ten times smaller. This result is expected because the input voltage noise source due to 1/f noise is inversely proportional to the equivalent transistor area, which in this case is $10\ WL$.

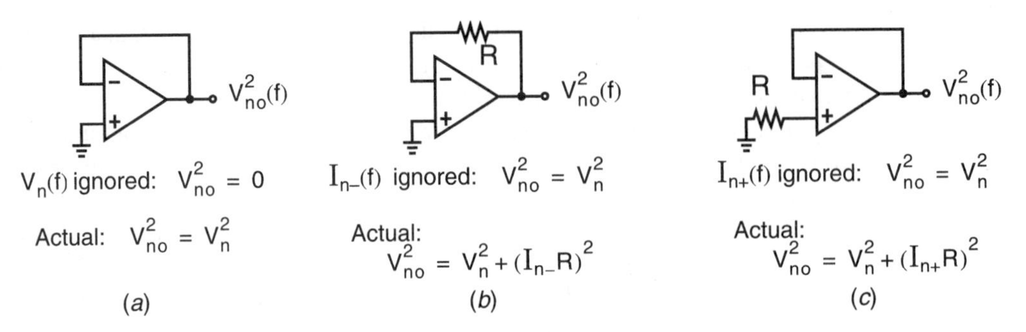

Fig. 4.12 Opamp circuits showing the need for three noise sources in an opamp noise model. Assume the resistance, R, is noiseless. Also, notation is simplified from $V_n(f)$ to V_n, and so on.

Opamps

Noise in opamps is modelled using three uncorrelated input-referred noise sources, as shown in Fig. 4.11. With an opamp that has a MOSFET input stage, the current noises can often be ignored at low frequencies since their values are small. However, for bipolar input stages, all three noise sources are typically required, as shown in Fig. 4.12. In Fig. 4.12(*a*), if $V_n(f)$ is not included in the model, a unity-gain buffer with no resistors indicates that the circuit output is noiseless. In fact, the voltage noise source, $V_n(f)$, dominates. If $I_{n-}(f)$ is not included in an opamp model, the circuit shown in Fig. 4.12(*b*) indicates that the output noise voltage equals $V_n(f)$. Here, the current noise may dominate if the resistance, R, is large. A similar conclusion can be drawn with $I_{n+}(f)$, as shown in Fig. 4.12(*c*).

Capacitors and Inductors

Capacitors and inductors do not generate any noise. However, they do accumulate noise generated by other noise sources. Here, we will see that the capacitor noise mean-squared value equals kT/C when it is connected to an arbitrary resistor value.

Consider a capacitance, C, in parallel with a resistor of arbitrary size, R, as shown in Fig. 4.13(*a*). The equivalent circuit for noise analysis is shown in Fig. 4.13(*b*). To determine the total noise mean-squared value across the capacitor, we note that $V_{no}(f)$ is simply a first-order, low-pass, filtered signal with $V_R(f)$ as the

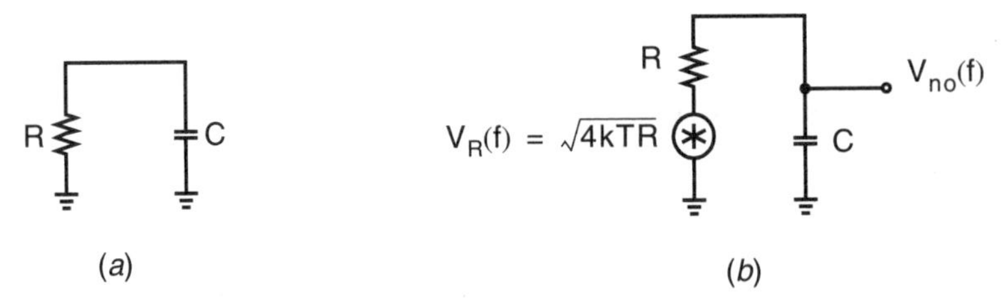

Fig. 4.13 (*a*) Capacitor, C, in parallel with a resistor, and (*b*) equivalent noise model circuit.

input. Therefore, we recognize that the noise bandwidth is given by $(\pi/2)f_0$ (see Section 4.2, Eq. (4.36), and since the input has a white spectral density, the total output mean-squared value is calculated as

$$V_{no(rms)}^2 = V_R^2(f)\left(\frac{\pi}{2}\right)f_0 = (4kTR)\left(\frac{\pi}{2}\right)\left(\frac{1}{2\pi RC}\right)$$
$$V_{no(rms)}^2 = \frac{kT}{C} \tag{4.57}$$

In other words, the rms voltage value across a capacitor is equal to $\sqrt{kT/C}$, regardless of the resistance seen across it. Such a result is due to fact that small resistances have less noise spectral density but result in a wide bandwidth, compared to large resistances, which have reduced bandwidth but larger noise spectral density.

Finally, it should be stated that this noise property for capacitors gives a fundamental limit on the minimum noise level across a capacitor.[6] Thus, to lower the noise level, either the temperature must be lowered or the capacitance value must be increased.

EXAMPLE 4.7

At a room temperature of 300 °K, what capacitor size is needed to achieve a 96-dB dynamic range in an analog circuit with maximum signal levels of 1 V rms?

Solution

The value of noise that can be tolerated here is 96 dB down from 1 V rms, which is

$$V_{n(rms)} = \frac{1V}{10^{96/20}} = 15.8\ \mu V\ rms \tag{4.58}$$

Using (4.57), we have

$$C = \frac{kT}{V_{n(rms)}^2} = 16.6\ pF \tag{4.59}$$

Thus, the minimum capacitor size that can be used (without oversampling) is 16.6 pF. The use of this minimum capacitor size determines the maximum resistance size allowed to achieve a given time constant.

Finally, it should be mentioned that the equivalent noise current mean-squared value in an inductor of value L is given by (see Problem 4.9)

$$I_{no(rms)}^2 = \frac{kT}{L} \tag{4.60}$$

6. Some feedback circuits can make the noise smaller but signal levels are also typically smaller (see Problem 4.8).

Sampled Signal Noise

In many cases, one should obtain a sampled value of an analog voltage. For example, switched-capacitor circuits make extensive use of sampling, and sample-and-hold circuits are commonly used in analog-to-digital and digital-to-analog conversion.

Consider a basic sample-and-hold circuit, as shown in Fig. 4.14. When ϕ_{clk} drops, the transistor turns off and, in an ideal noiseless world, the input voltage signal at that instance would be held on capacitance C. However, when thermal noise is present, the resistance when the transistor is switched on causes the capacitance voltage noise to be equal to $\sqrt{kT/C}$. Thus, when the switch is turned off, the noise as well as the desired signal is held on C. As a result, a fundamental limit occurs for sampled signals using a capacitance C—an rms noise voltage of $\sqrt{kT/C}$.

It should be noted that this noise voltage will not depend on the sampling rate and is independent from sample to sample. This fact suggests a method to reduce the noise level of a signal measurement. Specifically, in the case where V_{in} is a dc (or low-frequency) signal, taking only one sample results in a noise voltage of $\sqrt{kT/C}$. However, if many samples are taken (say, 1,000) and all samples are averaged, the averaged value will have a reduced noise level. The reason this technique improves the measurement accuracy is that, when individual sampled values are summed together, their signal values add linearly, whereas their noise values add as the root of the sum of squares. This technique is known as oversampling and will be discussed at length with respect to oversampling converters in Chapter 14.

4.4 NOISE ANALYSIS EXAMPLES

In this section, a variety of circuits are analyzed from a noise perspective. Although some useful design techniques for reducing noise are presented, the main purpose of this section is to give the reader some experience in analyzing circuits from a noise perspective.

Opamp Example

Consider an inverting amplifier and its equivalent noise model, as shown in Fig. 4.15. Here, $V_n(f)$, $I_{n-}(f)$, and $I_{n+}(f)$ represent the opamp's equivalent input noise, and the remaining noise sources are resistor thermal noise sources. Note that current noise sources are used in the models for R_1 and R_f, whereas a voltage noise source is used for R_2. As we will see, these choices of noise sources simplify the circuit analysis.

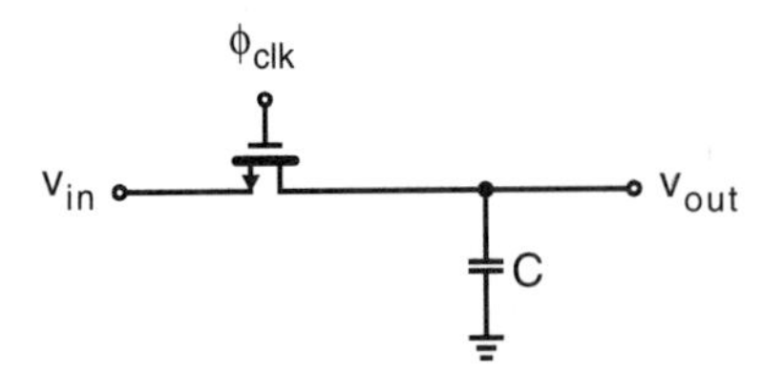

Fig. 4.14 A sample-and-hold circuit.

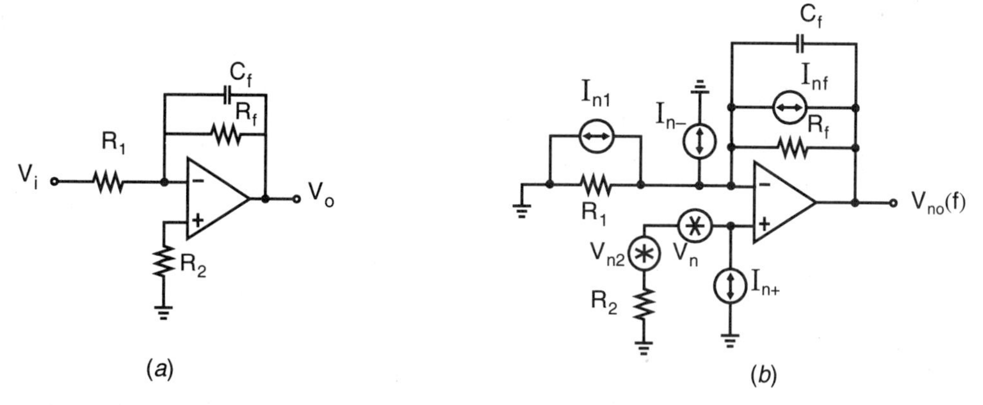

Fig. 4.15 (*a*) Low-pass filter, and (*b*) equivalent noise model.

First, using superposition and assuming all noise sources are uncorrelated, consider the output noise voltage, $V_{no1}^2(f)$, due only to I_{n1}, I_{nf}, and I_{n-}. These three noise currents add together, and their total current sum is fed into the parallel combination of C_f and R_f. Note that no current passes through R_1 since the voltage across it is zero due to the virtual ground at the negative opamp terminal (assuming a high-gain opamp). Thus, the output noise mean-squared value due to these three noise currents is given by

$$V_{no1}^2(f) = [I_{n1}^2(f) + I_{nf}^2(f) + I_{n-}^2(f)]\left|\frac{R_f}{1 + j2\pi f R_f C_f}\right|^2 \tag{4.61}$$

This equation indicates that this part of the output noise value equals the sum of the noise currents multiplied by R_f^2. This noise portion is then shaped by a low-pass filter with a 3-dB frequency equal to $f_0 = 1/(2\pi R_f C_f)$.

Using superposition further, the output mean-squared value, $V_{no2}^2(f)$, due to the three noise sources at the positive opamp terminal can be found as follows: By converting I_{n+} to a voltage source (by multiplying it by R_2), we see that the three noise voltages are summed and are applied to the positive opamp terminal. Since the gain from the positive opamp terminal to the output signal is easily found, the output noise mean-squared value due to these three noise sources is given by

$$V_{no2}^2(f) = [I_{n+}^2(f)R_2^2 + V_{n2}^2(f) + V_n^2(f)]\left|1 + \frac{R_f/R_1}{1 + j2\pi f C_f R_f}\right|^2 \tag{4.62}$$

This equation indicates that this part of the output noise mean-squared value equals the sum of the noise voltages, and this noise portion is then shaped by the shown transfer function. For this transfer function, if $R_f << R_1$, then its gain approximately equals unity for all frequencies. Thus, for an ideal opamp, the noise would exist up to infinite frequency, resulting in an infinite amount of mean-squared volts. However, for practical circuits, the gain drops off above the unity-gain frequency of the opamp, and thus the noise is effectively low-pass filtered. In the case where $R_f >> R_1$, the low-frequency gain is roughly R_f/R_1, and its 3-dB frequency is the same as in the case of the noise sources at the negative opamp terminal (i.e., $f_0 = 1/(2\pi R_f C_f)$). However, in this case, the gain only decreases to unity and then remains at that level. Treating the Bode plot as exact and noting that, above f_0, the gain drops off at –20 dB/decade, this transfer function reaches unity around $f_1 = (R_f/R_1)f_0$. Thus, one should also include the opamp's positive input noise (with a gain of one) integrated over the region between f_1 and the unity-gain frequency of the opamp.

Finally, the total output noise mean-squared value is simply the sum

$$V_{no}^2(f) = V_{no1}^2(f) + V_{no2}^2(f) \tag{4.63}$$

or, if rms values are found,

$$V_{no(rms)}^2 = V_{no1(rms)}^2 + V_{no2(rms)}^2 \tag{4.64}$$

EXAMPLE 4.8

Estimate the total output noise rms value for a 10-kHz low-pass filter, as shown in Fig. 4.15, when $C_f = 160$ pF, $R_f = 100$ k, $R_1 = 10$ k, and $R_2 = 9.1$ k. Also, find the SNR for an input signal equal to 100 mV rms. Assume that the noise voltage of the opamp is given by $V_n(f) = 20\ \text{nV}/\sqrt{\text{Hz}}$, both its noise currents are $I_n(f) = 0.6\ \text{pA}/\sqrt{\text{Hz}}$, and that its unity-gain frequency equals 5 MHz.

Solution

Assuming the device is at room temperature, the resistor noise sources are equal to

$$I_{nf} = 0.406\ \text{pA}/\sqrt{\text{Hz}} \tag{4.65}$$

$$I_{n1} = 1.28\ \text{pA}/\sqrt{\text{Hz}} \tag{4.66}$$

$$V_{n2} = 12.2\ \text{nV}/\sqrt{\text{Hz}} \tag{4.67}$$

The low-frequency value of $V_{no1}^2(f)$ is found by letting $f = 0$ in (4.61).

$$\begin{aligned} V_{no1}^2(0) &= [I_{n1}^2(0) + I_{nf}^2(0) + I_{n-}^2(0)]R_f^2 \\ &= (0.406^2 + 1.28^2 + 0.6^2)(1 \times 10^{-12})^2(100\ \text{k})^2 \\ &= (147\ \text{nV}/\sqrt{\text{Hz}})^2 \end{aligned} \tag{4.68}$$

Since (4.61) also indicates that this noise is low-pass filtered, the rms output noise value due to these three sources can be found using the concept of a noise equivalent bandwidth. Specifically, we multiply the spectral density by $(\pi f_0)/2$, where $f_0 = 1/(2\pi R_f C_f)$. Thus,

$$\begin{aligned} V_{no1(rms)}^2 &= (147\ \text{nV}/\sqrt{\text{Hz}})^2 \times \frac{1}{4(100\ \text{k}\Omega)(160\ \text{pF})} \\ &= (18.4\ \mu\text{V})^2 \end{aligned} \tag{4.69}$$

To estimate the output noise due to the sources at the positive opamp terminal, we find $V_{no2}^2(0)$ to be given by

$$\begin{aligned} V_{no2}^2(0) &= [I_{n+}^2(f)R_2^2 + V_{n2}^2(f) + V_n^2(f)](1 + R_f/R_1)^2 \\ &= (24.1 \text{ nV}/\sqrt{\text{Hz}})^2 \times 11^2 \\ &= (265 \text{ nV}/\sqrt{\text{Hz}})^2 \end{aligned} \tag{4.70}$$

This noise is also low-pass filtered at f_0 until $f_1 = (R_f/R_1)f_0$, where the noise gain reaches unity and it remains until $f_t = 5$ MHz (i.e, the unity-gain frequency of the opamp). Thus, breaking this noise into two portions, and using (4.38) to calculate the first portion, we have

$$\begin{aligned} V_{no2(rms)}^2 &= (265 \times 10^{-9})^2\left(\frac{1}{4R_fC_f}\right) + (24.1 \times 10^{-9})^2\left(\frac{\pi}{2}\right)(f_t - f_1) \\ &= (74.6\ \mu\text{V})^2 \end{aligned} \tag{4.71}$$

Thus, the total output noise is estimated to be

$$V_{no(rms)} = \sqrt{V_{no1(rms)}^2 + V_{no2(rms)}^2} = 77\ \mu\text{V rms} \tag{4.72}$$

It should be noted here that the major source of noise at low frequencies is due to the opamp's voltage noise, $V_n(f)$.

To obtain the SNR for this circuit with a 100-mV rms input level, one can find the output signal mean-squared value and relate that to the output noise value. Alternatively, one can find the equivalent input noise value by dividing the output noise value by the circuit's gain and then relate the input signal and noise mean-squared values. Taking the first approach results in an output signal level of 1 V rms, which gives an SNR of

$$\text{SNR} = 20\log\left(\frac{1\text{ V}}{77\ \mu\text{V}}\right) = 82\text{ dB} \tag{4.73}$$

Note here that using a lower-speed opamp would have reduced the total output noise, as would choosing an opamp with a lower noise voltage. Also note that R_2 contributes to the output noise through both its thermal noise and the noise current of the opamp's positive input. Sincc its only purpose is to improve the dc offset performance, R_2 should be eliminated in a low-noise circuit (assuming dc offset can be tolerated).

Bipolar Common-Emitter Example

In this example, we consider a bipolar common-emitter amplifier, as shown in Fig. 4.16. Here, we wish to find the optimum bias current to minimize the equivalent input noise of this amplifier (i.e., maximize the signal-to-noise ratio of the amplifier).

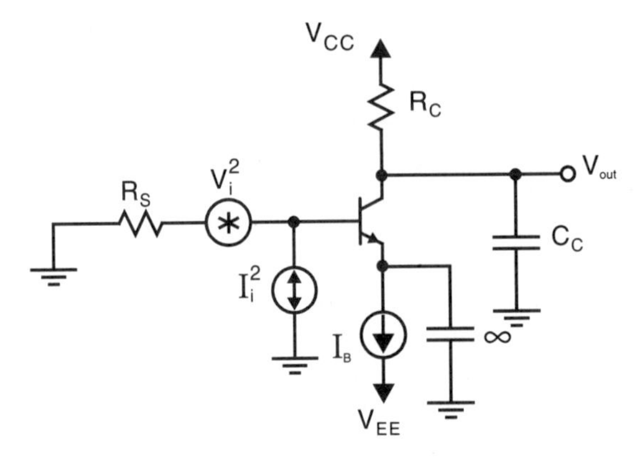

Fig. 4.16 A bipolar common-emitter amplifier.

We should state that it is assumed here that the collector-current shot noise dominates in the input voltage noise source and the base-current shot noise dominates in the input current noise source. Therefore, the input voltage noise due to the transistor alone is given by

$$V_i^2(f) = 2kT/g_m \tag{4.74}$$

and the input current noise due to the transistor alone is given by

$$I_i^2(f) = 2qI_B \tag{4.75}$$

These assumptions are reasonable since usually 1/f noise is not important for bipolar transistors, and the collector shot noise is usually not a major component of the input current noise source for wideband examples. Also, although the base resistance has temporarily been ignored, it will be taken into account shortly by simply modifying the size of the source resistance.

The first step is to replace the input current noise source by an input voltage noise source. First, note that the gain from the base to the collector (ignoring the transistor output impedance, r_o) is equal to $g_m R_C$, and the impedance looking into the base is r_π. Therefore, the output noise due to $I_i^2(f)$ is given by

$$V_{oi}^2(f) = I_i^2(f)[(R_S \| r_\pi)g_m R_C]^2 \tag{4.76}$$

Now the gain from the input voltage noise source to the output is given by

$$A_V = \frac{V_o}{V_i} = \frac{r_\pi}{r_\pi + R_S} g_m R_C \tag{4.77}$$

To represent the output noise due to the input current source by an equivalent input voltage source, we have

$$V_{ieq}^2(f) = \frac{V_{oi}^2(f)}{A_V^2} = \left(\frac{\dfrac{r_\pi R_S}{(r_\pi + R_S)} g_m R_C}{\dfrac{r_\pi}{(r_\pi + R_S)} g_m R_C} \right)^2 I_i^2(f) = I_i^2(f) R_S^2 \tag{4.78}$$

We can now replace all noise sources by a single voltage noise source, which includes the noise due to the input voltage noise source and the input current noise source of the transistor, as well as the noise of the source resistor. We have

$$V_{i=total}^2(f) = V_i^2(f) + V_{ieq}^2(f) + 4kTR_S \tag{4.79}$$

The first two terms represent the noise of the transistor and the third term represents the noise of the source resistance. Using (4.74), (4.75), and (4.78), we have

$$V_{i=total}^2(f) = \frac{2kT}{g_m} + 2qI_BR_S^2 + 4kTR_S \tag{4.80}$$

Substituting $g_m = qI_C/(kT)$ and $I_B = I_C/\beta$, we find

$$V_{i=total}^2(f) = \frac{2(kT)^2}{qI_C} + \frac{2qI_CR_S^2}{\beta} + 4kTR_S \tag{4.81}$$

The noise of the transistor base resistor can now be included by replacing R_S with $R_S + r_b$ to obtain

$$V_{i=total}^2(f) = \frac{2(kT)^2}{qI_C} + \frac{2qI_C(R_S + r_b)^2}{\beta} + 4kT(R_S + r_b) \tag{4.82}$$

Alternatively, (4.82) can be expressed in terms of $g_m = qI_C/(kT)$ as in [Buchwald, 1995]

$$V_{i=total}^2(f) = 4kT\left[R_s + r_b + \frac{1}{2g_m} + \frac{g_m(R_s + r_b)^2}{2\beta}\right] \tag{4.83}$$

These two equations are good approximations for most applications of bipolar common-emitter amplifiers. Notice in (4.82), the first term models the transistor base-current shot noise and increases with increased bias current. However, the second term (which models the transistor collector shot noise) decreases with increasing collector bias current. Finally, the terms modelling the noise of the source resistor and the transistor base resistor are independent of the transistor bias current. To find the optimum bias current, we differentiate (4.82) with respect to I_C and set the result to zero. This yields the rather simple result of

$$I_{C=opt} = \frac{kT}{q}\frac{\sqrt{\beta}}{R_S + r_b} \tag{4.84}$$

or, equivalently,

$$g_{m=opt} = \frac{\sqrt{\beta}}{R_S + r_b} \tag{4.85}$$

EXAMPLE 4.9

Consider the common-emitter amplifier shown in Fig. 4.16, where $R_C = 10\ k\Omega$, $C_C = 1\ pF$, $R_S = 500\ \Omega$, $\beta = 100$, and the base resistance is given by $r_b = 300\ \Omega$. At room temperature, (i.e. 300 °K), find the optimum bias current and the total equivalent input noise.

Solution

Using (4.84), we find

$$I_{C=opt} = 0.026 \times \frac{\sqrt{100}}{500+300} = 0.325 \text{ mA} \tag{4.86}$$

implying that

$$g_m = \frac{I_C}{V_T} = 12.5 \text{ mA/V} \tag{4.87}$$

To find the output noise spectral density, we now use (4.83) to find

$$V_{i=total}^2(f) = 4kT(500+300+40+40) = 1.46 \times 10^{-17}\ V^2/Hz \tag{4.88}$$

Notice that the noise due to the source resistance dominates, even though the source resistance is moderately small. Even for no source resistance, the thermal noise due to the base resistance would still dominate. Thus, for a very low source resistance, the major way to improve the noise is to minimize the base resistance by using larger transistors (or to combine a number of parallel transistors).

Assuming the RC time constant at the collector dominates the frequency response, the noise bandwidth of the amplifier is given by

$$f_x = \frac{1}{4R_C C_C} = 25 \text{ MHz} \tag{4.89}$$

resulting in the total input-referred voltage noise given by

$$V_{ni(rms)} = [f_x V_{i=total}^2(f)]^{1/2} = 19.1\ \mu V \tag{4.90}$$

CMOS Example

In this example, we look at the input circuitry of a traditional two-stage CMOS opamp, as shown in Fig. 4.17. Note that each of the transistors have been modelled using an equivalent voltage noise source, as presented in Fig. 4.11. Voltage noise sources are used here since we will be addressing the low-frequency noise performance of this stage. It should be mentioned here that in the following derivations, we have assumed matching between transistor pair Q_1 and Q_2 as well as in pair Q_3 and Q_4.

We start by finding the gains from each noise source to the output node, V_{no}. The gains from V_{n1} and V_{n2} are the same as the gains from the input signals, resulting in

$$\left|\frac{V_{no}}{V_{n1}}\right| = \left|\frac{V_{no}}{V_{n2}}\right| = g_{m1} R_o \tag{4.91}$$

where R_o is the output impedance seen at V_{no}.

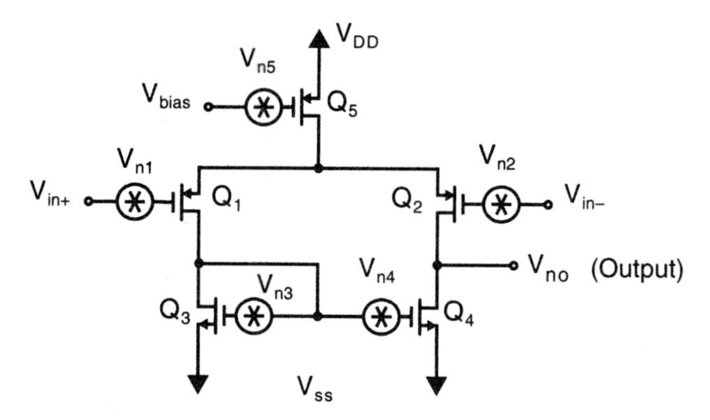

Fig. 4.17 A CMOS input stage for a traditional opamp with MOSFET noise sources shown.

Next, for V_{n3}, notice that the current through this voltage source must be zero because one side is connected to the gate of Q_3 only. Therefore, assuming all other sources are zero, the drain current is unaffected by V_{n3}. This implies that the gate voltage of Q_3 is also unaffected by V_{n3}. Therefore, in the small-signal model, V_{n3} is equal to v_{gs4}, and the gain from V_{n3} to the output is the same as the gain from V_{n4} to the output. Thus, we have

$$\left|\frac{V_{no}}{V_{n3}}\right| = \left|\frac{V_{no}}{V_{n4}}\right| = g_{m3}R_o \tag{4.92}$$

Finally, the noise gain from V_{n5} to the output can be found by noting that it modulates the bias current and the fact that, due to the symmetry in the circuit, the drain of Q_2 will track that of Q_1. As a result, the last gain factor is given by

$$\left|\frac{V_{no}}{V_{n5}}\right| = \frac{g_{m5}}{2g_{m3}} \tag{4.93}$$

Since this last gain factor is relatively small compared to the others, it will be ignored from here on.

Using the gain factors just shown, the output noise value is seen to be given by

$$V_{no}^2(f) = 2(g_{m1}R_o)^2V_{n1}^2(f) + 2(g_{m3}R_o)^2V_{n3}^2(f) \tag{4.94}$$

This output noise value can be related back to an equivalent input noise value, $V_{neq}(f)$, by dividing it by the gain, $g_{m1}R_o$, which results in

$$V_{neq}^2(f) = 2V_{n1}^2(f) + 2V_{n3}^2(f)\left(\frac{g_{m3}}{g_{m1}}\right)^2 \tag{4.95}$$

Thus, for the white noise portion of $V_{n1}(f)$ and $V_{n3}(f)$, we make the substitution

$$V_{ni}^2(f) = 4kT\left(\frac{2}{3}\right)\left(\frac{1}{g_{mi}}\right) \tag{4.96}$$

resulting in

$$V_{neq}^2(f) = \left(\frac{16}{3}\right)kT\left(\frac{1}{g_{m1}}\right) + \left(\frac{16}{3}\right)kT\left(\frac{g_{m3}}{g_{m1}}\right)^2\left(\frac{1}{g_{m3}}\right) \tag{4.97}$$

Assuming g_{m3}/g_{m1} is not far from unity, we see here that the two pairs of transistors contribute an approximately equal amount of noise, and this noise is inversely proportional to the transconductance of g_{m1}. *In other words,* g_{m1} *should be made as large as possible to minimize thermal noise contribution.*

However, to look at the effects of 1/f, or flicker, noise, which normally greatly dominates at low frequencies, we make the following substitution into (4.95),

$$g_{mi} = \sqrt{2\mu_i C_{ox}\left(\frac{W}{L}\right)_i I_{Di}} \tag{4.98}$$

resulting in

$$V_{ni}^2(f) = 2V_{n1}^2(f) + 2V_{n3}^2(f)\left[\frac{(W/L)_3\mu_n}{(W/L)_1\mu_p}\right] \tag{4.99}$$

Now, letting each of the noise sources have a spectral density given by

$$V_{ni}^2(f) = \frac{K_i}{W_i L_i C_{ox} f} \tag{4.100}$$

we have [Bertails, 1979]

$$V_{ni}^2(f) = \frac{2}{C_{ox}f}\left[\frac{K_1}{W_1 L_1} + \left(\frac{\mu_n}{\mu_p}\right)\left(\frac{K_3 L_1}{W_1 L_3^2}\right)\right] \tag{4.101}$$

Recall that the first term in (4.101) is due to the p-channel input transistors, Q_1 and Q_2, and the second term is due to the n-channel loads, Q_3 and Q_4. We note some points for 1/f noise here:

1. For $L_1 = L_3$, the noise of the n-channel loads dominate since $\mu_n > \mu_p$ and typically n-channel transistors have larger 1/f noise than p-channel transistors (i.e., $K_3 > K_1$).
2. Taking L_3 longer greatly helps due to the inverse squared relationship in the second term of (4.101). This limits the signal swings somewhat, but it may be a reasonable trade-off where low noise is important.
3. The input noise is independent of W_3, and therefore we can make it large to maximize signal swing at the output.
4. Taking W_1 wider also helps to minimize 1/f noise. (Recall that it helps white noise, as well.)
5. Taking L_1 longer increases the noise because the second term in (4.101) is dominant. Specifically, this decreases the input-referred noise of the p-channel drive transistors, which are not the dominant noise sources, but it also increases

the input-referred noise of the n-channel load transistors, which are the dominant noise sources!

Finally, we can integrate (4.101) from f_1 to f_2 to find the equivalent input noise value given by

$$V_{ni(rms)}^2 = 2\left[\frac{a_p}{W_1 L_1} + a_n\left(\frac{\mu_n}{\mu_p}\right)\left(\frac{L_1}{W_1 L_3^2}\right)\right] \tag{4.102}$$

where

$$a_i = \frac{K_i}{C_{ox}} \ln\left(\frac{f_2}{f_1}\right) \tag{4.103}$$

Fiber-Optic Preamp Example

The most popular means of detecting light from a fiber-optic cable is to use a transresistance amplifier, as shown in Fig. 4.18. The light from the fiber cable hits a photodetector, such as a reverse-biased diode. The light produces electron hole carriers in the depletion region of the reverse-biased diode, causing current to flow through the resistor, and therefore the output of the amplifier becomes positive. A popular choice for the first stage of the amplifier is to use a common-source amplifier with a resistor load. In low-noise applications, an active load would be too noisy. Assuming a CMOS transistor is used,[7] the preamp can be modelled as shown in Fig. 4.19. The photodetector is modelled as an input current source along with a parasitic capacitance, C_{in}. Also shown in Fig. 4.19 are the two major noise sources, namely the thermal current noise at the drain of Q_1 and the thermal noise from the feedback resistor, R_F. The 1/f noise of the transistor is ignored because it is assumed that the circuit is high speed, such that thermal noise dominates. The second (and perhaps subsequent) stage is modelled by an amplifier that has a gain of A_2. The noise due to this second stage is also ignored since the noise sources in the second stage are not amplified by the gain of the first stage.

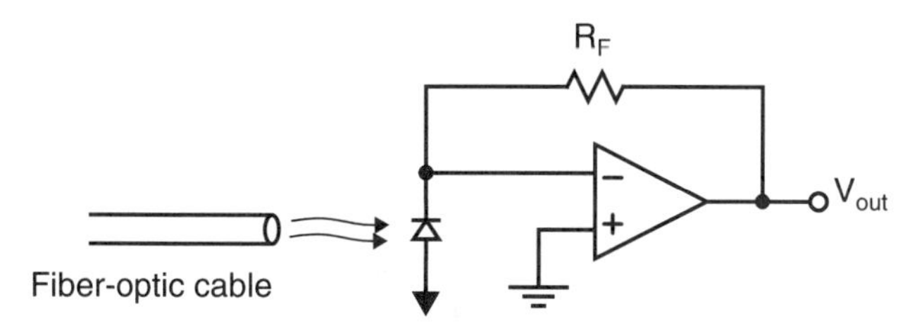

Fig. 4.18 A fiber-optic transresistance preamp.

7. Fiber-optic preamps are often realized using JFET transistors as well, but since their thermal noise model is identical to the noise model of the CMOS transistor, the analysis would be almost unchanged.

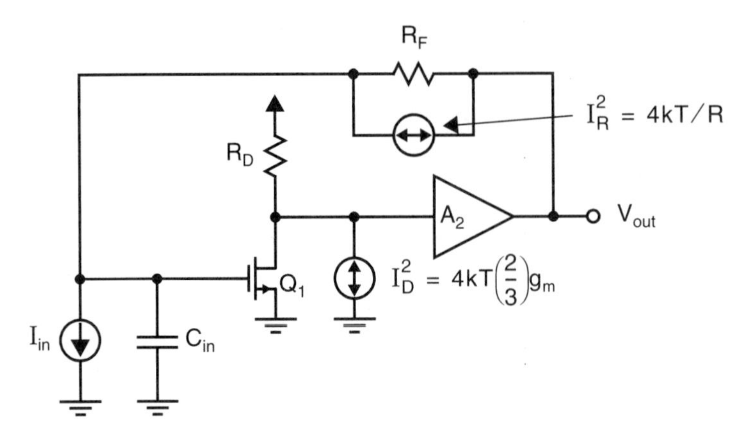

Fig. 4.19 A simplified model for a CMOS fiber-optic preamp.

A simplified small-signal model of this preamplifier, used for noise analysis, is shown in Fig. 4.20. The only parasitic capacitance considered in the transistor model is C_{gs}, since the gain of the first stage is only moderate, due to the resistor load, and therefore we can assume that C_{gd} and C_{db} can be ignored. The transfer function from i_{in} to the output is found by using nodal analysis to be

$$\frac{V_{out}(s)}{I_{in}(s)} = R_F\left(\frac{A_V}{1+A_V}\right)\frac{1}{1+s\left(\frac{R_F C_T}{1+A_V}\right)} \tag{4.104}$$

where $A_V = A_1 A_2$ is the total voltage gain of the preamp and $C_T = C_{gs} + C_{in}$. It is also found that this is the same transfer function from I_R^2 to the output. Thus, the noise current source I_R^2 can be replaced by an input current noise source having a spectral density of $4kT/R_F$. In a typical design, one would choose R_F as large as possible to limit this noise. However, this choice is constrained by the bandwidth requirements of passing the signal frequencies. From (4.104), we have the –3-dB frequency of the amplifier, given by

$$\omega_{-3\text{ dB}} = \frac{1+A_V}{R_F C_T} \tag{4.105}$$

For a given amplifier gain and detector capacitance, R_F is chosen using (4.105) to place the –3-dB frequency as small as possible without substantially attenuating the signals. This makes the dominant node for determining stability the input node. The bandwidth of the second amplifier must be substantially greater than that of the input node to guarantee stability. Unfortunately, this constraint greatly amplifies the thermal noise due to input transistor Q_1, as we will see next.

The gain from the noise source I_D^2 to the output is found by using nodal analysis to be

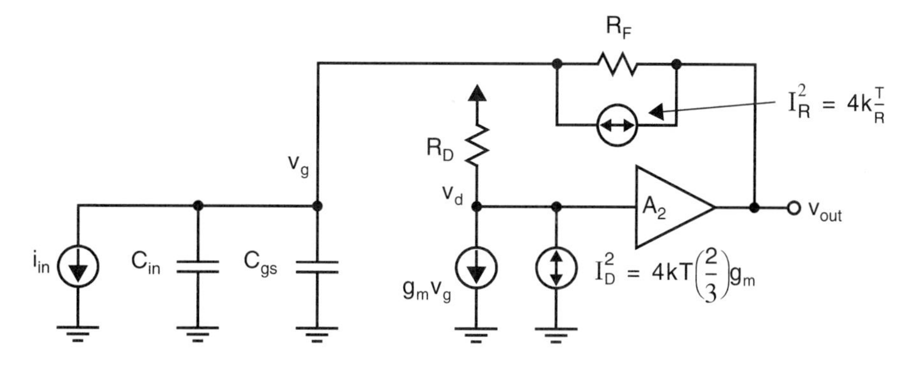

Fig. 4.20 A simplified small-signal model used for noise analysis.

$$\frac{V_{out}(s)}{I_D(s)} = \frac{1}{g_m}\frac{A_V}{1 + A_V}\frac{1 + sR_FC_T}{1 + s\left[\frac{R_FC_T}{(1 + A_V)}\right]} \tag{4.106}$$

At low frequencies, this gain is approximately given by $1/g_m$, whereas at high frequencies it is as much as $1 + A_V$ times greater. Thus, the high-frequency output noise due to Q_1 is much greater than the low-frequency noise due to Q_1. Furthermore, the only bandwidth limitation of this noise at high frequencies is due to the finite bandwidth of the second amplifier, A_2. From the discussion on compensation in Chapter 5, we know the second pole frequency of the amplifier must be almost four times greater than the closed-loop –3-dB frequency when lead compensation is not used.[8]

Continuing, in order to refer the noise due to Q_1 back to the input, we use (4.104) and (4.106) to obtain

$$I_{iD}^2(f) = I_D^2(f)\left|\frac{V_{out}(j\omega)}{I_D(jw)}\right|^2\left|\frac{I_{in}(j\omega)}{V_{out}(jw)}\right|^2 = I_D^2(f)\frac{1 + \omega^2(R_FC_T)^2}{(g_mR_F)^2} \tag{4.107}$$

We use

$$I_D^2(f) = 4kT\left(\frac{2}{3}\right)g_m \tag{4.108}$$

and we add to this the input noise source that models the noise of the feedback resistor to obtain the total input-referred noise, given by

$$I_i^2(f) = \frac{4kT}{R_F} + 4kT\left(\frac{2}{3}\right)\frac{1 + \omega^2(R_FC_T)^2}{g_mR_F^2} \tag{4.109}$$

Notice that the second term starts to quickly increase at a frequency of

8. Lead compensation could possibly be achieved by placing a capacitor in parallel with R_F. This is beyond the scope of this text.

$$\omega_z = \frac{1}{R_F C_T} = \frac{\omega_{-3\text{ dB}}}{1 + A_V} \tag{4.110}$$

which is a relatively low frequency.

Continuing, normally $2/(3g_m R_F) << 1$, and we have

$$\begin{aligned} I_i^2(f) &= \frac{4kT}{R_F}\left(1 + \frac{2}{3g_m R_F}\right) + \frac{8kT\omega^2 C_T^2}{3 \quad g_m} \\ &\cong \frac{4kT}{R_F} + \frac{8kT\omega^2 C_T^2}{3 \quad g_m} \end{aligned} \tag{4.111}$$

Using the facts that

$$g_m = \mu_n C_{ox} \frac{W}{L} V_{eff} \tag{4.112}$$

and that

$$C_{gs} = \frac{2}{3} C_{ox} WL \tag{4.113}$$

we can write (4.111) as

$$I_i^2(f) \cong \frac{4kT}{R_F} + \frac{16}{9} kT \frac{L^2}{\mu_n V_{eff}} \omega^2 \frac{(C_{gs} + C_{in})^2}{C_{gs}} \tag{4.114}$$

Normally, V_{eff} is taken as large as possible given power-supply-voltage and power-dissipation constraints. The only parameter left for the designer to choose is the width of the input transistor and, therefore, C_{gs}. It can be shown (by differentiating the second term of (4.114) with respect to C_{gs} and then setting the result to zero) that the second term of (4.114) is minimized by the choice $C_{gs} = C_{in}$, in which case we have

$$I_i^2(f) \cong 4kT\left(\frac{1}{R_F} + \frac{4}{9} \frac{L^2}{\mu_n V_{eff}} \omega^2 C_{in}\right) \tag{4.115}$$

4.5 REFERENCES

J. C. Bertails, "Low-Frequency Noise Considerations for MOS Amplifiers Design," *IEEE J. of Solid-State Circuits,* Vol. SC-14, No. 4, pp. 773–776, August, 1979.

A. Buchwald and K. Martin. *Integrated Fiber-Optic Receivers*. Kluwer Academic Publishers, Norwell, Massachusetts, 1995.

P. R. Gray and R. G. Meyer. *Analysis and Design of Analog Integrated Circuits.* John Wiley & Sons, New York, 1993.

J. B. Johnson, *Phys. Rev.*, Vol. 32, pp. 97–109, 1928.

E. J. Kennedy, *Operational Amplifier Circuits: Theory and Applications.* Holt, Rinehart, & Winston, New York, 1988.

A. Leon-Garcia, *Probability and Random Processes for Electrical Engineering*. Addison-Wesley, Reading, Massachusetts, 1989.

C. D. Motchenbacher and J. A. Connelly. *Low Noise Electronic System Design*. John Wiley & Sons, New York, 1993.

H. Nyquist, *Phys. Rev.*, Vol. 32, pp. 110–113, 1928.

W. Schottky, *Ann. Phys. (Leipzig)*, Vol. 57, pp. 541–567, 1918.

M. Steyaert, Z. Y. Chang, and W. Sansen. "Low-Noise Monolithic Amplifier Design: Bipolar versus CMOS," *Analog Integrated Circuits and Signal Processing 1*. Kluwer Academic Publishers, Boston, pp. 9–19, September 1991.

A. Van der Ziel, *Noise in Solid State Devices and Circuits*. John Wiley & Sons, New York, 1986.

4.6 PROBLEMS

Unless otherwise stated, assume dBm values are referenced to $50\ \Omega$.

4.1 If a signal is measured to have $V_{(rms)}$ volts, what is the difference in db if it is expressed in dBm referenced to $50\ \Omega$ as opposed to being referenced to $75\ \Omega$?

4.2 Consider the sum of two noise sources of values -20 dBm and -23 dBm. Find the total noise power in dBm for the cases in which the two noise sources are (*a*) uncorrelated, (*b*) $C = 0.3$, (*c*) $C = +1$, and (*d*) $C = -1$.

4.3 The output noise of a circuit is measured to be -40 dBm around 100 kHz when a resolution bandwidth of 30 Hz is used. What is the expected dBm measurement if a resolution bandwidth of 10 Hz is used? Find the root spectral density in $V/\sqrt{Hz}$.

4.4 At 0.1 Hz, a low-frequency measurement has a noise value of -60 dBm when a resolution bandwidth of 1 mHz is used. Assuming 1/f noise dominates, what would be the expected noise value (in dBm) over the band from 1 mHz to 1 Hz?

4.5 Show that, when two resistors of values R_1 and R_2 are in series, their noise model is the same as a single resistance of value $R_1 + R_2$. Repeat the problem for parallel resistances.

4.6 Sketch the spectral density of voltage noise across a 100-pF capacitor when it is in parallel with a $1\text{-k}\Omega$ resistor. Make another sketch for the same capacitor but with a $1\text{-M}\Omega$ resistance in parallel. What can you say about the area under the curves of the two sketches?

4.7 Consider the circuit shown in Fig. P4.7 on p. 218, where the opamp has a unity-gain frequency of 1 MHz and equivalent input voltage and current noise sources of $V_n(f) = 20\ nV/\sqrt{Hz}$ and $I_n(f) = 10\ pA/\sqrt{Hz}$, respectively. Estimate the total rms noise voltage at V_o.

4.8 Assuming an ideal noiseless opamp, find the noise level at V_x and V_o for the opamp circuit shown in Fig. P4.8. By what factor is the noise value at V_o larger (or smaller) than $kT/1\ nF$? How do you account for this increase (or decrease)? Also explain why the noise value at V_x is smaller than $kT/1\ nF$.

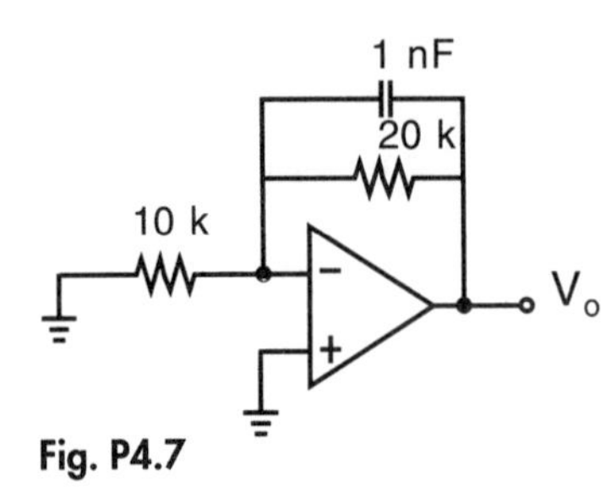

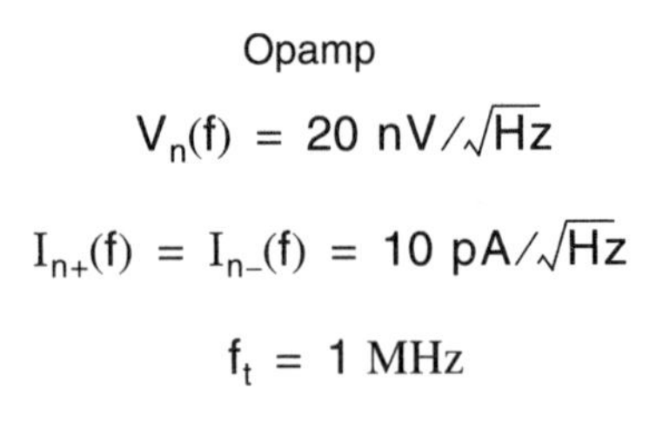

Fig. P4.7

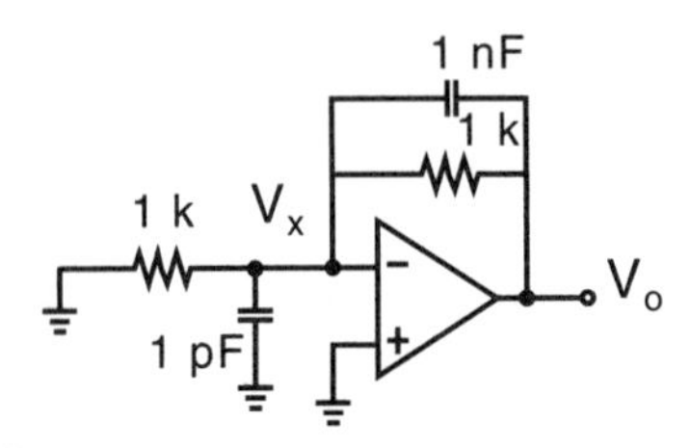

Fig. P4.8

4.9 Consider an inductor of value L and an arbitrary resistor in parallel, as shown in Fig. P4.9. Show that the current noise, $i_{no}(t)$, has a noise value given by

$$I_{no(rms)}^2 = \frac{kT}{L}$$

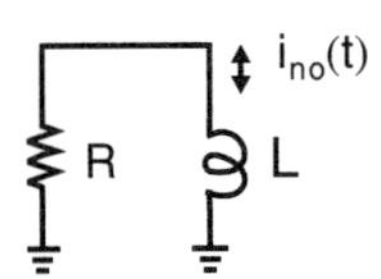

Fig. P4.9

4.10 The two circuits shown in Fig. P4.10 realize a first-order, low-pass filter. Assume the opamps are ideal and noiseless.

(*a*) Show that the two circuits have the same input-output transfer function.

(*b*) Estimate the total output noise for circuit I, using only dominant noise sources.

(*c*) Repeat (*b*) for circuit II.

4.11 Modify circuit I in Fig. P4.10 such that the new circuit has the same transfer function but uses an 80-pF capacitor instead of an 80-nF capacitor. If the opamp is ideal and noiseless, what is the new total output noise?

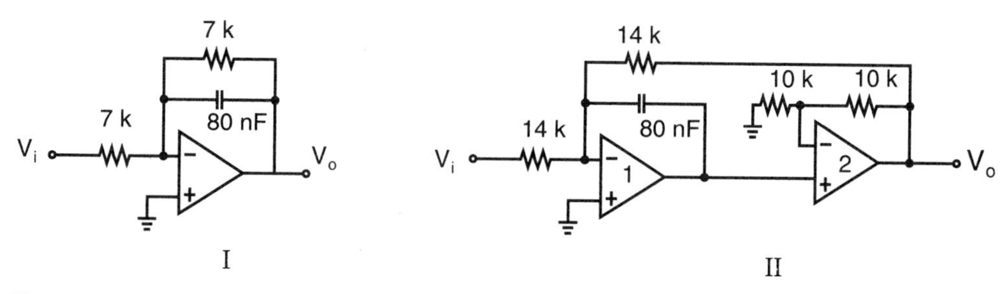

Fig. P4.10

4.12 Using the 1/f tangent principle, estimate the total noise above 0.1 Hz for the spectral density shown in Fig. 4.3.

4.13 Consider a bandpass amplifier that has equivalent input noise root spectral density and amplifier response, as shown in Fig. P4.13. Sketch the root spectral density for the output signal. Estimate the total output rms noise value by applying the 1/f tangent principle.

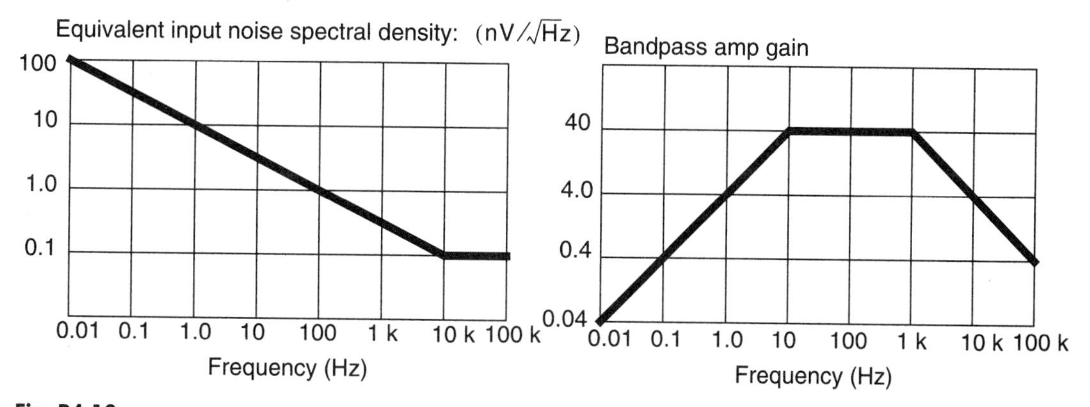

Fig. P4.13

4.14 Consider the noise root spectral density of the signal shown in Fig. P4.14. Find the total rms noise value using a graphical approach for 0.01 to ∞ Hz. Compare your result with that obtained when using the 1/f tangent principle.

4.15 We saw on page 193, Eq. (4.36), that the noise bandwidth of a first-order, low-pass filter is $(\pi/2)f_0$. Show that the noise bandwidth of a second-order, low-pass filter given by

$$A(s) = \frac{A_o}{\left(1 + \frac{s}{(2\pi f_0)}\right)^2}$$

is equal to $(\pi/4)f_o$.

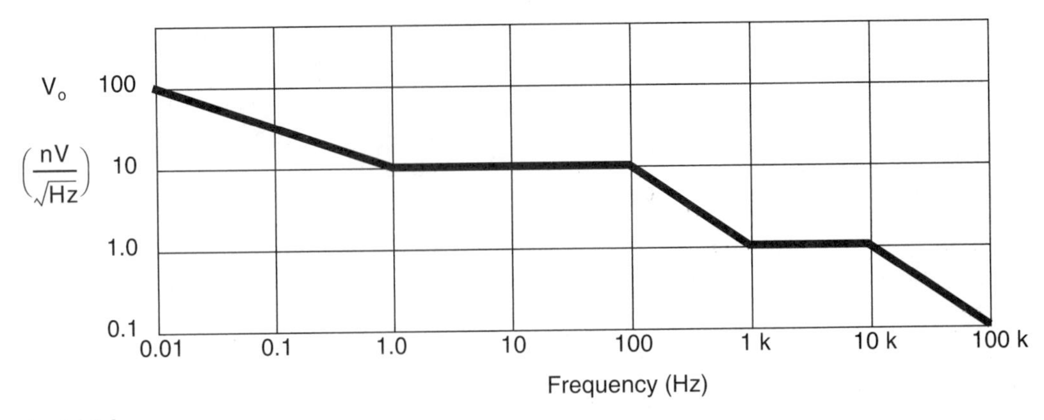

Fig. P4.14

4.16 Consider the two bipolar current mirrors shown in Fig. P4.16, where I_{in} is a 1-mA bias current plus a 100-μA(rms) signal current. Assuming that the base resistance for each transistor is $r_b = 330\ \Omega$ and dominates the output noise, estimate the resulting SNR (in dB) for the two current mirrors over a 50-MHz bandwidth (also assume the output noise is white).

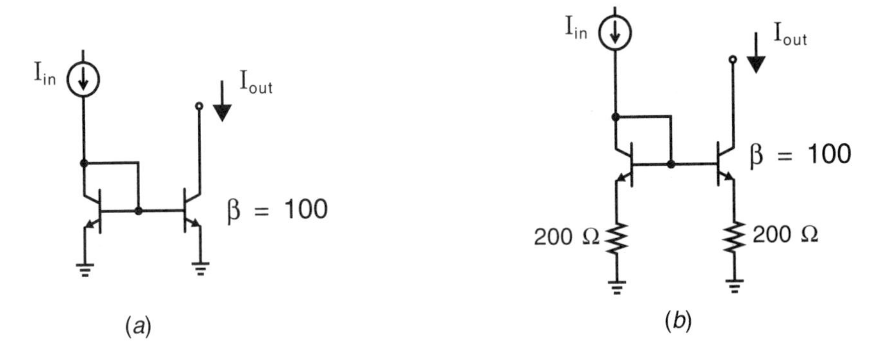

Fig. P4.16

4.17 Estimate the total output noise rms value for a low-pass filter, as shown in Fig. 4.15, when $C_f = 1\ \text{nF}$, $R_f = 16\ \text{k}$, $R_1 = 1.6\ \text{k}$, and $R_2 = 0$. Also, find the SNR for an input signal equal to 100 mV rms. Assume that the noise voltage of the opamp is given by $V_n(f) = 20\ \text{nV}/\sqrt{\text{Hz}}$, that both its noise currents are $I_n(f) = 0.6\ \text{pA}/\sqrt{\text{Hz}}$, and that its unity-gain frequency equals 2 MHz.

4.18 Consider the CMOS differential input stage shown in Fig. 4.17, where Q_5 supplies a bias current of $100\ \mu\text{A}$, resulting in $g_{m1} = g_{m2} = 1\ \text{mA/V}$ and $g_{m3} = g_{m4} = 0.5\ \text{mA/V}$. Find the equivalent input noise spectral density associated with thermal noise. If the bias current is doubled, how does the equivalent input noise density change?

CHAPTER

5 Basic Opamp Design and Compensation

This chapter describes the fundamental principles of basic opamp design. To illustrate many of these principles, the design of a traditional opamp—namely, the two-stage CMOS opamp—is used. This example illustrates compensation techniques needed to ensure stability in closed-loop amplifiers as well as to introduce a number of other important design techniques, such as ensuring zero systematic input-offset-voltage and process-insensitive lead compensation. Although the two-stage CMOS opamp is a classic circuit used in many integrated circuits, other more modern architectures have gained popularity recently. These more advanced architectures are described in Chapter 6. However, one should not overlook the potential of the two-stage opamp because fully-differential versions of this classic opamp are well-suited to low-voltage applications since they do not require cascode output stages. Fully-differential opamps are also described Chapter 6.

5.1 TWO-STAGE CMOS OPAMP

The two-stage circuit architecture has historically been the most popular approach for both bipolar and CMOS opamps, where a complementary process that has reasonable n-type and p-type devices is available. Although a CMOS version of this two-stage opamp is described here, a bipolar version is similar (but slightly more complicated). When properly designed, the two-stage opamp has a performance very close to more modern designs and is somewhat more suitable when resistive loads need to be driven.[1] Furthermore, it is an excellent example to illustrate many important design concepts that are also directly applicable to other more modern designs.

A block diagram of a typical two-stage CMOS opamp is shown in Fig. 5.1. "Two-stage" refers to the number of gain stages in the opamp. Figure 5.1 actually shows three stages—two gain stages and a unity-gain output stage. The output buffer is normally present only when resistive loads need to be driven. If the load is purely capacitive, then it is seldom included. The first gain stage is a differential-input single-ended output stage, often very similar to that shown in Fig. 3.19. The second gain stage is normally a common-source gain stage that has an active load, often very similar to that shown previously in Fig 3.4. Capacitor C_C is included to ensure stability

1. In a CMOS integrated circuit, opamp loads are often, but not always, purely capacitive.

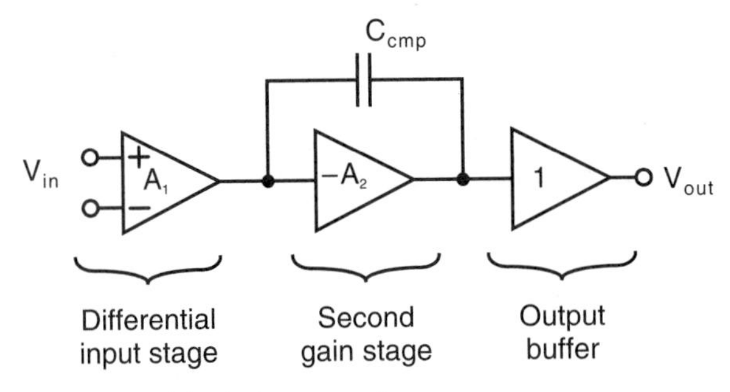

Fig. 5.1 A block diagram of a two-stage opamp.

when the opamp is used with feedback. Because C_C is between the input and the output of the high-gain second stage, it is often called a Miller capacitance since its effective capacitive load on the first stage is larger than its physical value.

An example of a practical CMOS version of the two-stage opamp is shown in Fig. 5.2. This example is used to illustrate many of the important design principles when realizing the two-stage amplifier.

It should be noted that the first stage has a p-channel differential input pair with an n-channel current-mirror active load. This is a complementary differential gain stage to that shown previously in Fig. 3.19. The trade-offs between having p-channel input transistors versus this stage and the alternative stage of Fig. 3.19 will be discussed later in this section. Also, the numbers next to the transistors represent reasonable transistor widths for a 1-μm process. Reasonable sizes for the lengths of the transistor might be somewhere between 1.5 and 2 times the minimum transistor length

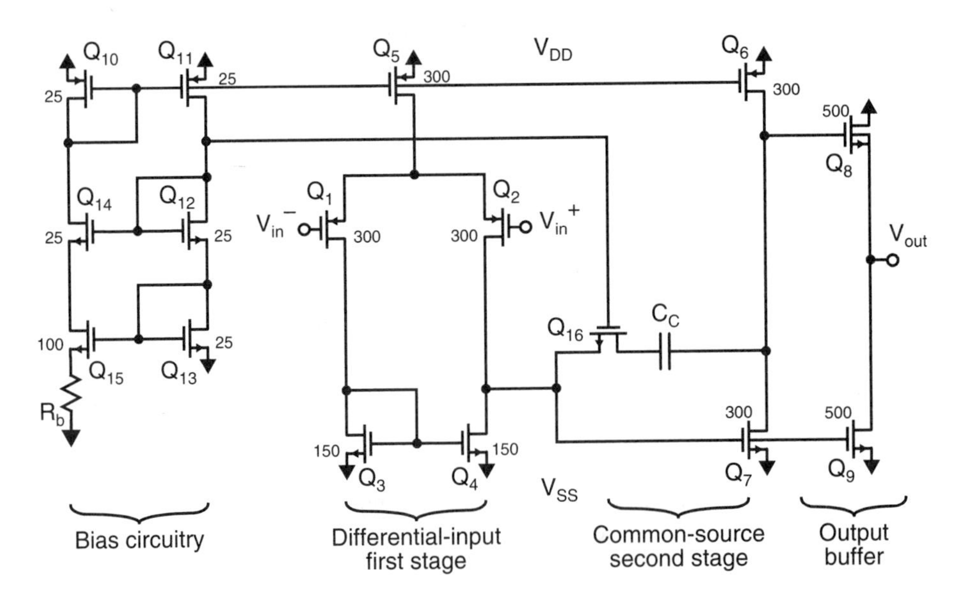

Fig. 5.2 A CMOS realization of a two-stage amplifier. All transistor lengths are 1.6 μm.

of a particular technology, whereas digital logic typically makes use of the minimum transistor length.

Opamp Gain

First we discuss the overall gain of the opamp. For low-frequency applications, this gain is one of the most critical parameters of an opamp.

The gain of the first stage has already been derived, resulting in (3.70), and is repeated here for convenience:

$$A_{v1} = g_{m1}(r_{ds2} \| r_{ds4}) \tag{5.1}$$

Recall from Chapter 1 that g_{m1} is given by

$$g_{m1} = \sqrt{2\mu_p C_{ox}\left(\frac{W}{L}\right)_1 I_{D1}} = \sqrt{2\mu_p C_{ox}\left(\frac{W}{L}\right)_1 \frac{I_{bias}}{2}} \tag{5.2}$$

Also, an approximation to the finite output impedance, r_{dsi}, of transistor Q_i is given by

$$r_{dsi} \approx \alpha \frac{L_i}{I_{Di}} \sqrt{V_{DGi} + V_{ti}} \tag{5.3}$$

where α is a technology-dependent parameter of around $5 \times 10^6\ \sqrt{V}/m$. This equation is at best approximate and ignores all short-channel effects that become more important for shorter channel lengths in modern technologies.

The second gain stage is simply a common-source gain stage with a p-channel active load, Q_6. Its gain is given by $(-g_{m7})(r_{ds6} \| r_{ds7})$. Thus, we have

$$A_{v2} = -g_{m7}(r_{ds6} \| r_{ds7}) \tag{5.4}$$

The third stage is a common-drain buffer stage. This stage is often called a *source follower*, because the source voltage follows the gate voltage of Q_8, except for a level shift. As shown, the substrate of Q_8 is at the same voltage as the source of Q_8. This connection is only possible when Q_8 is realized in a well that is isolated from the bulk. Thus, for this example opamp, one would require a p-well or twin-well process to realize this output stage. Tying the substrate of Q_8 to its source eliminates gain degradations due to the body effect. This connection also results in a smaller dc voltage drop from the gate to the source of Q_8, which is a major limitation on the maximum positive output voltage. As Chapter 3 shows, the gain of this source-follower stage is given by

$$A_{v3} \cong \frac{g_{m8}}{G_L + g_{m8} + g_{ds8} + g_{ds9}} \tag{5.5}$$

where G_L is the load conductance being driven by the buffer stage. When it is not possible to tie the substrate of Q_8 to its source, as is the case when an n-well process is used (currently, a popular process), then the gain of the buffer stage is given by

$$A_{v3} \cong \frac{g_{m8}}{G_L + g_{m8} + g_{s8} + g_{ds8} + g_{ds9}} \tag{5.6}$$

where g_s is a body-effect conductance and is given by

$$g_s = \frac{g_m \gamma}{2\sqrt{V_{SB} + 2\phi_F}} \tag{5.7}$$

Voltage V_{SB} is the source-to-substrate (or bulk) voltage, γ is the body-effect constant (around $0.5\ V^{1/2}$), and $2\phi_F$ is twice the difference between the Fermi level in the bulk and the Fermi level of intrinsic silicon (around 0.7 V). Thus, g_s is around $g_m/5$.

EXAMPLE 5.1

Find the gain of the opamp shown in Fig. 5.2 using the following assumptions. Assume the bias current of the input differential pair is given by $I_{D5} = 100\ \mu A$, the power supplies are $V_{DD} = -V_{SS} = 2.5\ V$, and $R_L = 10\ k\Omega$. Also assume the following process parameters: $\mu_n C_{ox} = 3\mu_p C_{ox} = 96\ \mu A/V^2$, $\gamma = 0.5\ V^{1/2}$, $\phi_F = 0.35\ V$, $\alpha = 5 \times 10^6\ \sqrt{V}/m$, and $V_{tn} = -V_{tp} = 0.8\ V$. To estimate output impedances, assume the drain-to-gate voltages of the first-stage transistors equal $0.5\ V$, whereas the transistors in the second and third stages have $V_{DGi} = 1\ V$. Finally, do not assume the substrate of Q_8 is connected to its source—rather, assume it is connected to the negative power supply.

Solution

First the bias currents are calculated. Since $I_{D5} = 100\ \mu A$, we have

$$I_{D1} = I_{D2} = I_{D3} = I_{D4} = I_{D5}/2 = 50\ \mu A$$

$$I_{D6} = I_{D7} = (W_6/W_5)I_{D5} = 100\ \mu A$$

and

$$I_{D8} = I_{D9} = (W_9/W_7)I_{D7} = 167\ \mu A$$

We can now calculate the transconductances of Q_1, Q_2, Q_7, and Q_8. Using (5.2), we have $g_{m1} = g_{m2} = 0.775\ mA/V$, $g_{m7} = 1.90\ mA/V$, and $g_{m8} = 3.16\ mA/V$.

Next, we need to estimate the output impedances of the transistors. To find these impedances, we use (5.3) together with the given approximations that the first-stage transistors have $V_{DGi} = 0.5\ V$ whereas the transistors in the other two stages have drain-to-gate voltages of 1 V. It should be mentioned here that these very rough drain-gate voltage approximations are reasonable since (5.3) is only of very moderate accuracy, perhaps 50 percent at best. These values for the transistor output impedances can be corrected later once a SPICE analysis is run. With the use of (5.3), we find that all of the transistors in the first stage have output impedances of

$$r_{ds1} = r_{ds2} = r_{ds3} = r_{ds4} = 182 \text{ k}\Omega$$

The output impedances of transistors in the second stage are

$$r_{ds6} = r_{ds7} = 107 \text{ k}\Omega$$

whereas the output impedances of transistors in the output stage are

$$r_{ds8} = r_{ds9} = 64 \text{ k}\Omega$$

The last parameter we need to calculate is the body-effect conductance of Q_8. Using (5.7) and assuming V_{out} is at ground results in $V_{SB8} = 2.5$ V, and thus we have $g_{s8} = 0.44$ mA/V.

Finally, using (5.1), we calculate that

$$A_{v1} = g_{m1}(r_{ds2} \| r_{ds4}) = 70.2 \text{ V/V}$$

Using (5.4),

$$A_{v2} = -g_{m7}(r_{ds6} \| r_{ds7}) = -102 \text{ V/V}$$

And using (5.6),

$$A_{v3} \cong \frac{g_{m8}}{G_L + g_{m8} + g_{s8} + g_{ds8} + g_{ds9}} = 0.85 \text{ V/V}$$

Thus, the total gain is equal to $A_{v1}A_{v2}A_{v3} = -6{,}090$ V/V. Once again, it should be mentioned here that this result is a rough approximation and should be verified using SPICE. The benefit of performing the hand calculations is to see how the gain is affected by different design parameters.

Frequency Response

We now wish to investigate the frequency response of the two-stage opamp at frequencies where the compensation capacitor, C_C, has caused the magnitude of the gain to begin to decrease, but still at frequencies well below the unity-gain frequency of the opamp. This corresponds to midband frequencies for many applications. This allows us to make a couple of simplifying assumptions. First, we will ignore all capacitors except the compensation capacitor, C_C, which normally dominates at all frequencies except around the unity-gain frequency of the opamp.[2] Second, we also assume that transistor Q_{16} is not present. This transistor operates as a resistor, which is included to achieve lead compensation and it has an effect only around the unity-gain frequency of the opamp, as we will see when we discuss compensation in Section 5.2. The simplified circuit used for analysis is shown in Fig. 5.3. It is worth mentioning that this simplified circuit is often used during system-level simulations when speed is more important than accuracy, but aspects such as slew-rate limiting (see next subsection) should be included in the simulation.

2. Recall that the unity-gain frequency of an opamp is that frequency where the magnitude of the open-loop opamp gain has decreased to one.

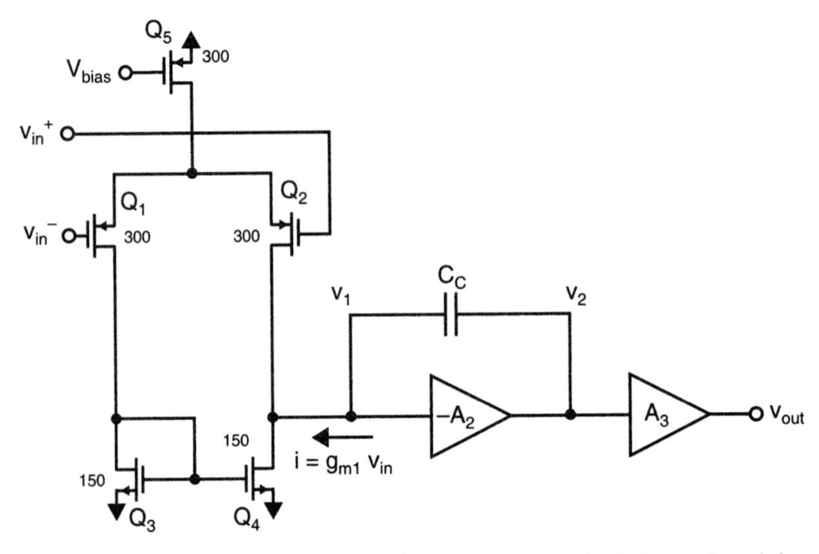

Fig. 5.3 A simplified model for the opamp used to find the midband frequency response.

The second stage introduces primarily a capacitive load on the first stage due to the compensation capacitor, C_C. Using Miller's Theorem [Sedra, 1991], one can show that the equivalent load capacitance, C_{eq}, at node v_1 is given by,

$$C_{eq} = C_C(1 + A_2) \approx C_C A_2 \tag{5.8}$$

The gain in the first stage can now be found using the small-signal model of Fig. 3.20, resulting in

$$A_1 = \frac{v_1}{v_{in}} = -g_{m1} Z_{out1} \tag{5.9}$$

where

$$Z_{out1} = r_{ds2} \parallel r_{ds4} \parallel \frac{1}{sC_{eq}} \tag{5.10}$$

For midband frequencies, the impedance of C_{eq} dominates, and we can write

$$Z_{out1} \cong \frac{1}{sC_{eq}} \cong \frac{1}{sC_C A_2} \tag{5.11}$$

For the overall gain, we have

$$A_v(s) \equiv \frac{v_{out}}{v_{in}} = A_3 A_2 A_1 \cong A_3 A_2 \frac{g_{m1}}{sC_C A_2} \tag{5.12}$$

using (5.9) and (5.11). If we further assume that $A_3 \cong 1$, then the overall gain, given in (5.12), simplifies to

$$A_v(s) = \frac{g_{m1}}{sC_C} \tag{5.13}$$

This simple equation can be used to find the approximate[3] unity-gain frequency. Specifically, to find the unity-gain frequency, ω_{ta}, we set $|A_v(j\omega_{ta})| = 1$, and solve for ω_{ta}. Performing such a procedure with (5.13), we obtain the following relationship:

$$\omega_{ta} = \frac{g_{m1}}{C_C} \tag{5.14}$$

Note here that the unity-gain frequency is directly proportional to g_{m1} and inversely proportional to C_C.

EXAMPLE 5.2

Using the same parameters as in Example 5.1, and assuming $C_C = 5$ pF, what is the unity-gain frequency in Hz?

Solution

Using $g_{m1} = 0.771$ mA/V and (5.14), we find that

$$\omega_{ta} = \frac{0.771 \times 10^{-3}}{5 \times 10^{-12}} = 154.2 \text{ Mrad/s}$$

Thus, we find that $f_{ta} = \omega_{ta}/(2\pi) = 24.5$ MHz.

Slew Rate

Another important high-frequency parameter of an opamp is its slew rate. The *slew rate* is the maximum rate at which the output changes when input signals are large. When the opamp of Fig. 5.2 is limited by its slew rate because a large input signal is present, all of the bias current of Q_5 goes into either Q_1 or Q_2, depending on whether v_{in} is negative or positive. When v_{in} is a large positive voltage, the bias current, I_{D5}, goes entirely through Q_1 and also goes into the current-mirror pair, Q_3, Q_4. Thus, the current coming out of the compensation capacitor, C_C, (i.e., I_{D4}) is simply equal to I_{D5} since Q_2 is off. When v_{in} is a large negative voltage, the current-mirror pair Q_3 and Q_4 is shut off because Q_1 is off, and now the bias current, I_{D5}, goes directly into C_C. In either case, the maximum current entering or leaving C_C is simply the total bias current, I_{D5}.

Defining the slew rate, SR, to be the maximum rate that v_2 can change, and recalling that $v_{out} \cong v_2$, we have

3. It is approximate because it is assumed that $A_3 \cong 1$ and that the higher-frequency poles are being ignored. Both of these effects will cause the unity-gain frequency to be slightly smaller than that calculated.

$$SR \equiv \left.\frac{dV_{out}}{dt}\right|_{max} = \frac{I_{C_C}|_{max}}{C_C} = \frac{I_{D5}}{C_C} \tag{5.15}$$

where we used the charge equation $q = CV$, which leads to $I = dq/dt = C(dV/dt)$. Since $I_{D5} = 2I_{D1}$, we can also write

$$SR = \frac{2I_{D1}}{C_C} \tag{5.16}$$

where I_{D1} is the original bias current of Q_1 with no signals present. Also, using (5.14), we have $C_C = g_{m1}/\omega_{ta}$, and substituting this into (5.16), we have

$$SR = \frac{2I_{D1}\omega_{ta}}{g_{m1}} \tag{5.17}$$

Recalling that

$$g_{m1} = \sqrt{2\mu_p C_{ox}\left(\frac{W}{L}\right)_1 I_{D1}} \tag{5.18}$$

we finally have another relationship for the slew-rate value.

$$SR = \frac{2I_{D1}}{\sqrt{2\mu_p C_{ox}(W/L)_1 I_{D1}}}\omega_{ta} = V_{eff1}\omega_{ta} \tag{5.19}$$

where

$$V_{eff1} = \sqrt{\frac{2I_{D1}}{\mu_p C_{ox}(W/L)_1}} \tag{5.20}$$

from Chapter 1. Normally one has little control over ω_{ta}, assuming a given maximum power dissipation is allowed. It is usually constrained to be less than two-thirds of the second-pole frequency, as we will see in Section 5.2, on compensation. As a result, *the only ways of improving the slew rate for the two-stage CMOS opamp is to increase* V_{eff1}, ω_{ta}, or both. As we will soon see, obtaining a high slew rate and unity-gain frequency are two of the major reasons for choosing p-channel input transistors rather than n-channel input transistors. It should be mentioned here that increasing V_{eff1} lowers the transconductance of the input stage. Although increasing V_{eff1} helps to minimize distortion, a lower transconductance in the first stage decreases the dc gain and increases the equivalent input thermal noise (see Chapter 4, Section 4.4).

EXAMPLE 5.3

Using the same parameters as in Example 5.1, and assuming $C_C = 5\text{ pF}$, what is the slew rate? What circuit changes could be made to double the slew rate but keep ω_{ta} and bias currents unchanged?

Solution

From (5.16), we have

$$\mathrm{SR} = \frac{2 \times 50\ \mu\mathrm{A}}{5\ \mathrm{pF}} = 20\ \mathrm{V}/\mu\mathrm{s} \tag{5.21}$$

To double the slew rate while maintaining the same bias currents, C_C should be set to 2.5 pF. To maintain the same unity-gain frequency, g_{m1} should be halved, which can be accomplished by decreasing the widths of Q_1 and Q_2 by 4 (i.e., one could let $W_1 = W_2 = 75\ \mu\mathrm{m}$).

Systematic Offset Voltage

When designing the two-stage CMOS opamp, if one is not careful, it is possible that the design will have an inherent (or systematic) input-offset voltage. Indeed, this was the case for many of the original designs used in production microcircuits. To see what is necessary to guarantee that no inherent input-offset voltage exists, consider the first two stages of the opamp, as shown in Fig. 5.4. To ensure that no systematic input-offset voltage exists, when the differential input voltage is zero (i.e, when $V_{in}^{+} = V_{in}^{-}$), the output voltage of the first stage, V_{GS7}, should be that which is required to make I_{D7} equal to its bias current, I_{D6}. Specifically, the value of V_{GS7} should be given by

$$V_{GS7} = \sqrt{\frac{2I_{D6}}{\mu_n C_{ox}(W/L)_7}} + V_{tn} \tag{5.22}$$

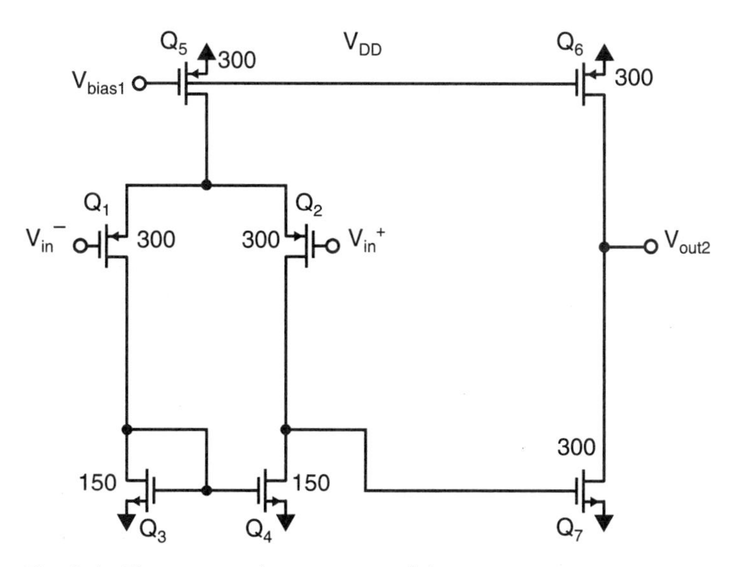

Fig. 5.4 The input and gain stages of the two-stage opamp.

When the differential input voltage is zero, the drain voltages of both Q_3 and Q_4 are equal by arguments of symmetry. Therefore, the output voltage of the first stage, V_{GS7}, is given by

$$V_{GS7} = V_{DS3} = V_{GS4} \tag{5.23}$$

This value is the voltage necessary to cause I_{D7} to be equal to I_{D6}, or, in other words, to satisfy (5.22). If this value is not achieved, then the output of the second stage (with Q_6, Q_7) would clip at either the negative or positive rail since this stage has such a high gain.[4] However, the gate-source voltage of Q_4 is given by

$$V_{GS4} = \sqrt{\frac{2I_{D4}}{\mu_n C_{ox}(W/L)_4}} + V_{tn} \tag{5.24}$$

so equating (5.22) and (5.24) to satisfy (5.23) results in

$$\sqrt{\frac{2I_{D4}}{\mu_n C_{ox}(W/L)_4}} = \sqrt{\frac{2I_{D6}}{\mu_n C_{ox}(W/L)_7}} \tag{5.25}$$

or, equivalently,

$$\frac{I_{D4}}{(W/L)_4} = \frac{I_{D6}}{(W/L)_7} \tag{5.26}$$

This equality, when the current density of Q_4 is equal to the current density of Q_7, guarantees that they both have the same effective gate-source voltages.

Since

$$\frac{I_{D6}}{I_{D4}} = \frac{I_{D6}}{I_{D5}/2} = \frac{(W/L)_6}{(W/L)_5/2} \tag{5.27}$$

we see that the necessary condition to ensure that no input-offset voltage is present is

$$\frac{(W/L)_7}{(W/L)_4} = 2\frac{(W/L)_6}{(W/L)_5} \tag{5.28}$$

which is satisfied for the sizes shown. Note that this analysis ignores the voltage drop of the level-shifter output stage (Q_8, Q_9) and any mismatches between the output impedance of p-channel and n-channel transistors. Fortunately, these effects cause only minor offset voltages, and by satisfying (5.28), offset voltages on the order of 5 mV or less can be obtained.

EXAMPLE 5.4

Consider the opamp of Fig. 5.2, where Q_3 and Q_4 are each changed to widths of 120 μm, and we want the output stage to have a bias current of 150 μA.

4. When the opamp is incorrectly designed to have an inherent input-offset voltage, feedback will cause the differential input voltage of the first stage to be equal to a nonzero voltage needed to ensure that the second stage does not clip.

Find the new sizes of Q_6 and Q_7, such that there is no systematic offset voltage.

Solution

Since I_{D6} determines the bias current of the output stage, and since it should have 50 percent more current than I_{D5}, its width should be 50 percent greater than W_5, resulting in

$$W_6 = 450\ \mu m \tag{5.29}$$

For Q_7, we use (5.28), which leads to

$$W_7 = \frac{2(450)(120)}{300} = 360\ \mu m \tag{5.30}$$

---

n-Channel or p-Channel Input Stage

The two-stage opamp discussed above has p-channel input transistors. It is also possible to realize a complementary opamp where the first stage has an n-channel differential pair and the second stage is a common-source amplifier having a p-channel input drive transistor. The choice of which configuration to use depends on a number of trade-offs that are discussed here.

First, the overall dc gain is largely unaffected by the choice since both designs have one stage with one or more n-channel driving transistors, and one stage with one or more p-channel driving transistors.

For a given power dissipation, and therefore bias current, having a p-channel input-pair stage maximizes the slew rate. This result is seen from (5.19) since p-channel input transistors for the first stage have a larger V_{eff} than would be the case for n-channel input transistors (assuming similar maximum widths have been chosen to maximize the gain). This slew-rate improvement can be one of the most important considerations, and thus most knowledgeable designers choose a p-channel input for the first stage.

Having a p-channel input first stage implies that the second stage has an n-channel input drive transistor. This arrangement maximizes the transconductance of the drive transistor of the second stage, which is critical when high-frequency operation is important. As we will see in the next section, the equivalent second pole, and therefore the unity-gain frequency as well, are both proportional to the transconductance of the second stage.

Another consideration is whether a p-channel or n-channel source-follower output stage is desired. Typically, an n-channel source follower is preferable because this will have less of a voltage drop. Also, since an n-channel transistor has a higher transconductance, the effect on the equivalent second pole due to its load capacitance is minimized, which is another important consideration. Finally, there is less degradation of the gain when small load resistances are being driven. The one disadvantage of having an n-channel source follower is that, for n-well processes, it is not possible to connect the source to the substrate, thereby minimizing the voltage drop. Finally, note that, for opamps that drive purely capacitive loads, the buffer stage should not be included, and for this case, the output stage is clearly not a consideration.

Noise is another important consideration when choosing which input stage to use. Perhaps the major noise source of MOS opamps is due to 1/f noise caused by carriers randomly entering and leaving traps introduced by defects near the semiconductor surface. This 1/f noise source can be especially troublesome unless special circuit design techniques are used.[5] Typically, p-channel transistors have less 1/f noise than n-channel transistors since their majority carriers (holes) have less potential to be trapped in surface states. Thus, having a first-stage with p-channel inputs minimizes the output noise due to the 1/f noise. The same is not true when thermal noise is considered. When thermal noise is referred to the input of the opamp, it is minimized by using input transistors that have large transconductances (Chapter 4, Section 4.4), which unfortunately degrades the slew rate. However, when thermal noise is a major consideration, then a more modern architecture, such as a folded-cascode opamp (discussed in Chapter 6), is normally used.

In summary, when using a two-stage opamp, a p-channel input transistor for the first stage is almost always the best choice because it optimizes slew rate, unity-gain frequency, and minimizes 1/f noise, with the major disadvantage being an increase in wideband thermal noise.

5.2 FEEDBACK AND OPAMP COMPENSATION

This section discusses using opamps in closed-loop configurations and how to compensate an opamp to ensure that the closed-loop configuration is not only stable, but also has good settling characteristics. Although the two-stage opamp is used as an example, almost all the material discussed here applies to most other opamps as well.

Optimum compensation of opamps is typically considered to be one of the most difficult parts of the opamp design procedure. However, if the systematic approach taken here is used, then a straightforward procedure can be used that almost always results in a near-optimum compensation network.

Before discussing compensation, some properties of feedback and closed-loop amplifiers will first be reviewed.

First-Order Model of Closed-Loop Amplifier

A simple first-order model for the transfer function of a dominant-pole compensated opamp, $A(s)$, is given by

$$A(s) = \frac{A_0}{(1 + s/\omega_{p1})} \tag{5.31}$$

5. Some useful circuit techniques to reduce the effects of 1/f noise are correlated double sampling and chopper stabilization.

where A_0 is the dc gain of the opamp and ω_{p1} is the (real-axis) dominant pole. At this point, it should be mentioned that all the poles and zeros in this chapter occur on the real axis and are represented by the notations ω_p and ω_z, respectively.[6]

Recall that the definition of the unity-gain frequency of an opamp, ω_{ta}, is the frequency at which $|A(j\omega_{ta})| = 1$. We then have the following approximation, since $\omega_{ta} >> \omega_{p1}$:

$$|A(j\omega_{ta})| = 1 \cong \frac{A_0}{\omega_{ta}/\omega_{p1}} \tag{5.32}$$

Thus, we have the following important relationship for this first-order model:

$$\omega_{ta} \cong A_0\omega_{p1} \tag{5.33}$$

From here on, we will define ω_{ta} *to be exactly equal to* $A_0\omega_{p1}$, *which is approximately equal to the unity-gain frequency of the opamp (assuming a first-order model for the opamp).*[7] Substituting (5.33) into (5.31) for the case in which $\omega_{p1} << \omega << \omega_{ta}$, we have at midband frequencies

$$A(s) \cong \frac{\omega_{ta}}{s} \tag{5.34}$$

This approximate relationship is often used to analyze a closed-loop circuit for the effects of the opamp's finite bandwidth at midband frequencies.

An opamp with feedback can be modelled by the block diagram shown in Fig. 5.5 [Sedra, 1991]. Here, the feedforward amplifier, $A(s)$, models the open-loop response of the opamp. The feedback term, β, represents the feedback factor, which is assumed to be frequency independent; this is typically the case for amplifiers, but may not be the case for applications such as damped integrators. It can be shown using signal-flow graph analysis that the closed-loop gain, $A_{CL}(s)$, for this model is given by

$$A_{CL}(s) = \frac{A(s)}{1 + \beta A(s)} \tag{5.35}$$

At midband frequencies, the transfer function of the closed-loop amplifier may be found by substituting (5.34) into (5.35), resulting in

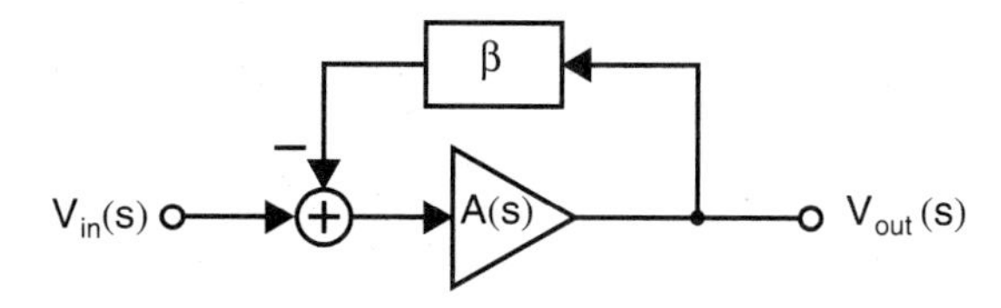

Fig. 5.5 A block diagram of a feedback circuit.

6. To be more precise, ω_p and ω_z are the inverses of the coefficients multiplying the s terms in the denominators and numerators, respectively. The actual poles and zeros are the negatives of these terms.

7. The unity-gain frequency of an optimally compensated opamp can be very different from ω_{ta} due to high-frequency poles and zeros that are ignored in the first-order model, especially when the closed-loop configuration has large gain. However, at midband frequencies (5.34) is still valid.

$$A_{CL}(s) \approx \frac{\omega_{ta}}{\beta\omega_{ta} + s} = \frac{1}{\beta}\frac{1}{(1 + s/\beta\omega_{ta})} \tag{5.36}$$

Thus, the closed-loop amplifier has a closed-loop gain at low frequencies approximately equal to $1/\beta$ [8] and it has a –3-dB frequency given by

$$\omega_{-3\text{ dB}} \cong \beta\omega_{ta} \tag{5.37}$$

It will be shown shortly that, for optimally compensated opamps, $\beta\omega_{ta} \cong \omega_t$, where ω_t is the unity-gain frequency of the open-loop transfer function. It will also be shown that, for optimally compensated amplifiers, ω_t is independent of the feedback factor, β, and dependent only on high-frequency poles and zeros. *Thus, the –3-dB frequency of the closed-loop gain of an optimally compensated amplifier is approximately equal to the unity-gain frequency of the loop gain, independent of the actual closed-loop gain.* However, the component sizes of the compensation network and ω_{ta} will be shown to depend on the desired closed-loop gain for an optimally compensated amplifier.

Linear Settling Time

The settling-time performance of integrated amplifiers is often an important design parameter. For example, in switched-capacitor circuits, the charge from one or more capacitors must be mostly transferred to a feedback capacitor within about half a clock period. This charge transfer is closely related to the opamp's step response. As a result, the *settling time* is defined to be the time it takes for an opamp to reach a specified percentage of its final value when a step input is applied.

This settling time consists of two distinct segments—linear and nonlinear settling time segments. The linear settling-time portion is due to the finite unity-gain frequency of the opamp, and thus it sets a minimum value for the overall settling time independent of the step size of the opamp's output. In contrast, the nonlinear settling time is due to slew-rate limiting, and thus this portion is strongly dependent on the output's step size. For example, for small step sizes in the output signal's level, the opamp may not reach a slew-rate limit at all, resulting in a nonlinear settling time of zero. Here, we discuss only the linear settling time by modelling the opamp as ideal but with a finite unity-gain frequency. Another simplification is to use the first-order opamp model, which has a 90-degree phase margin. Such a simplification results in a simple settling behavior that can be easily analyzed. Thus, the results here are used only to estimate the necessary unity-gain frequency for a circuit to settle within the linear settling-time segment. Simulations should be used in the latter parts of a design to determine more accurate settling-time estimates.

From (5.36), we recognize that the time constant of the closed-loop amplifier, τ, is given by

8. For inverting configurations, (5.35) should be modified to be $A_{CL}(s) = KA(s)/[1 + \beta A(s)]$, and the closed-loop gain at low frequencies is given by K/β. But these facts have no effect on the discussions to follow regarding optimum compensation because we are primarily interested in the loop gain, which is still given by $\beta A(s)$.

$$\tau = \frac{1}{\omega_{-3\text{ dB}}} = \frac{1}{\beta\omega_{ta}} \tag{5.38}$$

The important result here is that the –3-dB frequency determines the settling-time response for a step input. Recall that the transient response of any first-order circuit is given by

$$x(t) = x(\infty) - [x(\infty) - x(t)]e^{-t/\tau} \tag{5.39}$$

where τ is the time constant of the circuit. For the closed-loop amplifier, the step response is found using (5.39) to be given by

$$V_{out}(t) = V_{step}(1 - e^{-t/\tau}) \tag{5.40}$$

Here, V_{step} is the size of the voltage step. With this exponential relationship, the time required for a first-order circuit to settle to within a specified value can be found. For example, if 1 percent accuracy is required, then one must allow $e^{-t/\tau}$ to reach 0.01, which is achieved at a time of 4.6τ. For settling to within a 0.1 percent accuracy, the settling time needed becomes approximately 7τ. Also, note that just after the step input, the slope of the output will be at its maximum, given by

$$\left.\frac{d}{dt}V_{out}(t)\right|_{t=0} = \frac{V_{step}}{\tau} \tag{5.41}$$

If the slew rate of the opamp is larger than this value, no slew-rate limiting would occur.

EXAMPLE 5.5

One phase of a switched-capacitor circuit is shown in Fig. 5.6, where the input signal can be modelled as a voltage step. If 0.1 percent accuracy is needed in the linear settling-time portion corresponding to $0.1\ \mu s$, find the required unity-gain frequency in terms of the capacitance values, C_1 and C_2. For $C_2 = 10C_1$, what is the necessary unity-gain frequency of the opamp? What unity-gain frequency is needed in the case in which $C_2 = 0.2C_1$?

Solution

We first note that a capacitive feedback network is used rather than a resistive one. A difficulty with this network in a nonswitched circuit is that no bias current

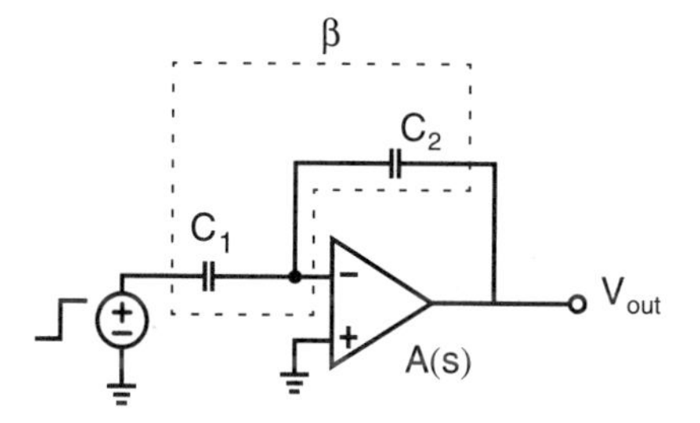

Fig. 5.6 One phase of a switched-capacitor circuit.

flows into the negative opamp input terminal. However, in a switched circuit using a CMOS opamp, this connection shown occurs only for a short time and does not cause any problems. Second, note that the feedback factor is simply a constant value since it is determined by a capacitive divider consisting of C_1 and C_2.

The feedback factor, β, is given by

$$\beta = \frac{1/(sC_1)}{1/(sC_1) + 1/(sC_2)} = \frac{C_2}{C_1 + C_2} \tag{5.42}$$

Now for 7τ settling within 0.1 μs, we see that τ must be less than 14.2 ns, and, since $\tau = 1/(\beta\omega_{ta})$, we see that

$$\omega_{ta} \geq \left(\frac{C_1 + C_2}{C_2}\right)\left(\frac{1}{14.2\text{ ns}}\right) \tag{5.43}$$

For the case in which $C_2 = 10C_1$, a unity-gain frequency of 12.3 MHz is needed, whereas in the case of $C_2 = 0.2C_1$, f_{ta} should be larger than 66.8 MHz.

Opamp Compensation

When one is compensating an opamp, the first-order model given by (5.31) is insufficient. It ignores high-frequency poles and zeros, which cause possible instability, even though they may be at frequencies greater than ω_t. It is important to *accurately* model the open-loop transfer function at higher frequencies where the loop-gain is approximately unity. Fortunately, when all poles and zeros are on the real axis, they can be modelled reasonably well by a single additional pole. Specifically, we can model $A(s)$ by

$$A(s) = \frac{A_0}{(1 + s/\omega_{p1})(1 + s/\omega_{eq})} \tag{5.44}$$

where ω_{p1} is the first dominant-pole frequency and ω_{eq} is the pole frequency that models higher-frequency poles. The relationship to approximate ω_{eq} given a set of real-axis poles, ω_{pi}, and zeros, ω_{zi}, is given by

$$\frac{1}{\omega_{eq}} \cong \sum_{i=2}^{m} \frac{1}{\omega_{pi}} - \sum_{i=1}^{n} \frac{1}{\omega_{zi}} \tag{5.45}$$

It should be noted here that the approximation in (5.45) is different than that given in [Sedra, 1991] since we are mostly interested in the phase shifts, rather than the attenuation, due to higher-frequency poles and zeros. In practice, ω_{eq} is found from simulation as the frequency at which the transfer function has a –135° phase shift (–90° due to the dominant pole and another –45° due to the higher-frequency poles and zeros).

At frequencies much greater than the dominant pole frequency, $\omega >> \omega_{p1}$, we see that $1 + j\omega/\omega_{p1} \cong j\omega/\omega_{p1}$, and so (5.44) can be accurately approximated by

$$A(s) \cong \frac{\omega_{ta}}{s(1 + s/\omega_{eq})} \tag{5.46}$$

Note that this approximation result is especially valid at the unity-gain frequency of the loop (which we are presently interested in) since, from (5.33), $\omega_t/\omega_{p1} \cong (1/\beta)A_0$, which is almost certainly much greater than 1.

When the opamp is included in a circuit that has a feedback factor, β, then, at frequencies around ω_{ta}, the loop gain, $LG(s)$, is given by

$$LG(s) = \beta A(s) = \frac{\beta\omega_{ta}}{s(1 + s/\omega_{eq})} \tag{5.47}$$

The unity-gain frequency, ω_t, of the loop gain, $LG(s)$, can now be found by setting the magnitude of (5.47) equal to unity after substituting $s = j\omega_t$. Once this is done and the equation is rearranged, one can write,

$$\beta\frac{\omega_{ta}}{\omega_{eq}} = \frac{\omega_t}{\omega_{eq}}\sqrt{1 + \left(\frac{\omega_t}{\omega_{eq}}\right)^2} \tag{5.48}$$

This equation will be used shortly to relate a specified phase margin to the Q factor of the closed-loop configuration.

Note also that from (5.48) we have

$$\omega_{ta} = \frac{\omega_t}{\beta}\sqrt{1 + \left(\frac{\omega_t}{\omega_{eq}}\right)^2} \cong \frac{\omega_t}{\beta} \tag{5.49}$$

for the special case in which the unity-gain frequency is much less than the equivalent nondominant pole frequency (i.e., $\omega_t << \omega_{eq}$).

The phase margin, PM, is an often-used measure for how far an opamp with feedback is from becoming unstable. PM is defined as the difference between the loop-gain phase shift and –180°. From (5.47), the phase shift, $\angle LG(j\omega)$, is found as

$$\angle LG(j\omega) = -90° - \tan^{-1}(\omega/\omega_{eq}) \tag{5.50}$$

This equation implies that at the unity-gain frequency, $s = j\omega_t$, we have

$$PM = \angle LG(j\omega_t) - (-180°) = 90° - \tan^{-1}(\omega_t/\omega_{eq}) \tag{5.51}$$

and, therefore,

$$\omega_t/\omega_{eq} = \tan(90° - PM) \tag{5.52}$$

which implies that

$$\omega_t = \tan(90° - PM)\omega_{eq} \tag{5.53}$$

Equation (5.52) can be used to derive ω_t/ω_{eq} for a specified phase margin. *Note that the unity-gain frequency of the loop is independent of the feedback factor, β, and therefore, in the case of an optimally compensated amplifier, it is independent of the closed-loop gain as well.*

EXAMPLE 5.6

A closed-loop amplifier is compensated to have a 75° phase margin for $\beta = 1$. What is ω_t if $f_{eq} = \omega_{eq}/(2\pi) = 50\text{ MHz}$? What is ω_{ta}?

Solution

Using (5.53), we have $\omega_t = 0.268\omega_{eq}$, which implies that the loop-gain unity-gain frequency, f_t, is given by $f_t = \omega_t/2\pi = 13.4\text{ MHz}$. Using (5.49), we also have, for $\beta = 1$, $\omega_{ta} = \omega_t\sqrt{1+0.268^2} = 1.035\omega_t$, which implies that $f_{ta} = \omega_{ta}/2\pi = 13.9\text{ MHz}$.

When considering optimum compensation, (5.34) is not accurate enough and one must use the more accurate relationship for $A(s)$ given in (5.46). Assuming β is frequency independent, substituting (5.44) into (5.35), and rearranging gives

$$A_{CL}(s) = \frac{A_{CL0}}{1 + \dfrac{s(1/\omega_{p1} + 1/\omega_{eq})}{1+\beta A_0} + \dfrac{s^2}{(1+\beta A_0)(\omega_{p1}\omega_{eq})}} \tag{5.54}$$

where

$$A_{CL0} = \frac{A_0}{1+\beta A_0} \cong \frac{1}{\beta} \tag{5.55}$$

Thus, the closed-loop response near the unity-gain frequency, ω_t, is closely approximated by a second-order transfer function. It should be noted here that this approximation is inaccurate in the case where the high-frequency poles and zeros are quite close to ω_t. However, in this case, it would be extremely difficult to adequately compensate the opamp, and thus, this case is ignored.

The result of (5.54) can be equated to the general equation for a second-order all-pole transfer function, written as

$$H_2(s) = \frac{K\omega_0^2}{s^2 + \left(\dfrac{\omega_0}{Q}\right)s + \omega_0^2} = \frac{K}{1 + \dfrac{s}{\omega_0 Q} + \dfrac{s^2}{\omega_0^2}} \tag{5.56}$$

Recall that parameter ω_0 is called the resonant frequency and parameter Q is called the Q factor[9] [Sedra, 1991]. It is well known that, if $Q = \sqrt{1/2} \cong 0.707$, then the magnitude of the transfer function will have the widest passband without any peaking. Furthermore, for $Q = \sqrt{1/2}$, the –3-dB frequency is equal to ω_0. When the step response is investigated, restrictions on the Q factor can also be found to guar-

9. The Q factor is 1/2 times the inverse of the damping factor. The damping factor is an alternative method of indicating the pole locations in second-order transfer functions.

antee no peaking. Specifically, for there to be no peaking in the step response, it is necessary that both poles be real, which is equivalent to the requirement that $Q \le 0.5$. In the case where $Q > 0.5$, the percentage overshoot of the output voltage is given by

$$\%\text{ overshoot} = 100e^{\frac{-\pi}{\sqrt{4Q^2 - 1}}} \tag{5.57}$$

Equating (5.54) with (5.56) and solving for ω_0 and Q results in

$$\omega_0 = \sqrt{(1 + \beta A_0)(\omega_{p1}\omega_{eq})} \cong \sqrt{\beta\omega_{ta}\omega_{eq}} \tag{5.58}$$

and

$$Q = \frac{\sqrt{(1 + \beta A_0)/\omega_{p1}\omega_{eq}}}{1/\omega_{p1} + 1/\omega_{eq}} \cong \sqrt{\frac{\beta A_0 \omega_{p1}}{\omega_{eq}}} = \sqrt{\frac{\beta\omega_{ta}}{\omega_{eq}}} \tag{5.59}$$

where the approximation of Q is valid since $\beta A_0 >> 1$ and $\omega_{p1} << \omega_{eq}$.

It is now possible to relate a specified phase margin to the Q factor. Equation (5.52) can be used to find ω_t/ω_{eq}. This result can be substituted into (5.48) to find $\beta(\omega_{ta}/\omega_{eq})$, which can then be substituted into (5.59) to find the equivalent Q factor. Finally, (5.57) can be used to find the corresponding percentage overshoot for a step input. This procedure gives us the information in Table 5.1.

Table 5.1 leads to some interesting observations. First, a frequency response with $Q \cong \sqrt{1/2}$ roughly corresponds to a phase margin of 65°. Therefore, one should design for a phase margin of at least 65°, given both process and temperature changes. Second, if one wants to ensure that there is no overshoot for a step input, then the phase margin should be at least 75°, again, given both process and temperature variations. Normally, values of 80° to 85° should be the nominal phase margin to account for these variations. These phase margins are much larger than what was traditionally thought to be necessary.

Finally, it is worth mentioning here that, when the feedback network is frequency independent and less than unity, (i.e., when $\beta \le 1$), the worst-case phase margin occurs for $\beta = 1$. Thus, for a general-purpose opamp where $0 < \beta \le 1$, if the opamp is compensated for $\beta = 1$, it is guaranteed to be stable for all other β, although it will not be optimally compensated and will be slower than necessary.

Table 5.1 The relationship between PM, ω_t/ω_{eq}, Q factor, and percentage overshoot

PM (Phase margin)	ω_t/ω_{eq}	Q factor	Percentage overshoot for a step input
55°	0.700	0.925	13.3%
60°	0.580	0.817	8.7%
65°	0.470	0.717	4.7%
70°	0.360	0.622	1.4%
75°	0.270	0.527	0.008%

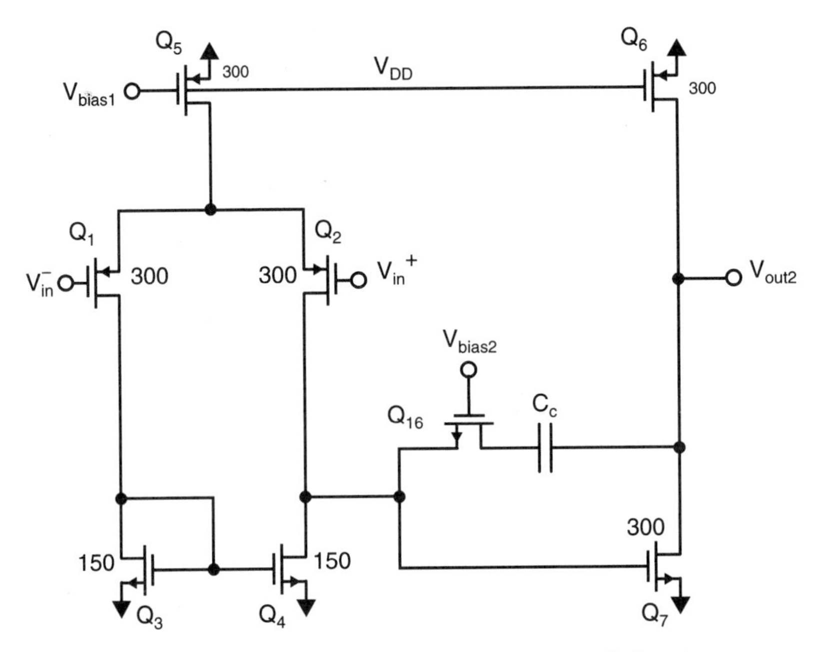

Fig. 5.7 The first two stages of the two-stage opamp, including the compensation network.

Compensating the Two-Stage Opamp

It is now possible to consider compensation of the two-stage opamp. To make the discussion easier to visualize, the first two stages of the opamp, including the compensation network, are shown in Fig. 5.7.

The capacitor, C_C, realizes what is commonly called dominant-pole compensation. It controls the dominant first pole, (i.e., ω_{p1}), and thereby the frequency, ω_{ta}, since (repeating (5.33) here for convenience),

$$\omega_{ta} = A_0 \omega_{p1} \tag{5.60}$$

Transistor Q_{16} has $V_{DS16} = 0$ since no dc bias current flows through it, and therefore Q_{16} is hard in the triode region. Thus, this transistor operates as a resistor, R_C, of value given by

$$R_C = r_{ds16} = \frac{1}{\mu_n C_{ox}\left(\frac{W}{L}\right)_{16} V_{eff16}} \tag{5.61}$$

This transistor is included in order to realize a left-half-plane zero at frequencies around or slightly above ω_t. Without Q_{16}, we have a right-half-plane zero, which makes compensation much more difficult. The addition of such an extra left-half-plane zero is what is commonly called *lead-compensation.* It should be noted here that r_{ds} indicates the drain-source resistance of Q_{16} when it is in the triode region as opposed to the finite-output impedance of Q_{16} when it is in the active mode. The same notation, r_{ds}, is used to indicate the drain-source resistance in both cases—whether the transis-

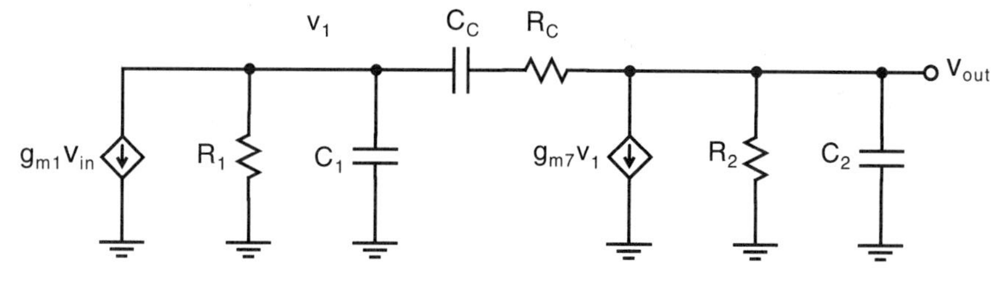

Fig. 5.8 A small-signal model of the two-stage opamp used for compensation analysis.

tor is in the active or the triode region. One simply has to check which region a transistor is operating in to ensure that the correct equation is used to determine r_{ds}.

A simplified small-signal model for this opamp is shown in Fig. 5.8, where the output buffer has again been ignored. Also, it is assumed that the first stage is much faster than the second stage and can therefore be modelled by a simple voltage-controlled current source. In the small-signal model, we have

$$R_1 = r_{ds4} \| r_{ds2} \tag{5.62}$$

$$C_1 = C_{db2} + C_{db4} + C_{gs7} \tag{5.63}$$

$$R_2 = r_{ds6} \| r_{ds7} \tag{5.64}$$

$$C_2 = C_{db7} + C_{db6} + C_{L2} \tag{5.65}$$

Note that if no output buffer is present, then C_{L2} is the output load capacitance. If an output buffer is present, then C_{L2} is the load capacitance introduced by the output buffer and C_1 also includes the capacitance needed to drive Q_9 in Fig. 5.2.

To show the need for R_C, we first assume $R_C = 0$ and perform nodal analysis at the nodes designated by v_1 and v_{out}. The following transfer function is obtained:

$$\frac{V_{out}}{V_{in}} = \frac{g_{m1}g_{m7}R_1R_2\left(1 - \frac{sC_C}{g_{m7}}\right)}{1 + sa + s^2b} \tag{5.66}$$

where

$$a = (C_2 + C_C)R_2 + (C_1 + C_C)R_1 + g_{m7}R_1R_2C_C \tag{5.67}$$

and

$$b = R_1R_2(C_1C_2 + C_1C_C + C_2C_C) \tag{5.68}$$

It is possible to find approximate equations for the two poles based on the assumption that the poles are real and widely separated.[10] This assumption allows us to express the denominator, $D(s)$, as[11]

10. If this assumption is not valid, then it is extremely difficult to properly compensate the opamp for unity-gain stability.

11. Note that the notation $[1 + (s/\omega_x)]$ implies that the root is in the left half plane if $\omega_x > 0$, whereas the root is in the right half plane if $\omega_x < 0$.

$$D(s) = \left(1 + \frac{s}{\omega_{p1}}\right)\left(1 + \frac{s}{\omega_{p2}}\right) \cong 1 + \frac{s}{\omega_{p1}} + \frac{s^2}{\omega_{p1}\omega_{p2}} \tag{5.69}$$

Setting the coefficients of (5.66) equal to the coefficients of (5.69) and solving for ω_{p1} and ω_{p2} results in the following relationships. The dominant pole, ω_{p1}, is given by

$$\begin{aligned} \omega_{p1} &\cong \frac{1}{R_1[C_1 + C_C(1 + g_{m7}R_2)] + R_2(C_2 + C_C)} \\ &\cong \frac{1}{R_1C_C(1 + g_{m7}R_2)} \\ &\cong \frac{1}{g_{m7}R_1R_2C_C} \end{aligned} \tag{5.70}$$

whereas the nondominant pole, ω_{p2}, is given by

$$\begin{aligned} \omega_{p2} &\cong \frac{g_{m7}C_C}{C_1C_2 + C_2C_C + C_1C_C} \\ &\cong \frac{g_{m7}}{C_1 + C_2} \end{aligned} \tag{5.71}$$

Note that, from (5.66), another zero, ω_z, is located in the right half plane and is given by

$$\omega_z = \frac{-g_{m7}}{C_C} \tag{5.72}$$

This zero should come as no surprise since there are two signal paths having opposite signs in the small-signal circuit of Fig. 5.8—one through the voltage-controlled current source, $g_{m7}v_1$, and the other through the compensation capacitor, C_C.

From the preceding relationships for ω_{p1} and ω_{p2}, we can see that, as g_{m7} increases, the separation between the first and second poles increases. This separation tends to make the circuit more stable; hence, the use of a Miller capacitance for compensation is often called *pole-splitting compensation.* Also note that increasing C_C moves the dominant pole, ω_{p1}, to a lower frequency without affecting the second pole, ω_{p2}. This effect also makes the opamp more stable.

However, a problem arises due to the right-half-plane zero, ω_z. Because the zero is in the right half plane, it introduces negative phase shift, or phase lag, in the transfer function of the opamp. This makes stability more difficult. Making C_C larger does not help matters because this decreases the frequencies of both the first pole and the zero without making them more widely separated. Indeed, because of this right-half-plane zero, it is often impossible to choose C_C such that the step response has no overshoot (assuming $R_C = 0$). Fortunately, all is not lost, since introducing R_C allows adequate compensation, as we discuss in the next subsection.

Lead Compensation

If the small-signal model of Fig. 5.8 is reanalyzed with a nonzero R_C, then a third-order denominator results. The first two poles are still approximately at the frequencies given by (5.70) and (5.71). The third pole is at a high frequency and has almost no effects. However, the zero is now determined by the relationship,

$$\omega_z = \frac{-1}{C_C(1/g_{m7} - R_C)} \tag{5.73}$$

This result allows the designer a number of possibilities. One could take

$$R_C = 1/g_{m7} \tag{5.74}$$

to eliminate the right-half-plane zero altogether. Alternatively, one could choose R_C to be even larger and thus move the right-half-plane zero into the left half plane to cancel the nondominant pole, ω_{p2}. Setting (5.73) equal to (5.71) and solving for R_C results in the following equation for R_C:

$$R_C = \frac{1}{g_{m7}}\left(1 + \frac{C_1 + C_2}{C_C}\right) \tag{5.75}$$

Unfortunately, C_2 is often not known a priori, especially when no output stage is present.

The third possibility (recommended by the authors) is to choose R_C even larger yet to move the now left-half-plane zero to a frequency slightly greater than the unity-gain frequency that would result if the lead resistor were not present—say, 20 percent larger [Roberge, 1975]. For this case, one should satisfy the following equation:

$$\omega_z = 1.2\omega_t \tag{5.76}$$

Assuming $R_C \gg (1/g_{m7})$, then $\omega_z \cong 1/(R_C C_C)$, and recalling from (5.14) that $\omega_t \cong g_{m1}/C_C$, then one should choose R_C according to

$$R_C \cong \frac{1}{1.2g_{m1}} \tag{5.77}$$

This approach leads to the following design procedure for lead compensation of a two-stage CMOS opamp:

1. Start by choosing, somewhat arbitrarily, $C'_C \cong 5$ pF.
2. Using SPICE, find the frequency at which a $-125°$ phase shift exists. Let the gain at this frequency be denoted A'. Also, let the frequency be denoted ω_t. This is the frequency that we would like to become the unity-gain frequency of the loop gain.
3. Choose a new C_C so that ω_t becomes the unity-gain frequency of the loop gain, thus resulting in a 55° phase margin. (Obtaining this phase margin is the reason we chose $-125°$ in step 2.) This can be achieved by taking C_C according to the equation

$$C_C = C'_C A' \tag{5.78}$$

It might be necessary to iterate on C_C a couple of times using SPICE.

4. Choose R_C according to

$$R_C = \frac{1}{1.2\omega_t C_C} \tag{5.79}$$

This choice will increase the unity-gain frequency by about 20 percent, leaving the zero near to the final resulting unity-gain frequency, which will end up about 15 percent below the equivalent second-pole frequency. The resulting phase margin is approximately 85°.[12] This allows a margin of 5° to account for processing variations without the poles of the closed-loop response becoming real. *This choice is also almost optimum lead compensation for almost any opamp when a resistor is placed in series with the compensation capacitor.* It might be necessary to iterate on R_C a couple of times to optimize the phase margin. However, one should check that the gain continues to steadily decrease at frequencies above the new unity-gain frequency; otherwise, the transient response can be poor. This situation sometimes occurs when unexpected zeros at frequencies only slightly greater than ω_t are present.

5. If, after step 4, or the phase margin is not adequate, then increase C_C while leaving R_C constant. This will move both ω_t and the lead zero to lower frequencies while keeping their ratio approximately constant, thus minimizing the effects of higher-frequency poles and zeros which, hopefully, do not also move to lower frequencies. In most cases, the higher-frequency poles and zeros (except for the lead zero) will not move to significantly lower frequencies when C_C is increased.

6. The final step is to replace R_C by a transistor. The size of the transistor can be chosen using equation (5.61), which is repeated here for convenience:

$$R_C = r_{ds16} = \frac{1}{\mu_n C_{ox}\left(\frac{W}{L}\right)_{16} V_{eff16}} \tag{5.80}$$

Finally, SPICE can be used again to fine-tune the device dimensions to optimize the phase margin to that obtained in steps 4 and 5.

Not only does the procedure just described apply to the two-stage opamp, but it (or a very similar procedure) has been found to be almost optimum for compensating most types of opamps.

EXAMPLE 5.7

An opamp has an open-loop transfer function given by

$$A(s) = \frac{A_0(1 + s/\omega_z)}{(1 + s/\omega_{p1})(1 + s/\omega_2)} \tag{5.81}$$

12. The first pole contributes –90° phase shift, the lead zero contributes approximately 45° phase shift, and the equivalent second pole contiguities approximately –40° phase shift.

Here, A_0 is the dc gain of the opamp and ω_z, ω_1, and ω_2 are the frequencies of a zero, the dominant pole, and the equivalent second pole, respectively. Assume that $\omega_2 = 2\pi \times 50$ MHz and that $A_0 = 10^4$.

(*a*) Assuming $\omega_z \to \infty$, find ω_{p1} and the unity-gain frequency, ω_t', so that the opamp has a unity-gain phase margin of 55°.

(*b*) Assuming $\omega_z = 1.2\omega_t'$ (where ω_t' is as found in part (*a*)), what is the new unity-gain frequency, ω_t? Also, find the new phase margin.

Solution

First, note that at frequencies much greater than the dominant-pole frequency (i.e., $\omega >> \omega_{p1}$), such as the unity-gain frequency, (5.81) can be approximated by

$$A(s) \cong \frac{A_0(1 + s/\omega_z)}{(s/\omega_{p1})(1 + s/\omega_2)} \tag{5.82}$$

(*a*) For $\omega_z \to \infty$ we use (5.82) to find the phase angle at ω_t':

$$\angle A(j\omega_t') = -90° - \tan^{-1}(\omega_t'/\omega_2) \tag{5.83}$$

For a phase margin of $PM = 55°$, we need

$$\angle A(j\omega_t') = -180° + PM = -125° \tag{5.84}$$

Setting (5.83) equal to (5.84) and solving for ω_t'/ω_2 results in

$$\begin{aligned} \tan^{-1}(\omega_t'/\omega_2) &= 35° \\ \Rightarrow \omega_t' &= 2.2 \times 10^8 = 2\pi \times 35 \text{ MHz} \end{aligned} \tag{5.85}$$

Next, setting $|A(j\omega_t')| = 1$ and again using (5.82) results in

$$\begin{aligned} \frac{A_0}{(\omega_t'/\omega_{p1})\sqrt{1 + (\omega_t'/\omega_2)^2}} &= 1 \\ \Rightarrow \omega_{p1} = \frac{\omega_t'\sqrt{1 + (\omega_t'/\omega_2)^2}}{A_0} &= 2\pi \times 4.28 \text{ kHz} \end{aligned} \tag{5.86}$$

(*b*) First, we set

$$\omega_z = 1.2\omega_t' = 2\pi \times 42 \text{ MHz} \tag{5.87}$$

To find the new unity-gain frequency, setting $|A(j\omega_t)| = 1$ in (5.82) now gives

$$\begin{aligned} \frac{A_0\sqrt{1 + (\omega_t/\omega_z)^2}}{(\omega_t/\omega_{p1})\sqrt{1 + (\omega_t/\omega_2)^2}} &= 1 \\ \Rightarrow \omega_t = \frac{A_0\omega_{p1}\sqrt{1 + (\omega_t/\omega_z)^2}}{\sqrt{1 + (\omega_t/\omega_2)^2}} & \end{aligned} \tag{5.88}$$

This equation can be solved for the new unity-gain frequency. Both sides can be squared, resulting in a quadratic equation in ω_t, which can be solved exactly. Alternatively, the value for ω_t' found in part (*a*) can be used as an initial guess, and (5.88) can be solved iteratively. After four iterations, one finds that $\omega_t = 2\pi \times 46.8 \text{ MHz}$. This unity-gain frequency is a 34 percent increase over that found in part (*a*). We can solve for the new phase shift by using

$$\angle A(j\omega_t) = -90° + \tan^{-1}(\omega_t/\omega_z) - \tan^{-1}(\omega_t/\omega_2) = -85.0° \quad (5.89)$$

Note that this phase value gives a phase margin of 95°, which is a 40° improvement! Normally, the improvement would not be this great due to additional high-frequency poles and zeros (which have been ignored here). These high-frequency poles and zeros will typically degrade the phase margin by an additional 5° or 10°. Regardless of this degradation, the improvement from using lead compensation is substantial.

Making Compensation Independent of Process and Temperature

This section shows how lead compensation can be made process and temperature insensitive. Repeating equations (5.14) and (5.71) here, we have

$$\omega_t = \frac{g_{m1}}{C_C} \quad (5.90)$$

and

$$\omega_{p2} \cong \frac{g_{m7}}{C_1 + C_2} \quad (5.91)$$

We see here that the second pole is proportional to the transconductance of the drive transistor of the second stage, g_{m7}. Also, the unity-gain frequency is proportional to the transconductance of the input transistor of the first stage, g_{m1}. Furthermore, the ratios of all of the transconductances remain relatively constant over process and temperature variations since the transconductances are all determined by the same biasing network.[13] Also, most of the capacitances also track each other since they are primarily determined by gate oxides. Repeating (5.73), when a resistor is used to realize lead compensation, the lead zero is at a frequency given by

$$\omega_z = \frac{-1}{C_C(1/g_{m7} - R_C)} \quad (5.92)$$

Thus, if R_C can also be made to track the inverse of transconductances, and in particular $1/g_{m7}$, then the lead zero will also be proportional to the transconductance of Q_7. As a result, the lead zero will remain at the same relative frequency with respect to ω_t and ω_{p2}, as well as all other high-frequency poles and zeros. In other words, the lead compensation will be mostly independent of process and temperature variations.

13. The ratio μ_n/μ_p is relatively constant for a given process (say, a 0.6-μm process) but often varies more significantly from process to process (say, from 0.6-μm to 0.4-μm processes).

It turns out that R_C can be made proportional to $1/g_{m7}$ as long as R_C is realized by a transistor in the triode region that has an effective gate-source voltage proportional to that of Q_7. To see this result, recall that R_C is actually realized by Q_{16}, and therefore we have

$$R_C = r_{ds16} = \frac{1}{\mu_n C_{ox}(W/L)_{16} V_{eff16}} \tag{5.93}$$

Also, g_{m7} is given by

$$g_{m7} = \mu_n C_{ox}(W/L)_7 V_{eff7} \tag{5.94}$$

Thus, the product $R_C g_{m7}$, which we want to be a constant, is given by

$$R_C g_{m7} = \frac{(W/L)_7 V_{eff7}}{(W/L)_{16} V_{eff16}} \tag{5.95}$$

Therefore, all that remains is to ensure that V_{eff16}/V_{eff7} is independent of process and temperature variations since clearly the remaining terms depend only on a geometric relationship. The ratio V_{eff16}/V_{eff7} can be made constant by deriving V_{GS16} from the same biasing circuit used to derive V_{GS7}. Specifically, consider the circuit shown in Fig. 5.9, which consists of a bias stage, the second stage, and the compensation network of the two-stage opamp. First we must make $V_a = V_b$, which is possible by making $V_{eff13} = V_{eff7}$. These two effective gate-source voltages can be made equal by taking

$$\sqrt{\frac{2I_{D7}}{\mu_n C_{ox}(W/L)_7}} = \sqrt{\frac{2I_{D13}}{\mu_n C_{ox}(W/L)_{13}}} \tag{5.96}$$

Squaring and simplifying, we see that the following condition must be satisfied:

$$\frac{I_{D7}}{I_{D13}} = \frac{(W/L)_7}{(W/L)_{13}} \tag{5.97}$$

However, the ratio I_{D7}/I_{D13} is set from the current mirror pair Q_6, Q_{11}, resulting in

$$\frac{I_{D7}}{I_{D13}} = \frac{(W/L)_6}{(W/L)_{11}} \tag{5.98}$$

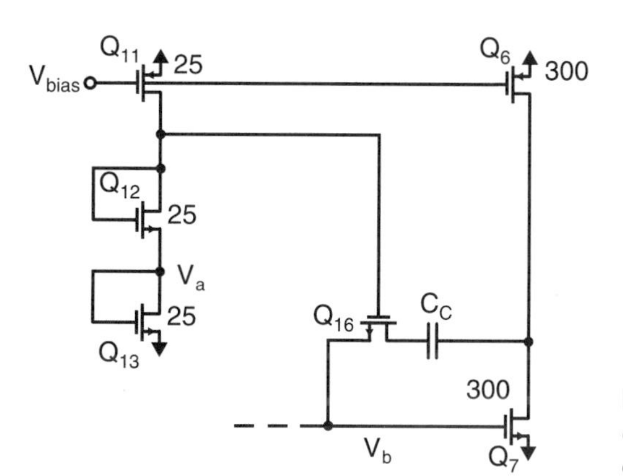

Fig. 5.9 The bias circuit, second stage, and compensation circuit of the two-stage opamp.

Thus, to make $V_{eff13} = V_{eff7}$, from (5.97) and (5.98), we need to satisfy the following relationship:

$$\frac{(W/L)_6}{(W/L)_7} = \frac{(W/L)_{11}}{(W/L)_{13}} \tag{5.99}$$

Assuming this condition is satisfied, we have $V_{eff13} = V_{eff7}$, and therefore we also have $V_a = V_b$. Since the gates of Q_{12} and Q_{16} are connected and their source voltages are the same, we also have $V_{eff12} = V_{eff16}$. Next, using these gate-source relationships and noting that $I_{D12} = I_{D13}$, we can write

$$\frac{V_{eff7}}{V_{eff16}} = \frac{V_{eff13}}{V_{eff12}} = \frac{\sqrt{\dfrac{2I_{D13}}{\mu_n C_{ox}(W/L)_{13}}}}{\sqrt{\dfrac{2I_{D12}}{\mu_n C_{ox}(W/L)_{12}}}} = \sqrt{\frac{(W/L)_{12}}{(W/L)_{13}}} \tag{5.100}$$

Finally, substituting (5.100) into (5.95), we have the product $R_C g_{m7}$, given by

$$R_C g_{m7} = \frac{(W/L)_7}{(W/L)_{16}} \sqrt{\frac{(W/L)_{12}}{(W/L)_{13}}} \tag{5.101}$$

which is only dependent on geometry and not on processing or temperature variations.

As a result, we have guaranteed that the drain-source resistance of a transistor in the triode region is inversely matched to the transconductance of a different transistor. This relationship can be very useful for many other applications as well. Indeed, in the next subsection, we will see that it's quite simple to make all of the transconductances of transistors in a microcircuit match the conductance of a single off-chip resistor. *This approach results in the possibility of on-chip "resistors," realized by using triode-region transistors that are accurately ratioed with respect to a single off-chip resistor.* This relationship can be very useful in modern circuit design.

Biasing an Opamp to Have Stable Transconductances

We have seen that transistor transconductances are perhaps the most important parameters in opamps that must be stabilized. This stabilization can be achieved by using a circuit approach first proposed in [Steininger, 1990], in which transistor transconductances are matched to the conductance of a resistor. As a result, to a first-order effect, the transistor transconductances are independent of power-supply voltage as well as process and temperature variations.

The bias circuit is shown in Fig. 5.10. First, it is assumed that $(W/L)_{10} = (W/L)_{11}$. This equality results in both sides of the circuit having the same current due to the current-mirror pair Q_{10}, Q_{11}. As a result, we also must have $I_{D15} = I_{D13}$. Now, around the loop consisting of Q_{13}, Q_{15}, and R_B, we have

$$V_{GS13} = V_{GS15} + I_{D15} R_B \tag{5.102}$$

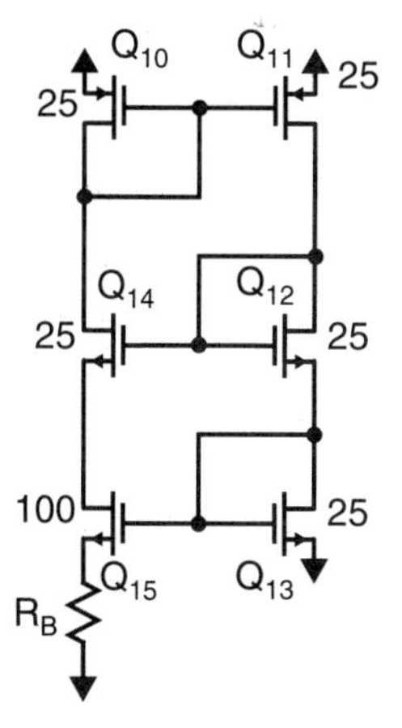

Fig. 5.10 A bias circuit that gives very predictable and stable transistor transconductances.

and recalling that $V_{effi} = V_{GSi} - V_t$, we can subtract the threshold voltage, V_t, from both sides, resulting in

$$V_{eff13} = V_{eff15} + I_{D15}R_B \tag{5.103}$$

This equation can also be written as

$$\sqrt{\frac{2I_{D13}}{\mu_n C_{ox}(W/L)_{13}}} = \sqrt{\frac{2I_{D15}}{\mu_n C_{ox}(W/L)_{15}}} + I_{D15}R_B \tag{5.104}$$

and since $I_{D13} = I_{D15}$, we can also write

$$\sqrt{\frac{2I_{D13}}{\mu_n C_{ox}(W/L)_{13}}} = \sqrt{\frac{2I_{D13}}{\mu_n C_{ox}(W/L)_{15}}} + I_{D13}R_B \tag{5.105}$$

Rearranging, we obtain

$$\frac{2}{\sqrt{2\mu_n C_{ox}(W/L)_{13}I_{D13}}}\left[1 - \sqrt{\frac{W/L_{13}}{W/L_{15}}}\right] = R_B \tag{5.106}$$

and recalling that $g_{m13} = \sqrt{2\mu_n C_{ox}(W/L)_{13}I_{D13}}$ results in the important relationship

$$g_{m13} = \frac{2\left[1 - \sqrt{\frac{(W/L)_{13}}{(W/L)_{15}}}\right]}{R_B} \tag{5.107}$$

Thus, the transconductance of Q_{13} is determined by geometric ratios only, independent of power-supply voltages, process parameters, temperature, or any other parameters with large variability. For the special case of $(W/L)_{15} = 4(W/L)_{13}$, we have simply

$$g_{m13} = \frac{1}{R_B} \tag{5.108}$$

Note that, not only is g_{m13} stabilized, but all other transconductances are also stabilized since all transistor currents are derived from the same biasing network, and, therefore, the ratios of the currents are mainly dependent on geometry. We thus have, for all n-channel transistors,

$$g_{mi} = \sqrt{\frac{(W/L)_i I_{Di}}{(W/L)_{13} I_{D13}}} \times g_{m13} \tag{5.109}$$

and for all p-channel transistors

$$g_{mi} = \sqrt{\frac{\mu_p}{\mu_n} \frac{(W/L)_i I_{Di}}{(W/L)_{13} I_{D13}}} \times g_{m13} \tag{5.110}$$

It should be noted here that the preceding analysis has ignored many second-order effects such as the transistor output impedance and the body effect. The body effect will modify the equation slightly, but the relationship will still depend primarily on geometry alone. The major limitation is due to the transistor output impedance. This effect can be made of little consequence by replacing the simple current mirrors with cascode mirrors. Although the use of cascode mirrors appears at first glance to require large power supplies, in Chapter 6 we will see that by using wide-swing current mirrors, this is not necessarily the case. A final limitation is that at high temperatures, the currents and effective gate-source voltages increase substantially to offset the mobility degradation and to keep the transconductances stable. In addition, the effective gate-source voltages also increase, thereby limiting signal swings more than is the case at normal temperatures. Since the carrier mobility is proportional to $T^{-3/2}$, this corresponds to a 27-percent reduction from room temperature of (300 °K) to 100 °C (373 °K). Thus, the effective gate-source voltages increase by 27 percent to keep the transistor transconductance unchanged, since

$$g_{mi} = \mu_i C_{ox} \left(\frac{W}{L}\right)_i V_{eff\text{-}i} \tag{5.111}$$

As long as the effective gate-source voltages have not initially been designed to be too large, this limitation is tolerable in most applications.[14] A typical value for effective gate-source voltages might be 0.2 V to 0.25 V at room temperature.

The bias circuit just presented is an example of a circuit having positive feedback. It is stable as long as the loop gain is less than unity, which is the case when $(W/L)_{15}/(W/L)_{13} > 1$. However, this circuit unfortunately can have a second stable state where all the currents are zero. To guarantee this condition doesn't happen, it is necessary to add a *start-up circuit* that only affects the operation if all the currents are zero at start up. Possible start-up circuits will be described in Chapter 6.

14. Also, if on-chip resistors, such as well or diffusion resistors, are used, this effect will be somewhat offset by their positive temperature-coefficient dependency.

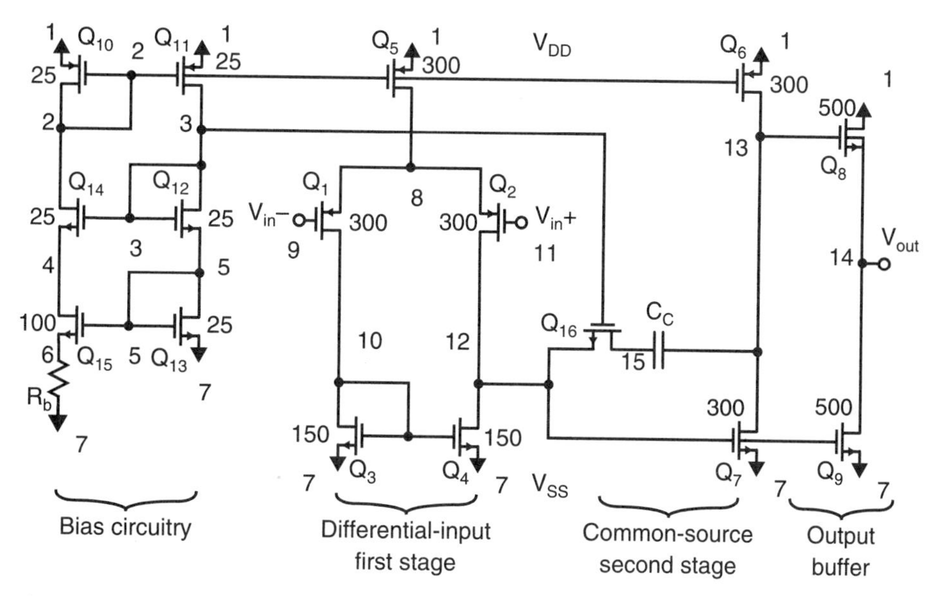

Fig. 5.11 All transistor lengths equal 1.6 μm.

5.3 SPICE SIMULATION EXAMPLES

In this section, SPICE simulation results are presented for Examples 5.1 and 5.2.

Simulations of Examples 5.1 and 5.2.

The circuit for Examples 5.1 and 5.2 is shown in Fig. 5.11. R_b is chosen such that Q_5 has a dc current of 100 μA. The size of Q_{16} is selected for proper lead compensation.

```
NETLIST:
vdd     1    0              dc 2.5
vss     7    0              dc –2.5
m10     2    2    1    1    PMOS w=25um l=1.6um
m11     3    2    1    1    PMOS w=25um l=1.6um
m14     2    3    4    7    NMOS w=25um l=1.6um
m12     3    3    5    7    NMOS w=25um l=1.6um
m15     4    5    6    7    NMOS w=100um l=1.6um
m13     5    5    7    7    NMOS w=25um l=1.6um
rb      6    7              8k
m5      8    2    1    1    PMOS w=300um l=1.6um
m1      10   9    8    1    PMOS w=300um l=1.6um
m2      12   11   8    1    PMOS w=300um l=1.6um
m3      10   10   7    7    NMOS w=150um l=1.6um
m4      12   10   7    7    NMOS w=150um l=1.6um
vin-    9    0              dc 0
```

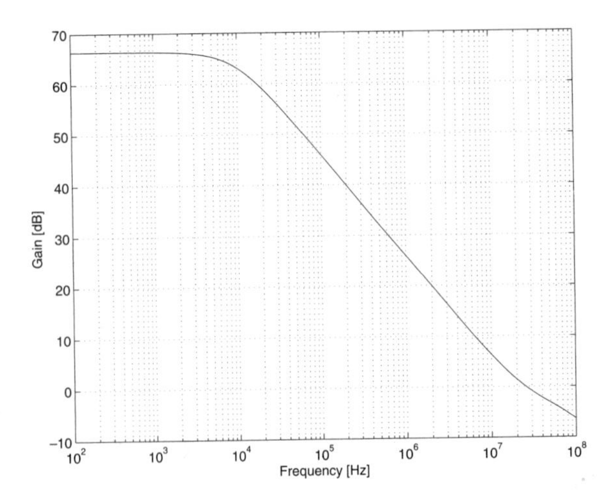

Fig. 5.12 A frequency plot of the two-stage opamp.

```
vin+   11   0                 dc 0 ac 1
m6     13   2    1    1       PMOS w=300um l=1.6um
m8     1    13   14   7       NMOS w=500um l=1.6um
m7     13   12   7    7       NMOS w=300um l=1.6um
m9     14   12   7    7       NMOS w=500um l=1.6um
cc     15   13                5pF
m16    15   3    12   7       NMOS w=100um l=1.6um

.op
.ac dec 20 0.1k 100Meg
.print vdb(14)
.option post
.options brief
.lib "../cmos_models" library
.end
```

The frequency plot result is shown in Fig. 5.12, where we see that the dc gain is 67 dB (i.e., 2240 V/V). The gain of the opamp is different from the hand-calculation result from Example 5.1 because the values for the transistor output impedances in the example are rough estimations. The unity-gain frequency is shown to be 25 MHz, which corresponds closely to the hand-calculated result since this calculation does not depend on poorly modelled variables.

5.4 REFERENCES

P. R. Gray and R. G. Meyer. *Analysis and Design of Analog Integrated Circuits*, 3rd ed. John Wiley & Sons, New York, 1993.

J. K. Roberge. *Operational Amplifiers*. John Wiley & Sons, New York, 1975.

A. S. Sedra and K. C. Smith. *Microelectronic Circuits*, 3rd ed. Holt, Rinehart & Winston, New York, 1991.

J. M. Steininger, "Understanding Wide-Band MOS Transistors," *IEEE Circuits and Devices*, Vol. 6, No. 3, pp. 26–31, May 1990.

5.5 PROBLEMS

Unless otherwise stated, assume the following:

- n-channel MOS transistors:

  $\mu_n C_{ox} = 92\ \mu A/V^2$
  $V_{tn} = 0.8\ V$
  $\gamma = 0.5\ V^{1/2}$
  $r_{ds}(\Omega) = 8{,}000\ L(\mu m)/I_D(mA)$ in active region
  $C_j = 2.4 \times 10^{-4}\ pF/(\mu m)^2$
  $C_{j\text{-}sw} = 2.0 \times 10^{-4}\ pF/\mu m$
  $C_{ox} = 1.9 \times 10^{-3}\ pF/(\mu m)^2$
  $C_{gs(overlap)} = C_{gd(overlap)} = 2.0 \times 10^{-4}\ pF/\mu m$
- p-channel MOS transistors:

  $\mu_p C_{ox} = 30\ \mu A/V^2$
  $V_{tp} = -0.9\ V$
  $\gamma = 0.8\ V^{1/2}$
  $r_{ds}(\Omega) = 12{,}000\ L(\mu m)/I_D(mA)$ in active region,
  $C_j = 4.5 \times 10^{-4}\ pF/(\mu m)^2$
  $C_{j\text{-}sw} = 2.5 \times 10^{-4}\ pF/\mu m$
  $C_{ox} = 1.9 \times 10^{-3}\ pF/(\mu m)^2$
  $C_{gs(overlap)} = C_{gd(overlap)} = 2.0 \times 10^{-4}\ pF/\mu m$

5.1 Assume that $I_{D5} = 100\ \mu A$, all transistor lengths are $1.2\ \mu m$, and $C_c = 10\ pF$. For the two-stage opamp of Fig. 5.2,

a. Estimate the –3-dB frequency of the first stage.

b. Estimate the unity-gain frequency of the opamp.

c. Find the slew rate.

5.2 Repeat Example 5.3, but let $C_C = 4\ pF$. What circuit changes could be made to double the slew rate while keeping ω_t and C_C unchanged?

5.3 How should the sizes of Q_6 and Q_7 change in Problem 5.2 to maintain zero systematic offset voltage?

5.4 Assume Q_3 and Q_4 of the two-stage opamp of Fig. 5.4 are both 50 μm wide, all transistor lengths are $1.6\ \mu m$, and $I_{D5} = 100\ \mu A$. Estimate the inherent input-offset voltage.

5.5 Ignoring the body effect, what is the output voltage range of the opamp of Fig. 5.2? What is the range of the common-mode input voltage, assuming ±5 V power supplies are used? Assume $I_{D5} = 100\ \mu A$.

5.6 Consider the opamp shown in Fig. 5.2, where Q_{16} is placed on the output side of C_C (rather than where it is shown). Explain why this circuit would then oscillate for large positive output values.

5.7 Repeat Problem 5.5, but take the body effect into account and find approximate answers. Assume the substrate of Q_8 is connected to V_{SS}.

5.8 Consider a system response, $H(s)$, with one dominant pole, other real-axis poles, and zeros such that

$$H(s) = \frac{A_0 \times \Pi\left(1 + \frac{s}{\omega_{zi}}\right)}{\left(1 + \frac{s}{\omega_{p1}}\right)\Pi\left(1 + \frac{s}{\omega_{pi}}\right)}$$

Also consider another approximating function, $H_{app}(s)$, given by

$$H_{app}(s) = \frac{A_o}{\left(1 + \frac{s}{\omega_{p1}}\right)\left(1 + \frac{s}{\omega_{eq}}\right)}$$

Show that, when $H(s)$ is approximated by $H_{app}(s)$ such that the phase of the two systems are approximately equal at ω_t of $H(s)$, ω_{eq} should be equal to

$$\frac{1}{\omega_{eq}} = \sum\frac{1}{\omega_{pi}} - \sum\frac{1}{\omega_{zi}}$$

5.9 An opamp has its first pole at 3 kHz and has high-frequency poles at 130 MHz, 160 MHz, and 180 MHz. Using iteration, find the frequency where the phase shift is $-135°$ and, therefore, find the equivalent time constant that models the high-frequency poles. How does this compare to the estimate given by equation (5.45)?

5.10 A two-stage opamp has a compensation capacitor connected between the input and the output of the second stage. It has an equivalent second-pole frequency of 60 MHz. Assume the input transistors of the first stage have a transconductance of 0.775 mA/V and the gain of the output buffer is exactly unity. What is the required size of the compensation capacitor if the phase margin is to be 55° for the feedback configuration shown in Fig. P5.10?

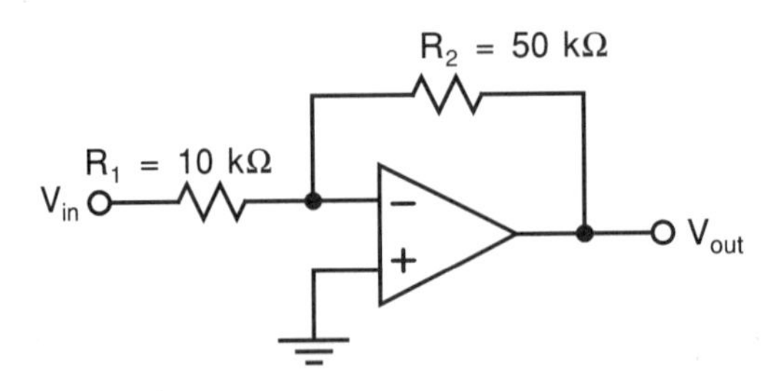

Fig. P5.10

5.11 For the result of Problem 5.10, what is a near-optimum size for a capacitor to be placed in parallel with R_2 in order to obtain lead compensation?

5.12 An opamp has an open-loop transfer function given by

$$H(s) = \frac{A_0}{\left(1 + \frac{s}{\omega_1}\right)\left(1 + \frac{s}{\omega_2}\right)}$$

Assume $A_0 = 10^4$ and $\omega_2 = 10^8$ rad/s. A zero, ω_z, is added such that the frequency of the zero is the same as the resulting open-loop unity-gain frequency, ω_t. Find ω_1 and ω_t so that the phase margin is 80°.

5.13 An opamp has an open-loop transfer function given by

$$A(s) \cong \frac{A_0(1 + s\tau_z)}{s\tau_1(1 + s\tau_2)}$$

Find the transfer function of the closed-loop amplifier, assuming a feedback factor β exists. Find approximate equations for the resonant frequency and the Q factor of the denominator of the transfer function of the closed-loop amplifier.

5.14 Prove that the frequencies of the first and second poles of the amplifier of Fig. 5.2 are still given by (5.70) and (5.71), even when lead compensation is used (i.e., when R_C is included in Fig. 5.8).

5.15 Design the bias circuit of Fig. 5.10 to have $V_{eff} = 0.25$ V for all transistors except Q_{15}. For $Q_{10} - Q_{14}$ having sizes of 10 μm/1.2 μm and Q_{15} having a size of 40 μm/1.2 μm, what is the required size of R? What would the effective gate-source voltage be at 70 °C assuming electron mobility varies proportional to $T sup - 3/2$? Confirm your answer using SPICE.

CHAPTER

Advanced Current Mirrors and Opamps

The classical two-stage opamp was discussed in Chapter 5. While this two-stage opamp has been used in many commercial integrated circuits, recently a number of alternate opamp designs have been gaining in popularity and are the topic of this chapter. Since many of these opamps make use of more advanced current mirrors, a discussion of advanced current mirrors is presented first. Following the current-mirror section, the design of two modern opamps having single-ended outputs is presented. Next, design principles are discussed for realizing opamps having fully differential outputs. Finally, the basics of "current-feedback opamps" are presented, where we see that these opamps can have large gain-bandwidth products without the need for adjusting the compensation capacitance.

6.1 ADVANCED CURRENT MIRRORS

Wide-Swing Current Mirrors

As newer technologies with shorter channel lengths are used, it becomes more difficult to achieve reasonable opamp gains due to transistor output-impedance degradation caused by short-channel effects. As a result, designers are often forced to use cascode current mirrors. Unfortunately, the use of conventional cascode current mirrors limits the signal swings available, which may not be tolerated in certain applications. Fortunately, circuits exist that do not limit the signal swings as much as the current mirrors discussed in Chapter 3. One such circuit is shown in Fig. 6.1 and is often called the "wide-swing cascode current mirror" [Sooch, 1985; Babanezhad, 1987].

The basic idea of this current mirror is to bias the drain-source voltages of transistors Q_2 and Q_3 to be close to the minimum possible without them going into the triode region. Specifically, if the sizes shown in Fig. 6.1 are used, and assuming the classical square-law equations for long channel-length devices are valid, transistors Q_2 and Q_3 will be biased right at the edge of the triode region. Before seeing how these bias voltages are created, note that the transistor pair Q_3, Q_4 acts like a single diode-connected transistor in creating the gate-source voltage for Q_3. These two transistors operate very similarly to how Q_3 alone would operate if its gate were con-

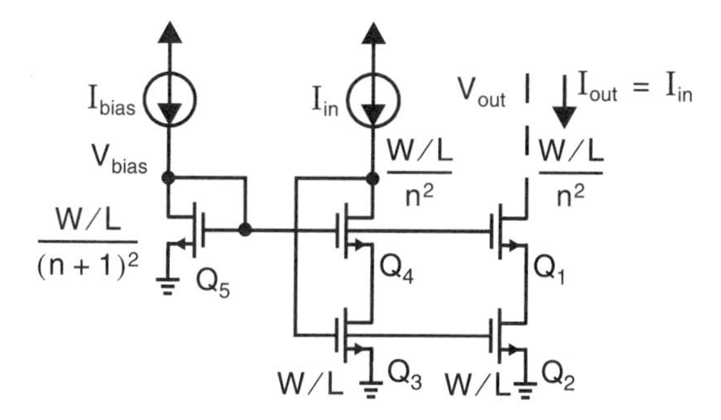

Fig. 6.1 The wide-swing cascode current mirror. I_{bias} typically is set to the nominal or maximum input current, I_{in}.

nected to its source. The reason for including Q_4 is to lower the drain-source voltage of Q_3 so that it is matched to the drain-source voltage of Q_2. This matching makes the output current, I_{out}, more accurately match the input current, I_{in}. If Q_4 were not included, then the output current would be a little smaller than the input current due to the finite output impedances of Q_2 and Q_3. Other than this, Q_4 has little effect on the circuit's operation.

To determine the bias voltages for this circuit, let V_{eff} be the effective gate-source voltage of Q_2 and Q_3, and assume all of the drain currents are equal. We therefore have

$$V_{eff} = V_{eff2} = V_{eff3} = \sqrt{\frac{2I_{D2}}{\mu_n C_{ox}(W/L)}} \tag{6.1}$$

Furthermore, since Q_5 has the same drain current but is $(n+1)^2$ times smaller, we have

$$V_{eff5} = (n+1)V_{eff} \tag{6.2}$$

Similar reasoning results in the effective gate-source voltages of Q_1 and Q_4 being given by

$$V_{eff1} = V_{eff4} = nV_{eff} \tag{6.3}$$

Thus,

$$V_{G5} = V_{G4} = V_{G1} = (n+1)V_{eff} + V_{tn} \tag{6.4}$$

Furthermore,

$$V_{DS2} = V_{DS3} = V_{G5} - V_{GS1} = V_{G5} - (nV_{eff} + V_{tn}) = V_{eff} \tag{6.5}$$

This drain-source voltage puts both Q_2 and Q_3 right at the edge of the triode region. Thus, the minimum allowable output voltage is now

$$V_{out} > V_{eff1} + V_{eff2} = (n+1)V_{eff} \tag{6.6}$$

A common choice for n might be simply unity, in which case the current mirror operates correctly as long as

$$V_{out} > 2V_{eff} \tag{6.7}$$

With a typical value of V_{eff} between 0.2 V and 0.25 V, the wide-swing current mirror can guarantee that all of the transistors are in the active (i.e., saturation) region even when the voltage drop across the mirror is as small as 0.4 V to 0.5 V.

However, there is one other requirement that must be met to ensure that all transistors are in the active region. Specifically, we need

$$V_{DS4} > V_{eff4} = nV_{eff} \tag{6.8}$$

to guarantee that Q_4 is in the active region. To find V_{DS4}, we note that the gate of Q_3 is connected to the drain of Q_4, resulting in

$$V_{DS4} = V_{G3} - V_{DS3} = (V_{eff} + V_{tn}) - V_{eff} = V_{tn} \tag{6.9}$$

As a result, one need only ensure that V_{tn} be greater than nV_{eff} for Q_4 to remain in the active region—not a difficult requirement.

It should be noted that this circuit was analyzed assuming the bias current, I_{bias}, equals the input current, I_{in}. Since, in general, I_{in} may be a varying current level, there is some choice as to what I_{bias} value should be chosen. One choice is to set I_{bias} to the largest expected value for I_{in}. Such a choice will ensure that none of the devices exit their active region, though the drain-source voltage of Q_2 and Q_3 will be larger than necessary except when the maximum I_{in} is applied. As a result, some voltage swing will be lost. Perhaps the more common choice in a wide-swing opamp is to let I_{bias} equal the nominal value of I_{in}. With this setting, some devices will enter triode and the output impedance will be reduced for larger I_{in} values (say, during slew-rate limiting), but such an effect during transient conditions can often be tolerated.

Before leaving this circuit, a few design comments are worth mentioning. In most applications, an experienced designer would take $(W/L)_5$ smaller than the size given in Fig. 6.1 to bias transistors Q_2 and Q_3 with slightly larger drain-source voltages than the minimum required (perhaps 0.1 V to 0.15 V larger). This increase accounts for the fact that practical transistors don't have a sharp transition between the triode and active regions. This increase would also help offset a second-order effect due to the body effect of Q_1 and Q_4, which causes their threshold voltages to increase, which tends to push Q_2 and Q_3 more into the triode region. In addition, to save on power dissipation, the bias branch consisting of Q_5 and I_{bias} might be scaled to have lower currents while keeping the same current densities and, therefore, the same effective gate-source voltages. A final common modification is to choose the lengths of Q_2 and Q_3 just a little larger than the minimum allowable gate length (as the drain-source voltage across them is quite small), but Q_1 and Q_4 would be chosen to have longer gate lengths since the output transistor (i.e., Q_1) often has larger voltages across it. A typical size might be twice the minimum allowable channel length. This choice of gate lengths helps eliminate detrimental short-channel effects, such as the drain-to-substrate leakage currents of Q_1 that might otherwise result. Minimizing the lengths of Q_2 and Q_3 maximizes the frequency response, as their gate-source capacitances are the most significant capacitances contributing to high-frequency poles.

Finally, it should be mentioned that this current mirror is presently becoming the most popular CMOS current mirror and is therefore very important in many analog designs besides opamps. The next subsection will show how it can be incorporated into the constant g_m bias circuit described in Chapter 5.

Wide-Swing Constant-Transconductance Bias Circuit

It is possible to incorporate wide-swing current mirrors into the constant-transconductance bias circuit described in Chapter 5. This modification greatly minimizes most of the detrimental second-order imperfections caused by the finite-output impedance of the transistors, without greatly restricting signal swings. The complete circuit is shown in Fig. 6.2. This circuit is a modification of the circuit described in Fig 5.10, and has both wide-swing current mirrors and a start-up circuit.

The n-channel wide-swing cascode current mirror consists of transistors Q_1–Q_4, along with the diode-connected biasing transistor Q_5. The pair Q_3, Q_4 acts similarly to a diode-connected transistor at the input side of the mirror. The output current comes from Q_1. The gate voltages of cascode transistors Q_1 and Q_4 are derived by the diode-connected transistor Q_5. The current for this biasing transistor is actually derived from the bias loop via Q_{10} and Q_{11}.

Similarly, the p-channel wide-swing cascode current mirror is realized by Q_6–Q_9. Transistors Q_8 and Q_9 operate as a diode-connected transistor at the input side of the mirror. The current-mirror output current is the drain current of Q_6. The cascode transistors Q_6 and Q_9 have gate voltages derived from diode-connected Q_{14}, which has a bias current derived from the bias loop via Q_{12} and Q_{13}.

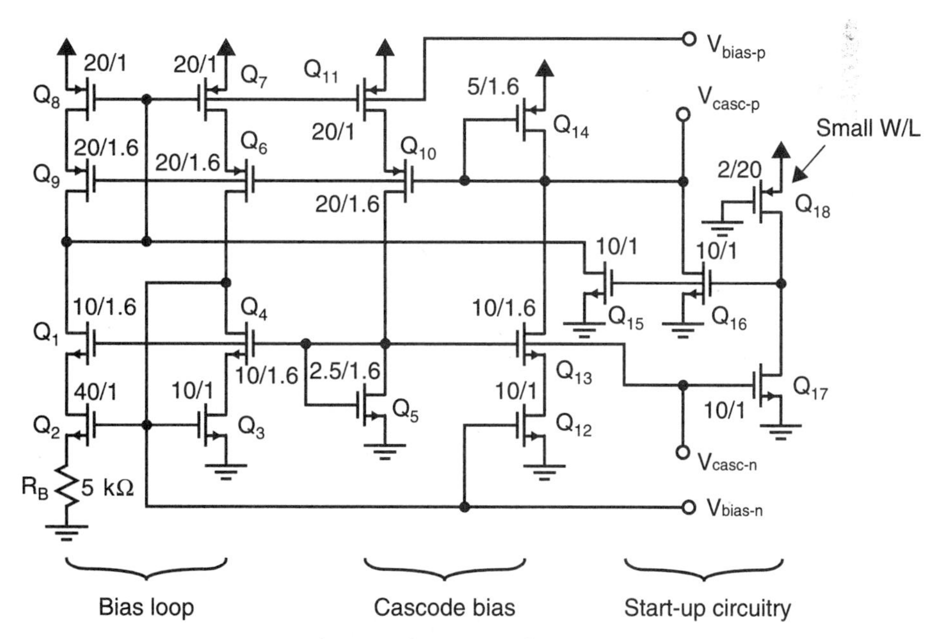

Fig. 6.2 A constant-transconductance bias circuit having wide-swing cascode current mirrors.

The bias loop does have the problem that at start-up it is possible for the current to be zero in all transistors, and the circuit will remain in this stable state forever. To ensure this condition does not happen, it is necessary to include start-up circuitry that affects only the bias-loop in the case that all currents in the loop are zero. An example of a start-up circuit consisting of transistors Q_{15}, Q_{16}, Q_{17}, and Q_{18} is also shown in Fig. 6.2. In the event that all currents in the bias loop are zero, Q_{17} will be off. Since Q_{18} operates as a high-impedance load that is always on, the gates of Q_{15} and Q_{16} will be pulled high. These transistors then will inject currents into the bias loop, which will start up the circuit. Once the loop starts up, Q_{17} will come on, sinking all of the current from Q_{18}, pulling the gates of Q_{15} and Q_{16} low, and thereby turning them off so they no longer affect the bias loop. This circuit is only one example of a start-up loop, and there are many other variations. For example, sometimes the p-channel transistor, Q_{18}, is replaced by an actual resistor (perhaps realized using a well resistor).

It is of interest to note that the bias circuit shown consists of four different loops—the main loop with positive feedback, the start-up loop that eventually gets disabled, and the two loops used for establishing the bias voltages for the cascode transistors. These latter two loops also constitute positive feedback but with very little gain.

Shown next to each transistor are reasonable W/L dimensions (in μm) of a possible realization that had its operation experimentally verified (it was realized in a 0.8-μm technology). Because this circuit allowed for accurately predictable transconductances, it also allows the performance of the realized opamp to be accurately predicted using moderately simple equations prior to realization.

Enhanced Output-Impedance Current Mirrors

Another variation on the cascode current mirror is often referred to as the *enhanced output-impedance current mirror.* A simplified form of this circuit is shown in Fig. 6.3, and it is used to increase the output impedance. The basic idea is to use a feedback amplifier to keep the drain-source voltage across Q_2 as stable as possible, irrespective of the output voltage. The addition of this amplifier ideally increases the

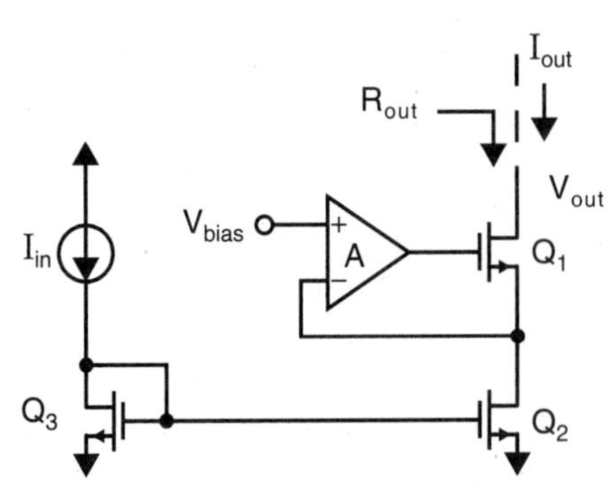

Fig. 6.3 The enhanced output-impedance current mirror.

output impedance by a factor equal to one plus the loop gain over that which would occur for a classical cascode current mirror. Specifically, using the results from Chapter 3,

$$R_{out} \cong g_{m1} r_{ds1} r_{ds1} (1 + A) \tag{6.10}$$

In practice, the output impedance might be limited by a parasitic conductance between the drain of Q_1 and its substrate due to short-channel effects. This parasitic conductance is a result of collisions between highly energized electrons resulting in electron-hole pairs to be generated with the holes escaping to the substrate. The generation of these electron-hole pairs is commonly called *impact ionization.*

It should be mentioned that the technique just described for output-impedance enhancement is not useful when bipolar transistors are used due to the finite base current of the transistors (see Problem 6.8).

EXAMPLE 6.1

Assume an amplifier, $A(s)$, has a transfer function given by

$$A(s) = \frac{A_0}{1 + s\tau'_1} \cong \frac{A_0}{s\tau'_1} \tag{6.11}$$

around its unity-gain frequency. Find approximate equations for any additional poles and/or zeros introduced by the gain enhancement circuit that are significant around the unity-gain frequency of the common-source amplifier shown in Fig. 6.4. Assume the gain of the common-source stage is given by

$$A_V(s) = \frac{V_{out}(s)}{V_{in}(s)} = -g_{m2}\left(R_{out}(s) \parallel \frac{1}{sC_L}\right) \tag{6.12}$$

where R_{out} is now a function of frequency and given by

$$R_{out}(s) = g_{m1} r_{ds1} r_{ds2} (1 + A(s)) \tag{6.13}$$

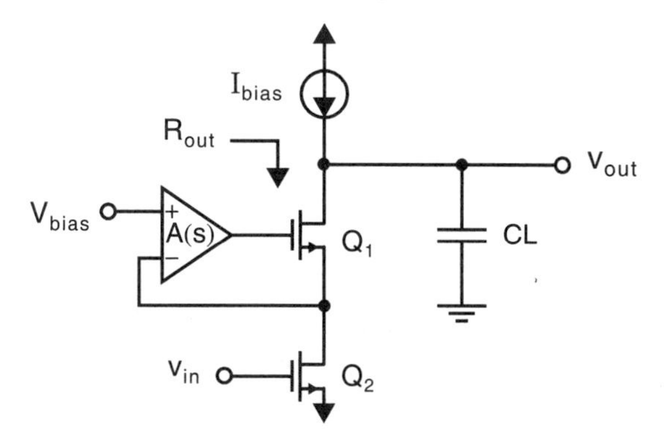

Fig. 6.4 A common-source amplifier with output-impedance enhancement.

Solution

Note that at high frequencies, it is not possible to assume that $|A(j\omega)| >> 1$. Working with conductances, we have

$$G_{out}(s) = \frac{1}{R_{out}(s)} = \frac{g_{ds1}g_{ds2}}{g_{m1}(1 + A(s))} \tag{6.14}$$

which can be inserted into (6.12) giving

$$\begin{aligned} A_V(s) &= \frac{-g_{m2}}{sC_L + G_{out}(s)} = \frac{-g_{m2}}{sC_L + \dfrac{g_{ds1}g_{ds2}}{g_{m1}(1 + A(s))}} \\ &= \frac{-g_{m2}(1 + A(s))}{sC_L(1 + A(s)) + \dfrac{g_{ds1}g_{ds2}}{g_{m1}}} \end{aligned} \tag{6.15}$$

Substituting (6.11) into (6.15) and rearranging gives

$$A_V(s) = \frac{-g_{m2}(A_0 + s\tau_1')}{s\left(A_0C_L + \dfrac{g_{ds1}g_{ds2}\tau_1'}{g_{m1}} + sC_L\tau_1'\right)} \tag{6.16}$$

From (6.16), we see that there is a left-hand-plane zero at a frequency given by

$$\omega_z = \frac{A_0}{\tau_1'} \tag{6.17}$$

which is the approximate unity-gain frequency of the amplifier, $A(s)$. Also, there are two poles, but the pole at dc is not accurate since we have made use of approximate equations that are valid only near the unity-gain frequency of the amplifier. However, the other pole location is a good approximation and it occurs in the left-half-plane given by (after some rearranging)

$$p_2 = \frac{A_0}{\tau_1'} + \frac{g_{ds1}g_{ds2}}{C_Lg_{m1}} = \frac{A_0}{\tau_1'} + \frac{1}{C_LR'} \tag{6.18}$$

Here, we have defined

$$R' = g_{m1}r_{ds1}r_{ds2} \tag{6.19}$$

which is approximately the output impedance of the cascode mirror without enhancement and is quite a large value. Furthermore, note that the term A_0/τ_1' is a value near the unity-gain frequency of the amplifier, and therefore, typically the first term in (6.18) dominates. Thus, the added pole frequency is approximately given by

$$p_2 \cong \frac{A_0}{\tau_1'} \cong \omega_z' \tag{6.20}$$

Therefore, like the new zero, the new pole also occurs around the unity-gain frequency of the amplifier, and their effects mostly cancel each other out. In addition, if the unity-gain frequency of the enhancement loop is greater than the unity-gain frequency of the opamp, the frequency response of the enhancement amplifier shouldn't have a large detrimental effect on the overall frequency response.

The enhanced output-impedance current mirror appears to have been originally proposed in [Hosticka, 1979] for single-ended input amplifiers and more recently described almost simultaneously in [Bult, 1990] and [Säckinger, 1990] for differential-input amplifiers. Bult proposed that complete opamps be used for each current mirror. In a modern fully differential opamp, this might increase the number of opamps by four, although these extra opamps can be scaled to have less power dissipation. The implementation proposed by Säckinger is shown in Fig. 6.5. The feedback amplifier in this case is realized by the common-source amplifier consisting of Q_3 and its current source I_{B1}. Assuming the output impedance of current source I_{B1} is approximately equal to r_{ds3}, the loop gain will be $(g_{m3}r_{ds3})/2$, and the final ideal output impedance will be given by

$$r_{out} \cong \frac{g_{m1}g_{m3}r_{ds1}r_{ds2}r_{ds3}}{2} \tag{6.21}$$

The circuit consisting of Q_4, Q_5, Q_6, I_{in}, and I_{B2} operates almost identically to a diode-connected transistor, but is used instead to guarantee that all transistor bias voltages are accurately matched to those of the output circuitry consisting of Q_1, Q_2, Q_3, and I_{B1}. As a result, I_{out} will very accurately match I_{in}.

The Säckinger realization is much simpler than that proposed by Bult, but has a major limitation in that the signal swing is significantly reduced. This reduction is a result of Q_2 and Q_5 being biased to have drain-source voltages much larger than the minimum required. Specifically, their drain-source voltages are given by

$$V_{DS2} = V_{DS5} = V_{eff3} + V_{tn} \tag{6.22}$$

rather than the minimum required, which would be equal to V_{eff2}. This limitation is especially harmful for modern technologies operating with power supply voltages of

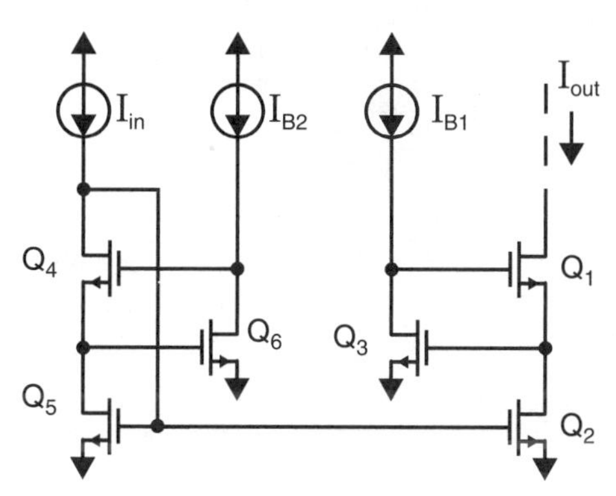

Fig. 6.5 The Säckinger implementation of the enhanced output-impedance current mirror.

3.3 V or lower. In the next subsection, an alternative realization is described that combines the wide-swing current mirror with the enhanced output-impedance circuit.

Wide-Swing Current Mirror with Enhanced Output Impedance

This current mirror was originally described in [Gatti, 1990] for use in current-mode continuous-time filters and was then used in the design of wide-signal-swing opamps [Coban, 1994; Martin, 1994]. It is very similar to the enhanced output-impedance current mirrors of [Säckinger, 1990], except that diode-connected transistors used as level shifters have been added in front of the common-source enhancement amplifiers, as shown in Fig. 6.6. At the output side, the level shifter is the diode-connected transistor, Q_4, biased with current I_{bias}. The circuitry at the input basically acts as a diode-connected transistor while ensuring that all bias voltages are matched to the output circuitry so that I_{out} accurately matches I_{in}. Shown next to each transistor is a reasonable width in μm. Note that for the case in which

$$I_{bias} = I_{in}/7 \tag{6.23}$$

all transistors are biased with nearly the same current density, except for Q_3 and Q_7. As a result, all transistors have the same effective gate-source voltages, V_{eff}, except for Q_3 and Q_7, which have gate-source voltages of $2V_{eff}$ because they are biased at four times the current density. Thus, we find that the gate voltage of Q_3 equals

$$V_{G3} = 2V_{eff} + V_{tn} \tag{6.24}$$

and the drain-source voltage of Q_2 is given by

$$V_{DS2} = V_{S4} = V_{G3} - V_{GS4} = (2V_{eff} + V_{tn}) - (V_{eff} + V_{tn}) = V_{eff} \tag{6.25}$$

Therefore, Q_2 is biased on the edge of the triode region and the minimum output voltage is given by

$$V_{out} > V_{DS2} + V_{eff1} = 2V_{eff} \tag{6.26}$$

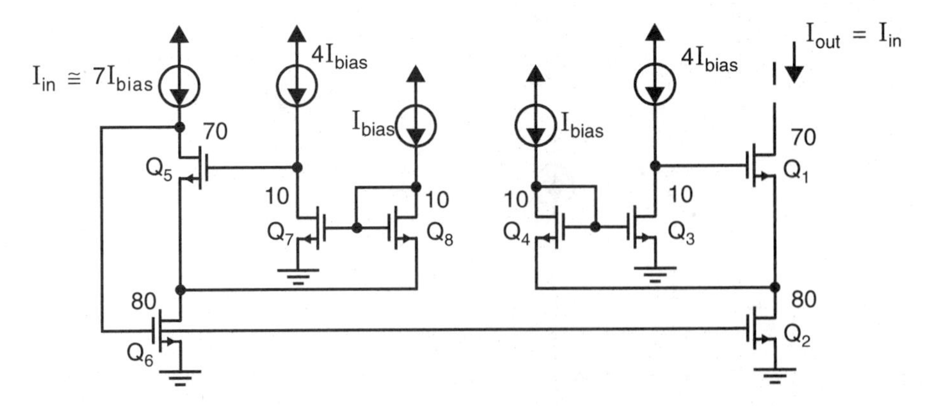

Fig. 6.6 A wide-swing current mirror with enhanced output impedance. Current source $7I_{bias}$ is typically set to the nominal or maximum input current, I_{in}.

With the shown values for the W/L ratios, the power dissipation would be almost doubled over that of a classical cascode current mirror, assuming I_{bias} is set to one-seventh the nominal or maximum input current value, I_{in}. However, it is possible to bias the enhancement circuitry at lower densities and thereby save on power dissipation, albeit at the expense of speed, as the additional poles introduced by the enhancement circuitry would then be at lower frequencies.

A slightly modified version of the wide-swing enhanced output-impedance mirror is shown in Fig. 6.7. This variation obtains the bias voltage of the cascode transistor of the input diode-connected branch from the bias-generation circuitry of Fig. 6.2. This circuit suffers slightly with regard to large-signal dc matching, but has less power dissipation and area than the circuit of Fig. 6.6. Notice also that Q_2 of Fig. 6.6 has now been separated into two transistors, Q_2 and Q_5, in Fig. 6.7. This change allows one to design an opamp without the output-impedance enhancement, but with wide-swing mirrors. Then, when output-impedance enhancement is desired, one simply includes the appropriate enhancement amplifiers and the only change required to the original amplifier is to connect the gates of the output cascode mirrors of the appropriate current mirrors to the enhancement amplifiers rather than the bias-generation circuit. Finally, it has been found through simulation that the modified circuit of Fig. 6.7 is less prone to instability than the circuit of Fig. 6.6.

It is predicted that this current mirror will become more important as the newer technologies force designers to use smaller power-supply voltages, which limit voltage swings. Also, the short-channel effects of newer technologies make the use of current mirrors with enhanced output impedance more desirable.

There are a couple of facts that designers should be aware of when they use output-impedance enhancement. The first fact is that sometimes it may be necessary to add local compensation capacitors to the enhancement loops to prevent ringing during transients. The second fact is that the inclusion of output-impedance enhancement can substantially slow down the settling times for large-signal transients. This is because during large-signal transients, the outputs of the enhancement loops can slew to voltages very different from those required after settling, and it can take a while for the outputs to

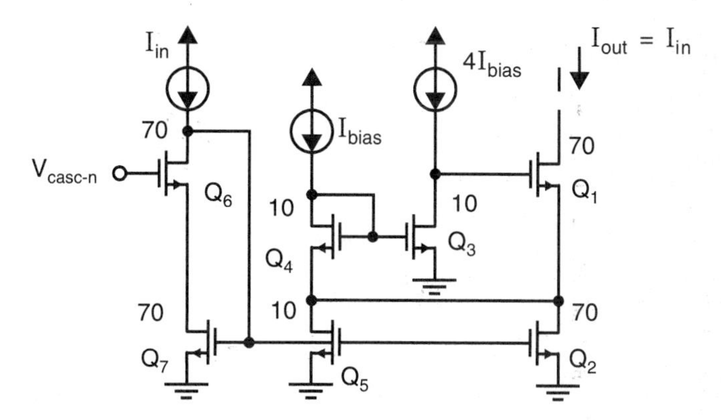

Fig. 6.7 A modified wide-swing current mirror with enhanced output impedance.

Fig. 6.8 (*a*) A symbol representing a current mirror. (*b*) Example of a simple current mirror.

return to the necessary voltages. A typical settling-time difference might be around a 50-percent increase. Whether this is worth the substantial increase in gain (perhaps as much as 30 dB or so) depends on the individual application.

Current-Mirror Symbol

It should now be clear that there are a number of different current-mirror circuits, each having various advantages and disadvantages, that can be used in a particular circuit application. Thus, from an architectural perspective, it is often desirable to describe a circuit without showing which particular current mirror is used. In these cases, we make use of the current-mirror symbol shown in Fig. 6.8(*a*). The arrow is on the input side of the mirror, which is the low-impedance side (with an impedance typically equal to $1/g_m$). The arrow also designates the direction of current flow on the input side. The ratio, $1:K$, represents the current gain of the mirror. For example, the current mirror shown in Fig. 6.8(*a*) might be realized by the simple current mirror shown in Fig. 6.8(*b*). Finally, it should be mentioned that for illustrative purposes, some circuits in this book will be shown with a particular current mirror (such as the wide-swing mirror) to realize a circuit architecture. However, it should be kept in mind that similar circuits using almost any of the current mirrors just described (or described elsewhere) are possible.

6.2 FOLDED-CASCODE OPAMP

Many modern integrated CMOS opamps are designed to drive only capacitive loads. With such capacitive-only loads, it is not necessary to use a voltage buffer to obtain a low output impedance for the opamp. As a result, it is possible to realize opamps having higher speeds and larger signal swings than those that must also drive resistive loads. These improvements are obtained by having only a single high-impedance node at the output of an opamp that drives only capacitive loads. The admittance seen at all other nodes in these opamps is on the order of a transistor's transconductance, and thus they have relatively low impedance. By having all internal nodes of relatively low impedance, the speed of the opamp is maximized. It should also be mentioned that these low node impedances result in reduced voltage signals at all nodes other than the output node; however, the current signals in the various transistors can be quite large.[1] With these opamps, we shall see that the compensation is usually achieved by the load capacitance. Thus, as the load capacitance gets larger, the opamp usually becomes more stable but also slower. One of the most important parameters of these modern opamps is their transconductance value (i.e., the ratio of the output current to the input voltage). Therefore, some designers refer to these modern opamps as transconductance opamps, or *Operational Transconductance Amplifiers* (OTAs).

An example of an opamp with a high-output impedance is the folded-cascode opamp, as shown in Fig. 6.9. The design shown is a differential-input single-ended output design. Note that all current mirrors in the circuit are wide-swing cascode current mirrors, discussed previously. The use of these mirrors results in high-output impedance for the mirrors (compared to simple current mirrors), thereby maximizing the dc gain of the opamp. The basic idea of the folded-cascode opamp is to apply cascode transistors to the input differential pair but using transistors opposite in type from those used in the input stage. For example, the differential-pair transistors consisting of Q_1 and Q_2 are n-channel transistors in Fig. 6.9, whereas the cascode transistors consisting of Q_5 and Q_6 are p-channel transistors. This arrangement of opposite-type transistors allows the output of this single gain-stage amplifier to be taken at the same bias-voltage levels as the input signals. It should be mentioned that even though a folded-cascode amplifier is basically a single gain stage, its gain can be quite reasonable, on the order of 700 to 3,000. Such a high gain occurs because the gain is determined by the product

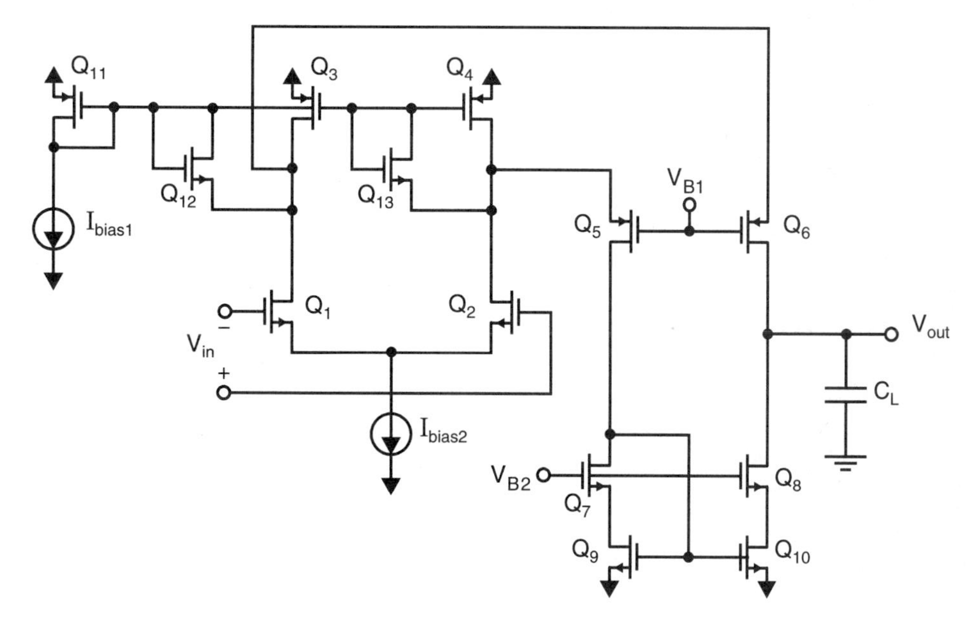

Fig. 6.9 A folded-cascode opamp.

1. Because of their reduced voltage signals but large current signals, these types of opamps are sometimes referred to as *current-mode opamps*.

of the input transconductance and the output impedance, and the output impedance is quite high due to the use of cascode techniques.

The shown differential-to-single-ended conversion is realized by the wide-swing current mirror composed of Q_7, Q_8, Q_9, and Q_{10}. In a differential-output design, these might be replaced by two wide-swing cascode current sinks, and common-mode feedback circuitry would be added, as is discussed in a later section in this chapter.

It should be mentioned that a simplified bias network is shown so as not to complicate matters unnecessarily when the intent is to show the basic architecture. In practical realizations, Q_{11} and I_{bias1} might be replaced by the constant-transconductance bias network of Fig. 6.2. In this case, V_{B1} and V_{B2} would be connected to $V_{casc\text{-}p}$ and $V_{casc\text{-}n}$, respectively, of Fig. 6.2 to maximize the output voltage signal swing.

An important addition of the folded-cascode opamp shown is the inclusion of two extra transistors, Q_{12} and Q_{13}. These two transistors serve two purposes. One is to increase the slew-rate performance of the opamp, as will be discussed in Example 6.2. However, more importantly, during times of slew-rate limiting, these transistors prevent the drain voltages of Q_1 and Q_2 from having large transients where they change from their small-signal voltages to voltages very close to the negative power-supply voltage. Thus the inclusion of Q_{12} and Q_{13} allows the opamp to recover more quickly following a slew-rate condition.

The compensation is realized by the load capacitor, C_L, and realizes dominant-pole compensation. In applications where the load capacitance is very small, it is necessary to add additional compensation capacitance in parallel with the load to guarantee stability. If lead compensation is desired, a resistor can be placed in series with C_L. While lead compensation may not be possible in some applications, such as when the compensation capacitance is mostly supplied by the load capacitance, it is more often possible than many designers realize (i.e., it is often possible to include a resistor in series with the load capacitance).

The bias currents for the input differential-pair transistors are equal to $I_{bias2}/2$. The bias current of one of the p-channel cascode transistors, Q_5 or Q_6, and hence the transistors in the output-summing current mirror as well, is equal to the drain current of Q_3 or Q_4 minus $I_{bias2}/2$. This drain current is established by I_{bias1} and the ratio of $(W/L)_3$, or $(W/L)_4$, to $(W/L)_{11}$. Since the bias current of one of the cascode transistors is derived by a current subtraction, for it to be accurately established, it is necessary that both I_{bias1} and I_{bias2} be derived from a single-bias network. In addition, any current mirrors used in deriving these currents should be composed of transistors realized as parallel combinations of unit-size transistors. This approach eliminates inaccuracies due to second-order effects caused by transistors having non-equal widths.

Small-Signal Analysis

In a small-signal analysis of the folded-cascode amplifier, it is assumed that the differential output current from the drains of the differential pair, Q_1, Q_2, is applied to the load capacitance, C_L. Specifically, the small-signal current from Q_1 passes directly from the source to the drain of Q_6 and thus C_L, while the current from Q_2 goes indirectly through Q_5 and the current mirror consisting of Q_7 to Q_{10}[2]. Although these two paths have slightly different transfer functions due to the poles caused by the current mirror, when an n-channel transistor current mirror is used, a pole-zero doublet occurs at frequencies quite a bit greater than the opamp unity-gain frequency and can usually be ignored. Hence, ignoring high-frequency poles and zero, the approximate small-signal transfer function for the folded-cascode opamp is given by

$$A_V = \frac{V_{out}(s)}{V_{in}(s)} = g_{m1}Z_L(s) \tag{6.27}$$

Here, g_{m1} is the transconductance of each of the transistors in the input differential pair and $Z_L(s)$ is the impedance to ground seen at the output node. This impedance consists of the parallel combination of the output load capacitance, the impedance of any additional network added for stability, and the output impedance of the opamp (i.e., looking into the drains of Q_6 and Q_8). Thus, when the compensation is realized by the output capacitance only, we have

$$A_V = \frac{g_{m1}r_{out}}{1 + sr_{out}C_L} \tag{6.28}$$

where r_{out} is the output impedance of the opamp. This impedance is quite high, on the order of $g_m r_{ds}^2/2$ or greater if output-impedance enhancement is used.

For mid-band and high frequencies, the load capacitance dominates, and we can ignore the unity term in the denominator and thus have

$$A_V \cong \frac{g_{m1}}{sC_L} \tag{6.29}$$

from which the unity-gain frequency of the opamp is found to be

$$\omega_t = \frac{g_{m1}}{C_L} \tag{6.30}$$

Therefore, for large load capacitances, maximizing the transconductance of the input transistors maximizes the bandwidth, assuming the load capacitance is large enough so that the unity-gain frequency is much less than the limit imposed by the second poles. The transconductance of the input transistors is maximized by using wide n-channel devices and ensuring that the input transistor pair's bias current is substantially larger than the bias current of the cascode transistors and current mirror. Note that this approach also maximizes the dc gain (i.e., $g_{m1}r_{out}$) since not only does it maximize g_{m1}, but it also maximizes r_{out} by resulting in all transistors connected to the output node being biased at lower current levels (for a given total power dissipation). A practical upper limit on the ratio of the bias currents of the input transistors to the currents of the cascode transistors might be around four or so. If too high a ratio is used, the bias currents of the cascode transistors cannot be well established since they

2. This is based on the assumption that g_{m5} and g_{m6} are much larger than g_{ds3} and g_{ds4}.

are derived by current subtractions. Another advantage of having very large transconductances for the input devices is that the thermal noise due to this input pair is reduced. Thus, since much of the bias current in folded-cascode opamps flows through the input differential pair, these opamps often have a better thermal noise performance than other opamp designs having the same power dissipation.

When the load is a series combination of a compensation capacitance, C_L, and a lead resistance, R_C, the small-signal transfer function is given by

$$A_V = \frac{g_{m1}}{\frac{1}{r_{out}} + \frac{1}{R_C + 1/sC_L}} \cong \frac{g_{m1}(1 + sR_CC_L)}{sC_L} \tag{6.31}$$

where the approximate term is valid at mid and high frequencies. Here we see that, as in Chapter 5, R_C can be chosen to place a zero at 1.2 times the unity-gain frequency.

The second poles of this opamp are primarily due to the time constants introduced by the impedance and parasitic capacitances at the sources of the p-channel cascode transistors, Q_5 and Q_6. The impedances at these nodes is one over the transconductances of the cascode transistors.[3] Since p-channel transistors are used here, possibly biased at lower currents, these impedances are typically substantially greater than the source impedances of most n-channel transistors in the signal path. As a consequence, when high-frequency operation is important, these impedances can be reduced by making the currents in the p-channel cascode transistors around the same level as the bias currents of the input transistors. The parasitic capacitance at the sources of the cascode transistors is primarily due to the gate-source capacitances of the cascode transistors as well as the drain-to-bulk and drain-to-gate capacitances of the input transistors and the current-source transistors Q_3 and Q_4. Therefore, minimizing junction areas and peripheries at these two nodes is important.

Slew Rate

The diode-connected transistors, Q_{12} and Q_{13}, are turned off during normal operation and have almost no effect on the opamp. However, they substantially improve the operation during times of slew-rate limiting [Law, 1983]. To appreciate their benefit, consider first what happens during times of slew-rate limiting when they are not present. Assume there is a large differential input voltage that causes Q_1 to be turned on hard and Q_2 to be turned off. Since Q_2 is off, all of the bias current of Q_4 will be directed through the cascode transistor Q_5, through the n-channel current mirror, and out of the load capacitance. Thus, the output voltage will decrease linearly with a slew-rate given by

3. This is true only at high frequencies, where the output impedance of the amplifier is small due to the load and/or compensation capacitance. This is the case of interest here. At low frequencies, the impedance looking into the source of Q_6 is given by $R_{s6} = 1/g_{m6} + R_L/(g_{m6}r_{ds6})$, where R_L is the resistance seen looking out the drain of Q_6.

$$SR = \frac{I_{D4}}{C_L} \tag{6.32}$$

Also, since all of I_{bias2} is being diverted through Q_1, and since this current is usually designed to be greater than I_{D3}, both Q_1 and the current source I_{bias2} will go into the triode region, causing I_{bias2} to decrease until it is equal to I_{D3}. As a result, the drain voltage of Q_1 approaches that of the negative power-supply voltage. When the opamp is coming out of slew-rate limiting, the drain voltage of Q_1 must slew back to a voltage close to the positive power supply before the opamp operates in its linear region again. This additional slewing time greatly increases the distortion and also increases the transient times during slew-rate limiting (which occurs often for opamps used in switched-capacitor applications).

Next, consider the case where the diode-connected transistors, Q_{12} and Q_{13}, are included. *Their main purpose is to clamp the drain voltages of* Q_1 *or* Q_2 *so they don't change as much during slew-rate limiting*. A second, more subtle effect dynamically increases the bias currents of both Q_3 and Q_4 during times of slew-rate limiting. This increased bias current results in a larger maximum current available for charging or discharging the load capacitance. To see this increase in bias current, consider a case similar to the one just described, where a large differential input causes Q_1 to be fully on while Q_2 is off. In this case, the diode-connected transistor Q_{12} conducts with the current through Q_{12} coming from the diode-connected transistor Q_{11}. Thus, the current in Q_{11} increases, causing the currents in bias transistors Q_3 and Q_4 to also increase, until the sum of the currents of Q_{12} and Q_3 are equal to the bias current I_{bias2}. Note that the current in Q_4 also increases since it is equal to the current in Q_3. This increase in bias current of Q_4 results in an increase of the maximum current available for discharging C_L. In summary, not only are the voltage excursions less, but the maximum available current for charging or discharging the load capacitance is also greater during times of slew-rate limiting.

EXAMPLE 6.2

Find reasonable transistor sizes for the folded-cascode opamp shown in Fig. 6.9 to satisfy the following design parameters. Also find the unity-gain frequency and slew rate, both without and with the clamp transistors. What would be a reasonable size for a compensation resistor in series with C_L, if lead compensation were desired?

- Assume ± 2.5-V power supplies are used and limit the power dissipation of the opamp to 2 mW.
- Set the ratio of the current in the input transistors to that of the cascode transistors to be 4:1. Also, set the bias current of Q_{11} to be 1/30th that of Q_3 (or Q_4) such that its current can be ignored in the power dissipation calculation.
- The maximum transistor width should be 300 μm and channel lengths of 1.6 μm should be used in all transistors.

- All transistors should have effective gate-source voltages of around 0.25 V except for the input transistors, whose widths should be set to the maximum value of 300 μm. Also, round all transistor widths to the closest multiple of 10 μm, keeping in mind that if a larger transistor is to be matched to a smaller one, the larger transistor should be built as a parallel combination of smaller transistors.
- Finally, assume the load capacitance is given by $C_L = 10\text{ pF}$, and that $\mu_n C_{ox} = 3\mu_p C_{ox} = 96\ \mu A/V^2$.

Solution

The total current in the opamp, I_{total}, excluding the current in the bias network, is equal to $I_{D3} + I_{D4}$, which is equal to $2(I_{D1} + I_{D6})$. Defining $I_B \equiv I_{D5} = I_{D6}$ and noting that we are asked to make $I_{D1} = 4I_{D6}$, we have

$$I_{total} = 2(I_{D1} + I_{D6}) = 2(4I_B + I_B) = 10I_B \tag{6.33}$$

Since the power dissipation is specified to be 2 mW, we have

$$I_B = I_{D5} = I_{D6} = \frac{I_{total}}{10} = \frac{P_{diss}/5\text{ V}}{10} = 40\ \mu A \tag{6.34}$$

This bias current for I_B implies that $I_{D3} = I_{D4} = 5I_{D5} = 200\ \mu A$, and $I_{D1} = I_{D2} = 4I_{D5} = 160\ \mu A$. Now, we let all transistor channel lengths be 1.6 μm. This choice allows us to immediately determine the sizes of most transistors using

$$\left(\frac{W}{L}\right)_i = \frac{2I_{Di}}{\mu_i C_{ox} V_{eff i}^2} \tag{6.35}$$

and then round to the closest multiple of 10 in transistor widths. When the widths found were larger than 300 μm, they were limited to 300 μm. Table 6.1 lists reasonable values for the resulting dimensions of all transistors. Note that the larger widths are exactly divisible by the smaller widths of a transistor of the same type and thus allows larger transistors to be realized as a parallel combination of smaller transistors. The width of Q_{11} was determined from the requirement that $I_{D11} = I_{D3}/30 = 6.6\ \mu A$. The widths of Q_{12} and Q_{13} were somewhat arbitrarily chosen to equal the width of Q_{11}. Note that, as requested, the widths of the input transistors were set to a value of 300 μm to maximize their transconductance. The transconductance of the input transistors

Table 6.1 The transistor dimensions (in μm) of the opamp of Fig. 6.9.

Q_1	300/1.6	Q_6	60/1.6	Q_{11}	10/1.6
Q_2	300/1.6	Q_7	20/1.6	Q_{12}	10/1.6
Q_3	300/1.6	Q_8	20/1.6	Q_{13}	10/1.6
Q_4	300/1.6	Q_9	20/1.6		
Q_5	60/1.6	Q_{10}	20/1.6		

is given by

$$g_{m1} = \sqrt{2I_{D1}\mu_n C_{ox}(W/L)_1} = 2.4 \text{ mA/V} \tag{6.36}$$

The unity-gain frequency of the opamp is given by

$$\omega_t = \frac{g_{m1}}{C_L} = 2.4\times10^8 \text{ rad/s} \Rightarrow f_t = 38 \text{ MHz} \tag{6.37}$$

A reasonable size for the lead resistor, R_C, in series with C_L, is given by

$$R_C = \frac{1}{1.2C_L\omega_t} = \frac{1}{1.2g_{m1}} = 347\ \Omega \tag{6.38}$$

The slew rate without the clamp transistors is given by

$$SR = \frac{I_{D4}}{C_L} = 20 \text{ V}/\mu\text{s} \tag{6.39}$$

When the clamp transistors are included, during slew-rate limiting, we have

$$I_{D12} + I_{D3} = I_{bias2} \tag{6.40}$$

But

$$I_{D3} = 30I_{D11} \tag{6.41}$$

and

$$I_{D11} = 6.6\ \mu\text{A} + I_{D12} \tag{6.42}$$

Substituting (6.41) and (6.42) into (6.40) and solving for I_{D11} gives

$$I_{D11} = \frac{I_{bias2} + 6.6\ \mu\text{A}}{31} \tag{6.43}$$

which implies that the value of I_{D11} during slew-rate limiting equals 10.53 μA and $I_{D3} = I_{D4} = 30I_{D11} = 0.32$ mA, which is substantially larger than the slew current available without the clamp transistors. This larger bias current will give a slew-rate value of

$$SR = \frac{I_{D4}}{C_L} = 32 \text{ V}/\mu\text{s} \tag{6.44}$$

More importantly, the time it takes to recover from slew-rate limiting will be substantially decreased.

6.3 CURRENT-MIRROR OPAMP

Another popular opamp often used when driving on-chip capacitive-only loads is shown in simplified form in Fig. 6.10. It is immediately seen that all nodes are low impedance except for the output node. By using good current mirrors having high

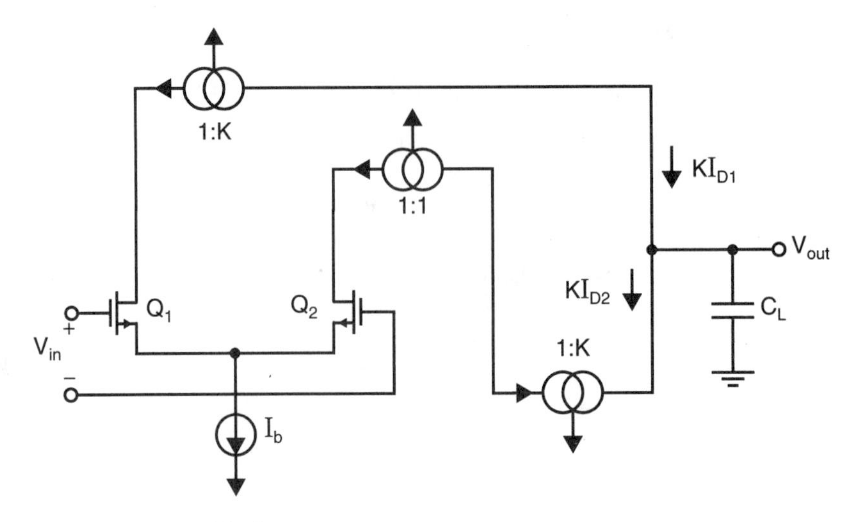

Fig. 6.10 A simplified current-mirror opamp.

output impedance, a reasonable overall gain can be achieved. A more detailed example of a current-mirror opamp is shown in Fig. 6.11.

The overall transfer function of this opamp will closely approximate a single-pole operation. In a similar analysis to that given for the folded-cascode opamp, we have

$$A_V = \frac{V_{out}(s)}{V_{in}(s)} = Kg_{m1}Z_L(s) = \frac{Kg_{m1}r_{out}}{1 + sr_{out}C_L} \cong \frac{Kg_{m1}}{sC_L} \tag{6.45}$$

The K factor is the current gain from the input transistors to the output sides of the current mirrors connected to the output node. Using (6.45), we can solve for the unity-gain frequency, resulting in

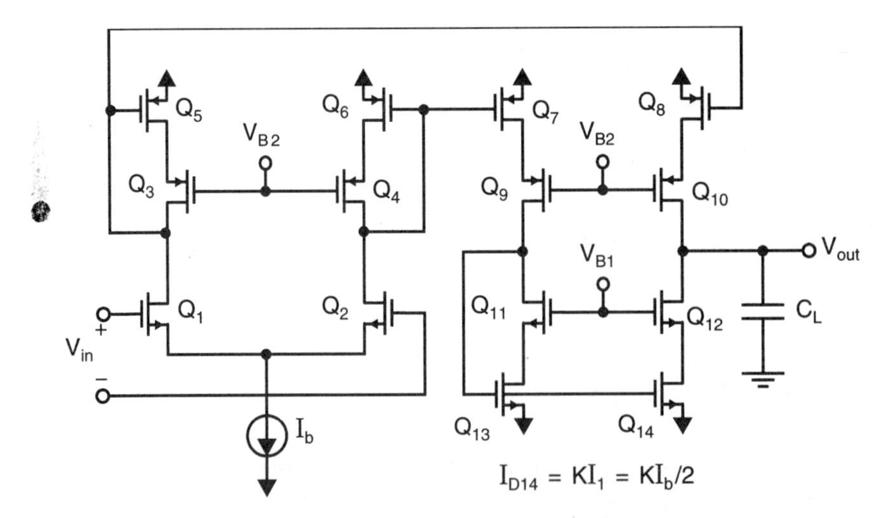

Fig. 6.11 A current-mirror opamp with wide-swing cascode current mirrors.

$$\omega_t = \frac{Kg_{m1}}{C_L} = \frac{K\sqrt{2I_{D1}\mu_n C_{ox}(W/L)_1}}{C_L} \tag{6.46}$$

If the power dissipation is specified, the total current,

$$I_{total} = (3 + K)I_{D1} \tag{6.47}$$

is known for a given power-supply voltage. Substituting (6.47) into (6.46), we obtain

$$\omega_t = \frac{K\sqrt{2\left(\frac{I_{total}}{3+K}\right)\mu_n C_{ox}(W/L)_1}}{C_L} = \frac{K}{\sqrt{3+K}}\,\frac{\sqrt{2I_{total}\mu_n C_{ox}(W/L)_1}}{C_L} \tag{6.48}$$

Obviously, for larger values of K, the opamp's transconductance is larger (i.e., Kg_{m1}), and therefore, the unity-gain frequency is also larger. This simple result assumes the unity-gain frequency is limited by the load capacitance rather than any high-frequency poles caused by the time constants of the internal nodes. A practical upper limit on K might be around five. The use of large K values also maximizes the gain for I_{total} fixed since r_{out} is roughly independent of K for large K.

In the circuit shown in Fig. 6.11, the important nodes for determining the non-dominant poles are the drain of Q_1, primarily, and the drains of Q_2 and Q_9, secondly. Increasing K increases the capacitances of these nodes while also increasing the equivalent resistances.[4] As a result, the equivalent second pole moves to lower frequencies. If K is increased too much, an increase in C_L will be required to keep ω_t below the frequency of the equivalent second pole to maintain stability. Thus, increasing K decreases the bandwidth when the equivalent second poles dominate. In the case where the load capacitance is small, the equivalent second pole will limit the unity-gain frequency of the opamp, and if it is very important that speed is maximized, K might be taken as small as one. From experience it has been found that a reasonable compromise for a general-purpose opamp might be to let $K = 2$.

The slew rate of the current-mirror opamp is found by assuming there is a very large input voltage. In this case, all of the bias current of the first stage will be diverted through either Q_1 or Q_2. This current will be amplified by the current gain from the input stage to the output stage, (i.e., K). Thus, during slew-rate limiting, the total current available to charge or discharge C_L is KI_b. The slew rate is therefore given by

$$SR = \frac{KI_b}{C_L} \tag{6.49}$$

For a given total power dissipation, this slew rate is maximized by choosing a large K value. For example, with $K = 4$ and during slew-rate limiting, 4/5 of the total bias current of the opamp will be available for charging or discharging C_L. This result gives a current-mirror opamp superior slew rates when compared to a folded-cascode

4. Increasing K implies that the currents at the input sides of the current mirrors are smaller, for a given total power dissipation. Also, the widths of transistors at the input sides of current mirrors will be smaller. Both of these effects cause the transconductances of transistors at the input side of current mirrors to be smaller, and therefore the impedances (in this case, 1/gm) will be larger.

opamp, even when the clamp transistors have been included in the folded-cascode opamp. Also, there are no problems with large voltage transients during slew-rate limiting for the current-mirror opamp.

In summary, due primarily to the larger bandwidth and slew rate, the current-mirror opamp is usually preferred over a folded-cascode opamp. However, it will suffer from larger thermal noise when compared to a folded-cascode amplifier because its input transistors are biased at a lower proportion of the total bias current and therefore have a smaller transconductance. As seen in the chapter on noise, a smaller transconductance results in larger thermal noise when the noise is referred back to the input of the opamp.

EXAMPLE 6.3

Assume the current-mirror opamp shown in Fig. 6.11 has all transistor lengths equal to 1.6 μm and transistor widths as given in Table 6.2. Notice that $K = 2$. Assume $C_L = 10\ \text{pF}$, $\mu_n C_{ox} = 3\mu_p C_{ox} = 96\ \mu\text{A/V}^2$, the total power dissipation is specified to be 2 mW (ignoring the power dissipation of the bias network, which would normally be small), and ±2.5 V power supplies are used.

Find the slew rate and the unity-gain frequency, assuming the equivalent second pole does not dominate. Estimate the equivalent second pole, assuming $C_{ox} = 1.92\ \text{fF}/\mu\text{m}^2$. Would it be necessary to increase C_L if a 75° phase margin were required without using lead compensation? What if lead compensation were used?

Solution

Since the total bias current is given by $(3 + K)I_b/2$, we have

$$I_b = \frac{2I_{total}}{(3+K)} = \frac{2P_{diss}/5\ \text{V}}{(3+K)} = 160\ \mu\text{A} \tag{6.50}$$

Thus, the bias currents of all transistors, except those in the output stage, are 80 μA. The bias currents of transistors in the output stage are twice that of the input stage, which is 160 μA. The transconductance of the input transistors is

$$g_{m1} = \sqrt{2I_{D1}\mu_n C_{ox}(W/L)_1} = 1.7\ \text{mA/V} \tag{6.51}$$

Table 6.2 The transistor sizes (in μm) for the opamp used in Example 6.3.

Q_1	300/1.6	Q_7	60/1.6	Q_{13}	30/1.6
Q_2	300/1.6	Q_8	120/1.6	Q_{14}	60/1.6
Q_3	60/1.6	Q_9	60/1.6		
Q_4	60/1.6	Q_{10}	120/1.6		
Q_5	60/1.6	Q_{11}	30/1.6		
Q_6	60/1.6	Q_{12}	60/1.6		

The transconductance of the opamp will be K times this or 3.4 mA/V. The unity-gain frequency is now given by

$$\omega_t = \frac{Kg_{m1}}{C_L} = 3.4\times10^8 \text{rad/s} \Rightarrow f_t = 54 \text{ MHz} \tag{6.52}$$

This result should be compared to the 38 MHz unity-gain frequency of the 2 mW folded-cascode opamp of the previous example, showing that current-mirror opamps are superior when driving large capacitive loads.

Continuing, the slew rate is given by

$$SR = \frac{KI_b}{C_L} = 32 \text{ V}/\mu\text{s} \tag{6.53}$$

which compares favorably to 20 $\text{V}/\mu\text{s}$ for the folded-cascode opamp without the clamp transistors (although it equals the value of 32 $\text{V}/\mu\text{s}$ for the folded-cascode opamp with the clamp transistors).

When estimating the equivalent second pole, the junction capacitances will be ignored to simplify matters. In reality, the drain-to-bulk capacitances of the input transistors could contribute a significant factor, and this amount can be determined using simulation. The dominant node almost certainly will occur at the drain of Q_1. The impedance at this node is given by

$$R_1 = 1/g_{m5} = 2.29 \text{ k}\Omega \tag{6.54}$$

In addition, the capacitance will be primarily due to the gate-source capacitances of Q_5 and Q_8. We therefore have

$$C_1 = C_{gs5} + C_{gs8} = (1+K)C_{gs5} \tag{6.55}$$

resulting in

$$C_1 = (1+K)C_{ox}W_5 L_5 = 0.553 \text{ pF} \tag{6.56}$$

With these values of impedance, the time constant for this node is given by

$$\tau_1 = R_1C_1 = 1.27 \text{ ns} \tag{6.57}$$

In a similar manner, we can calculate the impedances and, hence, the time constant for the drain of Q_2 to be $R_2 = 2.29 \text{ k}\Omega$, $C_2 = 0.369 \text{ pF}$, and $\tau_2 = R_2C_2 = 0.85 \text{ ns}$. The other important time constant comes from the parasitic capacitors at the drain of Q_9. Here, we have

$$R_3 = 1/g_{m13} = 1.86 \text{ k}\Omega \tag{6.58}$$

and

$$C_3 = C_{gs13} + C_{gs14} = (1+K)C_{ox}W_{13} L_{13} = 0.276 \text{ pF} \tag{6.59}$$

resulting in $\tau_3 = 0.51$ ns. The time constant of the equivalent second pole can now be estimated to be given by

$$\tau_{2eq} = \tau_1 + \tau_2 + \tau_3 = 2.6 \text{ ns} \tag{6.60}$$

This result gives an equivalent second pole at the approximate frequency of

$$p_{2eq} = \frac{1}{\tau_{2eq}} = 3.8\times10^{8}\ \text{rad/s} = 2\pi\times 61\ \text{MHz} \tag{6.61}$$

If lead compensation is not used and a 75° phase margin is required, the unity-gain frequency must be constrained to be less than 0.27 times the equivalent second pole from Table 5.1 of Chapter 5. Specifically, the unity-gain frequency must be less than 16.5 MHz. To achieve this unity-gain frequency, we need to increase C_L by the factor 54 MHz/16.5 MHz, resulting in C_L being 32.7 pF.

If lead compensation is to be used, the unity-gain frequency can be chosen so that only a 55° phase margin is achieved *before* the lead resistor is added. This approach would allow the unity-gain frequency to be as high as 0.7 times the equivalent second-pole frequency, which is 42.7 MHz. This unity-gain frequency can be realized by increasing C_L to 10 pF times 54 MHz/42.7 MHz, resulting in the new value of C_L being 12.6 pF. After the lead resistor is added, the unity-gain frequency increases by about 20 percent to 51 MHz, which is *over a three times improvement as compared to when lead compensation is not used.* The benefits of lead compensation are obvious.

As a final comment, notice that the delay through the signal path from the drain of Q_2 to the output (i.e., $\tau_2 + \tau_3 = 1.36$ ns) is approximately the same as the delay through the signal path from the drain of Q_1 to the output, which is $\tau_1 = 1.27$ ns. This near equivalence helps to minimize harmful effects caused by the pole-zero doublet introduced at high frequencies (i.e., around the frequency of the second pole) by the existence of two signal paths.

6.4 LINEAR SETTLING TIME REVISITED

We saw in Chapter 5 that the time constant for linear settling time was equal to the inverse of $\omega_{-3\ dB}$ for the closed-loop circuit gain. We also saw that $\omega_{-3\ dB}$ is given by the relationship

$$\omega_{-3\ dB} = \beta\omega_t \tag{6.62}$$

However, while for the classical two-stage CMOS opamp the unity-gain frequency remains relatively constant for varying load capacitances, the unity-gain frequencies of the folded-cascode and current-mirror amplifiers are strongly related to their load capacitance. As a result, their settling-time performance is affected by both the feedback factor as well as the effective load capacitance.

For the folded-cascode opamp, its unity-gain frequency is given by

$$\omega_t = \frac{g_{m1}}{C_L} \tag{6.63}$$

whereas for a current-mirror opamp, it is equal to

$$\omega_t = \frac{Kg_{m1}}{C_L} \tag{6.64}$$

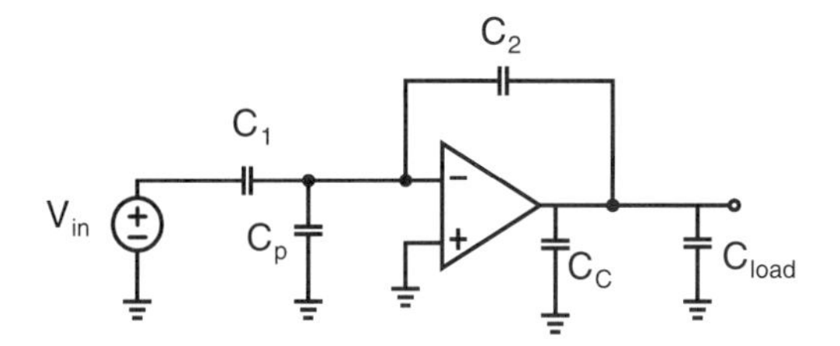

Fig. 6.12 A general circuit for determining the –3-dB frequency of a closed-loop amplifier.

Thus we see that for both these high-output impedance opamps, their unity-gain frequency is inversely proportional to the load capacitance. To determine the –3-dB frequency of a closed-loop opamp, consider the general case shown in Fig. 6.12. At the opamp output, C_{load} represents the capacitance of the next stage that the opamp must drive, while C_C is a compensation capacitance that might be added to maintain a sufficient phase margin. At the input side, C_p represents parasitic capacitance due to large transistors at the opamp input as well as any switch capacitance.

The feedback network is due to capacitances C_1, C_2, and C_p resulting in

$$\beta = \frac{1/[s(C_1 + C_p)]}{1/[s(C_1 + C_p)] + 1/(sC_2)} = \frac{C_2}{C_1 + C_p + C_2} \tag{6.65}$$

The effective load capacitance seen by the opamp output is found by analyzing this circuit from a series-shunt feedback perspective. Specifically, treating the negative opamp input as an open circuit, the effective load capacitance, C_L, is seen to be given by the parallel combination of C_C and C_{load}, as well as that seen looking into C_2. The capacitance seen looking into C_2 is equal to the series combination of C_2 together with $C_1 + C_p$. Combining all these capacitances together, we have

$$C_L = C_C + C_{load} + \frac{C_2(C_1 + C_p)}{C_1 + C_p + C_2} \tag{6.66}$$

EXAMPLE 6.4

Consider the current-mirror opamp with no lead compensation in Example 6.3 being used in the circuit shown in Fig. 6.12 with $C_1 = C_2 = C_C = C_{load} = 5\ \text{pF}$. What is the linear settling time required for 0.1 percent accuracy?

Solution

First, the gate-source capacitances of the input devices of the current-mirror opamp can be calculated to be

$$C_{gs1} = 300 \times 1.6 \times 1.92\ \text{fF}/\mu\text{m}^2 = 0.92\ \text{pF} \tag{6.67}$$

The capacitance seen looking into the inverting input of the opamp is one-half this value since the gate-source capacitances of the two input devices are in series. Thus, the parasitic capacitance, C_P, is 0.46 pF. Therefore, the effective load capacitance is given by

$$C_L = 5 + 5 + \frac{5(5 + 0.46)}{5 + 5 + 0.46} = 12.61 \text{ pF} \tag{6.68}$$

which results in

$$\omega_t = \frac{Kg_{m1}}{C_L} = \frac{2 \times 1.7 \text{ mA/V}}{12.61 \text{ pF}} = 2.70 \times 10^8 \text{ rad/s} \tag{6.69}$$

or equivalently,

$$f_t = 42.9 \text{ MHz} \tag{6.70}$$

Now, the feedback factor, β, is seen to be

$$\beta = \frac{5}{5 + 0.46 + 5} = 0.48 \tag{6.71}$$

resulting in

$$\tau = \frac{1}{\beta\omega_t} = 7.8 \text{ ns} \tag{6.72}$$

Finally, for 0.1 percent accuracy, we need a linear settling time of 7τ or 54 ns.

6.5 FULLY DIFFERENTIAL OPAMPS

Most modern high-performance analog integrated circuits make use of fully differential signal paths. With opamps, this technique results in differential outputs as well as inputs, and hence, they are referred to as fully differential opamps. One of the major driving forces behind the use of fully differential signals is to help reject noise from the substrate as well as from pass-transistor switches turning off in switched-capacitor applications. The reason for this noise rejection is that if the circuit is built in a symmetric manner (sometimes referred to as a balanced circuit), then ideally the noise will affect both signal paths identically, and will then be rejected, since only the difference between signals is of importance. In other words, the noise will have no effect on the differential signal, which is the signal of interest, since both sides of the differential signal see the same noise. In reality, this rejection only partially occurs since, unfortunately, the mechanisms introducing the noise are usually nonlinear with respect to voltage levels. For example, substrate noise will usually feed in through junction capacitances, which are nonlinear with voltage. Also, the clock feedthrough noise introduced by switches turning off usually has some voltage-dependent nonlinearities that can cause more noise to feed into one path than the other, thereby injecting a differential noise. However, almost certainly, the noise rejection of a fully differential design will be much better than that for a single-ended output design.

One drawback of using fully differential opamps is that a common-mode feedback (CMFB) circuit must be added. This extra circuitry is needed to establish the common-mode (i.e., average) output voltage. Ideally, it will keep this common-mode voltage immovable, preferably close to halfway between the power-supply voltages, even when large differential signals are present. Without it, the common-mode voltage is left to drift, since, although the opamp is placed in a feedback configuration, the common-mode loop gain is not typically large enough to control its value. Such is not the case with differential signals as the differential loop gain is typically quite large.

As we will see in the next section, the design of a good common-mode feedback circuit is nontrivial. First, it usually must have a speed performance comparable to the unity-gain frequency of the differential path. This speed requirement is necessary, otherwise noise on the power supplies may be significantly amplified such that the output signal becomes clipped (or distorted). Second, straightforward continuous-time designs often don't work when large differential signals are present, which may result in large common-mode signals being injected. Typically, these continuous-time CMFB circuits are the major limitation on maximum allowable signals, often limiting the signals substantially more than the differential signal-path circuitry does. Also, even when switched-capacitor CMFB circuits (described in the next section) are used, although the signal swings are not usually limited, the CMFB circuitry is often a source of noise injection and can also substantially increase the load capacitance that needs to be driven.

Another drawback of fully differential opamps is that in many designs, the single-ended slew-rate in one direction is substantially reduced when it is compared to that of equivalent single-ended output designs. This slew-rate reduction occurs because the maximum current for slewing is often limited by fixed-bias currents in the output stages. For example, in the fully differential amplifiers to be presented shortly, the negative-going current in the output stage is fixed. The unity-gain frequency, however, is usually increased because one of the current-mirrors is typically eliminated from the signal path.

Regardless of some of the limitations just described, a well-designed, fully differential amplifier works very well and can substantially improve the noise rejection. For this reason, differential designs are becoming more and more popular, perhaps constituting a majority in new microcircuit designs.

Fully Differential Folded-Cascode Opamp

A simplified schematic of a fully differential folded-cascode opamp is shown in Fig. 6.13. The n-channel current-mirror of Fig. 6.9 has been replaced by two cascode current sources, one composed of Q_7 and Q_8, the other composed of Q_9 and Q_{10}. In addition, a common-mode feedback (CMFB) circuit has been added. The gate voltage on the drive transistors of these current sources is determined by the output voltage, V_{cntrl}, of the common-mode feedback circuit. The inputs to the CMFB circuit are the two outputs of the fully differential amplifier. The CMFB circuit will detect the average of these two outputs and force it to be equal to a predetermined value, as we shall see in the next section.

Note that when the opamp output is slewing, the maximum current available for the negative slew rate is limited by the bias currents of Q_7 or Q_9. If the CMFB circuit

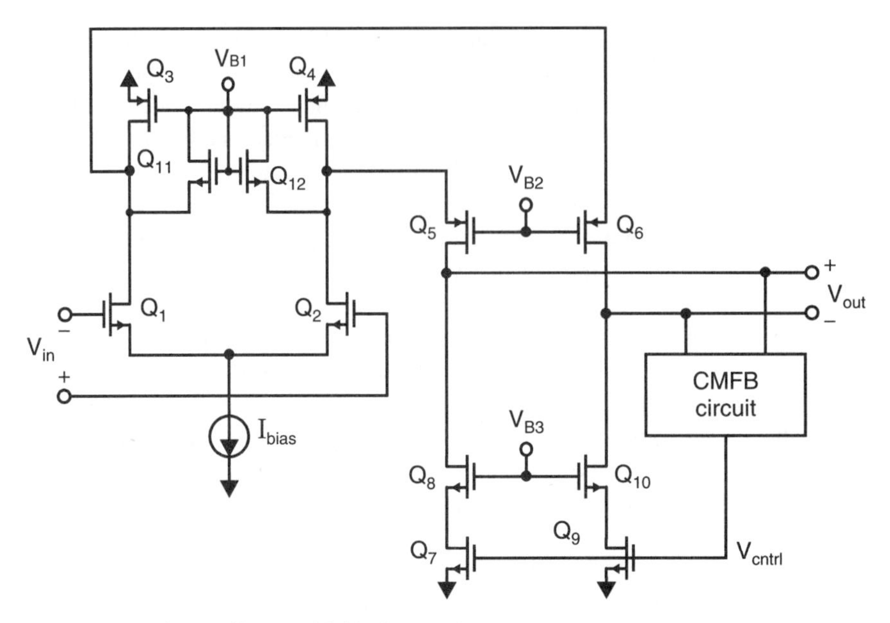

Fig. 6.13 A fully differential folded-cascode opamp.

is very fast, these will be increased dynamically to some degree during slewing, but seldom to the degree of a single-ended output fully differential opamp. For this reason, fully differential folded-cascode opamps are usually designed with the bias currents in the output stage equal to the bias currents in the input transistors. Also, to minimize transient voltage changes during slew-rate limiting, clamp transistors Q_{11} and Q_{12} have been added, as in the single-ended design presented earlier. If these transistors are not included (as has historically been the case), the time to recover from slew-rate limiting of the folded-cascode amplifier can be comparatively poor.

Each signal path now consists of only one node in addition to the output nodes, namely the drain nodes of the input devices. These nodes will most certainly be responsible for the equivalent second poles. In cases where the load capacitance is small, and it is important to maximize the bandwidth, a complementary design (i.e., n- and p-channel transistors and the power supplies interchanged) would be preferable to maximize the frequency of the equivalent second poles. This complementary circuit would result in the impedance at the drains of the input devices being the inverse of the transconductance of n-channel transistors, rather than p-channel transistors, resulting in smaller impedances and therefore faster time constants. The trade-off involved is that the opamp's transconductance, and therefore the opamp's dc gain, would become smaller due to the input transistors now being p-channel transistors. Also, the CMFB circuitry could possibly be slower, as the current sources being modulated would also have to be p-channel drive transistors. In any case, a complementary design is often a reasonable choice for high-speed fully differential designs.

Alternative Fully Differential Opamps

An alternative design is the fully differential current-mirror opamp, as shown in simplified form in Fig. 6.14. As with the folded-cascode design, this circuit can be realized as shown or in a complementary design, with p-channel input transistors, n-channel current mirrors, and p-channel bias current sources. Which design is preferable is primarily determined by whether the load capacitance or the equivalent second poles are limiting the bandwidth, and by whether maximizing the dc gain or the bandwidth is more important. n-channel inputs are preferable in the former case and p-channel inputs are preferable in the latter case, for reasons similar to those discussed for the folded-cascode opamp. Also, n-channel input transistors will give lower thermal noise (due to their larger transconductances), whereas p-channel inputs will result in less opamp input-referred 1/f noise.

If a general-purpose fully differential opamp is desired, then this design with large p-channel input transistors, a current gain of $K = 2$, and wide-swing enhanced output-impedance cascode mirrors and current sources is probably a good choice, compared to all other fully differential alternatives.

One of the limitations of the fully differential opamps seen so far is that the maximum current at the output for single-ended slewing in one direction is limited by fixed current sources. It is possible to modify the designs to get bidirectional drive capability at the output, as is shown in Fig. 6.15. This opamp is similar to the current mirror design of Fig. 6.14, but now the current mirrors at the top (i.e., the p-channel current mirrors) have been replaced by current mirrors having two outputs. The first

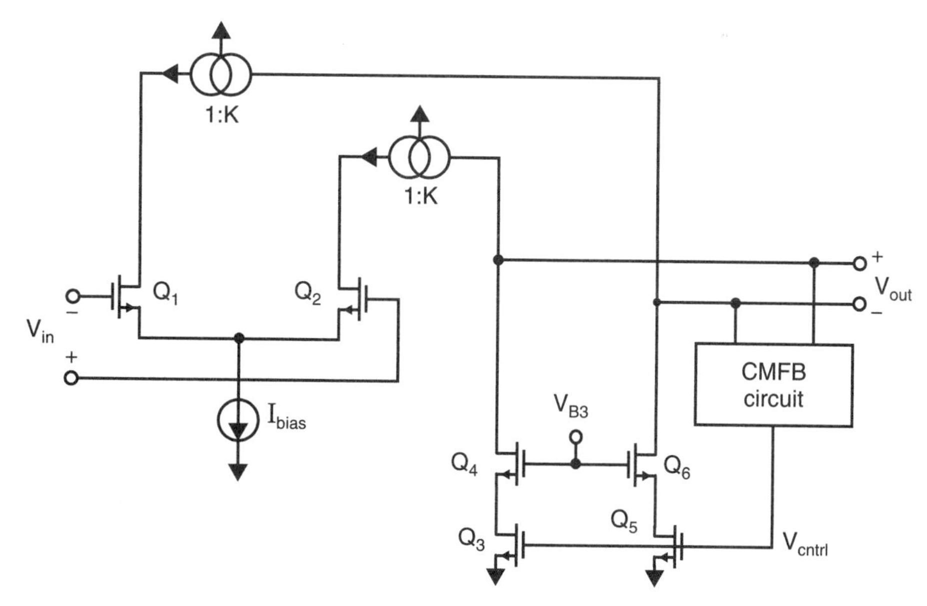

Fig. 6.14 A fully differential current-mirror opamp.

output has a current gain of K and goes to the output of the opamp, as before. The second output has a gain of one and goes to a new n-channel current mirror that has a current gain of K, where it is mirrored a second time and then goes to the opamp's opposite output, as shown. Assume now the differential output is slewing in the positive direction. For this case, the differential input will be very positive and the slewing current going into V_{out+} will be KI_{bias}, but due to the additional n-channel mirrors, the current being sinked from V_{out-} will also be KI_{bias}. This circuit has an improved slew rate at the expense of slower small-signal response due to the addition of the extra outputs on the p-channel mirrors and the additional n-channel mirrors, although in many applications this trade-off may be well merited. Of course, common-mode feedback must be added to the circuit of Fig. 6.15.

A third alternative, which makes use of a class-AB input stage, is shown in Fig. 6.16 [Castello, 1985]. Note here that for simplicity, the CMFB circuit is not shown, although it is required. This design has two differential-pair input stages connected in parallel. Each differential pair is composed of an n-channel and a p-channel transistor, and some level-shift transistors. Specifically, one input stage consists of a level-shift circuit composed of source follower Q_1 and diode-connected Q_2, and a differential pair consisting of p-channel transistor Q_3 and n-channel transistor Q_4. The other input stage consists of level shifter Q_5 and Q_6, and differential pair Q_7 and Q_8. In addition, there are four current mirrors to sum the drain currents of all differential pair transistors at the output nodes. At each output node, currents originally coming from the n-channel transistor of one differential pair and from the p-channel transistor of the other differential pair are summed.

The advantage of the input stage in this opamp is that during slew-rate limiting, one differential pair will turn off, but the total current in the other differential pair will dynamically increase substantially. For example, assuming a large positive input

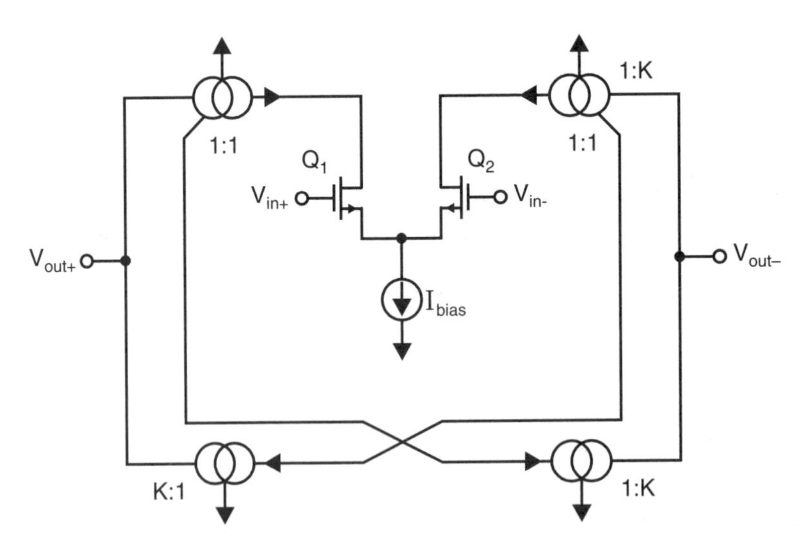

Fig. 6.15 A fully differential opamp with bidirectional output drive.

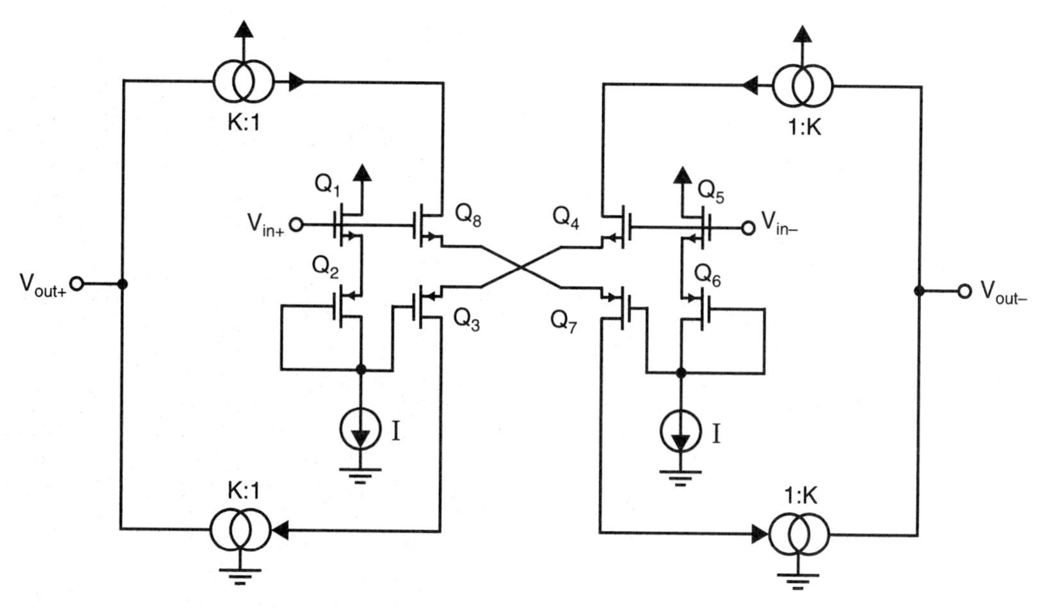

Fig. 6.16 A class AB fully differential opamp. CMFB circuit not shown.

voltage, the pair Q_3, Q_4 turns off, while the current through the pair Q_7, Q_8 increases. As a result, the increased current through Q_8 goes through a current mirror to the V_{out+} node, causing it to slew in a positive direction, whereas the current from Q_7 goes through a current mirror to the V_{out-} node, causing it to slew in a negative direction. The currents at this time of slew-rate limiting will be much larger than the small-signal bias currents, giving this opamp a very large slew-rate performance.

The disadvantage of this design is that the level-shift circuitry required at the input increases the noise and adds additional parasitics, which contribute to the equivalent second pole. In addition, the common-mode range of the input must remain at least $2V_t + 3V_{eff}$ above the lower power supply (and typically higher for the slew-rate performance to be maintained). This is a major problem when 5-V power supplies are being used, and it effectively eliminates this design from consideration for use with 3.3-V power supply voltages. However, for applications where the power-supply voltages are large, the load capacitances are large, and the slew rate is very important, this approach is quite reasonable.

Another alternative for a fully differential opamp is to use two single-ended output opamps with their inputs connected in parallel and each of their outputs being one output side of the fully differential circuit. An example of such an approach using current-mirror subcircuits is shown in Fig. 6.17. This design also has a fairly large slew rate when compared to the simpler fully differential current-mirror opamp of Fig. 6.14, but has additional current mirrors and complexity. Another advantage of this design is that the noise voltages due to the input transistors sum in a squared

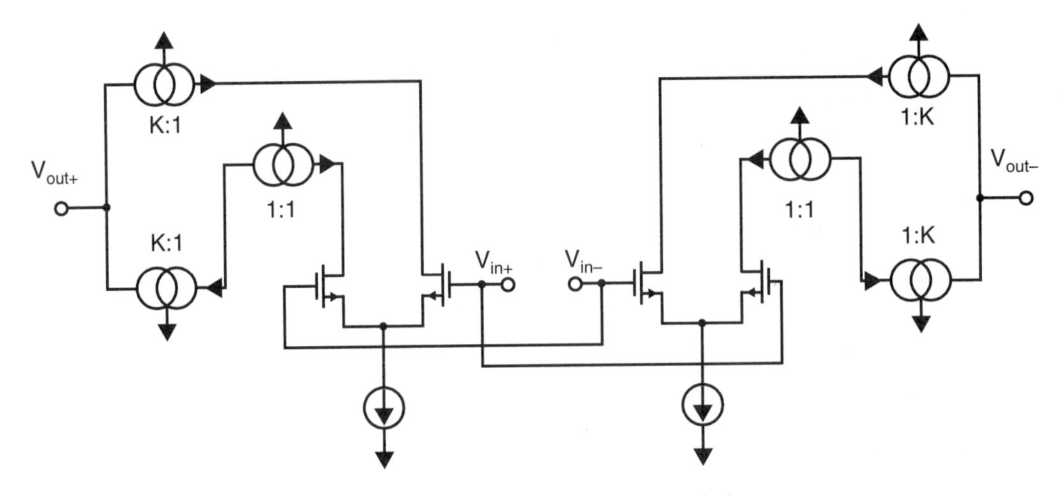

Fig. 6.17 A fully differential opamp composed of two single-ended output current-mirror opamps. CMFB circuit not shown.

fashion, while the two signal paths sum linearly.[5] As a result, this design has an improvement in signal-to-noise ratio of the input-referred gain by a factor of $\sqrt{2}$, or 3 dB. Also, the increase in total power dissipation is not significant when K is moderately large. As for the previous design, the compensation of the common-mode feedback loop is more difficult than for designs having fixed current sources biasing the output stages.

Another alternative design is shown in somewhat simplified form in Fig. 6.18. The advantage of this circuit is that due to the use of both n-channel and p-channel transistors in the two differential input pairs, the input common-mode voltage range is increased [Babanezhad, 1988; Hogervorst, 1992; Coban, 1994]. This feature can be particularly important when low power-supply voltages are being used. When the input common-mode voltage range is close to one of the power-supply voltages, one of the input differential pairs will turn off, but the other one will remain active. In an effort to keep the opamp gain relatively constant during this time, the bias currents of the still-active differential pair are dynamically increased. For example, if the input common-mode voltage range was close to the positive power-supply voltage, Q_3 and Q_4 would turn off, and Q_6 would conduct all of I_2. This current would go through current mirror M_1 and increase the bias current of the differential pair consisting of Q_1 and Q_2, which is still active. A similar situation occurs if the input common-mode voltage is near the negative power-supply rail. With careful design, it has been reported that the transconductance of the input stage can be held constant to within 15 percent of its nominal value with an input common-mode voltage

5. See Chapter 4 regarding noise.

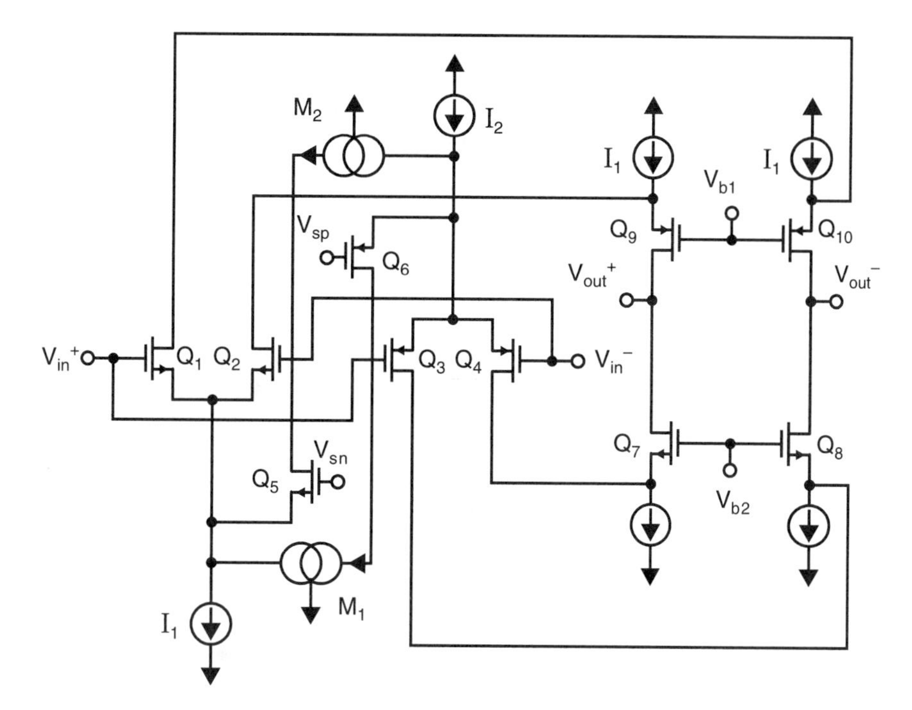

Fig. 6.18 An opamp having rail-to-rail input common-mode voltage range. CMFB circuit not shown.

range as large as the difference between the power-supply voltages [Coban, 1994].

6.6 COMMON-MODE FEEDBACK CIRCUITS

Typically, when using fully-differential opamps in a feedback application, the applied feedback determines the differential signal voltages, but does not affect the common-mode voltages. It is therefore necessary to add additional circuitry to determine the output common-mode voltage and to control it to be equal to some specified voltage, usually about halfway between the power-supply voltages. This circuitry, referred to as the *common-mode feedback* (CMFB) *circuitry,* is often the most difficult part of the opamp to design.

There are two typical approaches to designing CMFB circuits—a continuous-time approach and a switched-capacitor approach. The former approach is often the limiting factor on maximizing the signal swings, and, if nonlinear, may actually introduce common-mode signals. The latter approach is typically only used in switched-capacitor circuits, since in continuous-time applications it introduces clock-feedthrough glitches.

An example of a continuous-time CMFB circuit is shown in Fig. 6.19 [Martin, 1985; Whatly, 1986]. To illustrate its operation, assume a common-mode output voltage of zero and that V_{out+} is equal in magnitude, but opposite in sign, to V_{out-}. Furthermore, assume the two differential pairs have infinite common-mode input

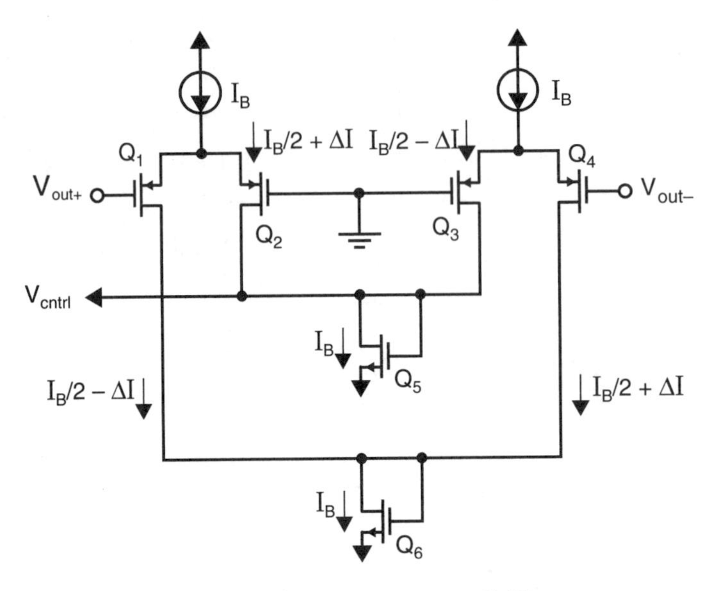

Fig. 6.19 An example of a continuous-time CMFB circuit.

rejection, which implies that the large-signal output currents of the differential pairs depend only on their input differential voltages. Since the two pairs have the same differential voltages being applied, the current in Q_1 will be equal to the current in Q_3, while the current in Q_2 will equal the current in Q_4. This result is valid independent of the nonlinear relationship between a differential pair's input voltage and its large-signal differential drain currents. Now, letting the current in Q_2 be denoted as $I_{D2} = I_B/2 + \Delta I$, where I_B is the bias current of the differential pair and ΔI is the large-signal current change in I_{D2}, the current in Q_3 is given by $I_{D3} = (I_B/2) - \Delta I$, and the current in Q_5 is given by

$$I_{D5} = I_{D2} + I_{D3} = (I_B/2 + \Delta I) + ((I_B/2) - \Delta I) = I_B \tag{6.73}$$

Thus, as long as the voltage V_{out+} is equal to the negative value of V_{out-}, the current through diode-connected Q_5 will not change even when large differential signal voltages are present. Since the voltage across diode-connected Q_5 is used to control the bias voltages of the output stage of the opamps, this means that when no common-mode voltage is present, the bias currents in the output stage will be the same regardless of whether a signal is present or not. Note, however, that the above result does not remain valid if the output voltage is so large that transistors in the differential pairs turn off.

Next consider what happens when a common-mode voltage other than zero[6] is present. For example, assume a positive common-mode signal is present. This positive voltage will cause the currents in both Q_2 and Q_3 to increase, which

6. For this example, it is assumed that positive and negative power supplies are present and that the desired common-mode voltage is zero. In a single-supply application, the gates of Q_2 and Q_3 would be connected to a bias voltage equal to the desired output common-mode voltage.

causes the current in diode-connected Q_5 to increase, which in turn causes its voltage to increase. This voltage is the bias voltage that sets the current levels in the n-channel current sources at the output of the opamp. Thus, both current sources will have larger currents pulling down to the negative rail, which will cause the common-mode voltage to decrease, bringing the common-mode voltage back to zero (or almost zero). Thus, as long as the common-mode loop gain is large enough, and the differential signals are not so large as to cause transistors in the differential pairs to turn off, the common-mode output voltage will be kept very close to ground.

The size of the differential signals that can be processed without one of the differential-pair signals turning off is maximized if the differential-pair transistors are designed to have large effective gate-source voltages. Alternatively, source degeneration can be used to allow them to have larger input signals without all of the current being directed to one side of the differential pair. However, even when this maximization is performed, the CMFB circuit still limits the differential signals to be less than what can be processed by the rest of the opamp.

Finally, it should be noted that when realizing the CMFB circuit of Fig. 6.19, the current sources I_B should be high-output impedance cascode current sources to ensure good common-mode rejection of the two differential pairs.

A modified version of the CMFB circuit of Fig. 6.19 is shown in Fig. 6.20 [Duque-Carillo, 1993]. This version adds an additional transistor (Q_7) and an additional current source to make use of the other sides of the differential pairs, which were previously unused in Fig. 6.19. This change effectively doubles the common-mode gain of the circuit for little additional complexity. As reported in [Duque-Carillo, 1993], it has very good linearity (better than 0.01%).

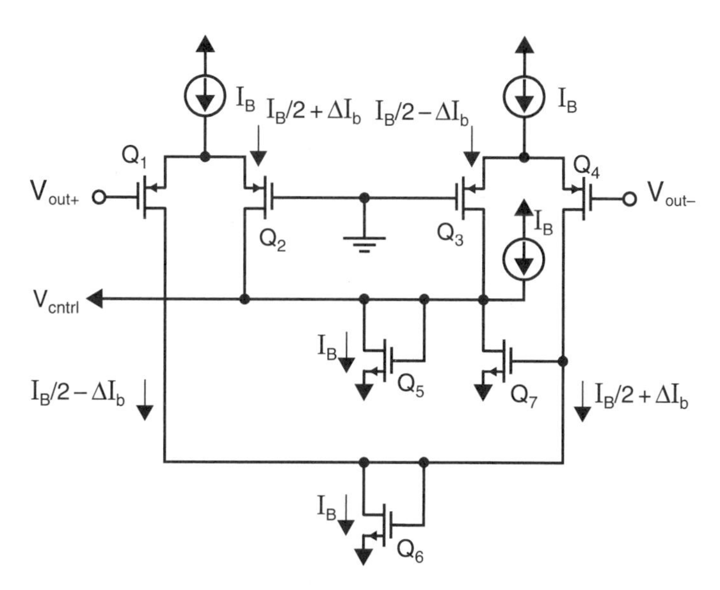

Fig. 6.20 A modified CMFB circuit having twice the common-mode gain as that of Fig. 6.19.

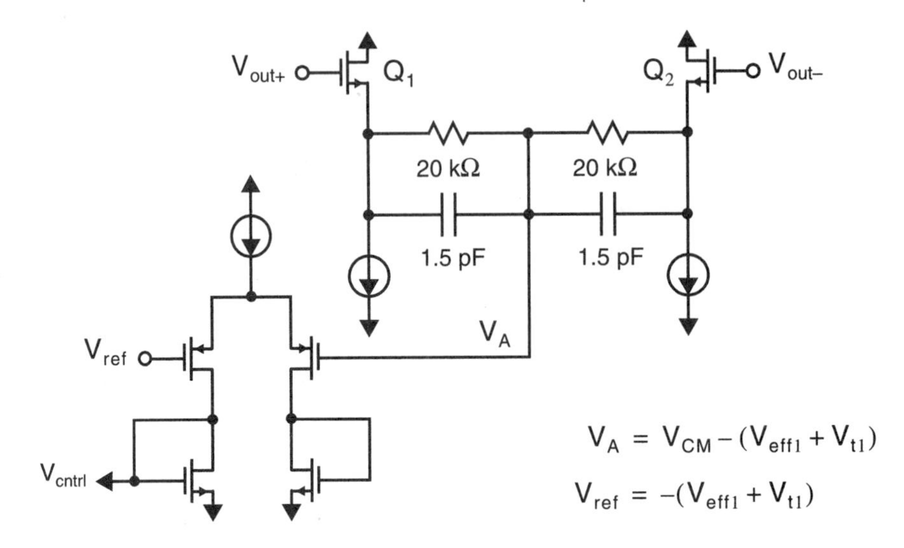

Fig. 6.21 An alternative continuous-time CMFB circuit.

An alternative approach for realizing CMFB circuits is shown in Fig. 6.21 [Banu, 1988]. This circuit generates the common-mode voltage of the output signals (minus a dc level shift) at node V_A. This voltage is then compared to a reference voltage, V_{ref}, using a separate amplifier. Although this approach works well, it has a major limitation in that the voltage drop across the source-follower transistors Q_1, Q_2 severely limits the differential signals that can be processed (unless transistors with native threshold voltages, such as 0.3 volts, are available). This limitation is particularly important when small power-supply voltages are used. In addition, the additional nodes of the common-mode feedback circuitry make the circuit slightly more difficult to compensate. When bipolar transistors are available, as is the case in a BiCMOS process, this approach is much more desirable.

An important consideration when designing CMFB circuits is that they be well compensated. Otherwise, the injection of common-mode signals can cause them to ring or even possibly become unstable. Thus, when the circuit is being designed, the phase margin and step-response of the common-mode loop should be found and verified by simulation. Often, the compensation of the common-mode loop can be realized using the same compensation capacitors used to stabilize the differential loop. This multipurpose compensation is achieved by connecting two compensation (or loading) capacitors between the opamp outputs and ground (or some other reference voltage). The other approach would be to use a single compensation capacitor connected directly between the two outputs, but this method only compensates the differential loop. It should be mentioned that by having as few nodes in the common-mode loop as is possible, compensation is simplified without having to severely limit the speed of the CMFB circuit. For this reason, the CMFB circuit is usually used to control current sources in the output stage of the opamp, as opposed to the current sources in the input stage of the opamp. The high speed of the CMFB circuit is neces-

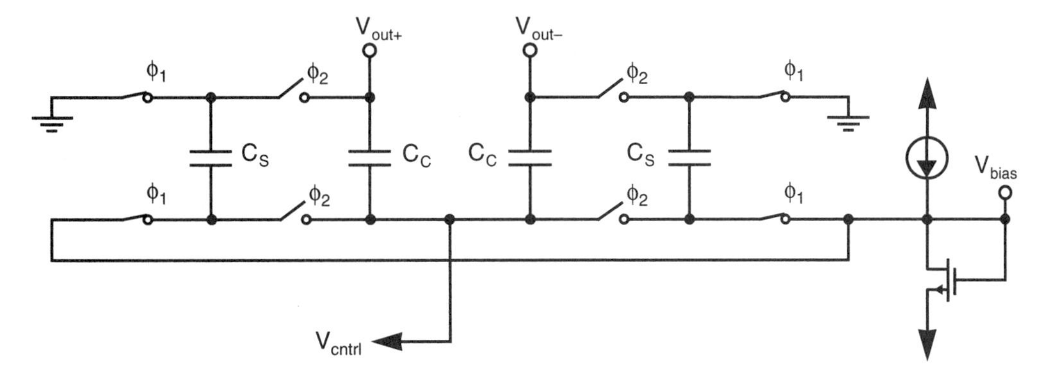

Fig. 6.22 A switched-capacitor CMFB circuit.

sary to minimize the effects of high-frequency common-mode noise, which could be amplified, causing the opamp outputs to saturate. Finally, it should be stated that designing continuous-time CMFB circuits that are both linear and operate with low power-supply voltages is an area of continuing research.

A third approach for realizing CMFB circuits is based on the use of switched-capacitor circuits. An example of this approach is shown in Fig. 6.22 [Senderowicz, 1982; Castello, 1985]. In this approach, capacitors labelled C_C generate the average of the output voltages, which is used to create control voltages for the opamp current sources. The dc voltage across C_C is determined by capacitors C_S, which are switched between bias voltages and between being in parallel with C_C. This circuit acts much like a simple switched-capacitor low-pass filter having a dc input signal. The bias voltages are designed to be equal to the difference between the desired common-mode voltage and the desired control voltage used for the opamp current sources.

The capacitors being switched, C_S, might be between one-quarter and one-tenth the sizes of the nonswitched capacitors, C_C. Using larger capacitance values overloads the opamp more than is necessary during the phase ϕ_2, and their size is not critical to circuit performance. Reducing the capacitors too much causes common-mode offset voltages due to charge injection of the switches. Normally, all of the switches would be realized by minimum-size n-channel transistors only, except for the switches connected to the outputs, which might be realized by transmission gates (i.e., parallel n-channel and p-channel transistors both having minimum size) to accommodate a wider signal swing.

In applications where the opamp is being used to realize switched-capacitor circuits, switched-capacitor CMFB circuits are generally preferred over their continuous-time counterparts since they allow a larger output signal swing.

6.7 CURRENT-FEEDBACK OPAMPS

The last type of opamp described in this chapter is referred to as a *current-feedback opamp*. This type of opamp has become popular recently, particularly in applications where high gain and high speed are required [Comlinear, 1985; Bowers, 1990]. One of this opamp's advantages is that its closed-loop gain can be changed, when used in a feedback application, without significantly affecting its loop gain. Because of this feature, a single compensation capacitor of one size can be used irrespective of the gain selected. As will be seen, the opamp is similar to many of the high-speed CMOS opamps described above, in that all nodes are low impedance except for the node where the compensation capacitor is located. Currently, the most popular technology used for realizing current-feedback opamps is a high-speed complementary-bipolar technology having vertical pnp transistors that have speeds comparable to vertical npn transistors.

A simplified example of a current-feedback opamp is shown in Fig. 6.23, where R_o is the impedance at the output of the current mirrors. The input signal, v_{in}, is applied to a high-impedance input, while a feedback current connects to a low-impedance node labelled v_n. The impedance seen looking into this low-impedance node is on the order of $1/(2g_{m1})$.

Ignoring stability issues, the basic operation of this current-feedback opamp can be understood as follows. The voltage at v_n is equal to the input voltage, v_{in}, due to the class-AB unity-gain buffer stage consisting of Q_1, Q_2, and the two diodes. Now assuming R_o is very large and ignoring C_C (which is included for compensation), a small feedback current, i_f, results in a large output voltage, v_{out}. Thus, during stable

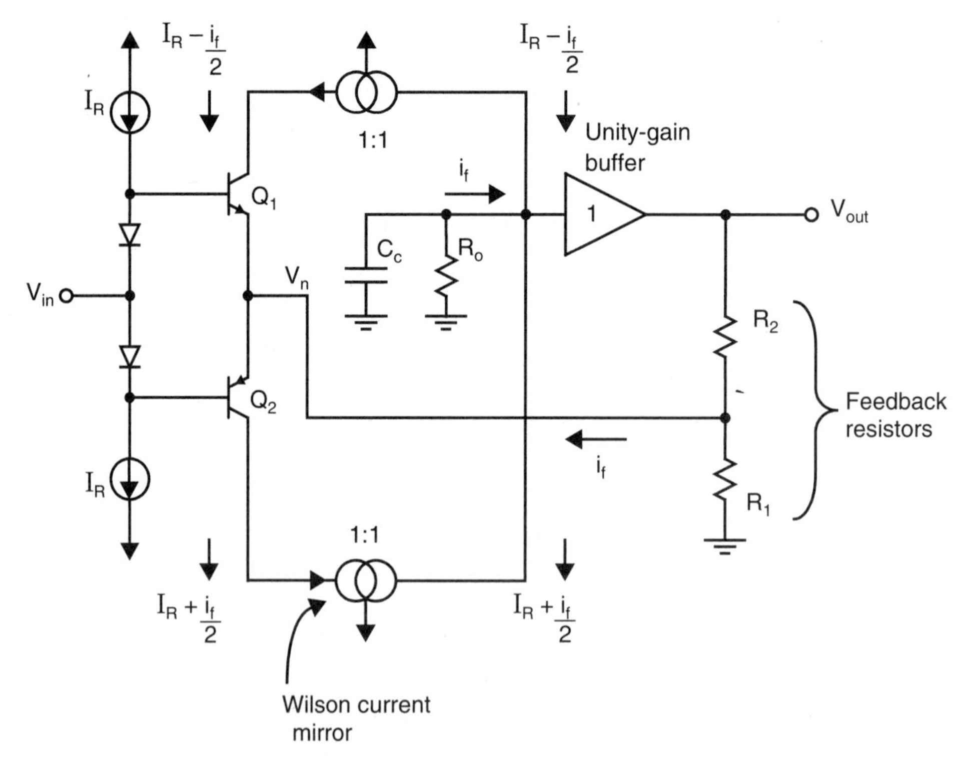

Fig. 6.23 A current-feedback opamp.

operation, $i_f \approx 0$ such that v_{out} goes to a finite voltage value. Finally, v_{out} is seen to be related to $v_n = v_{in}$ due to the two resistors, R_1 and R_2. Specifically, noting that $i_f = 0$, we have

$$\frac{v_{in}}{R_1} = \frac{v_{out} - v_{in}}{R_2} \tag{6.74}$$

which can be rearranged to give the following input/output relationship:

$$\frac{v_{out}}{v_{in}} = \frac{R_1 + R_2}{R_1} \tag{6.75}$$

To understand the feedback nature of the current-feedback opamp, we make some simplifying assumptions here. First, assume that the transistor current gain, β, is infinite. Also assume that the impedance looking into the emitters of Q_1 and Q_2, which is $1/(g_{m1} + g_{m2})$, is much smaller than $R_1 \| R_2$. Finally, assume the feedback current, i_f, splits equally between the two transistors Q_1 and Q_2. Although this equal current splitting assumption is unnecessary (since the two currents will add to i_f in any case), it does make the explanation simpler.

The signal being fed back in the opamp is the signal current i_f coming from the feedback network of R_1 and R_2 and going into the emitters of Q_1 and Q_2. Recalling that $v_n = v_{in}$, this current is given by

$$i_f = i_{R_2} - i_{R_1} = \frac{v_{out} - v_{in}}{R_2} - \frac{v_{in}}{R_1} = \frac{v_{out}}{R_2} - v_{in}\left(\frac{1}{R_1} + \frac{1}{R_2}\right) \tag{6.76}$$

Now, with Q_1 and Q_2 having infinite current gains, β, the collector currents of Q_1 and Q_2 are both equal to their bias currents, I_B, added to their fed-back signal currents, $i_f/2$. This current is mirrored by current mirrors whose outputs are summed at the high-impedance node. Note that this high-impedance node sees an impedance to ground equal to the compensation capacitor, C_C, in parallel with a high resistance, R_o. This resistance, R_o, is large, typically on the order of $(g_m r_o^2)/4$. The loop gain for this circuit can be found by breaking the loop at the top of R_2 and injecting a signal, v_x, at the top of R_2. By breaking the loop at the low-impedance output of the unity-gain buffer, there is only a small error in ignoring the loading effect of the feedback network. The loop gain is then equal to the ratio of V_{out} to v_x, while zeroing the input signal, v_{in}. The loop gain, $LG(s)$, is seen to be

$$LG(s) = \left(\frac{-I_f}{V_x}\right)\left(\frac{V_{out}}{I_f}\right) \tag{6.77}$$

The first term in (6.77) can be found using (6.76) for $v_{in} = 0$ and letting $V_x = V_{out}$. We have

$$\frac{I_f}{V_x} = \frac{1}{R_2} \tag{6.78}$$

assuming again that $V_x = V_{in}$. For the second term, note from Fig. 6.23 that the small-signal portions of the currents from the current mirrors sum together and both flow out of the parallel combination of C_C and R_o, resulting in

$$V_{out} = \frac{-I_f}{sC_C + 1/R_o} \tag{6.79}$$

Substituting (6.78) and (6.79) into (6.77), we have

$$LG(s) = \frac{R_o/R_2}{1 + sC_C R_o} \cong \frac{1}{sC_C R_2} \tag{6.80}$$

where the approximation is valid at high frequencies. Thus, we see here that the high-frequency behavior of the loop gain is independent of R_1. As a result, one can change R_1 to realize various closed-loop gains without affecting the unity-gain frequency or the closed-loop stability. Using (6.80), we see the unity-gain frequency, ω_t, is given by

$$\omega_t \cong \frac{1}{C_C R_2} \tag{6.81}$$

To find the closed-loop transfer function, we can substitute (6.79) into (6.76) and simplify to obtain

$$A_{cl}(s) = \frac{V_{out}}{V_{in}} = \frac{1/R_1 + 1/R_2}{1/R_1 + 1/R_2 + sC_c} = \frac{R_o(R_1 + R_2)}{(R_o + R_2)R_1}\left(\frac{1}{1 + sC_C(R_o \| R_2)}\right) \tag{6.82}$$

For $R_o >> R_2$, which is typically a good assumption, we have

$$A_{cl}(s) \cong \frac{R_1 + R_2}{R_1}\left(\frac{1}{1 + sC_C R_2}\right) \tag{6.83}$$

Thus, the –3-dB frequency of the closed-loop transfer function is given by

$$\omega_{-3\ dB} = \frac{1}{C_C R_2} \tag{6.84}$$

which is the same as the unity-gain frequency of the open-loop response, and is independent of R_1 and the low-frequency closed-loop gain. Therefore C_C can be chosen once, for a given R_2, and will remain optimum to a first-order approximation, irrespective of R_1 and the closed-loop gain. This independence is one of the main benefits of using current-feedback amplifiers. Specifically, they can attain very high 3-dB frequency values for large closed-loop gains, or in other words, large gain-bandwidth products. For example, if a current-feedback amplifier has a closed-loop gain of unity and a –3-dB frequency of 100 MHz, it has a gain-bandwidth product of 100 MHz. Now, by simply reducing R_1, the same amplifier might have a closed-loop gain of 20, but its –3-dB frequency remains near 100 MHz, and thus its gain-bandwidth product is near 2 GHz! It should be noted here, though, that above 100 MHz the gain would most likely fall off quite a bit faster than –20 dB/decade, since the nondominant poles will be only somewhat greater than 100 MHz.

This compensation independence does not occur for the more common voltage-feedback amplifiers. With voltage-feedback amplifiers, as the closed-loop gain is increased, the closed-loop –3-dB frequency is decreased by the same amount if the compensation capacitor remains the same. As a result, the gain-bandwidth product for a voltage-feedback amplifier remains constant for a given compensation capacitor. For example, if a voltage-feedback amplifier has a closed-loop gain of unity and a –3-dB frequency of 100 MHz, when the closed-loop gain is changed to 20, the –3-dB frequency lowers to 5 MHz. However, in this case, the gain above 5 MHz will fall off at –20 dB/decade until somewhat greater than 100 MHz (when the nondominant poles come into effect). In addition, it should be noted that for a given fixed closed-loop gain, voltage-feedback amplifiers can achieve similar gain-bandwidth products to their current-feedback counterparts if one is allowed to change their compensation capacitor. Specifically, for larger closed-loop gains, the feedback factor β, and thus loop gain, is decreased; as a result, a smaller compensation capacitor can be used to obtain the same phase margin.

In most practical realizations of current-feedback amplifiers, Darlington-pair transistors are used in the input stage to decrease the input bias currents required. Unfortunately, this approach makes the opamp somewhat noisier and increases the opamp input offset voltages somewhat. Another limitation is the requirement that R_1 be taken much larger than the impedances looking into the emitters of Q_1 and Q_2, namely $1/(g_{m1} + g_{m2})$. As a result, R_1, and therefore R_2 as well, need to be larger than would be desired for optimum high-frequency performance. This limitation is more severe in high-gain applications where $R_1 << R_2$. Additionally, these current-feedback opamps typically make use of a purely resistive feedback network and may become difficult to compensate if reactive components (such as a capacitor) are used in the feedback network. Regardless of these limitations, current-feedback opamps exhibit excellent high-frequency characteristics and are quite popular in many video and telecommunications applications.

6.8 SPICE SIMULATION EXAMPLES

In this section, simulation results for Example 6.2 and Example 6.3 are presented.

Simulation of Example 6.2

The circuit for Example 6.2 is shown in Fig. 6.24. First, the clamp transistors, Q_{12} and Q_{13}, are excluded. Bias voltages V_{B1} and V_{B2} are set such that all transistors are active, and an ac analysis is performed to determine the unity-gain frequency. Next, the opamp is placed in unity-gain configuration, and a step response is performed to determine the slew rate. Finally, the clamp transistors are included, and the resulting slew rate is determined.

```
NETLIST:
vdd    1    0               dc 2.5
vss    2    0               dc -2.5
```

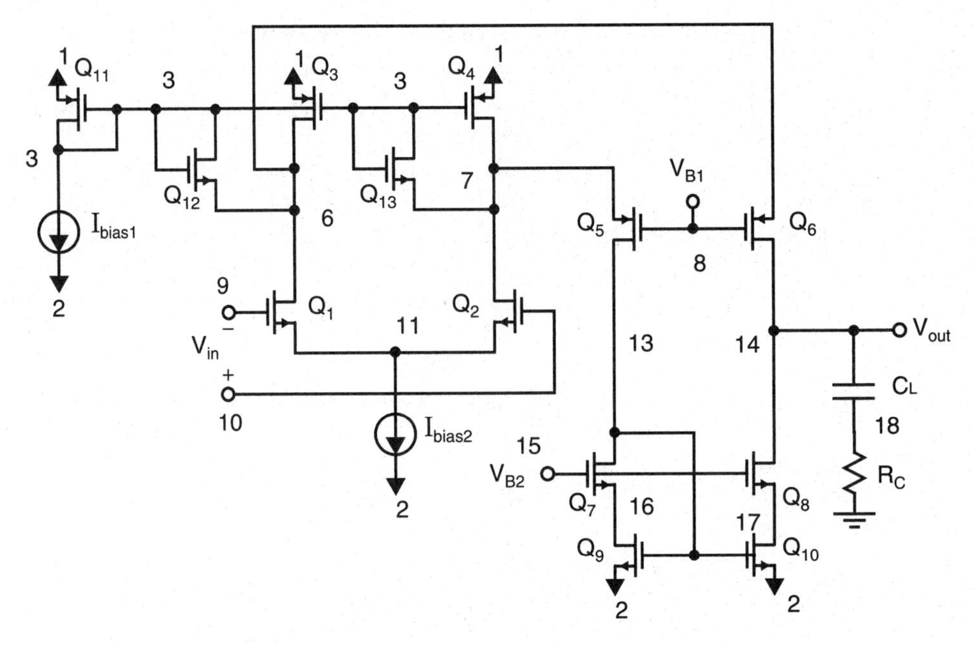

Fig. 6.24 A folded-cascode opamp.

m11	3	3	1	1	PMOS w = 10um l = 1.6um
ibias1	3	2			dc 6.6μ
m3	6	3	1	1	PMOS w = 300um l = 1.6um
m4	7	3	1	1	PMOS w = 300um l = 1.6um
m1	6	9	11	2	NMOS w = 300um l = 1.6um
m2	7	10	11	2	NMOS w = 300um l = 1.6um
ibias2	11	2			dc 320μ
m5	13	8	7	1	PMOS w = 60um l = 1.6um
m6	14	8	6	1	PMOS w = 60um l = 1.6um
m7	13	15	16	2	NMOS w = 20um l = 1.6um
m8	14	15	17	2	NMOS w = 20um l = 1.6um
m9	16	13	2	2	NMOS w = 20um l = 1.6um
m10	17	13	2	2	NMOS w = 20um l = 1.6um
Cl	14	18			10p
Rc	18	0			347
vb1	8	0			dc 0.75
vb2	15	0			dc −1
vin10	10	0			dc 0 ac 1
vin9	9	0			dc 0

The frequency plot of the folded-cascode opamp is shown in Fig. 6.25, where we find that the unity-gain frequency occurs at about 40 MHz. The step response of the folded-cascode opamp without clamp transistors is shown in Fig. 6.26(*a*), where we

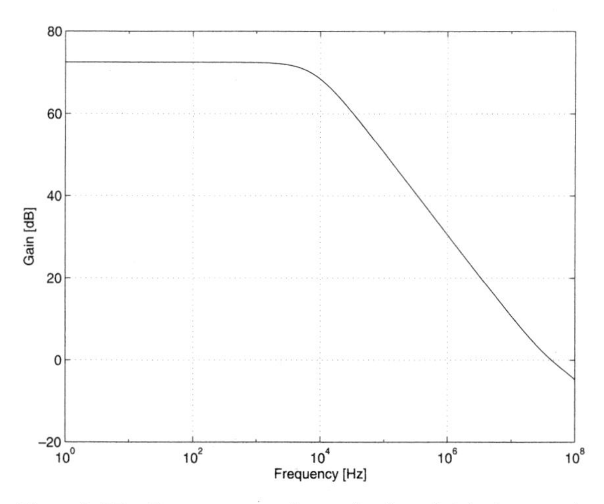

Fig. 6.25 Frequency plot of the folded-cascode opamp.

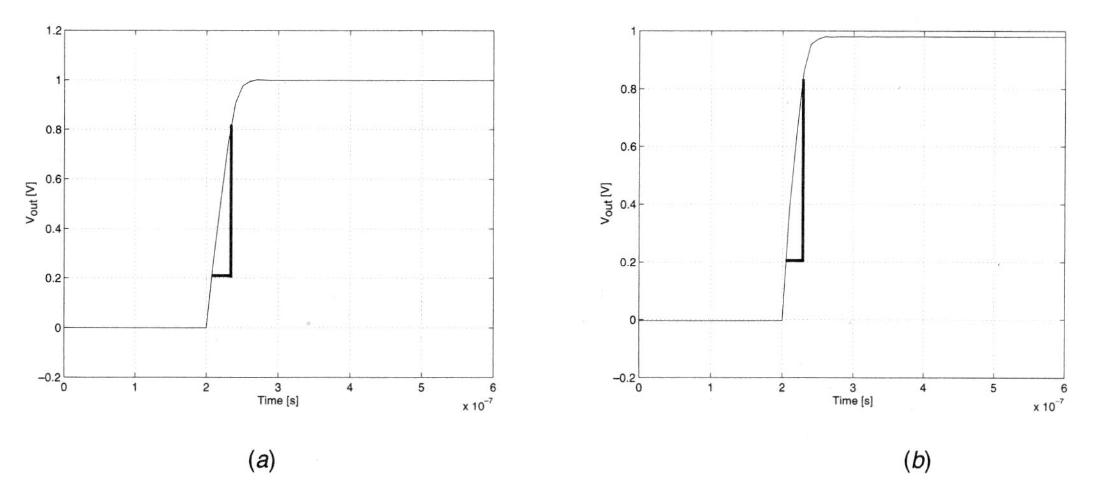

Fig. 6.26 The step response of the folded-cascode opamp (*a*) without clamp transistors and (*b*) with clamp transistors.

find the slew rate is about 23 V/μs. The step response of the folded-cascode opamp with clamp transistors is shown in Fig. 6.26(*b*), where we find that the slew rate is now 27 V/μs. While this slew-rate improvement is not that substantial, it should be noted that the main advantage of including the clamp transistors is to help the opamp recover from slew-rate conditions more quickly.

Simulation of Example 6.3

The circuit for Example 6.3 is shown in Fig. 6.27. First, an ac analysis is performed to determine the unity-gain frequency. Then the opamp is placed in unity-gain

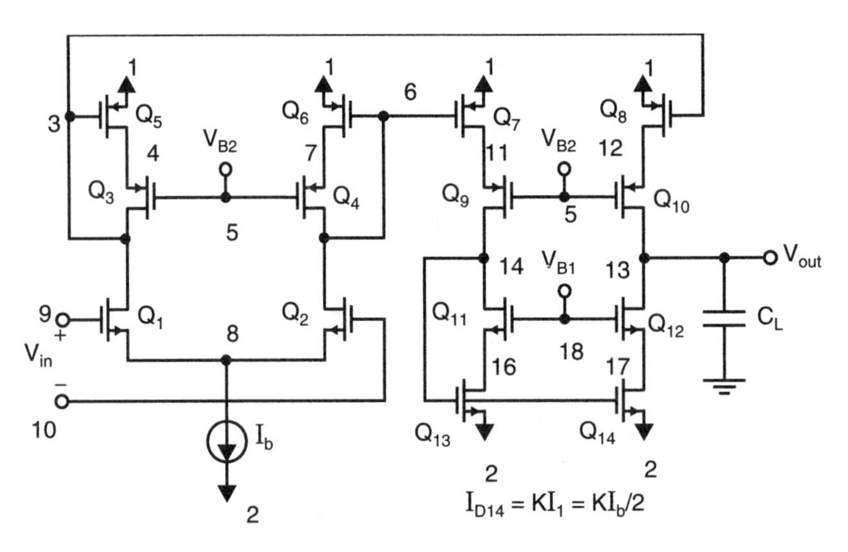

Fig. 6.27 A current-mirror opamp.

configuration, and a step response is performed to determine the slew rate.

```
NETLIST:
vdd    1    0              dc 2.5
vss    2    0              dc -2.5
m5     4    3    1    1    PMOS w = 60um l = 1.6um
m6     7    6    1    1    PMOS w = 60um l = 1.6um
m3     3    5    4    1    PMOS w = 60um l = 1.6um
m4     6    5    7    1    PMOS w = 60um l = 1.6um
m1     3    9    8    2    NMOS w = 300um l = 1.6um
m2     6    10   8    2    NMOS w = 300um l = 1.6um
ib     8    2              160um
m7     11   6    1    1    PMOS w = 60um l = 1.6um
m8     12   3    1    1    PMOS w = 120um l = 1.6um
m9     14   5    11   1    PMOS w = 60um l = 1.6um
m10    13   5    12   1    PMOS w = 120um l = 1.6um
m11    14   18   16   2    NMOS w = 30um l = 1.6um
m12    13   18   17   2    NMOS w = 60um l = 1.6um
m13    16   14   2    2    NMOS w = 30um l = 1.6um
m14    17   14   2    2    NMOS w = 60um l = 1.6um
cl     13   0              10p
vb2    5    0              dc 0.6
vb1    18   0              dc -1.24
vin1   9    0              dc 0 ac 1
vin2   10   0              dc 0
```

The frequency plot of the opamp is shown in Fig. 6.28(*a*), where it is seen that unity-gain frequency is about 50 MHz. The step response of the opamp is shown in Fig. 6.28(*b*), where it can be determined that the slew rate is about 30 V/μs.

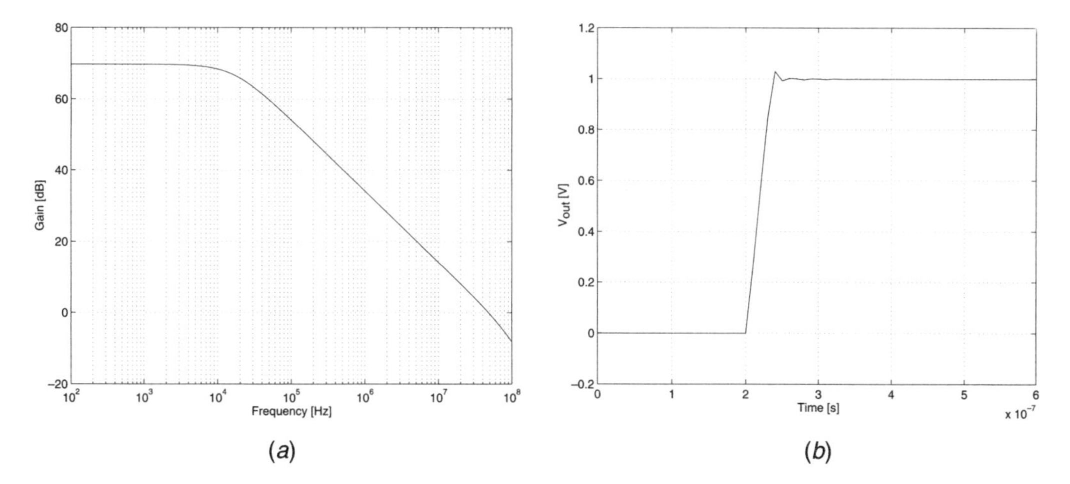

Fig. 6.28 Simulation results for a current-mirror opamp. (*a*) Frequency response. (*b*) Step response.

6.9 REFERENCES

J. N. Babanezhad, "A Rail-to-Rail CMOS Opamp," *IEEE J. of Solid-State Circuits,* Vol. 23, no. 6, pp. 1414–1417, December 1988.

J. N. Babanezhad and R. Gregorian, "A Programmable Gain/Loss Circuit," *IEEE J. of Solid-State Circuits,* Vol. 22, no. 6, pp. 1082–1090, December 1987.

M. Banu, J. M. Khoury, and Y. Tsividis, "Fully Differential Operational Amplifiers with Accurate Output Balancing," *IEEE J. of Solid-State Circuits,* Vol. 23, no. 6, pp. 1410–1414, December 1988.

D. F. Bowers, "Applying 'Current-Feedback' to Voltage Amplifiers," *Analog IC Design: The Current-Mode Approach.* C. Toumazou, F. J. Lidgey, and D. G. Haigh, eds., Peter Peregrinus, London, 1990.

K. Bult and G. J. G. M. Geelen, "A Fast-Settling CMOS Opamp for SC Circuits with 90-dB DC Gain," *IEEE J. of Solid-State Circuits,* Vol. 25, no. 6, pp. 1379–1384, December 1990.

R. Castello and P. R. Gray, "A High-Performance Micropower Switched-Capacitor Filter," *IEEE J. of Solid-State Circuits,* Vol. 20, no. 6, pp. 1122–1132, December 1985.

A. Coban and P. Allen, "A 1.75-V Rail-to-Rail CMOS Opamp," *Proceedings of the IEEE Int. Symp. on Circuits and Systems,* Vol. 5, pp. 5.497–5.500, London, June 1994.

Comlinear Corporation, "A New Approach to Opamp Design," *Comlinear Corporation Application Note* 300-1, March 1985.

J. F. Duque-Carillo, "Control of the Common-Mode Component in CMOS Continuous-Time Fully Differential Signal Processing," *Analog Integrated Circuits and Signal Processing, An International Journal,* Kluwer Academic Publishers, September 1993.

U. Gatti, F. Maloberti, and G. Torelli, "A Novel CMOS Linear Transconductance Cell for Continuous-Time Filters," proceedings of the *IEEE Int. Symp. on Circuits and Systems,* pp. 1173–1176, New Orleans, May 1990.

R. Hogervorst, et al., "CMOS Low-Voltage Operational Amplifiers with Constant-G_m Rail-to-Rail Input Stage," proceedings of the *IEEE Int. Symp. on Circuits and Systems,* pp. 2876–2879, San Diego, May 1992.

B. Hosticka, "Improvement of the Gain of MOS Amplifiers," *IEEE J. of Solid-State Circuits,* Vol. SC-14, no. 14, pp. 1111–1114, December 1979.

S. Law, Private conversation, Xerox Corp., 1983.

K. Martin, Class notes, UCLA, 1985.

K. Martin, Laboratory notes (independently derived, albeit after [Coban 1994]), 1994.

E. Säckinger and W. Guggenbühl, "A High-Swing, High-Impedance MOS Cascode Circuit," *IEEE J. of Solid-State Circuits,* Vol. 25, no. 1, pp. 289–298, February 1990.

D. Senderowicz, et al., "A Family of Differential NMOS Analog Circuits for a PCM Codec Filter Chip," *IEEE J. of Solid-State Circuits,* Vol. 17, no. 6, pp. 1014–1023, December 1982.

C. C. Shih and P. R. Gray, "Reference Refreshing Cyclic Analog-to-Digital and Digital-to-Analog Converters," *IEEE J. of Solid-State Circuits,* Vol. 21, no. 4, pp. 544–554, August 1986.

N. S. Sooch, "MOS Cascode Current Mirror," U.S. patent no. 4,550,284, October 1985.

R. A. Whatly, "Fully Differential Operational Amplifier with DC Common-Mode Feedback," U.S. patent no. 4573020, February 1986.

6.10 PROBLEMS

Unless otherwise stated, assume the following:

- n-channel MOS transistors:
  $\mu_n C_{ox} = 92\ \mu A/V^2$
  $V_{tn} = 0.8\ V$
  $\gamma = 0.5\ V^{1/2}$
  $r_{ds}(\Omega) = 8{,}000 L(\mu m)/I_D(mA)$ in active region
  $C_j = 2.4 \times 10^{-4}\ pF/(\mu m)^2$
  $C_{j\text{-}sw} = 2.0 \times 10^{-4}\ pF/\mu m$
  $C_{ox} = 1.9 \times 10^{-3}\ pF/(\mu m)^2$
  $C_{gs(overlap)} = C_{gd(overlap)} = 2.0 \times 10^{-4}\ pF/\mu m$
- p-channel MOS transistors:
  $\mu_p C_{ox} = 30\ \mu A/V^2$
  $V_{tp} = -0.9\ V$
  $\gamma = 0.8\ V^{1/2}$
  $r_{ds}(\Omega) = 12{,}000 L(\mu m)/I_D(mA)$ in active region
  $C_j = 4.5 \times 10^{-4}\ pF/(\mu m)^2$
  $C_{j\text{-}sw} = 2.5 \times 10^{-4}\ pF/\mu m$
  $C_{ox} = 1.9 \times 10^{-3}\ pF/(\mu m)^2$
  $C_{gs(overlap)} = C_{gd(overlap)} = 2.0 \times 10^{-4}\ pF/\mu m$

6.1 Calculate the output impedance of the two-transistor diode-connected circuit shown in Fig. P6.1 using small-signal analysis. Assume both transistors are in the active region, ignore the body effect, and assume $g_{m1} = g_{m2}$, $r_{ds1} = r_{ds2}$, and $g_m r_{ds} >> 1$.

6.2 Assume n is chosen to be 1 in Fig. 6.1 and all transistors are taken to be equal sizes, except $(W/L)_5$ is chosen so that $V_{DS3} = V_{DS2} = V_{eff3} + 0.15\ V$, where V_{eff} is chosen to be 0.2 V for all transistors except Q_5. Ignoring the body effect, find the required size for all transistors assuming they all have lengths equal to 1.6 μm. Assume that $I_{bias} = 50\ \mu A$.

6.3 What are the drain-source voltages of Q_3 and Q_4 from Problem 6.2 if the lengths of Q_3 and Q_4 are decreased to 1 μm to maximize speed?

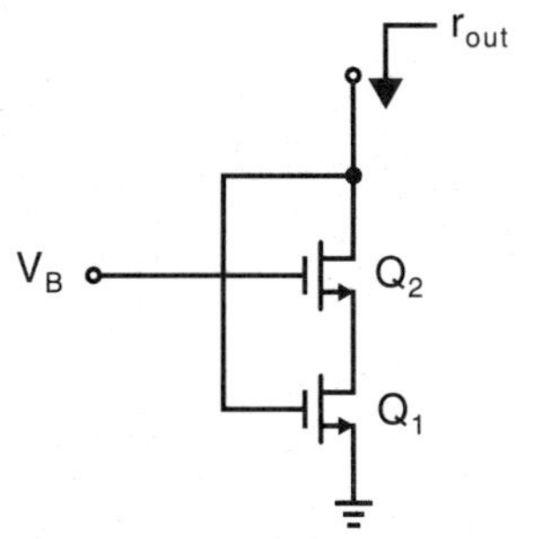

Fig. P6.1

6.4 What is the value required for R_B in Fig. 6.2 to give $V_{eff3} = 0.2$ V?

6.5 Assuming that μ_n is proportional to $T^{-3/2}$, what would V_{eff3} be at 100 °C if it was 0.2 V at 20 °C?

6.6 Assume that R_B from Fig. 6.2 has a temperature dependence of +0.3%/°C. What would V_{eff3} be at 100 °C if it was 0.2 V at 20 °C? Assume μ_n varies as in Problem 6.5.

6.7 Using small-signal analysis and ignoring the body effect, show that the circuit of Fig. 6.3 has an output impedance given by $r_{out} \cong g_{m1}r_{ds1}r_{ds2}(1 + A)$.

6.8 For the circuit shown in Fig. P6.8, using small-signal analysis, find the output impedance. What is an upper limit on the output impedance assuming $R_E >> r_\pi$? Compare this to the upper limit on the output impedance of a simple cascode mirror (i.e., $A = 0$) for the same case.

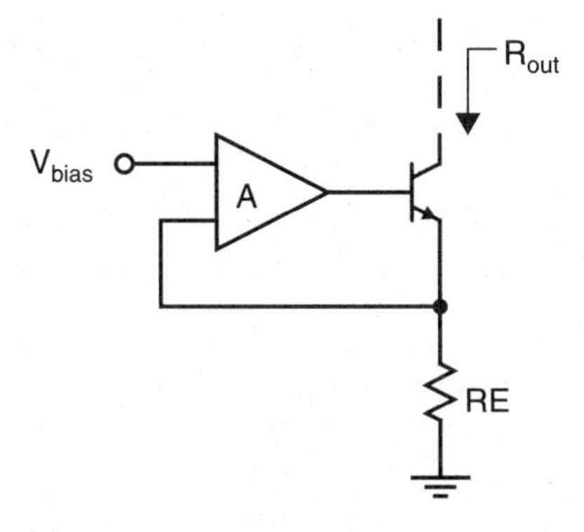

Fig. P6.8

6.9 For the circuit shown in Fig. 6.6 with the W/L ratios shown and assuming all lengths are 1.6 μm, estimate the output impedance. Assume that all current sources are ideal and that $I_{in} = 7I_{bias}$. Compare this to the output impedance of a wide-swing cascode current mirror where the gate of Q_1 was connected to $V_{casc\text{-}n}$ from the bias-generation circuitry of Fig. 6.2. Assume $I_{bias} = 50\ \mu A$.

6.10 For the folded-cascode amplifier shown in Fig. 6.9, with the transistor sizes given in Table 6.1, find the unity-gain frequency and the slew rate. Assume

± 2 V power supplies are used, that C_L is 10 pF, and that the power dissipation is 1 mW.

6.11 For the folded-cascode amplifier shown in Fig. 6.9, with the transistor sizes given in Table 6.1, estimate the approximate frequency of the second pole caused by the parasitic capacitances at the drains of Q_1 and Q_2. Assume ± 2 V power supplies are used, that C_L is 10 pF, and that the power dissipation is 1 mW. To simplify matters, junction capacitances can be ignored. What should C_L be to achieve 70° phase margin? What would the corresponding unity-gain frequency and slew rate be?

6.12 For the amplifier of Problem 6.11 with lead compensation, find the load capacitance, C_L, for a 70° phase margin. What size resistor should be used? What is the final unity-gain frequency and slew rate? Assume lead compensation with $\omega_z = 1.2\omega_t$ improves the initial unity-gain frequency by 20 percent and the phase margin by 30 degrees.

6.13 For the folded-cascode amplifier shown in Fig. 6.9, assume the bias currents of Q_1 and Q_2 are K times greater than the bias currents in Q_5 and Q_6. Assume the total bias current in the opamp, I_{total}, is determined by a specified power dissipation. Derive an equation for the unity-gain frequency in terms of K and I_{total}. Then show that the unity-gain frequency is maximized by taking K large.

6.14 Repeat Problem 6.13, but this time show that the dc gain is optimized by taking K large.

6.15 Derive an equation for the ratio of the unity-gain frequency of the folded-cascode amplifier of Fig. 6.9 to the unity-gain frequency of the current-mirror opamp of Fig. 6.11 in terms of K and I_{total} assuming both amplifiers have the same size input transistors, total power dissipation, and load capacitances. For the folded-cascode opamp, define K to be I_{D2}/I_{D7}. What is this ratio for K equal to 1, 2, and 4?

6.16 Consider the current-mirror opamp described in Example 6.3, in a feedback configuration shown in Fig. 6.12, with $C_1 = 1$ pF and $C_2 = C_C = C_{load} = 5$ pF. What is the linear settling time for 1 percent accuracy?

6.17 For the case described in Example 6.4, and assuming a current-mirror opamp described in Example 6.3, also assume the input voltage is a 1-V step change. Would the opamp slew rate limit? If so, how close must the output voltage come to its final value before its output voltage rate of change is less than the slew rate? How long would this take? How much longer would be required for linear settling to 1 percent of the total voltage change at the output?

6.18 What are the single-ended slew rates of the fully differential folded-cascode and current-mirror opamps, assuming $K = 2$ and the current densities are the same as those of Example 6.2 and Example 6.3, respectively? Calculate the slew rates in both the positive and the negative directions. Assume the load capacitances are 10 pF between each output and ground. For the folded-cascode design, assume the clamp transistors Q_{11} and Q_{12} of Fig. 6.13 have not been included. Also, assume the current sources biasing the output stages do not change during transients.

6.19 Derive the slew rate of the fully differential opamp of Fig. 6.15 in terms of C_L, K, and I_{bias}.

6.20 Derive an equation for the unity-gain frequency of the fully differential opamp shown in Fig. 6.16 in terms of K and I_{total}.

6.21 Assume the CMFB circuit of Fig. 6.19 has ±2.5 V power supply voltages, and that the current sources require 0.5 V across them in order to have all transistors remain in the active regions. Ignoring the body effect, what V_{eff} bias voltage should be used for the p-channel transistors to maximize signal swing? What is the maximum single-ended signal swing before the gain of the common-mode feedback circuitry goes to zero and why?

6.22 Repeat Problem 6.21, but do not ignore the body effect. Assume the n-wells of the p-channel transistors are connected to V_{DD}.

6.23 Derive the low-frequency gain of the current-feedback amplifier of Fig. 6.23 without assuming that the impedance looking into the emitters of Q_1 and Q_2 is zero.

6.24 Consider the circuit shown in Fig. P6.24, where a high-output-impedance opamp is used. Capacitor C_p is the parasitic capacitance at the opamp input, while C_o is the output capacitance. Assuming linear settling, show that the optimal value for minimizing the time constant of this circuit is given by

$$C_{1,opt} = \sqrt{MC_oC_p}$$

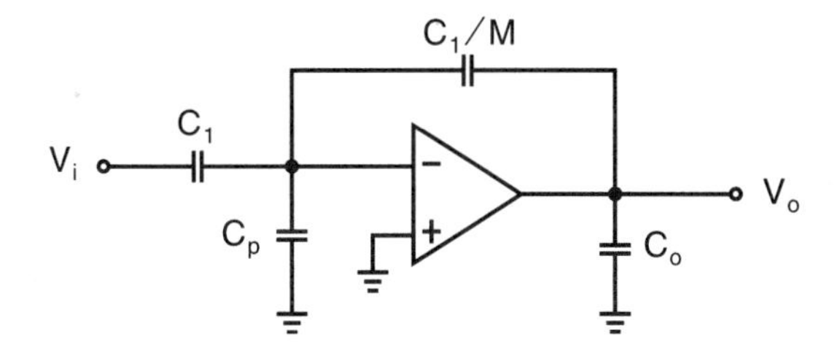

Fig. P6.24

6.25 Based on the results of Problem 6.24, derive τ_{min} for $M = 1$, $C_o = 1$ pF, and $C_p = 0.05$ pF. Sketch τ for other values of C_1 between 0.1 pF and 1 pF.

CHAPTER

7 Comparators

Perhaps the second most widely used electronic components (after amplifiers) are comparators. A comparator is used to detect whether a signal is greater or smaller than zero, or to compare the size of one signal to another. As we will see in Chapter 13, comparators are used in large abundance in A/D converters. They also find widespread use in many other applications, such as data transmission, switching power regulators, and others. In this chapter, we look at comparator design and practical limitations, where a number of different approaches are discussed. First, we examine a simplistic approach of using an open-loop opamp for a comparator. Although this approach is too slow for practical applications, it is a good example to use when we discuss several design principles for minimizing input-offset voltage and charge-injection errors. A number of other approaches are also described, such as multiple-stage comparators, positive-feedback track-and-latch comparators, and fully differential comparators. Finally, recent examples of both CMOS and bipolar comparator circuits are also described.

7.1 USING AN OPAMP FOR A COMPARATOR

A simplistic approach for realizing a comparator is to use an open-loop opamp, as shown in Fig. 7.1. The main drawback to this approach is the slow response time since the opamp output has to slew a large amount of output voltage and settles too slowly. However, temporarily ignoring this slow response time, we will first investigate another problem due to input-offset voltage.

The simplistic opamp approach shown in Fig. 7.1 has a resolution limited to the input-offset voltage of the opamp. This offset might be on the order of 2 mV to 5 mV for typical MOS processes, which is inadequate for many applications. An alternative architecture that can resolve signals with accuracies much less than the input-offset voltages of opamps is shown in Fig. 7.2 [McCreary, 1975; Yee, 1978]. Although this circuit has been used many times in early analog-to-digital converters, as we will see,

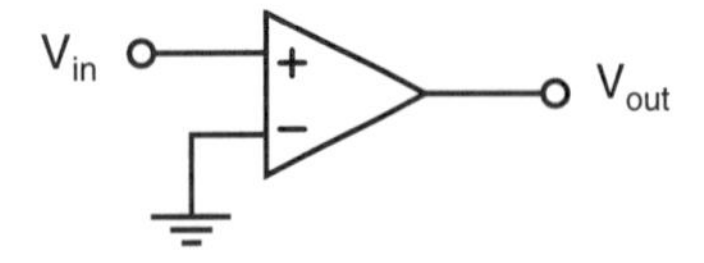

Fig. 7.1 A simplistic approach of using an open-loop opamp for a comparator.

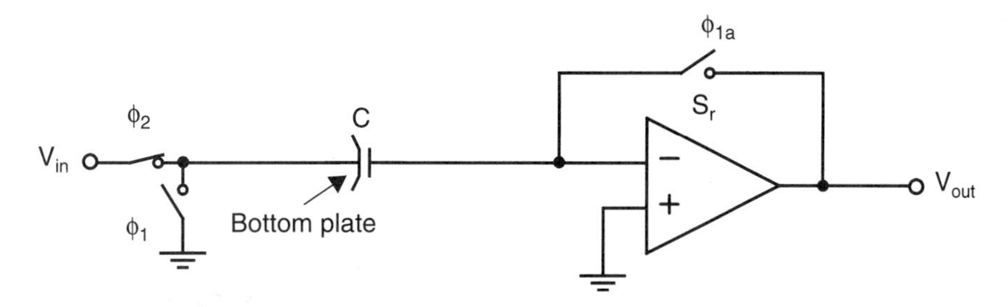

Fig. 7.2 Cancelling the offset voltage of a comparator—the comparator here must be stable, with unity-gain feedback during ϕ_{1a}. (ϕ_{1a} is a slightly advanced version of ϕ_1 so that charge-injection effects are reduced.)

it is not preferable nowadays. However, it is a simple example that can be used to illustrate many important design principles. The circuit of Fig. 7.2 operates as follows: During ϕ_1, known as the *reset phase,* the bottom plate[1] of the capacitor C (i.e., the left side of capacitor C) is connected to ground, and the top plate is connected to the inverting input of the opamp. At the same time, the output of the opamp is also connected to the inverting input of the opamp by closing switch S_r. Assuming the opamp is ideal, this connection causes the capacitor to be charged to zero volts. Next, during the *comparison phase*, the reset switch, S_r, is turned off, and the bottom plate of the capacitor is connected to the input voltage. The opamp is now in an open-loop configuration. If the input signal is greater than zero, the output of the opamp swings to a large negative voltage. If the input signal is less than zero, the output of the opamp swings to a large positive voltage. These two cases are easily resolved and the decision can be stored using a simple digital latch.

The limitations of this approach become apparent when one considers nonideal opamps, which have finite gains and require compensation to be stable during the reset phase.

EXAMPLE 7.1

Consider the case in which a 0.5-mV signal must be resolved using the circuit shown in Fig. 7.2, where the opamp's output should be 5 V. Assuming the opamp's unity-gain frequency is 10 MHz, find the maximum clocking rate of the comparator circuit if reset and comparison phases are equal and if six time constants are allowed for settling.

Solution

After the comparison phase, the output of the opamp should have a 5-V difference between the cases in which the input signal is either –0.25 mV or

1. The bottom plate of a capacitor has significant parasitic capacitance between it and ground. Therefore, it is almost always connected to the less sensitive input side rather than to the critical amplifier side (see Chapter 10).

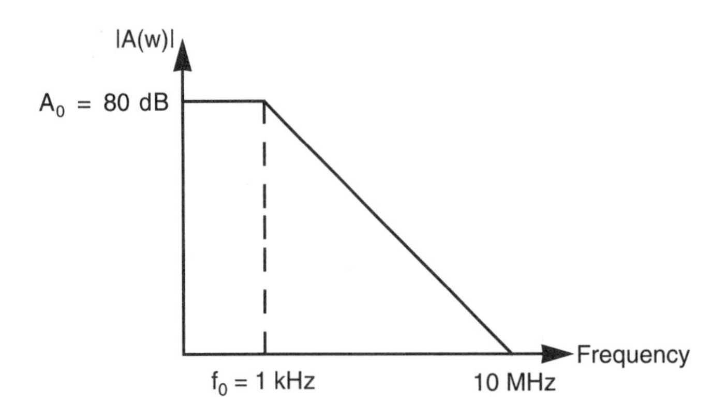

Fig. 7.3 The open-loop transfer function of the opamp used to realize the comparator.

+0.25 mV.[2] As a result, the opamp gain must be at least 10,000 V/V. By assuming that the dominant-pole compensation is used to guarantee stability during the reset phase, we obtain an open-loop transfer function for the opamp similar to that shown in Fig. 7.3. Here, the –3-dB frequency of the opamp is given by

$$f_{-3\text{ dB}} = \frac{f_t}{A_0} = 1\text{ kHz} \tag{7.1}$$

During the comparison phase, the output of the opamp will have a transient response similar to that of a first-order system that has a time constant given by

$$\tau = \frac{1}{2\pi f_{-3\text{ dB}}} \cong 0.16\text{ ms} \tag{7.2}$$

If six time constants are allowed for settling during the comparison phase, then approximately 1 ms is needed for the comparison phase. Assuming the reset time is the same as the comparison time, the clock frequency can be no greater than 500 Hz—a frequency that is much too slow for most applications.

One possibility for speeding up the comparison time is to disconnect the compensation capacitor during the comparison phase. For example, a simplified opamp schematic is shown in Fig. 7.4. In this opamp, transistor Q_1 is used to achieve lead compensation when it is on during the reset phase, ϕ_1. During the comparison phase, Q_1 is turned off, which disconnects the compensation capacitor, C_C, thereby greatly speeding up the opamp during this phase. Using this technique, it is possible to use

2. This assumes that ±2.5-V power supplies are being used and that digital output values are the same as the power-supply voltages, as is typical for CMOS. If a single power supply voltage of 5 V is used, then analog ground can be considered to be at 2.5 V.

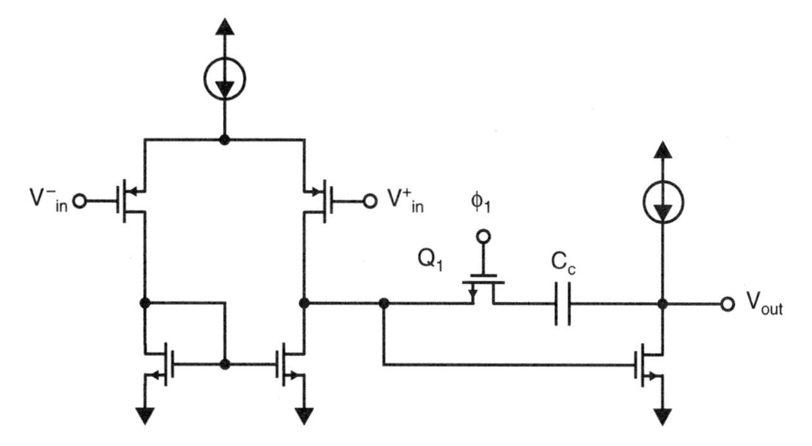

Fig. 7.4 An opamp that has its compensation capacitor disconnected during the comparison phase.

clock frequencies ten to fifty times greater than would otherwise be the case—perhaps as high as 25 or 50 kHz in our example. If this is adequate, then the approach is reasonable. Unfortunately, this often isn't adequate, and other approaches, to be described shortly, are necessary.

One superior aspect of the approach just described is that the input capacitor, C in Fig. 7.2, is never charged or discharged during operation. Specifically, in phase ϕ_1, C is always charged to 0 V. In phase ϕ_2, the top plate, connected to the inverting input of the opamp, is open-circuited at that time (assuming the parasitic capacitors are ignored), and the voltage across capacitor C remains at 0 V (in other words, the top-plate voltage follows V_{in}). This approach greatly minimizes the charge required from the input when V_{in} changes. If the switches attached to the bottom plate had their phases interchanged, then the comparison operation would be noninverting, but now capacitor C must be charged or discharged during the reset phase since the bottom plate follows V_{in}, whereas the top plate remains at virtual ground. Normally, we like to use a reasonably large input capacitor to minimize clock-feedthrough effects (described in Section 7.2). This charging/discharging requirement puts severe constraints on the input signal source, and, thus, the clock phasing shown in Fig. 7.2 should be used. Besides, it is usually possible to tolerate an inverting comparison since it can be made noninverting through the use of a digital inversion somewhere else in the system.

Input-Offset Voltage Errors

One source of error is due to the input-offset voltage of the opamp. This input offset might be caused by device mismatches or might be inherent in the design of the comparator. In switched-capacitor comparators, such as that shown in Fig. 7.2, input offset is not a problem since it is stored across the capacitor during the reset phase, and then the error is cancelled during the comparison phase. To appreciate this cancella-

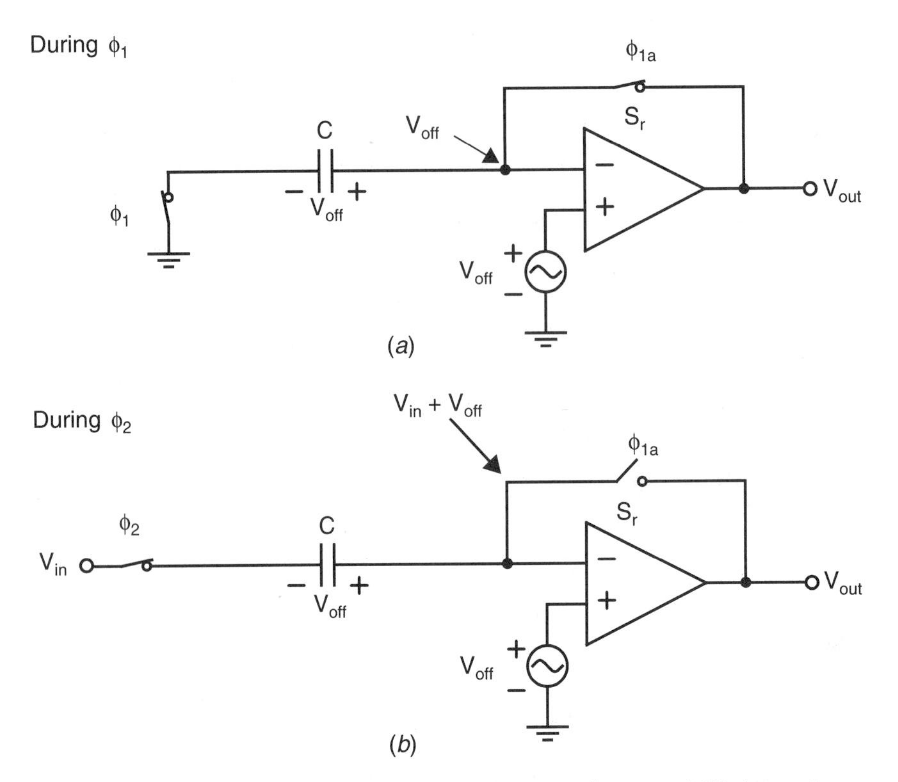

Fig. 7.5 The circuit configuration (*a*) during the reset phase, and (*b*) during the comparison phase, assuming the opamp has an input-offset voltage given by V_{off}.

tion, assume the opamp has an input-offset voltage error that is modelled as a voltage source in series with one of the opamp's inputs. The circuit configuration during the reset phase is shown in Fig. 7.5(*a*). Assuming the opamp gain is very large, then the inverting input of the opamp is at the voltage V_{off}, which implies that the input capacitor is charged to V_{off} during this phase. Next, during ϕ_2, as shown in Fig. 7.5(*b*), the left side of capacitor C is connected to the input voltage. The right side of the capacitor has a voltage given by $V_{in} + V_{off}$, which results in the comparator output becoming negative if V_{in} is greater than zero or positive if V_{in} is less than zero, regardless of the value of V_{off}. Not only does this technique eliminate input-offset voltage errors, but it also minimizes errors caused by low-frequency 1/f noise, which can be large in CMOS microcircuits.

7.2 CHARGE-INJECTION ERRORS

Perhaps the major limitation on the resolution of comparators is due to what is referred to as *charge injection,* also commonly called *clock feedthrough*. This error is due to unwanted charges being injected into the circuit when the transistors turn off. For the comparator in Fig. 7.2, the switches are normally realized by either n-channel transistors alone or CMOS transmission gates (which consist of n-channel transistors in parallel with p-channel transistors, both of which must turn off). When MOS switches are on, they operate in the triode region and have zero volts between their drain and their source. When MOS switches turn off, charge errors occur by two mechanisms. The first is due to the channel charge, which must flow out from the channel region of the transistor to the drain and the source junctions.[3] The channel charge of a transistor that has zero V_{DS} is given by

$$Q_{CH} = WLC_{ox}V_{eff} = WLC_{ox}(V_{GS} - V_t) \tag{7.3}$$

This charge often dominates. The second charge (typically smaller, unless V_{eff} is very small) is due to the overlap capacitance between the gate and the junctions.

Figure 7.6 shows the comparator of Fig. 7.2, where the switches have been realized as n-channel transistors. Also shown in Fig. 7.6 are the parasitic capacitances due to gate-drain and gate-source overlap capacitors. Finally, although it is not shown in Fig. 7.6, one should remember to include the channel charge dispersion of any transistors that change from being on to off.[4] In the circuit shown, all transistors are n-channel, implying that the channel charge is negative.

Consider first when Q_3 turns off. If the clock waveform is very fast, the channel charge due to Q_3 will flow equally out through both junctions [Shieh, 1987]. The $Q_{ch}/2$ charge that goes to the output node of the opamp will have very little effect other than causing a temporary glitch. However, the $Q_{ch}/2$ charge that goes to the inverting-input node of the opamp will cause the voltage across C to change, which introduces an error. Since this charge is negative, for an n-channel transistor, the node

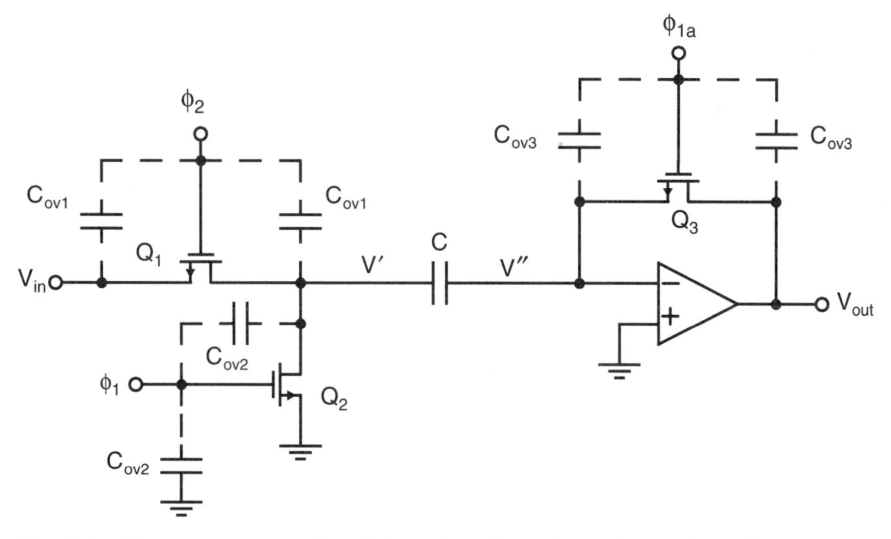

Fig. 7.6 The comparator in Fig. 7.2, with n-channel switches and overlap parasitic capacitances shown. Transistors turned on also have channel charge.

3. For a transistor in the triode region that has zero V_{DS}, it doesn't matter which junctions are called the drain and the source, since both junctions are at the same potential.
4. Transistors also accumulate channel charge when turning on, but this does not typically affect circuit performance since the charge comes from low-impedance nodes.

voltage V'' will become negative. The voltage change due to the channel charge is given by

$$\Delta V'' = \frac{(Q_{ch}/2)}{C} = -\frac{V_{eff3}C_{ox}W_3L_3}{2C} = -\frac{(V_{DD}-V_{tn})C_{ox}W_3L_3}{2C} \tag{7.4}$$

since the effective gate-source voltage of Q_3 is given by $V_{eff3} = V_{GS3} - V_{tn} = V_{DD} - V_{tn}$. The preceding voltage change in V'' is based on the assumption that Q_2 turns off slightly after Q_3 turns off. More will be said about this assumption shortly.

To calculate the change in voltage due to the overlap capacitance, it is first necessary to introduce the capacitor-divider formula. This formula is used to calculate the voltage change at the internal node of two series capacitors, when the voltage at one of the end terminals changes. This situation is shown in Fig. 7.7, where it is assumed that V_{in} is changing and we want to calculate the change in $V_{out} = V_{C2}$. The series combination of C_1 and C_2 is equal to a single capacitor, C_{eq}, given by

$$C_{eq} = \frac{C_1C_2}{C_1+C_2} \tag{7.5}$$

When V_{in} changes, the charge flow into this equivalent capacitor is given by

$$\Delta Q_{eq} = \Delta V_{in}C_{eq} = \Delta V_{in}\frac{C_1C_2}{C_1+C_2} \tag{7.6}$$

All of the charge that flows into C_{eq} is equal to the charge flow into C_1, which is also equal to the charge flow into C_2. Thus, we have

$$\Delta V_{out} = \Delta V_{C2} = \frac{\Delta Q_{C2}}{C_2} = \frac{\Delta V_{in}C_1}{C_1+C_2} \tag{7.7}$$

This formula is often useful when calculating charge flow in integrated circuits. It can be applied to the circuit of Fig. 7.6 to calculate the change in V'' due to the overlap capacitance of Q_3 when it turns off. For this case, we have $C_1 = C_{ov}$, $C_2 = C$, and $\Delta V_{in} = -(V_{DD} - V_{SS})$. This assumes the clock signals change from V_{DD} to V_{SS}. The change in V'' due to the overlap capacitance is now found to be given by

$$\Delta V'' = \frac{-(V_{DD}-V_{SS})C_{ov}}{C_{ov}+C} \tag{7.8}$$

This change is normally less than that due to the change caused by the channel charge since C_{ov} is small.

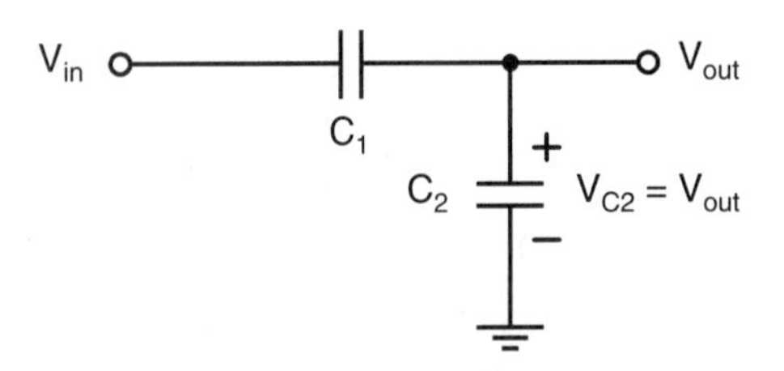

Fig. 7.7 A capacitor divider.

EXAMPLE 7.2

Assume that a resolution of ± 2.5 mV is required from the circuit of Fig. 7.6. The following values are given: $C = 1$ pF, $C_{ox} = 1.92$ fF/(μm)^2, $(W/L)_3 = 10$ μm/0.8 μm, $L_{ov} = 0.1$ μm, $V_{tn} = 0.8$ V, $V_{DD} = 2.5$ V, and $V_{SS} = -2.5$ V. What is the change in $\Delta V''$ when Q_3 turns off?

Solution

Using (7.4), we have $\Delta V'' = -13.1$ mV. The overlap capacitance is given by $C_{ov} = W_3 L_{ov} C_{ox} = 1.92$ fF. Using (7.8), this gives a voltage change in $\Delta V''$ of $\Delta V'' = -9.6$ mV, which is smaller than the change due to the channel charge, but not insignificant. The total change is found by adding the two effects so that $\Delta V'' = -22.7$ mV.

This result is on the order of 10 times greater than the value of ± 2.5 mV that should be resolved. Clearly, additional measures should be taken to minimize the effects of charge injection if this resolution is required.

Making Charge-Injection Signal Independent

The charge injection due to transistors Q_1 and Q_2 may cause temporary glitches, but it will have much less effect than the charge injection due to Q_3 if we assume that Q_2 turns off slightly after Q_3 does [Haigh, 1983]. This time difference is the reason why the clock voltage of Q_2 is denoted ϕ_1, whereas the clock voltage of Q_3 is denoted ϕ_{1a} (ϕ_1 advanced). The argument for this arrangement is slightly complicated, but goes as follows: When Q_2 turns off, its charge injection causes a negative glitch at V', but this will not cause any change in the charge stored in C since the right side of C is connected to an effective open circuit, assuming that Q_3 has already turned off. Later, when Q_1 turns on, the voltage V' will settle to V_{in} regardless of the previous charge injection of Q_2. Thus, the voltage at the left of C is unaffected by the charge injection of Q_2, and the voltage across C is also unaffected by the charge injection of Q_2. Therefore, the voltage V'' is unaffected by the charge injection of Q_2. This is not the case if Q_2 is turned off at the same time or before Q_3 is turned off. The charge injection of Q_1 has no effect for a simpler reason—the comparison has already been made when it turns off. In addition, its charge injection has no effect when it turns on because the right side of C is connected to an open circuit at that time, assuming the clocks do not overlap. *In summary, by turning off* ϕ_{1a} *first, the circuit is affected only by the charge injection of* Q_3 *and not by any of the other switches.*

Fig. 7.8 shows a simple digital circuit, originally developed by one of the authors [Martin, 1980], that is capable of generating the desired clock waveforms. The waveforms ϕ_1 and ϕ_{1a} do not overlap with ϕ_2 and ϕ_{2a}. Also, ϕ_{1a} will be advanced slightly (by two inverter delays), compared to ϕ_1.

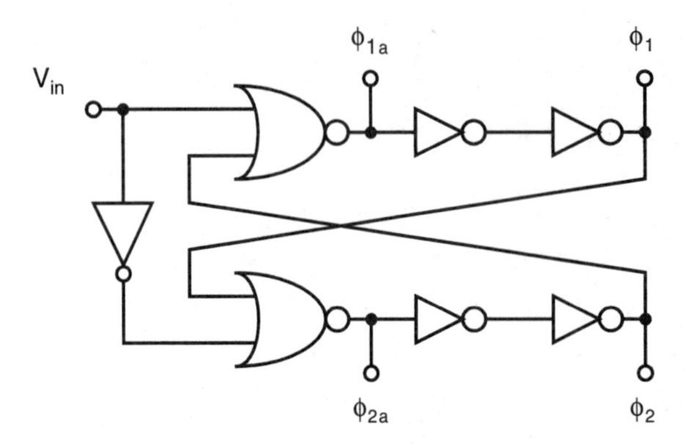

Fig. 7.8 A clock generator suitable for generating the desired clock waveforms for the comparator in Fig. 7.2.

Minimizing Errors Due to Charge Injection

The simplest way to reduce errors due to charge injection is to use larger capacitors. For our previous example, capacitors on the order of about 100 pF guarantee that the clock-feedthrough errors are less than 0.5 mV. Unfortunately, this large amount of capacitance would require a large amount of silicon area. Also, integrated capacitors have parasitic capacitances between the bottom plate and the substrate that might be about 20 percent of the size of the realized capacitor. Thus, a bottom plate capacitance of about 20 pF would have to be driven by the input circuits, which would greatly slow down the circuits.

An alternative approach for minimizing errors due to charge injection is to use fully differential design techniques for comparators, similar to what is often done for opamps. A simple example of a one-stage, switched-capacitor, fully differential comparator is shown in Fig. 7.9. Ideally, when the comparator is taken out of reset mode,

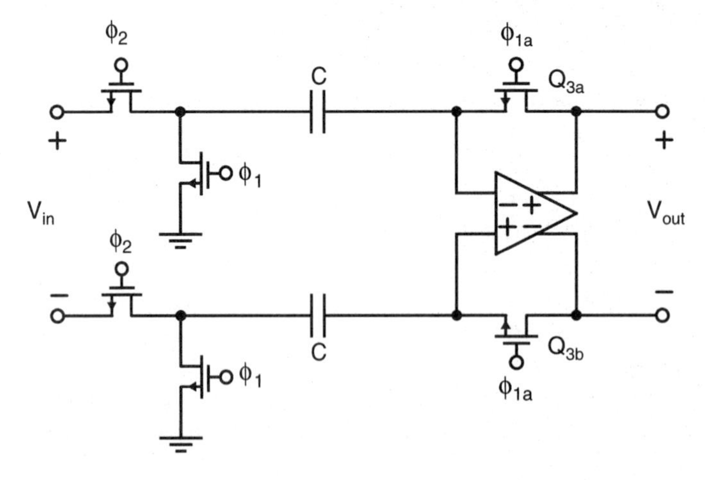

Fig. 7.9 A fully differential, single-stage, switched-capacitor comparator.

the clock feedthrough of reset switch Q_{3a} matches the clock feedthrough of Q_{3b}. In this case, the common-mode voltage may be slightly affected, but the differential input voltage is unaffected. The only errors now are due to mismatches in the clock feedthrough of the two switches, which will typically be at least ten times smaller than in the single-ended case. For this reason, virtually all modern integrated comparators utilize fully differential design techniques.

A third alternative that can be used along with fully differential design techniques is to realize a multi-stage comparator [Poujois, 1978; Vittoz, 1985], where the clock feedthrough of the first stage is stored on coupling capacitors between the first and second stage, thereby eliminating its effect. The clock feedthrough of the second stage can be stored on the coupling capacitors between the second and the third stages, and so on. Although this technique is almost always used along with fully differential configurations, we will explain the technique using single-ended configurations, for simplicity. Consider the three-stage comparator shown in Fig. 7.10(*a*), along with the appropriate clock waveforms shown in Fig. 7.10(*b*). Consider the time when ϕ_1' drops and the switch of the first stage has charge injection. Figure 7.11 illustrates this case, along with the parasitic capacitances of the first-stage switch. When ϕ_1' drops, Q_1 injects charge into both the inverting input and the output of the first stage through parasitic capacitors C_{p1} and C_{p2}, respectively. The charge injected at the first stage

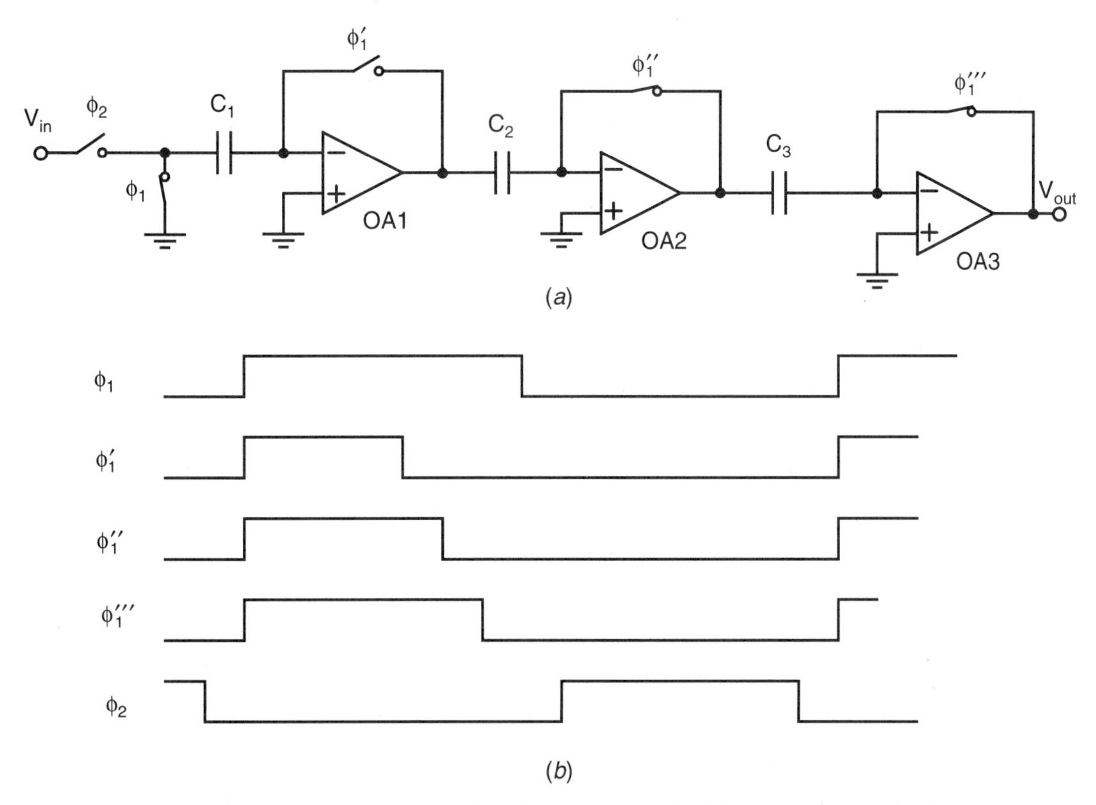

Fig. 7.10 (*a*) A multi-stage comparator used to eliminate clock feedthrough errors, with (*b*) the appropriate clock waveforms.

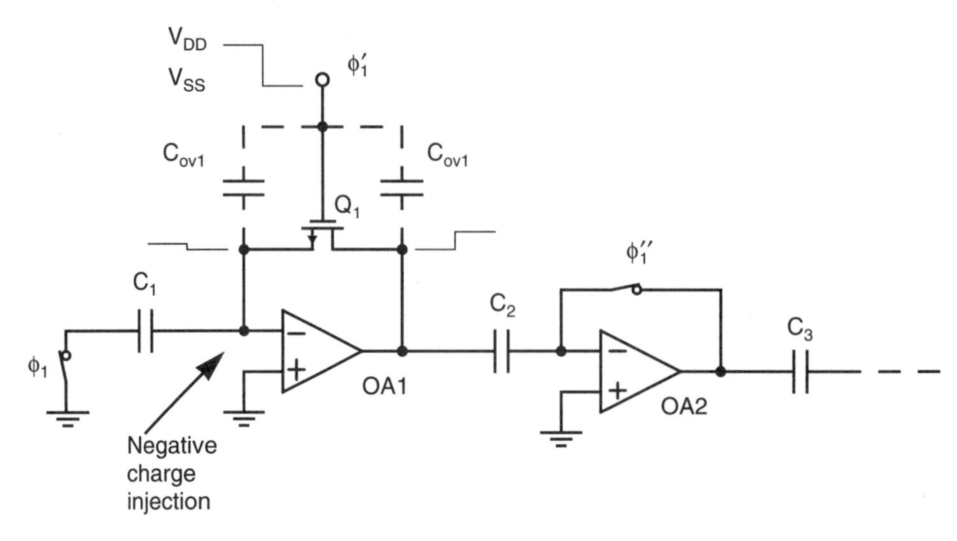

Fig. 7.11 The first stage of the comparator in Fig. 7.10, when the first stage is injecting charge.

output causes only a temporary glitch there. The charge injected at the inverting input causes this node to become negative. The amount by which this node voltage becomes negative is calculated using the analysis method described previously. In any practical case, this amount will be in the range of tens of millivolts, assuming that coupling capacitors of around 1 pF are used. After the inverting input becomes negative, the output of the first stage becomes positive by an amount equal to the negative transition of the inverting input multiplied by the first stage's gain. However, at this time, ϕ_1'' is still high. Therefore, the second stage is still being reset and C_2 is charged up to the output error caused by the clock feedthrough of the first stage, thereby eliminating its effect. In a similar manner, when ϕ_1'' turns off and the second stage goes from closed-loop reset mode to open-loop comparison mode, the third stage is still in reset mode and the clock feedthrough of the second stage is stored on coupling capacitor C_3. Finally, when the third stage turns off, its charge injection is not cancelled. However, the error it causes in resolving an input voltage to all three stages is equal to the voltage transition at the inverting input of the third stage divided by the negative of the gains of the first two stages (i.e., this is the input voltage needed to cancel the effect of the clock feedthrough of the third stage coming out of reset mode).

In a variation on this multi-stage approach [Poujois, 1978], the first stage was not reset. However, the input-offset voltage of the first stage was still cancelled by storing it on coupling capacitors between the first and second stage during a reset mode. This approach is described Section 7.4.

EXAMPLE 7.3

Assume all capacitors are 1 pF, transistor sizes are $W/L = 10\ \mu m/0.8\ \mu m$, $C_{ox} = 1.92\ fF/(\mu m)^2$, $L_{ov} = 0.1\ \mu m$, $V_{tn} = 0.8\ V$, $V_{DD} = 2.5\ V$, and

$V_{SS} = -2.5$ V, and each stage has a gain of 20. What is the input-offset voltage error caused by the clock feedthrough of the third stage when the third stage comes out of reset?

Solution

The values used are identical to those used in Example 7.2. Therefore, the charge injection at the inverting input of the third stage is the same as that found in the solution to Example 7.2, or –22.7 mV. The input signal that can overcome this 22.7 mV value is given by

$$\Delta V_{in} = \frac{22.7\ \text{mV}}{A_1 A_2} = 57\ \mu\text{V} \tag{7.9}$$

Thus, the equivalent input-offset voltage is 57 μV, which is much less than the resolution found in Example 7.2 of 22.7 mV.

Speed of Multi-Stage Comparators

The approach just described can be used to realize very-high-resolution comparators, especially when it is combined with fully differential circuit design techniques. However, this approach does have the limitation that it requires multiple-phase clock waveforms, which slow down the circuits.

Although the multi-stage approach is limited in speed due to the need for the signal to propagate through all the stages, it can still be reasonably fast because each of the individual stages can be made to operate fast. Typically, each stage consists of a single-stage amplifier that has only a 90° phase shift and therefore does not need compensation capacitors (i.e., each stage has a 90° phase margin without compensation). To see why this multi-stage approach can be fast, consider the simplified case of a cascade of first-order, uncompensated inverters, as shown in Fig. 7.12. This circuit is a very rough approximation of a multi-stage comparator when it is not being reset. The parasitic load capacitance at the output of the ith stage is approximately given by

$$C_{pi} \cong C_{0\text{-}i} + C_{gs\text{-}i+1} \tag{7.10}$$

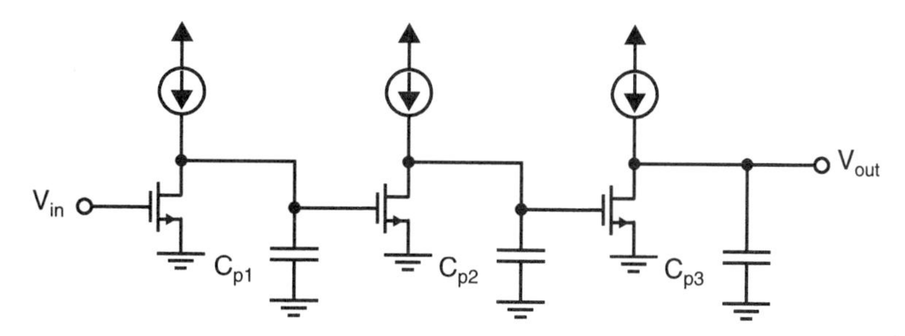

Fig. 7.12 Realizing a comparator by using a cascade of first-order gain stages.

where $C_{0\text{-}i}$ is the output capacitance of the ith stage, which would normally be due to junction capacitance, and $C_{gs\text{-}i+1}$ is the gate-source capacitance of the input transistor of the succeeding stage. Equation (7.10) will not necessarily be true for the last stage, but the load of the last stage should be of the same magnitude. If one assumes the stages are matched, then normally $C_{0\text{-}i} < C_{gs\text{-}i+1}$ since junction capacitances are usually less than gate-source capacitances, implying that one can usually assume

$$C_{pi} < 2C_{gs\text{-}i} \tag{7.11}$$

The unity-gain frequency of a single stage is then on the order of

$$\omega_{ti} \sim \frac{g_{mi}}{2C_{gs\text{-}i}} \tag{7.12}$$

or larger, where g_{mi} is the transconductance of the input capacitor of the ith stage. Thus, the unity-gain frequency of a single gain-stage is on the order of one-half the unity-gain frequency of a single transistor. If one assumes the ith stage has a dc gain $A_{0\text{-}i}$ and that the ith stage is well described by a first-order transfer function, then the transfer function of a single stage is approximately given by

$$A_i(s) = \frac{A_{0\text{-}i}}{1 + s/\omega_{p\text{-}i}} \tag{7.13}$$

where

$$\omega_{p\text{-}i} \cong \frac{\omega_{t\text{-}i}}{A_{0\text{-}i}} \sim \frac{g_{mi}}{2A_{0\text{-}i}C_{gs\text{-}i}} \tag{7.14}$$

Thus, the –3-dB frequency of a single stage is on the order of one-half the unity-gain frequency of a transistor divided by the gain of the stage. If one has a cascade of n stages, then the overall transfer function is given by

$$A_{total}(s) = \prod A_i(s) \tag{7.15}$$

This result can be approximated by a first-order transfer function, where all higher-order terms are ignored, resulting in

$$A_{total}(s) \cong \frac{\prod A_{0\text{-}i}}{1 + s\sum 1/\omega_{p\text{-}i}} \cong \frac{A_0^n}{1 + ns/\omega_{p\text{-}i}} \tag{7.16}$$

Thus, a cascade of n stages has a time constant approximately given by

$$\tau_{total} \cong \frac{2nA_0C_{gs}}{g_m} \cong 2nA_0\tau_T \tag{7.17}$$

where $\tau_T = C_{gs}/g_m$ is the approximate transit time of a single transistor (i.e., 1 over the unity-gain frequency of a single transistor). In other words, the time constant of the cascade of first-order stages is approximately equal to n times the time constant of a single stage, which is roughly given by $2A_0$ times the transit time of a single transistor. This result should be compared to using a single opamp, which, for the same overall gain, will have a time constant considerably greater than $2A_0^n$ times the transit time of a single transistor. Although the preceding speed estimate is valid for simple single-ended stages only, the general principles apply to more complicated, fully differential stages as long as each stage is first order and doesn't require compensation.

Equation (7.17) can be simplified further. Recall from Chapter 2 that

$$C_{gs} = \frac{2}{3}C_{ox}WL \tag{7.18}$$

for a transistor in the active region, and that

$$g_m = \sqrt{2\mu_n C_{ox}(W/L)I_D} \tag{7.19}$$

We can substitute (7.18) and (7.19) into (7.17), and, after simple manipulation, arrive at

$$\tau_{total} \cong \frac{2nA_0C_{gs}}{g_m} \cong \frac{4nA_0L^2}{3\mu_n V_{eff}} \tag{7.20}$$

where, again from Chapter 2,

$$V_{eff} = \sqrt{\frac{2I_D}{\mu_n C_{ox}(W/L)}} \tag{7.21}$$

Equation (7.20) is very useful in quickly accessing the speed and resolution capabilities of a given technology. It also gives some insight into designing comparators. First, the effective gate-source voltages of input drivers of each stage should be as large as possible. Second, the widths of the transistors are relatively unimportant, assuming they are large enough so that C_{gs} dominates parasitic capacitances due to interconnect and external loading.

EXAMPLE 7.4

Assume for a 0.8-μm technology that $A_0 = 20$, $n = 3$, $V_{eff} = 0.5$ V, and $\mu_n = 0.05\ M^2/V \cdot s$. What is the maximum clocking frequency of the comparator?

Solution

Using (7.20), we have $\tau_{total} = 4.1$ ns. If we assume that 3τ is required for each half period, then a complete period requires 6τ, which corresponds to a clocking frequency of 40.1 MHz. The resolution is on the order of $5\ V/A_0^n = 0.625$ mV. This corresponds fairly closely to state-of-the-art realized comparators in a 0.8-μm technology.

7.3 LATCHED COMPARATORS

Modern high-speed comparators typically have one or two stages of preamplification followed by a *track-and-latch stage*, as Fig. 7.13 [Yukawa, 1985] shows in simplified

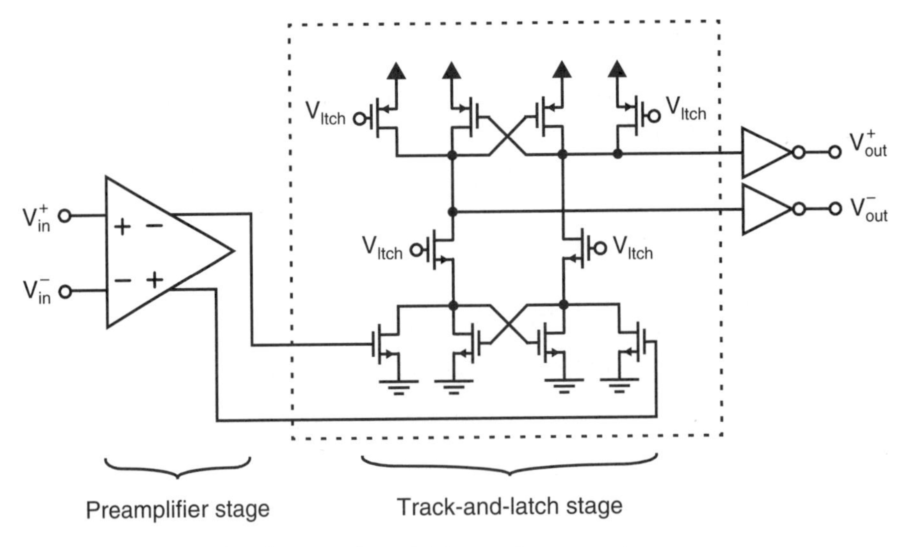

Fig. 7.13 A typical architecture for a high-speed comparator.

form for a CMOS implementation. The rationale behind this architecture is as follows: The preamplifier or preamplifiers are used to obtain higher resolution and to minimize the effects of kickback (which will be explained shortly). The output of the preamplifier, although larger than the comparator input, is still much smaller than the voltage levels needed to drive digital circuitry. The track-and-latch stage then amplifies this signal further during the track phase, and then amplifies it again during the latch-phase, when positive feedback is enabled. The positive feedback regenerates the analog signal into a full-scale digital signal. The track-and-latch stage minimizes the total number of gain stages required, even when good resolution is needed, and, thus, is faster than the multi-stage approach just described.

The preamplifier typically has some gain, perhaps 4 to 10, although sometimes it may simply be a unity-gain buffer, if very high speed but only moderate resolution is required. The preamplifier usually does not have gains much greater than 10; otherwise its time constant is too large and its speed is limited. It is not good practice to eliminate the preamplifier altogether since kickback into the driving circuitry will then result in very limited accuracy. *Kickback* denotes the charge transfer either into or out of the inputs when the track-and-latch stage goes from track mode to latch mode. This charge transfer is caused by the charge needed to turn the transistors in the positive-feedback circuitry on and by the charge that must be removed to turn transistors in the tracking circuitry off. Without a preamplifier or buffer, this kickback will enter the driving circuitry and cause very large glitches, especially in the case when the impedances seen looking into the two inputs are not perfectly matched.

In high-resolution applications, capacitive coupling and reset switches are also typically included to eliminate any input-offset-voltage and clock-feedthrough errors, in a manner similar to that described in Sections 7.1 and 7.2.

One very important consideration for comparators is to ensure that no memory is transferred from one decision cycle to the next. For example, if a comparator toggles in one direction, it might have a tendency to stay in that direction. This tendency is sometimes called *hysteresis*. In order to eliminate it, one can reset the different stages before entering track mode. This might be achieved by connecting internal nodes to one of the power supplies or by connecting differential nodes together using switches before entering track mode. For example, the comparator shown in Fig. 7.13 has internal nodes of the latch reset to both V_{DD} and ground when the V_{ltch} signal is low. Not only does this eliminate memory, but it also sets the comparator to its *trip point,* which speeds up operation when the comparator resolves small input signals.

Another consideration for comparators is that the gain should not be too large; otherwise the time constants are too large, which limits the speed, especially in track mode.

The track-and-latch stage has many variations. The circuitry shown in Fig. 7.13 should be considered only symbolic, although it will suffice in many applications.

Latch-Mode Time Constant

The time constant of the latch, when it is in its positive-feedback (i.e., latch) phase, can be found by analyzing a simplified circuit consisting of two back-to-back inverters, such as those shown in Fig. 7.14. If we assume that the output voltages of the inverters are close to each other at the beginning of the latch phase, and that the inverters are in their linear range, then each of the inverters can be modelled as a voltage-controlled current source driving an RC load, as shown in Fig. 7.15, where A_V is the low-frequency gain of each inverter, which has a transconductance given by $G_m = A_V/R_L$. For this linearized model, we have

$$\frac{A_V}{R_L}V_y = -C_L\left(\frac{dV_x}{dt}\right) - \left(\frac{V_x}{R_L}\right) \tag{7.22}$$

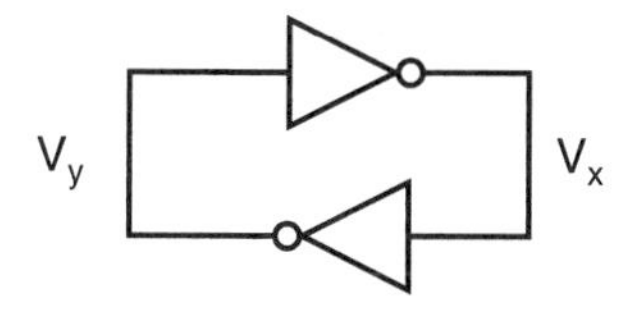

Fig. 7.14 Two back-to-back inverters used as a simplified model of a track-and-latch stage in its latch phase.

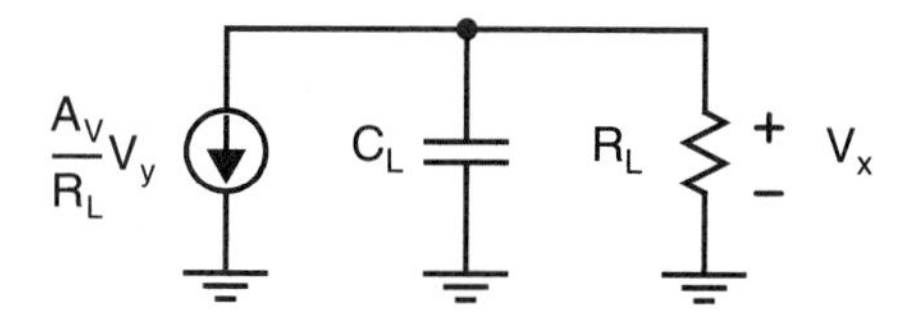

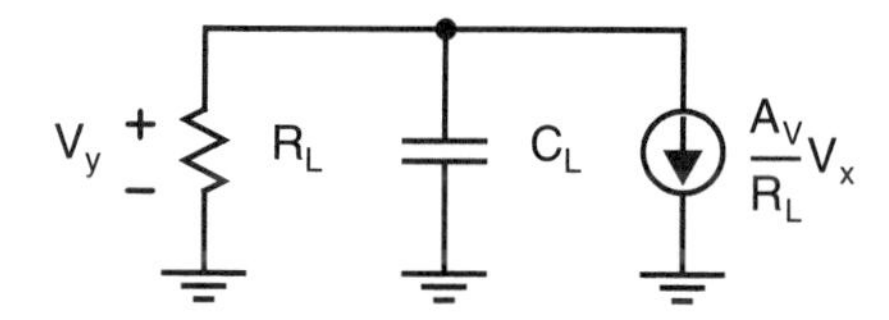

Fig. 7.15 A linearized model of the track-and-latch stage when it is in its latch phase.

and

$$\frac{A_V}{R_L}V_x = -C_L\left(\frac{dV_y}{dt}\right) - \left(\frac{V_y}{R_L}\right) \quad (7.23)$$

Multiplying (7.22) and (7.23) by R_L and rearranging gives

$$\tau\left(\frac{dV_x}{dt}\right) + V_x = -A_V V_y \quad (7.24)$$

and

$$\tau\left(\frac{dV_y}{dt}\right) + V_y = -A_V V_x \quad (7.25)$$

where $\tau = R_L C_L$ is the time constant at the output node of each inverter. Subtracting (7.25) from (7.24) and rearranging terms gives

$$\left(\frac{\tau}{A_V - 1}\right)\left(\frac{d\Delta V}{dt}\right) \cong \Delta V \quad (7.26)$$

where $\Delta V = V_x - V_y$ is the voltage difference between the output voltages of the inverters. Equation (7.26) is a first-order differential equation with no forcing function. Its solution is given by

$$\Delta V = \Delta V_0 e^{(A_V - 1)t/\tau} \quad (7.27)$$

where ΔV_0 is the initial voltage difference at the beginning of the latch phase. Thus, the voltage difference increases exponentially in time with a time constant given by

$$\tau_{ltch} = \frac{\tau}{A_V - 1} \cong \frac{R_L C_L}{A_V} = \frac{C_L}{G_m} \quad (7.28)$$

where, again, $G_m = A_V / R_L$ is the transconductance of each inverter. Note that τ_{ltch} is roughly equal to the inverse of the unity-gain frequency of each inverter.

In the case of MOS devices, normally the output load is proportional to the gate-source capacitance of a single transistor, or specifically,

$$C_L = K_1 W L C_{ox} \quad (7.29)$$

Here, K_1 is a proportionality constant between 1 and 2. Also, the inverter transconductance is proportional to the transconductance of a single transistor, and is given by

$$G_m = K_2 g_m = K_2 \mu_n C_{ox} \frac{W}{L} V_{eff} \quad (7.30)$$

where K_2 might be between 0.5 and 1. Substituting (7.29) and (7.30) into (7.28) gives

$$\tau_{ltch} = \frac{K_1}{K_2} \frac{L^2}{\mu_n V_{eff}} = K_3 \frac{L^2}{\mu_n V_{eff}} \quad (7.31)$$

where K_3 might be between 2 and 4. Note that (7.31) implies that τ_{ltch} depends primarily on the technology and not on the design (assuming a reasonable design is used that maximizes V_{eff} and minimizes C_L). Note also the similarity between (7.31) and (7.20), the equation for the time constant of a cascade of gain stages. For a given technology, (7.31) is very useful in determining a rough estimate for the maximum clock frequency of a latch-and-track comparator.

If it is necessary for a voltage difference of ΔV_{logic} to be obtained in order for succeeding logic circuitry to safely recognize the correct output value, then, by using (7.27), we find the time necessary for this to happen is given by

$$T_{ltch} = \frac{C_L}{G_m} \ln\left(\frac{\Delta V_{logic}}{\Delta V_0}\right) = K_3 \frac{L^2}{\mu_n V_{eff}} \ln\left(\frac{\Delta V_{logic}}{\Delta V_0}\right) \quad (7.32)$$

If ΔV_0 is small, this latch time can be large, perhaps larger than the allowed time for the latch phase. Such an occurrence is often referred to as *metastability*. In other words, because its initial value is too small, the differential output voltage of the latch does not increase enough to be recognized as the correct logic value by succeeding circuitry. Sometimes, even when the initial voltage difference is large enough, it is possible that circuit noise can cause the initial voltage difference to become small enough to cause metastability.

EXAMPLE 7.5

Assume for a 0.8-μm technology that $K_3 = 2.5$, $\Delta V_0 = 10$ mV, $\Delta V_{logic} = 2$ V, $V_{eff} = 0.5$ V, and $\mu_n = 0.05\ M^2/V \cdot s$. What is the maximum clocking frequency of the comparator?

Solution

Using (7.32), we have $T_{ltch} = 0.34$ ns. Assuming this is half the total period (which might be quite an optimistic assumption), the smallest possible period for a track-and-latch might be 0.68 ns, which corresponds to a clocking frequency of 1.5 GHz. In a typical comparator, the maximum clock rate would probably be limited by the frequency response of the preamplifiers and the track-and-latch during the track phase, rather than by the speed of the latch during the latch phase.

7.4 EXAMPLES OF CMOS AND BiCMOS COMPARATORS

This section describes a number of high-speed comparators that are currently popular.

The literature on integrated circuit technology has many examples of latched comparators. For example, Fig. 7.16 shows a comparator presented in [Song, 1990]. This comparator has the positive feedback of the second stage always enabled. In track mode, when the two diode-connected transistors of the gain stage are enabled, the gain around the positive-feedback loop is less than one, and the circuit is stable. The combination of the diode-connected transistors of the gain stage and the transistors of the

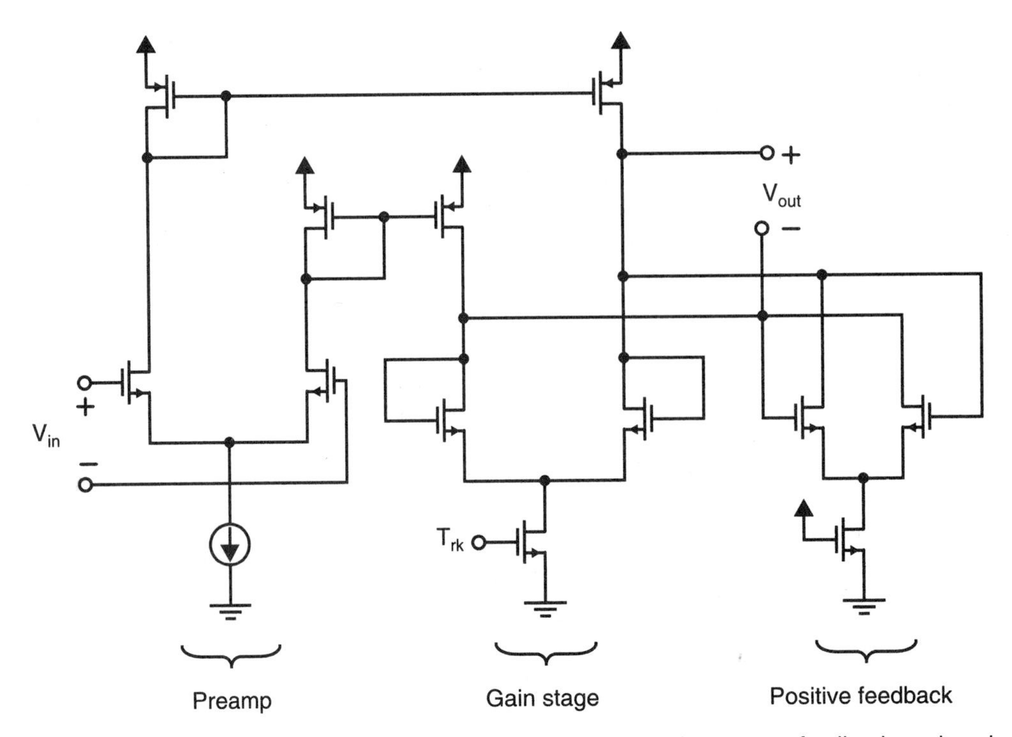

Fig. 7.16 A two-stage comparator that has a preamplifier and a positive-feedback track-and-latch stage.

positive-feedback loop acts as a moderately large impedance and gives gain from the preamplifier stage to the track-and-latch stage. The diode-connected loads of the preamplifier stage give a limited amount of gain in order to maximize speed, while still buffering the kickback from the input circuitry.

A second comparator is shown in Fig. 7.17 [Norsworthy, 1989]. This design also uses diode-connected loads to keep all nodes at relatively low impedance (similar to current-mode circuit-design techniques), thereby keeping all node time constants small and giving fast operation. This design also uses precharging to eliminate any memory from the previous decision. For example, the positive-feedback stage is precharged low, whereas the digital-restoration stage is precharged high, somewhat similar to what is done in Domino CMOS logic [Krambeck, 1982].

A third comparator was designed, fabricated, and characterized by K. Martin in 1984 (unpublished) and is shown in Fig. 7.18, with appropriate clock waveforms shown in Fig. 7.19. This design eliminates any input-offset voltages from both the first and second stages by using capacitive coupling. It also has common-mode feedback circuitry for the first preamplifier stage, which allows for input signals that have large common-mode signals. Unlike in fully differential opamps, in fully differential comparators, the linearity of the common-mode feedback circuitry is noncritical since, whenever large signals are present (and the common-mode feedback circuitry becomes nonlinear), there is no ambiguity in resolving the sign of the input signal. This allows a simple differential pair to be used for CMFB circuitry. Measured (but

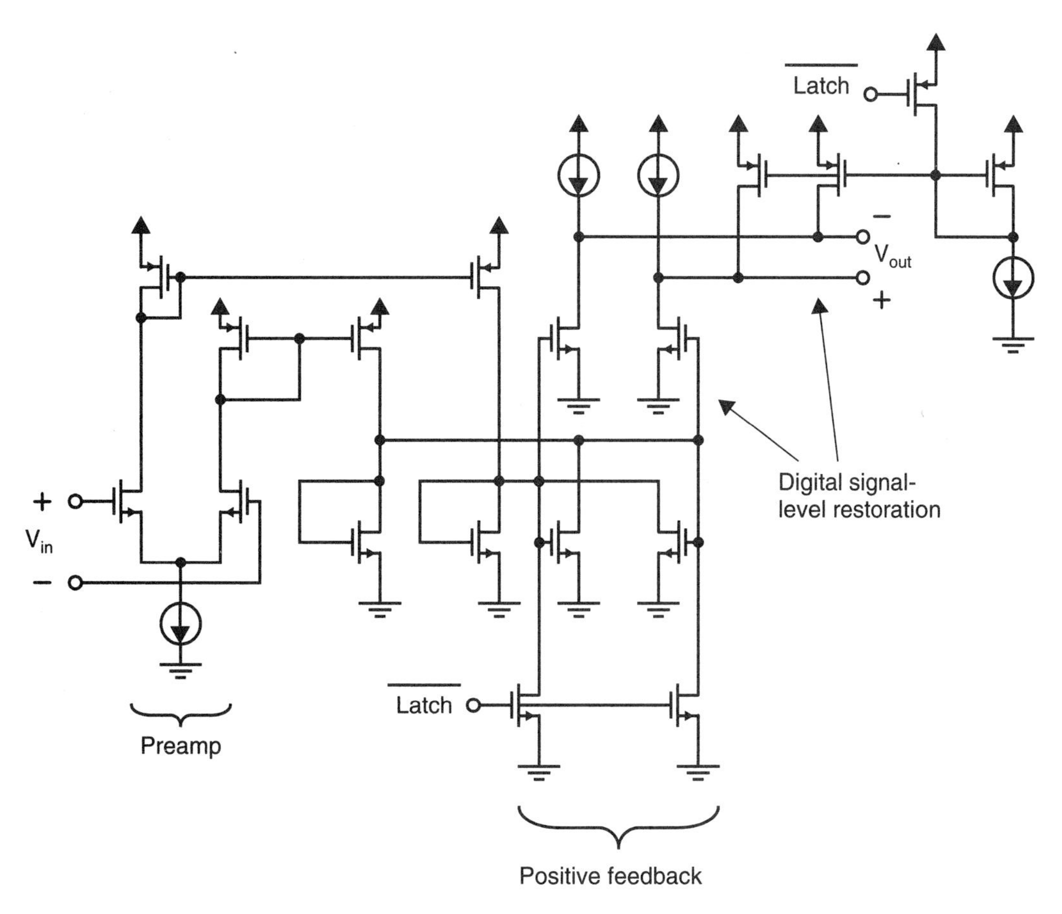

Fig. 7.17 A two-stage comparator.

unpublished) performance of this circuit resulted in a 0.1-mV resolution at a 2-MHz clock frequency, despite a very old 5-μm technology. The performance measured was limited by the test set-up rather than by the circuitry.

A fourth comparator that is realized in a BiCMOS technology and that exhibits very good performance, is described in [Razavi, 1992]. Indeed, this comparator is one of the better BiCMOS designs to date. It is based on the idea that it is not necessary to reset the first stage; the input-offset errors of the first stage can still be eliminated by resetting the right side of the coupling capacitors between the first stage and the second stage, assuming that the gain of the first stage is not too large [Poujois, 1978], [Vittoz, 1985]. A simplified schematic of this comparator is shown in Fig. 7.20. During the reset phase, the inputs to the preamplifier are connected directly to ground (or to a reference voltage), and the outputs of the coupling capacitors are connected to ground as well. This stores any offset voltages of the first stage on the capacitors. When the comparator is taken out of reset phase, then the effect of the clock feedthrough of S_5 and S_6 on the input resolution is divided by the gain of the first stage. In the realization described in [Razavi, 1992], the first stage is a BiCMOS

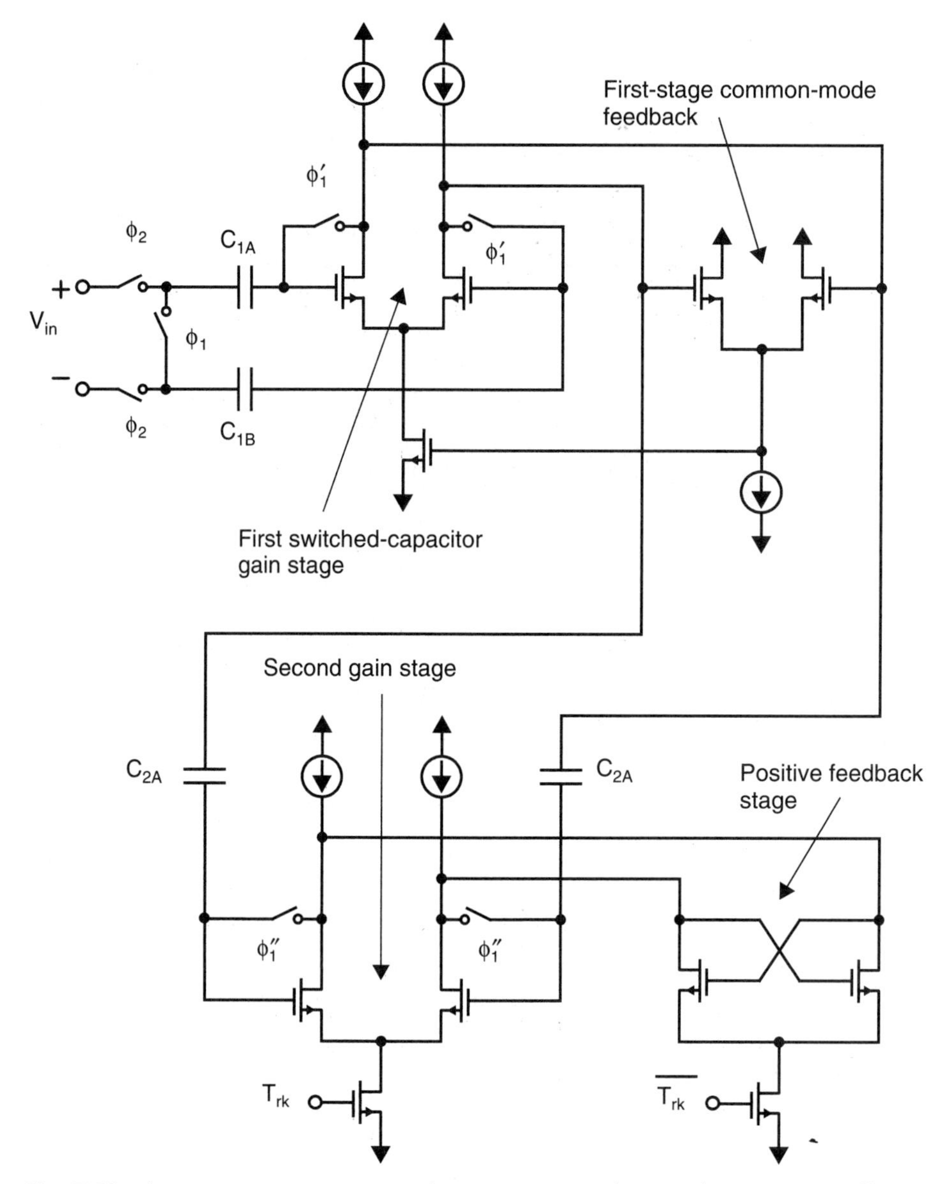

Fig. 7.18 A two-stage comparator with capacitive coupling to eliminate input-offset voltage and clock-feedthrough errors, along with positive feedback for fast operation.

preamplifier, consisting of MOS source followers followed by a bipolar differential amplifier and emitter-follower output buffers, as shown in Fig. 7.21. Notice that the circuit operates between ground and a negative voltage supply. Note also that the switches are realized using p-channel transistors. The track-and-latch stage has both a bipolar latch and a CMOS latch, as shown in Fig. 7.22. During the reset phase, ϕ_1 is low and ϕ_2 is high, X_1 and Y_1 are connected to ground through switches M_1 and M_2, C_3 is discharged to the minus supply through M_{11}, and X_2 and Y_2 are discharged to

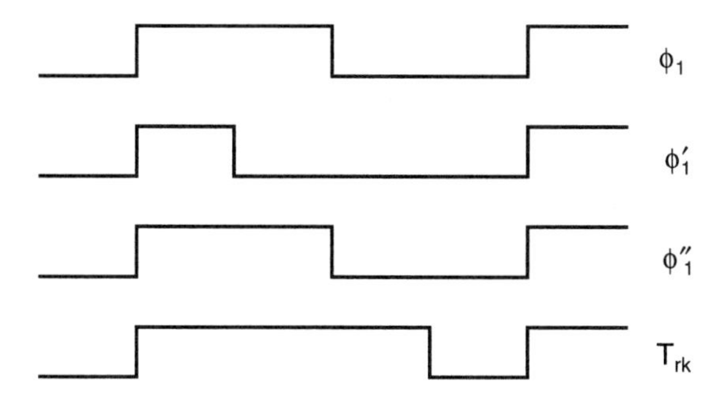

Fig. 7.19 The clock waveforms required by the comparator in Fig. 7.18.

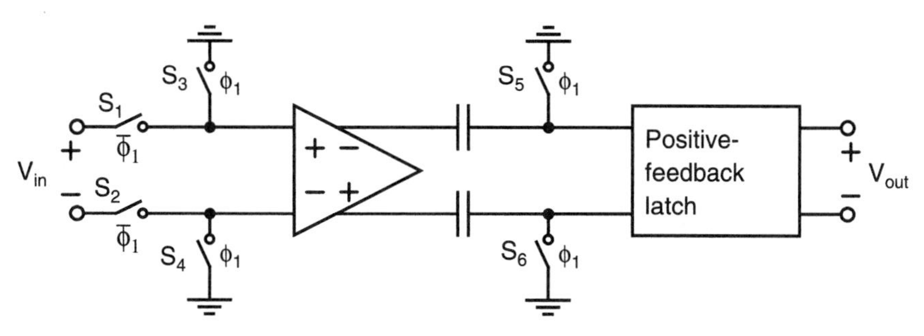

Fig. 7.20 A simplified schematic of the comparator described in [Razavi, 1992].

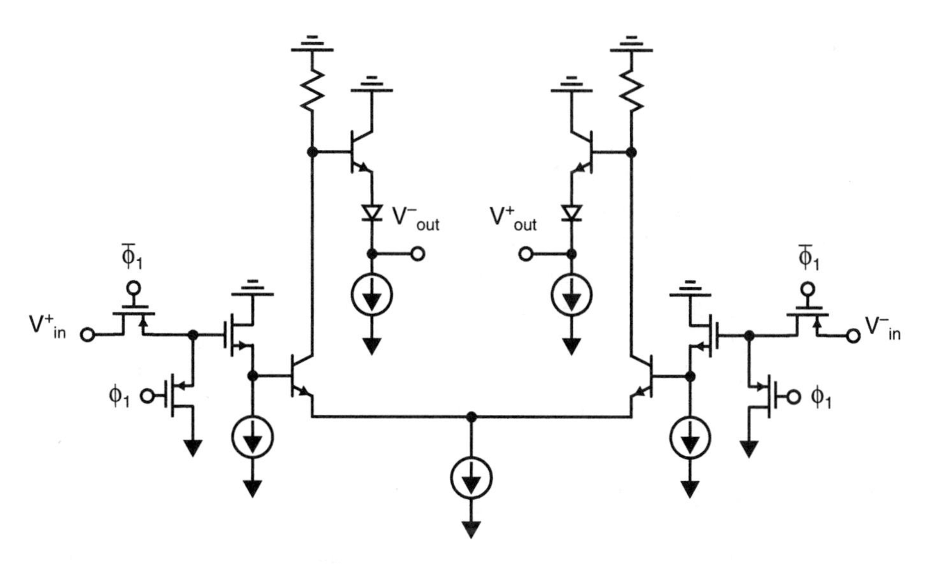

Fig. 7.21 A BiCMOS preamplifier.

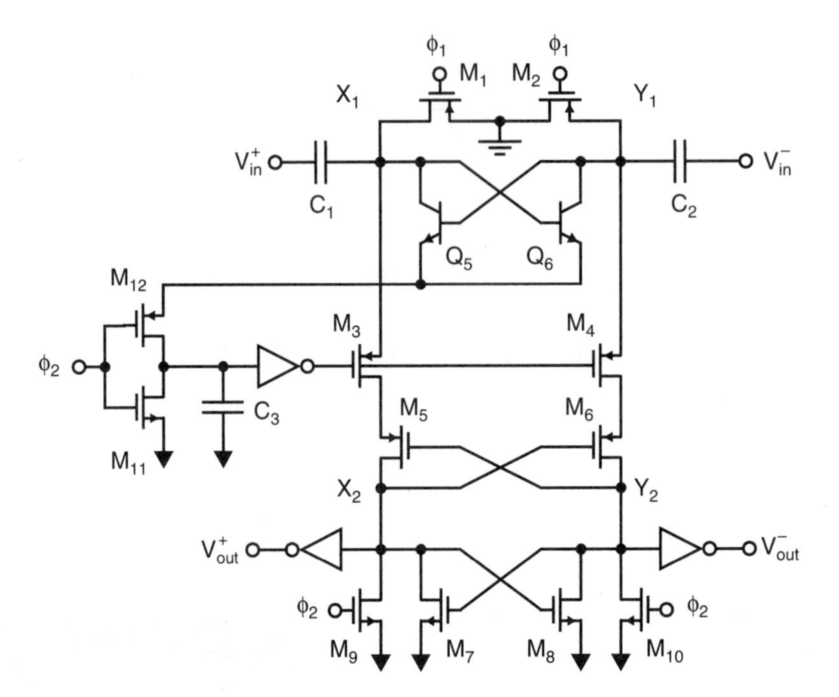

Fig. 7.22 The BiCMOS track-and-latch stage in [Razavi, 1992].

the minus supply through M_9 and M_{10}. Also, at this time, M_3 and M_4 are off. Next, ϕ_1 becomes high, which leaves X_1 and Y_1 floating and connects the outputs of the preamplifier to its input signal. After a short delay, which is needed for the transient response of the preamplifier, ϕ_2 becomes low, which turns on M_{12}, thereby activating the positive feedback action of Q_5 and Q_6. This develops a differential voltage between X_1 and Y_1 of about 200 mV, since C_3 is about one-fifth the size of C_1 and C_2. The offset of this bipolar latch is quite small, typically on the order of one millivolt or less. Also note that, when activated, the only power dissipated is that required to charge C_3—there is no dc power dissipation. A short time after the bipolar latch is activated, the inverter connected to C_3 changes state, and its output voltage will drop. This low voltage turns on M_3 and M_4, which activates the CMOS part of the latch consisting of cross-coupled M_5 and M_6 and cross-coupled M_7 and M_8. The reasons for including cross-coupled M_5 and M_6 are to prevent nodes X_1 and Y_1 from being discharged very far from ground toward the minus power supply and at the same time to amplify the 200-mV difference signal developed by the bipolar latch to almost a full-level CMOS voltage change. Thus, the latch is accurate, relatively fast, and low power. Also described in [Razavi, 1992] is an interesting CMOS-only comparator, which is not described here, but which interested readers may investigate.

Input-Transistor Charge Trapping

When realizing very accurate CMOS comparators, an error mechanism that must be considered is charge trapping in the gate oxide of the input transistors of the MOS

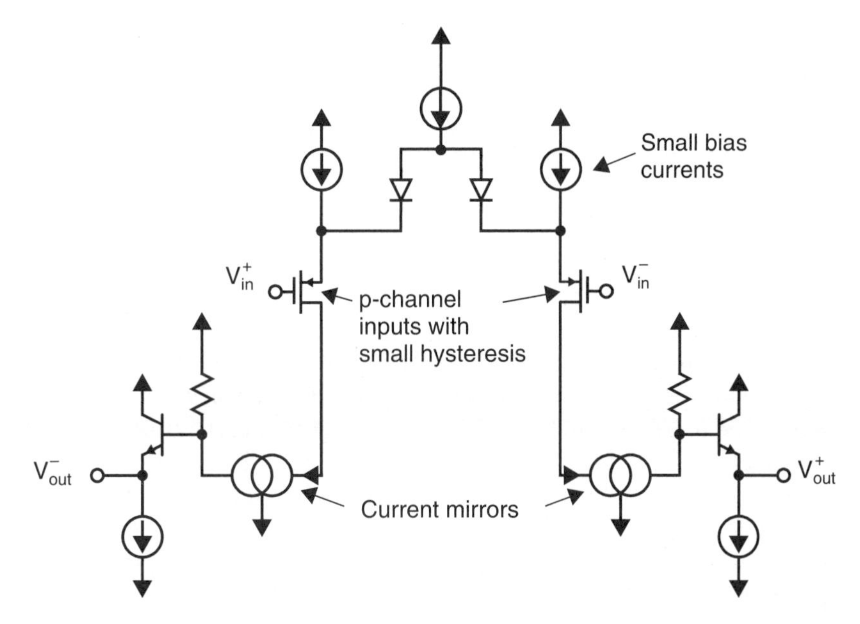

Fig. 7.23 A comparator with little hysteresis. (Hysteresis is caused by charge trapping.)

comparators [Tewksbury, 1989]. When n -channel transistors are stressed with large positive gate voltages, electrons can become trapped via a tunneling mechanism in which electrons tunnel to oxide traps close to the conduction band. The time constant for the release of these trapped electrons is on the order of milliseconds and is much longer than the time it takes for them to become trapped. During the time they are trapped, the effective transistor threshold voltage is increased. This leads to a comparator hysteresis on the order of 0.1 to 1 mV. This effect correlates well with transistor 1/f noise and is much smaller in p -channel transistors. When very accurate comparators are needed, charge trapping must be considered and minimized. One possible means of minimizing this effect, assuming we have a BiCMOS technology, is described in [Miller, 1990] and is illustrated in Fig. 7.23. A BiCMOS technology allows diodes to be used in the input stage, which allows the inclusion of two additional small-current sources in the input stage; this guarantees that the p-channel input transistors never turn off.[5] Also, by using p-channel input transistors, the charge trapping is greatly minimized because p-channel transistors exhibit much less hysteresis than n-channel transistors.

An alternative approach is to flush the input transistors after each use where the junctions and wells of n-channel transistors are connected to a positive power supply whereas the gates are connected to a negative power supply. This effectively eliminates the trapped electrons [Swanson, 1993]. Alternatively, or in addition, two input stages can be used for the comparator—a rough stage can be used during times when

5. p-channel transistors exhibit trapping for positive gate voltages since only electrons, and not holes, are trapped. Therefore, to prevent trapping, they should never be turned off.

large signals with overloads are possible, whereas a fine stage can be used during times when accurate comparisons are necessary and it can be guaranteed that no large signals are present [Swanson, 1993; Tan, 1990].

7.5 EXAMPLES OF BIPOLAR COMPARATORS

High-speed bipolar comparators are typically latched comparators, as shown previously in Fig. 7.13, where preamplifier and track-and-latch stages are used. A typical example is shown in Fig. 7.24 [Wakimoto, 1988]. The preamplifier is a simple differential amplifier that might have a gain of around 5. Besides increasing the resolution, this preamplifier helps to eliminate kickback from the track-and-latch stage. The track-and-latch stage is very similar to a current-mode digital latch. During track phase, the differential pair consisting of transistors Q_3 and Q_4 is enabled and operates as a second differential amplifier. When latch mode is enabled, I_2 is diverted to the differential pair consisting of Q_5 and Q_6. The inputs to this third differential pair come from the outputs of the second differential pair through emitter-follower buffers consisting of Q_9 and Q_{10}. The outputs of the third differential pair are connected back to the high-impedance output nodes in a cross-coupled manner so that positive feedback results. When this positive feedback is enabled, it takes the initially-small differential voltage and quickly amplifies it to between 0.4 and 0.6 V (depending on the logic levels being used). There are two reasons for using the emitter-follower buffers. These buffers isolate the

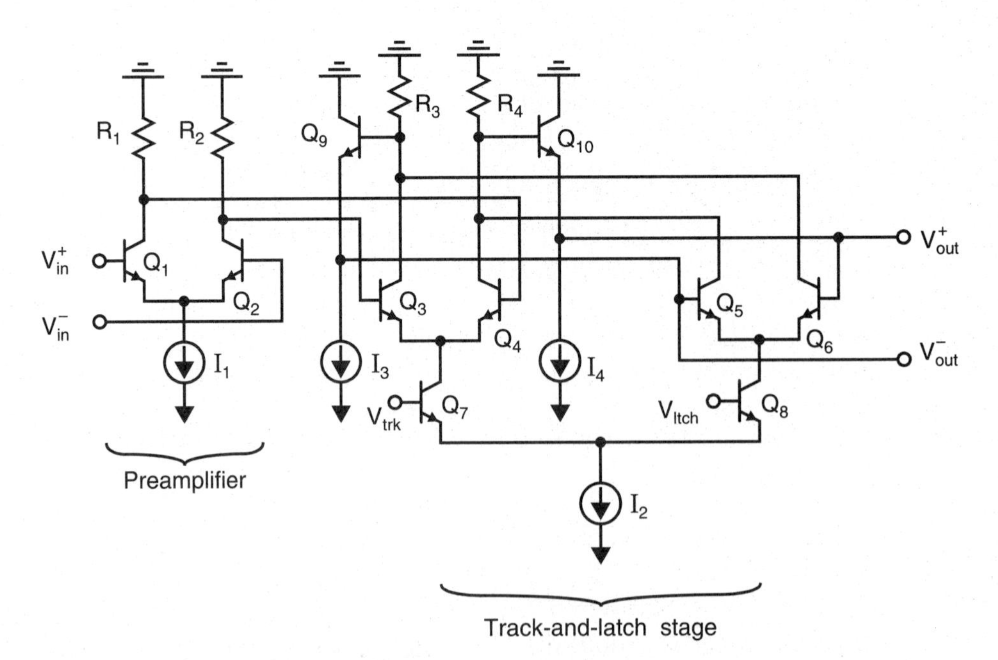

Fig. 7.24 A typical bipolar comparator.

high-impedance nodes at the collectors of the second and third differential pairs from the base-emitter capacitances of the third differential pair. They also isolate the same high impedance nodes from the output load capacitance.

Another architecture for a bipolar comparator is shown in Fig. 7.25 [Van de Plassche, 1988]. In this comparator, the input signals go to a differential pair that is always turned on (even during the latch phase). During the track phase, Q_3 and Q_4 are on. They operate as common-base transistors, and therefore the comparator operates as a cascode amplifier during this phase. When the comparator goes into latch mode, the collector current from the input transistors is redirected from the common-base transistors to transistors Q_7 and Q_8, which are connected in a positive-feedback loop. Since the bias current of the input transistors is never turned off, the kickback from them is greatly reduced without adding additional preamplifier stages. This maximizes the bandwidth during the tracking phase.

A third bipolar comparator is shown in Fig. 7.26. Although this comparator has never been experimentally tested, it is used as an example to illustrate many comparator design principles. A small current source is added to the tracking differential pair so that this differential pair doesn't turn off completely when the comparator is in hold mode. This makes the comparator faster when it goes from hold mode into track mode. Also, a cascode common-base stage is added just below the output voltages. This stage is used to isolate the outputs from the positive-feedback loop. This in turn allows for much smaller resistors to be used in the positive-feedback loop, which speeds this loop up and eliminates the need for emitter-follower buffers in the loop. Finally, a larger resistor is added between the emitters of the cascode transistors. This resistor prevents the cascode transistors from turning off completely during the latch phase, which also speeds up the transition from latch mode to tracking mode.

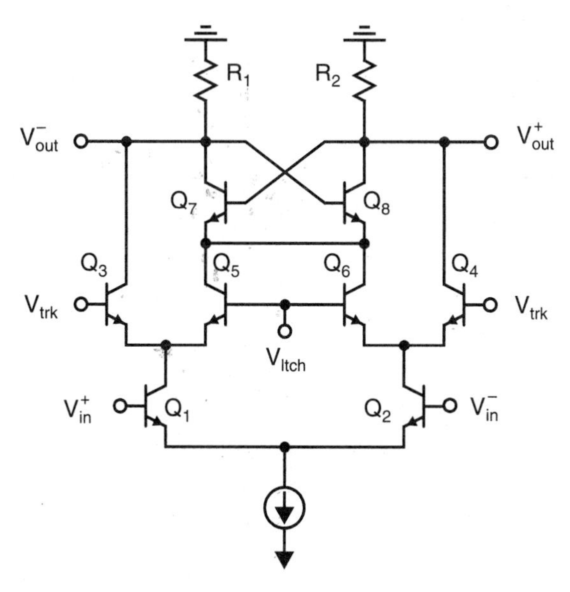

Fig. 7.25 A second bipolar comparator.

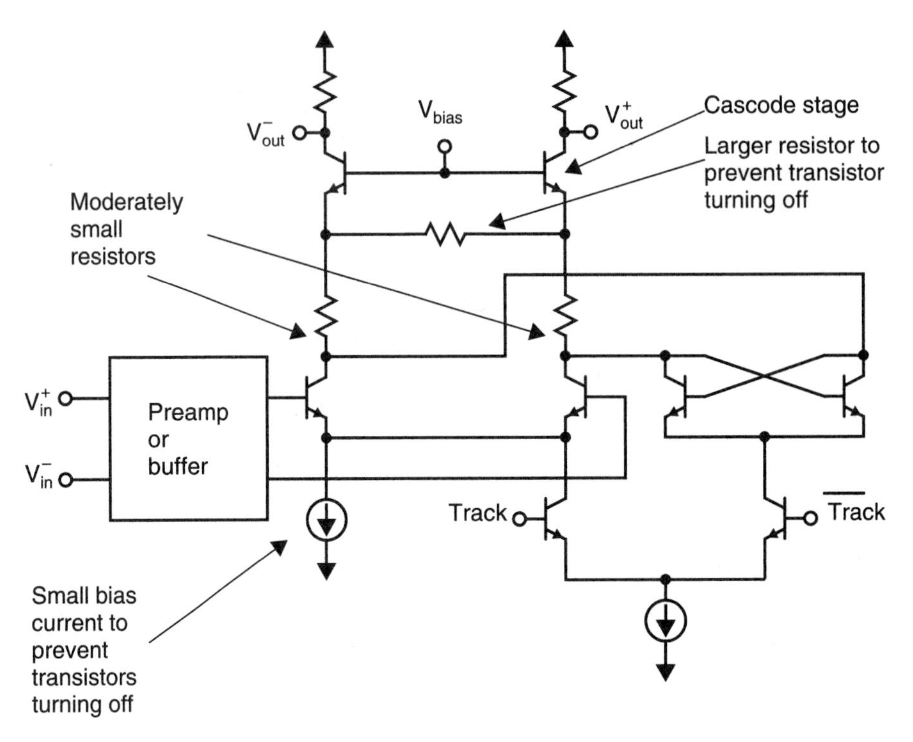

Fig. 7.26 A third bipolar comparator.

7.6 REFERENCES

D. G. Haigh and B. Singh. "A Switching Scheme for Switched-Capacitor Filters, Which Reduces Effect of Parasitic Capacitances Associated with Control Terminals," *Proc. IEEE Int. Symp. on Circuits and Systems,* Vol. 2, pp. 586–589, June 1983.

R. H. Krambeck, C. Lee, and H. S. Law. "High-Speed Compact Circuits with CMOS," *IEEE J. of Solid-State Circuits,* Vol. 17, no. 3, pp. 614–619, June 1982.

K. Martin, "Improved Circuits for the Realization of Switched-Capacitor Filters," *IEEE Trans. Circuits and Systems*, Vol. CAS-27, no. 4, pp. 237–244, April 1980.

———. "New Clock-Feedthrough Cancellation Technique for Analog MOS Switched Capacitor," *Electron. Lett.*, no. 18, 1992.

K. Martin, L. Ozcolak, Y. S. Lee, and G. C. Temes. "A Differential Switched-Capacitor Amplifier," *IEEE J. of Solid-State Circuits*, Vol. SC-22, no. 1, pp. 104–106, February 1987.

L. L. McCreary and P. R. Gray. "All-MOS Charge Redistribution Analog-to-Digital Conversion Techniques—Part I, "*IEEE J. of Solid-State Circuits*, Vol. SC-10, pp. 371–379, December 1975.

G. Miller, M. Timko, H. S. Lee, E. Nestler, M. Mueck, and P. Ferguson. "An 18-b 10-μs Self-Calibrating ADC," *IEEE Intern. Solid-State Circuits Conf.*, pp. 168–169, February 1990.

S. Norsworthy, I. Post, and S. Fetterman. "A 14-bit 80-KHz Sigma-Delta A/D Converter: Modeling, Design, and Performance Evaluation," *IEEE J. of Solid-State Circuits*, Vol. 24, no. 2, pp. 256–267, April 1989.

R. Poujois and J. Borel. "A Low-Drift Fully Integrated MOSFET Operational Amplifier," *IEEE J. of Solid-State Circuits*, Vol. SC-13, pp. 499–503, August 1978.

B. Razavi, *Principles of Data Conversion System Design*, IEEE Press, Piscataway, New Jersey, 1995.
B. Razavi and B. Wooley. "Design Techniques for High-Speed, High-Resolution Comparators," *IEEE J. of Solid-State Circuits*, Vol. 27, no. 12, pp. 1916–1926, December 1992.
J. Shieh, M. Patil, and B. L. Sheu. "Measurement and Analysis of Charge Injection in MOS Analog Switches," *IEEE J. of Solid-State Circuits*, Vol. 22, no. 2, pp. 277–281, April 1987.
B. Song, H. S. Lee, and M. Tompsett. "A 10-b 15-MHz CMOS Recycling Two-Step A/D Converter," *IEEE J. of Solid-State Circuits*, Vol. 25, no. 6, pp. 1328–1338, December 1990.
E. J. Swanson, "Method and Circuitry of Decreasing the Recovery Time of a MOS Differential Voltage Comparator," U.S. patent no. 5,247,210, September 1993.
K. Tan, et al. "Error-Correction Techniques for High-Performance Differential A/D Converters," *IEEE J. of Solid-State Circuits,* Vol. 25, no. 6, pp. 1318–1327, December 1990.
T. L. Tewksbury, H. S. Lee, and G. Miller. "The Effects of Oxide Traps on the Large-Signal Transient Response of Analog MOS Circuits," *IEEE J. of Solid-State Circuits,* Vol. 24, no. 2, pp. 542–543, April 1989.
R. Van de Plassche and P. Baltus. "An 8-bit 100-MHz Full-Nyquist Analog-to-Digital Converter," *IEEE J. of Solid-State Circuits*, Vol. 23, no. 6, pp. 1334–1344, December 1988.
E. A. Vittoz, "Dynamic Analog Techniques," in *Design of MOS VLSI Circuits for Telecommunications*, ed. Y. Tsividis and P. Antognetti, Prentice Hall, Englewood Cliffs, New Jersey, 1985.
T. Wakimoto, Y. Akazawa, and S. Konaka. "Si Bipolar 2-GHz 6-bit Flash A/D Conversion LSI," *IEEE J. of Solid-State Circuits,* Vol. 23, no. 6, pp. 1345–1350, December 1988.
Y. S. Yee, L. M. Terman, and L. G. Heller. "A 1-mV MOS Comparator," *IEEE J. of Solid-State Circuits*, Vol. SC-13, pp. 294–297, June 1978.
Yubawa, A. "A CMOS 8-bit High-Speed A/D Converter," *IEEE J. of Solid-State Circuits*, Vol. SC-20, no. 3, pp. 775–779, June 1985.

7.7 PROBLEMS

Unless otherwise stated, assume the following:

- n-channel MOS transistors:
  $\mu_n C_{ox} = 92\ \mu A/V^2$
  $V_{tn} = 0.8\ V$
  $\gamma = 0.5\ V^{1/2}$
  $r_{ds}\ (\Omega) = 8000\ L\ (\mu m)/I_D\ (mA)$ in active region
  $C_j = 2.4 \times 10^{-4}\ pF/(\mu m)^2$
  $C_{j\text{-}sw} = 2.0 \times 10^{-4}\ pF/\mu m$
  $C_{ox} = 1.9 \times 10^{-3}\ pF/(\mu m)^2$
  $C_{gs(overlap)} = C_{gd(overlap)} = 2.0 \times 10^{-4}\ pF/\mu m$
- p-channel MOS transistors:
  $\mu_p C_{ox} = 30\ \mu A/V^2$
  $V_{tp} = -0.9\ V$
  $\gamma = 0.8\ V^{1/2}$
  $r_{ds}\ (\Omega) = 12{,}000\ L\ (\mu m)/I_D\ (mA)$ in active region
  $C_j = 4.5 \times 10^{-4}\ pF/(\mu m)^2$
  $C_{j\text{-}sw} = 2.5 \times 10^{-4}\ pF/\mu m$
  $C_{ox} = 1.9 \times 10^{-3}\ pF/(\mu m)^2$
  $C_{gs(overlap)} = C_{gd(overlap)} = 2.0 \times 10^{-4}\ pF/\mu m$

7.1 A simple CMOS inverter is used as a comparator, as shown in Fig. P7.1. Assume ϕ_{1a} drops before ϕ_1 drops, so that only the charge injection from Q_1 need be considered. What is the change in V_{out} due to this charge injection?

Assume the transistors in the inverter have sizes chosen such that the inverter threshold voltage is at $V_{DD}/2 = 2.5$ V (i.e., $W_5/W_4 = \mu_n/\mu_p$), and that all transistor lengths are 0.8 μm. Also assume the inverter gain is –24 at $V_{out} = 2.5$ V, and ignore overlap capacitance. Do not ignore the body effect on the threshold voltage of Q_1.

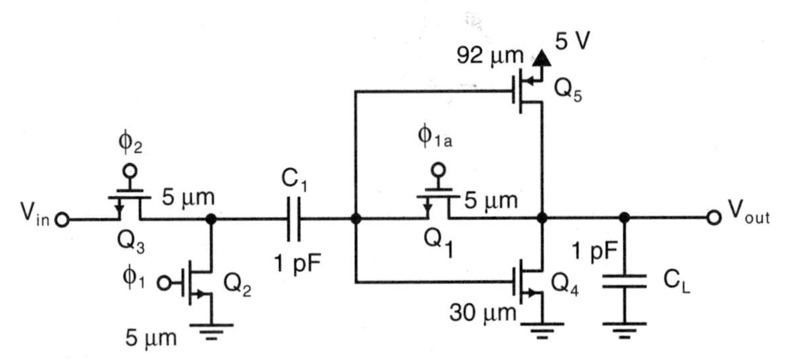

Fig. P7.1

7.2 Assume that three stages of inverters identical to the inverter used in Problem 7.1 are cascaded as shown in Fig. 7.10. What is the input-voltage offset error due to the charge injection of the reset switch of the last stage only?

7.3 Estimate the time constant for the inverter used in Problem 7.1 during its reset phase and also during its comparison phase.

7.4 For the circuit shown in Fig. 7.10, find the voltages at each of the opamp inputs at the end of ϕ_1', ϕ_1'', ϕ_1''', ϕ_1, and ϕ_2, where V_i is the input and $V_{err\text{-}i}$ is the error due to the reset switch of the ith opamp. Assume opamps are offset free and that they remain in their linear region after their reset switch is opened. Show that the overall input-referred clock feedthrough is given by $|V_{err\text{-}3}|/(A_1A_2)$.

7.5 Repeat Problem 7.4 where offsets of the opamps are not ignored. What is the input-referred offset and clock feedthrough?

7.6 Assume a MOS transistor's unity-gain frequency is approximately given by $\omega_u = g_m/C_{gs}$. Find an expression for ω_t in terms of V_{eff} and C_{ox}. Assume a single gain stage without compensation has a unity-gain frequency given by $\omega_t = \omega_u/2$. Further assume that a comparator is realized by a cascade of three stages, each having a gain of –15. If three time constants are required for settling during the open-loop comparison phase, what is the maximum frequency of operation for the comparator? What is the resolution? You may assume that $V_{eff} = 0.25$ V, $\mu_n = 0.05\ M^2/V \cdot s$, and that there are only n-channel transistors in the signal path.

7.7 Give reasonable device widths of the comparator in Fig. 7.16 so that the first stage has a gain of 5 and so that the latch-and-track stage has a gain of 10 when in track mode and a loop gain of 4 when in latch mode. Assume the input transistors are 25-μm wide and that the bias current of the first stage is 0.1 mA.

7.8 Assume the comparator of Fig. 7.16 has the device sizes found in Problem 7.7. Estimate the equivalent time constant when the comparator is in track mode. If the input signal changes from 0 to 5 mV at time 0, how long would it take the differential output voltage to change by 50 mV?

7.9 Consider the comparator in Fig. 7.16 with the device sizes found in Problem 7.7. Estimate the equivalent time constant of the latch-and-track stage when the comparator is in latch mode. If the differential output voltage is 50 mV at the beginning of the latch phase, how long would it take for the output voltage to change to a differential voltage of 2 V?

7.10 What is the minimum power-supply voltage that can be used for the comparator in Fig. 7.25 if the current source and the resistive loads both require 0.4 V across them? You may assume that for the bipolar transistors $V_{CE\text{-}sat} = 0.3$ V.

7.11 The CMOS transmission gate shown has an input voltage of 2.5 V when it turns off. The W/L of the n-channel is 5 μm/0.8 μm and the W/L of the p-channel is 15 μm/0.8 μm. Assume the total parasitic capacitance between the output node and ground is 50 fF, $V_{DD} = 5$ V, and that the clock signals change very fast. Also ignore changes due to overlap capacitance. Estimate the change in output voltage due to charge injection. What will the final output voltage be?

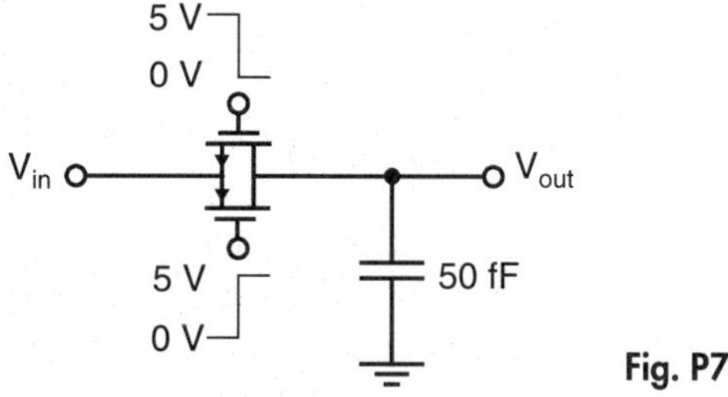

Fig. P7.11

CHAPTER

8 Sample and Holds, Voltage References, and Translinear Circuits

In this chapter, some common analog building blocks are described other than those already presented. Specifically, circuits to realize sample and holds, bandgap voltage references, and translinear gain cells and multipliers are discussed. A *sample and hold* (S/H) is used to sample an analog signal and to store its value for some length of time. A *bandgap voltage reference* is often used to bias circuits or to supply a reference to which other voltages are compared. A *translinear gain cell* is commonly used to create an amplifier whose gain can be adjusted through the use of a controlling input current. Finally, a *translinear multiplier* is realized by interconnecting two gain cells and finds a variety of uses in applications such as a modulator (in a communication system) or a phase detector (in a phase-locked loop).

8.1 PERFORMANCE OF SAMPLE-AND-HOLD CIRCUITS

An important analog building block, especially in data-converter systems, is the sample-and-hold-circuit. Before proceeding, it is worthwhile to mention that sample-and-hold circuits are also often referred to as "track-and-hold" circuits. Normally, these two terms are synonymous except for a few particular switched-capacitor sample-and-hold circuits that do not have a phase where the output signal is tracking the input signal. Sample-and-hold circuits are necessary components in many data-acquisition systems such as A/D converters. In many cases, the use of a sample and hold (at the front of the data converter) can greatly minimize errors due to slightly different delay times in the internal operation of the converter.

Before discussing the basic principles of sample-and-hold circuits, it is necessary to mention some performance parameters used in characterization.

1. The first of these parameters is a *sampling pedestal* or a *hold step*. This is an error that occurs each time a sample and hold goes from sample mode to hold mode. During this change in operation, there is always a small error in the voltage being held that makes it different from the input voltage at the time of sampling. Obviously, this error should be as small as possible. *Perhaps more importantly, this error should be signal independent; otherwise it can introduce nonlinear distortion.*

2. Another parameter is a measure of how isolated the sampled signal is from the input signal during hold mode. Ideally, the output voltage will no longer be affected by changes in the input voltage. In reality, there is always some signal feedthrough, usually through parasitic capacitive coupling from the input to the output. In well-designed sample and holds, this signal feedthrough can be greatly minimized.
3. A third important parameter is the speed at which a sample and hold can track an input signal, when in sample mode. In this mode, a sample and hold will have both small-signal and large-signal limitations due to its –3-dB bandwidth and finite slew rate, respectively. Both the –3-dB bandwidth and slew rate should be maximized for high-speed operation.
4. Yet another limitation (somewhat less important in high-speed designs) is the *droop rate* in hold mode. This error is a slow change in output voltage, when in hold mode, caused by effects such as leakage currents due to the finite base currents of bipolar transistors and reverse-biased junctions. In most CMOS designs, this droop rate is so small it can often be ignored.
5. A fifth limitation is *aperture jitter* or *aperture uncertainty*. This error is the result of the effective sampling time changing from one sampling instance to the next and becomes more pronounced for high-speed signals. Specifically, when high-speed signals are being sampled, the input signal changes rapidly, resulting in small amounts of aperture uncertainty causing the held voltage to be significantly different from the ideal held voltage.

Other performance parameters are also important when realizing sample and holds. These include parameters such as dynamic range, linearity, gain, and offset error. Some of the mechanisms whereby these errors arise in particular sample and holds are described in the next section using specific S/H examples. In addition, a number of design principles whereby these errors can be minimized are explained.

Testing Sample and Holds

Before describing various S/H architectures, it is appropriate to first describe a popular method for testing S/Hs, referred to as the *beat test*. This test consists of clocking the S/H at its maximum clock frequency and applying a sinusoidal input signal that has a frequency slightly different than the clock frequency.[1] The output of this system is demodulated to a low frequency equal to the difference between the frequency of the clock signal and that of the input signal. This low-frequency signal is then characterized using a spectrum analyzer or by digitizing it using a high-accuracy A/D converter clocked at the difference frequency and then analyzed using a computer. The test setup for a beat test is shown in Fig. 8.1. Example waveforms for the input signal, the sampling signal, and the output signal of the sample and hold are shown in Fig. 8.2.

1. If the circuit being tested is, in fact, a track and hold, then the output of the track and hold would be resampled by a second track and hold clocked on the opposite phase as the original.

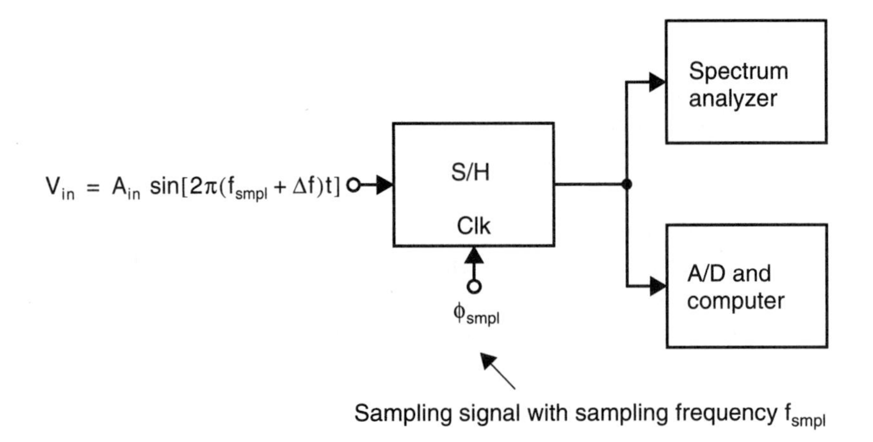

Fig. 8.1 The test setup for characterizing a sample and hold using a beat test.

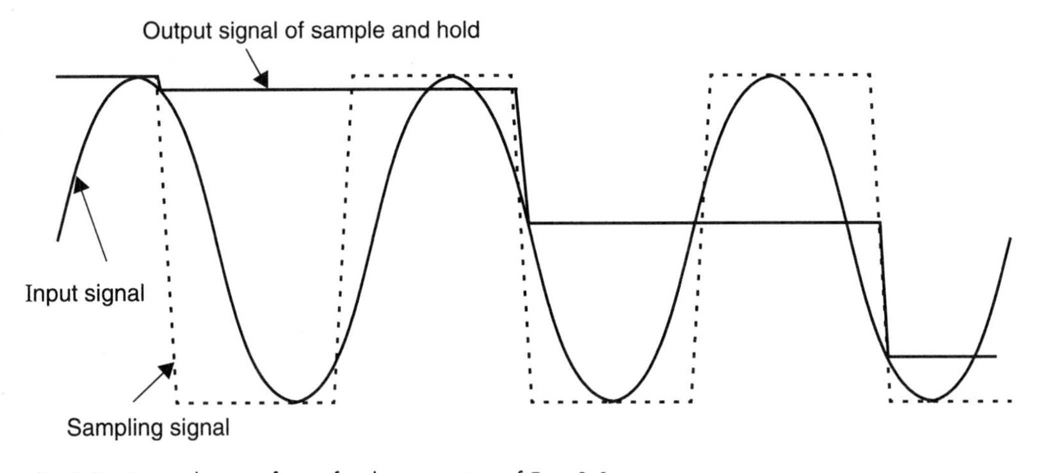

Fig. 8.2 Example waveforms for the test setup of Fig. 8.1.

If a computer is used for analysis, the signal would normally be curve-fitted to a sinusoidal wave at the beat (i.e., difference) frequency. This ideal closest-fit sinusoid would then be subtracted from the measured signal, and the error signal would be analyzed for RMS content and spectral components using a fast-Fourier transform.

It might be mentioned that a very similar test setup can be used for analog-to-digital converters, but now the output of the A/D converter would be loaded into the computer at a sampling rate equal to the beat frequency.

8.2 MOS SAMPLE-AND-HOLD BASICS

Perhaps the simplest sample and hold that can be realized using a CMOS technology is shown in Fig. 8.3. When ϕ_{clk} is high, V' follows V_{in}. When ϕ_{clk} goes low, V' will ideally stay constant from then on, having a value equal to V_{in} at the instance ϕ_{clk}

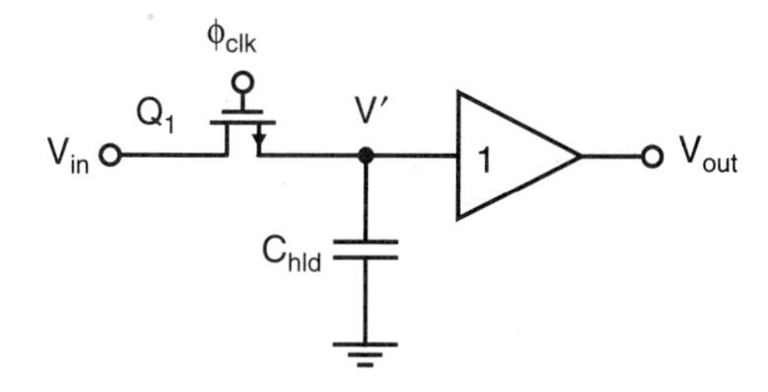

Fig. 8.3 An open-loop track and hold realized using MOS technology.

went low. Unfortunately, V' will have a negative going hold step at this time caused by the channel charge of Q_1. When Q_1 turns off, its channel charge must flow out from under its gate into its junctions. Since this charge is negative, it will cause the junction voltages to have negative glitches. If one assumes the source impedance at the node V_{in} is very low, then the glitch at this node will be small and have a very short duration. However, the negative charge that goes to the node with C_{hld} connected to it (i.e., node V') will cause a negative voltage change that is long lasting (until the next time Q_1 turns on again). If clock ϕ_{clk} turns off fast, then the channel charge, Q_{CH}, will flow equally into both junctions [Shieh, 1987] since the channel becomes pinched off at both ends (i.e., next to the junctions) while the charge is flowing out to the two junctions. The charge flowing to the junction labelled V' is therefore given by

$$\Delta Q_{C_{hld}} = \frac{Q_{CH}}{2} = \frac{C_{ox}WLV_{eff\text{-}1}}{2} \tag{8.1}$$

where $V_{eff\text{-}1}$ is given by

$$V_{eff\text{-}1} = V_{GS1} - V_{tn} = V_{DD} - V_{tn} - V_{in} \tag{8.2}$$

Here, V_{in} is the input voltage at the instance Q_1 turns off. It should be noted here that this result assumes that the clock signal, ϕ_{clk}, goes between V_{DD} and the most negative voltage in the circuit.

The change in the voltage V' is found by using the relationship $Q = CV$, resulting in

$$\Delta V' = \frac{\Delta Q_{C\text{-}hld}}{C_{hld}} = -\frac{C_{ox}WLV_{eff\text{-}1}}{2C_{hld}} = -\frac{C_{ox}WL(V_{DD} - V_{tn} - V_{in})}{2C_{hld}} \tag{8.3}$$

Notice that $\Delta V'$ is linearly related to V_{in}, which results in a gain error for the overall sample-and-hold circuit. However, more importantly, $\Delta V'$ is also linearly related to V_{tn}, which is nonlinearly related to the input signal, V_{in}, due to variations in the source-substrate voltage (assuming the substrate is tied to one of the voltage rails). This nonlinear relationship with V_{in} results in distortion for the overall sample-and-hold circuit.

There is also an additional change in V' due to the gate overlap capacitance. Using a derivation similar to that used to find (7.8), we have

$$\Delta V' \cong -\frac{C_{ox}WL_{ov}(V_{DD} - V_{SS})}{C_{hld}} \tag{8.4}$$

where V_{SS} is the most negative voltage in the circuit. This component is usually smaller than that due to the channel charge, and appears simply as an offset since it is signal independent. Therefore, this error component is not typically important since signal-independent offsets can often be removed in most systems. However, it may cause noise problems if care is not taken to ensure that the clock signal, ϕ_{clk}, is relatively noise free. For example, if the clock signal has power-supply noise on it due to being realized with a simple inverter tied to digital power supplies, the power-supply rejection ratio of this circuit might be poor.

EXAMPLE 8.1

Consider the sample and hold of Fig. 8.3 with $C_{hld} = 1$ pF, $C_{ox} = 1.92$ fF/(μm)^2, $V_{tn} = 0.8$ V, and $(W/L)_1 = (5\ \mu m/0.8\ \mu m)$. Assume the power supply voltages are ±2.5 V and the input signal is 1 V peak to peak. Find the hold step for V_{in} equal to 1 V, and then repeat for V_{in} equal to –1 V. Use the estimates for errors at ±1 V to estimate the dc offset.

Solution

At $V_{in} = 1$ V, using (8.3), we find $\Delta V'(1\text{ V}) = -2.69$ mV. At $V_{in} = -1$ V, again using (8.3), we find $\Delta V'(-1\text{ V}) = -10.26$ mV. The average offset is given by

$$V_{offset\text{-}avg} = \frac{\Delta V'(1\text{ V}) + \Delta V'(-1\text{ V})}{2} = -6.48\text{ mV} \tag{8.5}$$

which is a reasonable estimate of the amount of dc offset injected in this example.

There have been a number of changes proposed to minimize the signal-dependent hold step. One approach is to replace the n-channel switch by a CMOS transmission gate, as shown in Fig. 8.4. The idea behind this approach is that if the size of the p-channel transistor is taken the same as the n-channel transistor, then the charge injection due to each transistor will cancel when the transmission gate turns off. This result is somewhat true when V_{in} is in the middle region between the power supplies, assuming the clock waveforms are fast and exactly complementary. Unfortunately, these conditions are seldom possible to achieve in practice. When the finite slopes of the clock waveforms are taken into account, it is seen (as is explained shortly) that the

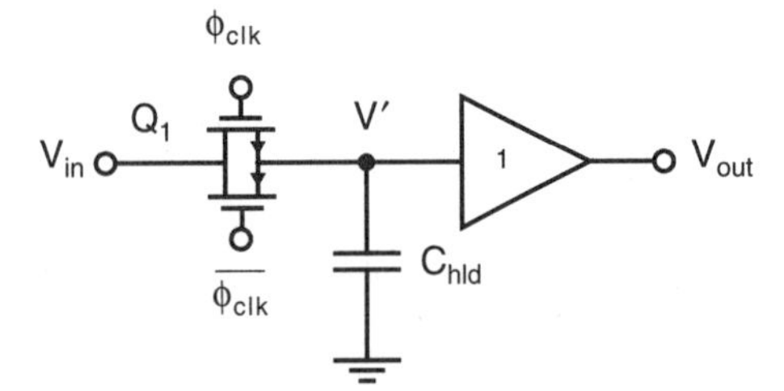

Fig. 8.4 An open-loop track and hold realized using a CMOS transmission gate.

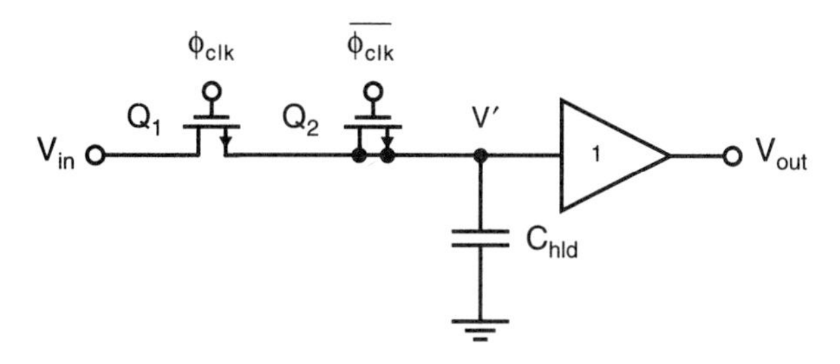

Fig. 8.5 An open-loop track and hold realized using an n-channel switch along with a dummy switch for clock-feedthrough cancellation.

turn-off times of the transistors are signal dependent and this signal dependence causes the n-channel transistor to turn off at different times than the p-channel transistor. Even if we ignore the errors caused by nonperfect clock waveforms, when V_{in} is closer to V_{DD}, the charge from the p-channel transistor is greater than that from the n-channel transistor because of its larger effective gate-source voltage resulting in a positive hold step. The opposite happens when the input signal is closer to the negative power supply. Seldom will these two effects have the same magnitude in practice.

Another modification, often proposed to minimize clock-feedthrough errors, is to add a dummy switch as shown in Fig. 8.5 [McCreary, 1975]. The theory behind this technique is that if the width of Q_2 is taken exactly one-half that of Q_1, and if the clock waveforms are fast, then the charges will cancel. In practice, it is seldom possible to have the clock waveforms change fast enough so that the ideal ratio of widths is exactly one-half. When the ideal ratio is not one-half, it is difficult to make the ratio equal to the optimum required for perfect cancellation. However, when the clock waveforms are fast, this technique usually can minimize the hold pedestal to less than about one-fifth the value it would have without it. For this to be the case, however, it is necessary

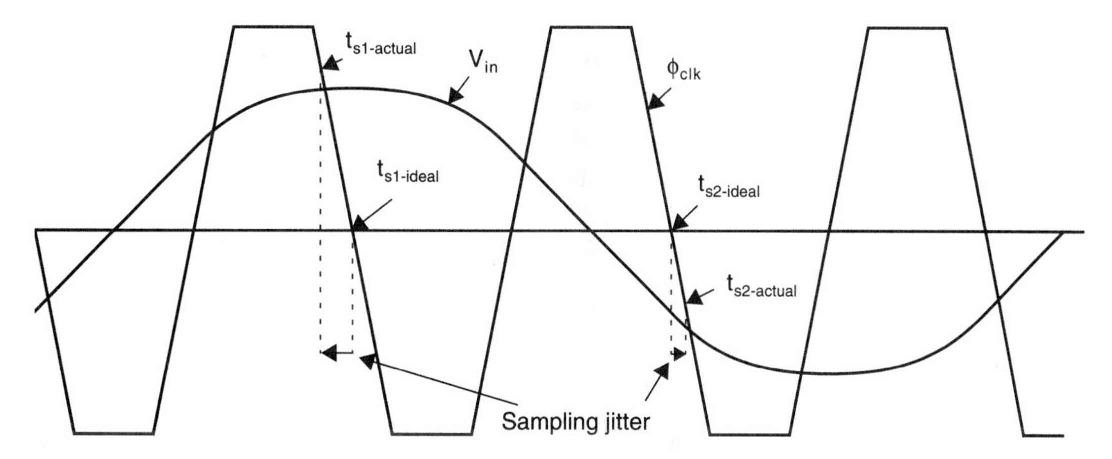

Fig. 8.6 The clock waveforms for V_{in} and ϕ_{clk} used to illustrate how a finite slope for the sampling clock introduces sampling-time jitter.

EXAMPLE 8.2

Consider the S/H circuit of Fig. 8.3, where V_{in} is a 20-MHz band-limited signal with a 2-V_{pp} amplitude. Assume that ϕ_{clk} is a 100-MHz square wave having a peak amplitude of ± 2.5 V with linear rise and fall times of 1.5 ns. What is the maximum uncertainty of the sampling time? Assume V_{tn} is 0.8 V.

Solution

First, we note that the slope of ϕ_{clk} is $(5\ \text{V})/(1.5\ \text{ns}) = 3.33\ \text{V/ns}$, and the true sampling time is when the clock waveform is 0.8 V greater than the input signal.

For V_{in} equal to 1 V, the sampling transistor will turn off when ϕ_{clk} is 1.8 V, which is $(0.7\ \text{V})/(3.33\ \text{V/ns}) = 0.21\ \text{ns}$ after the clock first starts to move down.

When V_{in} is –1 V, the sampling time occurs when ϕ_{clk} is –0.2 V, which is 0.81 ns after the clock first starts to move upward.

Therefore, the sampling-time uncertainty is $0.81 - 0.21 = 0.6$ ns. Also, assuming the ideal sampling time is 0.75 ns after the clock first starts to move, the sampling jitter is from –0.54 ns to +0.06 ns from the ideal sampling time.

---

that the clock of Q_2 changes slightly after that of Q_1. This clock arrangement guarantees that the cancelling charge of Q_2 cannot escape through Q_1 while it is still on.

Another source of error for the simple sample and hold of Fig. 8.3 is caused by clock waveforms having finite slopes. To understand this error source consider the waveforms for V_{in} and ϕ_{clk} shown in Fig. 8.6. Assume the ideal sampling time is defined to be the negative-going zero-crossing of ϕ_{clk}. The true sampling time is that when the sampling-clock voltage passes through the value when it is one transistor threshold voltage drop above the input voltage. In other words, transistor Q_1 of Fig. 8.3 turns off when ϕ_{clk} is V_{tn} above V_{in}, as shown in Fig. 8.6. Thus, when V_{in} is above ground, the true sampling time is earlier than the ideal sampling time, whereas when V_{in} is less than 0 V, the true sampling time is late.

A more elaborate sample-and-hold circuit is to include an opamp in a feedback loop, as shown in Fig. 8.7. When the clock, ϕ_{clk}, is high, the complete circuit responds similarly to an opamp in a unity-gain feedback configuration. When ϕ_{clk} goes low, the input voltage at that time is stored on C_{hld}, similarly to a simple sample and hold. By including an opamp in the feedback loop, the input impedance of the sample and hold is greatly increased. Another advantage of this configuration is that even if the unity-gain buffer at the output has an offset voltage, the dc error due to this buffer will be divided by the gain of the input opamp (although the input-offset of the input opamp will remain). Thus, very simple source followers can be used for the output buffer.

A disadvantage of the configuration shown is that the speed of operation can be seriously degraded due to the necessity of guaranteeing that the loop is stable when it

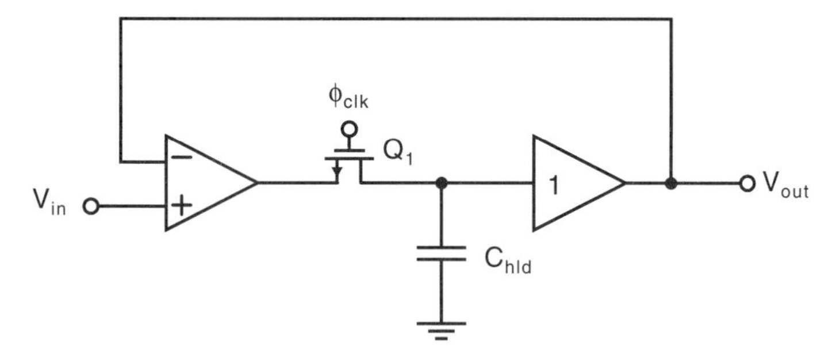

Fig. 8.7 Including an opamp in a feedback loop of a sample and hold to increase the input impedance.

is closed. Another source of speed degradation is that when in hold mode, the opamp is open loop, resulting in its output almost certainly saturating at one of the power-supply voltages. When the sample and hold next goes back into track mode, it will take some time for the opamp output voltage to slew back to its correct closed-loop value. This slewing time can be greatly minimized by adding two additional transistors as shown in Fig. 8.8. During hold mode, switch Q_2 keeps the output of the first opamp close to the voltage it will need to be at when the T/H goes into track mode. It should also be noted that this track-and-hold configuration still has errors due to charge injection from the switch, Q_1, similar to the simple S/H of Fig. 8.3. The errors due to finite clock rise and fall times are also similar.

A similar, but improved, configuration is shown in Fig. 8.9 [Stafford, 1974; Lim, 1991]. In this configuration, the holding capacitor is not going to ground, but rather is placed in the feedback path of a second opamp. This configuration has a number of desirable features. Perhaps the most important of these is due to the fact that, assuming the second opamp has a large gain, the voltages on both sides of switch Q_1 are very nearly signal independent. Thus, when Q_1 turns off, there will still be charge injection to the left side of C_{hld}, which will cause the output voltage of opamp 2 to

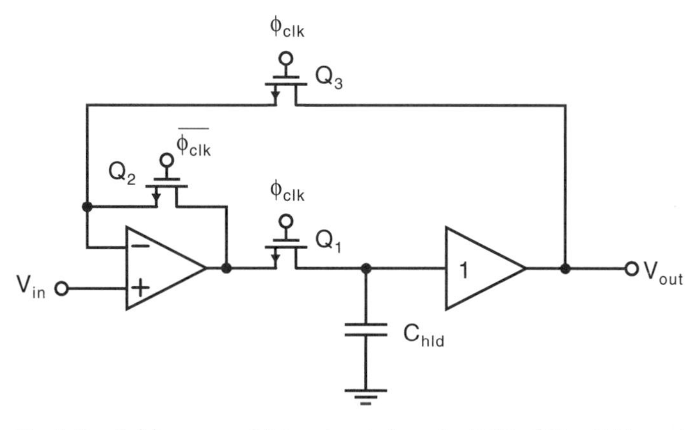

Fig. 8.8 Adding an additional switch to the S/H of Fig. 8.7 to minimize slewing time.

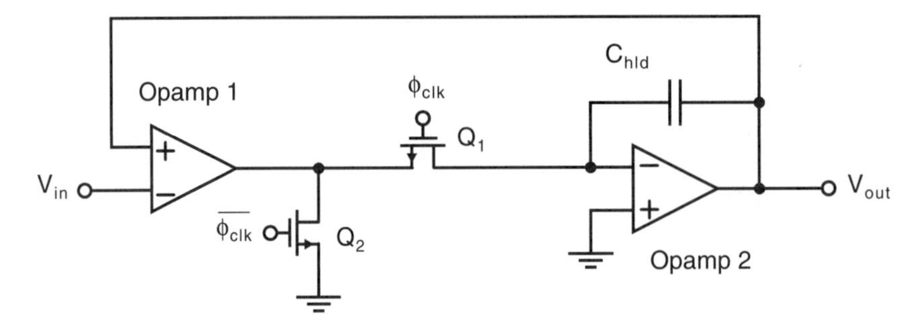

Fig. 8.9 An improved configuration for an S/H as compared to that of Fig. 8.8.

have a positive hold step, *but this hold step will be just a dc offset and will be signal independent*. In other words, the charge injection due to Q_1 will cause some dc offset but no distortion. In addition, the sampling time will not change because of the finite slopes of the sampling-clock waveform.

Another advantage is due to the inclusion of Q_2. This switch grounds the output of the first opamp during hold mode. This grounding keeps the output of the first opamp close to the voltage it must change to when the S/H goes back into track mode. This approach greatly speeds up the time it takes the S/H to return to track mode. The switch also greatly minimizes signal feedthrough when the S/H is in hold mode by grounding the signal path.

A major limitation of this configuration is that the speed will be degraded because of the necessity to guarantee stability in the track mode. This limitation is worsened since there are now two opamps in the loop along with the on resistance of the sampling switch, Q_1, during the closed-loop phase.

Figure 8.10 shows an interesting modification of Fig. 8.9 by including some additional circuitry intended to minimize the hold pedestal and thereby minimize the dc offset [Martin, 1987; Nayebi, 1989]. The basic idea is to match the charge injection into C_{hld} with a similar charge injection into C'_{hld}. Since these capacitors are chosen to have the same size, their voltage changes will match, and the common-

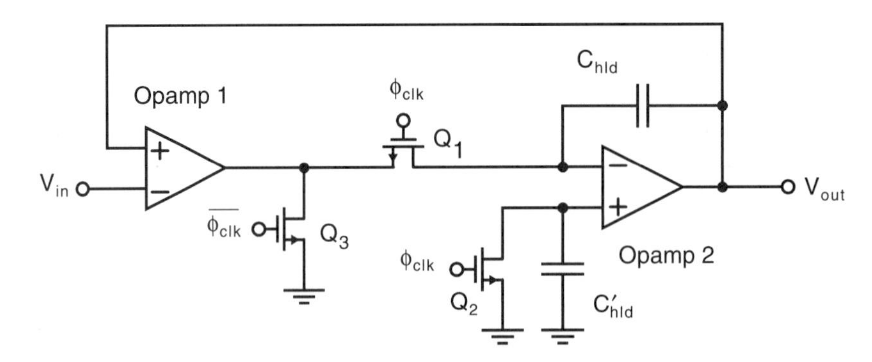

Fig. 8.10 An S/H similar to that of Fig. 8.9, but with clock-feedthrough cancellation circuitry added.

mode rejection of the opamp will eliminate the effects of these voltage changes on the output voltage of the S/H. The major limitation of this approach is a second-order effect caused by a mismatch in impedance levels at the left of Q_1 and the bottom of Q_2. Errors due to this mismatch can be minimized by including small capacitors (0.5 pf to 1 pf) between each node and ground. These extra capacitors help keep these nodes at constant voltages while the clocks are turning off, assuming the clock waveforms are fast. It should be mentioned that the configuration of [Nayebi, 1989] was a fully differential design and is a very reasonable choice for many S/H applications.

8.3 EXAMPLES OF CMOS S/H CIRCUITS

In this section, a number of CMOS S/H circuits are described other than those already mentioned. Unfortunately, the optimum choice for a sample-and-hold circuit is highly dependent on the application's requirements and the technology available. Therefore, choosing the best circuit is nontrivial and a matter of ongoing research. For these reasons, it is difficult to definitively state which configurations are superior or inferior.

Before proceeding, it should be noted that many of these circuits could also be realized in a BiCMOS technology, where a speed advantage would be obtained by increasing the unity-gain frequency of any opamps used. For even higher speeds, a BiCMOS circuit can make use of a bipolar diode bridge, as described in Section 8.4.

Another structure for an S/H circuit is shown in Fig. 8.11 [Ishikawa, 1988]. This configuration is intended to operate at higher speeds, assuming a high-speed opamp capable of driving resistive loads is available. Unfortunately, in CMOS technologies, such opamps are difficult to obtain. Furthermore, the required output buffer would not only limit speed, but could also limit the signal swing. In bipolar and BiCMOS technologies, this configuration could be a very viable alternative. When in track mode, Q_1 is on and Q_2 is off, resulting in the sample and hold acting as an inverting low-pass circuit having a –3-dB frequency given by $\omega_{-3\text{ dB}} = 1/(RC)$. When Q_1 turns off, the output voltage will remain constant from then on. Since the junctions of Q_1 are always at voltages very close to ground, the clock feedthrough of this sample and

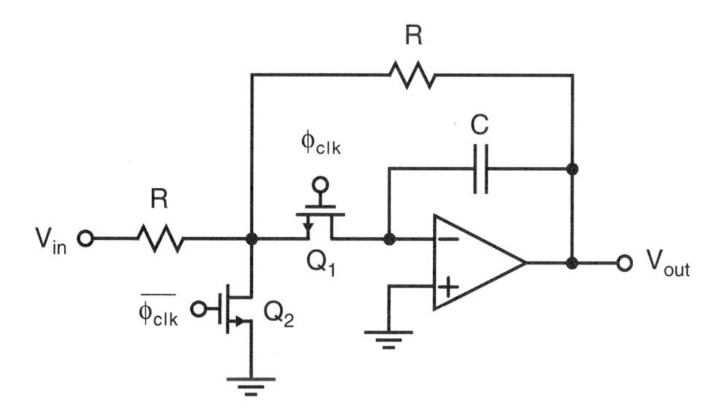

Fig. 8.11 An inverting track and hold.

hold is signal independent. Also, finite clock rise and fall times do not make the sampling time a function of the signal, either. The function of Q_2 is to minimize signal feedthrough when in hold mode and to keep the common node of the resistive network close to the voltage required when the S/H goes back into track mode. In the implementation described in [Ishikawa, 1988], an additional small bypass capacitor was included in parallel to the input resistor. It was reported, and supported by analysis, that this capacitor improved the speed of the S/H.

Another alternative is shown in Fig. 8.12 [Sone, 1993]. This track and hold places the opamp into a unity-gain follower during track mode. Also, the positive input terminal of the opamp is connected to V_{in} at this time. When the T/H goes into hold mode, the input signal is stored across C_1, since Q_1 is turned off. At the same time, transistor Q_2 is also turned off. Ideally, the charge feedthrough from Q_1, although signal dependent, will be matched by the charge feedthrough from Q_2, since both are at the same voltage and have the same clock waveforms. For this to be true, the opamp must have a low-impedance output.[2] Furthermore, the signal dependence of the clock signals can be minimized by having clock signals that change above and below the input signal by fixed amounts. As will be seen in the next section, this is not too difficult to realize using diode clamps in a bipolar or BiCMOS technology. In a CMOS technology, it is more difficult. Regardless, the topology of Fig. 8.12 is attractive for high-speed CMOS applications that are not too demanding.

Yet another S/H architecture is shown in Fig. 8.13 [Lim, 1991]. Shown in Fig. 8.14(*a*) is the circuit configuration when the S/H is in sample mode. During this mode, the opamp is being reset, and both capacitors are connected between the input voltage and the virtual input of the opamp. Thus the total capacitance that needs to be charged or discharged during this mode is $C_1 + C_2$. During hold mode, the configuration is as shown in Fig. 8.14(*b*). During this mode, the effective hold capacitor is the

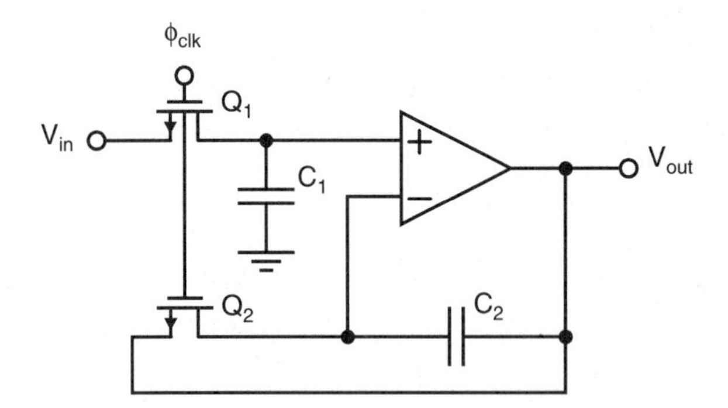

Fig. 8.12 A simple noninverting sample and hold with clock-feedthrough cancellation.

2. This can be partially realized using a high-output impedance OTA by having a load capacitor connected to ground, but without having a resistor included in series with the load capacitor (as would be required if lead compensation was desired).

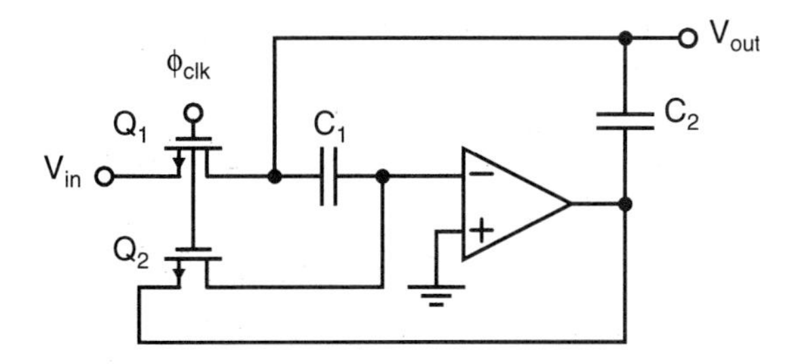

Fig. 8.13 An open-loop architecture but with a Miller holding capacitor.

Miller capacitor given by

$$C_{hld-eff} = (1 + A)\left(\frac{C_1 C_2}{C_1 + C_2}\right) \tag{8.6}$$

which is typically much larger than the capacitance that needs to be charged during the sampling mode. This allows smaller capacitances and therefore smaller switches to be used than would otherwise be the case. Since the voltage changes at the output

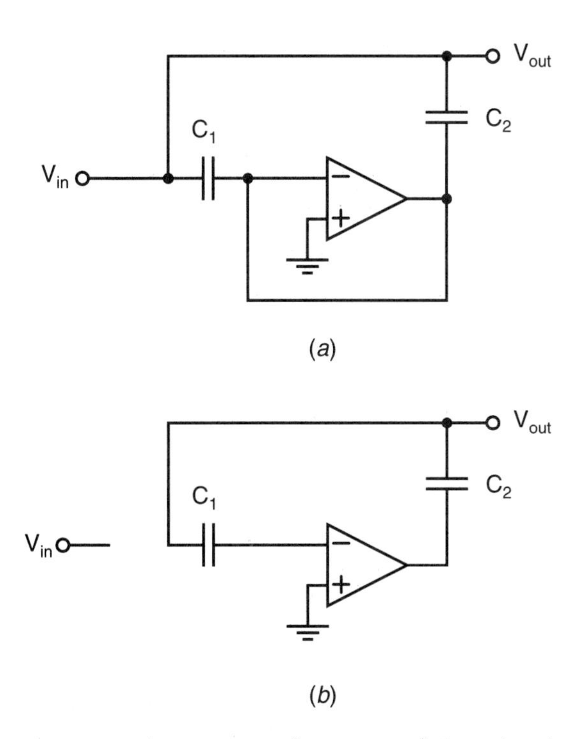

Fig. 8.14 The circuit configurations of the S/H of Fig. 8.13 during (*a*) sample mode and (*b*) hold mode.

of the amplifier are very small, it is easy to design the amplifier for very high speed. The sampling switch, Q_1, does inject signal-dependent charge as in other open-loop architectures, but again this is minimized by the large size of the Miller hold capacitance, especially if Q_2 is turned off slightly before Q_1.

At lower frequencies, there are a number of sample-and-hold architectures based on switched-capacitor technology that are often used. These S/Hs tend to be quite accurate but not necessarily fast. A simple example is shown in Fig. 8.15. Note how similar this S/H is to the S/H of Fig. 8.13. During ϕ_1, capacitor C_H is connected between the input signal source and the inverting input of the opamp. At the same time, the inverting input and the output of the opamp are connected together. This causes the voltages at both of these nodes to be equal to the input-offset voltage of the opamp, V_{off}. Thus, C_H is charged up to $V_{in} - V_{off}$. Next, during ϕ_2, the opamp is taken out of reset mode, and the capacitor C_H is connected between the opamp output and its inverting input. This will cause the output voltage to be equal to V_{in}, irrespective of the input-offset voltage of the opamp. Note that this circuit is an example of a sample and hold that is not thought of as a track-and-hold circuit. Specifically, during the sample phase, the opamp output voltage will be very close to 0 V rather than tracking the input voltage. This places a requirement on the opamp to have a very large slew rate, or the speed is seriously degraded. Also, obviously, the opamp output is invalid during ϕ_1, the sample phase.

An interesting variation somewhat similar to the S/Hs of Figs. 8.13 and 8.15 adds two unity-gain buffers and two more switches to help improve accuracy. The S/H is shown in Fig. 8.16 [Miller, 1990; Real, 1991]. During sampling, all switches are on except Q_2. This connects C_1 and C_2 together and charges them to the input voltage. During this time, the input impedance is quite large due to the inclusion of buffer B_1. At the same time, the opamp is being reset and the positive terminal of the opamp is charged to 0 V. Next, Q_4 and Q_5 are turned off, and shortly thereafter Q_1 and Q_3 are turned off. Finally, Q_2 is turned on. This configures the circuit as shown in Fig. 8.17. Note the similarity between this configuration and that of Fig. 8.14(*b*). The unity-gain feedback holds the output during this phase, and also the output impedance of the S/H is very low. Because Q_4 and Q_5 turn off slightly before Q_1 and Q_3, their clock feedthrough is not only signal

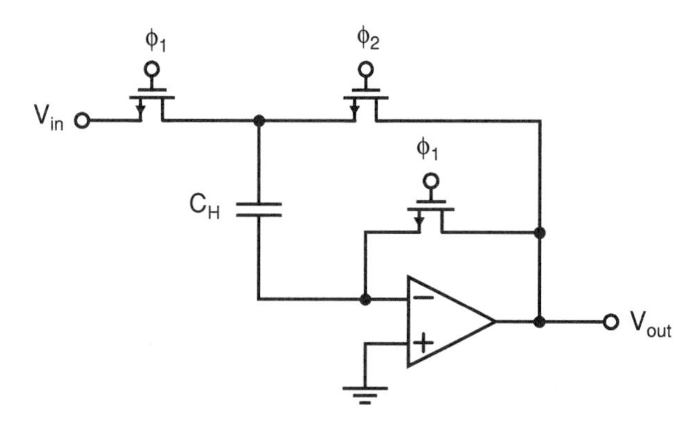

Fig. 8.15 A simple switched-capacitor S/H.

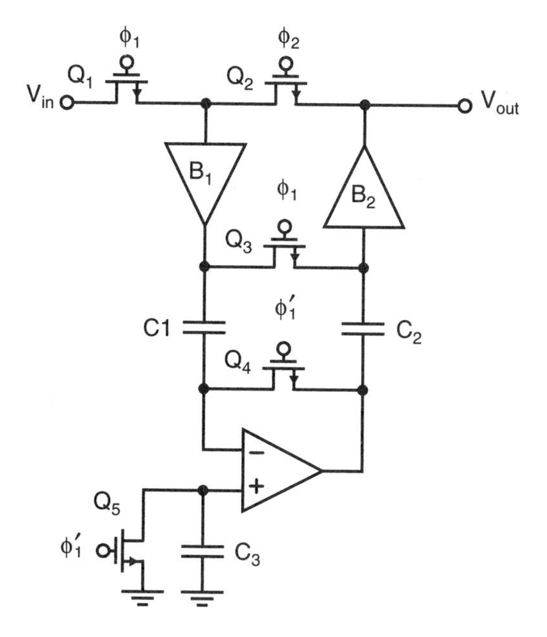

Fig. 8.16 A recycling S/H.

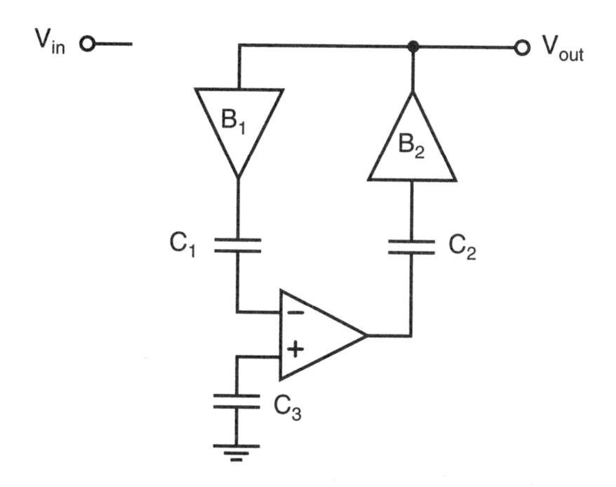

Fig. 8.17 The S/H of Fig. 8.16 when in hold mode.

independent but largely cancels due to the common-mode rejection of the input stage of the amplifier [Martin, 1987]. The charge injection of Q_1 and Q_2 does not affect the output voltage. The charge injection of Q_3 is signal dependent and does affect the output voltage, but its effect is minimized by the loop gain similar to the S/H of Fig. 8.13.

An interesting variation on the S/Hs of Figs. 8.10 and 8.15 is the S/H shown in Fig. 8.18 from [Gatti, 1992]. This T/H has an output that is always valid and incorporates a number of features. During sample mode, the input-offset voltage of the first opamp is stored across C_{OF}. At the same time, C_S is sampling the input voltage, and the second opamp is holding the previously-sampled input voltage. During the next

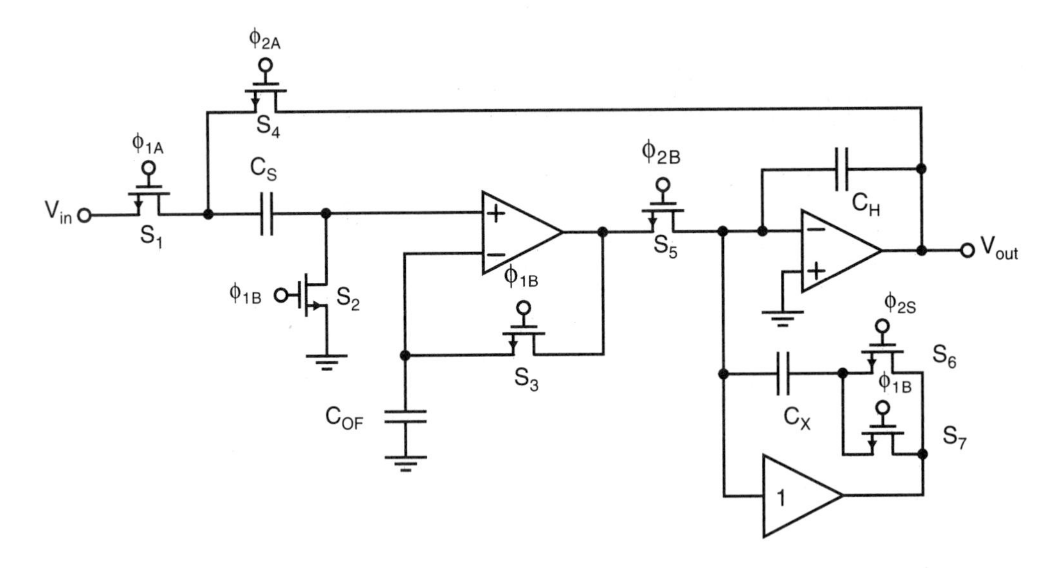

Fig. 8.18 A switched-capacitor S/H.

phase, the input-offset voltage of the first opamp is eliminated, and C_S is connected to the output while the feedback loop is enabled. This arrangement causes the output voltage to be equal to the just-sampled input voltage. The clock feedthrough of switch S_2 is cancelled by the clock feedthrough of switch S_3. The clock feedthrough of switch S_5 is cancelled by the clock feedthrough of switch S_6 and the added buffer network. The other switches ideally do not cause appreciable errors due to clock feedthrough. The details concerning the operation of this T/H and the required clocking waveforms are left for the interested reader to consult [Gatti, 1992].

Yet another example of a switched-capacitor S/H is shown in Fig. 8.19 [Sooch, 1991]. This circuit combines a sample and hold with a low-pass filter and was used to convert a signal from the sampled-data domain to the continuous-time domain at

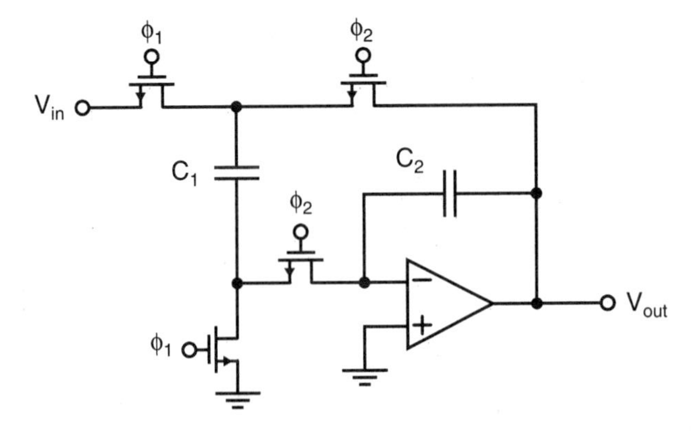

Fig. 8.19 A switched-capacitor sample and hold and low-pass filter.

the output of a high-quality audio digital-to-analog converter based on oversampling techniques. The low-pass filtering was used to help convert from a single-bit high-frequency signal to a decimated lower-frequency multi-bit signal. During ϕ_1, C_1 is connected between the input voltage and ground. Next, during ϕ_2, it is connected between the output and the inverting input of the opamp. At this time, its charge is shared with the charge being stored across C_2, which came from previous samples. For low-frequency signals, the output voltage will equal the input voltage in the steady state. Higher frequency changes in the input signal will be low-pass filtered. For C_2 greater than C_1 by at least a few times, it can be shown using switched-capacitor analysis techniques [Martin, 1980] that the –3-dB frequency of the low-pass filter is approximately given by

$$f_{-3\text{ dB}} \cong \frac{1}{2\pi}\frac{C_1}{C_2}f_{clk} \tag{8.7}$$

where f_{clk} is the clock frequency. While this example does not eliminate the effects of opamp finite-input voltage offsets, unlike the previous two examples, it does have the advantage that its output is always valid.

The switched-capacitor sample-hold circuits just shown (except for that of Fig. 8.18) do have dc offsets due to clock feedthrough from the switches connected to the inverting inputs of the opamps. However, these offsets are largely signal independent assuming the opamp has a reasonable gain. Also, it is possible to minimize these offsets by including a clock-feedthrough cancellation network connected to the positive input of the opamp, similar to what is described in [Martin, 1987]. Alternatively, fully differential sample and holds can be used, which also helps minimize the effects of clock feedthrough.

8.4 BIPOLAR AND BICMOS SAMPLE AND HOLDS

In this section, bipolar and BiCMOS sample-and-hold circuits are described. It should be noted that BiCMOS sample-and-hold circuits are often realized using the CMOS techniques described above. Here, we look at BiCMOS circuits more closely related to their bipolar counterparts.

Most of the early integrated bipolar sample and holds were based on using a *diode-bridge* switch, as shown in Fig. 8.20 [Erdi, 1978]. In track mode the bias current, I_B, is turned on and evenly divides between D_1 and D_2, as well as between D_3 and D_4. This current division results in a low-impedance path between the input and the output with a small-signal impedance of r_d, where

$$r_d = \frac{V_T}{I_{D1}} = \frac{V_T}{I_B/2} \tag{8.8}$$

In hold mode, both current sources labelled I_B are turned off, leaving the hold capacitor, C_{hld}, open circuited.

A more practical realization of a diode-bridge track and hold is shown in Fig. 8.21 [Matsuzawa, 1990]. During track mode, V_{trk} is at a higher voltage than V_{hld},

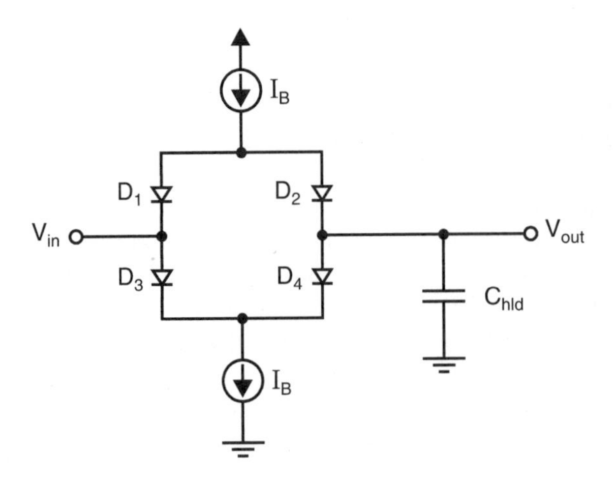

Fig. 8.20 A bipolar track and hold based on using a diode bridge.

and Q_2 will be on and conducting $2I_B$. Half of this current comes from the current source connected to the node V_2. The other half comes from the current source connected to node V_1 after going through the diode bridge. Thus, during track mode, the diode bridge is low impedance, and the node V_3 will track the input signal. The output voltage will also track this node through the unity-gain output buffer. Also, during this phase, the diodes D_5 and D_6 are reverse biased.

When the track and hold is placed in hold mode, Q_2 is turned off and Q_1 is turned on. Since Q_2 is turned off, the current from the source connected to the node V_2 now

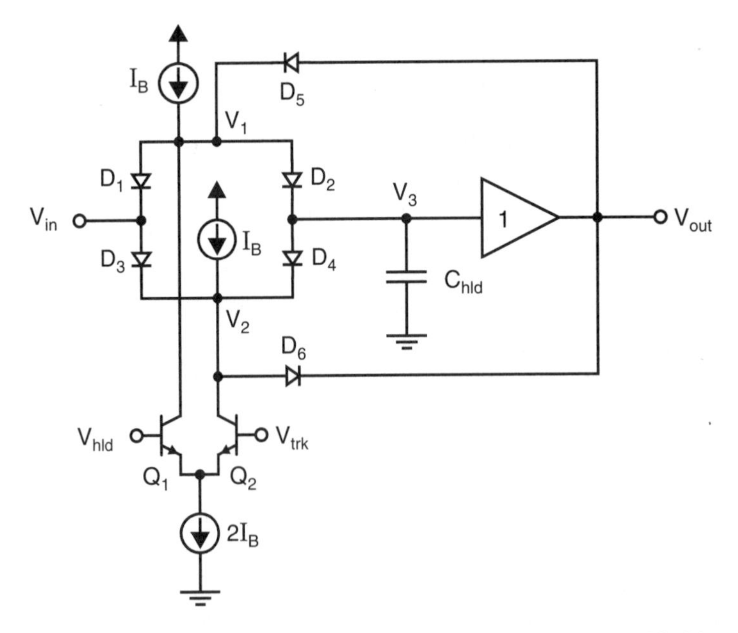

Fig. 8.21 An improved example of a diode-bridge track and hold.

flows through the diode D_6, turning it on. This causes V_2 to be one diode drop above V_{out}, which causes diodes D_3 and D_4 to become reverse biased. At the same time, Q_1 will be conducting current $2I_B$. Half of this will come from the current source connected to node V_1. The other half comes from V_{out} through diode D_5, causing it to become forward biased. Therefore, V_1 will be one diode drop below V_{out}, and diodes D_1 and D_2 will be reverse biased. Thus, all the diodes in the diode bridge are reverse biased, and the node V_3 is disconnected from the input signal.

The circuit of Fig. 8.21 has a number of advantages compared to the simple circuit of Fig. 8.20. Perhaps the most important of these is that the clock feedthrough is signal independent. This independence occurs because when the diode bridge turns off, the voltage changes of nodes V_1 and V_2 are from a voltage closely equal to V_{in}, to a voltage one diode drop below V_{in}, and a voltage one diode drop above V_{in}, respectively, regardless of the value of V_{in}. Thus, the charge from turning the diode bridge off, and changing the reverse-bias voltage of the diodes in the bridge, is signal independent. Furthermore, the charge from D_1 and D_2 will (to a first-order degree) cancel the charge from D_3 and D_4, assuming fast clock waveforms, which also minimizes the effects of charge injection. It should be mentioned that in many implementations, the diodes would all be Schottky diodes, which don't have minority carrier charge storage when they are on, and, thus, turn off faster than regular diodes and have smaller hold steps. A second advantage of Fig. 8.21 is that the sample-time uncertainty caused by the finite slope of the clock waveforms is greatly minimized. Finally, a third advantage of Fig. 8.21 is that when in hold mode, the signal feedthrough is substantially smaller. When the simple circuit of Fig. 8.20 is in hold mode, the diodes are turned off but are floating. The input signal is then capacitively coupled through the junction capacitance of these reverse-biased diodes to the hold capacitor. The magnitude of this signal feedthrough is easily calculated using the capacitor-divider formula. This signal feedthrough is minimized in the improved circuit of Fig. 8.21, because during hold mode, nodes V_1 and V_2 are connected to V_{out} through low-impedance paths consisting of conducting diodes D_5 and D_6, respectively. This effectively shields the node V_3 from the input signal during hold mode.

In the BiCMOS realization presented in [Matsuzawa, 1990], the input transistors to the unity-gain buffer were p-channel MOS transistors that minimized the droop on the hold capacitor during hold-mode, which would normally be caused by finite-based currents of bipolar input transistors.

It should be noted in Fig. 8.21 that the maximum input-signal excursions during hold mode are limited to less than two diode drops above or below the input signal voltage at the time the track and hold went into hold mode. Otherwise, one of the diodes D_1 or D_3 will become forward biased, which will tend to clamp the input signal. For high-speed track and holds, this input amplitude restriction is not usually a serious limitation as peak signals are usually limited to be less than one volt. Furthermore, the input signal range can be increased by replacing each of D_5 and D_6 with two series diodes. This would increase the maximum allowable signal-swing excursion to three diode drops.

One of the major limitations of bipolar S/Hs is a droop in the voltage being stored across a capacitor during hold mode, caused by the finite base currents of bipolar transistors. One of the major ways of minimizing this droop is to use fully differential approaches where the droop is the same in both paths, and the differential signal is, to a first-order approximation, unaffected. An example of a fully differential diode-bridge-based T/H is given in [Colleran, 1993]. In this example, the current sources were realized using resistors, and additional circuitry was also included to minimize errors introduced by these resistors.

An interesting BiCMOS S/H is shown in Fig. 8.22 [Wakayama, 1992]. This S/H uses bipolar switches that appear similar to the diode bridge switches, but operate differently. Note that the direction of diodes D_2 and D_4 have been reversed compared to the circuit of Fig. 8.20. When V_{trk1} is greater than V_{hld1}, then Q_3 will be conducting $3I_B$. I_B of this will come from M_5 through Q_1, I_B of this will come from M_6, and I_B of this will come from M_7 through D_2 and D_1. Since both Q_1 and D_1 are conducting, V_{out1} will equal V_{in}. Diodes D_3 and D_4 are reverse biased and V_{out2} is isolated from V_{in}. The ratios of the currents in M_2, M_3, and M_4 are accurately determined by the feedback loop including M_1, M_2, M_5, and the unity-gain buffer. When V_{trk1} is less than V_{hld1}, then Q_4 will be conducting $3I_B$. This will cause Q_1, D_1, and D_2 to turn off and Q_2, D_3, and D_4 to turn on. In this phase, V_{out1} will be isolated from V_{in}, and V_{out2} will track V_{in}. Notice that the input is buffered from the holding capacitors by emitter followers (either Q_1 or Q_2) and that the load current on the input source is constant (and equal to the base current of the emitter followers). Also, when one output is in track mode, the other is in hold mode. This arrangement can possibly be used to effectively double a data rate. Finally, as is explained in [Wakayama, 1992], during slewing the current available to charge or discharge the holding capacitors is dynamically increased to $3I_B$ rather than the quiescent I_B.

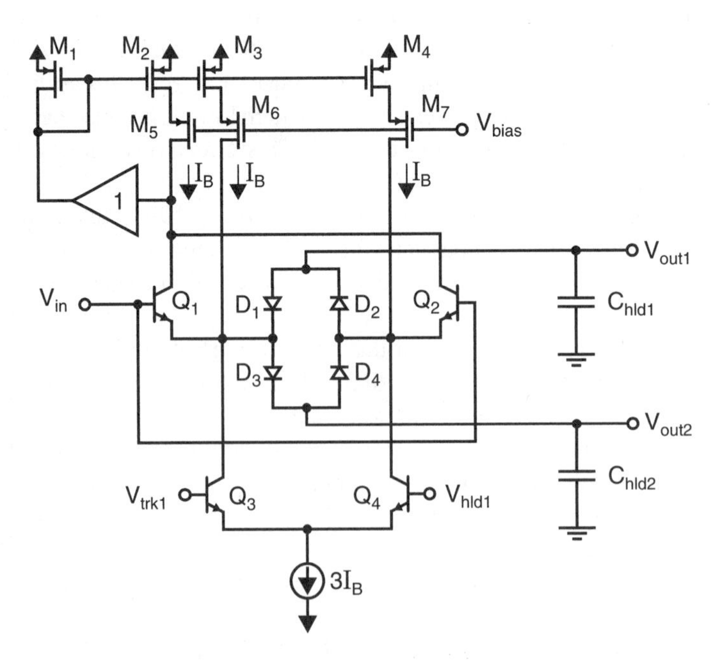

Fig. 8.22 A BiCMOS sample and hold.

An alternate BiCMOS S/H is described in [Sone, 1993], which makes use of the basic architecture shown in Fig. 8.12, except the switches are realized using bipolar switches. There have been a number of other alternative bipolar and BiCMOS track and holds that have been reported in the literature. Some alternative and interesting BiCMOS examples include [Fernandes, 1989; Robertson, 1990; and Real, 1991]. Some alternative and interesting bipolar examples include [Moraveji, 1991; Petschacher, 1990; and Vorenkamp, 1992]. The interested reader is referred to the applicable references.

8.5 BANDGAP VOLTAGE REFERENCE BASICS

Another important analog building block, especially in data acquisition systems, is a voltage reference. Ideally, this block will supply a fixed dc voltage of known amplitude that does not change with temperature. There have been a number of approaches that have been taken to realize voltage references in integrated circuits. These include

1. Making use of a zener diode that breaks down at a known voltage when reverse biased.
2. Making use of the difference in the threshold voltage between an enhancement transistor and a depletion transistor.
3. Cancelling the negative temperature dependence of a pn junction with a positive temperature dependence from a PTAT (proportional-to-absolute-temperature) circuit.

The first approach is not popular nowadays because the breakdown voltage of a zener diode is typically larger than the power supplies used in modern circuits. The second approach cannot be used in most CMOS circuits because depletion transistors are not typically available. In addition, although it can be used to make quite stable references, the actual value of the reference is difficult to determine accurately because of the process sensitivity of the difference between the threshold voltage of an enhancement device and a depletion device. For these reasons, the first two approaches are not covered here. Rather, the last approach, which is currently the most popular for both bipolar and CMOS technologies, will be discussed. Voltage references based on the last approach are commonly called "bandgap" voltage references for reasons that will become apparent shortly.

Bandgap Voltage Reference

As just mentioned, a bandgap voltage reference is based on subtracting the voltage of a forward-biased diode (or base-emitter junction) having a negative temperature coefficient from a voltage proportional to absolute temperature (PTAT). As we shall see, this PTAT voltage is realized by amplifying the voltage difference of two forward-biased base-emitter (or diode) junctions. A bandgap voltage reference system is shown symbolically in Fig. 8.23.

A forward-biased base-emitter junction of a bipolar transistor has an I-V relationship given by

$$I_C = I_S e^{qV_{BE}/kT} \tag{8.9}$$

where I_S is the transistor scale current and, although not shown, has a strong dependence on temperature.

Writing the base-emitter voltage as a function of collector current and temperature, it can be shown that [Brugler, 1967; Tsividis, 1980]

$$V_{BE} = V_{G0}\left(1 - \frac{T}{T_0}\right) + V_{BE0}\frac{T}{T_0} + \frac{mkT}{q}\ln\left(\frac{T_0}{T}\right) + \frac{kT}{q}\ln\left(\frac{J_C}{J_{C0}}\right) \tag{8.10}$$

Here, V_{G0} is the bandgap voltage of silicon extrapolated to 0 °K (approximately 1.206 V), k is Boltzmann's constant, and m is a temperature constant approximately equal to 2.3. Also, J_C and T are the collector current density and temperature, respectively, while the subscript 0 designates an appropriate quantity at a reference temperature, T_0. Specifically, J_{C0} is the collector current density at the reference temperature, T_0, whereas J_C is the collector current density at the true temperature, T. Also, V_{BE0} is the junction voltage at the reference temperature, T_0, whereas V_{BE} is the base-emitter junction voltage at the true temperature, T. Note that the junction current is related to the junction current density according to the relationship

$$I_C = A_E J_C \tag{8.11}$$

where A_E is the effective area of the base-emitter junction.

For I_C constant, V_{BE} will have approximately a –2 mV/°K temperature dependence around room temperature. This negative temperature dependence is cancelled by a PTAT temperature dependence of the amplified difference of two base-emitter junctions biased at fixed but different current densities. Using (8.10), it is seen that if there are two base-emitter junctions biased at currents J_2 and J_1, then the difference in their junction voltages is given by

$$\Delta V_{BE} = V_2 - V_1 = \frac{kT}{q}\ln\left(\frac{J_2}{J_1}\right) \tag{8.12}$$

Thus, the difference in the junction voltages is proportional to absolute temperature. This proportionality is quite accurate and holds even when the collector currents are temperature dependent, as long as their ratio remains fixed.

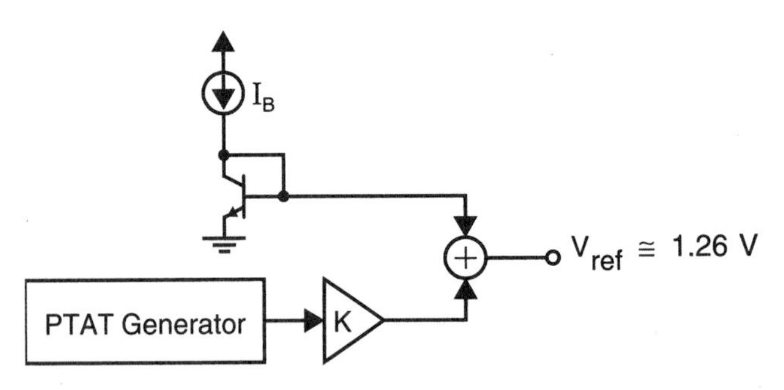

Fig. 8.23 A simplified circuit of a bandgap voltage reference.

EXAMPLE 8.3

Assume two transistors are biased at a current-density ratio of 10:1 at $T = 300\ °K$. What is the difference in their base-emitter voltages and what is its temperature dependence?

Solution

Using (8.12), we have

$$\Delta V_{BE} = \frac{kT}{q}\ln\left(\frac{J_2}{J_1}\right) = \frac{1.38 \times 10^{-23}(300)}{1.602 \times 10^{-19}}\ln(10) = 59.5\ \text{mV} \tag{8.13}$$

Since this voltage is proportional to absolute temperature, after a $1\ °K$ temperature increase, the voltage difference will be

$$\Delta V_{BE} = 59.5\ \text{mV}\frac{301}{300} = 59.7\ \text{mV} \tag{8.14}$$

Thus, the voltage dependence is $59.5\ \text{mV}/300\ °K$ or $0.198\ \text{mV}/°K$. Since the temperature dependence of a single V_{BE} is $-2\ \text{mV}/°K$, if it is desired to cancel the temperature dependence of a single V_{BE}, then ΔV_{BE} should be amplified by about a factor of 10, as explained next.

It will be seen shortly that when realizing a bandgap voltage reference, although the output voltage is temperature independent, the junction currents turn out to be proportional to absolute temperature (assuming the resistors used are temperature independent). Thus, to simplify derivations, we will first assume the junction currents are proportional to absolute temperature. Later, it will be verified that this proportionality relationship is true when circuit realizations are described. We therefore first assume

$$\frac{J_i}{J_{i0}} = \frac{T}{T_0} \tag{8.15}$$

where J_i is the current density of the collector current of the ith transistor, whereas J_{i0} is the same current density at the reference temperature.

Now, assume that the difference between two base-emitter voltages is multiplied by a factor of K and added to the base-emitter voltage of the junction with the larger current density. Using (8.12) and (8.15) along with (8.10), we have

$$\begin{aligned} V_{ref} &= V_{BE2} + K\Delta V_{BE} \\ &= V_{G0} + \frac{T}{T_0}(V_{BE0\text{-}2} - V_{G0}) + (m-1)\frac{kT}{q}\ln\left(\frac{T_0}{T}\right) + K\frac{kT}{q}\ln\left(\frac{J_2}{J_1}\right) \end{aligned} \tag{8.16}$$

Equation (8.16) is the fundamental equation giving the relationship between the output voltage of a bandgap voltage reference and temperature. If we want this relationship to be zero at a particular temperature, we can differentiate (8.16) with respect to temperature and set the derivative to zero at the desired reference temperature. From (8.16), we have

$$\frac{\partial V_{ref}}{\partial T} = \frac{1}{T_0}(V_{BE0\text{-}2} - V_{G0}) + K\frac{k}{q}\ln\left(\frac{J_2}{J_1}\right) + (m-1)\frac{k}{q}\left[\ln\left(\frac{T_0}{T}\right) - 1\right] \tag{8.17}$$

Setting (8.17) equal to zero at $T = T_0$, we see that for zero temperature dependence at the reference temperature, we need

$$V_{BE0\text{-}2} + K\frac{kT_0}{q}\ln\left(\frac{J_2}{J_1}\right) = V_{G0} + (m-1)\frac{kT_0}{q} \tag{8.18}$$

The left side of (8.18) is the output voltage V_{ref} at $T = T_0$ from (8.16). Thus for zero temperature dependence at $T = T_0$, we need

$$V_{ref\text{-}0} = V_{G0} + (m-1)\frac{kT_0}{q} \tag{8.19}$$

For the special case of $T_0 = 300\ °K$ and $m = 2.3$, (8.19) implies that

$$V_{ref\text{-}0} = 1.24\ V \tag{8.20}$$

for zero temperature dependence. Notice that this value is independent of the current densities chosen. Thus, if a larger current density is chosen, then K must be taken appropriately smaller to achieve the correct reference output voltage. In many integrated voltage references, this correct output voltage will be achieved by trimming at the time the wafer is being tested. From (8.18), the required value for K is

$$K = \frac{V_{G0} + (m-1)\frac{kT_0}{q} - V_{BE0\text{-}2}}{\frac{kT_0}{q}\ln\left(\frac{J_2}{J_1}\right)} = \frac{1.24 - V_{BE0\text{-}2}}{0.0258\ \ln\left(\frac{J_2}{J_1}\right)} \tag{8.21}$$

at 300 °K.

The reason for the name of the bandgap voltage should now be apparent. Specifically, for zero temperature dependence, the output of a bandgap voltage reference is given by the bandgap voltage plus a small correction term to account for second-order effects.

The output voltage of the reference for temperatures different from the reference is found after back-substituting (8.18) and (8.19) into (8.16). After some manipulations, the result is

$$V_{ref} = V_{G0} + (m-1)\frac{kT}{q}\left[1 + \ln\left(\frac{T_0}{T}\right)\right] \tag{8.22}$$

and

$$\frac{\partial V_{ref}}{\partial T} = (m-1)\frac{k}{q}\ln\left(\frac{T_0}{T}\right) \tag{8.23}$$

These equations can be used to estimate the temperature dependence at temperatures different from the reference temperature. In the next section, a practical bipolar realization of a bandgap reference will be described.

EXAMPLE 8.4

Estimate the temperature dependence at $0\ ^\circ C$ for a bandgap voltage reference that was designed to have zero temperature dependence at $20\ ^\circ C$. Present the result as $ppm/^\circ K$.

Solution

Recalling that $0\ ^\circ K$ corresponds to $-273\ ^\circ C$, we can write $T_0 = 293\ ^\circ K$ and $T = 273\ ^\circ K$. Substituting these values into (8.23), we have

$$\frac{\partial V_{ref}}{\partial T} = (2.3-1)\frac{1.38\times 10^{-23}}{1.6\times 10^{-19}}\ln\left(\frac{293}{273}\right) = 8\ \mu V/^\circ K \tag{8.24}$$

For a reference voltage of $1.24\ V$, a dependency of $8\ \mu V/^\circ K$ results in

$$\frac{8\ \mu V/^\circ K}{1.24\ V} = 6.5\times 10^{-6}\ parts/^\circ K = 6.5\ ppm/^\circ K \tag{8.25}$$

where ppm represents parts per million. It should be mentioned here that practical effects result in voltage references with typically 4 to 10 times larger values than this small amount of temperature dependency. It should also be noted that the ideal first-order temperature of this bandgap voltage circuit is $0\ ppm/^\circ K$ at the reference temperature of $20\ ^\circ C$.

8.6 CIRCUITS FOR BANDGAP REFERENCES

Bipolar Bandgap References

A voltage reference originally proposed in [Brokaw,1974] has been the basis for many bipolar bandgap references. A simplified schematic of the circuit is shown in Fig. 8.24. The amplifier in the feedback loop keeps the collector voltages of Q_1 and Q_2 equal. Since $R_3 = R_4$, this guarantees that both transistors have the same collector currents and collector-emitter voltages. Also, notice that the emitter area of Q_1 has been taken eight times larger than the emitter area of Q_2. Therefore, Q_2 has eight times the current density of Q_1, resulting in

$$\frac{J_2}{J_1} = 8 \tag{8.26}$$

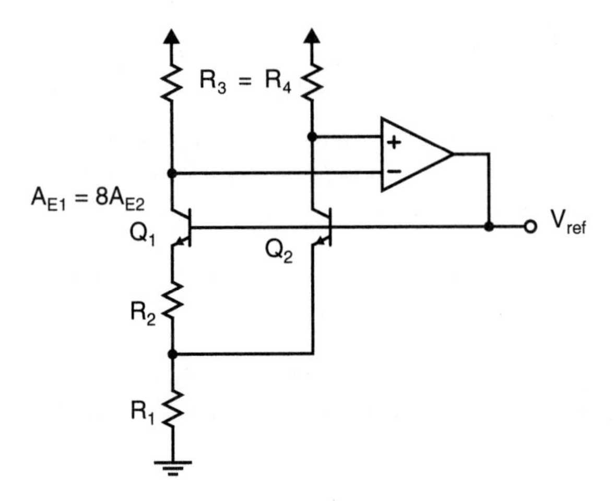

Fig. 8.24 A simplified schematic of a bipolar band-gap voltage reference.

We have for the circuit

$$V_{ref} = V_{BE2} + V_{R1} \tag{8.27}$$

Also

$$\begin{aligned} V_{R1} &= I_{R1}R_1 \\ &= 2I_{R2}R_1 \end{aligned} \tag{8.28}$$

But

$$I_{R2} = \frac{V_{R2}}{R_2} = \frac{V_{BE2} - V_{BE1}}{R_2} = \frac{\Delta V_{BE}}{R_2} \tag{8.29}$$

Substituting (8.28) and (8.29) into (8.27) gives

$$V_{ref} = V_{BE2} + \frac{2R_1}{R_2}\Delta V_{BE} \tag{8.30}$$

which is of the form desired to realize a bandgap reference. It is immediately recognizable that

$$K = \frac{2R_1}{R_2} \tag{8.31}$$

Assuming $V_{BE2\text{-}0} = 0.65\ \text{V}$, from (8.21)

$$\frac{R_1}{R_2} = \frac{1}{2} \times \frac{1.24 - 0.65}{0.0258 \times \ln(8)} = 5.5 \tag{8.32}$$

In an integrated implementation, R_1 or R_2 would be trimmed while monitoring V_{ref} to force it equal to the desired reference voltage. Furthermore, the optimum value for

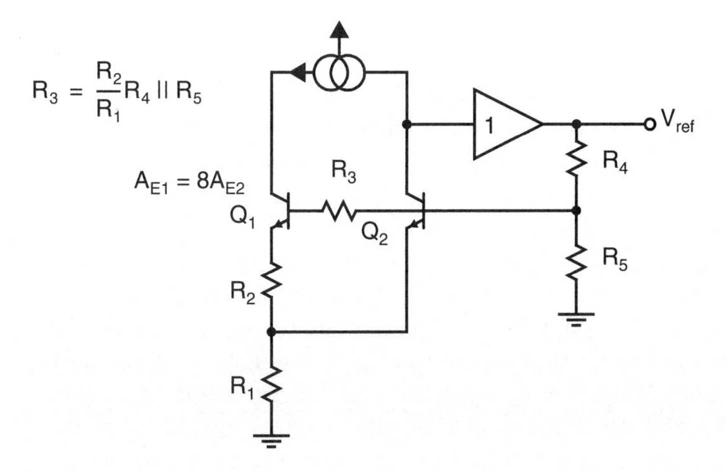

Fig. 8.25 A bipolar bandgap with output voltages greater than 1.24 V.

this voltage might be determined empirically during the prototype phase of the design cycle.

Notice also, from (8.12) and (8.29), we have

$$I_{E1} = I_{E2} = I_{R2} = \frac{\Delta V_{BE}}{R_2} = \frac{\frac{kT}{q}\ln\left(\frac{J_2}{J_1}\right)}{R_2} \tag{8.33}$$

implying that all currents are proportional to absolute temperature (assuming that resistor R_2 is temperature independent). Thus, as assumed earlier, all currents are indeed PTAT. It is worth mentioning here that PTAT currents are often used to bias many bipolar circuits, as they result in transistor transconductances being independent of temperature. This transconductance independence has the desirable feature that circuit speed is relatively independent of temperature, but, unfortunately, has the undesirable feature that the circuit power dissipation goes up considerably at high temperatures, which makes it more difficult to dissipate the heat.

In applications where it is desirable to have reference voltages larger than 1.24 V, a modified bandgap reference as shown in Fig. 8.25 can be used. It is not difficult to show that the output voltage is now given by

$$V_{ref\text{-}0} = \left(1 + \frac{R_4}{R_5}\right)\left[V_{G0} + (m-1)\frac{kT_0}{q}\right] \cong \left(1 + \frac{R_4}{R_5}\right)1.24 \text{ V} \tag{8.34}$$

Resistor R_3 has been added to cancel the effects of the finite base currents going through R_4 and should be chosen according to the formula in the figure. The interested reader is referred to [Brokaw, 1974] for additional details concerning this realization.

CMOS Bandgap References

The most popular method for realizing CMOS voltage references also makes use of a bandgap voltage reference despite the fact that *independent* bipolar transistors are not available. These CMOS circuits rely on using what are commonly called *well transistors.* These devices are vertical bipolar transistors that use wells as their bases and the substrate as their collectors. In an n-well process (the most common modern process), these vertical bipolar transistors are pnp types with their collectors connected to ground, as shown in Fig. 8.26(*a*). In a p-well process, they would be npn transistors with their collectors connected to the positive power supply, as shown in Fig. 8.26(*b*). These transistors have reasonable current gains, but their main limitation is the series base resistance, which can be high due to the large lateral dimensions between the base contact and the effective emitter region. To minimize errors due to this base resistance, the maximum collector currents through the transistors are usually constrained to be less than 0.1 mA. It is possible to use these transistors to implement bandgap voltage references using configurations similar to those shown in Fig. 8.27(*a*) for n-well processes [Kujik, 1973] or Fig. 8.27(*b*) for p-well processes [Ye, 1982].

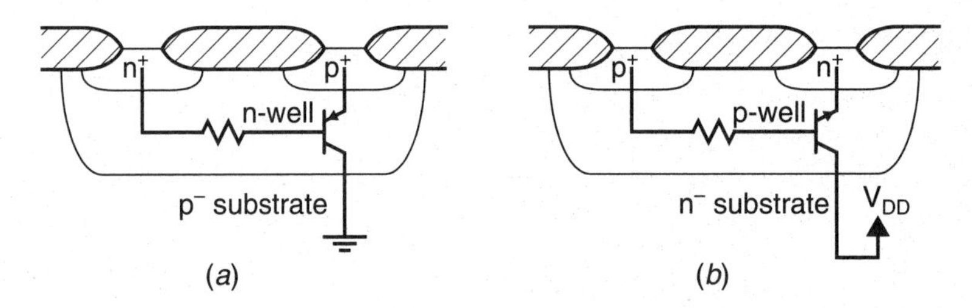

Fig. 8.26 Vertical CMOS well transistors realized in (*a*) an n-well process and (*b*) a p-well process.

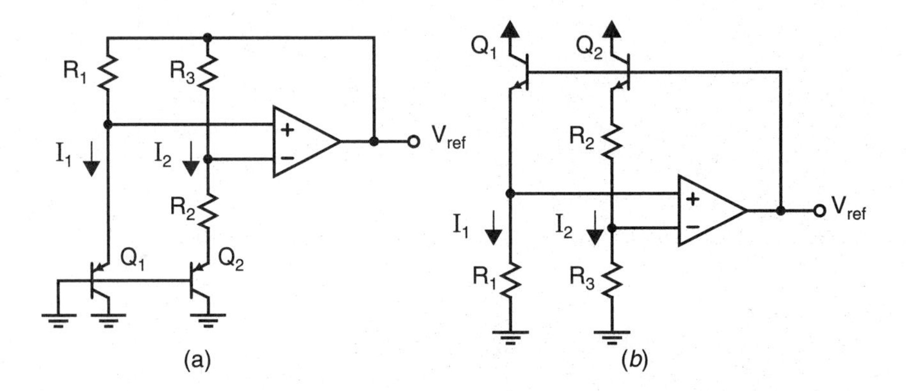

Fig. 8.27 Bandgap voltage references implemented with well transistors in (a) an n-well CMOS process, and (b) a p-well CMOS process.

With respect to the n-well implementation of Fig. 8.27(*a*), we have

$$V_{ref} = V_{EB1} + V_{R1} \tag{8.35}$$

Also, assuming the opamp has large gain and that its input terminals are at the same voltage, then

$$V_{R2} = V_{EB1} - V_{EB2} = \Delta V_{EB} \tag{8.36}$$

Now, since the current through R_3 is the same as the current through R_2, we have

$$V_{R3} = \frac{R_3}{R_2} V_{R2} = \frac{R_3}{R_2} \Delta V_{EB} \tag{8.37}$$

using (8.36). The opamp feedback also makes the voltage across R_1 equal to the voltage across R_3. Using this fact and substituting (8.37) into (8.35) results in

$$V_{ref} = V_{EB1} + \frac{R_3}{R_2} \Delta V_{EB} \tag{8.38}$$

which is in the required form to realize a bandgap reference. In integrated realizations of this reference, the bipolar transistors are often taken the same size, and the different current-densities are realized by taking R_3 greater than R_1, which causes I_1 to be larger than I_2. In this case, we would have

$$\frac{J_1}{J_2} = \frac{R_3}{R_1} \tag{8.39}$$

since R_1 and R_3 have the same voltage across them. Also, recalling from (8.12) that

$$\Delta V_{EB} = V_{EB1} - V_{EB2} = \frac{kT}{q} \ln\left(\frac{J_1}{J_2}\right) \tag{8.40}$$

and using (8.39) in (8.38) gives

$$V_{ref} = V_{EB1} + \frac{R_3 kT}{R_2 q} \ln\left(\frac{R_3}{R_1}\right) \tag{8.41}$$

It is immediately recognizable that

$$K = \frac{R_3}{R_2} \tag{8.42}$$

EXAMPLE 8.5

Find the resistances of a bandgap voltage reference based on Fig. 8.27(*a*), where $I_1 = 80\ \mu A$, $I_2 = 8\ \mu A$, and $V_{EB1\text{-}0} = 0.65\ V$ at $T = 300\ °K$.

Solution

Recalling from (8.20) that

$$V_{ref\text{-}0} = 1.24\text{ V} \tag{8.43}$$

therefore, from (8.35), we require

$$V_{R1} = V_{ref\text{-}0} - V_{EB1\text{-}0} = 0.59\text{ V} \tag{8.44}$$

Also, since $V_{R3} = V_{R1}$, we have

$$R_3 = \frac{V_{R3}}{I_2} = \frac{0.59\text{ V}}{8\ \mu\text{A}} = 73.8\text{ k}\Omega \tag{8.45}$$

and

$$R_1 = R_3\frac{I_2}{I_1} = 7.38\text{ k}\Omega \tag{8.46}$$

Now, recalling from (8.22) that

$$K = \frac{1.24 - 0.65\text{ V}}{0.0258 \times \ln(10)} = 9.93 \tag{8.47}$$

therefore

$$R_2 = \frac{R_3}{K} = 7.43\text{ k}\Omega \tag{8.48}$$

It is of interest to note here that using (8.40) and noting that $J_1/J_2 = I_1/I_2$ (since the sizes of Q_1 and Q_2 are assumed to be the same), we find that

$$\Delta V_{EB} = \frac{kT_0}{q}\ln\left(\frac{I_1}{I_2}\right) = 59\text{ mV} \tag{8.49}$$

gives the temperature dependence value of 198 mV/°K, as found in Example 8.3, which requires approximately a gain of 10 to cancel temperature dependency (actually 9.93 in this case).

The design equations for a voltage reference that is suitable for p-well processes and shown in Fig. 8.27(*b*) are essentially identical to those just given for the n-well reference.

In CMOS realizations of the references just described, the large value resistors are often realized by well resistors. Unfortunately, these resistors have a temperature dependence approximately given by [Michejda, 1984]

$$R(T) = R_0\frac{T^{\eta}}{T_0^{\eta}} \tag{8.50}$$

where $\eta = 2.2$. The errors caused by this temperature dependence can be eliminated by offsetting $V_{ref\text{-}0}$ slightly positive from the value given by (8.19) to

$$V_{ref\text{-}0} = V_{G0} + (m + \eta - 1)\frac{kT_0}{q} \tag{8.51}$$

Assuming the effects of the temperature coefficient of the resistors have been minimized, the next major source of error is often due to the input-offset voltage of the opamp [Michejda, 1984]. This results in an error term in the equation for ΔV_{BE} that is roughly equal to K times the input-offset voltage of the opamp. For example, a 1mV offset error that is temperature independent causes a temperature coefficient (TC), error approximately given by [Song, 1983]

$$TC_{error} \cong 26 \text{ ppm}/°\text{C} \tag{8.52}$$

One means of eliminating this source of error is to use switched-capacitor (SC) amplifiers that have input-offset compensation circuits [Song, 1983]. One possible SC-based voltage reference is shown in Fig. 8.28, where it makes use of an amplifier described in [Martin, 1987]. The amplifier shown results in a circuit having its output valid at all times and is less sensitive to finite opamp gain. A detailed explanation of how the amplifier shown in Fig. 8.28 operates is deferred until Chapter 10, where switched-capacitor circuits are discussed.

Assuming errors due to the input-offset voltage of the opamp have been minimized, there is still an inherent temperature dependence of the bandgap voltage reference, as we saw in (8.23). In addition, a further temperature dependence occurs because V_{G0} varies slightly with temperature (which has been ignored in the above analysis). Together, these two error sources limit the best achievable temperature coefficient to about 25 ppm/°K. Minimizing these second-order effects is beyond the scope of this book, but the interested reader is referred to [Palmer, 1981; Meijer, 1982; Song, 1983] to see examples of how errors due to second-order effects have been minimized.

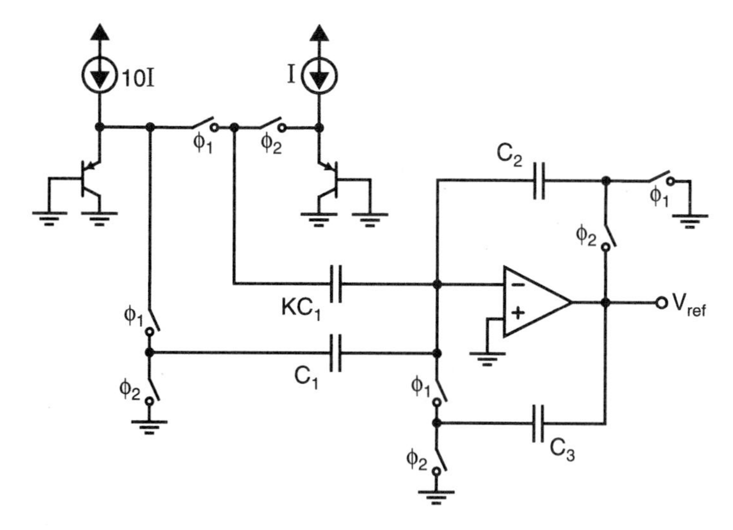

Fig. 8.28 An SC-based voltage reference that is insensitive to opamp input-offset voltages.

An alternative realization of a CMOS bandgap reference was reported in [Degrauwe, 1985], where lateral npn well transistors were used. Also, in [Tzanateas, 1979] a voltage reference was reported that was based on a realized PTAT voltage from the difference of the source-gate voltages of two MOS transistors biased in weak inversion. The interested reader can consult the references for details on these circuits.

8.7 TRANSLINEAR GAIN CELL

An important analog building block for bipolar circuits is the translinear gain cell[3] [Gilbert, 1968a; Gilbert, 1990]. These gain cells are useful when building circuits requiring varying gain, such as in adjustable continuous-time filters and voltage-controlled amplifiers. They are also useful in realizing four-quadrant analog multipliers, as will be seen in the next section.

A translinear gain cell is shown in Fig. 8.29. It has differential inputs and outputs that are current signals. The first stage simply consists of pn junction loads, which in this case are realized by the base-emitter junctions of Q_1 and Q_2. The second stage is a differential pair. To analyze this cell, we recall the relationship between the base-emitter voltage and collector current,

$$I_C = I_S e^{(V_{be}/V_T)} \quad \text{or} \quad V_{be} = V_T \ln\left(\frac{I_C}{I_S}\right) \tag{8.53}$$

Thus, ignoring the base currents of Q_3 and Q_4, we see that the voltage at the emitter of Q_1 is given by

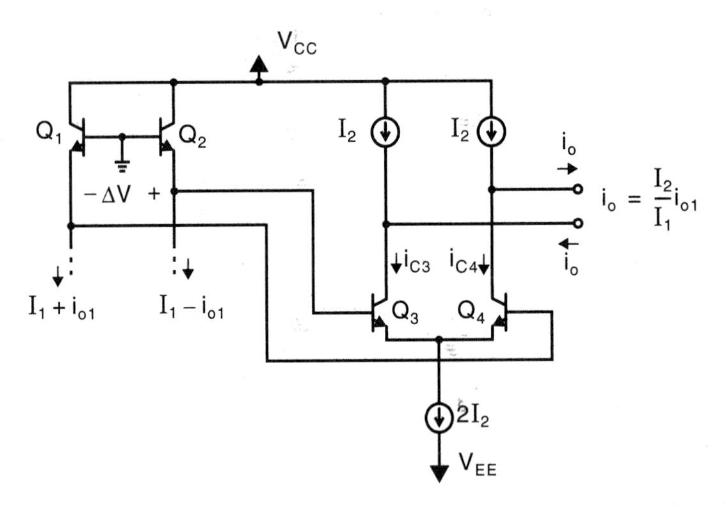

Fig. 8.29 A four-transistor gain cell. The output current, i_o, is a scaled version of i_{o1}.

3. These four transistor gain cells are also commonly called two-quadrant multipliers or Gilbert gain cells.

$$v_{e1} = (-V_T)\ln\left(\frac{I_1 + i_{o1}}{I_{S1}/\alpha}\right) \tag{8.54}$$

where $\alpha = \beta/(\beta+1)$. For Q_2, we have

$$v_{e2} = (-V_T)\ln\left(\frac{I_1 - i_{o1}}{I_{S2}/\alpha}\right) \tag{8.55}$$

Now, assuming Q_1 and Q_2 are matched such that $I_{S1} = I_{S2}$, and defining ΔV to be the difference between v_{e2} and v_{e1}, we have

$$\Delta V \equiv v_{e2} - v_{e1} = V_T \ln\left(\frac{I_1 + i_{o1}}{I_1 - i_{o1}}\right) \tag{8.56}$$

Note that this voltage difference, ΔV, does not depend on the scale current, I_S.

To analyze the current output from the differential pair of Q_3 and Q_4, we recall from Section 3.10 that the large-signal emitter currents through a differential pair are given by

$$i_{E3} = \frac{2I_2}{1 + e^{(-\Delta V)/V_T}} \quad \text{and} \quad i_{E4} = \frac{2I_2}{1 + e^{(\Delta V)/V_T}} \tag{8.57}$$

Therefore, the collector current, i_{C3}, is equal to

$$i_{C3} = \alpha i_{E3} = \frac{\alpha 2I_2}{1 + \left(\frac{I_1 - i_{o1}}{I_1 + i_{o1}}\right)} = \alpha I_2 + \alpha\frac{I_2}{I_1}i_{o1} \tag{8.58}$$

and for $\alpha = 1$, we have

$$i_{C3} = I_2 + \frac{I_2}{I_1}i_{o1} \tag{8.59}$$

In a similar manner, i_{C4} is equal to

$$i_{C4} = I_2 - \frac{I_2}{I_1}i_{o1} \tag{8.60}$$

Finally, the output current, i_o, is given by

$$i_o = i_{C3} - I_2 = I_2 - i_{C4} = \frac{I_2}{I_1}i_{o1} \tag{8.61}$$

In other words, the output current is equal to a scaled version of the input current where the scaling factor is determined by the ratio of two dc bias currents, I_2 and I_1.

An important point to note here is that not only is the difference between i_{C3} and i_{C4} linearly related to the input current, *but also each of* i_{C3} *and* i_{C4} *is also linearly related* (except for the dc bias current). This result is important since it implies that even if the difference between these two currents is not taken precisely, it does not affect the linearity of the overall circuit. Furthermore, note that the gain and linearity are not dependent on any transistor parameters such as β or I_S. However, one should be aware that there are mechanisms that will create distortion in a gain cell, such as mismatches between transistors and the occurrence of linear resistances in the pairs Q_1, Q_2 and Q_3, Q_4.

EXAMPLE 8.6

Consider the gain cell shown in Fig. 8.29 where I_1 is fixed to a bias current of 200 μA, while I_2 varies such that the gain changes from 0.5 to 4. What are the extreme values of I_2 needed? Also, what is the maximum peak current that should be used for i_{o1}, assuming all transistors should remain in the active region with at least 20 μA of emitter current?

Solution

Making use of (8.61), when a gain of 0.5 is needed, I_2 should be one-half of I_1, which is 100 μA. When a gain of 4 is needed, I_2 should be four times that of I_1, which is 800 μA. Therefore, the extreme values of I_2 are 100 μA and 800 μA.

The maximum peak current of i_{o1} is found by noting that the emitter currents of either transistors Q_1 or Q_2 should not go below 20 μA. Since they are biased with 200 μA each, the peak current of i_{o1} should be less than 180 μA. Note that if I_1 were varied while I_2 remained fixed, the maximum peak level of i_{o1} would change as I_1 was varied.

8.8 TRANSLINEAR MULTIPLIER

A translinear multiplier[4] is one approach for realizing a four-quadrant multiplier circuit. Multiplier circuits are useful in a variety of applications, such as frequency modulators in communication systems or phase detectors in phase-locked loops. A translinear multiplier can be realized using two translinear gain cells, as seen in the six-transistor translinear multiplier shown in Fig. 8.30 [Gilbert, 1968b]. We can analyze this multiplier by making use of the results for the gain-cell circuit just discussed preceding. Note that the preceding multiplier circuit is effectively two gain-cell circuits with their output signals cross-coupled. Specifically, one gain cell consists of the two differential pairs Q_1, Q_2 and Q_5, Q_6, while the other gain cell consists of the differential pair Q_3, Q_4 and makes use of the same voltage generating pair, Q_5, Q_6. Thus, recognizing the similarity of this circuit with that of Fig. 8.29 and making use of (8.59) and (8.60), we find the following four collector currents:

4. A translinear multiplier is also commonly called a Gilbert multiplier or an analog multiplier.

$$i_{C1} = \frac{(I_2 + i_2)}{2} + \frac{(I_2 + i_2)}{2I_1} i_1 \tag{8.62}$$

$$i_{C2} = \frac{(I_2 + i_2)}{2} - \frac{(I_2 + i_2)}{2I_1} i_1 \tag{8.63}$$

$$i_{C3} = \frac{(I_2 - i_2)}{2} + \frac{(I_2 - i_2)}{2I_1} i_1 \tag{8.64}$$

$$i_{C4} = \frac{(I_2 - i_2)}{2} - \frac{(I_2 - i_2)}{2I_1} i_1 \tag{8.65}$$

Note here that we have made the substitutions of signal currents, $I_2 \pm i_2$, in what was simply a bias current of $2I_2$ in Fig. 8.29. In addition, once again these four collector currents are linear in i_2 if i_1 remains a constant value and linear in i_1 if i_2 remains constant. To find the output currents, we combine collector currents to obtain

$$I_2 + i_o \equiv i_{C1} + i_{C4} = I_2 + \frac{i_1 i_2}{I_1} \tag{8.66}$$

and

$$I_2 - i_o \equiv i_{C2} + i_{C3} = I_2 - \frac{i_1 i_2}{I_1} \tag{8.67}$$

and taking the difference of these two equations results in

$$i_o = \frac{i_1 i_2}{I_1} \tag{8.68}$$

Thus, we see that the output signals are linear with respect to i_2 and can be scaled by the ratio i_1/I_1. Also, note that i_1 can be made positive or negative (as it is the amount of the bias current deviation from I_1), and so the output current, i_o, is a scaled positive or negative version of i_2. A similar result occurs with respect to the output signal and i_1. Thus, this multiplier is called a four-quadrant multiplier. Recall that a gain cell can only scale i_1 with respect to the ratio of two positive bias currents and is therefore only a two-quadrant multiplier.

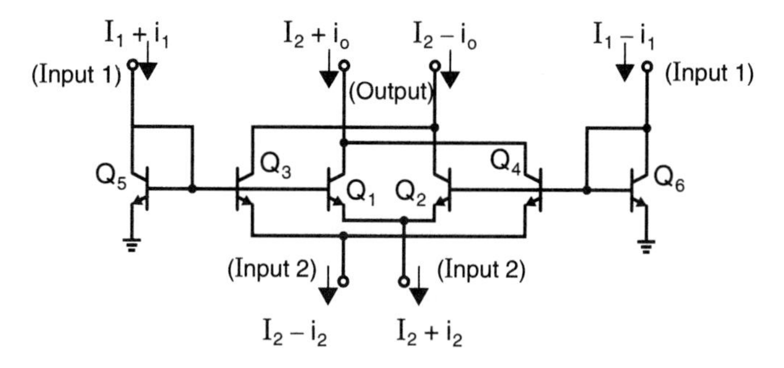

Fig. 8.30 A six-transistor translinear multiplier.

If voltage inputs, rather than current inputs, are desired, then one can precede the multiplier with voltage-to-current converters. The simplest of these are differential pairs having emitter degeneration. More linear, but more complicated, voltage-to-current converters include local feedback to eliminate distortions due to the modulation of the base-emitter voltages. Examples of these circuits are given in Chapter 15, where continuous-time filters are discussed.

EXAMPLE 8.7

Consider the translinear multiplier circuit shown in Fig. 8.30, where I_1 and I_2 are fixed bias currents of 100 μA and 200 μA, respectively. If i_1 is the gain-controlling current, what is that maximum gain for i_o/i_2, assuming all transistors should remain in the active region with at least 20 μA of emitter current?

Solution

To maintain 20 μA of current through Q_5 or Q_6, the magnitude of i_1 should be limited to 80 μA. As seen from (8.68), such a limit results in the peak gains being ±0.8. A gain of zero would occur if collector currents of Q_5 and Q_6 were equal.

Note that this multiplier circuit cannot get gain from i_2 to i_o, while gain is possible if the input signal is taken to be i_1.

8.9 REFERENCES

P. Brokaw, "A Simple Three-Terminal IC Bandgap Reference," *IEEE J. of Solid-State Circuits,* Vol. SC-9, pp. 388–393, December 1974.

J. Brugler, "Silicon Transistor Biasing for Linear Collector Current Temperature Dependence," *IEEE J. of Solid-State Circuits,* Vol. SC-2, pp. 57–58, June 1967.

W. Colleran and A. Abidi, "A 10-b, 75-MHz Two-Stage Pipelined Bipolar A/D Converter," *IEEE J. of Solid-State Circuits,* Vol. 28, no. 12, pp. 1187–1199, December 1993.

M. Degrauwe, O. Leuthold, E. Vittoz, H. Oguey, and A. Descombes, "CMOS Voltage References Using Lateral Bipolar Transistors," *IEEE J. of Solid-State Circuits,* Vol. 20, pp. 1151–1157, December 1985.

G. Erdi and P. R. Henneuse, "A Precision FET-Less Sample-and-Hold with High Charge-to-Droop Current Ratio," *IEEE J. of Solid-State Circuits,* Vol. 13, no. 6, pp. 864–873, December 1978.

J. Fernandes, S. Miller, M. Mallison, and G. Miller, "A 14-Bit 10-μs Subranging A/D Converter with S/H," *IEEE J. of Solid-State Circuits*, Vol. 23, no. 6, pp. 1309–1315, December 1988.

U. Gatti, F. Maloberti, and G. Palmisano, "An Accurate CMOS Sample-and-Hold Circuit," *IEEE J. of Solid-State Circuits,* Vol. 27, no. 1, pp. 120–122, January 1992.

B. Gilbert, "A New Wide-Band Amplifier Technique," *IEEE J. of Solid-State Circuits,* Vol. 3, pp. 353–365, December 1968a.

B. Gilbert, "A Precise Four-Quadrant Multiplier with Subnanosecond Response," *IEEE J. of Solid-State Circuits,* Vol. 3, pp. 365–373, December 1968b.

B. Gilbert, "Current-Mode Circuits from a Translinear Viewpoint: A Tutorial," *Analog IC Design: The Current-Mode Approach,* ed. C. Toumazou, F. J. Lidgey, and D. G. Haigh, Peter Peregrinus, London, 1990.

M. Ishikawa and T. Tsukahara, "An 8-bit 50-MHz CMOS Subranging A/D Converter with Pipelined Wide-Band S/H," *IEEE J. of Solid-State Circuits*, Vol. 23, no. 6, pp. 1309–1315, December 1988.

K. Kujik, "A Precision Reference Voltage Source," *IEEE J. of Solid-State Circuits*, Vol. SC-8, pp. 222–226, June 1973.

P. Lim and B. Wooley, "A High-Speed Sample-and-Hold Technique Using a Miller Hold Capacitance," *IEEE J. of Solid-State Circuits,* Vol. 26, no. 4, pp. 643–651, April 1991.

K. Martin, "Improved Circuits for the Realization of Switched-Capacitor Filters," *IEEE Trans. Circuits and Systems,* Vol. CAS-27, no. 4, pp. 237–244, April 1980.

K. Martin, "New Clock-Feedthrough Cancellation Technique for Analog MOS Switched-Capacitor," *Elctron. Lett.,* no. 18, 1992.

K. Martin, L. Ozcolak, Y. S. Lee, and G. C. Temes, "A Differential Switched-Capacitor Amplifier," *IEEE J. of Solid-State Circuits,* Vol. SC-22, no. 1, pp. 104–106, February 1987.

A. Matsuzawa, et al., "A 10-b 30-MHz Two-Step Parallel BiCMOS ADC with Internal S/H," *IEEE Int. Solid-State Circuits Conf.*, pp. 162–163, February 1990.

J. McCreary and P. R. Gray, "All-MOS Charge Redistribution Analog-to-Digital Conversion Techniques—Part 1," *IEEE J. of Solid-State Circuits*, Vol. SC-10, pp. 371–379, December 1975.

G. Meijer, P. Schmale, and K. van Zalinge, "A New Curvature-Corrected Bandgap Reference," *IEEE J. of Solid-State Circuits,* Vol. SC-17, pp. 1139–1143, December 1982.

J. Michejda and S. Kim, "A Precision CMOS Bandgap Reference," *IEEE J. of Solid-State Circuits,* Vol. SC-19, no. 6, pp. 1014–1021, December 1984.

G. Miller and C. O'Connor, "Sample-and-Hold Amplifier Circuit," United States Patent No. 4,962,325, issued October 9, 1990.

F. Moraveji, "A 14-b, 150-ns Sample-and-Hold Amplifier with Low Hold Step," *IEEE Int. Solid-State Circuits Conf.*, pp. 166–167, February 1991.

M. Nayebi and B. Wooley, "A 10-bit Video BiCMOS Track-and-Hold Amplifier," *IEEE J. of Solid-State Circuits*, Vol. 24, no. 6, pp 1507–1516, December 1989.

G. Nicollini and D. Senderowicz, "A CMOS Bandgap Reference for Differential Signal Processing," *IEEE J. of Solid-State Circuits,* Vol. 26, no. 1, pp. 41–50, January 1991.

C. Palmer and R. Dobkin, "A Curvature-Corrected Micropower Voltage Reference," *IEEE Int. Solid-State Circuits Conf.*, pp. 58–59, February 1981.

R. Petschacher, B. Zojer, B. Astegher, H. Jessner, and A. Lechner, "A 10-b 75MSPS subranging A/D converter with integrated sample and hold," *IEEE J. of Solid-State Circuits*, Vol. 25, no. 6, pp. 1339–1346, December 1990.

P. Real and D. Mercer, "A 14-b Linear, 250ns Sample-and-Hold Subsystem with Self-Correction," *IEEE Int. Solid-State Circuits Conf.*, pp. 164–165, February 1991.

D. Robertson, P. Real, and C. Mangelsdorf, "A Wideband 10-bit 20Msps Pipelined ADC Using Current-Mode Signals," *IEEE Int. Solid-State Circuits Conf.*, pp. 160–161, February 1990.

J. Shieh, M. Patil, and B. L. Sheu, "Measurement and Analysis of Charge Injection in MOS Analog Switches," *IEEE J. of Solid-State Circuits*, Vol. 22, no. 2, pp. 277–281, April 1987.

K. Sone, Y. Nishida, and N. Nakadai, "A 10-b 100-Msample/s Pipelined Subranging BiCMOS ADC," *IEEE J. of Solid-State Circuits,* Vol. 28, no. 12, pp. 1180–1186, December 1993.

B. Song and P. Gray, "A Precision Curvature-Compensated CMOS Bandgap Reference," *IEEE J. of Solid-State Circuits,* Vol. SC-18, no. 6, pp. 634–643, December 1983.

N. Sooch, et al., "18-Bit Stereo D/A Converter with Integrated Digital and Analog Filters," *91st Audio Engineering Society Conv.,* Reprint # 3113(y-1), October 1991.

K. R. Stafford, P. R. Gray, and R. A. Blanchard, "A Complete Monolithic Sample/Hold Amplifier," *IEEE J. of Solid-State Circuits,* Vol. 9, no. 6, pp. 381–387, December 1974.

Y. Tsividis, "Accurate Analysis of Temperature Effects in I_C-V_{BE} Characteristics with Application to Bandgap Reference Sources," *IEEE J. of Solid-State Circuits,* Vol. 15, no. 6, pp. 1076–1084, December 1980.

G. Tzanateas, C. Salama, and Y. Tsividis, "A CMOS Bandgap Voltage Reference," *IEEE J. of Solid-State Circuits,* Vol. SC-14, pp. 655–657, June 1979.

P. Vorenkamp and J. Verdaasdonk, "A 10b 50MS/s Pipelined ADC," *IEEE Int. Solid-State Circuits Conf.*, pp. 32–33, February 1992.

M. Wakayama, H. Tanimoto, T. Tasai, and Y. Yoshida, "A 1.2-μm BiCMOS Sample-and-Hold Circuit with a Constant-Impedance Slew-Enhanced Sampling Gate," *IEEE J. of Solid-State Circuits,* Vol. 27, no. 12, pp. 1697–1708, December 1992.

R. Widlar, "New Developments in IC Voltage Regulators," *IEEE J. of Solid-State Circuits,* Vol. 23, no. SC-6, pp. 2–7, February 1971.

R. Ye and Y. Tsividis, "Bandgap Voltage Reference Sources in CMOS Technology," *Electron. Lett.*, Vol. 18, no. 1, January 1982.

8.10 PROBLEMS

Unless otherwise stated, assume the following:

- n-channel MOS transistors:

  $\mu_n C_{ox} = 92\ \mu A/V^2$
  $V_{tn} = 0.8\ V$
  $\gamma = 0.5\ V^{1/2}$
  $r_{ds}\ (\Omega) = 8{,}000\ L\ (\mu m)/I_D\ (mA)$ in active region,
  $C_j = 2.4 \times 10^{-4}\ pF/(\mu m)^2$
  $C_{j\text{-}sw} = 2.0 \times 10^{-4}\ pF/\mu m$
  $C_{ox} = 1.9 \times 10^{-3}\ pF/(\mu m)^2$
  $C_{gs(overlap)} = C_{gd(overlap)} = 2.0 \times 10^{-4}\ pF/\mu m$

- p-channel MOS transistors:

  $\mu_p C_{ox} = 30\ \mu A/V^2$
  $V_{tp} = -0.9\ V$
  $\gamma = 0.8\ V^{1/2}$
  $r_{ds}\ (\Omega) = 12{,}000\ L\ (\mu m)/I_D\ (mA)$ in active region,
  $C_j = 4.5 \times 10^{-4}\ pF/(\mu m)^2$
  $C_{j\text{-}sw} = 2.5 \times 10^{-4}\ pF/\mu m$
  $C_{ox} = 1.9 \times 10^{-3}\ pF/(\mu m)^2$
  $C_{gs(overlap)} = C_{gd(overlap)} = 2.0 \times 10^{-4}\ pF/\mu m$

8.1 Using the transient analysis of SPICE with a sinusoidal input signal of 2 V peak-to-peak, find the offset, gain, and linearity error of the simple S/H of Fig. 8.3.

8.2 For the T/H of Fig. 8.4, assume V_{in} is a 20-MHz sinusoid with a 2 $V_{p\text{-}p}$ amplitude. Also assume that ϕ_{clk} is a 100-MHz square wave having a peak amplitude of $\pm 2.5\ V$ with rise and fall times of 1.5 ns. What is the maximum time difference between the turn-off times of the n-channel and p-channel transistors? Ignore the body effect.

8.3 The T/H of Fig. 8.5 has transistors that are $10\ \mu m/0.8\ \mu m$ and a 1-pF hold capacitor. Assume the clock waveforms are fast enough that the channel charge of the transistors is evenly distributed between the two junctions. Compare the final hold pedestal between the case when the dummy switch turns on substantially before the sampling switch turns off, and the case when it turns on substantially after the sampling switch.

8.4 Assume the second stage of the S/H of Fig. 8.10 has a gain of –30 and a unity-gain frequency of 100 MHz. Also assume the input signal is a 5-MHz sinusoid with a 2 $V_{p\text{-}p}$ amplitude and the clock signals are at 20 MHz and go between +2.5 V and –2.5 V, with rise and fall times of 1.5 ns. What is the sampling time uncertainty?

8.5 Assume the S/H of Fig. 8.12 is being realized in a BiCMOS process. This means that diodes and BJT transistors are available. Design a clock-driver circuit so that the clock voltages go from two diode drops above the input voltage (in track mode) to two diode drops below the output voltage (in hold mode).

8.6 Assume the opamp in the S/H of Fig. 8.15 has a finite gain of A_0. Derive the output voltage in terms of V_{in} and A_0 during hold mode (i.e., when ϕ_2 is high).

8.7 Derive the output voltage of the S/H of Fig. 8.19 at the end of ϕ_2 in terms of the input voltage at the end of ϕ_1, the output voltage at the end of ϕ_1 from the previous period, and the capacitor ratio C_1/C_2. Take the z-transform of this difference equation, and substitute $z = e^{j\omega T}$ to find the frequency-domain transfer function of the S/H. Making the assumption that $e^{j\omega T} \cong 1 + j\omega T$ for $\omega << (1/T) = f_{clk}$, where f_{clk} is the sampling frequency, show that

$$f_{-3\text{ dB}} \cong \frac{1}{2\pi}\frac{C_1}{C_2}f_{clk}$$

8.8 Assume the S/H of Fig. 8.21 has each of D_5 and D_6 replaced by two series diodes. Show the voltages at all nodes for the cases of sampling a 1 V input and a –1 V input for before as well as after track mode.

8.9 Assume two transistors in a PTAT are biased at a current-density ratio of 8 : 1 at $T = 320$ °K. What is the difference in their base-emitter voltages and what is its temperature dependence?

8.10 Assuming $V_{BE0\text{-}2} = 0.65$ V, what is the value required for K in order to get zero temperature dependence at $T = 320$ °K for a current-density ratio of 8:1 in the two transistors?

8.11 Prove equations (8.22) and (8.23).

8.12 Find values for the bipolar reference of Fig. 8.25, where $V_{ref} = 2.5$ V.

8.13 Prove that equation (8.41) holds for the voltage reference of Fig. 8.27(*b*).

8.14 Consider the gain cell shown in Fig. 8.29, where I_2 is fixed to a bias current of 200 μA while I_1 varies such that the gain changes from 0.5 to 4. What are the extreme values of I_1 needed? Also, what is the maximum peak current that should be used for i_{o1}, assuming all transistors should remain in the active region with at least 20 μA of emitter current?

8.15 Consider the translinear multiplier shown in Fig. 8.30, where $I_1 = 100$ μA, $I_2 = 200$ μA, and i_1 is set such that the gain $i_o/i_2 = 0.5$. If a common-mode noise signal of 5 μA appears on I_1, what is the size of the output signal due to that noise signal? Repeat for a common-mode noise signal on I_2.

8.16 Repeat Problem 8.15 when the β's of the transistors' worst-case mismatch is 5 percent.

8.17 Design a four-quadrant translinear multiplier having differential voltage-current conversion stages based on emitter degeneration at both inputs and a differential voltage output. All transistors should be biased at 0.2 mA. Design the multiplier so that when both differential inputs are 1 V the differential output should be 1 V. Assume a 5 V power supply is available.

CHAPTER

Discrete-Time Signals

A basic understanding of discrete-time signal processing is essential in the design of most modern analog systems. For example, discrete-time signal processing is heavily used in the design and analysis of oversampling analog-to-digital (A/D) and digital-to-analog (D/A) converters used in digital audio and instrumentation applications. Also, a discrete-time filtering technique known as switched-capacitor filtering is probably the most popular approach for realizing fully integrated analog filters. Switched-capacitor filters are in the class of analog filters since voltage levels in these filters remain continuous. In other words, switched-capacitor filters operate and are analyzed using discrete-time steps but involve no A/D or D/A converters. This chapter presents some basic concepts of discrete-time signals and filters.

9.1 OVERVIEW OF SOME SIGNAL SPECTRA

Consider the spectra of sampled and continuous-time signals in the block diagram systems shown in Fig. 9.1, where it is assumed that the continuous-time signal, $x_c(t)$, is band limited through the use of an anti-aliasing filter (not shown). *DSP* refers to *Discrete-time signal processing,* which may be accomplished using fully digital processing or discrete-time analog circuits such as switched-capacitor filters. Some

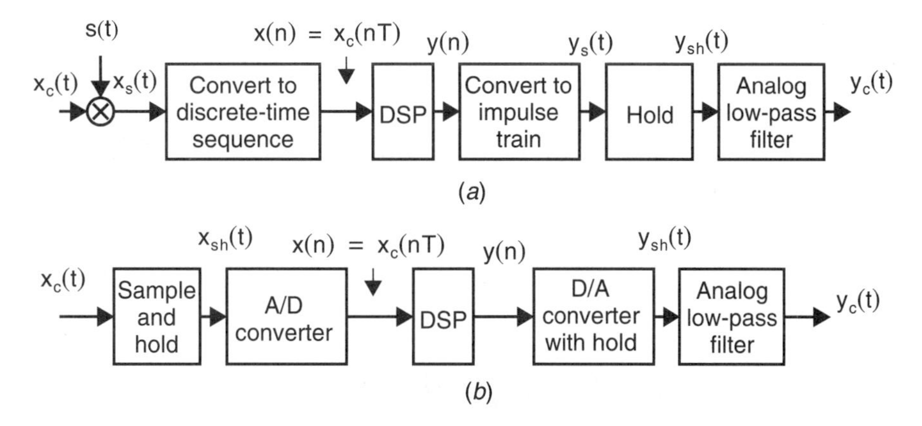

Fig. 9.1 Performing DSP on analog signals. (*a*) Conceptual realization, and (*b*) typical physical realization.

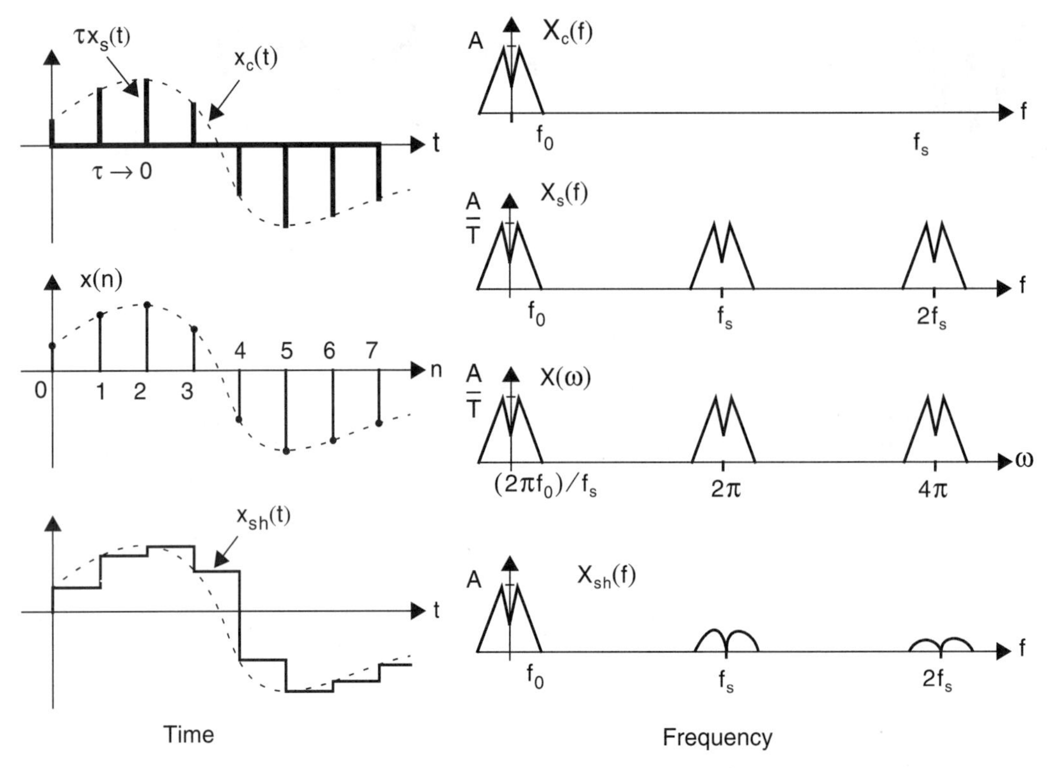

Fig. 9.2 Some time signals and frequency spectra.

example time signals and frequency spectra for this system are shown in Fig. 9.2. Here, $s(t)$ is a periodic impulse train in time with a period of T, where T equals the inverse of the sampling frequency, f_s. Some relationships for the signals in Figs. 9.1 and 9.2 are as follows:

1. $x_s(t)$ has the same frequency spectrum as $x_c(t)$, but the baseband spectrum repeats every f_s (assuming that no aliasing occurs and hence, an anti-aliasing filter is needed).
2. $x(n)$ has the same frequency spectrum as $x_s(t)$, but the sampling frequency is normalized to 1.
3. The frequency spectrum for $X_{sh}(t)$ equals that of $X_s(t)$ multiplied by a response of $(\sin x)/x$ (in effect, multiplying $X_s(f)$ by this response helps to remove high-frequency images).

The remainder of this chapter confirms these spectral relationships and introduces other basic discrete-time concepts.

9.2 LAPLACE TRANSFORMS OF DISCRETE-TIME SIGNALS

Consider the sampled signal, $x_s(t)$, related to the continuous-time signal, $x_c(t)$, as shown in Fig. 9.3. Here, $x_s(t)$ has been scaled by τ such that the area under the pulse

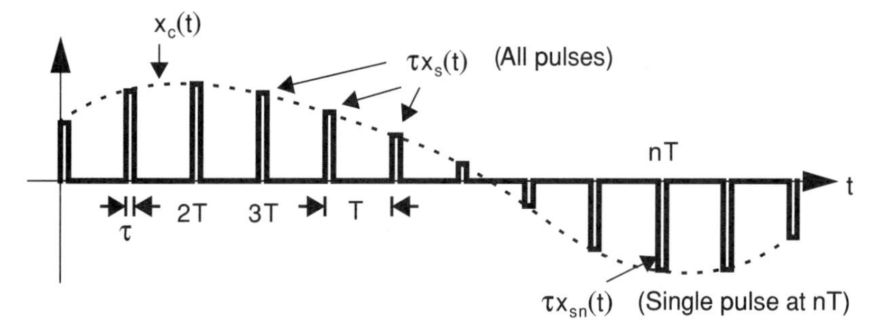

Fig. 9.3 Sampled and continuous-time signals.

at nT equals the value of $x_c(nT)$. In other words, at $t = nT$, we have

$$x_s(nT) = \frac{x_c(nT)}{\tau} \tag{9.1}$$

such that the area under the pulse, $\tau x_s(nT)$, equals $x_c(nT)$. Thus, as $\tau \to 0$, the height of $x_s(t)$ at time nT goes to ∞, and so we plot $\tau x_s(t)$ instead of $x_s(t)$.

We define $\vartheta(t)$ to be the step function given by

$$\vartheta(t) \equiv \begin{cases} 1 & (t \geq 0) \\ 0 & (t < 0) \end{cases} \tag{9.2}$$

Then $x_s(t)$ can be represented as a linear combination of a series of pulses, $x_{sn}(t)$, where $x_{sn}(t)$ is zero everywhere except for a single pulse at nT. The single-pulse signal, $x_{sn}(t)$, can be written as

$$x_{sn}(t) = \frac{x_c(nT)}{\tau}[\vartheta(t - nT) - \vartheta(t - nT - \tau)] \tag{9.3}$$

so that we can now write $x_s(t)$ as

$$x_s(t) = \sum_{n=-\infty}^{\infty} x_{sn}(t) \tag{9.4}$$

Note that these signals are defined for *all time,* so we can find the Laplace transform of $x_s(t)$ in terms of $x_c(t)$. Using the notation that $X_s(s)$ is the Laplace transform of $x_s(t)$, we find the Laplace transform $X_{sn}(s)$ for $x_{sn}(t)$ to be given by

$$X_{sn}(s) = \frac{1}{\tau}\left(\frac{1 - e^{-s\tau}}{s}\right)x_c(nT)e^{-snT} \tag{9.5}$$

Since $x_s(t)$ is merely a linear combination of $x_{sn}(t)$, we also have (for $\tau \neq 0$)

$$X_s(s) = \frac{1}{\tau}\left(\frac{1 - e^{-s\tau}}{s}\right)\sum_{n=-\infty}^{\infty} x_c(nT)e^{-snT} \tag{9.6}$$

Using the expansion $e^x = 1 + x + (x^2/2!) + \cdots$, when $\tau \to 0$, the term before the summation in (9.6) goes to unity. Therefore, in the limiting case as $\tau \to 0$, we have

$$X_s(s) = \sum_{n=-\infty}^{\infty} x_c(nT)e^{-snT} \tag{9.7}$$

Spectra of Discrete-Time Signals

The spectrum of the sampled signal, $x_s(t)$, can be found by replacing s by $j\omega$ in (9.7). However, a more intuitive approach to find the spectrum of $x_s(t)$ is to recall that multiplication in the time domain is equivalent to convolution in the frequency domain. To use this fact, note that, for $\tau \to 0$, $x_s(t)$ can be written as the product

$$x_s(t) = x_c(t)s(t) \tag{9.8}$$

where $s(t)$ is a periodic pulse train, or mathematically,

$$s(t) = \sum_{n=-\infty}^{\infty} \delta(t - nT) \tag{9.9}$$

where $\delta(t)$ is the unit impulse function, also called the *Dirac delta function.* It is well known that the Fourier transform of a periodic impulse train is another periodic impulse train. Specifically, the spectrum of $s(t)$, $S(j\omega)$, is given by

$$S(j\omega) = \frac{2\pi}{T}\sum_{k=-\infty}^{\infty} \delta\left(\omega - k\frac{2\pi}{T}\right) \tag{9.10}$$

Now writing (9.8) in the frequency domain, we have

$$X_s(j\omega) = \frac{1}{2\pi}X_c(j\omega) \otimes S(j\omega) \tag{9.11}$$

where $\otimes$ denotes convolution. Finally, by performing this convolution either mathematically or graphically, the spectrum of $X_s(j\omega)$ can be seen to be given by

$$X_s(j\omega) = \frac{1}{T}\sum_{k=-\infty}^{\infty} X_c\left(j\omega - \frac{jk2\pi}{T}\right) \tag{9.12}$$

or, equivalently,

$$X_s(f) = \frac{1}{T}\sum_{k=-\infty}^{\infty} X_c(j2\pi f - jk2\pi f_s) \tag{9.13}$$

Equations (9.12) and (9.13) show that the spectrum for the sampled signal, $x_s(t)$, equals a sum of shifted spectra of $x_c(t)$, and therefore no aliasing occurs if $X_c(j\omega)$ is band limited to $f_s/2$. The relation in (9.13) also confirms the example spectrum for $X_s(f)$, shown in Fig. 9.2. Note that, for a discrete-time signal, $X_s(f) = X_s(f \pm kf_s)$, where k is an arbitrary integer as seen by substitution in (9.13).

Finally, note that the signal $x_s(t)$ cannot exist in practice when $\tau \to 0$ since an infinite amount of power would be required to create it. (Integrating $X_s(f)$ over all frequencies illustrates this remark.)

9.3 z-TRANSFORM

For our purposes, the z-transform is merely a shorthand notation for (9.7). Specifically, defining

$$z \equiv e^{sT} \tag{9.14}$$

we can write

$$X(z) \equiv \sum_{n=-\infty}^{\infty} x_c(nT) z^{-n} \tag{9.15}$$

where $X(z)$ is called the z- transform of the samples $x_c(nT)$.

Two properties of the z-transform that can be deduced from Laplace transform properties are as follows:

1. If $x(n) \leftrightarrow X(z)$, then $x(n-k) \leftrightarrow z^{-k}X(z)$
2. Convolution in the time domain is equivalent to multiplication in the frequency domain. Specifically, if $y(n) = h(n) \otimes x(n)$, where $\otimes$ denotes convolution, then $Y(z) = H(z)X(z)$. Similarly, multiplication in the time domain is equivalent to convolution in the frequency domain.

Note that $X(z)$ is not a function of the sampling rate but is related only to the numbers $x_c(nT)$, whereas $X_s(s)$ is the Laplace transform of the signal $x_s(t)$ as $\tau \to 0$. In other words, the signal $x(n)$ is simply a series of numbers that may (or may not) have been obtained by sampling a continuous-time signal. One way of thinking about this series of numbers as they relate to the samples of a possible continuous-time signal is that the original sample time, T, has been effectively normalized to one (i.e., $f_s' = 1$ Hz). Such a normalization of the sample time, T, in both time and frequency, justifies the spectral relation between $X_s(f)$ and $X(\omega)$ shown in Fig. 9.2. Specifically, the relationship between $X_s(f)$ and $X(\omega)$ is given by

$$X_s(f) = X\left(\frac{2\pi f}{f_s}\right) \tag{9.16}$$

or, equivalently, the following frequency scaling has been applied:

$$\omega = \frac{2\pi f}{f_s} \tag{9.17}$$

This normalization results in discrete-time signals having ω in units of radians/sample, whereas the original continuous-time signals have frequency units of cycles/second (hertz) or radians/second. For example, a continuous-time sinusoidal signal of 1 kHz when sampled at 4 kHz will change by $\pi/2$ radians between each sample.

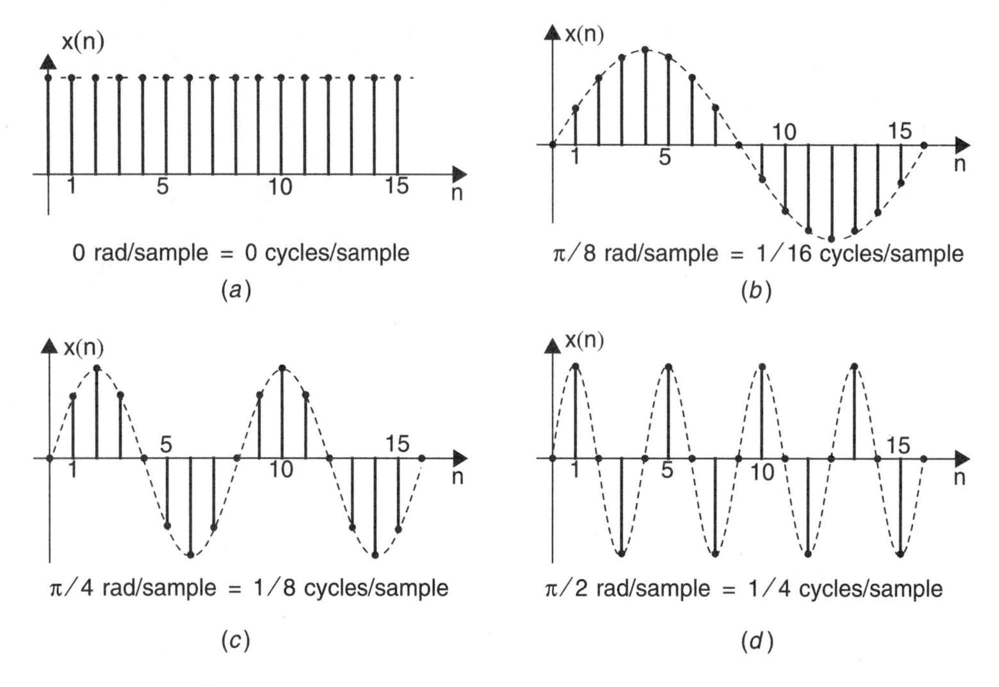

Fig. 9.4 Some discrete-time sinusoidal signals.

Therefore, such a discrete-time signal is defined to have a frequency of $\pi/2$ rad/sample. Other examples of discrete-time sinusoidal signals are shown in Fig. 9.4. It should be noted here that discrete-time signals are not unique since the addition of 2π results in the same signal. For example, a discrete-time signal that has a frequency of $\pi/4$ rad/sample is identical to that of $9\pi/4$ rad/sample. Thus, normally discrete-time signals are defined to have frequency components only between $-\pi$ and π rad/sample. For a more detailed discussion of this unit topic, see [Proakis, 1992].

EXAMPLE 9.1

Consider the spectra of $X_c(f)$ and $X_s(f)$, shown in Fig. 9.2, where f_0 is 1 Hz and f_s is 4 Hz. Compare the time and spectrum plots of $X_s(f)$ and $X_{s2}(f)$, where $X_{s2}(f)$ is sampled at 12 Hz. How does $X(\omega)$ differ between the two sampling rates?

Solution

By sampling at 12 Hz, the spectrum of $X_c(f)$ repeats every 12 Hz, resulting in the signals shown in Fig. 9.5.

Note that, for $X(\omega)$, 4 Hz is normalized to 2π rad/sample, whereas for $X_2(\omega)$, 12 Hz is normalized to 2π rad/sample.

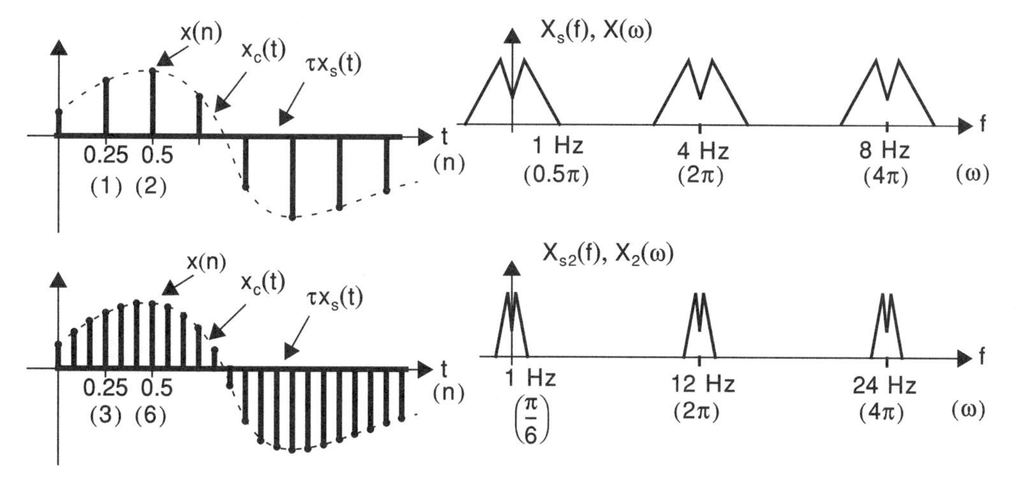

Fig. 9.5 Comparing time and frequency of two sampling rates.

9.4 DOWNSAMPLING AND UPSAMPLING

Two operations that are quite popular in discrete-time signal processing are downsampling and upsampling. Downsampling is used to reduce the sample rate (hopefully, without information loss), whereas upsampling is used to increase the sample rate. Although noninteger downsampling and upsampling rates can be achieved, here we consider only the case in which L is an integer value.

Downsampling is achieved by keeping every Lth sample and discarding the others. As Fig. 9.6 shows, the result of downsampling is to expand the original spectra by L. Thus, to avoid digital aliasing, the spectra of the original signal must be band limited to π/L before downsampling is done. In other words, the signal must be sampled L times above its minimum sampling rate so that no information is lost during downsampling.

Upsampling is accomplished by inserting $L-1$ zero values between samples, as shown in Fig. 9.7. In this case, one can show that the spectra of the resulting

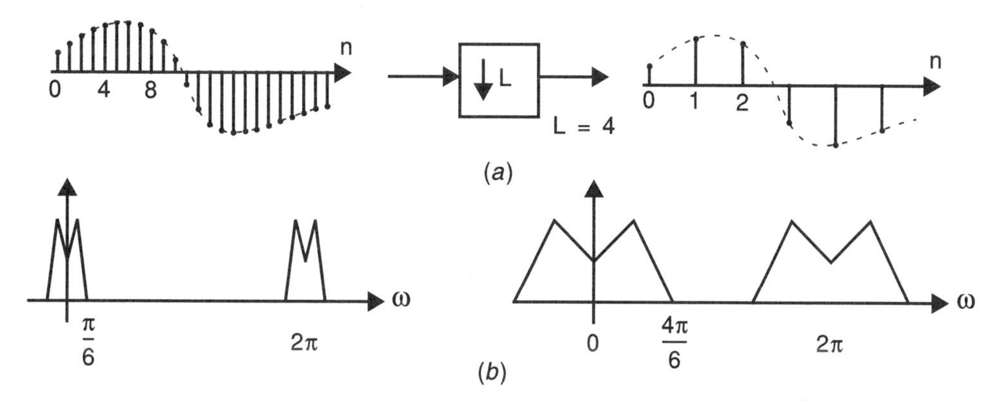

Fig. 9.6 Downsampling by 4: (*a*) time domain, and (*b*) frequency domain.

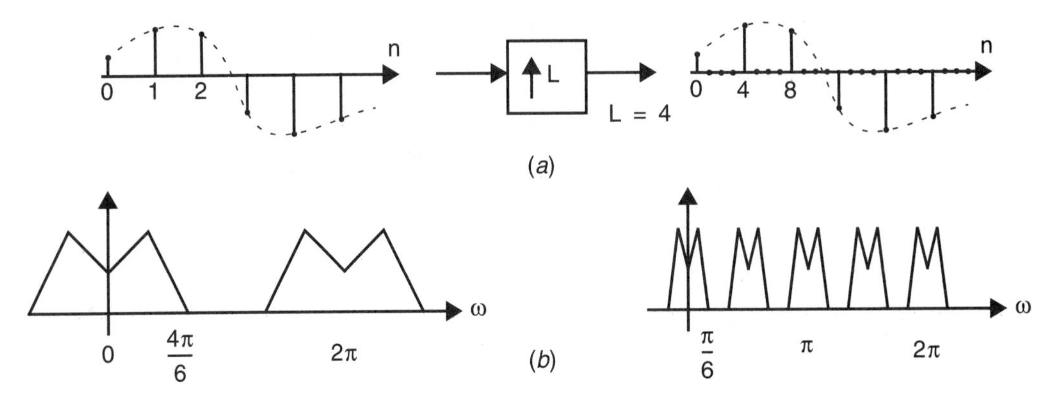

Fig. 9.7 Upsampling by 4: (*a*) time domain, and (*b*) frequency domain.

upsampled signal are identical to the original signal but with a renormalization along the frequency axis. Specifically, when a signal is upsampled by L, the frequency axis is scaled by L such that 2π now occurs where $L2\pi$ occurred in the original signal. This operation is useful when one wishes to increase the effective sampling rate of a signal, particularly if postfiltering is then applied.

Follow the next two examples carefully—they help explain the spectral changes due to downsampling and upsampling.

EXAMPLE 9.2

As an example of downsampling of the signal $x_2(n)$ in Example 9.1, find the new spectrum, $X_3(\omega)$, when every third sample is kept and the others are discarded (i.e., when we downsample by 3). In other words, if the original series of numbers is $x_1, x_2, x_3, x_4, \ldots$, the new series is $x_1, x_4, x_7, x_{10}, \ldots$.

Solution

Clearly the solution here is that the new sequence is equivalent to $x(n)$ in Example 9.1, and thus its spectrum is simply $X_3(\omega) = X(\omega)$.

Note that no information was lost in this example because the initial images of the signal $x_2(n)$ were far enough apart. However, if the downsampling operation results in the spectra overlapping one another, then an aliasing phenomenon occurs that is similar to that which occurs when sampling an analog signal lower than the Nyquist rate. In fact, it can be shown that, to avoid any aliasing during downsampling by the factor L, the original signal should be band limited to π/L.

EXAMPLE 9.3

As an example of upsampling of the signal $x(n)$ in Example 9.1, find the new spectrum, $X_4(\omega)$, if two zeros are inserted between each sample (i.e., if we upsample by 3). In other words, if the original series of numbers was $x(n) = x_1, x_2, x_3, \ldots$ and had a spectrum $X(\omega)$, then the new series of numbers is now $x_4(n) = (x_1, 0, 0, x_2, 0, 0, x_3, \ldots)$ and has a spectrum $X_4(\omega)$.

Solution

Perhaps the simplest way to conceptualize the insertion of these extra 0 samples in a series is to look at $X_{s4}(f)$ in time and frequency when we let $f_{s4} = 12$ Hz. Recall that $x_{s4}(t)$ is defined for all time, where it is zero between its impulses, and the Laplace transform is used to observe its frequency response. With such a sampling frequency, we note that $x_{s4}(t)$ is equal to the signal $x_s(t)$, and thus $X_{s4}(f) = X_s(f)$. To find $X_4(\omega)$, recall that it is simply a frequency normalization of $X_{s4}(f)$, where the sampling frequency is normalized to 2π. In other words, the series $x(n)$ is simply a normalization of the time between impulses to 1. However, by inserting extra zeros between samples, the normalization is effectively being done for a time period smaller than that between nonzero impulses. Thus, the images remain the same while the normalization along the frequency axis is different.

In this example, since two zeros are inserted, the effective sampling rate might be thought of as 12 Hz, but now the images at 4 Hz and 8 Hz remain in contrast to $X_2(\omega)$ in Example 9.1. The resulting signals are shown in Fig. 9.8.

Finally, note that if signal processing is used on the samples $x_4(n)$ to eliminate the two images at $(2\pi)/3$ and $(4\pi)/3$, then the resulting signal will equal $x_2(n)$. In other words, as long as an analog signal is originally sampled higher than the Nyquist rate, we can use upsampling and digital-signal processing to derive signals that are the same as if the analog signal were sampled at a much higher rate.

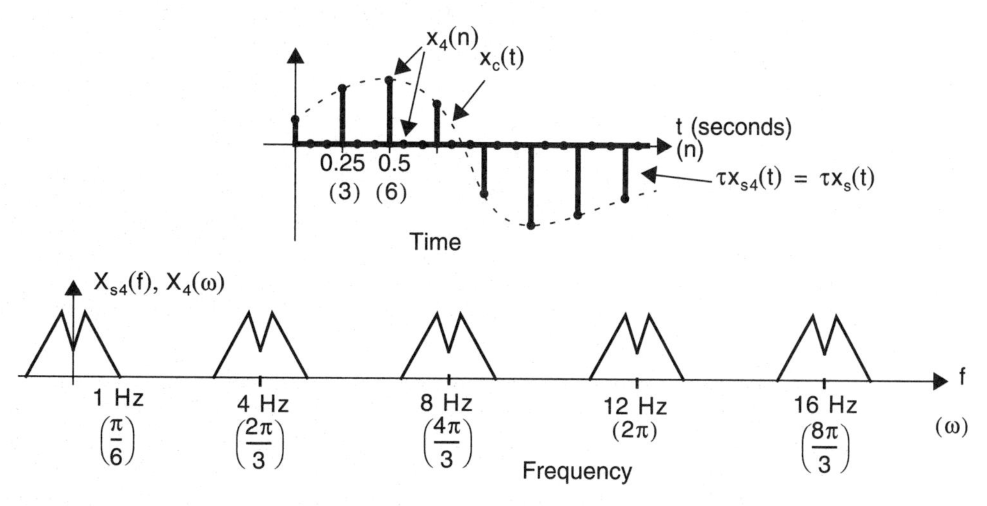

Fig. 9.8 Spectrum of a discrete-time signal upsampled by 3.

9.5 DISCRETE-TIME FILTERS

Thus far, we have seen the relationship between continuous-time and discrete-time signals in the time and frequency domain. However, often one wishes to perform filtering on a discrete-time signal to produce another discrete-time signal. In other words, an input series of numbers is applied to a discrete-time filter to create an output series of numbers. This filtering of discrete-time signals is most easily visualized with the shorthand notation of z-transforms.

Consider the system shown in Fig. 9.9, where the output signal is defined to be the impulse response, $h(n)$, when the input, $u(n)$, is an impulse (i.e., 1 for $n = 0$ and 0 otherwise). The transfer function of the filter is said to be given by $H(z)$, which is the z-transform of the impulse response, $h(n)$.

Frequency Response of Discrete-Time Filters

The transfer functions for discrete-time filters appear similar to those for continuous-time filters, except that, instead of polynomials in s, polynomials in z are obtained. For example, the transfer function of a low-pass, continuous-time filter, $H_c(s)$, might appear as

$$H_c(s) = \frac{4}{s^2 + 2s + 4} \tag{9.18}$$

The poles for this filter are determined by finding the roots of the denominator polynomial, which are $-1.0 \pm 1.7321j$ for this example. This continuous-time filter is also defined to have two zeros at ∞ since the denominator polynomial is two orders higher than the numerator polynomial. To find the frequency response of $H_c(s)$, the poles and zeros can be plotted in the s-plane (Fig. 9.10(*a*)), and the substitution $s = j\omega$ is equivalent to finding the magnitude and phase of vectors from a point along the $j\omega$ axis to all the poles and zeros.

An example of a discrete-time, low-pass transfer function is given by the following equation:

$$H(z) = \frac{0.05}{z^2 - 1.6z + 0.65} \tag{9.19}$$

Here, the poles now occur at $0.8 \pm 0.1j$ in the z-plane, and two zeros are again at ∞. To find the transfer function of $H(z)$, the poles and zeros can be plotted in the z-plane; however, instead of going along the vertical $j\omega$ axis as in the s-plane,

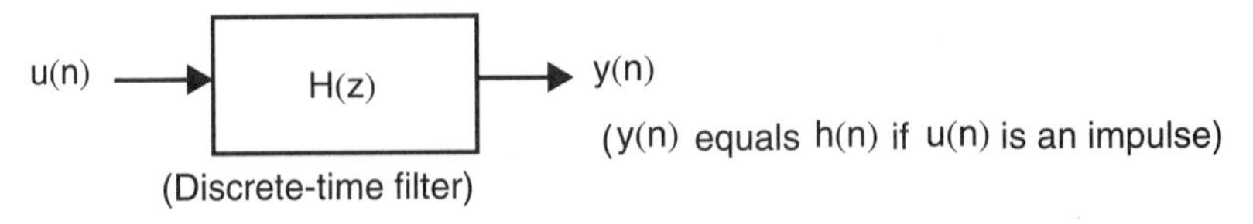

Fig. 9.9 Discrete-time filter system.

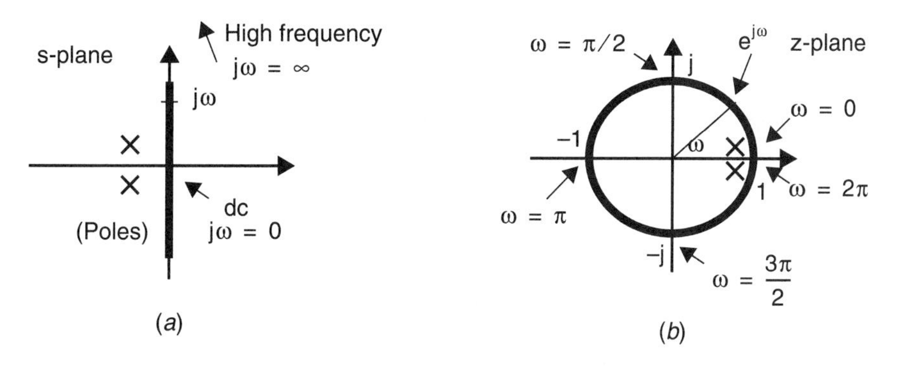

Fig. 9.10 Transfer function response. (*a*) Continuous-time, and (*b*) discrete-time.

the unit circle contour is used, so that $z = e^{j\omega}$, as shown in Fig. 9.10(*b*). Note that substituting $z = e^{j\omega}$ in the z-domain transfer function, $H(z)$, is simply a result of substituting $s = j\omega$ into (9.14), where T has been normalized to 1, as discussed in Section 9.3. Also note that poles or zeros occurring at $z = 0$ do not affect the magnitude response of $H(z)$ since a vector from the origin to the unit circle always has a length of unity. However, they would affect the phase response.

We see here that in the discrete-time domain, $z = 1$ corresponds to the frequency response at both dc (i.e., $\omega = 0$) and at $\omega = 2\pi$. Also, the time normalization of setting T to unity implies that $\omega = 2\pi$ is equivalent to the sampling-rate speed (i.e., $f = f_s$) for $X_s(f)$. In addition, note that the frequency response of a filter need be plotted only for $0 \le \omega \le \pi$ (i.e., $0 \le \omega \le f_s/2$) since, for filters with real coefficients, the poles and zeros always occur in complex-conjugate pairs (or on the real axis), so the magnitude response of the filter is equal to that for $\pi \le \omega \le 2\pi$ (the filter's phase is antisymmetric). Going around the circle again gives the same result as the first time, implying that the frequency response repeats every 2π.

Before we leave this section, a word of caution is in order. To simplify notation, the same variables, f and ω, are used in both the continuous-time and discrete-time domains in Fig. 9.10. However, these variables are not equal in the two domains, and care should be taken not to confuse values from the two domains. The continuous-time domain is used here for illustrative reasons only since the reader should already be quite familiar with transfer function analysis in the s-domain. In summary, the unit circle, $e^{j\omega}$, is used to determine the frequency response of a system that has its input and output as a series of numbers, whereas the $j\omega$-axis is used for a system that has continuous-time inputs and outputs. However, ω is different for the two domains.

EXAMPLE 9.4

Assuming a sample rate of $f_s = 100$ kHz, find the magnitude of the transfer function in (9.19) for 0 Hz, 100 Hz, 1 kHz, 10 kHz, 50 kHz, 90 kHz, and 100 kHz.

Table 9.1 Finding the gain of H(z) at some example frequencies

Frequency (kHz)	z-plane locations		$\lvert H(z) \rvert$
	$e^{j\omega}$	$z = x + jy$	
0	e^{j0}	$1.0 + j0.0$	1.0
0.1	$e^{j0.002\pi}$	$0.9999803 + j0.00628314$	0.9997
1	$e^{j0.02\pi}$	$0.9980267 + j0.06279052$	0.968
10	$e^{j0.2\pi}$	$0.809 + j0.5878$	0.149
50	$e^{j\pi}$	$-1.0 + j0.0$	0.0154
90	$e^{j1.8\pi}$	$0.809 - j0.5878$	0.149
100	$e^{j2\pi}$	$1.0 + j0.0$	1.0

Solution

To find the magnitude of $H(z)$ at these frequencies, first we find their equivalent z-domain locations by normalizing f_s to 2π. Next, we find the gain of $H(z)$ by putting these z-domain values into (9.19) and finding the magnitude of the resulting complex value. Table 9.1 summarizes the results.

Note that the gain of $H(z)$ is the same at both 0 Hz and 100 kHz, as expected, since their z-plane locations are the same. Also, the gain is the same at both 10 kHz and 90 kHz since their z-plane locations are complex conjugates of each other. Finally, note that the minimum gain for this transfer function occurs at $z = -1$, or equivalently, $f_s/2 = 50$ kHz.

EXAMPLE 9.5

Consider a first-order $H(z)$ having its zero at ∞ and its pole on the real axis, where $0 < a < 1$. Mathematically, the transfer function is represented by $H(z) = b/(z-a)$. Find the value of ω, where the magnitude of $H(z)$ is 3 dB lower than its dc value. What is the 3-dB value of ω for a real pole at 0.8? What fraction of the sampling rate, f_s, does it correspond to?

Solution

Consider the pole-zero plot shown in Fig. 9.11, where the zero is not shown since it is at ∞. Since the zero is at ∞, we need be concerned only with the magnitude of the denominator of $H(z)$ and when it becomes $\sqrt{2}$ larger than its dc value. The magnitude of the denominator of $H(z)$ for $z = 1$ (i.e., dc) is shown as the vector I_1 in Fig. 9.11. This vector changes in size as z goes around the unit circle and is shown as I_2 when it becomes $\sqrt{2}$ larger than I_1. Thus we can write,

$$I_2 = e^{j\omega} - a \equiv \sqrt{2}(1-a) \tag{9.20}$$

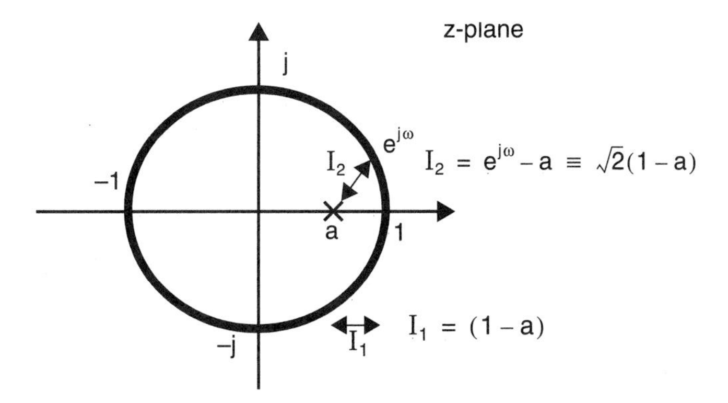

Fig. 9.11 Pole-zero plot used to determine the 3-dB frequency of a first-order, discrete-time filter.

Writing $e^{j\omega} = \cos(\omega) + j\sin(\omega)$, we have

$$|(\cos(\omega) - a) + j\sin(\omega)|^2 = 2(1-a)^2 \tag{9.21}$$

$$\cos^2(\omega) - (2a)\cos(\omega) + a^2 + \sin^2(\omega) = 2(1-a)^2 \tag{9.22}$$

$$1 - (2a)\cos(\omega) + a^2 = 2(1 - 2a + a^2) \tag{9.23}$$

which is rearranged to give the final result.

$$\omega = \cos^{-1}\left(2 - \frac{a}{2} - \frac{1}{2a}\right) \tag{9.24}$$

For $a = 0.8$, $\omega = 0.2241$ rad, or 12.84 degrees. Such a location on the unit circle corresponds to $0.2241/(2\pi)$ times f_s or, equivalently, $f_s/28.04$.

Stability of Discrete-Time Filters

To realize rational polynomials in z, discrete-time filters use delay elements (i.e., z^{-1} building blocks) much the same way that analog filters can be formed using integrators (i.e., s^{-1} building blocks). The result is that finite difference equations represent discrete-time filters rather than the differential equations used to describe continuous-time filters.

Consider the block diagram of a first-order, discrete-time filter shown in Fig. 9.12. A finite difference equation describing this block diagram can be written as

$$y(n+1) = bx(n) + ay(n) \tag{9.25}$$

In the z-domain, this equation is written as

$$zY(z) = bX(z) + aY(z) \tag{9.26}$$

where the z-domain property of delayed signals is used. We define the transfer function of this system to be $H(z)$ given by

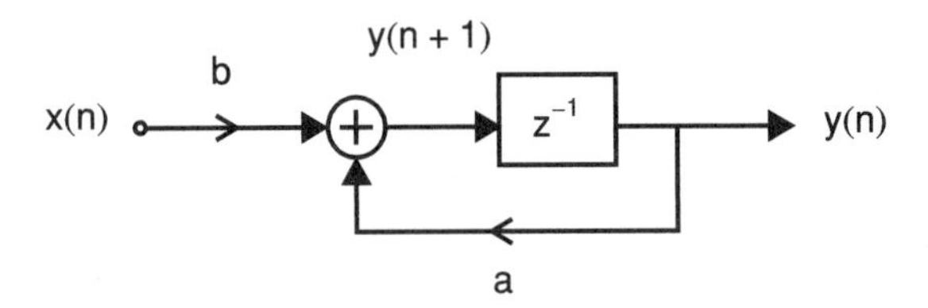

Fig. 9.12 A first-order, discrete-time filter.

$$H(z) \equiv \frac{Y(z)}{X(z)} = \frac{b}{z-a} \tag{9.27}$$

which has a pole on the real axis at $z = a$.

To test for stability, we let the input, $x(n)$, be an impulse signal (i.e., 1 for $n = 0$ and 0 otherwise), which gives the following output signal, according to (9.25),

$$y(0) = k$$

where k is some arbitrary initial state value for y.

$$y(1) = b + ak$$
$$y(2) = ab + a^2k$$
$$y(3) = a^2b + a^3k$$
$$y(4) = a^3b + a^4k$$
$$\vdots$$

More concisely, the response, $h(n)$, is given by

$$h(n) = \begin{cases} 0 & (n<0) \\ k & (n=0) \\ (a^{n-1}b + a^nk) & (n \geq 1) \end{cases} \tag{9.28}$$

Clearly, this response remains bounded only when $|a| \leq 1$ for this first-order filter and is unbounded otherwise.

Although this stability result is shown only for first-order systems, in general, an arbitrary, linear, time-invariant, discrete-time filter, $H(z)$, is stable if and only if all its poles are located within the unit circle. In other words, if z_{pi} are the poles, then $|z_{pi}| < 1$ for all i. Locating some poles on the unit circle is similar to poles being on the imaginary $j\omega$-axis for continuous-time systems. For example, in the preceding first-order example, if $a = 1$, the pole is at $z = 1$, and the system is marginally stable (in fact, it is a discrete-time integrator). If we let $a = -1$, this places the pole at $z = -1$, and one can show that the system oscillates at $f_s/2$, as expected.

IIR and FIR Filters

Infinite-impulse-response (IIR) filters are discrete-time filters that, when excited by an impulse, their outputs remain nonzero, assuming infinite precision arithmetic. For example, the filter given in (9.27) is an IIR filter (for a not equal to zero) since, although its impulse response decays to zero (as all stable filters should), it remains nonzero for a finite value of n.

Finite-impulse-response (FIR) filters are discrete-time filters that, when excited by an impulse, their outputs go precisely to zero (and remain zero) after a finite value of n. As an example of an FIR filter, consider the following filter,

$$y(n) = \frac{1}{3}[x(n) + x(n-1) + x(n-2)] \tag{9.29}$$

Defining the transfer function for this filter to be $H(z)$, we can easily show that

$$H(z) = \frac{1}{3}\sum_{i=0}^{2} z^{-i} \tag{9.30}$$

This filter is essentially a running average filter since its output is equal to the average value of its input over the last three samples. Applying an impulse signal to this filter results in an output that is nonzero for only three samples and, therefore, this is an FIR type filter. Note that this FIR filter has poles, but they all occur at $z = 0$.

Some advantages of FIR filters are that stability is never an issue (they are always stable) and exact linear phase filters can be realized (a topic beyond the scope of this chapter). However, for many specifications, an IIR filter can meet the same specifications as an FIR filter, but with a much lower order, particularly in narrowband filters in which the poles of an IIR filter are placed close to the unit circle (i.e., has a slowly decaying impulse response).

Bilinear Transform

With modern filter design software, desired discrete-time transfer functions that meet specifications can be obtained entirely within the discrete-time domain. However, another approach draws on the wealth of knowledge of continuous-time transfer-function approximation and uses the bilinear transform. Assuming that $H_c(p)$ is a continuous-time transfer function (where p is the complex variable equal to $\sigma_p + j\Omega$), the bilinear transform is defined to be given by[1]

$$p = \frac{z-1}{z+1} \tag{9.31}$$

The inverse transformation is given by

$$z = \frac{1+p}{1-p} \tag{9.32}$$

1. It should be noted here that in many other textbooks, the bilinear transform is defined as $s = (2/T)[(z-1)/(z+1)]$ where T is the sampling period. Here, we have normalized T to 2 since we use the bilinear transform only as a temporary transformation to a continuous-time equivalent, and then we inverse transform the result back to discrete time. Thus, the value of T can be chosen arbitrarily as long as the same value is used in each transformation.

A couple of points of interest about this bilinear transform are that the z-plane locations of 1 and –1 (i.e., dc and $f_s/2$) are mapped to p-plane locations of 0 and ∞, respectively. However, with a little analysis, we will see that this bilinear transformation also maps the unit circle, $z = e^{j\omega}$, in the z-plane to the entire $j\Omega$-axis in the p-plane. To see this mapping, we substitute $z = e^{j\omega}$ into (9.31),

$$p = \frac{e^{j\omega}-1}{e^{j\omega}+1} = \frac{e^{j(\omega/2)}\left(e^{j(\omega/2)} - e^{-j(\omega/2)}\right)}{e^{j(\omega/2)}\left(e^{j(\omega/2)} + e^{-j(\omega/2)}\right)} \tag{9.33}$$

$$= \frac{2j\sin(\omega/2)}{2\cos(\omega/2)} = j\tan(\omega/2)$$

Thus, we see that points on the unit circle in the z-plane are mapped to locations on the $j\Omega$-axis in the p-plane, and we have

$$\Omega = \tan(\omega/2) \tag{9.34}$$

As a check, note that the z-plane locations of 1 and –1, which correspond to ω equal to 0 and π, respectively, map to Ω equal to 0 and ∞.

One way to use this transform is to design a continuous-time transfer function, $H_c(p)$, and choose the discrete-time transfer function, $H(z)$, such that

$$H(z) \equiv H_c[(z-1)/(z+1)] \tag{9.35}$$

With such an arrangement, one can show that,

$$H(e^{j\omega}) = H_c[j\tan(\omega/2)] \tag{9.36}$$

and so the response of $H(z)$ is shown to be equal to the response of $H_c(p)$, except with a frequency warping according to (9.34). Note that the order of $H(z)$ equals that of $H_c(p)$ since, according to (9.35), each p term is replaced by another first-order function.

EXAMPLE 9.6

Using the bilinear transform, find the 3-dB frequency of a first-order discrete-time filter, $H(z)$, that has a pole at 0.8 and a zero at –1.

Solution

Using (9.31), we see that the z-plane pole of 0.8 and zero of –1 are mapped to a pole at –0.11111 and to a zero at ∞ in the p-plane, respectively. Such a continuous-time filter has a 3-dB frequency at $\Omega = 0.11111$ rad/s. Using (9.34), we find that the equivalent 3-dB frequency value in the z-plane is given by $\omega = 0.2213$ rad/sample, or $f_s/28.4$.

Note that this result is very close to that in Example 9.5. However, the two results are not in exact agreement since a zero at –1 is also present in this example.

EXAMPLE 9.7

Using the bilinear transform, find a first-order $H(z)$ that has a 3-dB frequency at $f_s/20$, a zero at –1, and a dc gain of 1.

Solution

Using (9.34), the frequency value, $f_s/20$, or equivalently, $\omega = (2\pi)/20 = 0.314159$ is mapped to $\Omega = 0.1584$ rad/s. Thus, $H_c(p)$ should have a 3-dB frequency value of 0.1584 rad/s. Such a 3-dB frequency value is obtained by having a p-plane zero equal to ∞ and a pole equal to –0.1584. Transforming these continuous-time pole and zero back into the z-plane using (9.32) results in a z-plane zero at –1 and a pole at 0.7265. Therefore, $H(z)$ appears as

$$H(z) = \frac{k(z+1)}{z-0.7265}$$

The constant k can be determined by setting the dc gain to 1, or equivalently, $|H(1)| = 1$, which results in $k = 0.1368$.

9.6 SAMPLE-AND-HOLD RESPONSE

In this section, we look at the frequency response that occurs when we change a discrete-time signal back into an analog signal with the use of a sample-and-hold circuit. Note that here we plot a frequency response for all frequencies (as opposed to only up to $f_s/2$) since the output signal is a continuous-time signal rather than a discrete-time one.

A sample-and-hold signal, $x_{sh}(t)$, is related to its sampled signal by the mathematical relationship

$$x_{sh}(t) = \sum_{n=-\infty}^{\infty} x_c(nT)[\vartheta(t-nT) - \vartheta(t-nT-T)] \tag{9.37}$$

Note that, once again, $x_{sh}(t)$ is well defined for all time, and thus the Laplace transform can be found to be equal to

$$\begin{aligned} X_{sh}(s) &= \frac{1-e^{-sT}}{s} \sum_{n=-\infty}^{\infty} x_c(nT)e^{-snT} \\ &= \frac{1-e^{sT}}{s} X_s(s) \end{aligned} \tag{9.38}$$

This result implies that the hold transfer function, $H_{sh}(s)$, is equal to

$$H_{sh}(s) = \frac{1-e^{-sT}}{s} \tag{9.39}$$

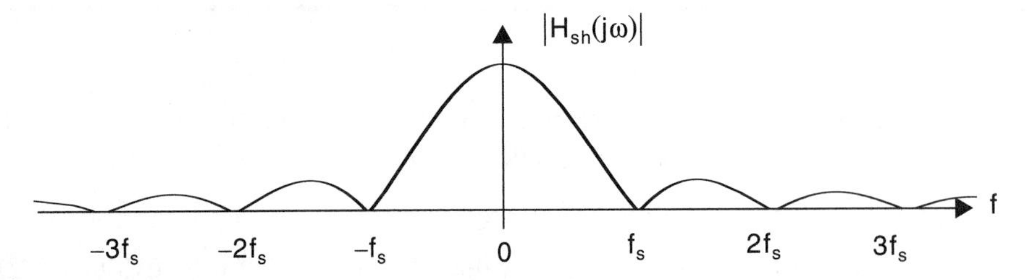

Fig. 9.13 Sample-and-hold response (also called the sinc response).

It should be mentioned here that this transfer function is usually referred to as the *sample-and-hold response* although, in fact, it only accounts for the hold portion.

The spectrum for $H_{sh}(s)$ is found by substituting $s = j\omega$ into (9.39), resulting in

$$H_{sh}(j\omega) = \frac{1 - e^{-j\omega T}}{j\omega} = T \times e^{-j\omega T/2} \times \frac{\sin(\omega T/2)}{(\omega T/2)} \tag{9.40}$$

The magnitude of this response is given by

$$|H_{sh}(j\omega)| = T\frac{|\sin(\omega T/2)|}{|\omega T/2|}$$

or

$$|H_{sh}(f)| = T\frac{|\sin(\pi f/f_s)|}{|(\pi f)/f_s|} \tag{9.41}$$

and is often referred to as the $(\sin x)/x$ or *sinc response*. This magnitude response is illustrated in Fig. 9.13.

Multiplying this sinc response by $X_s(f)$ in Fig. 9.2 confirms the spectral relationship for $X_{sh}(f)$. It should be noted here that this frequency shaping of a sample and hold occurs only for a continuous-time signal. Specifically, although the sample and hold before the A/D converter shown in Fig. 9.1(*b*) would result in $x_{sh}(t)$ having smaller images at higher frequencies (due to the sinc response), the images of $x(n)$ are all of the same height (i.e., they are not multiplied by the sinc response) since it is a discrete-time signal. In other words, a sample and hold before an A/D converter simply allows the converter to have a constant input value during one conversion and does not aid in any anti-aliasing requirement.

EXAMPLE 9.8

Consider the discrete-time signal processing system shown in Fig. 9.1(*b*), where a sample (and clock rate) of 50 kHz is used and the digital filter has the response

$$H(z) = \frac{0.2}{z - 0.8}$$

For a 10-kHz input sinusoidal signal of 1 V rms, find the magnitude of the output signal, $y_{sh}(t)$, at 10 kHz, and at the images 40 kHz and 60 kHz. Assume the quantization effects of the A/D and D/A converters are so small that they can be ignored. Also assume the converters complement each other such that a 1-V dc signal into the A/D converter results in a 1-V dc signal from the D/A converter if the converters are directly connected together.

Solution

First, the magnitude of $H(z)$ is found for a 10-kHz signal when the sampling rate is 50 kHz by noting that

$$e^{j(2\pi \times 10\text{ kHz}/50\text{ kHz})} = e^{j0.4\pi} = \cos(0.4\pi) + j\sin(0.4\pi)$$
$$= 0.309017 + j0.951057$$

and therefore the gain of $H(z)$ for a 10-kHz signal is found to be given by,

$$|H(e^{j0.4\pi})| = \frac{0.2}{|(0.309017 - 0.8) + j0.951057|} = 0.186864$$

To determine the magnitude of $y_{sh}(t)$ at various frequencies, we need only multiply the filter's gain by the sinc response shown in Fig. 9.13. Specifically, the magnitude of the sample-and-hold response for a 10-kHz signal when using a 50-kHz clock is equal to

$$|H_{sh}(10\text{ kHz})| = T\frac{\sin(\pi 10/50)}{(\pi 10/50)} = 0.9355T$$

and therefore the magnitude of $y_{sh}(t)$ at 10 kHz is equal to $0.18686 \times 0.9355 = 175\text{ mV}_{rms}$ (note that the T term cancels with a similar T term due to the creation of a discrete-time signal—see (9.13) in Section 9.2). Similarly, the magnitude of $H_{sh}(f)$ at 40 kHz and 60 kHz is 0.2399 and 0.1559, respectively, resulting in the magnitude of $y_{sh}(t)$ at 40 kHz and 60 kHz to be 43.7 and 29.1 mV_{rms}, respectively.

9.7 REFERENCES

R. Gregorian and G. C. Temes, *Analog MOS Integrated Circuits for Signal Processing*. John Wiley & Sons, New York, 1986.

A. V. Oppenheim and R. W. Schafer, *Discrete-Time Signal Processing*. Prentice Hall, Englewood Cliffs, New Jersey, 1989.

J. G. Proakis and D. G. Manolakis, *Digital Signal Processing: Principles, Algorithms, and Applications*. Macmillan, New York, 1992.

P. P. Vaidyanathan, *Multirate Systems and Filter Banks*. Prentice Hall, Englewood Cliffs, New Jersey, 1993.

9.8 PROBLEMS

9.1 Show that if $x(n) \leftrightarrow X(z)$, then $x(n-k) \leftrightarrow z^{-k}X(z)$.

9.2 What is the maximum amount of downsampling that can be applied to $x_{s2}(n)$ in Example 9.1 before aliasing occurs?

9.3 Is downsampling a time-invariant operation? If so, explain. If not, give a simple counter example. Is downsampling a linear operation?

9.4 Repeat Example 9.3, but upsample by 8 (i.e., insert 7 zeros between samples).

9.5 Upsampling and downsampling are often used for converting between two sample rates. Give a block diagram showing how one might convert 0 – 20 kHz band-limited audio signal from a sampling rate of 50 kHz to 40 kHz.

9.6 Find the location of the poles and zeros of the transfer function given by $H(z)$ in (9.19).

9.7 Find the location of the pole required in a first-order filter that has its zero at ∞ such that its 3-dB frequency is at $f_s/10$.

9.8 Sketch the magnitude response (in dB) of the following transfer function, $H(z)$, by finding the gains at dc and at $\omega = \pi$, as well as the location of the 3-dB frequency.

$$H(z) = \frac{0.05z}{z - 0.95}$$

9.9 Consider the following difference equation, which describes a filter with input $u(n)$ and output $y(n)$.

$$y(n) = u(n) - 0.3y(n-1)$$

Determine $y(n)$ when $u(n)$ is an impulse (i.e., when the impulse is 1 at $n = 1$ and 0 otherwise) assuming $y(0) = 0$.

9.10 Find a first-order $H(z)$ that has a 3-dB frequency at $f_s/30$, a zero at –1, and a dc gain of 1. Sketch the magnitude response from $\omega = 0$ to $\omega = 2\pi$.

9.11 Using a bilinear transform, find an $H(z)$ to realize a second-order transfer function with a dc gain of one, a maximally-flat passband (i.e., $Q = 1/\sqrt{2}$ in the continuous-time domain), and a –3-dB frequency at 1 kHz when the sampling rate is 100 kHz.

9.12 Find the output sequence of a discrete-time integrator, $H(z) = 1/(z-1)$, when its input is a sinusoidal waveform of peak height 1 at $f_s/4$ and its initial state is zero. In other words, the input waveform is the series of numbers $\{1, 0, -1, 0, 1, 0, -1, \ldots\}$. Using a frequency analysis, what is the magnitude and phase response for $H(z)$ at $f_s/4$? Check the frequency domain result with the time domain result.

9.13 A 100-Hz sinusoidal signal, $x(n)$, is digitally created using a 400-Hz sampling rate producing a sample-and-hold signal, $x_{sh}(t)$. Before any smoothing filter is applied and assuming the magnitude of $x_{sh}(t)$ at 100 Hz is 1 V_{rms}, what is the magnitude of $x_{sh}(t)$ at 300 Hz? Also, find the magnitude of $x_{sh}(t)$ at 500 Hz and at 700 Hz.

9.14 Repeat Problem 9.13, but assume the 100-Hz signal is digitally created using a 10-kHz signal, and find the magnitude of $x_{sh}(t)$ at 9.9 kHz and at 10.1 kHz.

9.15 Repeat Example 9.8, but assume an input signal frequency of 100 Hz and find the magnitude of the output signal, $y_{sh}(t)$, at 100 Hz, 99.9 kHz, and 100.1 kHz.

9.16 A 0–5-kHz voiceband microphone is modelled as having a low-pass, first-order frequency response with a pole at 10 kHz. If the microphone output is sampled at 100 kHz, how much attenuation is required in the anti-aliasing filter for 80-dB noise rejection around 100 kHz?

CHAPTER

10 Switched-Capacitor Circuits

Perhaps the most popular approach for realizing analog signal processing in MOS (or BiCMOS) integrated circuits is through the use of switched-capacitor circuits [Poschenrieder, 1966; Fried, 1972; Caves, 1977; Hosticka, 1977; Young, 1977]. A switched-capacitor circuit operates as a discrete-time signal processor (although without the use of A/D or D/A converters). As a result, these circuits are most easily analyzed with the use of z-transform techniques and typically require anti-aliasing and smoothing filters.

As a filtering technique, switched-capacitor circuits have become extremely popular due to their accurate frequency response as well as good linearity and dynamic range. Accurate discrete-time frequency responses are obtained since filter coefficients are determined by capacitance ratios which can be set quite precisely in an integrated circuit (on the order of 0.1 percent). Such an accuracy is orders of magnitude better than that which occurs for integrated RC time constants (which can vary by as much as 20 percent). Once the coefficients of a switched-capacitor discrete-time filter are accurately determined, its overall frequency response remains a function of the clock (or sampling) frequency. Fortunately, clock frequencies can also be set very precisely through the use of a crystal oscillator.

In addition to creating filtering functions, switched-capacitor circuit techniques can be used to realize a variety of other signal-processing blocks such as gain-stages, voltage-controlled oscillators, and modulators. This chapter will introduce the basic principles of switched-capacitor filters as well as some other nonfiltering functions realized using switched-capacitor circuits.

10.1 BASIC BUILDING BLOCKS

A switched-capacitor circuit is realized with the use of some basic building blocks such as opamps, capacitors, switches, and nonoverlapping clocks. Brief descriptions of each of these blocks and their important nonidealities with respect to switched-capacitor circuits are given in this section.

Opamps

The basic principles of switched-capacitor circuits can be well understood assuming ideal opamps. However, some important opamp nonidealities in practical switched-capacitor circuits are dc gain, unity-gain frequency and phase-margin, slew-rate, and dc offset. Of less concern is opamp input impedance as it is essentially capacitive assuming MOSFET input stages are used.[1]

The *dc gain* of opamps in a MOS technology intended for switched-capacitor circuits is typically on the order of 40 to 80 dB. Low dc gains affect the coefficient accuracy of the discrete-time transfer function of a switched-capacitor filter.

The *unity-gain frequency and phase margin* of an opamp gives an indication of the small signal settling behavior of an opamp. A general rule of thumb is that the clock frequency should be at least five times lower in frequency than the unity-gain frequency of the opamp assuming little slew-rate behavior occurs and the phase margin is greater than 70 degrees. Modern SC circuits are often realized using high-frequency single-stage opamps having very large output impedances (on the order of 100 kΩ or much larger). Since the loads of these opamps are purely capacitive (never resistive), their dc gains remain high despite the lack of an output buffer stage. It should be noted here that their unity-gain frequency and phase margin are determined by the load capacitance, which also serves as the compensation capacitor. Thus, in these single-stage opamps, doubling the load capacitance would halve the unity-gain frequency and *improve* the phase margin.

The finite *slew rate* of an opamp can limit the upper clock rate in a switched-capacitor circuit as these circuits rely on charge being quickly transferred from one capacitor to another. Thus, at the instance of charge transfer, it is not uncommon for opamps to slew-rate limit.

A nonzero *dc offset* can result in a high output dc offset for the circuit depending on the topology chosen. Fortunately, a technique known as correlated double sampling can significantly reduce this output offset and at the same time reduce low frequency opamp input noise (known as 1/f noise).

Capacitors

A highly linear capacitance in an integrated circuit is typically constructed between two polysilicon layers, as shown in Fig. 10.1(*a*).[2] These capacitors are known as double-poly capacitors. The thin oxide is an insulating layer separating the two relatively conductive polysilicon layers, while the thick oxide is also an insulating layer but of much thicker width. The desired capacitance, C_1, is formed as the intersection of area between the two polysilicon layers, poly1 and poly2. However, since the substrate below the poly2 layer is an ac ground (the substrate is connected to one of the power supplies or analog ground), there also exists a substantial parasitic capacitance, C_{p2}, which may be as large as 20 percent of C_1. This large parasitic capacitance, C_{p2}, is known as the bottom plate capacitance. A top plate capacitance, C_{p1},

---

1. BJT input stages are rarely used as they would result in charge leakage due to their nonzero input bias current.

2. It is also possible to realize capacitors between other layers such as polysilicon and heavily-doped silicon, or between two metal layers.

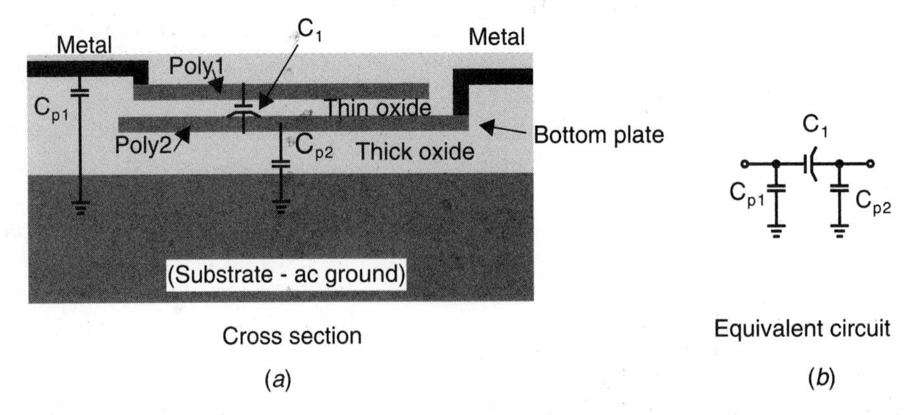

Fig. 10.1 A double-poly capacitor, C_1. Note that a large parasitic capacitance, C_{p2}, to ground exists on the bottom plate of C_1 while a smaller parasitic capacitance, C_{p1}, exists on the top plate.

also exists due primarily to the interconnect capacitance, but it is typically much smaller (on the order of 1 to 5 percent of C_1). In summary, the equivalent model for a single integrated capacitor is three capacitors, as shown in Fig. 10.1(*b*), where the bottom plate is often explicitly shown to indicate that a larger parasitic capacitance occurs on that node.

Switches

The requirements for switches used in switched-capacitor circuits are that they have a very high off resistance (so little charge leakage occurs), a relatively low on resistance (so that the circuit can settle in less than half the clock period), and introduce no offset voltage when turned on (as does a bipolar switch whose on voltage equals $V_{CE(sat)}$). The use of MOSFET transistors as switches satisfies these requirements, as MOSFET switches have off resistances in the $G\Omega$ range,[3] have no offset on voltages, and have on resistances in the 100 Ω to 5 $k\Omega$ range depending on transistor sizing.

The symbol for a switch and some MOSFET circuits that realize a switch are shown in Fig. 10.2. Here, the signal ϕ is one of two logic levels typically corresponding to the maximum and minimum power supply levels. As a convention, when the clock signal ϕ is high, the switch is assumed to be on (i.e., shorted). Although a switch can be implemented using a single transistor, as shown in Fig. 10.2(*b*) and Fig. 10.2(*c*), the signal range of the switch is reduced. For example, consider the ideal n-channel transistor switch shown in Fig. 10.2(*b*) being used in a circuit having power supplies of 0 to 5 V and a threshold voltage of $V_{tn} = 0.8$ V. The switch can

3. Leakage current is typically dominated by reverse biased diodes rather than the MOSFET channel which is assumed off.

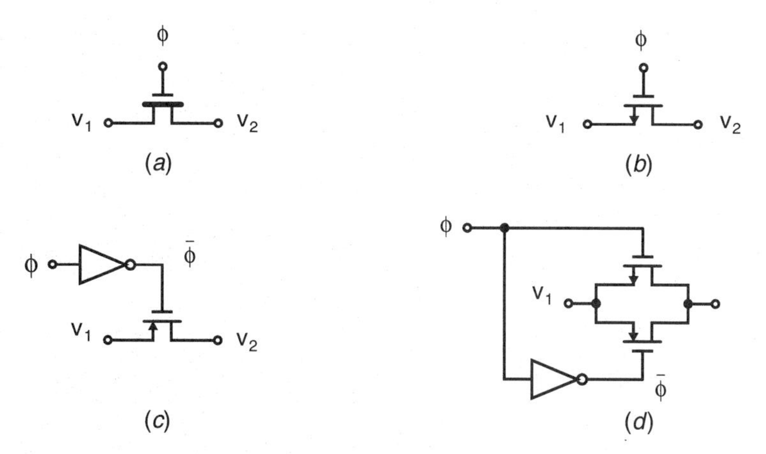

Fig. 10.2 Switch symbol and some transistor circuits: (*a*) symbol, (*b*) n-channel switch, (*c*) p-channel switch, (*d*) transmission gate.

always be turned off irrespective of the input voltage by putting 0 V on its gate. However, when the switch is on, its gate voltage will be V_{DD} or 5 V; in this case, the switch can only equalize the two voltages if V_1 and/or V_2 are lower than $V_{DD} - V_{tn}$ or around 4.0 V taking the body effect into account. Thus, the signal range for this switch is limited from 0 to 4.0 V. A similar argument can be made for the p-channel switch shown in Fig. 10.2(*c*) except that its signal range would be limited from 1.0 to 5 V. While this signal range is acceptable in many situations, the full signal range of 0 to 5 V can be achieved using two transistors in parallel, as shown in Fig. 10.2(*d*). This combination is commonly called either a CMOS switch (as opposed to an NMOS switch of Fig. 10.2(*a*)) or a CMOS transmission gate.

Some of the important nonideal switch effects in switched-capacitor circuits are the nonlinear capacitance on each side of the switch, channel charge injection, and capacitive coupling from the logic signal, ϕ, to each side of the switch. Nonlinear capacitance effects are alleviated through the use of parasitic insensitive structures, as discussed in Section 10.2, while channel charge injection is reduced by careful switch timing, as seen in Section 10.5.

Nonoverlapping Clocks

At least one pair of nonoverlapping clocks is essential in switched-capacitor circuits. These clocks determine when charge transfers occur and they must be nonoverlapping in order to guarantee charge is not inadvertently lost. As seen in Fig. 10.3(*a*), the term *nonoverlapping clocks* refers to two logic signals running at the same frequency and arranged in such a way that at no time are both signals high. Note that the time axis in Fig. 10.3(*a*) has been normalized with respect to the clock period, T. Such a normalization illustrates the location of the sample numbers of the discrete-time signals that occur in switched-capacitor filters. As a convention, we denote the sampling numbers

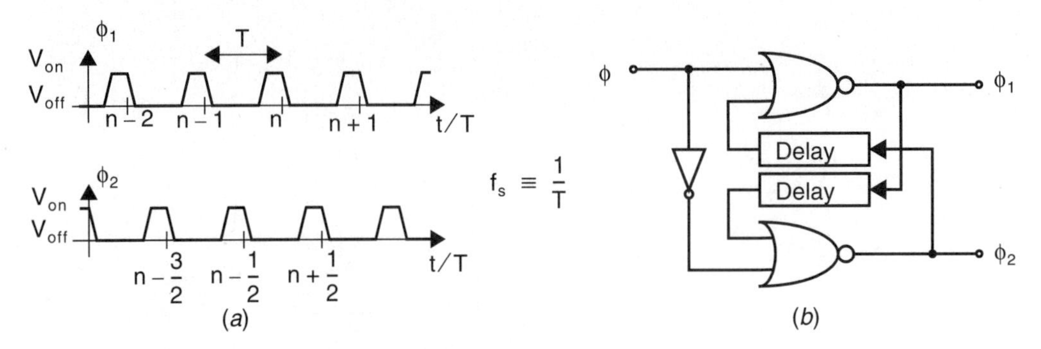

Fig. 10.3 Nonoverlapping clocks. (*a*) Clock signals, ϕ_1 and ϕ_2. (*b*) A possible circuit implementation of nonoverlapping clocks from a single clock.

to be integer values (i.e., $(n-1)$, (n), $(n+1)$, etc.) just before the end of clock phase ϕ_1, while the end of clock phase ϕ_2 is deemed to be $1/2$ sample off the integer values as shown (i.e., $(n-3/2)$, $(n-1/2)$, etc.). However, it should be noted that it is not important that the falling clock edge of ϕ_2 occur precisely one-half a clock period earlier than the falling edge of ϕ_1. In general, the locations of the clock edges of ϕ_1 and ϕ_2 need only be moderately controlled to allow for complete charge settling (assuming separate low-jitter sample and holds are used at the input and output of the overall circuit).

One simple method for generating nonoverlapping clocks is shown in Fig. 10.3(*b*) [Martin, 1981]. Here, delay blocks are used to ensure that the clocks remain nonoverlapping. These delays could be implemented as a cascade of an even number of inverters or perhaps an RC network.

10.2 BASIC OPERATION AND ANALYSIS

Resistor Equivalence of a Switched Capacitor

Consider the switched-capacitor circuit shown in Fig. 10.4(*a*), where V_1 and V_2 are two dc voltage sources. To analyze this circuit's behavior, we analyze the circuit from a charge perspective. Recall that the charge on a capacitor, Q_x, is equal to the capacitance value, C_x, times the voltage across it, V_x. In mathematical terms, we have

$$Q_x = C_x V_x \tag{10.1}$$

Now, since ϕ_1 and ϕ_2 are assumed to be a pair of nonoverlapping clocks, C_1 is charged to V_1 and then V_2 during each clock period. Therefore, the change in charge (in coulombs) over one clock period, ΔQ_1, that is transferred from node V_1 into node V_2 is given by

$$\Delta Q_1 = C_1(V_1 - V_2) \tag{10.2}$$

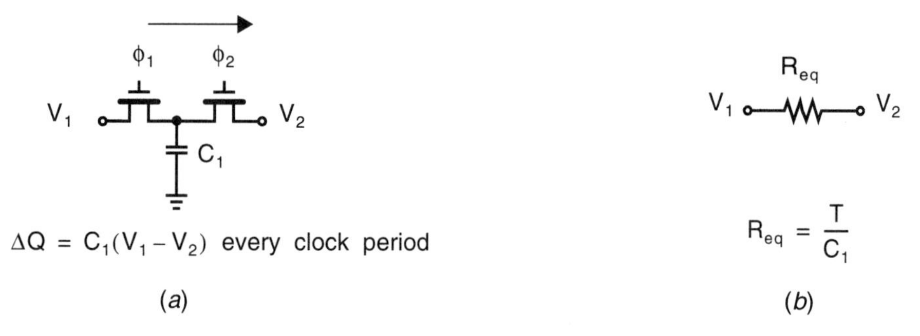

Fig. 10.4 Resistor equivalence of a switched capacitor. (*a*) Switched-capacitor circuit (*b*) Resistor equivalent.

Since this charge transfer is repeated every clock period, we can find the equivalent average current due to this charge transfer by dividing by the clock period. In other words, the average current, I_{avg} (in amps), is equal to the average number of coulombs/second and is given by

$$I_{avg} = \frac{C_1(V_1 - V_2)}{T} \qquad (10.3)$$

where T is the clock period (i.e., the sampling frequency $f_s = 1/T$). Relating (10.3) to an equivalent resistor circuit shown in Fig. 10.4(*b*) where

$$I_{eq} = \frac{V_1 - V_2}{R_{eq}} \qquad (10.4)$$

we see that the average current through the switched-capacitor circuit of Fig. 10.4(*a*) will equal that for the resistor circuit of Fig. 10.4(*b*) if

$$R_{eq} = \frac{T}{C_1} = \frac{1}{C_1 f_s} \qquad (10.5)$$

Note that the above relation makes intuitive sense. For example, as the clock frequency is increased, the same charge is transferred each period but at a faster rate, so the average current is higher. If C_1 is increased, a larger amount of charge is transferred each period, which would also increase the average current. In addition, increasing the average current is equivalent to decreasing the equivalent resistance seen between the two voltage nodes, and therefore the equivalent resistance is inversely related to the capacitance and clock frequency.

Finally, it should be noted that this resistor approximation is mainly useful for looking at the low-frequency behavior of a switched-capacitor circuit. For moderate input frequencies (in relation to the sampling frequency), an accurate discrete-time analysis is required, as discussed in the next subsection. Discrete-time analysis can be used in switched-capacitor circuits because the charge transfers are dependent on values of node voltages at particular instances in time and are ideally independent of transients during nonsampled times.

EXAMPLE 10.1

What is the equivalent resistance of a 5 pF capacitance sampled at a clock frequency of 100 kHz?

Solution

Using (10.5), we have

$$R_{eq} = \frac{1}{(5 \times 10^{-12})(100 \times 10^{3})} = 2\ \text{M}\Omega$$

What is interesting here is that a very large impedance of 2 MΩ can be realized on an integrated circuit through the use of two transistors, a clock, and a relatively small capacitance. Such a large impedance typically requires a large amount of CMOS silicon area if it is realized as a resistor without any special processing fabrication steps.

Parasitic-Sensitive Integrator

An example of one of the first switched-capacitor discrete-time integrators [Hosticka, 1977] is shown in Fig. 10.5. Here, an extra switch is shown near the output voltage, $v_{co}(t)$, to indicate to the reader that the output signal is valid at the end of ϕ_1 in keeping with the sampling-time convention shown in Fig. 10.3(*a*). In other words, another circuit that uses the output of this discrete-time integrator should sample $v_{co}(t)$ on ϕ_1. Note however that no assumptions are made about when $v_{ci}(t)$ changes its value (it could be either ϕ_1 or ϕ_2); it is not important because $v_{ci}(t)$ is sampled at the end of ϕ_1 by the circuit shown.

To analyze this circuit, we again look at the charge behavior and note that a virtual ground appears at the opamp's negative input. Assuming an initial integrator output voltage of $v_{co}(nT - T)$ implies that the charge on C_2 is equal to $C_2 v_{co}(nT - T)$ at time $(nT - T)$. Now also at time $(nT - T)$, ϕ_1 is just turning off (and ϕ_2 is off) so the input signal $v_{ci}(t)$ is sampled, resulting in the charge on C_1 being equal to

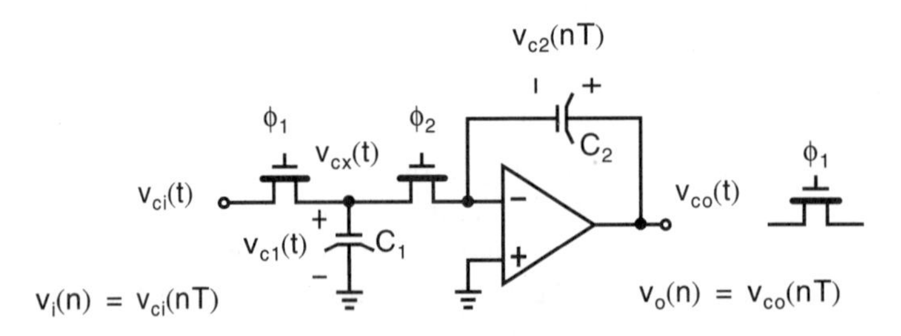

Fig. 10.5 A discrete-time integrator. This structure is sensitive to parasitic capacitances (not shown).

$C_1 v_{ci}(nT - T)$, as shown in Fig. 10.6(*a*). When ϕ_2 goes high, its switch forces C_1 to discharge since it places a virtual ground on the top plate of C_1, as shown in Fig. 10.6(*b*). However, this discharging current must pass through C_2 and hence the charge on C_1 is added to the charge already present on C_2. Note that a positive input voltage will result in a negative voltage across C_2 and therefore a negative output voltage; thus, the integrator is an inverting integrator. At the end of ϕ_2 (once things have settled out), we can write the charge equation

$$C_2 v_{co}(nT - T/2) = C_2 v_{co}(nT - T) - C_1 v_{ci}(nT - T) \tag{10.6}$$

The negative sign in (10.6) reflects the fact that the integrator is an inverting integrator.

Although (10.6) is correct, we would like to find the charge on C_2 at the end of ϕ_1 (or equivalently, nT) as indicated by the switch shown near $v_{co}(t)$ in Fig. 10.5. To find this charge, we note that once ϕ_2 turns off, the charge on C_2 will remain the same during the next ϕ_1, until ϕ_2 turns on again in its next cycle. Therefore, the charge on C_2 at time (nT) at the end of the next ϕ_1 is equal to that at time $(nT - T/2)$, or mathematically,

$$C_2 v_{co}(nT) = C_2 v_{co}(nT - T/2) \tag{10.7}$$

and we can combine (10.6) and (10.7) to write

$$C_2 v_{co}(nT) = C_2 v_{co}(nT - T) - C_1 v_{ci}(nT - T) \tag{10.8}$$

Dividing by C_2 and using the discrete-time variables $v_i(n) = v_{ci}(nT)$ and $v_o(n) = v_{co}(nT)$, we have the following discrete-time relationship for the circuit of Fig. 10.5:

$$v_o(n) = v_o(n-1) - \frac{C_1}{C_2} v_i(n-1) \tag{10.9}$$

Taking the z-transform of (10.9), we also have

$$V_o(z) = z^{-1} V_o(z) - \frac{C_1}{C_2} z^{-1} V_i(z) \tag{10.10}$$

from which we can finally find the transfer function, $H(z)$, to be given by

$$H(z) \equiv \frac{V_o(z)}{V_i(z)} = -\left(\frac{C_1}{C_2}\right) \frac{z^{-1}}{1 - z^{-1}} \tag{10.11}$$

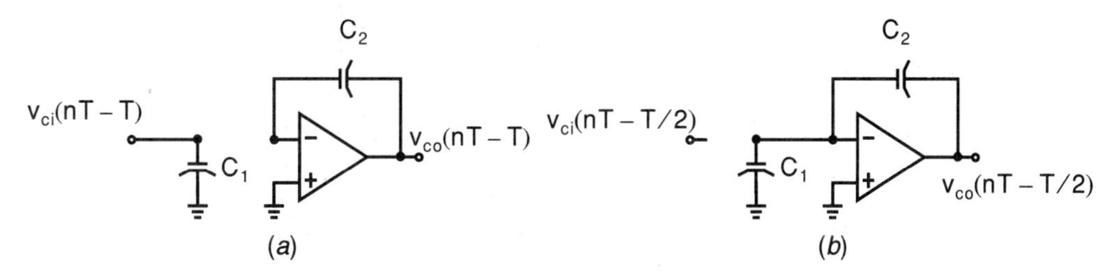

Fig. 10.6 The parasitic-sensitive integrator circuit for the two clock phases: (*a*) ϕ_1, (*b*) ϕ_2.

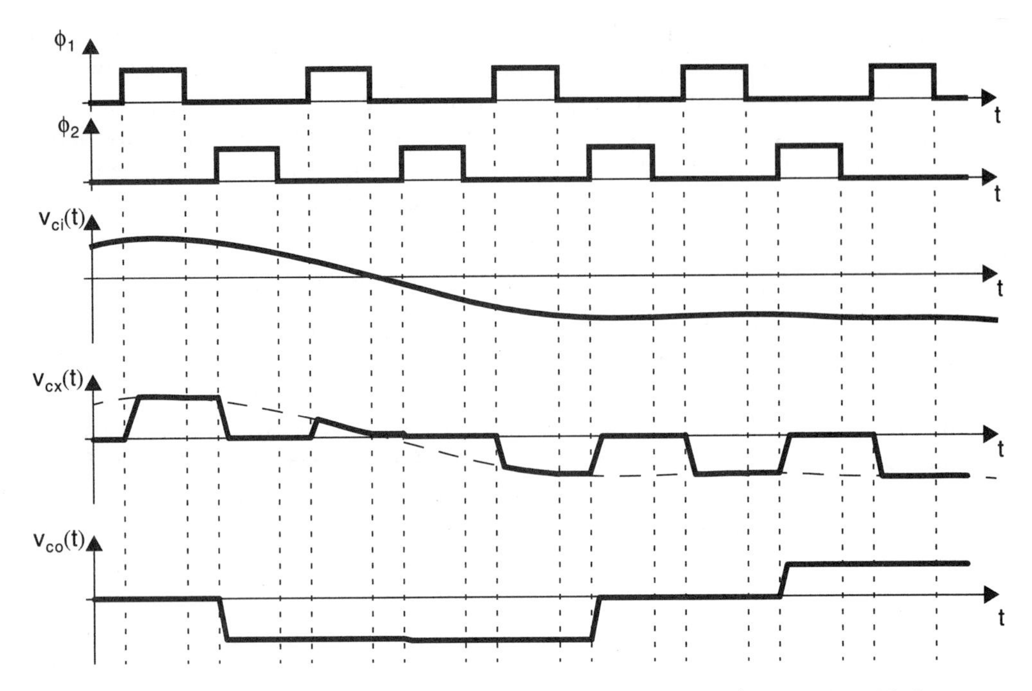

Fig. 10.7 Typical voltage waveforms for the discrete-time integrator shown in Fig. 10.5.

Note that this transfer function realizes its gain coefficient as a ratio of two capacitances, and thus the transfer function can be accurately defined in an integrated circuit. Often, the transfer function would be rewritten to eliminate terms of z having negative powers to give

$$H(z) \equiv \frac{V_o(z)}{V_i(z)} = -\left(\frac{C_1}{C_2}\right)\frac{1}{z-1} \tag{10.12}$$

Some typical voltage waveforms for this discrete-time integrator are shown in Fig. 10.7. Note here that the discrete-time equation derived only represents the relationship of the voltage signals $v_{ci}(t)$ and $v_{co}(t)$ at the time (nT), which occurs just before the end of ϕ_1. It does not say anything about what the voltages are at other times. In actual fact, the voltages are constant at other times (ignoring second-order effects), which is equivalent to taking the sampled voltages and then using them as inputs to a sample and hold for this particular SC integrator.

EXAMPLE 10.2

Show that for frequencies much less than the sampling frequency, equation (10.11) approximates the transfer function of an ideal continuous-time integrator.

Solution

Equation (10.11) can be rewritten as

$$H(z) = -\left(\frac{C_1}{C_2}\right)\frac{z^{-1/2}}{z^{1/2} - z^{-1/2}} \tag{10.13}$$

To find the frequency response, we make use of the fact that

$$z = e^{j\omega T} = \cos(\omega T) + j\sin(\omega T) \tag{10.14}$$

where T, the period, is one over the sampling frequency. Therefore,

$$z^{1/2} = \cos\left(\frac{\omega T}{2}\right) + j\sin\left(\frac{\omega T}{2}\right) \tag{10.15}$$

and

$$z^{-1/2} = \cos\left(\frac{\omega T}{2}\right) - j\sin\left(\frac{\omega T}{2}\right) \tag{10.16}$$

Substituting (10.15) and (10.16) into (10.13) gives

$$H(z) = -\left(\frac{C_1}{C_2}\right)\frac{z^{-1/2}}{j2\sin\left(\frac{\omega T}{2}\right)} \cong -\left(\frac{C_1}{C_2}\right)\frac{z^{-1/2}}{j\omega T} \tag{10.17}$$

for $\omega T \ll 1$. The $z^{-1/2}$ in the numerator represents a simple delay and can be ignored. Thus, we see that the transfer function is approximately that of a continuous-time integrator having a gain constant of

$$K_I \cong -\left(\frac{C_1}{C_2 T}\right) \tag{10.18}$$

which is a function of the integrator capacitor ratio and clock frequency only.

Thus far we have ignored the effect of the parasitic capacitances due to the creation of C_1 and C_2 as well as the nonlinear capacitances associated with the switches. The addition of such parasitic capacitances results in the circuit shown in Fig. 10.8. Here, C_{p1} represents the parasitic capacitance of the top plate of C_1 as well as the nonlinear capacitances associated with the two switches; C_{p2} represents the parasitic capacitance of the bottom plate of C_1; while C_{p3} represents the parasitic capacitance associated with the top plate of C_2, the input capacitance of the opamp, and that of the ϕ_2 switch. Finally, C_{p4} accounts for the bottom plate parasitic capacitance of C_2 as well as any extra capacitance that the opamp output must drive.

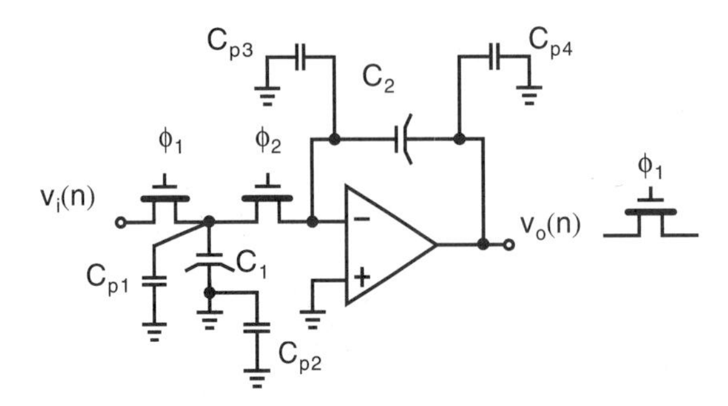

Fig. 10.8 A discrete-time integrator with parasitic capacitances shown.

In accounting for these parasitic capacitances, we can immediately discard the effect of C_{p2} since it always remains connected to ground. Similarly, the effect of C_{p3} on the transfer function is small since it is always connected to the virtual ground of the opamp. Also, C_{p4} is connected to the opamp output; therefore, although it may affect the speed of the opamp, it would not affect the final settling point of the opamp output. Finally, we note that C_{p1} is in parallel with C_1 and therefore the transfer function of this discrete-time integrator including the effects of parasitic capacitance is given by

$$H(z) = -\left(\frac{C_1 + C_{p1}}{C_2}\right)\frac{1}{z-1} \tag{10.19}$$

From (10.19), we see that the gain coefficient is related to the parasitic coefficient C_{p1}, which is not well controlled and would be partially nonlinear due to the input capacitances of the switches and top-plate parasitic capacitance. To overcome this deficiency, ingenious circuits known as parasitic-insensitive structures were developed [Martin, 1978; Allstot, 1978; Jacobs, 1978].

Parasitic-Insensitive Integrators

The parasitic-insensitive integrator was a critical development that allowed the realization of high-accuracy integrated circuits. In the first integrated integrator-based filters, the noninverting integrators (Fig. 10.9) were parasitic insensitive, whereas the inverting integrators (Fig. 10.8) were parasitic sensitive [Allstot, 1978]. The development of the parasitic-insensitive inverting integrator [Martin, 1978; Jacobs, 1978] allowed the complete filter to be insensitive, which greatly decreased second-order errors. To realize a parasitic-insensitive discrete-time integrator, two extra switches are used, as shown in Fig. 10.9. Analyzing this circuit as before, note that when ϕ_1 is on, C_1 is charged to $C_1 v_{ci}(nT - T)$, as shown in Fig. 10.10(*a*). When ϕ_2 turns on, C_1 is effectively "hooked" up backwards, and the discharge now occurs through what was the ground node rather than the input signal side, as shown in Fig. 10.10(*b*). Such

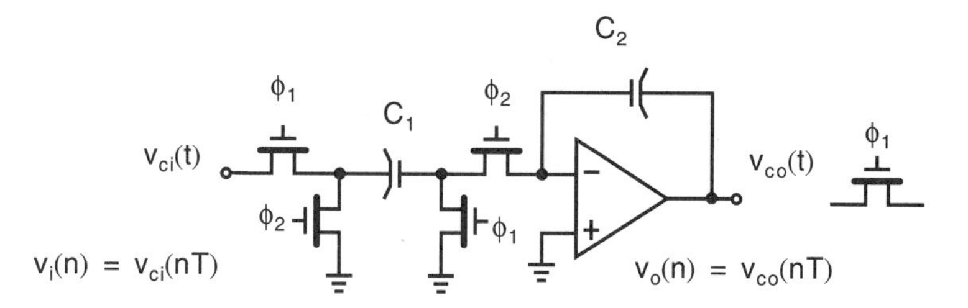

Fig. 10.9 A noninverting delaying discrete-time integrator (not sensitive to parasitic capacitances).

a reverse connection results in $v_{co}(t)$ rising for a positive $v_{ci}(nT - T)$, and therefore the integrator is noninverting. With the rest of the analysis being performed as before, the transfer function, $H(z)$, for this discrete-time integrator is given as

$$H(z) \equiv \frac{V_o(z)}{V_i(z)} = \left(\frac{C_1}{C_2}\right)\frac{z^{-1}}{1 - z^{-1}} \tag{10.20}$$

Note that (10.20) represents a positive discrete-time integrator with a full delay from input to output (due to the z^{-1} in the numerator, which represents a one-period delay) so we refer to the circuit of Fig. 10.9 as a delaying discrete-time integrator. As was done for the stray sensitive inverting integrator of Fig. 10.8, one often rewrites the transfer function to eliminate negative powers of z to give

$$H(z) \equiv \frac{V_o(z)}{V_i(z)} = \left(\frac{C_1}{C_2}\right)\frac{1}{z - 1} \tag{10.21}$$

To investigate the behavior of this noninverting integrator with respect to parasitic capacitances, consider the same circuit drawn with parasitic capacitances, as shown in Fig. 10.11. Here, C_{p3} and C_{p4} do not affect the operation of the circuit as before. In addition, C_{p2} is either connected to ground through the ϕ_1 switch or to virtual ground through the ϕ_2 switch. Therefore, since C_{p2} always remains discharged (after settling), it also does not affect the operation of the circuit. Finally, C_{p1} is continuously being charged to $v_i(n)$ and discharged to ground. However during ϕ_1 on, the fact that C_{p1} is also charged to $v_i(n-1)$ does not affect the charge that is placed on

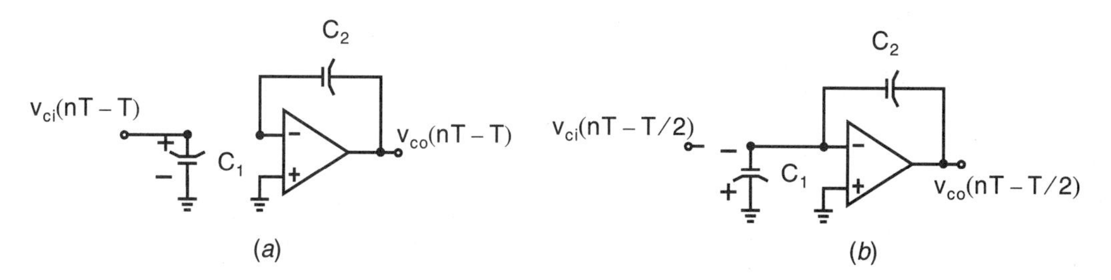

Fig. 10.10 The noninverting discrete-time integrator on the two clock phases: (a) ϕ_1, (b) ϕ_2.

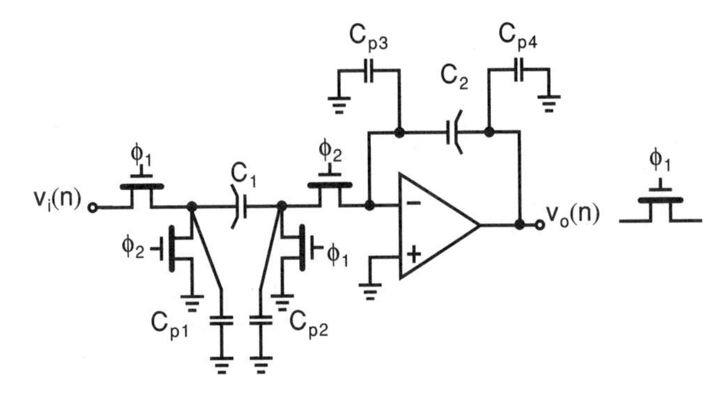

Fig. 10.11 A parasitic-insensitive integrator with parasitic capacitances shown.

C_1. When ϕ_2 is turned on, C_{p1} is discharged through the ϕ_2 switch attached to its node and none of its discharging current passes through C_1 to affect the charge accumulating on C_2. Therefore, it also does not affect the circuit operation. In summary, while the parasitic capacitances may slow down settling time behavior, they do not affect the discrete-time difference equation that occurs in the integrator shown in Fig. 10.9.

To obtain an inverting discrete-time integrator that is also parasitic insensitive, the same circuit as the parasitic-insensitive noninverting integrator can be used, but with the two switch phases on the switches near the opamp input side of C_1 (that is, the top-plate of C_1) interchanged as shown in Fig. 10.12 [Martin, 1978]. With this arrangement, note that the charge on C_2 does not change when ϕ_2 turns on (and ϕ_1 is off). Therefore, we can write

$$C_2 v_{co}(nT - T/2) = C_2 v_{co}(nT - T) \tag{10.22}$$

Also note that C_1 is fully discharged at the end of ϕ_2 on. Now when ϕ_1 is turned on, C_1 is charged up to $v_{ci}(t)$. However, the current needed to charge C_1 passes through C_2, in effect changing the charge on C_2 by that amount. At the end of ϕ_1 on, the charge left on C_2 is equal to its old charge (before ϕ_1 was on) subtracted from the charge needed to charge up C_1 to $v_{ci}(nT)$. Thus, we have the following charge relationship:

$$C_2 v_{co}(nT) = C_2 v_{co}(nT - T/2) - C_1 v_{ci}(nT) \tag{10.23}$$

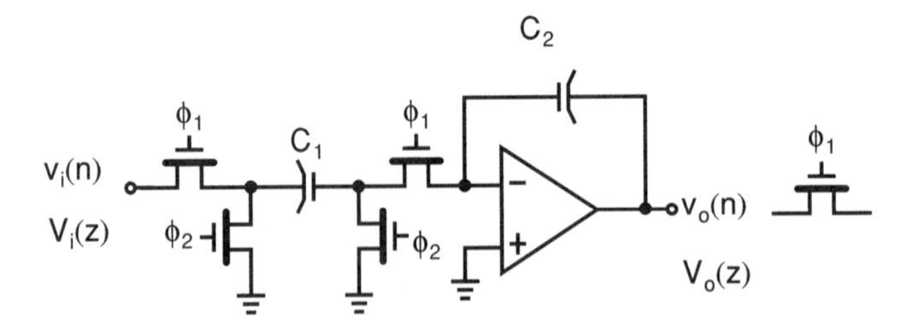

Fig. 10.12 A delay-free discrete-time integrator (not sensitive to parasitic capacitances).

Note that $v_{ci}(nT)$ occurs in the above equation rather than $v_{ci}(nT - T)$, since the charge on C_2 at the end of ϕ_1 is related to the value of $v_{ci}(nT)$ at the same time, and therefore the integrator is considered to be delay free. Substituting in (10.22), dividing by C_2, and switching to discrete-time variables, we have

$$v_o(n) = v_o(n-1) - \frac{C_1}{C_2}v_i(n) \tag{10.24}$$

Taking the z-transform of both sides, the transfer function, $H(z)$, for this delay-free integrator is found to be

$$H(z) \equiv \frac{V_o(z)}{V_i(z)} = -\left(\frac{C_1}{C_2}\right)\frac{1}{1-z^{-1}} \tag{10.25}$$

where the fact that the numerator has a purely real term is indicative of the integrator being delay free. As before, (10.25) is often rewritten as

$$H(z) \equiv \frac{V_o(z)}{V_i(z)} = -\left(\frac{C_1}{C_2}\right)\frac{z}{z-1} \tag{10.26}$$

Finally, note that the above delay-free integrator has a negative gain, while the delaying integrator has a positive gain. While these sign changes in the gain can be quite useful in single-ended switched-capacitor designs, many modern integrated circuits make use of fully differential signals for improved noise performance, and sign changes can be achieved by simply cross-coupling output wires.

Signal-Flow-Graph Analysis

Clearly, applying the charge equations just described would be quite tedious for larger circuits. So instead, a few rules can be developed to create and analyze such circuits. Consider the three-input integrator shown in Fig. 10.13. Using the principle of superposition, it is clear that the input-output relationships for $V_2(z)$ and $V_3(z)$ are given by

$$\frac{V_o(z)}{V_2(z)} = \left(\frac{C_2}{C_A}\right)\left(\frac{z^{-1}}{1-z^{-1}}\right) \tag{10.27}$$

$$\frac{V_o(z)}{V_3(z)} = -\left(\frac{C_3}{C_A}\right)\left(\frac{1}{1-z^{-1}}\right) \tag{10.28}$$

For the input $V_1(z)$, we note that its relationship to $V_o(z)$ is simply an inverting gain stage given by

$$\frac{V_o(z)}{V_1(z)} = -\left(\frac{C_1}{C_A}\right) \tag{10.29}$$

with the output once again being sampled at the end of ϕ_1 as shown.

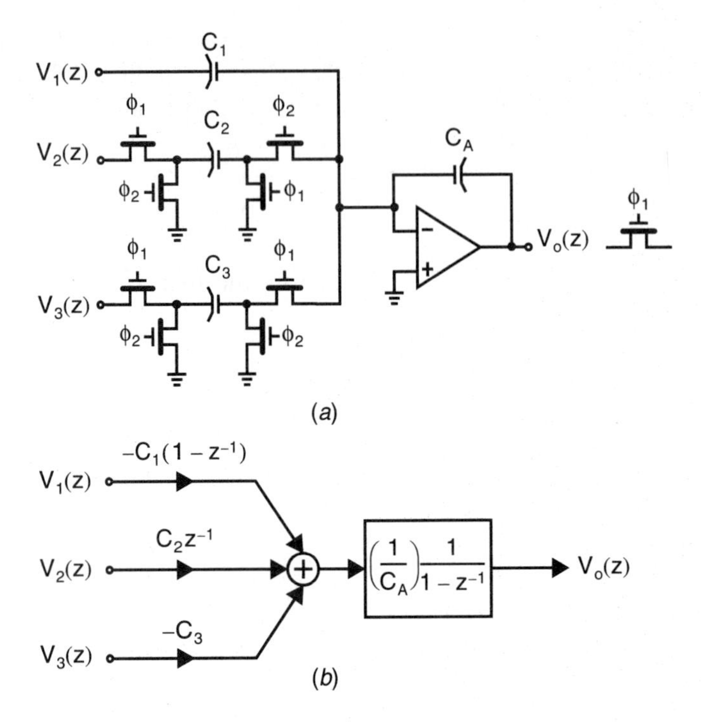

Fig. 10.13 Three-input switched capacitor summing/integrator stage. (*a*) Circuit. (*b*) Equivalent signal flow graph.

Combining (10.27), (10.28), and (10.29), the input-output relationship for the circuit of Fig. 10.13(*a*) is equal to

$$V_o(z) = -\left(\frac{C_1}{C_A}\right)V_1(z) + \left(\frac{C_2}{C_A}\right)\left(\frac{z^{-1}}{1-z^{-1}}\right)V_2(z) - \left(\frac{C_3}{C_A}\right)\left(\frac{1}{1-z^{-1}}\right)V_3(z) \qquad (10.30)$$

Such a relationship can also be represented by the signal flow graph shown in Fig. 10.13(*b*), where the input stages have been separated from the opamp stage. Thus, we see that to obtain an equivalent signal flow graph, the block $(1/C_A)[1/(1-z^{-1})]$ is used to represent the opamp stage, while three different gain factors are used to represent the possible input stages. For a nonswitched capacitor input stage, a gain factor of $-C_1(1-z^{-1})$ is used. If a delaying switched capacitor is used, it is represented by C_2z^{-1}. Finally, for a delay-free switched-capacitor input, a gain factor of $-C_3$ is used. Note, however, that the output, $V_o(z)$, should be sampled on ϕ_1 when the switched-capacitor inputs sample their input signals on ϕ_1.

As before, this transfer function might often be rewritten in powers of z rather than z^{-1}, resulting in

$$V_o(z) = -\left(\frac{C_1}{C_A}\right)V_1(z) + \left(\frac{C_2}{C_A}\right)\left(\frac{1}{z-1}\right)V_2(z) - \left(\frac{C_3}{C_A}\right)\left(\frac{z}{z-1}\right)V_3(z) \qquad (10.31)$$

10.3 FIRST-ORDER FILTERS

We saw on page 398 that at low frequencies, a switched capacitor is equivalent to a resistor. Such an equivalence is often used to derive good switched-capacitor filter structures from their active-RC counterparts. However, although such structures would behave very similarly for low-frequency input signals (when compared to the clock frequency), for signal frequencies moderately close to the clock frequency, the behavior of the switched-capacitor circuit is precisely determined through the use of the z-domain signal-flow-graph approach just discussed.

A general first-order active-RC filter is shown in Fig. 10.14. To obtain a switched-capacitor filter having the same low-frequency behavior, the resistors are replaced with delay-free switched capacitors (delay-free inputs are used since they create negative integration, as do resistive inputs), while the nonswitched capacitor feed in is left unchanged. The resulting first-order switched-capacitor filter and its equivalent signal flow graph are shown in Fig. 10.15. An equation that describes this signal flow graph is

$$C_A(1-z^{-1})V_o(z) = -C_3V_o(z) - C_2V_i(z) - C_1(1-z^{-1})V_i(z) \tag{10.32}$$

from which the transfer function for this first-order filter is found to be

$$\begin{aligned} H(z) \equiv \frac{V_o(z)}{V_i(z)} &= -\frac{\left(\frac{C_1}{C_A}\right)(1-z^{-1}) + \left(\frac{C_2}{C_A}\right)}{1-z^{-1}+\frac{C_3}{C_A}} \\ &= -\frac{\left(\frac{C_1+C_2}{C_A}\right)z - \frac{C_1}{C_A}}{\left(1+\frac{C_3}{C_A}\right)z-1} \end{aligned} \tag{10.33}$$

The pole of (10.33) is found by equating the denominator to zero, which results in a pole location, z_p, given by

$$z_p = \frac{C_A}{C_A + C_3} \tag{10.34}$$

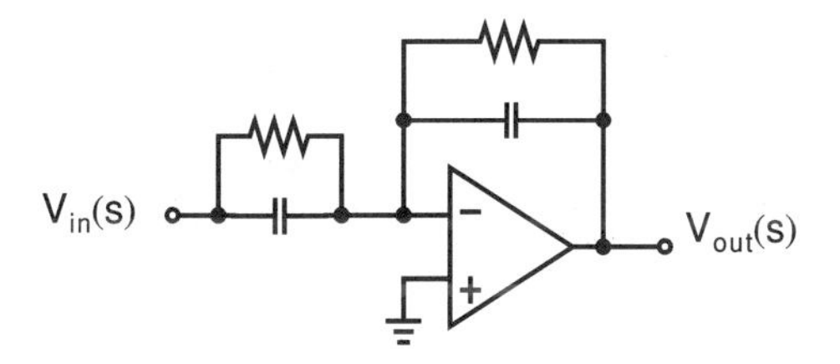

Fig. 10.14 A first-order active-RC filter.

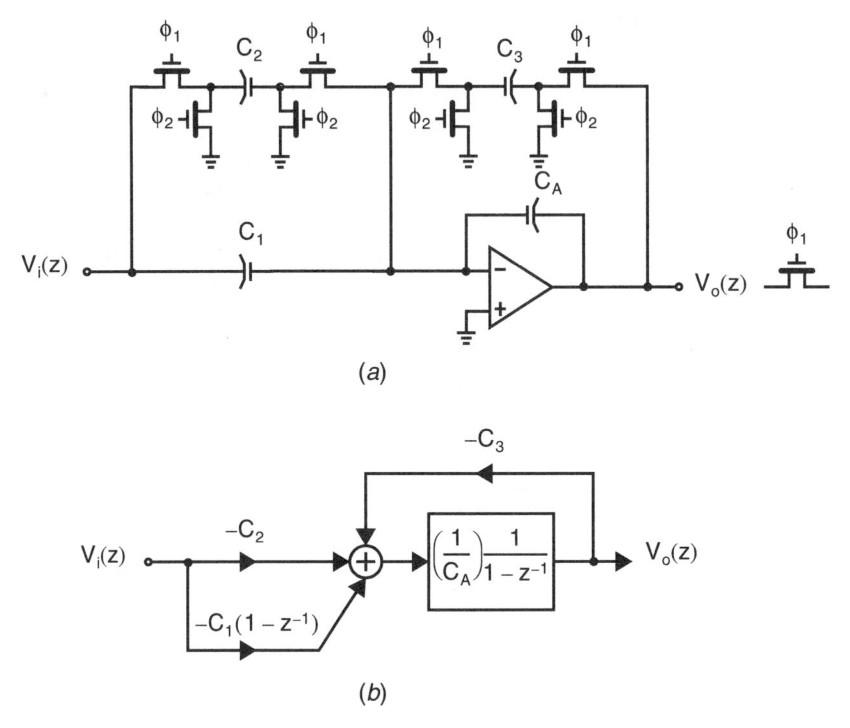

Fig. 10.15 A first-order switched-capacitor filter: (*a*) circuit (no switch sharing), (*b*) equivalent signal flow graph.

Note that for positive capacitance values, this pole is restricted to the real axis between zero and one, and therefore the circuit is always stable. In fact, with $C_3 = 0$ the circuit's pole is at $z = 1$ or, in other words, it becomes a discrete-time integrator, as expected. The zero of (10.33) is found by equating the numerator to zero, which results in a zero location, z_Z, given by

$$z_Z = \frac{C_1}{C_1 + C_2} \tag{10.35}$$

Therefore, for positive capacitance values, this zero location is also restricted to the real axis between zero and one. Additionally, the dc gain for this circuit is found by setting $z = 1$, which results in the dc gain being given by

$$H(1) = \frac{-C_2}{C_3} \tag{10.36}$$

Finally, it should be noted that in a fully differential implementation, effective negative capacitances for C_1, C_2, and C_3 can be achieved by simply interchanging the input wires. In this way, a zero at $z = -1$ could be realized by setting

$$C_1 = -0.5C_2 \tag{10.37}$$

In other words, the pair of capacitors representing C_1 would be half the size of those representing C_2 and the input wires interchanged into the C_1 pair such that $-V_i(z)$ is applied to that input branch.

EXAMPLE 10.3

Find the capacitance values needed for a first-order switched-capacitor circuit such that its 3-dB point is at 10 kHz when a clock frequency of 100 kHz is used. It is also desired that the filter have zero gain at 50 kHz and the dc gain be unity. Assume $C_A = 10\text{ pF}$.

Solution

First note that having a zero gain at 50 kHz is equivalent to placing a zero at –1. Now, making use of the bilinear transform $p = (z-1)/(z+1)$ as described in Section 9.5, the zero at –1 is mapped to $\Omega = \infty$. In addition, the frequency warping between the two domains, $\Omega = \tan(\omega/2)$, maps the –3-dB frequency of 10 kHz (or 0.2π rad/sample) in the z-domain to

$$\Omega = \tan\left(\frac{0.2\pi}{2}\right) = 0.3249$$

in the continuous-time domain. Therefore, our specification is now simplified to find the pole location for a continuous-time filter having a –3-dB frequency equal to 0.3249 rad/s and a zero at ∞. Clearly, the continuous-time pole, p_p, required here is

$$p_p = -0.3249$$

This pole is mapped back to a z-domain pole, z_p, of value

$$z_p = \frac{1+p_p}{1-p_p} = 0.5095$$

Therefore, the desired discrete-time transfer function, $H(z)$, is given by

$$H(z) = \frac{k(z+1)}{z-0.5095}$$

where k is determined by setting the dc gain to one (i.e., $H(1) = 1$) resulting in

$$H(z) = \frac{0.24525(z+1)}{z-0.5095}$$

or equivalently,

$$H(z) = \frac{0.4814z+0.4814}{1.9627z-1}$$

Equating these coefficients with those of (10.33) (and assuming $C_A = 10\text{ pF}$) results in

$$C_1 = 4.814 \text{ pF}$$
$$C_2 = -9.628 \text{ pF}$$
$$C_3 = 9.628 \text{ pF}$$

where a differential input can realize the negative capacitance.

For designs where the zero and pole frequencies are substantially less than the clock frequency (unlike the previous example, which had a zero at one-half the clock frequency), it is not necessary to pre-warp the specifications using the bilinear transform. Rather, some approximations can be used to derive equations for the capacitor ratios given the desired zero and pole, which are both assumed to be real and positive. As in Example 10.2, using (10.33), we have

$$H(z) \equiv \frac{V_o(z)}{V_i(z)} = -\frac{\left(\frac{C_1}{C_A}\right)(z^{1/2} - z^{-1/2}) + \left(\frac{C_2}{C_A}\right)z^{1/2}}{z^{1/2} - z^{-1/2} + \frac{C_3}{C_A}z^{1/2}} \tag{10.38}$$

Substituting (10.15) and (10.16) into (10.38) gives

$$H(e^{j\omega T}) \equiv \frac{V_o(e^{j\omega T})}{V_i(e^{j\omega T})} = -\frac{j\frac{2C_1 + C_2}{C_A}\sin\left(\frac{\omega T}{2}\right) + \frac{C_2}{C_A}\cos\left(\frac{\omega T}{2}\right)}{j\left(2 + \frac{C_3}{C_A}\right)\sin\left(\frac{\omega T}{2}\right) + \frac{C_3}{C_A}\cos\left(\frac{\omega T}{2}\right)} \tag{10.39}$$

This transfer function is exact. Making the approximation $\omega T \ll 1$ allows (10.39) to be simplified as

$$H(e^{j\omega T}) \equiv \frac{V_o(e^{j\omega T})}{V_i(e^{j\omega T})} = -\frac{j\left(\frac{C_1 + C_2/2}{C_A}\right)\omega T + \frac{C_2}{C_A}}{j\left(1 + \frac{C_3}{2C_A}\right)\omega T + \frac{C_3}{C_A}} \tag{10.40}$$

The procedure just used is quite useful when finding a continuous-time transfer function that approximates a discrete-time transfer function. Setting the numerator of (10.40) to zero gives the following approximate equation for the zero frequency:

$$j\omega_z T = \frac{-C_2/C_1}{1 + \frac{C_2}{2C_1}} \tag{10.41}$$

Similarly, setting the denominator to zero gives the following approximate equation for the pole frequency:

$$j\omega_p T = \frac{-C_3/C_A}{1 + \frac{C_3}{2C_A}} \tag{10.42}$$

EXAMPLE 10.4

Find the capacitance values C_3 needed for a first-order low-pass switched-capacitor filter that has $C_1 = 0$ and a pole at the 1/64th sampling frequency using the approximate equations. The low-frequency gain should be unity.

Solution

Using (10.42) and solving for C_3/C_A, we have

$$\frac{C_3}{C_A} = \frac{\omega_p T}{1 - \omega_p T/2} = \frac{2\pi/64}{1 - 2\pi/128} = 0.1032 \tag{10.43}$$

If we choose $C_A = 10$ pF, then we have $C_3 = 1.032$ pF. From (10.40) with $C_1 = 0$, we see that the low-frequency gain is given by C_2/C_3. For unity gain, this means we need $C_2 = 1.032$ pF as well.

Switch Sharing

Careful examination of the switched-capacitor filter shown in Fig. 10.15(*a*) reveals that some of the switches are redundant. Specifically, the top plate of both C_2 and C_3 are always switched to the opamp's virtual ground and true ground at the same time. Therefore, one pair of these switches can be eliminated and the two top plates of C_2 and C_3 can be connected together, as shown in Fig. 10.16.

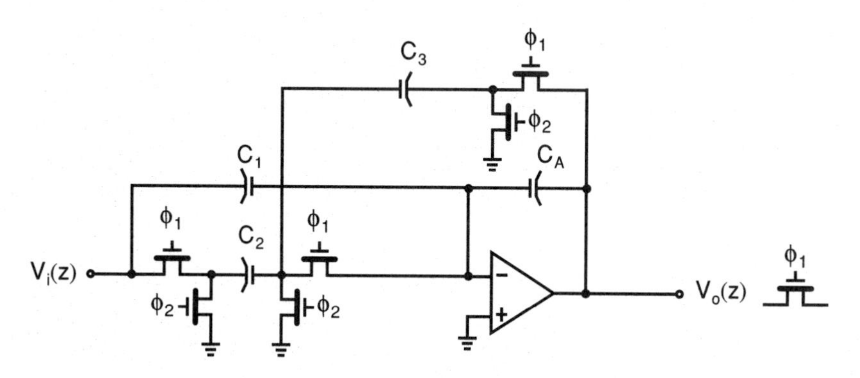

Fig. 10.16 A first-order switched-capacitor filter with switch sharing.

Fully Differential Filters

While the circuits seen so far have all been shown with single-ended signals, in most analog applications it is desirable to keep the signals fully differential. Fully differential signals imply that the *difference between two lines represents the signal component,* and thus any noise which appears as a common-mode signal on those two lines does not affect the signal. Fully differential circuits should also be balanced, implying that the differential signals operate symmetrically around a dc common-mode voltage (typically, analog ground or halfway between the power-supply voltages). Having balanced circuits implies that it is more likely noise will occur as a common-mode signal due to the symmetry of the circuit. Fully differential circuits have the additional benefit that if each single-ended signal is distorted symmetrically around the common-mode voltage, the differential signal will have only odd-order distortion terms (which are often much smaller). To see this distortion improvement, consider the block diagram shown in Fig. 10.17, where the two nonlinear elements are identical and, for simplicity, the common-mode voltage is assumed to be zero. If the nonlinearity is memoryless, then the outputs can be found as a Taylor series expansion given by

$$v_o = k_1 v_i + k_2 v_i^2 + k_3 v_i^3 + k_4 v_i^4 + \cdots \qquad (10.44)$$

where k_i are constant terms. In this case, the differential output signal, v_{diff}, consists of only the linear term, $2k_1v_1$, and odd-order distortion terms (which are typically smaller than the second-order term). With these two important advantages, most modern switched-capacitor circuits are realized using fully differential structures.

A fully differential realization of the first-order filter is shown in Fig. 10.18. Note here that the fully differential version is essentially two copies of the single-ended version, which might lead one to believe that it would consume twice the amount of integrated area. Fortunately, the increased area penalty is not that high. First, we see that only one opamp is needed although it does require extra common-mode feedback circuitry. Secondly, note that the input and output signal swings have now been doubled in size. For example, if single-ended signals are restricted to ±1 V, then the maximum peak-to-peak signal swing of single-ended signals is 2 V, while for differential signals it is 4 V. Thus, to maintain the same dynamic range due to $(kT)/C$ noise, the

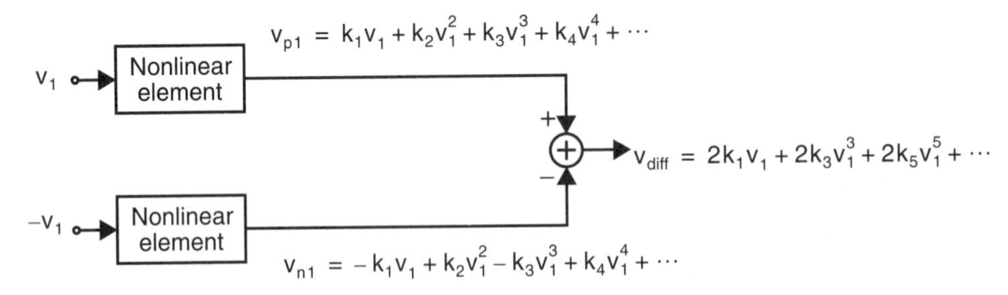

Fig. 10.17 Demonstrating that even-order distortion terms cancel in fully differential circuits if the distortion is symmetrical around the common-mode voltage. Here, the common-mode voltage is assumed to be zero.

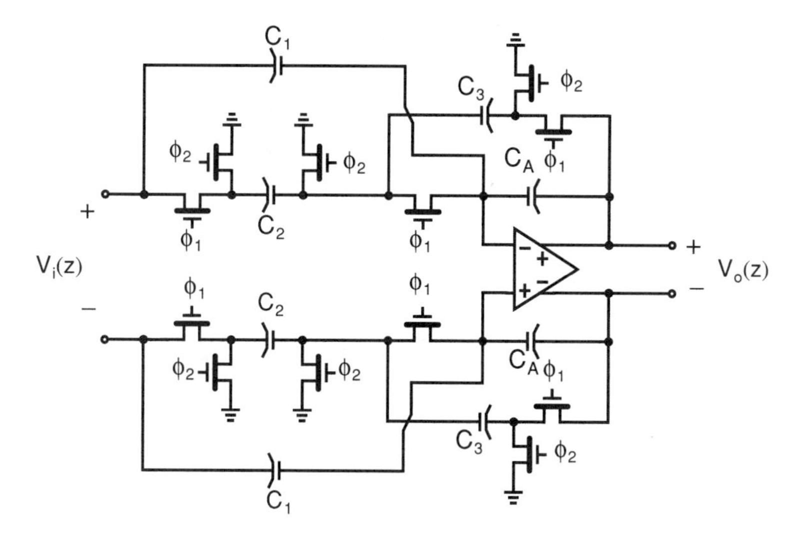

Fig. 10.18 A fully differential first-order switched-capacitor filter.

capacitors in the fully differential version can be half the size of those in the single-ended case.[4] Since smaller capacitors can be used, the switch size widths may also be reduced to meet the same settling-time requirement. However, the fully differential circuit has more switches and wiring so this circuit will be somewhat larger than its single-ended counterpart. However, recall that the fully differential circuit has the advantages of rejecting much more common-mode noise signals as well as having better distortion performance.

10.4 BIQUAD FILTERS

Similar to the first-order case, good switched-capacitor biquad filter structures can be obtained by emulating well-known continuous-time filter structures. However, also as in the first-order case, once a filter structure is obtained, its precise frequency response is determined through the use of discrete-time analysis using the signal-flow-graph technique discussed in Section 10.2. This exact transfer function, or a close approximation to it, is then used when determining capacitor ratios during the design phase.

Low-Q Biquad Filter

A direct-form continuous-time biquad structure can be obtained by rewriting the general biquad transfer function, $H_a(s)$, as follows:

4. When the signal voltage doubles, the signal power goes up by a factor of four. Fully differential circuits would have twice the noise power of single-ended circuits having the same size capacitors due to there being two capacitive networks generating noise. Halving the capacitor size of the differential circuit would result in another doubling in noise power, resulting in a signal-to-noise ratio equal to that of a single-ended circuit having capacitors twice as large.

$$H_a(s) \equiv \frac{V_{out}(s)}{V_{in}(s)} = -\frac{k_2 s^2 + k_1 s + k_0}{s^2 + \left(\frac{\omega_0}{Q}\right)s + \omega_0^2} \tag{10.45}$$

Here, ω_0 and Q are the pole frequency and pole Q, respectively, whereas k_0, k_1, and k_2 are arbitrary coefficients that place this biquad's zeros (i.e., the numerator's roots). Multiplying through by the denominators and dividing by s^2, (10.45) can be written as

$$V_{out}(s) = \left[-k_2 - \frac{k_1}{s}\right]V_{in}(s) - \frac{\omega_0}{Qs}V_{out}(s) - \frac{k_0}{s^2}V_{in}(s) - \frac{\omega_0^2}{s^2}V_{out}(s) \tag{10.46}$$

Finally, (10.46) can be expressed as two integrator-based equations. Specifically,

$$V_{c1}(s) \equiv -\frac{1}{s}\left[\frac{k_0}{\omega_0}V_{in}(s) + \omega_0 V_{out}(s)\right] \tag{10.47}$$

and

$$V_{out}(s) = -\frac{1}{s}\left[(k_1 + k_2 s)V_{in}(s) + \frac{\omega_0}{Q}V_{out}(s) - \omega_0 V_{c1}(s)\right] \tag{10.48}$$

A signal flow graph describing the preceding two equations is shown in Fig. 10.19.

Now, allowing negative resistors to be used, an active-RC realization of this general biquad signal flow graph is shown in Fig. 10.20 [Brackett, 1978]. Note that for convenience the integrating capacitors, C_A and C_B, have been set to unity.

A switched-capacitor biquad based on this active-RC circuit can be realized as shown in Fig. 10.21 [Martin, 1979]. Here, all positive resistors were replaced with delay-free feed-in switched-capacitor stages while the negative resistor was replaced with a delaying feed-in stage. Note that this switched-capacitor biquad circuit has redundant switches, as switch sharing has not yet been applied.

The input capacitor K_1C_1 is the major signal path when realizing low-pass filters; the input capacitor K_2C_2 is the major signal path when realizing bandpass filters; whereas, the input capacitor K_3C_2 is the major signal path when realizing high-pass filters.

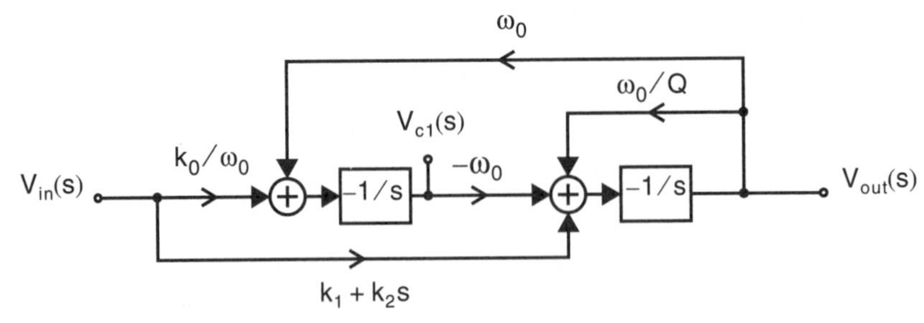

Fig. 10.19 A signal-flow-graph representation of a general continuous-time biquad filter.

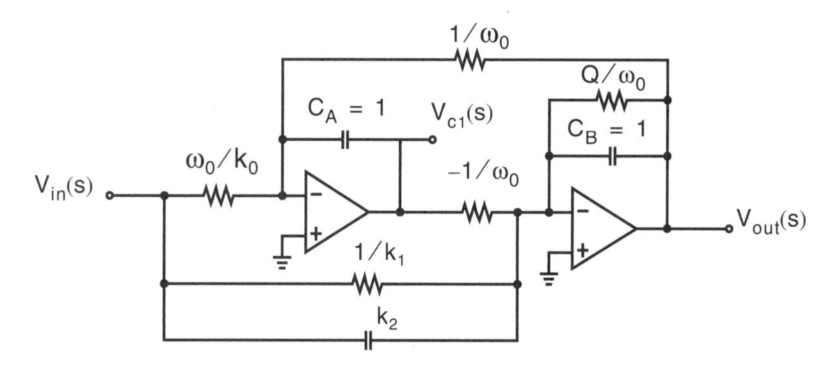

Fig. 10.20 An active-RC realization of a general continuous-time biquad filter.

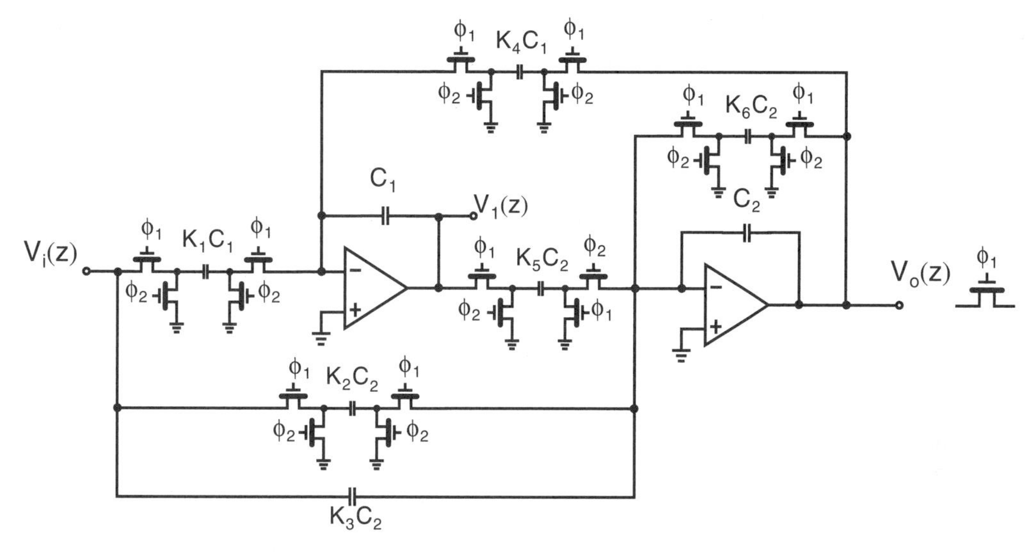

Fig. 10.21 A low-Q switched-capacitor biquad filter (without switch sharing).

Before proceeding, it is worth mentioning that while this switched-capacitor filter is a natural result when using single-ended circuits, the use of fully differential circuits could result in quite a few other circuits, some of which could have poor performance properties. For example, all the positive resistors could be replaced with delay-free feed-in stages as before, but in addition, the negative resistor could also be replaced with a delay-free feed-in stage where the differential input wires are interchanged. Such a circuit would have a delay-free loop around the two integrators and may have an excessive settling time behavior. As another example, all resistors might be replaced with delaying feed-in stages where positive resistors are realized by interchanging the input wires of those stages. In this case, settling time behavior would not suffer as there would be two delays around the two-integrator loop, but coefficient sensitivity would be worse for filters having high-Q poles. In summary, what is important is that the two-integrator loop have a single delay around the loop. Such an arrangement is referred to as using lossless discrete integrators (LDI) [Bruton, 1975].

Using the approach presented in Section 10.2, the signal flow graph for the biquad circuit shown in Fig. 10.21 can be found and is shown in Fig. 10.22. The transfer function, H(z), for this signal flow graph is found to be given by

$$H(z) \equiv \frac{V_o(z)}{V_i(z)} = -\frac{(K_2 + K_3)z^2 + (K_1K_5 - K_2 - 2K_3)z + K_3}{(1 + K_6)z^2 + (K_4K_5 - K_6 - 2)z + 1} \tag{10.49}$$

While the preceding relation is useful for analyzing the switched-capacitor circuit shown, this relation can also be used during design of a desired transfer function. Assuming the desired discrete-time biquad transfer function has been found using a digital-filter design program, and is given by

$$H(z) = -\frac{a_2z^2 + a_1z + a_0}{b_2z^2 + b_1z + 1} \tag{10.50}$$

we can equate the individual coefficients of z in (10.49) and (10.50), resulting in the following design equations:

$$K_3 = a_0 \tag{10.51}$$

$$K_2 = a_2 - a_0 \tag{10.52}$$

$$K_1K_5 = a_0 + a_1 + a_2 \tag{10.53}$$

$$K_6 = b_2 - 1 \tag{10.54}$$

$$K_4K_5 = b_1 + b_2 + 1 \tag{10.55}$$

Note that there is some flexibility in choosing the values of K_1, K_4, and K_5. This single degree of freedom occurs because specifying the overall transfer function, H(z), places no constraint on the size of the signal at the internal node, $V_1(z)$. Thus, one way to choose appropriate capacitor values is to initially let $K_5 = 1$ and determine other initial component values, find the signal levels that occur at nodes $V_1(z)$

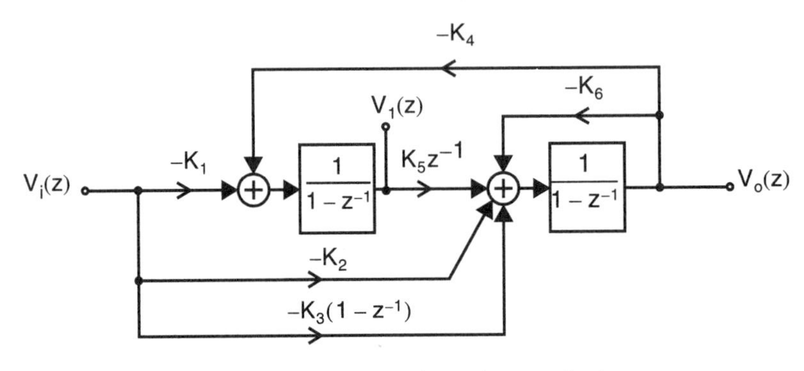

Fig. 10.22 A signal flow graph describing the switched-capacitor circuit in Fig. 10.21.

and $V_o(z)$, then using that signal level information, perform dynamic range scaling to find the final component values. Alternatively, the time constants of the two discrete-time integrators can be set equal by choosing

$$K_4 = K_5 = \sqrt{b_1 + b_2 + 1} \tag{10.56}$$

Such a choice usually results in a near optimal dynamically range-scaled circuit.

If it is desired to design the SC biquad so that its transfer function matches that of a continuous-time transfer function as closely as possible, then the approach used in Example 10.2 can be taken. Specifically, (10.49) can be rewritten as

$$H(z) \equiv \frac{V_o(z)}{V_i(z)} = -\frac{K_1K_5 + K_2(z^{1/2} - z^{-1/2})z^{1/2} + K_3(z^{1/2} - z^{-1/2})^2}{K_4K_5 + K_6(z^{1/2} - z^{-1/2})z^{1/2} + (z^{1/2} - z^{-1/2})^2} \tag{10.57}$$

Rewriting (10.15) and (10.16) here for convenience, we have

$$z^{1/2} = \cos\left(\frac{\omega T}{2}\right) + j\sin\left(\frac{\omega T}{2}\right) \tag{10.58}$$

and

$$z^{-1/2} = \cos\left(\frac{\omega T}{2}\right) - j\sin\left(\frac{\omega T}{2}\right) \tag{10.59}$$

Substituting these equations into (10.57) and rearranging gives

$$H(e^{j\omega T}) \equiv \frac{V_o(e^{j\omega T})}{V_i(e^{j\omega T})} = -\frac{K_1K_5 + jK_2\ \sin(\omega T) + (4K_3 + 2K_2)\sin^2\left(\frac{\omega T}{2}\right)}{K_4K_5 + jK_6\ \sin(\omega T) + (4 + 2K_6)\sin^2\left(\frac{\omega T}{2}\right)} \tag{10.60}$$

For $\omega T \ll 1$, we further have

$$H(e^{j\omega T}) \equiv \frac{V_o(e^{j\omega T})}{V_i(e^{j\omega T})} \approx -\frac{K_1K_5 + jK_2(\omega T) + (K_3 + K_2/2)(\omega T)^2}{K_4K_5 + jK_6(\omega T) + (1 + K_6/2)(\omega T)^2} \tag{10.61}$$

The coefficients of (10.61) can be matched to the desired coefficients of the continuous-time transfer function. A closer matching between transfer functions can be obtained if one chooses capacitor ratios to match the real and imaginary parts of both the numerators and denominators of (10.60) and the desired continuous-time transfer function at the resonant frequency, as is outlined in [Martin, 1980].

Finally, an estimate of the largest to smallest capacitance ratio can be made by using the resistor approximation (10.5) for this switched-capacitor circuit. Comparing Fig. 10.20 and Fig. 10.21 and using (10.5), we see the following approximations are valid for the feedback capacitor ratios, K_4, K_5, and K_6 (the capacitor ratios that determine the pole locations).

$$K_4 \cong K_5 \cong \omega_0 T \tag{10.62}$$

$$K_6 \cong \frac{\omega_0 T}{Q} \tag{10.63}$$

However, the sampling rate, $1/T$, is typically much larger that the approximated pole frequency, ω_0, resulting in

$$\omega_0 T \ll 1 \tag{10.64}$$

Thus, the largest capacitors (that determine the pole locations) in the switched-capacitor biquad circuit of Fig. 10.21 are the integrating capacitors, C_1 and C_2. Additionally, if $Q < 1$, then the smallest capacitors are K_4C_1 and K_5C_2, resulting in an approximate capacitance spread of $1/(\omega_0 T)$. However, if $Q > 1$, then from (10.63) the smallest capacitor would be K_6C_2, resulting in an approximate capacitance spread of $Q/(\omega_0 T)$.

High-Q Biquad Filter

The resultant high-capacitance spread in the preceding low-Q circuit, when $Q \gg 1$, was a direct result of having a large damping resistance value of Q/ω_0 in Fig. 10.20. One way to obtain a high-Q switched-capacitor circuit that does not have such a high capacitance spread is to eliminate this damping resistance in the continuous-time filter counterpart. Such an elimination can be accomplished by rewriting (10.46) as

$$V_{out}(s) = -\frac{1}{s}[k_2 s V_{in}(s) - \omega_0 V_{c1}(s)] \tag{10.65}$$

and

$$V_{c1}(s) = -\frac{1}{s}\left[\left(\frac{k_0}{\omega_0} + \frac{k_1 s}{\omega_0}\right)V_{in}(s) + \omega_0 V_{out}(s) + \frac{s}{Q}V_{out}(s)\right] \tag{10.66}$$

The signal-flow-graph representation of these two equations is shown in Fig. 10.23. Note that here the damping path, s/Q coefficient, passes through the first integrator stage and can be realized through the use of a capacitor, as shown in Fig. 10.24.

Once again, replacing resistors with appropriate switched-capacitor feed-in stages, the final high-Q switched-capacitor biquad circuit is shown in Fig. 10.25 [Martin, 1980].

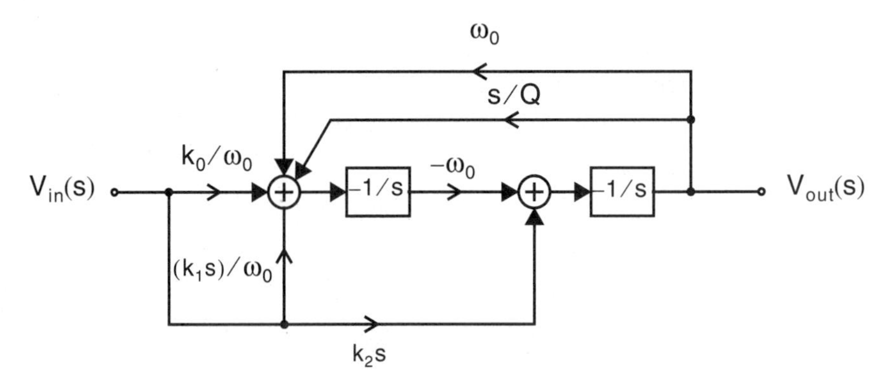

Fig. 10.23 An alternate realization for a general continuous-time biquad filter.

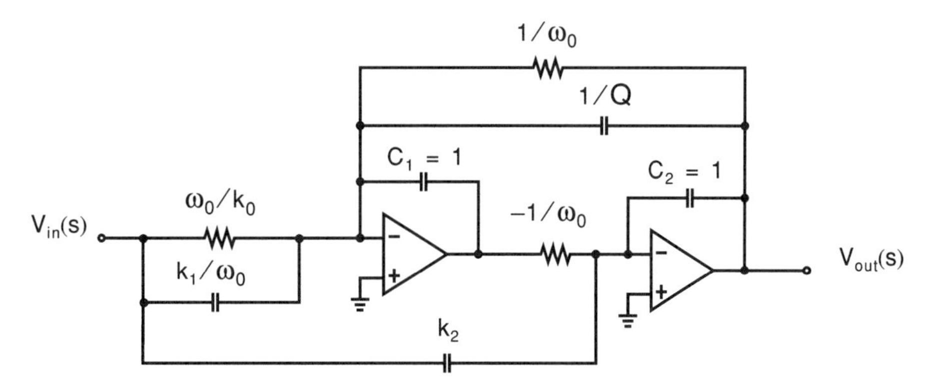

Fig. 10.24 An alternate realization of a general active-RC biquad filter.

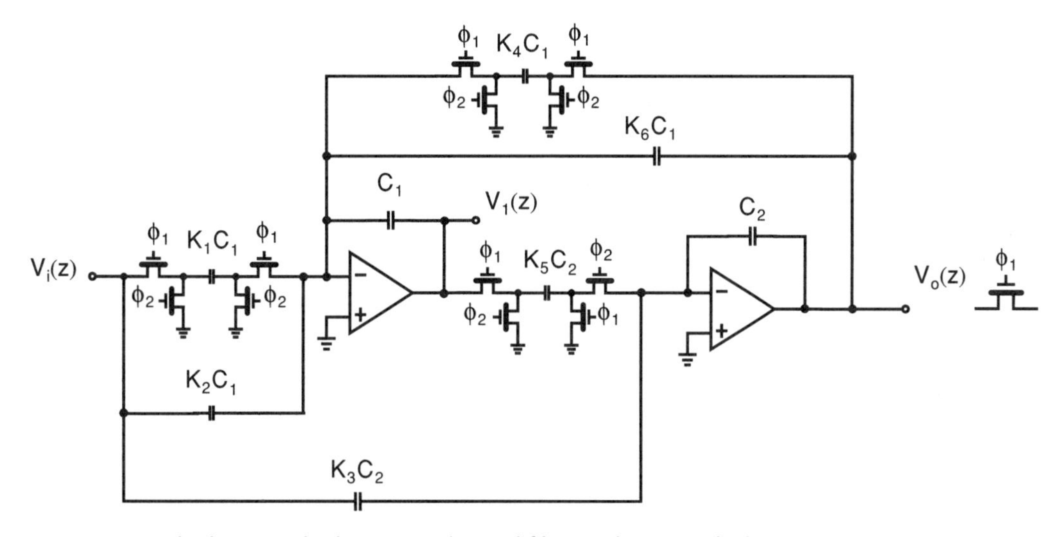

Fig. 10.25 A high-Q switched-capacitor biquad filter (without switch sharing).

The input capacitor K_1C_1 is the major signal path when realizing low-pass filters, the input capacitor K_2C_1 is the major signal path when realizing bandpass filters, and the input capacitor K_3C_2 is the major signal path when realizing high-pass filters. Other possibilities for realizing different types of functions exist. For example, a non-delayed switched capacitor going from the input to the second integrator could also be used to realize an inverting bandpass function. If the phases on the input switches of such a switched capacitor were interchanged, then a noninverting bandpass function would result, but with an additional period delay.

Using the signal-flow-graph approach described in Section 10.2, the transfer function for this circuit is found to be given by

$$H(z) \equiv \frac{V_o(z)}{V_i(z)} = -\frac{K_3z^2 + (K_1K_5 + K_2K_5 - 2K_3)z + (K_3 - K_2K_5)}{z^2 + (K_4K_5 + K_5K_6 - 2)z + (1 - K_5K_6)} \tag{10.67}$$

Matching the coefficients of (10.67) to the coefficients of a discrete-time transfer function given by[5]

$$H(z) = -\frac{a_2 z^2 + a_1 z + a_0}{z^2 + b_1 z + b_0} \tag{10.68}$$

the following equations result:

$$K_1 K_5 = a_0 + a_1 + a_2 \tag{10.69}$$

$$K_2 K_5 = a_2 - a_0 \tag{10.70}$$

$$K_3 = a_2 \tag{10.71}$$

$$K_4 K_5 = 1 + b_0 + b_1 \tag{10.72}$$

$$K_5 K_6 = 1 - b_0 \tag{10.73}$$

As for the low-Q biquad, we see that there is some freedom in determining the coefficients as there is one less equation than the number of coefficients. A reasonable first choice might be to take

$$K_4 = K_5 = \sqrt{1 + b_0 + b_1} \tag{10.74}$$

after which all other ratios follow. This will give a good dynamic range but not necessarily optimum. After the biquad has been designed, it can be analyzed and the ratios adjusted to equalize the maximum signal strength at the two opamp outputs.

EXAMPLE 10.5

The following transfer function describes a bandpass filter having a peak gain of 5 near $f_s/10$ and a Q of about 10.

$$H(z) = -\frac{0.288(z-1)}{z^2 - 1.572z + 0.9429} \tag{10.75}$$

Find the largest to smallest capacitor ratio if this transfer function is realized using the high-Q biquad circuit. Compare the results to those which would be obtained if the low-Q circuit were used instead. Let $C_1 = C_2 = 1$ in both cases.

Solution

For the high-Q biquad circuit, making use of (10.69) to (10.72) and assuming $K_4 = K_5$ results in the following capacitor values.

$$K_4 = K_5 = 0.6090$$

$$K_1 = 0$$

$$K_2 = 0.4729$$

5. Note that (10.68) is written in slightly different form than (10.50) to simplify the resulting equations.

$$K_3 = 0$$

$$K_6 = 0.0938$$

Recalling that $C_1 = C_2 = 1$, in this high-Q circuit the smallest capacitor is K_6C_1, which makes the capacitance spread about 11 to 1.

For the low-Q biquad circuit design, we first rewrite (10.75) as

$$H(z) = -\frac{0.3054(z-1)z}{1.0606z^2 - 1.667z + 1} \tag{10.76}$$

which makes applying equations (10.51) to (10.56) easier. In addition, an extra zero at $z = 0$ was added to avoid negative capacitances. This extra zero does not change the magnitude response.

We obtain the following ratios assuming $K_4 = K_5$:

$$K_4 = K_5 = 0.6274 \tag{10.77}$$

$$K_1 = 0 \tag{10.78}$$

$$K_2 = 0.3054 \tag{10.79}$$

$$K_3 = 0 \tag{10.80}$$

$$K_6 = 0.0606 \tag{10.81}$$

Thus the capacitance ratio for this circuit is determined by K_6, which primarily sets the accuracy of the pole-Q factor, resulting in a spread of near 17. It should be noted here that worse capacitance spreads will occur for this low-Q circuit if the pole Q is higher and/or the sampling frequency is increased relative to the passband edge.

There are many other biquadratic stages than the two presented here. For example, a general stage capable of realizing any z-transform is described in [Laker, 1994]. However, the authors have found that for the great majority of filtering applications, the choice of the circuits of either Fig. 10.21 or Fig. 10.25 (with switch sharing) is nearly optimum. The exception to this is for phase equalizers, where often feed-ins approximating negative resistors are needed. To realize higher-order switched-capacitor filters, one can cascade a number of biquad and/or first-order stages. Alternatively, one can use lower-sensitivity techniques such as signal-flow-graph simulations of the relationships between the state variables of passive doubly terminated LC ladder filters. The interested reader is referred to [Gregorian, 1986] for detailed descriptions on the design of high-order SC filters.

10.5 CHARGE INJECTION

As seen in Chapter 7, comparators that make use of switches suffer from charge-injection errors. Fortunately, by turning off certain switches first, the charge-injection errors of the overall circuit are due to only those early switches. This same result applies to switched-capacitor circuits, and thus charge-injection effects can be minimized by having some clock signals slightly advanced with respect to the remaining signals. Which clock signals should be advanced is most easily explained with the following example.

Consider the first-order switched-capacitor filter shown in Fig. 10.26, where switch sharing is used. Here, ϕ_{1a} and ϕ_{2a} are slightly advanced with respect to the remaining clock signals for the following reasons. First, since ϕ_{2a} is always connected to ground while ϕ_{1a} is always connected to virtual ground, when these switches are turned on they need only pass a signal near the ground node (not rail-to-rail signals as might occur at opamp outputs). Thus, these two switches can be realized using single n-channel transistors rather than full CMOS transmission gates.[6] A second, more important reason for using this clock arrangement is that the charge injections due to Q_3, Q_4 are not signal dependent, while those of Q_1, Q_6 are signal dependent. Specifically, we saw in Chapter 7 that the channel charge of an NMOS transistor in triode is given by

$$Q_{CH} = -WLC_{ox}V_{eff} = -WLC_{ox}(V_{GS} - V_t) \tag{10.82}$$

Here, we see that the charge is related to the gate-source voltage as well as to the threshold voltage. When Q_3 and Q_4 are on, $V_{GS} = V_{DD}$, and since their source remains at zero volts, their threshold voltages also remain constant. As a result, *the amount of charge injected by* Q_3, Q_4 *is the same from one clock cycle to the next and can be considered as a dc offset.* Unfortunately, the same cannot be said of Q_1 and Q_6. For example, when Q_1 is on, its channel charge is found from (10.82) to be

$$Q_{CH1} = -W_1L_1C_{ox}(V_{DD} - V_i - V_{tn}) \tag{10.83}$$

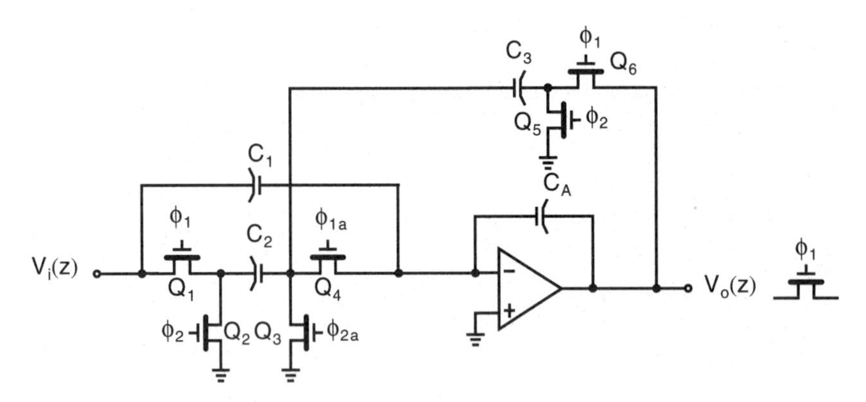

Fig. 10.26 A first-order switched-capacitor circuit with a clock arrangement to reduce charge-injection effects. ϕ_{1a} and ϕ_{2a} turn off Q_3 and Q_4 slightly early.

6. Typically, switches Q_1, Q_6 would be realized as CMOS transmission gates since they may need to pass large-swing signals.

Thus, one portion of the channel charge is linearly related to V_i. However, its source voltage settles to V_i and thus the transistor's threshold voltage changes in a nonlinear relationship (due to the bulk effect, assuming its substrate is set to a fixed voltage). As a result, Q_{CH1} has a linear and nonlinear relationship to V_i and thus would cause a gain error and distortion if Q_1 were turned off early. Finally, while Q_2 and Q_5 would also inject constant charge from one cycle to the next, it is best not to add any unnecessary charge into the circuit so these switches are also not advanced.

In summary, to reduce the effects of charge injection in switched-capacitor circuits, realize all switches connected to ground or virtual ground as n-channel switches only, and turn off the switches near the virtual ground node of the opamps first. Such an approach will minimize distortion and gain error as well as keeping dc offset low.

EXAMPLE 10.6

Assuming an ideal opamp, estimate the amount of dc offset in the output of the circuit shown in Fig. 10.26 due to channel-charge injection (ignore overlap capacitance charge injection) when $C_1 = 0$ and $C_2 = C_A = 10C_3 = 10\text{ pF}$. Assume that switches Q_3, Q_4 are n-channel devices with a threshold voltage of $V_{tn} = 0.8\text{ V}$, width of $W = 30\ \mu\text{m}$, length of $L = 0.8\ \mu\text{m}$, $C_{ox} = 1.9 \times 10^{-3}\text{ pF}/(\mu\text{m})^2$, and that power supplies of $\pm 2.5\text{ V}$ are used.

Solution

From (10.82), we can calculate the amount of channel charge of Q_3, Q_4 (when on) to be

$$\begin{aligned} Q_{CH3} &= Q_{CH4} \\ &= -(30)(0.8)(0.0019)(2.5-0.8) = -77.5 \times 10^{-3}\text{ pC} \end{aligned} \tag{10.84}$$

Note that the dc feedback around the opamp will keep the virtual input of the opamp at zero volts. Also, all of the dc feedback current is charge transferred through the switched capacitor C_3. Thus, the charge being transferred through this capacitor must equal the charge injected by both Q_3, Q_4 (assuming the input is at zero volts and therefore does not contribute charge). The charge transfer into C_3 is given by

$$Q_{C_3} = -C_3 V_{out} \tag{10.85}$$

Now, we estimate that half the channel charges of Q_3 and Q_4 are injected to the virtual ground node by the following reasoning. When ϕ_{1a} turns off, half the channel charge of Q_4 goes into the virtual ground while half of its charge is placed at the node between Q_3 and Q_4. When ϕ_{2a} goes high, the second charge escapes to ground, but when ϕ_{2a} goes low, half of its channel charge is left between Q_3 and Q_4 and is passed into the virtual ground when ϕ_{1a} goes high again. As a result, we can write the following charge equation,

$$\frac{1}{2}(Q_{CH3} + Q_{CH4}) = Q_{C_3} \tag{10.86}$$

which can be used to find Q_{C_3}, and hence the output voltage from (10.85)

$$V_{out} = \frac{77.5 \times 10^{-3}\ \text{pC}}{1\,\text{pF}} = 78\ \text{mV} \tag{10.87}$$

Thus, we see that this output dc offset value is affected by the size of capacitors used (in particular, C_3) as well as the switch sizes and power-supply voltage. It should be mentioned here that if the dc input offset voltage of the opamp was included in the analysis, it would be multiplied by the dc gain of the circuit similar to that which occurs in an active-RC circuit.

Charge-injection is especially troublesome at higher frequencies. At higher frequencies, the time constant of the switch-on resistance and the capacitor being charged or discharged must be smaller; this necessitates larger switches for a given size capacitor and therefore greater charge-injection. It is possible to derive a very simple formula that gives an approximate upper bound on the frequency of operation of an SC circuit for a specified maximum voltage change due to charge injection. This upper bound takes into account switch channel charge only and ignores charge injection due to overlap capacitance.

Most SC circuits will have two series switches for each capacitor. If one assumes that, for good settling, the sampling clock half-period must be greater than five time constants, then we have

$$\frac{T}{2} > 5R_{on}C \tag{10.88}$$

where T is the sampling period, R_{on} is the on resistance of the n-channel switch, and C is the capacitor. Equation (10.88) may be rewritten

$$f_{clk} < \frac{1}{10R_{on}C} \tag{10.89}$$

where $f_{clk} = 1/T$ is the clock frequency. Recalling (1.108) from Chapter 1, we have

$$R_{on} = \frac{1}{\mu_n C_{ox} \frac{W}{L} V_{eff}} \tag{10.90}$$

Also, using (10.83), the charge change due to the channel charge only caused by turning an n-channel switch off is approximated by

$$|\Delta V| = \frac{Q_{CH}}{2C} = \frac{WLC_{ox}V_{eff}}{2C} \tag{10.91}$$

For a specified $|\Delta V|_{max}$, (10.92) implies that

$$C = \frac{WLC_{ox}V_{eff}}{2|\Delta V|_{max}} \tag{10.92}$$

Substituting (10.91) and (10.93) into (10.90) gives

$$f_{clk} = \frac{\mu_n|\Delta V|_{max}}{5L^2} \tag{10.93}$$

a very simple expression. Thus, the upper frequency limit to charge injection for SC circuits is inversely proportional to L^2 and is approximately independent of the size of capacitors or the power-supply voltages, assuming the switch sizes are chosen optimally; as technology improves, much higher switching frequencies are attainable. Finally, (10.93) ignores overlap capacitance, so it is somewhat optimistic and should be therefore considered an upper bound on the maximum clocking frequency.

EXAMPLE 10.7

Assuming the maximum voltage change due to clock feedthrough is 1 mV, what are the maximum clocking frequencies considering only channel charge injection for technologies having minimum channel lengths of 0.8 μm, 0.5 μm, and 0.3 μm?

Solution

Using (10.94) and $\mu_n = 0.05\ M^2/V \cdot s$ gives f_{clk}, which must be less than 15.6 MHz, 40 MHz, and 111 MHz, respectively, for the three different technologies. Notice the substantial increase in speed for SC circuits as the technology improves.

10.6 SWITCHED-CAPACITOR GAIN CIRCUITS

Perhaps the most common nonfiltering analog function is a gain circuit where the output signal is a scaled version of the input. In this section, we shall see that it is possible to realize accurate gain circuits by making use of switched-capacitor techniques.

Parallel Resistor-Capacitor Circuit

In active-RC circuits, a gain circuit can be realized as parallel combinations of resistors and capacitors in both the feedback and feed-in paths, as shown in Fig. 10.27(*a*). One of the first switched-capacitor gain circuits uses a similar approach, where the two resistors have been replaced by their switched-capacitor equivalents, as shown in Fig. 10.27(*b*) [Foxall, 1980]. Using the signal-flow-graph technique of Section 10.2, straightforward analysis results in the transfer function for this switched-capacitor gain circuit being given by

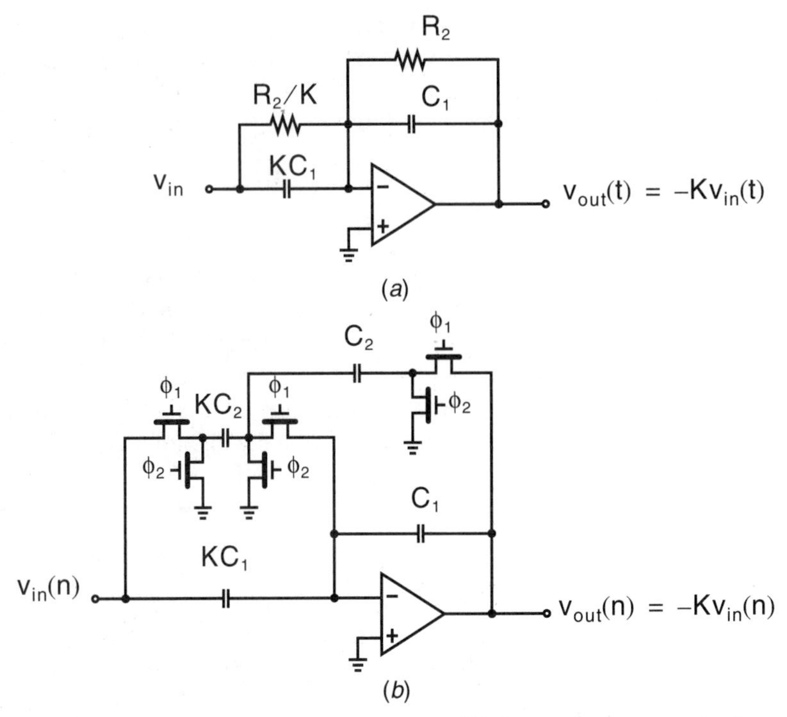

Fig. 10.27 (*a*) An active-RC gain circuit. (*b*) An equivalent switched-capacitor gain circuit.

$$H(z) = \frac{V_{out}(z)}{V_{in}(z)} = -K \tag{10.94}$$

Unfortunately, this design will also amplify the 1/f noise and offset voltage of the shown opamp by K. However, it does have the advantage that its output is a continuous waveform that does not incur any large slew-rate requirement (as occurs in the resettable gain circuit discussed next).

Resettable Gain Circuit

The next configuration continuously resets an integrating capacitor on each clock cycle, as shown in Fig. 10.28 [Gregorian, 1981]. Here, the voltage across the integrating capacitor, C_2, is cleared on each ϕ_2, while on ϕ_1 the input voltage charges C_1 and the charging current flows across C_2 at the same time. In this way, the change in charge across C_2, ΔQ_{C2}, equals the change in charge across C_1, ΔQ_{C1}, and therefore, at the end of ϕ_1, the output voltage is related to the input voltage by $V_{out}/V_{in} = -C_1/C_2$.

In addition, this circuit stores any opamp input-offset voltage across the input and feedback capacitors. When the output is sampled by the succeeding circuit (during ϕ_1), the effects of the input-offset voltage of the opamp are cancelled from the output

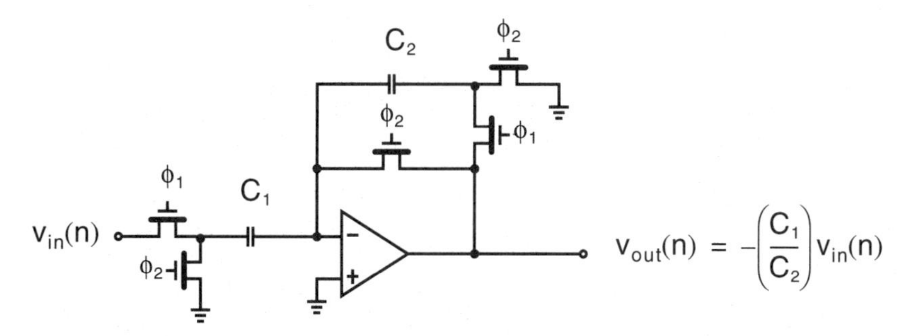

Fig. 10.28 A resettable gain circuit where opamp offset voltage is cancelled.

voltage. The elimination of the opamp's input-offset voltage is important for two reasons. One reason is that if the offset voltage is not cancelled, it will also be amplified when the input signal is amplified and therefore can be troublesome. A second reason (and often more important) is that when the offset voltage is eliminated, 1/f noise is also reduced. As we saw in Chapter 4, 1/f noise is large at low frequencies and can be a dominant noise source in many MOS circuits. However, when a circuit cancels dc offset voltages, the circuit has, in effect, a high-pass response from the opamp's input terminals to the output voltage. Thus, the 1/f noise is also high-pass filtered and hence reduced. It should be noted here that although there is a high-pass response from the opamp's input terminals to the output voltage, the response from the overall circuits does not necessarily have a high-pass response.

To see this offset cancellation, consider the gain circuit during ϕ_2 (i.e., when it is being reset) as shown in Fig. 10.29(*a*). The effect of the input offset voltage is being modelled as a voltage source, V_{off}, which is placed in series with one of the opamp inputs. In this case, it is placed in series with the positive input, which results in the analysis being marginally simpler. During ϕ_2, both the voltages across C_1 and C_2 are equal to the opamp offset voltage, V_{off}. Next, during ϕ_1 the circuit is configured as shown in Fig. 10.29(*b*), and at the end of ϕ_1 the voltage across C_1 is given by

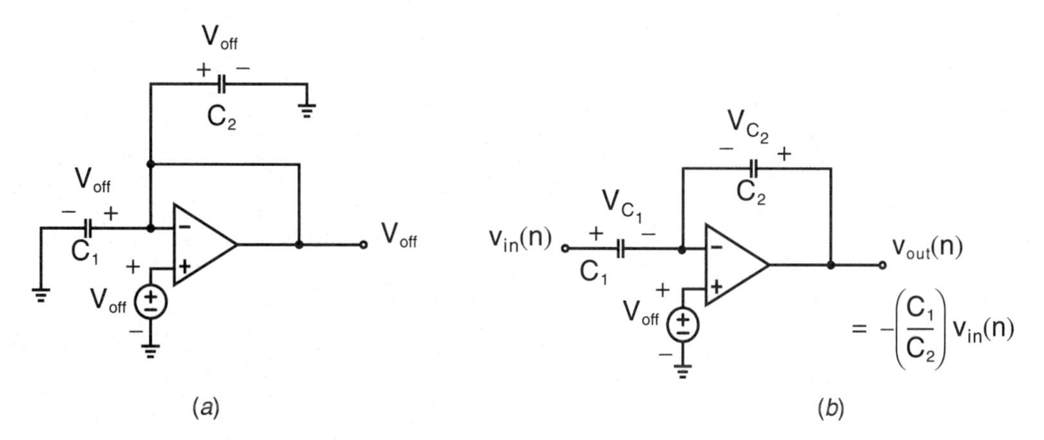

Fig. 10.29 The resettable gain circuit (*a*) during reset (ϕ_2), and (*b*) during valid output (ϕ_1).

$V_{C_1}(n) = v_{in}(n) - V_{off}$, while that for C_2 is given by $V_{C_2}(n) = v_{out}(n) - V_{off}$. Therefore we can write,

$$\Delta Q_{C1} = C_1[V_{C_1}(n) - (-V_{off})] = C_1 v_{in}(n) \tag{10.95}$$

and

$$V_{C_2}(n) = V_{C_2}(n - 1/2) - \frac{\Delta Q_{C2}}{C_2} \tag{10.96}$$

Since the voltage across C_2 one-half period earlier was $-V_{off}$ and $\Delta Q_{C2} = \Delta Q_{C1}$, we have

$$V_{C_2}(n) = -V_{off} - \frac{C_1 v_{in}(n)}{C_2} \tag{10.97}$$

Finally, since one side of C_2 is connected to the virtual ground of the opamp, which is at V_{off}, the output voltage is given by

$$\begin{aligned} v_{out}(n) &= V_{off} + V_{C_2}(n) \\ &= V_{off} + \left(-V_{off} - \frac{C_1 v_{in}(n)}{C_2}\right) \\ &= -\left(\frac{C_1}{C_2}\right) v_{in}(n) \end{aligned} \tag{10.98}$$

Thus, the output voltage is independent of the opamp offset voltage.

Note, however, that the output voltage is only valid during ϕ_1, and it is equal to V_{off} during ϕ_2, as shown in Fig. 10.30. The difficulty in realizing this waveform is that the opamp output must slew between $(-C_1/C_2)v_{in}$ and a voltage near 0 V each time the clock changes. Clearly, such a waveform requires a very high slew-rate opamp for proper operation.

Finally, note that a programmable gain circuit is easily realized by replacing C_1 with a programmable capacitor array[7] (PCA) where the capacitor size is determined by digital signals.

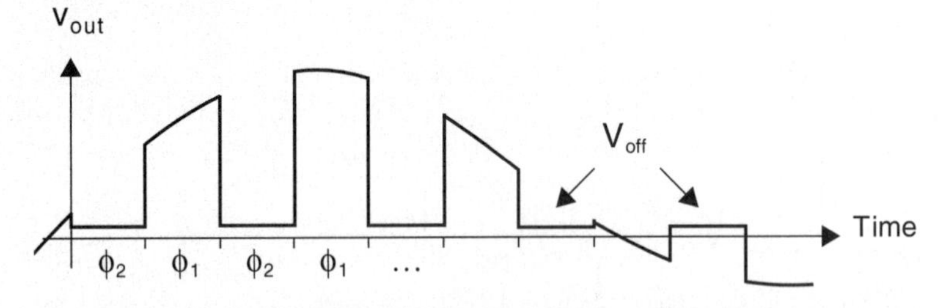

Fig. 10.30 An example output waveform for the resettable gain circuit.

7. A programmable capacitor array is a capacitor whose size can be digitally controlled. It normally consists of a number of binarily-weighted capacitors having one plate common for all capacitors and the other plate of each capacitor connected through digitally-controlled switches to either a second common node or ground.

Capacitive-Reset Gain Circuit

To eliminate the need for the opamp output to slew to approximately 0 V each clock period yet still cancel the opamp's offset voltage, a capacitive-reset gain circuit can be used. The basic idea of the gain circuit is to couple the opamp's output to the inverting input during the reset phase with a capacitor that has been previously charged to the output voltage. Thus, we shall see that one property of this gain circuit is that the opamp's output need only change by the opamp's offset voltage between clock phases. In addition, since the circuit is insensitive to the opamp's input offset voltage, it also reduces the effect of the opamp's 1/f noise. Finally, it can also be shown (see [Martin, 1987]) that for low-frequency inputs (or for inputs that are constant for at least two clock periods), the errors due to finite gain, A, of the opamp are proportional to $1/A^2$ (rather than the usual $1/A$). Such a result often allows the use of single-stage opamps that are very fast opamps.

The capacitive-reset gain circuit is shown in Fig. 10.31. Capacitor C_4 is an optional "deglitching" capacitor [Matsumoto, 1987] used to provide continuous-time feedback during the nonoverlap clock times when all the switches are open. This deglitching technique works well and should be used on almost all switched-capacitor circuits that would otherwise have no feedback connection at some instant of time. This capacitor would normally be small (around 0.5 pF or less).

This gain circuit can be either inverting or noninverting depending on the clock phases of the input stage. While the inverting circuit creates its output as a delay-free version of the input, the output for the noninverting case would be one-half clock cycle behind the input signal.

To see how this capacitive-reset gain circuit operates, consider the inverting circuit during the reset phase of ϕ_2, as shown in Fig. 10.32(*a*). We have assumed capacitor C_3 was charged to the output voltage during the previous ϕ_1 clock phase. Here we see that capacitors C_1 and C_2 are charged to the opamp's input-offset voltage, V_{off}, in the same manner as in the resettable gain circuit of Fig. 10.28. The next clock phase of ϕ_1 is shown in Fig. 10.32(*b*), where we see that the output voltage is

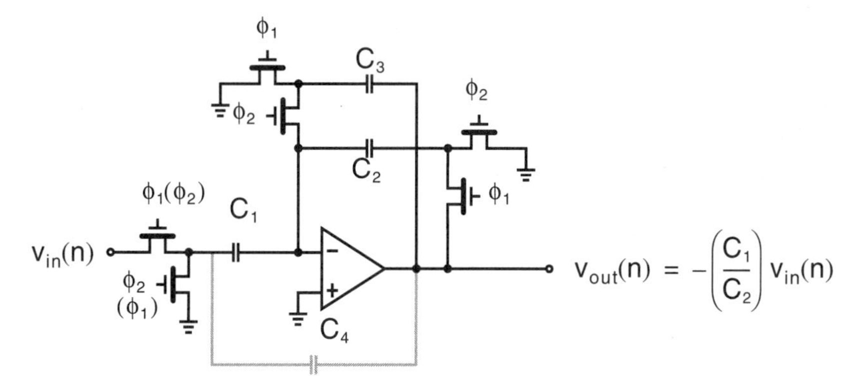

Fig. 10.31 A gain circuit using capacitive reset (C_4 is an optional deglitching capacitor). Depending on the input-stage clock signals, the gain can be either inverting (as shown) or noninverting (input-stage clocks shown in parentheses).

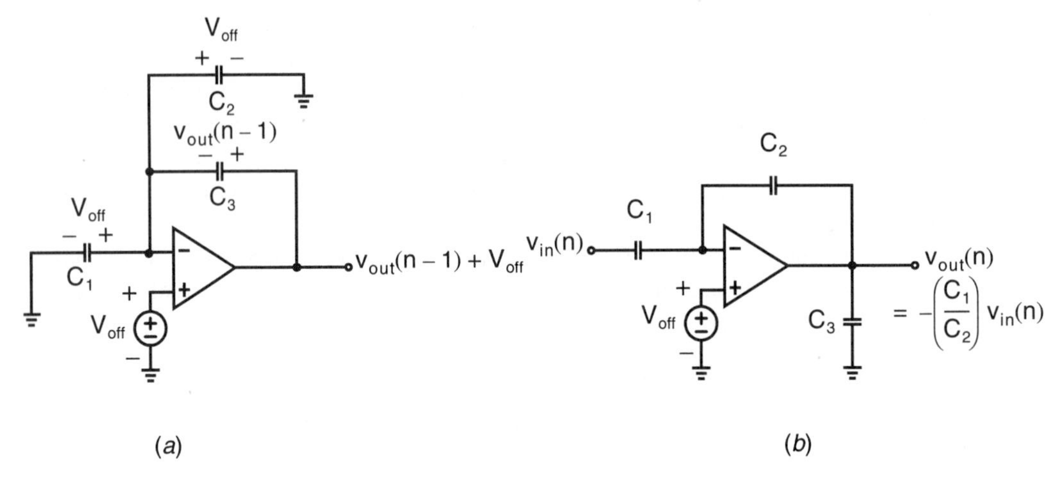

Fig. 10.32 Capacitive-reset gain circuit. (*a*) During reset (ϕ_2) when the output remains near previous output level. (*b*) During valid output (ϕ_1) when no offset error is present.

independent of the offset voltage in the same manner as the resettable gain circuit. In addition, we see that during ϕ_1, capacitor C_3 is charged to the output voltage. Before proceeding, it is of interest to note what happens to the charges on C_1 and C_2 when going from phase ϕ_1 to ϕ_2 (i.e., circuit (*b*) to (*a*) of Fig. 10.32). Since the charges on C_1 and C_2 are equal (recall that the charge on C_2 was obtained during the charging of C_1), when one side of each of C_1 and C_2 are grounded, their charges cancel each other and no charge is dumped onto C_3. Therefore, in the circuit of Fig. 10.32(*a*), the voltage on C_3 remains equal to the previous output voltage as shown. It should be

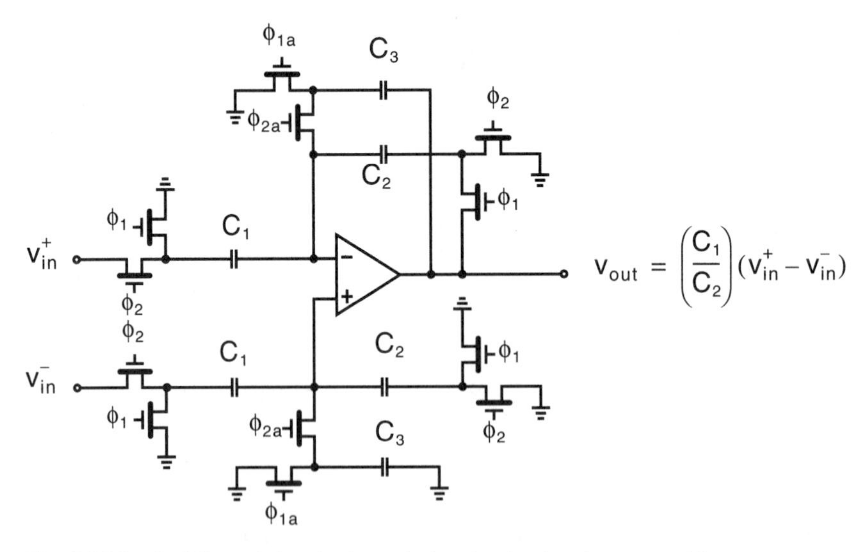

Fig. 10.33 A differential-to-single-ended gain circuit using capacitive reset.

noted here that even if the two charges did not precisely cancel, it would only result in the output voltage moving slightly further away from the previous output voltage during ϕ_2 but would not affect the output voltage during ϕ_1.

Finally, this capacitive-reset gain circuit can also realize a differential-to-single-ended gain stage, as shown in Fig. 10.33. Note that the shown circuit indicates that the switches connected around the virtual grounds are disconnected slightly sooner than other switches to reduce nonlinearities due to charge injection. This circuit has a switched-capacitor circuit connected to the positive input of the opamp that not only accepts differential inputs but also, to a first-order approximation, cancels the clock feedthrough of the switches [Martin, 1982]. The measured common-mode input rejection ratio for this circuit was 50 dB where a 5 μm technology was used together with a clock rate of 5 MHz [Martin, 1987].

10.7 CORRELATED DOUBLE-SAMPLING TECHNIQUES

The preceding SC gain amplifier is an example of using correlated double sampling (CDS) to minimize errors due to finite offset voltages, 1/f noise, and finite opamp gain. This technique is generally applicable to SC circuits of many different types [Temes, 1996]. To date, it has been used to realize highly accurate gain amplifiers, sample and holds, and integrators. The basic methodology in all cases is similar: During a calibration phase, the finite input voltage of the opamp is sampled and stored across capacitors; during the operation phase (when the output is being sampled), this error voltage is subtracted from the signal voltage by appropriate switching of the capacitors. A detailed description of CDS techniques is beyond the scope of this text; rather, a couple of examples will be briefly described and the interested reader can consult the tutorial [Temes, 1996] for more information.

A wide-band amplifier that uses CDS, similar to the SC amplifier of Fig. 10.31, but with superior high-frequency operation, is shown in Fig. 10.34. During ϕ_2, C_1' and C_2' are used to have

$$V_{out} \cong -\frac{C_1'}{C_2'} V_{in} \tag{10.99}$$

but with errors due to finite input-offset voltage, 1/f noise, and gain. At the same time, the finite opamp input voltage caused by these errors is sampled and stored across C_1 and C_2. Next, during ϕ_1, this input error voltage is subtracted from the signal (applied to the opamp input) at that time. Assuming that the input voltage and the opamp input error voltages did not change appreciably from ϕ_2 to ϕ_1, errors due to them will be significantly reduced.

The technique can also be used to realize accurate integrators. For example, the SC integrator shown in Fig. 10.35 [Nagaraj, 1986; Ki, 1991] uses an additional capacitor, C_2', to sample the opamp input error voltages during ϕ_1. Next, during ϕ_2, C_2' is connected in series with the opamp's inverting input and greatly minimizes the effects of these errors (by a factor of the inverse of the opamp's gain over what would otherwise occur at frequencies substantially less than the

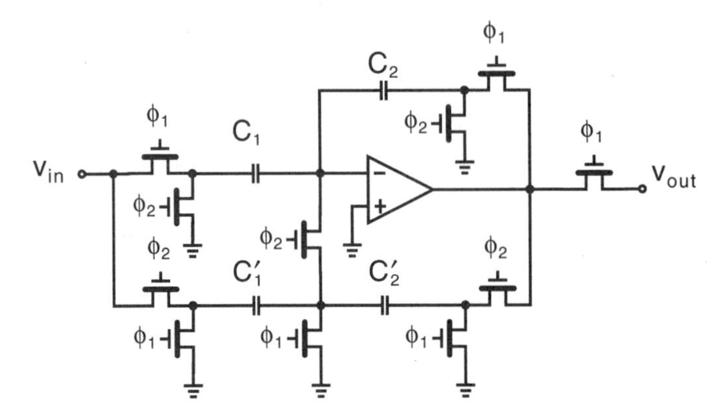

Fig. 10.34 An SC amplifier with CDS to minimize errors due to input- offset voltages, 1/f noise, and finite gain.

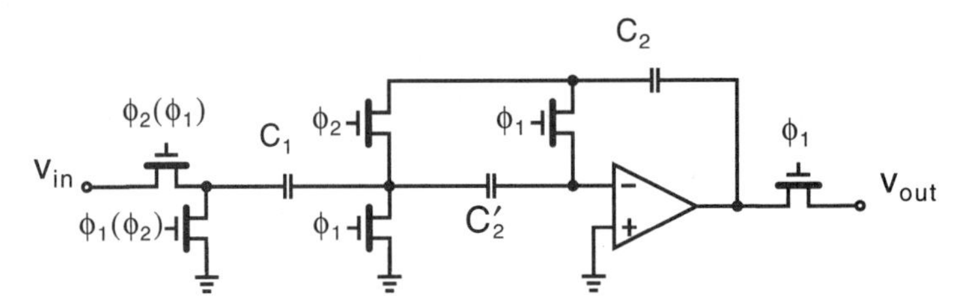

Fig. 10.35 An SC integrator with CDS to minimize errors due to input-offset voltages, 1/f noise, and finite gain.

sampling frequency). Similar architectures, with an open-loop capacitor in series with the opamp input during the phase when the output is sampled, can also be used for gain amplifiers.

When CDS sampling is used, the opamps should be designed to minimize thermal noise rather than 1/f noise. This might mean the use of n-channel input transistors rather than p-channel input transistors, for example.

When this technique is used in high-order SC filters, often only a couple of stages will have low-frequency gain from the opamp input terminals to the filter outputs and will require CDS circuitry. Other integrators can be more simply realized. This technique has proven to be very useful in applications such as oversampling A/D converters, where accurate integrators are required in the first stage [Hurst, 1989; Rebeschini, 1989] to reduce input-offset voltages and especially 1/f noise.

10.8 OTHER SWITCHED-CAPACITOR CIRCUITS

In this section, we present a variety of other switched-capacitor circuits useful for nonlinear applications. Specifically, we look at an amplitude modulator, full-wave rectifier, peak detector, voltage-controlled oscillator, and sinusoidal oscillator.

Amplitude Modulator

Amplitude modulators are used to shift a signal along the frequency axis. For example, to shift an information signal, $m(t)$, by a carrier frequency, ω_{ca}, we multiply $m(t)$ by a sinusoidal signal of frequency, ω_{ca}, or mathematically,

$$y(t) = m(t) \times \cos(\omega_{ca} t) \tag{10.100}$$

Since the Fourier transform of $\cos(\omega_{ca} t)$ is $\pi\delta(\omega - \omega_{ca}) + \pi\delta(\omega + \omega_{ca})$, and multiplication in the time domain is equivalent to convolution in the frequency domain, one can show that the spectrum of $y(t)$ is equal to

$$Y(\omega) = \frac{1}{2}M(\omega + \omega_{ca}) + \frac{1}{2}M(\omega - \omega_{ca}) \tag{10.101}$$

Thus, the output spectrum only has power around $\pm\omega_{ca}$.

However, in many cases it is difficult to realize a sinusoidal signal as well as a linear multiplier. As a result, many modulators make use of a square-wave carrier signal, in which case the output signal is simply

$$y(t) = m(t) \times s_q(t) = \pm m(t) \tag{10.102}$$

where $s_q(t)$ is a square wave operating at the carrier frequency whose amplitude has been normalized to 1. Thus, this approach simply requires a square-wave signal at the carrier frequency determining whether $m(t)$ or $-m(t)$ is passed to the output. The penalty paid here is that rather than $Y(\omega)$ having power located only around $\pm\omega_{ca}$, there will also be significant power around odd multiples of $\pm\omega_{ca}$ since a square-wave signal has power at odd harmonics of the fundamental. These extra images can typically be tolerated through the use of additional filtering stages.

A square-wave modulator can be realized using switched-capacitor techniques, as shown in Fig. 10.36. Here, the modulator is realized by making use of the capacitive-reset gain stage of Fig. 10.31. Note that the polarity of the output signal is determined by changing the clock phases of the input-transistor pair through the use of ϕ_A and ϕ_B, which are controlled by the square-wave carrier clock signal ϕ_{ca}. When ϕ_{ca} is high, $\phi_A = \phi_2$ and $\phi_B = \phi_1$, which results in a noninverting output of $(C_1/C_2)v_{in}$. When ϕ_{ca} goes low, the clock phases of ϕ_A, ϕ_B switch, and the circuit produces an inverting output of $-(C_1/C_2)v_{in}$.

Full-Wave Rectifier

A full-wave rectifier produces as its output the absolute value of the input. In other words, if the input is positive the output equals the input, but if the input is negative the output equals an inverted version of the input.

It is possible to realize a full-wave rectifier by making use of an amplitude modulator where, instead of a square-wave carrier signal being used, the output of a comparator controls the polarity of the output signal, as shown in Fig. 10.37. If the input signal is above ground, ϕ_{ca} goes high and the output is equal to a scaled version of the input. If the input signal is below ground, ϕ_{ca} goes low and the output equals an

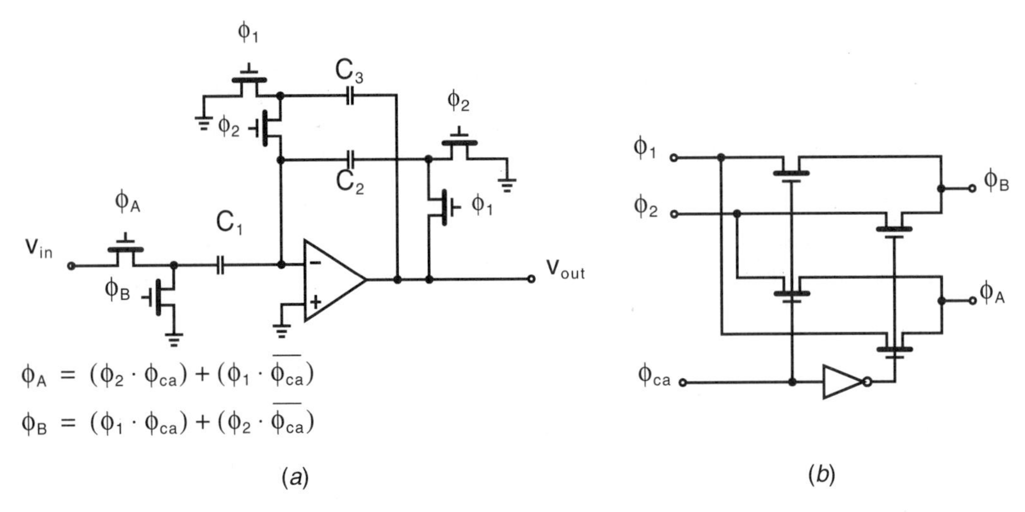

Fig. 10.36 (*a*) A switched-capacitor square-wave modulator where the input clock phases are controlled by the modulating square wave ϕ_M. (*b*) A possible circuit realization for ϕ_A and ϕ_B.

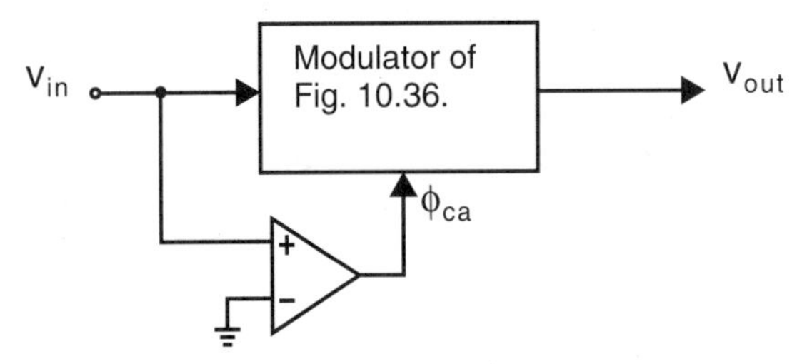

Fig. 10.37 A full-wave detector based on the square-wave modulator circuit of Fig. 10.36.

inverted scaled version of input. For proper operation, the comparator output must change synchronously with the sampling instances. Circuits to achieve this operation are given in [Gregorian, 1986].

Peak Detectors

Peak detectors produce as their output the maximum value of an input signal over some time period. While peak detectors will normally have some decaying circuitry associated with them (or a circuit to reset the peak occasionally), here we shall just look at the peak detector circuit itself. Two methods for realizing peak detectors are shown in Fig. 10.38.

In Fig. 10.38(*a*), a latched comparator is used to determine when $V_{in} > V_{out}$. When such a condition occurs, the comparator output goes high and the switch Q_1 turns on,

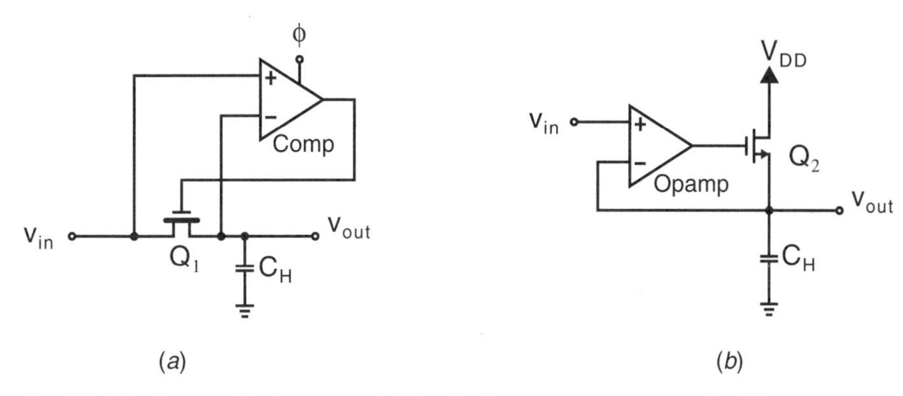

Fig. 10.38 Two-peak detectors: (*a*) latched-comparator approach; (*b*) continuous-time approach.

forcing $V_{out} = V_{in}$. The comparator is latched here to maintain its speed as well as ensuring that the comparator output is either high or low. Note that this circuit requires that the input signal directly drives C_H through Q_1, and hence buffering may be required. However, this circuit can be reasonably fast since latched comparators can be made to operate quickly and there is no opamp settling involved (only the RC time constant of Q_1 and C_H).

The circuit of Fig. 10.38(*b*) requires an opamp (with proper compensation), since it places Q_2 in a feedback connection. When $V_{in} > V_{out}$, the opamp output will start to go high, thus turning on Q_2 and hence turning on the feedback loop. While the feedback loop is on, the high loop gain will force $V_{out} = V_{in}$ with a high accuracy. Note that this circuit has a high input impedance as the input signal is applied to the noninverting terminal of an opamp. However, this opamp-based circuit will typically be slower than the comparator-based circuit of Fig. 10.38(*a*) since one needs to wait for the opamp output to slew and then settle. This slewing time can be minimized by adding circuitry to ensure that the gate of Q_2 never becomes lower than the output voltage when Q_2 is not conducting.

Voltage-Controlled Oscillator

A voltage-controlled oscillator (VCO) is an oscillator whose frequency can be adjusted through the use of a controlling voltage. As we shall see, the VCO described here generates an output square wave that places the oscillator alternatively in one of two states and is referred to as a relaxation oscillator. As we shall see in Chapter 16, a VCO is an integral part of a phase-locked loop (PLL).

The basic principle and typical waveforms of the relaxation oscillator are shown in Fig. 10.39 [Martin, 1981] and operate as follows. Assume that V_{out} has just toggled low to $-V_{SS}$ and node V_x is now at a level quite a bit higher than zero. On ϕ_1, k_2C will be charged up to a voltage of $-V_{SS}$, resulting in a charge value given by

$$Q_{k_2C} = -k_2CV_{SS} \tag{10.103}$$

Note that the voltage across the unswitched capacitor k_1C remains unchanged at $-V_{SS}$. Next, as ϕ_2 goes high, the charge on k_2C will be transferred to C (while the voltage across unswitched k_1C again remains unchanged) and the output will fall by an amount equal to

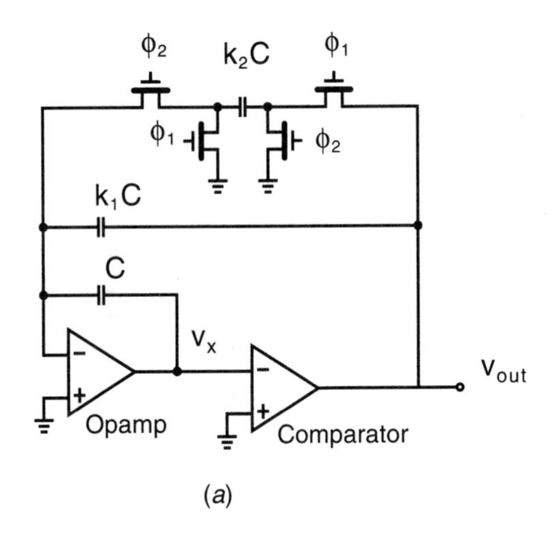

(a)

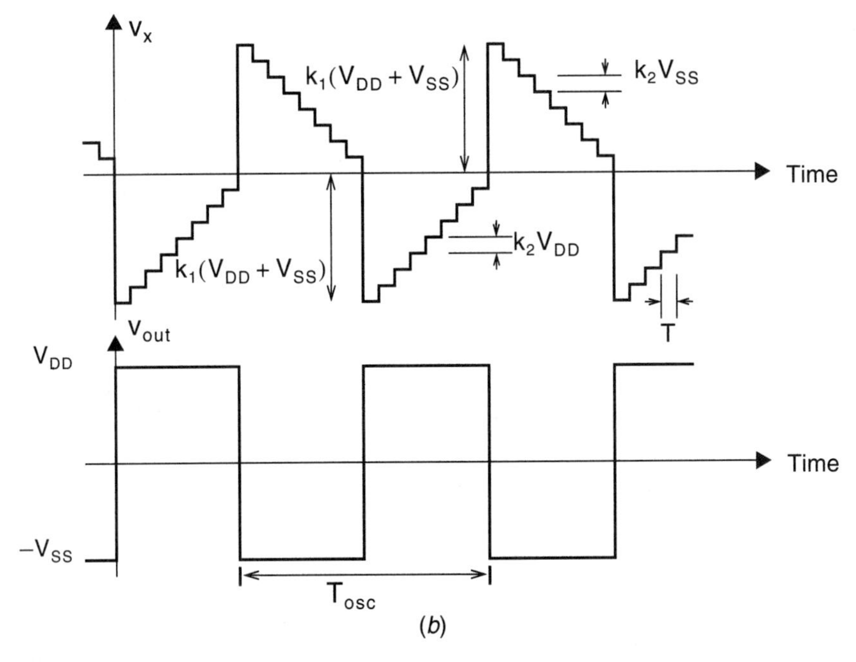

(b)

Fig. 10.39 A switched-capacitor relaxation oscillator.

$$\Delta v_x = \frac{\Delta Q_C}{C} = \frac{Q_{k_2C}}{C} = -k_2 V_{SS} \tag{10.104}$$

This operation will continue until v_x reaches zero volts. In effect, k_2C and C have formed a noninverting discrete-time integrator, while capacitor k_1C does not affect the operation of the circuit. Once v_x goes below zero, the comparator output will toggle to V_{DD}, and at this point a large amount of charge will pass through k_1C while the voltage across it changes from $-V_{SS}$ to V_{DD}. This large charge value is equal to

$$\Delta Q_{k_1C} = k_1 C(V_{DD} - (-V_{SS})) \tag{10.105}$$

and since this charge is drawn across C, then $\Delta Q_C = \Delta Q_{k_1C}$ and node v_x drops by an amount equal to

$$\Delta v_x = -\frac{\Delta Q_C}{C} = -k_1(V_{DD} + V_{SS}) \tag{10.106}$$

as shown in Fig. 10.39(*b*). After this time, v_x will step up in increments of k_2V_{DD} (since v_{out} now equals V_{DD}) until $v_x = 0$, at which point a large positive voltage step will occur. Just after this positive step, we will be back at our starting point, and the entire operation will repeat indefinitely.

To derive a formula for the oscillation period T_{osc}, we make the assumption that $k_1 >> k_2$ such that many small steps occur over one-half T_{osc} period. With this assumption, we can now write that the total number of clock cycles, T_{osc}/T, equals the number of steps during negative sloping v_x plus the number of steps during positive sloping v_x. Mathematically, we have

$$\frac{T_{osc}}{T} = \frac{k_1(V_{DD} + V_{SS})}{k_2V_{SS}} + \frac{k_1(V_{DD} + V_{SS})}{k_2V_{DD}} \tag{10.107}$$

Assuming $V_{DD} = V_{SS}$, we have a 50 percent duty cycle with

$$T_{osc} = 4\left(\frac{k_1}{k_2}\right)T \tag{10.108}$$

or equivalently,

$$f_{osc} = \frac{1}{4}\left(\frac{k_2}{k_1}\right)f \tag{10.109}$$

This fixed-frequency oscillator can be made voltage controlled by adding an extra feed-in capacitor to the oscillator, $k_{in}C$, as shown in Fig. 10.40. Here, when V_{in} is positive, extra charge packets are added into the integrator with the same sign as those for k_2C, and the circuit reaches $v_x = 0$ sooner. If V_{in} is negative, then extra charge packets are added into the integrator with the opposite sign as those for k_2C, and the circuit reaches $v_x = 0$ later. Assuming $k_1 >> k_2$, k_{in} and $V_{DD} = V_{SS}$, it is left as an exercise to the reader (Problem 10.12) to show that the frequency of oscillation for the

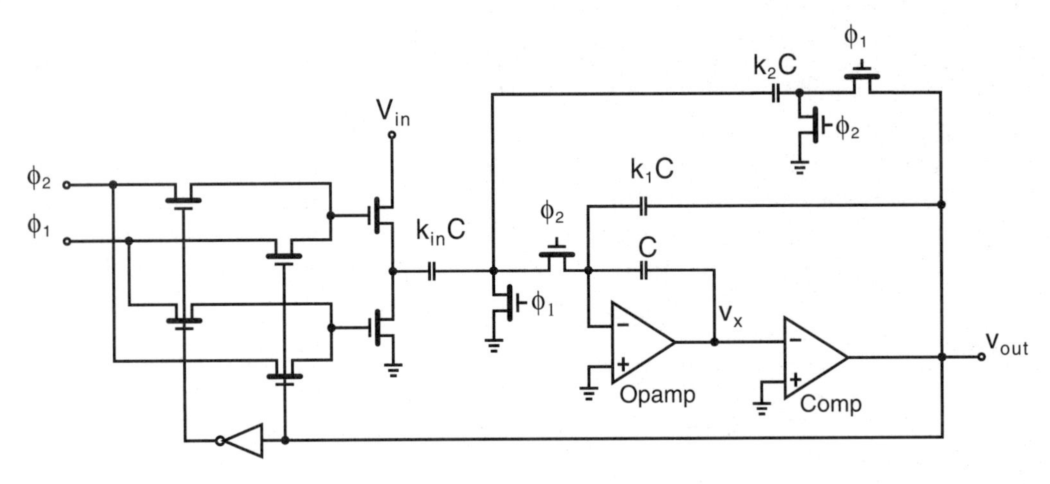

Fig. 10.40 A voltage-controlled oscillator (VCO) based on the relaxation oscillator of Fig. 10.39.

voltage-controlled oscillator of Fig. 10.40 is given by

$$f_{osc} = \frac{1}{4}\left(\frac{k_2}{k_1} + \frac{k_{in}V_{in}}{k_1 V_{DD}}\right)f \tag{10.110}$$

Sinusoidal Oscillator

In many applications, it is useful to realize an oscillator which produces a nearly sinusoidal output. For example, when a button is pressed on a touch-tone telephone, the signal sent down the telephone line is simply the sum of two sinusoidal signals at certain frequencies (the signal is known as a dual-tone multi-frequency (DTMF) signal). Thus, switched-capacitor sinusoidal oscillators could be used to generate the necessary touch-tone signals.

One method of generating a sinusoidal oscillator is shown in Fig. 10.41. Here, a high-Q bandpass filter with center frequency f_0 is used together with a hard limiter. Assuming some start-up circuit is used to force oscillations to exist, we see that a sinusoidal wave at v_{out} results in a square wave at v_1 which has a large component at f_0. The signal v_1 is filtered by the bandpass filter and thus maintains the sinusoidal signal, v_{out}. The oscillation frequency is set by the center frequency of the bandpass filter, while the sinusoidal amplitude is set by the center frequency gain of the filter and the peak values of the square-wave signal.

A switched-capacitor implementation of this sinusoidal oscillator is shown in Fig. 10.42 [Fleischer, 1985], where the high-Q bandpass filter has been realized using the circuit of Fig. 10.25. The comparator generates signals X and $\overline{X}$ with the result that the switched-input associated with V_{ref} is either positive or negative. The start-up circuit is used to sense when an ac signal is present at v_{out}. When no ac signal is detected, V_{st} is low and the extra capacitors C_5 and C_6 form a positive feedback

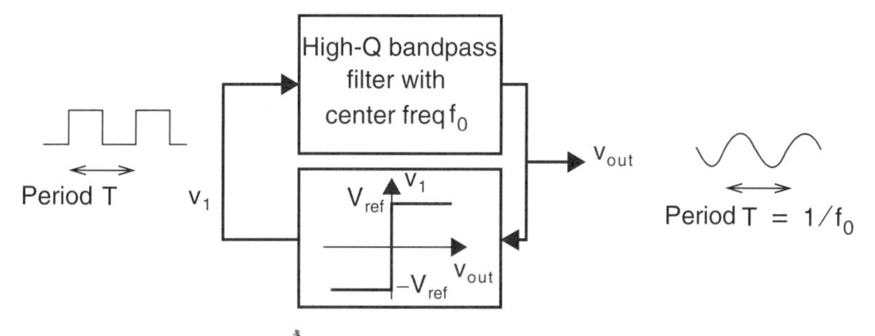

Fig. 10.41 A sinusoidal oscillator that makes use of a bandpass filter.

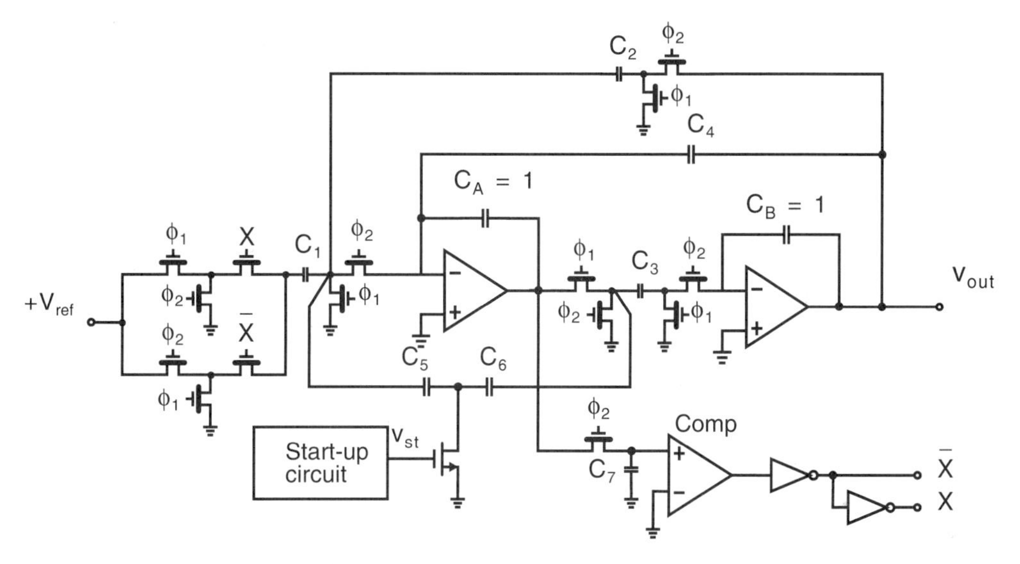

Fig. 10.42 A switched-capacitor implementation of a sinusoidal oscillator.

loop which causes oscillations to build up. Once an ac signal is detected, v_{st} goes high and the positive feedback loop is broken, thereby allowing oscillations to continue, as discussed previously.

For examples of other nonfiltering uses of switched-capacitor circuits, the interested reader is referred to [Gregorian, 1986].

10.9 REFERENCES

D. Allstot, R. Brodersen, and P. R. Gray. "MOS Switched-Capacitor Ladder Filters," *IEEE J. of Solid-State Circuits,* Vol. SC-13, no. 6, pp. 806–814, December 1978.

P. O. Brackett and A. S. Sedra. *Filter Theory and Design: Active and Passive.* Matrix Publishers, Champaign, Illinois, p. 575, 1978.

L. T. Bruton. "Low-Sensitivity Digital Ladder Filters," *IEEE Trans. Circuits and Systems,* Vol. CAS-22, no. 3, pp. 168–176, March 1975.

J. T. Caves, M. A. Copeland, C. F. Rahim, and S. D. Rosenbaum. "Sampled-Data Filters Using Switched Capacitors as Resistor Equivalents," *IEEE J. of Solid-State Circuits,* Vol. SC-12, pp. 592–600, December 1977.

P. E. Fleischer, A. Ganesan, and K. R. Laker, "A Switched-Capacitor Oscillator with Precision Amplitude Control and Guaranteed Start-up," *IEEE J. of Solid-State Circuits,* Vol. SC-20, pp. 641–647, April 1985.

D. L. Fried. "Analog Sample-Data Filters," *IEEE J. of Solid-State Circuits,* Vol. SC-7, pp. 302–304, August 1972.

T. Foxall, R. Whitbread, L. Sellars, A. Aitken, and J. Morris. "A Switched-Capacitor Bandsplit Filter Using Double Polysilicon Oxide-Isolated CMOS," *1980 ISSCC Dig. of Technical Papers,* pp. 90–91, February 1980.

R. Gregorian. "High-Resolution Switched-Capacitor D/A Converter," *Microelectronics J.,* Vol. 12, no. 2, pp. 10–13, March/April 1981.

R. Gregorian, K. Martin, and G. C. Temes. "Switched-Capacitor Circuit Design," *IEEE Proceedings,* Vol. 71, no. 8, pp. 941–966, August 1983.

R. Gregorian and G. C. Temes. *Analog MOS Integrated Circuits for Signal Processing,* John Wiley & Sons, New York, 1986.

B. J. Hosticka, R. W. Broderson, and P. R. Gray. "MOS Sampled Data Recursive Filters Using Switched-Capacitor Integrators," *IEEE J. of Solid-State Circuits,* Vol. SC-12, no. 6, pp. 600–608, December 1977.

P. J. Hurst and R. A. Levinson. "Delta-Sigma A/D with Reduced Sensitivity to Op-Amp Noise and Gain," *Proc. of IEEE Int. Symp. on Circuits and Systems,* pp. 254–257, 1989.

G. M. Jacobs, D. J. Allstot, R. W. Brodersen, and P. R. Gray. "MOS Switched-Capacitor Ladder Filters," *IEEE Trans. Circuits and Systems,* Vol. CAS-25, pp. 1014–1021, December 1978.

W. H. Ki and G. C. Temes. "Offset- and Clock-Feedthrough-Compensated SC Filters," *Proc. of IEEE Int. Symp. on Circuits and Systems,* pp. 1561–1564, 1991.

K. R. Laker and W. M. C. Sansen. *Design of Analog Integrated Circuits and Systems*. McGraw Hill, New York, 1994.

L. E. Larson and G. C. Temes. "SC Building Blocks with Reduced Sensitivity to Finite Amplifier Gain, Bandwidth, and Offset Voltage," *Proc. of IEEE Int. Symp. on Circuits and Systems,* pp. 334–338, 1987.

K. Martin. "Improved Circuits for the Realization of Switched-Capacitor Filters," *IEEE Trans. Circuits and Systems*, Vol. CAS-27, no. 4, pp. 237–244, April 1980; Also published as BNR Tech. Rep. TR1E 81-78-06, March 1978.

K. Martin. "Improved Circuits for the Realization of Switched Capacitor Filters," *Proc. 1979 Int. Symp. on Circuits and Systems (IEEE & IECE),* Tokyo, pp. 756–760, July 1979.

K. Martin. "A Voltage-Controlled Switched-Capacitor Relaxation Oscillator," *IEEE J. of Solid State Circuits,* Vol. SC-16, no. 4, pp. 412–414, August 1981.

K. Martin. "New Clock-Feedthrough Cancellation Technique for Analogue MOS Switched-Capacitor Circuits," *Electron. Lett.,* Vol. 18, no. 1, pp. 39–40, January 1982.

K. Martin. L. Ozcolak, Y. S. Lee, and G. C. Temes. "A Differential Switched-Capacitor Amplifier," *IEEE J. of Solid-State Circuits,* Vol. SC-22, no. 1, pp. 104–106, February 1987.

K. Martin and S. Rosenbaum. "Improvements in Sampled Analog Filter," filed March 1978, in Canada, the United States and Japan. Issued in Canada October 21, 1980, patent no. 1088161. Issued in England June 10, 1982, no. CYB 2019151B.

K. Martin and A. S. Sedra. "Strays-Insensitive Switched-Capacitor Filters Based on Bilinear z-Transform," *Electron. Lett.,* Vol. 15, no. 13, pp. 365–366, June 1979.

K. Martin and A. S. Sedra. "Exact Design of Switched-Capacitor Bandpass Filters Using Coupled-Biquad Structures," *IEEE Trans. Circuits and Systems*, Vol. CAS-27, no. 6, pp. 469–478, June 1980.

K. Martin and A. S. Sedra. "Switched-Capacitor Building Blocks for Adaptive Systems," *IEEE Trans. Circuits and Systems,* Vol. CAS-28, no. 6, pp. 576–584, June 1981.

K. Nagaraj et al. "SC Circuits with Reduced Sensitivity to Amplifier Gain," *Proc. of IEEE Int. Symp. on Circuits and Systems*, pp. 618–621, 1986.

H. Matsumoto and K. Watanabe. "Spike-Free Switched-Capacitor Circuits," *Electron. Lett.,* Vol. 23, no. 8, pp. 428–429, April 1987.

W. Poschenrieder, "Frequencz Filterung durch Netzwerke mit Periodis Gesteurten Schaltern," *Analys und Synthese von Netzwerken,* pp. 220–237, Tanungsheft, Stuttgart, 1966.

M. Rebeschini et al. "A High-Resolution CMOS Sigma-Delta A/D Converter with 320 kHz Output Rate," *Proc. of IEEE Int. Symp. on Circuits and Systems,* pp. 246–249, 1989.

R. Schaumann, M. S. Ghausi, and K. R. Laker. *Design of Analog Filters.* Prentice Hall, Englewood Cliffs, New Jersey, 1990.

G. C. Temes and C. Enz. "Autozeroing, Correlated Double Sampling, and Chopper Stabilization," *IEEE Proceedings,* to be published, 1996.

B. J. White, G. M. Jacobs, and G. F. Landsburg. "A Monolithic Dual-Tone Multifrequency Receiver," *IEEE J. of Solid-State Circuits,* Vol. SC-14, no. 6, pp. 991–997, December 1978.

I. A. Young, P. R. Gray, and D. A. Hodges. "Analog NMOS Sampled-Data Recursive Filters," *IEEE Int. Symp. on Solid-State Circuits,* pp. 156–157, February 1977.

10.10 PROBLEMS

10.1 Assuming logic signals have infinitely fast fall and rise times but are delayed by 1 ns through a logic gate, draw the output waveforms of the two phase clock generators shown in Fig. 10.3(*b*). Assume the delay blocks add delays of 10 ns each and a clock frequency of 10 MHz is applied to the input.

10.2 Ignoring the effect of parasitic capacitances, find the discrete-time transfer function of the switched-capacitor circuit shown in Fig. P10.2.

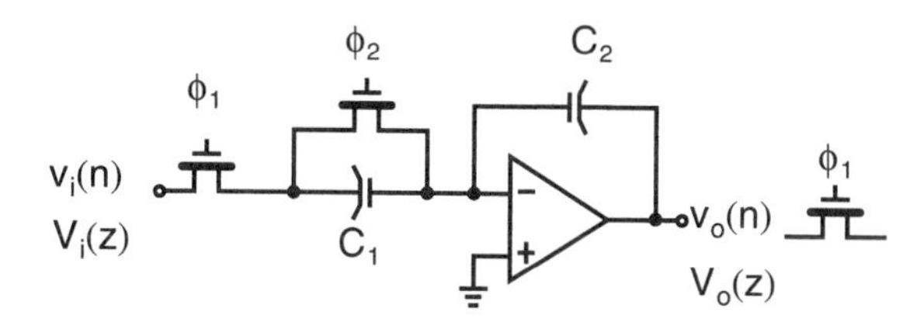

Fig. P10.2

10.3 Repeat Problem 10.2 but include the effect of parasitic capacitances.

10.4 Show that when an opamp has a finite gain of A, the transfer function for the discrete-time integrator of Fig. 10.5 is given by

$$\frac{V_o(z)}{V_i(z)} = -\left(\frac{C_1}{C_2}\right)\left(\frac{1}{z\left(1+\dfrac{C_1}{C_2 A}\right)-1}\right)$$

Also show that this transfer function has a dc gain of $-A$ and a pole that is located slightly to the left of 1 by an amount approximately equal to $(C_1/C_2)(1/A)$.

10.5 Repeat Example 10.3 but assume the discrete-time zero is at $z = 0$ and the pole remains at $z = 0.53327$. Note that with this change, C_1 is no longer required at the sacrifice of a nonzero gain being obtained at 50 kHz. What is the new gain at 50 kHz?

10.6 Find the capacitances needed for the first-order filter shown in Fig. 10.15(*a*) to realize a discrete-time transfer function having a dc gain of one, a zero at 0, and a 3-dB frequency of 1 kHz when a clock frequency of 50 kHz is used. Assume that $C_A = 50\ pF$. What is the gain (in dB) at 25 kHz?

10.7 Find the transfer function of the circuit shown in Fig. 10.16 when $C_1 = 0\ pF$, $C_2 = 2\ pF$, $C_3 = 2\ pF$, and $C_A = 20\ pF$. What is the magnitude and phase of the gain at dc, $f_s/4$ and $f_s/2$?

10.8 Find the discrete-time transfer function of the switched capacitor circuit shown in Fig. P10.8.

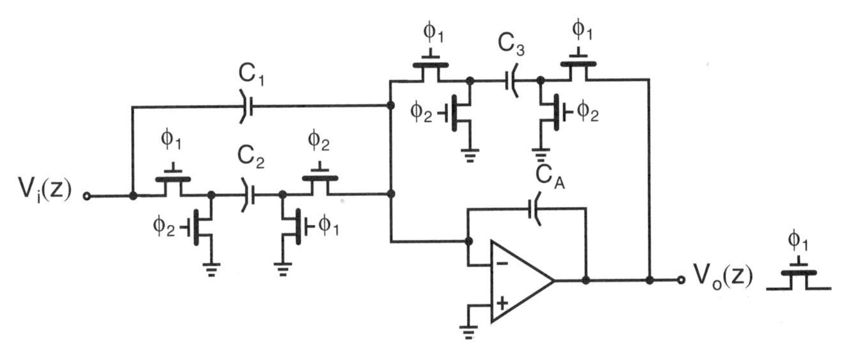

Fig. P10.8

10.9 Show that when $K_6 = 0$ in the biquad circuit of Fig. 10.21, if the poles are complex, they lie precisely on the unit circle (i.e., the circuit is a resonator).

10.10 Using the bilinear transform, design a bandpass filter with a $Q = 20$ (in the continuous-time domain) and a peak gain of one near $f_s/100$. Find the largest to smallest capacitor ratio if this transfer function is realized using the high-Q biquad circuit, and compare it to that which would be obtained if the low-Q circuit were used instead. Let $C_1 = C_2 = 1$ in both cases.

10.11 Sketch a typical output waveform for the capacitive-reset gain circuit of Fig. 10.31 if the opamp has an input offset of 100 mV and a gain of 10 is desired.

10.12 Verify that Eq. (10.104) gives the frequency of oscillation for the VCO of Fig. 10.40.

CHAPTER

11 Data Converter Fundamentals

In this chapter, fundamental aspects of analog-to-digital (A/D) and digital-to-analog (D/A) converters are presented without regard for their internal architecture or circuit design. In effect, converters are treated in this chapter as black boxes such that only their input-output relationships are discussed. Internal architectures and circuits techniques for realizing data converters are discussed in following chapters.

Before proceeding, it is useful to make the distinction between two main types of data converters.

Nyquist-Rate Converters We loosely define *Nyquist-rate data converters* as those converters that generate a series of output values in which each value has a one-to-one correspondence with a single input value. For example, a Nyquist-rate D/A converter would generate a series of analog output levels, where each level is a result of a single B-bit input word. However, it should be noted that Nyquist-rate converters are seldom used at the Nyquist rate due to the difficulty in realizing practical anti-aliasing and reconstruction filters. In most cases, Nyquist rate converters operate at 1.5 to 10 times the Nyquist rate (i.e., 3 to 20 times the input signal's bandwidth).

Oversampling Converters *Oversampling converters* are those converters that operate much faster than the input signal's Nyquist rate (typically 20 to 512 times faster) and increase the output's signal-to-noise ratio (SNR) by filtering out quantization noise that is not in the signal's bandwidth. In A/D converters, this filtering is performed digitally, whereas in D/A converters, analog filtering is used. Most often, oversampling converters use *noise shaping* to place much of the quantization noise outside the input signal's bandwidth (see Chapter 14).

This distinction is made here since much of the descriptions that follow refer more closely to Nyquist-rate converters than to oversampling converters.

11.1 IDEAL D/A CONVERTER

Consider the block diagram of an N-bit D/A converter shown in Fig. 11.1. Here, B_{in} is defined to be an N-bit digital signal (or word) such that

$$B_{in} = b_1 2^{-1} + b_2 2^{-2} + \cdots + b_N 2^{-N} \tag{11.1}$$

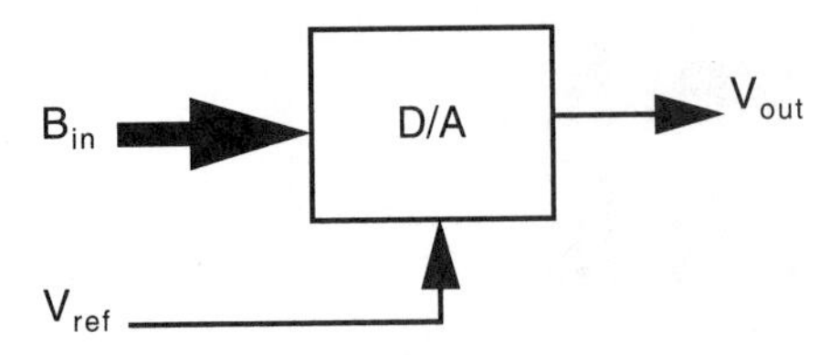

Fig. 11.1 A block diagram representing a D/A converter.

where b_i equals 1 or 0 (i.e., b_i is a binary digit). We also define b_1 as the most significant bit (MSB) and b_N as the least significant bit (LSB). Furthermore, we have assumed here that B_{in} represents a positive value, resulting in a unipolar D/A converter. A *unipolar D/A converter* produces an output signal of only one polarity. In contrast, *signed converters* produce output signals of either positive or negative polarity, depending on a sign bit (usually b_1). Extending the following concepts to the signed case is straightforward but requires knowledge of the type of digital representation used (i.e., sign magnitude, offset binary, or 2's complement). Signed codes are discussed in Section 11.4.

The analog output signal, V_{out}, is related to the digital signal, B_{in}, through an analog reference signal, V_{ref}. For simplicity, we assume that both V_{out} and V_{ref} are voltage signals, although, in general, they may be other physical quantities, such as current or charge. The relationship between these three signals for a unipolar D/A converter is given by

$$V_{out} = V_{ref}(b_1 2^{-1} + b_2 2^{-2} + \cdots + b_N 2^{-N}) = V_{ref} B_{in} \tag{11.2}$$

It is useful to define V_{LSB} to be the voltage change when one LSB changes, or, mathematically,

$$V_{LSB} \equiv \frac{V_{ref}}{2^N} \tag{11.3}$$

Also useful (particularly in measuring errors) is the definition of a new "unit," namely, LSB units, which are in fact unitless.

$$1 \text{ LSB} = \frac{1}{2^N} \tag{11.4}$$

The transfer curve for an ideal 2-bit D/A converter is shown in Fig. 11.2. Note here that, although only a finite number of analog values occur at the output, for an ideal D/A converter, the output signals are well-defined values. Also note that the maximum value of V_{out} is not V_{ref} but rather $V_{ref}(1 - 2^{-N})$, or equivalently, $V_{ref} - V_{LSB}$. Finally, as we can see from (11.2), a multiplying D/A converter is realized by simply allowing the reference signal, V_{ref}, to be a varying input signal along with the digital input, B_{in}. Such an arrangement results in V_{out} being proportional to the multiplication of the input signals, B_{in} and V_{ref}.

EXAMPLE 11.1

An 8-bit D/A converter has $V_{ref} = 5\ \text{V}$. What is the output voltage when $B_{in} = 10110100$? Also, find V_{LSB}.

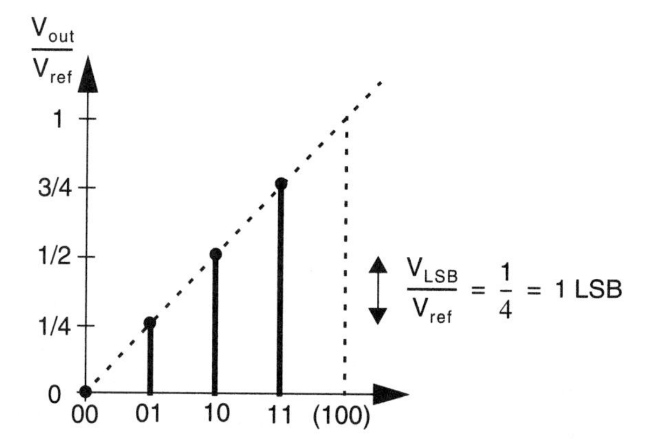

Fig. 11.2 Input-output transfer curve for an ideal 2-bit D/A converter.

Solution

We can find the decimal equivalent of B_{in} using (11.1)

$$B_{in} = 2^{-1} + 2^{-3} + 2^{-4} + 2^{-6} = 0.703125 \tag{11.5}$$

Then, using (11.2), we find

$$V_{out} = V_{ref}B_{in} = 3.516 \text{ V} \tag{11.6}$$

and

$$V_{LSB} = 5/256 = 19.5 \text{ mV} \tag{11.7}$$

11.2 IDEAL A/D CONVERTER

The block diagram representation for an A/D converter is shown in Fig. 11.3, where B_{out} is the digital output word while V_{in} and V_{ref} are the analog input and reference signals, respectively. Also, we define V_{LSB} to be the signal change corresponding to a single LSB change as in the D/A case.

For an A/D converter, the following equation relates these signals,

$$V_{ref}(b_1 2^{-1} + b_2 2^{-2} + \cdots + b_N 2^{-N}) = V_{in} \pm V_x$$

where

$$-\frac{1}{2}V_{LSB} \le V_x < \frac{1}{2}V_{LSB} \tag{11.8}$$

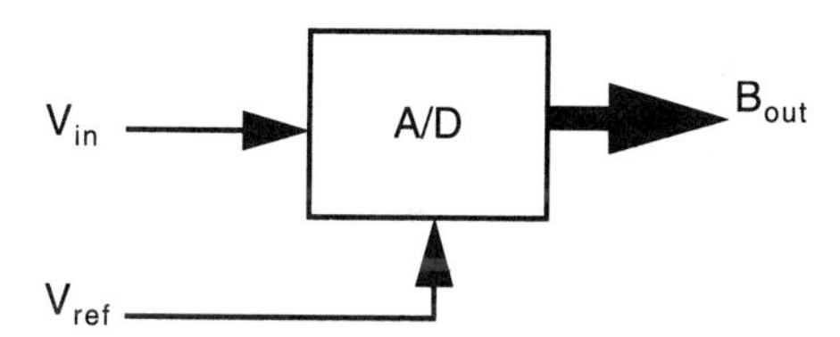

Fig. 11.3 A block diagram representing an A/D converter.

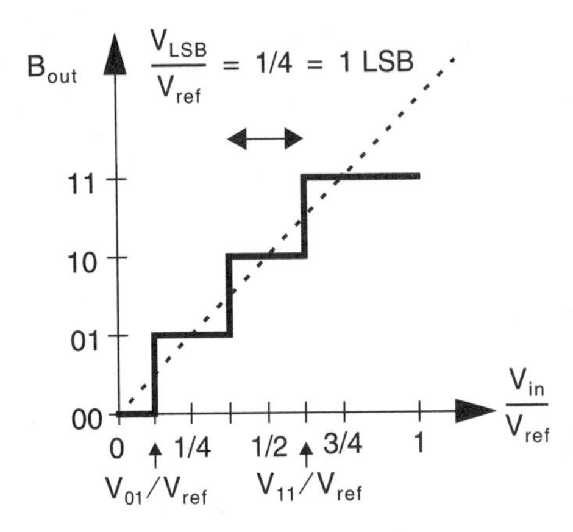

Fig. 11.4 Input-output transfer curve for a 2-bit A/D converter.

Note that there is now a range of valid input values that produce the same digital output word. This signal ambiguity produces what is known as *quantization error.* Also, note that no quantization error occurs in the case of a D/A converter since the output signals are well defined.[1]

As in the D/A case, a transfer curve for an A/D converter can be sketched as shown in Fig. 11.4 for a 2-bit converter. Note that the transitions along the V_{in} axis are offset by $1/2\ V_{LSB}$, so that the midpoints of the staircase curve fall precisely on the equivalent D/A transfer curve. Here, we define the transition voltages at V_{ij}, where the subscript ij indicates the upper B_{out} value of the transition. For example, V_{01} is shown in Fig. 11.4 as the voltage (normalized with respect to V_{ref}) for the transition from 00 to 01.

Finally, it should be noted that the relation shown in (11.8) holds only if the input signal remains within 1 LSB of the two last transition voltages. Specifically, for the 2-bit transfer curve shown in Fig. 11.4, V_{in} should remain less than $7/8\ V_{ref}$ and greater than $-1/8\ V_{ref}$. Otherwise, the quantizer is said to be *overloaded* since the magnitude of the quantization error would be larger than $V_{LSB}/2$.

11.3 QUANTIZATION NOISE

As mentioned in Section 11.2, quantization errors occur even in ideal A/D converters. In this section, we model these errors as being equivalent to an additive noise source and then find the power of this noise source. Consider the setup shown in Fig. 11.5, where both N-bit converters are ideal.

Since we have

$$V_Q = V_1 - V_{in} \tag{11.9}$$

---

1. Of course, quantization errors will occur if a 6-bit D/A converter is used to convert a 10-bit digital signal, but in this case, the quantization error occurs in the conversion from a 10-bit digital signal to a 6-bit digital signal.

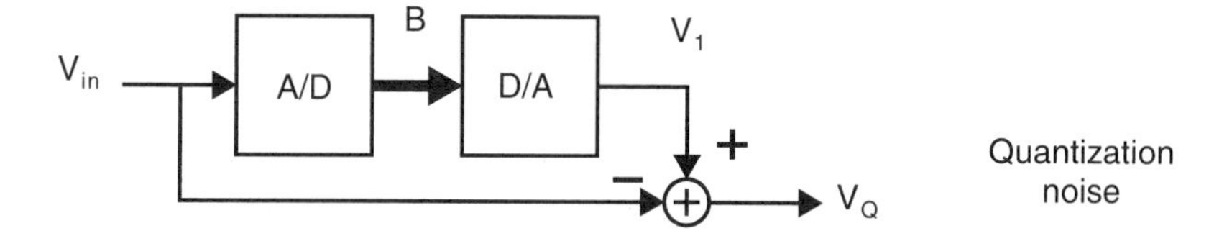

Fig. 11.5 A circuit to investigate quantization noise behavior.

we can rearrange this equation as

$$V_1 = V_{in} + V_Q \tag{11.10}$$

Although this rearrangement is trivial, it has important implications because (11.10) shows that the quantized signal, V_1, can be modelled as the input signal, V_{in}, plus some additive quantization noise signal, V_Q. Note that (11.10) is exact because no approximations have been made here. The quantization noise modelling becomes approximate once some assumptions are made about the statistical properties of V_Q.

Deterministic Approach

To gain an understanding of some of the properties of the quantization noise signal, V_Q, we will investigate its behavior when the input is a particular function. Specifically, let us assume the input signal, V_{in}, in Fig. 11.5 is a ramp. Such an input signal results in the output from the D/A, V_1, appearing as a staircase, as shown in Fig. 11.6, assuming no overloading occurs. Taking the difference between these two signals gives us the noise signal, V_Q, which is a result of quantization error. Note that the quantization signal, V_Q, is limited to $\pm V_{LSB}/2$ and will be so limited for all input signals (not just ramps). Clearly, the average of V_Q is zero. However, the rms value of the noise signal, $V_{Q(rms)}$, is given by

$$\begin{aligned} V_{Q(rms)} &= \left[\frac{1}{T}\int_{-T/2}^{T/2} V_Q^2\, dt\right]^{1/2} = \left[\frac{1}{T}\int_{-T/2}^{T/2} V_{LSB}^2\left(\frac{-t}{T}\right)^2 dt\right]^{1/2} \\ &= \left[\frac{V_{LSB}^2}{T^3}\left(\frac{t^3}{3}\bigg|_{-T/2}^{T/2}\right)\right]^{1/2} \end{aligned} \tag{11.11}$$

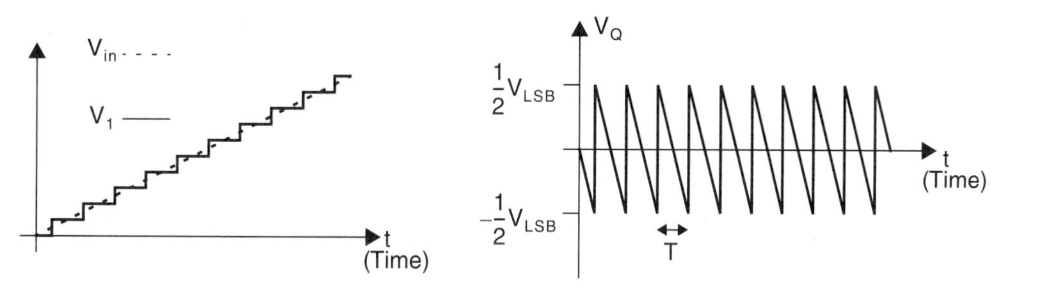

Fig. 11.6 Applying a ramp signal to the circuit in Fig. 11.5.

$$V_{Q(rms)} = \frac{V_{LSB}}{\sqrt{12}} \tag{11.12}$$

Thus, we see that the rms power of the quantization noise source is proportional to the size of V_{LSB}, which is determined by the number of bits, N, in the converter.

Stochastic Approach

The preceding deterministic approach was presented as a simple example to see some properties of the quantization noise signal. However, to deal with the more general input case, a stochastic approach is typically used. In a stochastic approach, we assume that the input signal is varying rapidly such that the quantization error signal, V_Q, is a random variable uniformly distributed between $\pm V_{LSB}/2$. The probability density function for such an error signal, $f_Q(x)$, will be a constant value, as shown in Fig. 11.7.

The average value of the quantization error, $V_{Q(avg)}$, is found to be zero as follows:

$$V_{Q(avg)} = \int_{-\infty}^{\infty} x f_Q(x)\, dx = \frac{1}{V_{LSB}}\left(\int_{-V_{LSB}/2}^{V_{LSB}/2} x\, dx\right) = 0 \tag{11.13}$$

In a similar fashion, the rms value of the quantization error is given by

$$V_{Q(rms)} = \left[\int_{-\infty}^{\infty} x^2 f_e(x)\, dx\right]^{1/2} = \left[\frac{1}{V_{LSB}}\left(\int_{-V_{LSB}/2}^{V_{LSB}/2} x^2\, dx\right)\right]^{1/2} = \frac{V_{LSB}}{\sqrt{12}} \tag{11.14}$$

which is the same result as (11.12), which is found using the deterministic ramp input signal. The fact that these two results are identical should come as no surprise since randomly chosen samples from the sawtooth waveform in the deterministic case would also have a uniformly distributed probability density function. In general, the rms quantization noise power equals $V_{LSB}/\sqrt{12}$ when the quantization noise signal is uniformly distributed over the interval $\pm V_{LSB}/2$.

Recalling that the size of V_{LSB} is halved for each additional bit and assuming that V_{ref} remains constant, we see from (11.14) that the noise power decreases by 6 dB for each additional bit in the A/D converter. Thus, given an input signal waveform, a

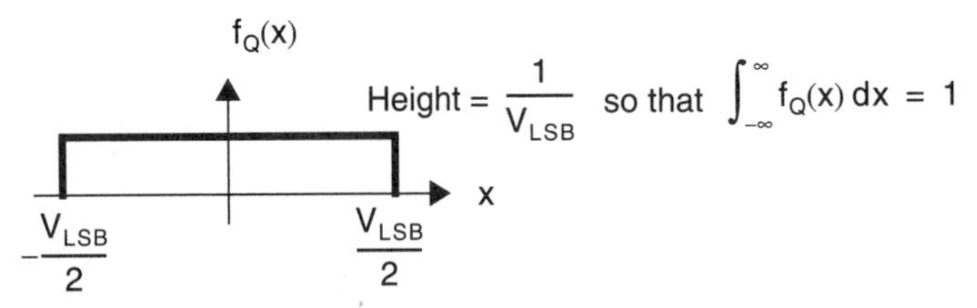

Fig. 11.7 Assumed probability density function for the quantization error, V_Q.

formula can be derived giving the best possible signal-to-noise ratio (SNR) for a given number of bits in an ideal A/D converter.

For example, assuming V_{in} is a sawtooth of height V_{ref} (or equivalently, a random signal uniformly distributed between 0 and V_{ref}) and considering only the ac power of the signal, the SNR is given by

$$\begin{aligned} SNR &= 20\log\left(\frac{V_{in(rms)}}{V_{Q(rms)}}\right) = 20\log\left(\frac{V_{ref}/\sqrt{12}}{V_{LSB}/\sqrt{12}}\right) \\ &= 20\log(2^N) = 6.02N \text{ dB} \end{aligned} \tag{11.15}$$

For example, a 10-bit A/D converter has a best possible SNR of about 60 dB.

Alternatively, a more common SNR formula is to assume V_{in} is a sinusoidal waveform between 0 and V_{ref}. Thus, the ac power of the sinusoidal wave is $V_{ref}/(2\sqrt{2})$, which results in

$$\begin{aligned} SNR &= 20\log\left(\frac{V_{in(rms)}}{V_{Q(rms)}}\right) \\ &= 20\log\left(\frac{V_{ref}/(2\sqrt{2})}{V_{LSB}/(\sqrt{12})}\right) \\ &= 20\log\left(\sqrt{\frac{3}{2}}2^N\right) \\ SNR &= 6.02N + 1.76 \text{ dB} \end{aligned} \tag{11.16}$$

In other words, a sinusoidal signal has 1.76 dB more ac power than a random signal uniformly distributed between the same peak levels.

Note that (11.16) gives the best possible SNR for an N-bit A/D converter. However, the idealized SNR decreases from this best possible value for reduced input signal levels. For example, Fig. 11.8 shows a plot of the idealized SNR for a 10-bit A/D converter versus the sinusoidal input signal amplitude. However, it should be noted that these SNR values could be improved through the use of oversampling techniques if the input signal's bandwidth is lower than the Nyquist rate. Oversampling will be discussed in detail in Chapter 14, where the design of oversampling converters is presented.

EXAMPLE 11.2

A 100-mV_{pp} sinusoidal signal is applied to an ideal 12-bit A/D converter for which $V_{ref} = 5$ V. Find the SNR of the digitized output signal.

Solution

First, we use (11.16) to find the maximum SNR if a full-scale sinusoidal waveform of ± 2.5 V were applied to the input.

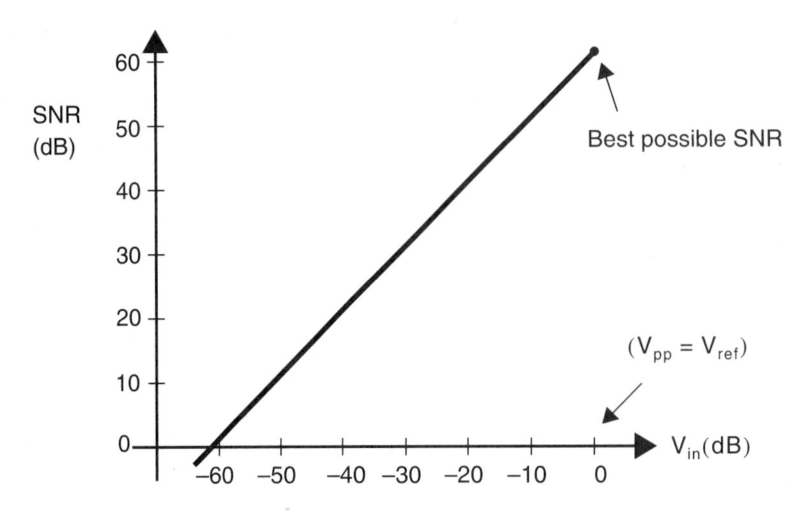

Fig. 11.8 Idealized SNR versus sinusoidal input signal amplitude for a 10-bit A/D converter. The 0-dB input signal amplitude corresponds to a peak-to-peak voltage equaling V_{ref}.

$$SNR_{max} = 6.02 \times 12 + 1.76 = 74 \text{ dB} \tag{11.17}$$

However, since the input is only a ±100-mV sinusoidal waveform that is 28 dB below full scale, the SNR of the digitized output is

$$SNR = 74 - 28 = 46 \text{ dB} \tag{11.18}$$

11.4 SIGNED CODES

In many applications, it is necessary to create a converter that operates with both positive and negative analog signals, resulting in the need for both positive and negative digital representations. Typically, the analog signal is bounded by $\pm 0.5V_{ref}$, such that its full-scale range is the same magnitude as in the unipolar case. Some common signed digital representations are sign magnitude, 1's complement, offset binary, and 2's complement, as shown in Table 11.1 for the 4-bit case. Note from Table 11.1 that all positive number representations are the same except for the offset-binary case, where the MSB is complemented.

Sign Magnitude

For negative numbers in the sign-magnitude case, all the bits are the same as for the positive number representation, except that the MSB is complemented. For example, the sign-magnitude representation for 5 is 0101, whereas for –5 it is 1101. However, note that this approach results in two representations for the number 0, and thus only $2^N - 1$ numbers are represented.

Table 11.1 Some 4-bit signed digital representations

Number	Normalized number	Sign magnitude	1's complement	Offset binary	2's complement
+7	+7/8	0111	0111	1111	0111
+6	+6/8	0110	0110	1110	0110
+5	+5/8	0101	0101	1101	0101
+4	+4/8	0100	0100	1100	0100
+3	+3/8	0011	0011	1011	0011
+2	+2/8	0010	0010	1010	0010
+1	+1/8	0001	0001	1001	0001
+0	+0	0000	0000	1000	0000
(–0)	(–0)	(1000)	(1111)		
–1	–1/8	1001	1110	0111	1111
–2	–2/8	1010	1101	0110	1110
–3	–3/8	1011	1100	0101	1101
–4	–4/8	1100	1011	0100	1100
–5	–5/8	1101	1010	0011	1011
–6	–6/8	1110	1001	0010	1010
–7	–7/8	1111	1000	0001	1001
–8	–8/8			0000	1000

1's Complement

In 1's-complement representation, negative numbers are represented as the complement of all the bits for the equivalent positive number. For example, here, 5 is once again 0101, whereas –5 is now 1010. It should be noted that the 1's-complement case also has two representations for 0, and thus only $2^N - 1$ numbers are represented.

Offset Binary

The offset-binary representation is obtained by assigning 0000 to the most negative number and then counting up, as in the unipolar case. In other words, this system can be thought of as simply a unipolar representation counting from 0 to 2^N, but where the decimal numbers represented are offset by 2^{N-1}, or equivalently, the decimal counts are from -2^{N-1} to 2^{N-1}. For example, the offset-binary code for the number 5 in the 4-bit case is the same as the unipolar code for the number 13, which is 1101, since $13 = 5 + 2^{4-1}$. The offset-binary code for –5 is the same as the unipolar code for 3, which is 0011. Note that the offset-binary code does not suffer from redundancy, and all sixteen numbers are uniquely represented. Also, this code has the advantage that it is closely related to the unipolar case through a simple offset.

Finally, the unipolar relationship for a D/A converter given in (11.2) is easily modified to the signed case, with offset-binary representation as

$$V_{out} = V_{ref}(b_1 2^{-1} + b_2 2^{-2} + \cdots + b_N 2^{-N}) - 0.5V_{ref} \tag{11.19}$$

Note here that the output signal is now bounded by $\pm 0.5V_{ref}$.

2's Complement

Finally, the 2's-complement representation is obtained from the offset-binary number by simply complementing the MSB. For the 4-bit example, 5 becomes 0101 in 2's complement (the same as in sign magnitude and in 1's complement), whereas –5 is now 1011. It should be mentioned that the 2's-complement code for negative numbers can also be obtained by adding 1 LSB to the equivalent 1's-complement code.

The main advantage of 2's-complement coding is that addition of both positive and negative numbers is performed using straightforward addition, and no extra hardware is required. Also, the subtraction of two numbers, $A - B$, can be easily performed by complementing all the bits for B (i.e., forming the 1's-complement equivalent) and then adding this result to A at the same time as adding a single LSB in order to create the 2's-complement equivalent of $-B$. Adding a single LSB is easily accomplished in hardware by setting the *carry-in bit* high in the overall adder. Thus, subtraction requires only a small amount of extra hardware. Finally, if many numbers are being added using 2's-complement codes, no overflow hardware is required as long as the final result is within the digital code range (even if intermediate results go well out of range). For these reasons, 2's-complement codes are the most popular representation for signed numbers when arithmetic operations are to be performed.

EXAMPLE 11.3

Show that, when adding $2/8$, $7/8$, and $-3/8$ together as 2's-complement numbers, the final result is correct although an intermediate result is incorrect.

Solution

The addition of $2/8$ and $7/8$ is given by

$$0010 + 0111 = 1001 \tag{11.20}$$

which corresponds to $-7/8$ (note that this temporary result is two steps past $7/8$ when thinking of a 2's-complement code as being circular). Although this intermediate result is incorrect since overflow occurred, we simply carry on and add $-3/8$ to find the final answer.

$$1001 + 1101 = 0110 \tag{11.21}$$

Thus, the final result of $6/8$ is correct, although a temporary overflow occurred, as we just saw.

11.5 PERFORMANCE LIMITATIONS

In this section, some commonly used terms describing the performance of data converters are defined. Before proceeding, definitions are required for determining the transfer responses of both D/A and A/D converters. The transfer response of a D/A converter is defined to be the analog levels that occur for each of the digital input words. Similarly, the transfer response of an A/D converter can be defined as the midpoints of the quantization intervals for each of the digital output words. However, since transitions are easier to measure than midpoint values, A/D converter errors are often measured in terms of the analog transition point values, V_{ij}, and we use that approach here.

Resolution The *resolution* of a converter is defined to be the number of distinct analog levels corresponding to the different digital words. Thus, an N-bit resolution implies that the converter can resolve 2^N distinct analog levels. Resolution is not necessarily an indication of the accuracy of the converter, but instead it usually refers to the number of digital input or output bits.

Offset and Gain Error In a D/A converter, the *offset error*, E_{off}, is defined to be the output that occurs for the input code that should produce zero output, or mathematically,

$$E_{off(D/A)} = \left.\frac{V_{out}}{V_{LSB}}\right|_{0...0} \tag{11.22}$$

where the offset error is in units of LSBs. Similarly, for an A/D converter, the offset error is defined as the deviation of $V_{0...01}$ from $1/2$ LSB, or mathematically,

$$E_{off(A/D)} = \frac{V_{0...01}}{V_{LSB}} - \frac{1}{2}\text{ LSB} \tag{11.23}$$

The gain error is defined to be the difference at the full-scale value between the ideal and actual curves when the offset error has been reduced to zero. For a D/A converter, the gain error, $E_{gain(D/A)}$, in units of LSBs, is given by

$$E_{gain(D/A)} = \left(\left.\frac{V_{out}}{V_{LSB}}\right|_{1...1} - \left.\frac{V_{out}}{V_{LSB}}\right|_{0...0}\right) - (2^N - 1) \tag{11.24}$$

For an A/D converter, the equivalent gain error, $E_{gain(A/D)}$ (in units of LSBs), is given by

$$E_{gain(A/D)} = \left(\frac{V_{1...1}}{V_{LSB}} - \frac{V_{0...01}}{V_{LSB}}\right) - (2^N - 2) \tag{11.25}$$

Graphical illustrations of gain and offset errors are shown in Fig. 11.9.

Accuracy

The *absolute accuracy* of a converter is defined to be the difference between the expected and actual transfer responses. The absolute accuracy includes the offset, gain, and linearity errors.

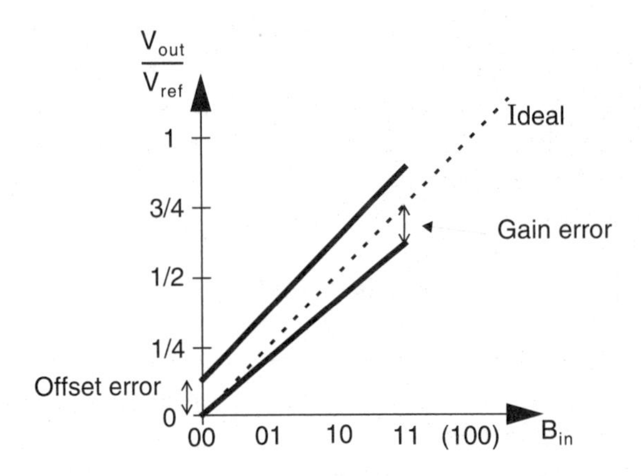

Fig. 11.9 Illustrating offset and gain errors for a 2-bit D/A converter.

The term *relative accuracy* is sometimes used and is defined to be the accuracy after the offset and gain errors have been removed. It is also referred to as the *maximum integral nonlinearity error* (described shortly) and we will refer to it as such.

Accuracy can be expressed as a percentage error of full-scale value, as the effective number of bits, or as a fraction of an LSB. For example, a 12-bit accuracy implies that the converter's error is less than the full-scale value divided by 2^{12}.

Note that a converter may have 12-bit resolution with only 10-bit accuracy, or 10-bit resolution with 12-bit accuracy. An accuracy greater than the resolution means that the converter's transfer response is very precisely controlled (better than the number of bits of resolution).

Integral Nonlinearity (INL) Error After both the offset and gain errors have been removed, the *integral nonlinearity (INL) error* is defined to be the deviation from a straight line. However, what straight line should be used? A conservative measure of nonlinearity is to use the endpoints of the converter's transfer response to define the straight line. An alternative definition is to find the best-fit straight line such that the maximum difference (or perhaps the mean squared error) is minimized. These two definitions are illustrated in Fig. 11.10. One should be aware that, in this book, we define INL values for each digital word (and thus these values can be plotted for a single converter), whereas others sometimes define the term "INL" as the maximum magnitude of the INL values (or equivalently, as the relative accuracy).

Differential Nonlinearity (DNL) Error In an ideal converter, each analog step size is equal to 1 LSB. In other words, in a D/A converter, each output level is 1 LSB from adjacent levels, whereas in an A/D, the transition values are precisely 1 LSB apart. *Differential nonlinearity* (DNL) is defined as the variation in analog step sizes away from 1 LSB (typically, once gain and offset errors have been removed). Thus, an ideal converter has its maximum differential nonlinearity of 0 for all digital values, whereas a converter with a maximum differential nonlinearity of 0.5 LSB has its step

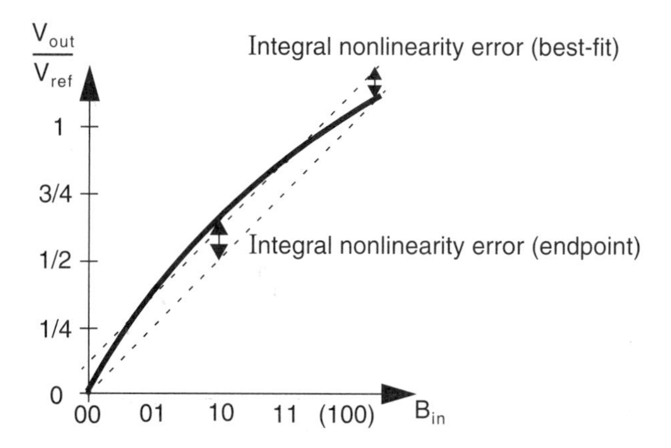

Fig. 11.10 Integral nonlinearity error in a 2-bit D/A converter.

sizes varying from 0.5 LSB to 1.5 LSB. Once again, as in the INL case, we define DNL values for each digital word, whereas others sometimes refer to DNL as the maximum magnitude of the DNL values.

Monotonicity A *monotonic* D/A converter is one in which the output always increases as the input increases. In other words, the slope of the D/A converter's transfer response is of only one sign. If the maximum DNL error is less than 1 LSB, then a D/A converter is guaranteed to be monotonic. However, many monotonic converters may have a maximum DNL greater than 1 LSB. Similarly, a converter is guaranteed to be monotonic if the maximum INL is less than 0.5 LSB.

Missing Codes Although monotonicity is appropriate for D/A converters, the equivalent term for A/D converters is *missing codes.* An A/D converter is guaranteed not to have any missing codes if the maximum DNL error is less than 1 LSB or if the maximum INL error is less than 0.5 LSB.

A/D Conversion Time and Sampling Rate In an A/D converter, the *conversion time* is the time taken for the converter to complete a single measurement including acquisition time of the input signal. On the other hand, the maximum sampling rate is the speed at which samples can be continuously converted and is typically the inverse of the conversion time. However, one should be aware that some converters have a large latency between the input and the output due to pipelining or multiplexing, yet they still maintain a high sampling rate. For example, a pipelined 12-bit A/D converter may have a conversion time of 2 μs (i.e., a sampling rate of 500 kHz) yet a latency from input to output of 24 μs.

D/A Settling Time and Sampling Rate In a D/A converter, the *settling time* is defined as the time it takes for the converter to settle to within some specified amount of the final value (usually 0.5 LSB). The *sampling rate* is the rate at which samples can be continuously converted and is typically the inverse of the settling time.

Sampling-Time Uncertainty Both A/D and D/A converters have limited accuracy when their sampling instances are ill defined. To quantify this *sampling time uncertainty,* also known as *aperture jitter,* for sinusoidal waveforms, consider a full-scale signal, V_{in}, applied to an N-bit, signed, A/D converter with frequency f_{in}. Mathematically,

$$V_{in} = \frac{V_{ref}}{2} \sin(2\pi f_{in} t) \tag{11.26}$$

Since the rate of change (or slope) of V_{in} at the peak of a sinusoidal waveform is small, sampling time uncertainty is less of a problem near the peak values. However, the maximum rate of change for this waveform occurs at the zero crossing and can be found by differentiating V_{in} with respect to time and setting $t = 0$. At the zero crossing, we find that

$$\left.\frac{\Delta V}{\Delta t}\right|_{max} = \pi f_{in} V_{ref} \tag{11.27}$$

If Δt represents some sampling-time uncertainty, and if we want to keep ΔV less than 1 V_{LSB}, we see that

$$\Delta t < \frac{V_{LSB}}{\pi f_{in} V_{ref}} = \frac{1}{2^N \pi f_{in}} \tag{11.28}$$

For example, an 8-bit converter sampling a full-scale 250-MHz sinusoidal signal must keep its sampling-time uncertainty under 5 ps to maintain 8-bit accuracy. Also, the same 5-ps time accuracy is required for a 16-bit converter operating on a full-scale, 1-MHz signal.

Dynamic Range The *dynamic range* of a converter is usually specified as the ratio of the rms value of the maximum amplitude input sinusoidal signal to the rms output noise plus the distortion measured when the same sinusoid is present at the output. The rms output noise plus distortion is obtained by first eliminating the sinusoid from the measured output. In a D/A converter, the output sinusoid can be eliminated by using a spectrum analyzer and ignoring the power at that particular frequency. For an A/D converter, a similar approach can be taken by using an FFT and eliminating the fundamental of the output, or a least-mean-squared fit can be used to find the amplitude and phase of a sinusoid at the input signal's frequency and then subtracting the best-fit sinusoid from the output signal. Dynamic range can also be expressed as an effective number of bits using the relationship presented in (11.16).

It should be noted that this approach often results in a dynamic range measurement that is a function of the frequency of the sinusoidal input. Hence, it is a more realistic way of measuring the performance of a converter than extrapolating the nonlinearity performance found using dc inputs. For example, if an 8-bit, 200-MHz, A/D converter has a band-limited preamplifier or a slew-rate limited sample and hold, dc inputs may show full 8-bit performance even at the maximum sample rate of 200 sample/s. However, a high-frequency, sinusoidal input, at, say, 40 MHz, will require the input stage to track a rapidly varying signal and may result in only 6-bit performance.

Finally, it should be mentioned that the distortion level (or nonlinearity performance) of many converters remains at a fixed level and is not a function of the input signal level. Thus, as the input signal level is decreased, the signal-to-distortion ratio decreases. This behavior occurs because the distortion level is often determined by component matching and thus is fixed once the converter is realized. However, in some converters, the distortion level decreases as the input signal level is decreased, which is often a desirable property to have. For example, most 1-bit oversampling converters have this desirable property since they do not rely on component matching and their distortion level is often a result of weak nonlinear effects at the input stage.

EXAMPLE 11.4

Consider a 3-bit D/A converter in which $V_{ref} = 4$ V, with the following measured voltage values:

$$\{0.011 : 0.507 : 1.002 : 1.501 : 1.996 : 2.495 : 2.996 : 3.491\}$$

1. Find the offset and gain errors in units of LSBs.
2. Find the INL (endpoint) and DNL errors (in units of LSBs).
3. Find the effective number of bits of absolute accuracy.
4. Find the effective number of bits of relative accuracy.

Solution

We first note that 1 LSB corresponds to $V_{ref}/2^3 = 0.5$ V.

1. Since the offset voltage is 11 mV, and since 0.5 V corresponds to 1 LSB, we see that the offset error is given by

$$E_{off(D/A)} = \frac{0.011}{0.5} = 0.022 \text{ LSB} \tag{11.29}$$

For the gain error, from (11.25) we have

$$E_{gain(D/A)} = \left(\frac{3.491 - 0.011}{0.5}\right) - (2^3 - 1) = -0.04 \text{ LSB} \tag{11.30}$$

2. For INL and DNL errors, we first need to remove both offset and gain errors in the measured D/A values. The offset error is removed by subtracting 0.022 LSB off each value, whereas the gain error is eliminated by subtracting off scaled values of the gain error. For example, the new value for 1.002 (scaled to 1 LSB) is given by

$$\frac{1.002}{0.5} - 0.022 + \left(\frac{2}{7}\right)(0.04) = 1.993 \tag{11.31}$$

Thus, the offset-free, gain-free, scaled values are given by

$$\{0.0 : 0.998 : 1.993 : 2.997 : 3.993 : 4.997 : 6.004 : 7.0\}$$

Since these results are in units of LSBs, we calculate the INL errors as the difference between these values and the ideal values, giving us

INL errors: $\{0 : -0.002 : -0.007 : -0.003 : -0.007 : -0.003 : 0.004 : 0\}$

For DNL errors, we find the difference between adjacent offset-free, gain-free, scaled values to give

DNL errors: $\{-0.002 : -0.005 : 0.004 : -0.004 : 0.004 : 0.007 : -0.004\}$

3. For absolute accuracy, we find the largest deviation between the measured values and ideal values, which, in this case, occurs at 0 V and is 11 mV. To relate this 11-mV value to effective bits, 11 mV should correspond to 1 LSB when $V_{ref} = 4$ V. In other words, we have the relationship

$$\frac{4\ \text{V}}{2^{N_{eff}}} = 11\ \text{mV} \tag{11.32}$$

which results in an absolute accuracy of $N_{abs} = 8.5$ bits.

4. For relative accuracy, we use the INL errors found in part 2, whose maximum magnitude is 0.007 LSB, or equivalently, 3.5 mV. We relate this 3.5-mV value to effective bits in the same manner as in part 3, resulting in a relative accuracy of $N_{rel} = 10.2$ bits.

EXAMPLE 11.5

A full-scale sinusoidal waveform is applied to a 12-bit A/D converter, and the output is digitally analyzed. If the fundamental has a normalized power of 1 W while the remaining power is 0.5 μW, what is the effective number of bits for the converter?

Solution

Using (11.16), we have

$$\text{SNR} = 6.02N_{eff} + 1.76 \tag{11.33}$$

In this case, the signal-to-noise ratio is found to be

$$\text{SNR} = 10\log\left(\frac{1}{0.5 \times 10^{-6}}\right) = 63\ \text{dB} \tag{11.34}$$

Substituting this SNR value into (11.33), we find

$$N_{eff} = \frac{63 - 1.76}{6.02} = 10.2 \text{ effective bits} \tag{11.35}$$

11.6 REFERENCES

Analog Devices. *Analog-Digital Conversion Handbook.* Prentice Hall, Englewood Cliffs, New Jersey, 1986.

A. B. Grebene. *Bipolar and MOS Analog Integrated Circuit Design.* John Wiley & Sons, New York, 1984.

R. van de Plassche. *Integrated Analog-to-Digital and Digital-to-Analog Converters.* Kluwer Academic Publishers, Dordrecht, the Netherlands, 1994.

11.7 PROBLEMS

11.1 For an ideal 10-bit, unipolar D/A converter with $V_{LSB} = 1$ mV, what is the largest output voltage?

11.2 What is the SNR for an ideal 12-bit unipolar A/D converter with $V_{ref} = 3$ V, when a sinusoidal input of 1 V_{pp} is applied? What size input would result in an SNR of 0 dB?

11.3 Find the equivalent of (11.19) for 2's-complement coding.

11.4 Show that, by using 4-bit 2's-complement words, the correct final sum is obtained when the numbers +5, +5, and –7 are added together while any overflow effects are ignored.

11.5 Starting with two 4-bit 2's-complement words, we want to add +5 and +7 to obtain the correct answer of +12 with a 5-bit word. Show how an extra bit can be added at the left of each of the 4-bit words such that numbers up to ±15 can be represented. This approach is called *sign extension* and can be used to increase the word size of any number.

11.6 What is the representation of +8 and –8 in 5-bit 2's complement? Assuming a circuit added only one of these two numbers to an arbitrary 5-bit word, show a simple logic circuit (i.e., nand, nor, xnor, etc.) that would accomplish such an addition. (*Hint:* Note that the 4 LSBs in +8 and –8 are all 0 and thus do not need to be added.)

11.7 The following measurements are found from a 3-bit unipolar D/A converter with $V_{ref} = 8$ V: (–0.01, 1.03, 2.02, 2.96, 3.95, 5.02, 6.00, 7.08). In units of LSBs, find the offset error, gain error, maximum DNL, and maximum INL.

11.8 How many bits of absolute accuracy does the converter in Problem 11.7 have? How many bits of relative accuracy does it have?

11.9 A 10-bit A/D converter has a reference voltage, V_{ref}, tuned to 10.24 V at $25\ °C$. Find the maximum allowable temperature coefficient in terms of $(\mu V)/°C$ for the reference voltage if the reference voltage is allowed to cause a maximum error of $(\pm 1/2)$ LSB over the temperature range of 0 to $50\ °C$.

11.10 Consider the following measured voltage values for a 2-bit D/A with a reference voltage of 4 V:

$$\{00 \leftrightarrow 0.01\ V\} \quad \{01 \leftrightarrow 1.02\ V\} \quad \{10 \leftrightarrow 1.97\ V\} \quad \{11 \leftrightarrow 3.02\ V\}$$

In units of LSB, find the offset error, gain error, worst absolute and relative accuracies, and worst differential nonlinearity. Restate the relative accuracy in terms of an N-bit accuracy.

11.11 Find the maximum magnitude of quantization error for a 12-bit A/D converter having V_{ref} equal to 5 V and 0.5-LSB absolute accuracy.

11.12 What sampling-time uncertainty can be tolerated for a 16-bit A/D converter operating on an input signal from 0–20 kHz ?

CHAPTER

12 Nyquist-Rate D/A Converters

In this chapter, a variety of methods are discussed for realizing integrated Nyquist-rate digital-to-analog converters (DAC). Nyquist-rate D/A converters can be roughly categorized into four main types: decoder-based, binary-weighted, thermometer-code, and hybrid. Oversampling D/A converters are discussed separately in Chapter 14 due to their importance and because they are best described using many signal processing concepts.

12.1 DECODER-BASED CONVERTERS

Perhaps the most straightforward approach for realizing an N-bit D/A converter is to create 2^N reference signals and pass the appropriate signal to the output, depending on the digital input word. We refer to such D/A converters as *decoder-based converters.*

Resistor String Converters

One of the first integrated MOS 8-bit D/A converters was based on selecting one tap of a segmented resistor string by a switch network [Hamadé, 1978].[1] The switch network was connected in a tree-like decoder, as in the 3-bit D/A converter shown in Fig. 12.1. Notice that there will be one, and only one, low-impedance path between the resistor string and the input of the buffer, and that path is determined by the digital input word, B_{in}. In a CMOS implementation, transmission gates might be used rather than n-channel switches. However, when only n-channel pass transistors are used, the transistor-tree decoder can be laid out quite compactly since no contacts are required in the tree. Also, an n-channel-only approach is not much different in speed than a CMOS transmission-gate implementation; a transmission-gate approach has extra drain and source capacitance to ground, but this extra capacitance is offset by the reduced switch resistance, which is due to the parallel combination of p-channel and n-channel transistors. In addition, a transmission-gate implementation can operate closer to the positive voltage supply.

1. In fact, the resistor-string D/A was used to realize an A/D converter in the reference.

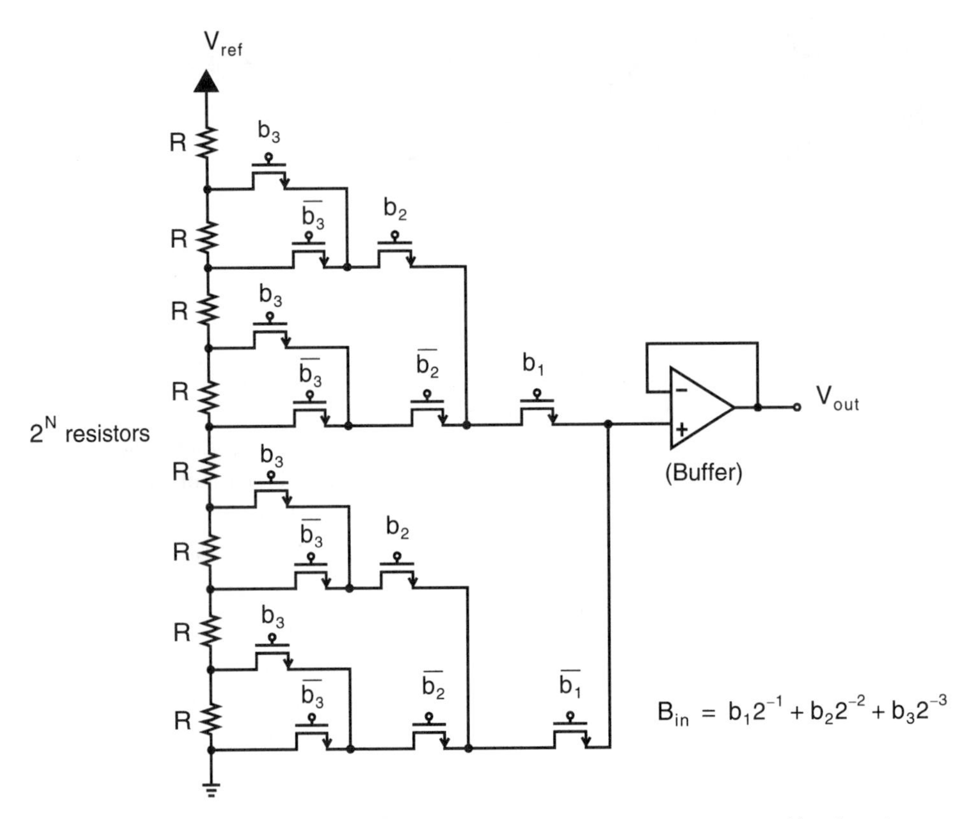

Fig. 12.1 Resistor-string 3-bit D/A converter with a transmission-gate, tree-like decoder.

With a resistor-string approach, if we assume the buffer's offset voltage does not depend on its input voltage, the D/A converter has guaranteed monotonicity since any tap on the resistor string must have a lower voltage than its upper, neighbor tap. Also, the accuracy of this D/A depends on the matching precision of R in the resistor string. Although this resistor-matching precision depends on the type of resistors used, the use of polysilicon resistors that have a resistivity of around 20–30 (ohms per square) can result in up to 10 bits of accuracy.

The delay through the switch network is the major limitation on speed. However, in a multiplying D/A, the delay through the resistor string would also be a major source of delay since V_{ref} would become a second input signal. A useful technique for estimating the settling-time behavior in RC type circuits (i.e., circuits that have only real-axis poles) is the *open-circuit time-constant* approach [Sedra, 1991]. Specifically, the dominant high-frequency time constant is estimated as the sum of the individual time constants due to each of the capacitances when all other capacitances are set to zero (i.e., replaced with open circuits). To find the individual time constant for a given capacitance, independent voltage sources are replaced with ground (independent current sources are opened), and the resistance seen by that capacitor is determined.

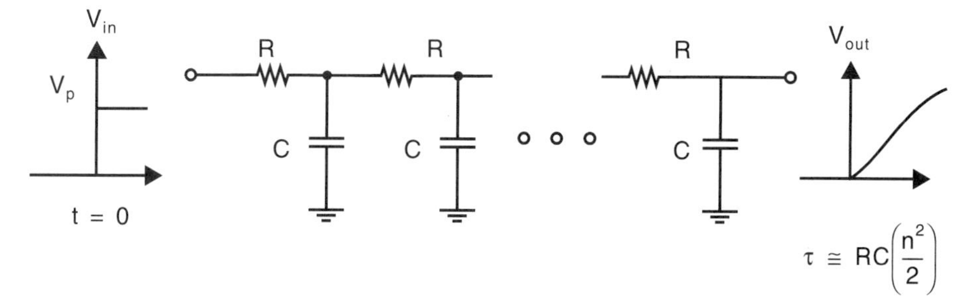

Fig. 12.2 Estimating the time constant for n resistors and capacitors in series.

EXAMPLE 12.1

Show that an estimate of the time constant for a network of n resistors, each of size R in series, with capacitive loading C at each node (see Fig. 12.2), is given by $\tau = RC(n^2/2)$. How much settling time is required for the output to settle to 0.1 percent of its final value?

Solution

The open-circuit time constant due to the first capacitor on the left is simply RC. The second capacitor from the left has an individual time constant of $2RC$ and so on. Thus, the dominant high-frequency time constant, τ, is estimated as

$$\tau \cong RC(1 + 2 + \cdots + n)$$

The sum from 1 to n can be shown to be equal to $n(n+1)/2$ and, thus, for a large n, the dominant time constant can be estimated by

$$\tau \cong RC\left(\frac{n^2}{2}\right)$$

Using this time constant to estimate the output voltage charging behavior, we have

$$V_{out} \cong (1 - e^{-(t/\tau)})V_p$$

Thus, for V_{out} to equal 0.999 V_p, we find that a time of about 7τ is needed.

In a higher-speed implementation, logic can be used for the decoder, and a single bus is connected to a single resistor-string node, as shown in Fig. 12.3. This approach takes more area for the decoder and also results in a large capacitive loading on the single bus because the 2^N transistors' junctions are connected to the bus. However, if the digital decoder is pipelined, the D/A can be moderately fast.

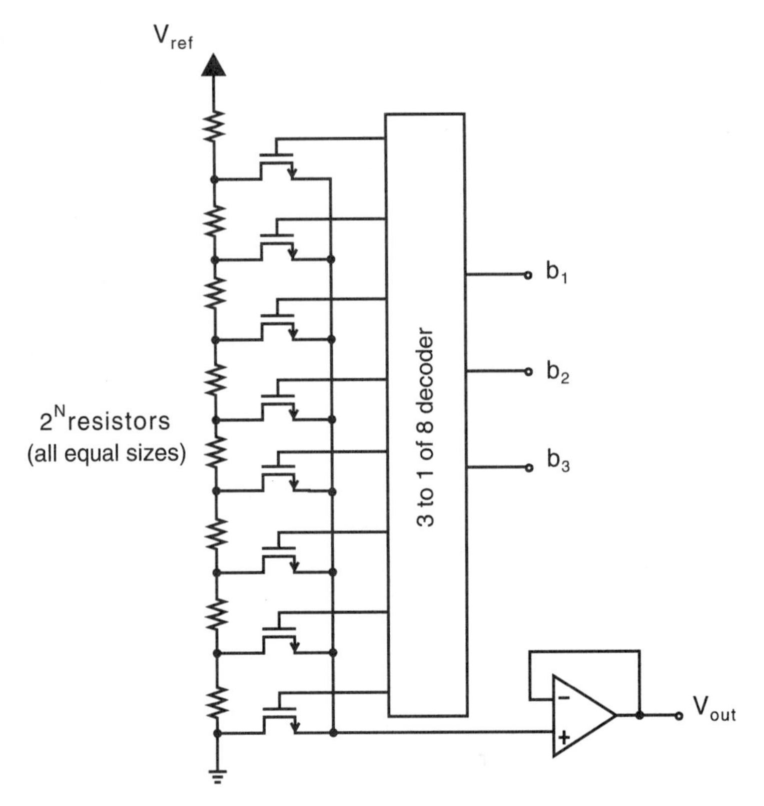

Fig. 12.3 Resistor-string 3-bit D/A converter with digital decoding.

Folded Resistor-String Converters

To reduce the amount of digital decoding and large capacitive loading, a *folded resistor-string D/A* can be used, as shown in Fig. 12.4 [Abrial, 1988]. This approach makes the decoding very similar to that for a digital memory, which reduces the total decoding area. To convert a digital input in this 4-bit example, the most significant bits, $\{b_1, b_2\}$, determine the single word line to be selected (all others will remain low). This operation connects a block of four adjacent resistor nodes to the four-bit lines. One of these bit lines is then connected to the output buffer by the bit-line decoder. Notice that the total number of transistor junctions on the output line is now $2\sqrt{2^N}$ because a set of transistors is connected directly to the output line and another set is connected to the chosen bit line. Thus, for a 10-bit converter, this approach would have a capacitive load of 64 junctions, as opposed to 1,024 junctions when we use the digital-decoding approach shown in Fig. 12.3. Unfortunately, the increase in speed is not equal to this large ratio since, when a word line goes high, all the bit lines must be pulled to new voltage levels—not just the one bit line connected to the output buffer.

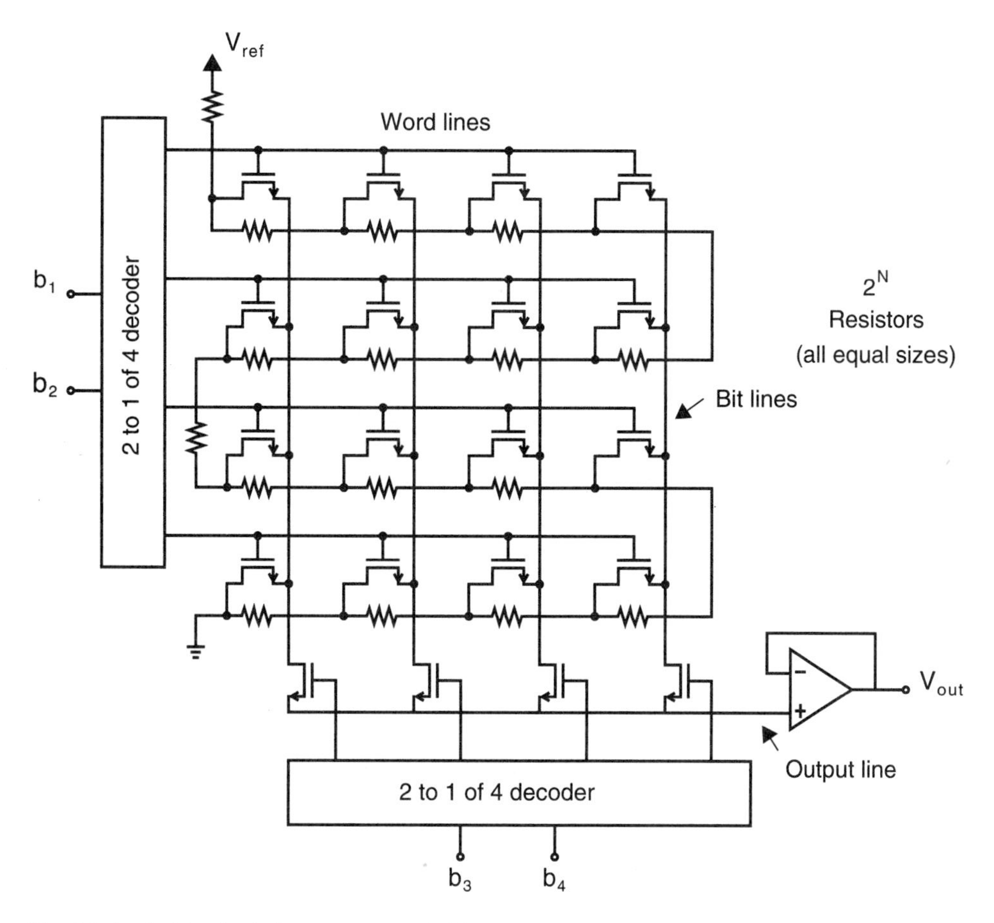

Fig. 12.4 A 4-bit folded resistor-string D/A converter.

Multiple R-String Converters

In this variation, a second tapped resistor string is connected between buffers whose inputs are two adjacent nodes of the first resistor string, as shown in Fig. 12.5 [Holloway, 1984]. In the shown 6-bit example, the three MSBs determine which two adjacent nodes of the first resistor string are connected to the two intermediate buffers. The second resistor string linearly interpolates between the two adjacent voltages from the first resistor string. Finally, the output is determined by the lower LSBs where extra logic must take into account that sometimes the top intermediate buffer has the higher voltage, whereas at other times it has the lower voltage. This approach requires only $2 \times 2^{N/2}$ resistors, making it suitable for higher-resolution, low-power applications. This approach also has guaranteed monotonicity, assuming the opamps have matched, voltage-insensitive offset voltages. However, the opamps must be fast and low noise, which can be achieved using a

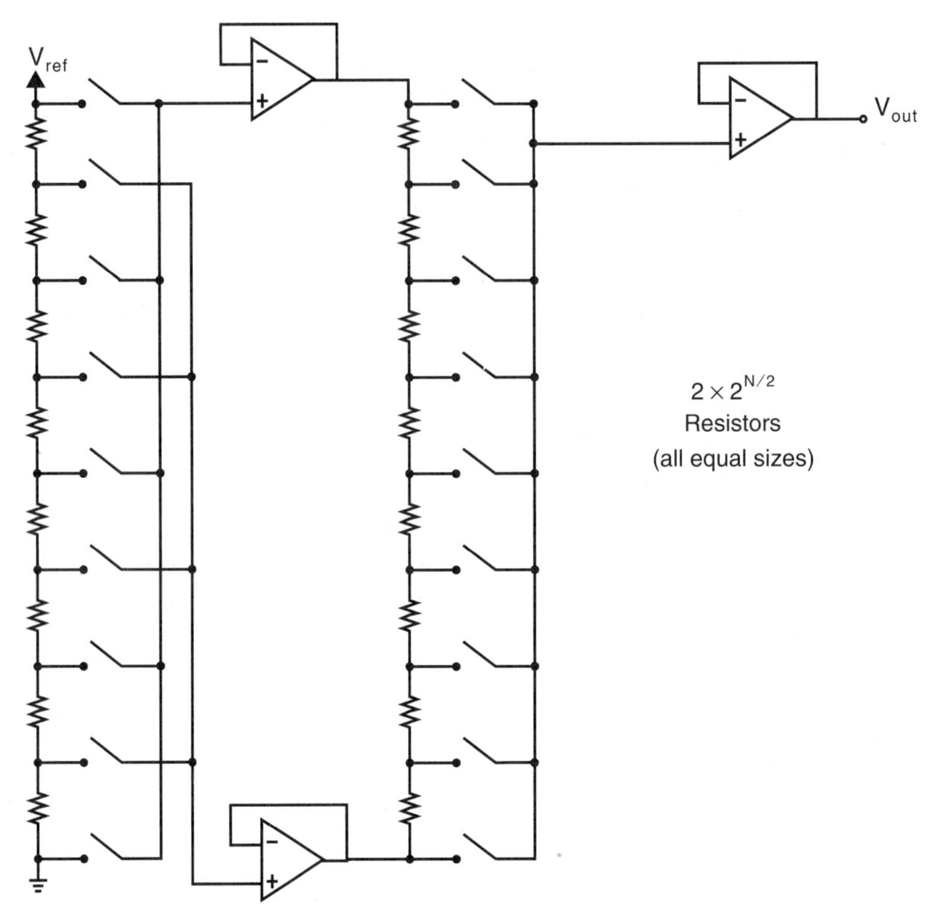

Fig. 12.5 Multiple R-string 6-bit D/A converter.

BiCMOS process. Another point to note here is that, since the second resistor string is used to decode only the lower-order bits, the matching requirements of the second resistor string are not nearly as severe as those for the first string.

EXAMPLE 12.2

Assume that the first resistor string of a 10-bit, multiple-R-string, D/A converter must match to 0.1 percent, and that the first string realizes the top 4 bits. What is the matching requirement of the second resistor string, which realizes the lower 6 bits?

Solution

Errors in the first resistor string correspond directly to errors in the overall D/A output. However, since the second resistor string forms the lower LSB bits (6 bits, in this case), errors in the matching of these resistors cause an error only in the LSB portion of the output voltage. As a result, the second resistor string need match only to

$$2^4 \times 0.1\% = 1.6\% \tag{12.1}$$

Signed Outputs

In applications where negative output voltages are required, the bottom of the resistor string can be connected to $-V_{ref}$. This requires a negative power supply, and the circuit needed to realize a dual power supply with exactly matched voltages is nontrivial since any error in matching will result in an offset error. Many papers on D/A converters assume that $-V_{ref}$ is available but do not explain how it was obtained. If it is obtained off chip, the cost is significantly higher. One possibility is to use a switched-capacitor gain amplifier, as shown in Fig. 12.6, where a negative output can be realized by changing the clock phases of the input switches so an inverting amplifier is realized [Martin, 1987]. Another possibility is to sense the center tap of the resistor string and adjust $-V_{ref}$ to get the center tap voltage equal to zero volts.

12.2 BINARY-SCALED CONVERTERS

The most popular approach for realizing at least some portion of D/A converters is to combine an appropriate set of signals that are all related in a binary fashion. This binary array of signals might be currents (in resistor or current approaches), but binary-weighted arrays of charge are also commonly used. In this section, resistor approaches are first discussed, followed by charge redistribution and current mode.

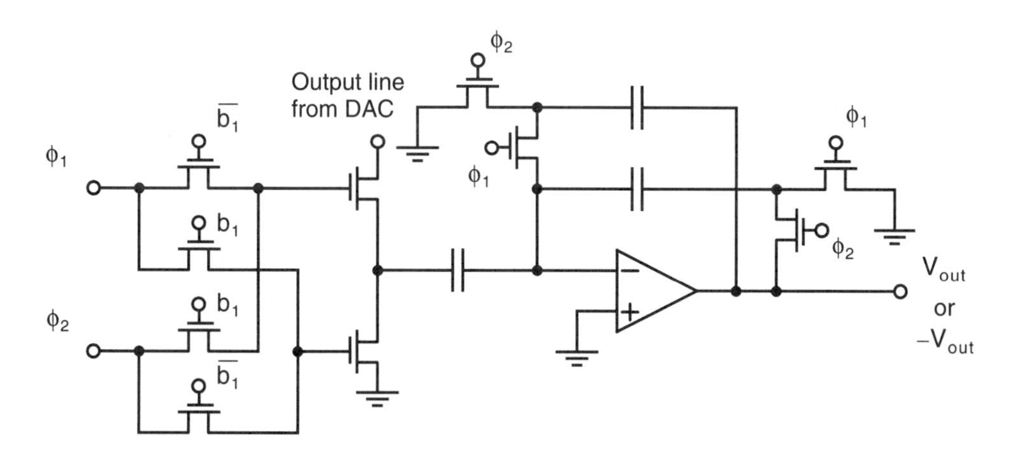

Fig. 12.6 Using an SC gain amplifier to realize a signed output from a unipolar (positive) DAC output. b_1 high causes a negative output.

Binary-Weighted Resistor Converters

Binary-weighted resistor converters are popular for a bipolar technology so that bipolar differential pairs can be used for current switches. The basic architecture for a 4-bit converter is shown in Fig. 12.7.

If b_i is a 1, then the current to the ith resistor comes from the virtual ground of the opamp; otherwise, it comes from ground. Therefore, we have

$$\begin{aligned} V_{out} &= -R_F V_{ref}\left(-\frac{b_1}{2R} - \frac{b_2}{4R} - \frac{b_3}{8R} - \dots\right) \\ &= \left(\frac{R_F}{R} V_{ref}\right) B_{in} \end{aligned} \tag{12.2}$$

where

$$B_{in} = b_1 2^{-1} + b_2 2^{-2} + b_3 2^{-3} + \cdots \tag{12.3}$$

Although this approach does not require many resistors or switches, it does have some disadvantages. The resistor and current ratios are on the order of 2^N, which may be large, depending on N. This large current ratio requires that the switches also be scaled so that equal voltage drops appear across them for widely varying current levels. Also, monotonicity is not guaranteed. Finally, this approach is prone to glitches for high-speed operation, as discussed on page 474.

Reduced-Resistance-Ratio Ladders

To reduce the large resistor ratios in a binary-weighted array, signals in portions of the array can be scaled by introducing a series resistor, as shown in Fig. 12.8. Here, note that the voltage node, V_A, is equal to one-fourth the reference voltage, V_{ref}, as a result of inserting the series resistor of value $3R$. Also note that an additional $4R$ resistor was added (to ground) such that resistance seen to the right of the $3R$ resistor equals

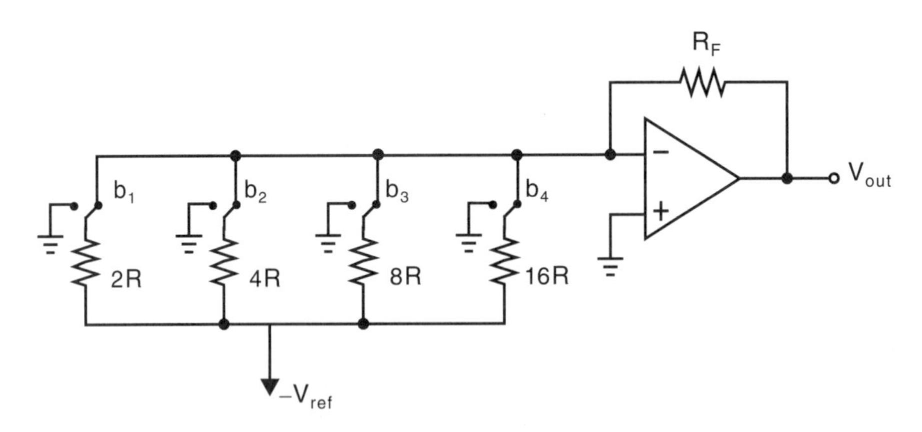

Fig. 12.7 Binary-weighted 4-bit resistor D/A converter.

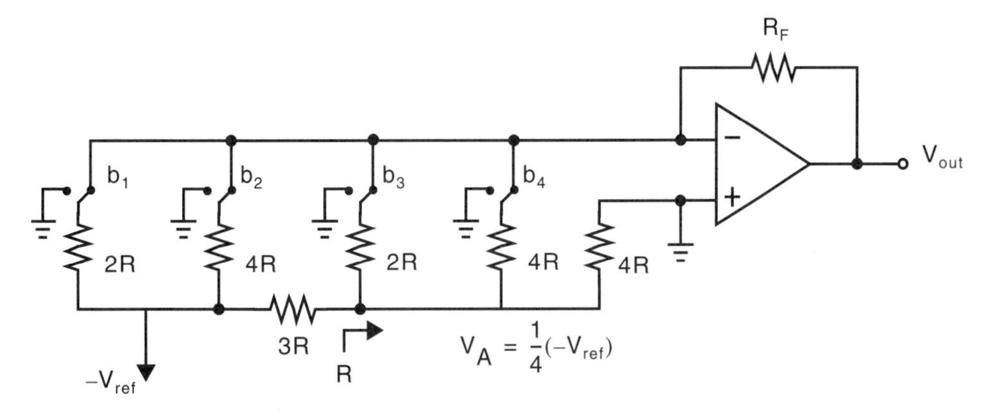

Fig. 12.8 Reduced-resistance-ratio 4-bit D/A converter.

R. Straightforward analysis shows that this converter has the same relationship to the binary digital signals as in the previous binary-weighted case, but with one-fourth the resistance ratio. Note, however, the current ratio has remained unchanged. Finally, note that, by repeating this procedure recursively to a binary-weighted ladder, one arrives at a structure commonly referred to as an *R-2R ladder,* described next.

R-2R-Based Converters

A very popular architecture for D/A converters uses R-2R ladders. These ladders are useful for realizing binary-weighted currents with a small number of components and with a resistance ratio of only 2, independent of the number of bits, N.

Consider the R-2R ladder network shown in Fig. 12.9. Analysis gives

$$\begin{aligned} R_4' &= 2R \\ R_4 &= 2R \,\|\, 2R = R \\ R_3' &= R + R_4 = 2R \\ R_3 &= 2R \,\|\, R_3' = R \end{aligned} \tag{12.4}$$

and so on. Thus, $R_i' = 2R$ for all i. This result gives the following current relationships:

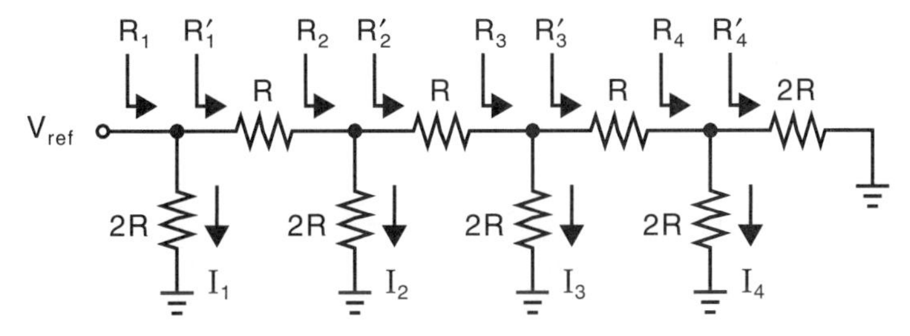

Fig. 12.9 R-2R resistance ladder.

$$I_1 = \frac{V_{ref}}{2R} \tag{12.5}$$

Also, the voltage at node 2 is one-half the voltage at node 1, giving

$$I_2 = \frac{V_{ref}}{4R} \tag{12.6}$$

At node 3, the voltage divides in half once again, therefore

$$I_3 = \frac{V_{ref}}{8R} \tag{12.7}$$

and so on.

Thus, the R-2R ladder can be used to obtain binary-weighted currents while using only a single-size resistor. (The resistors of size 2R are made out of two resistors of size R, to improve matching properties.) As a result, this R-2R approach usually gives both a smaller size and a better accuracy than a binary-sized approach.

A 4-bit D/A converter that uses an R-2R ladder is shown in Fig. 12.10. For this R-2R -based circuit, we see that

$$I_r = V_{ref}/(2R) \tag{12.8}$$

and

$$V_{out} = R_F \sum_{i=1}^{N} \frac{b_i I_r}{2^{i-1}} = V_{ref}\left(\frac{R_F}{R}\right) \sum_{i=1}^{N} \frac{b_i}{2^i} \tag{12.9}$$

However, as already mentioned, although the resistance ratio has been reduced, the current ratio through the switches is still large, and thus the switch sizes are usually scaled in size to accommodate the widely varying current levels. One approach to reduce this current ratio is shown in Fig. 12.11, where equal currents flow through all the switches. However, this configuration is typically slower since the internal nodes

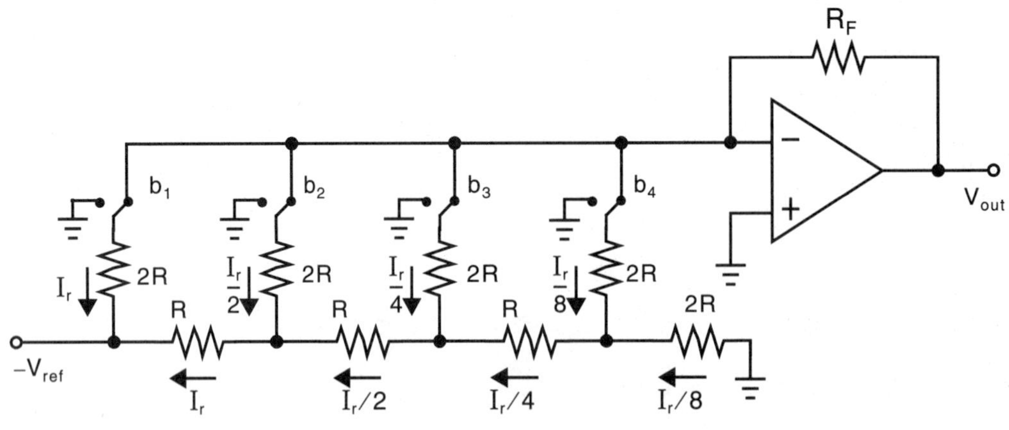

Fig. 12.10 4-bit R-2R based D/A converter.

of the R-2R ladder now exhibit some voltage swings (as opposed to the configuration in Fig. 12.10 where internal nodes all remain at fixed voltages).

Charge-Redistribution Switched-Capacitor Converters

The basic idea here is to simply replace the input capacitor of an SC gain amplifier by a *programmable capacitor array* (PCA) of binary-weighted capacitors, as shown in Fig. 12.12. As in the SC gain amplifier, the shown circuit is insensitive to opamp input-offset voltage, 1/f noise, and finite-amplifier gain. Also, an additional sign bit can be realized by interchanging the clock phases (shown in parentheses) for the input switches.

It should be mentioned here that, as in the SC gain amplifier, carefully generated clock waveforms are required to minimize the voltage dependency of clock feed-through, and a deglitching capacitor should be used. Also, the digital codes should be

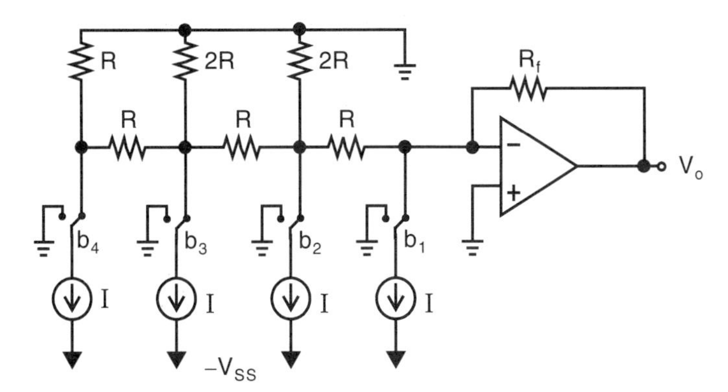

Fig. 12.11 R-2R ladder D/A converter with equal currents through the switches.

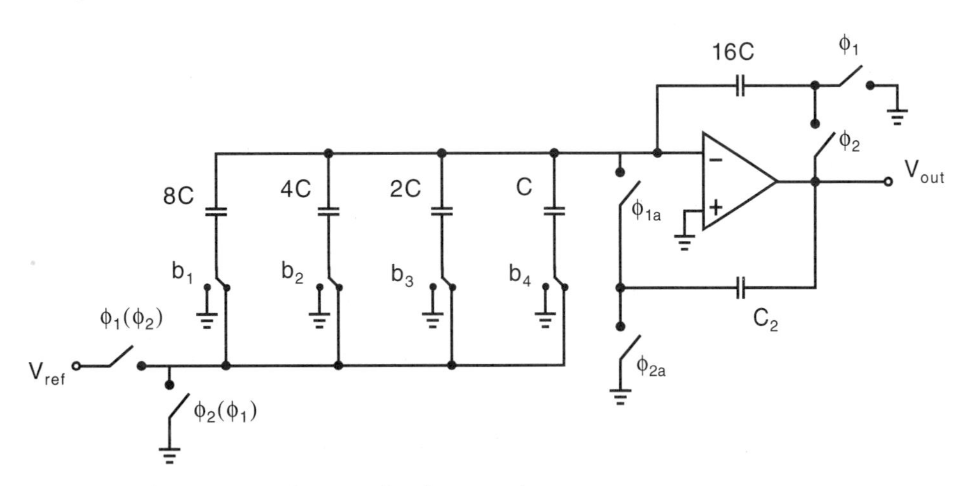

Fig. 12.12 Binary-array charge-redistribution D/A converter.

changed only when the input side of the capacitors are connected to ground, and thus the switching time is dependent on the sign bit, which requires some extra digital complexity.

Current-Mode Converters

Current-mode D/A converters are very similar to resistor-based converters, but are intended for higher-speed applications. The basic idea is to switch currents to either the output or to ground, as shown in Fig. 12.13. Here, the output current is converted to a voltage through the use of R_F, and the upper portion of each current source always remains at ground potential. Implementations of current-mode D/A converters are more fully discussed in Section 12.3, where thermometer codes are introduced.

Glitches

Glitches are a major limitation during high-speed operation for converters that have digital logic, $\{b_1, b_2, \ldots, b_N\}$, directly related to switching signals. Glitches are mainly the result of different delays occurring when switching different signals. For example, when the digital input code changes from 0111 . . . 1 to 1000 . . . 0, all of the $N-1$ LSBs turn off and the MSB turns on. However, it is possible that the currents due to the LSB switches will turn off slightly before the MSB current, causing the current to temporarily fall to zero. Alternatively, if the LSB switches turn off slightly after the MSB current, the current will temporarily go to its maximum value. In either case, a glitch occurs at the output unless these two delays are exactly matched, which is highly unlikely since the branches have different currents. This glitch phenomenon is illustrated in Fig. 12.14.

The glitch disturbance can be reduced by limiting the bandwidth (by placing a capacitor across the resistor R_f in Fig. 12.14), but this approach slows down the circuit. Another approach is to use a sample and hold on the output signal. Finally, the most popular way to reduce glitches is to modify some or all of the digital word from a binary code to a thermometer code, as discussed next.

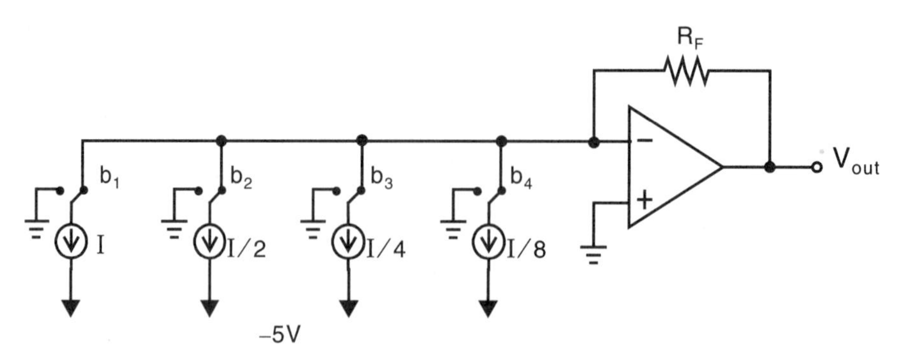

Fig. 12.13 Binary-weighted current-mode D/A converter.

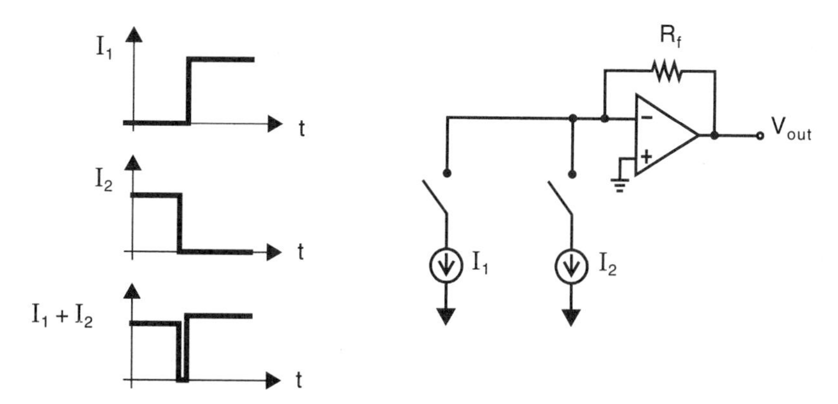

Fig. 12.14 Glitches. I_1 represents the MSB current, and I_2 represents the sum of the $N-1$ LSB currents. Here, the MSB current turns off slightly early, causing a glitch of zero current.

12.3 THERMOMETER-CODE CONVERTERS

Another method for realizing a D/A converter is to digitally recode the input value to a thermometer-code equivalent. A *thermometer code* differs from a binary one in that a thermometer code has $2^N - 1$ digital inputs to represent 2^N different digital values. Clearly, a thermometer code is not a minimal representation since a binary code requires only N digital inputs to represent 2^N input values. However, as we will see, a thermometer-based converter does have advantages over its binary counterpart, such as low DNL errors, guaranteed monotonicity, and reduced glitching noise.

Typically, in a thermometer-code representation, the number of 1s represents the decimal value. For example, with a 3-bit binary input, the decimal value 4 is binary coded as 100 while its thermometer equivalent is 0001111. The codes for the remaining values in this 3-bit example are shown in Table 12.1.

One method to realize a D/A converter with the use of a thermometer-code input is to build $2^N - 1$ equal-sized resistors and switches attached to the virtual ground of an opamp, as shown in Fig. 12.15. Note that monotonicity is guaranteed here since,

Table 12.1 Thermometer-code representations for 3-bit binary values

	Binary			Thermometer Code						
Decimal	b_1	b_2	b_3	d_1	d_2	d_3	d_4	d_5	d_6	d_7
0	0	0	0	0	0	0	0	0	0	0
1	0	0	1	0	0	0	0	0	0	1
2	0	1	0	0	0	0	0	0	1	1
3	0	1	1	0	0	0	0	1	1	1
4	1	0	0	0	0	0	1	1	1	1
5	1	0	1	0	0	1	1	1	1	1
6	1	1	0	0	1	1	1	1	1	1
7	1	1	1	1	1	1	1	1	1	1

when the binary input changes to the next higher number, one more digital value in the thermometer code goes high, causing additional current to be drawn out of the virtual ground and forces the opamp output to go some amount higher (never lower). This is not necessarily the case for a binary-array D/A converter since mismatches between elements may cause the output to go lower even though the digital input value is increased. It should be noted that this mismatch effect is usually the largest in a binary-array converter when the MSB is changed, resulting in the largest DNL at this location.

Perhaps more importantly, a D/A converter based on a thermometer code greatly minimizes glitches, as compared to binary-array approaches, since banks of resistors are never exchanged at slightly different times when the output should change by only 1 LSB. It should also be mentioned here that latches can be used in the binary-to-thermometer code conversion such that no glitches occur in the digital thermometer-code words and pipelining can be also used to maintain a high throughput speed.

It is also of interest to note that the use of a thermometer code does not increase the size of the analog circuitry compared to a binary-weighted approach. In a 3-bit binary-weighted approach, the resistor values of R, $2R$, and $4R$ are needed for a total resistance of $7R$. This total value is the same as for the 3-bit thermometer-code approach shown in Fig. 12.15, and since resistors are created on an integrated circuit using area that is proportional to their size, each approach requires the same area (ignoring interconnect). The same argument can be used to show that the total area required by the transistor switches is the same since transistors are usually size-scaled in binary-weighted designs to account for the various current densities. All transistor switches in a thermometer-code approach are of equal sizes since they all pass equal currents. Finally, it should be mentioned that a thermometer-code charge-redistribution D/A can also be realized, as shown in Fig. 12.16.

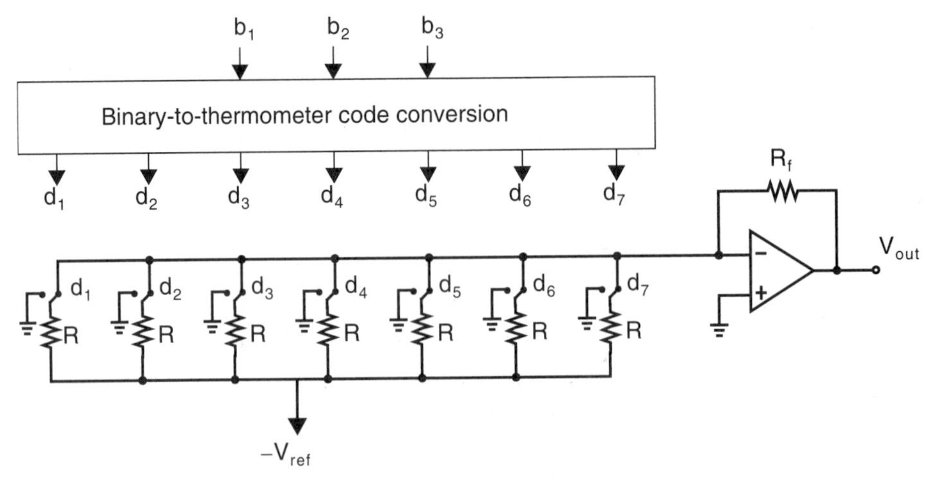

Fig. 12.15 A 3-bit thermometer-based D/A converter.

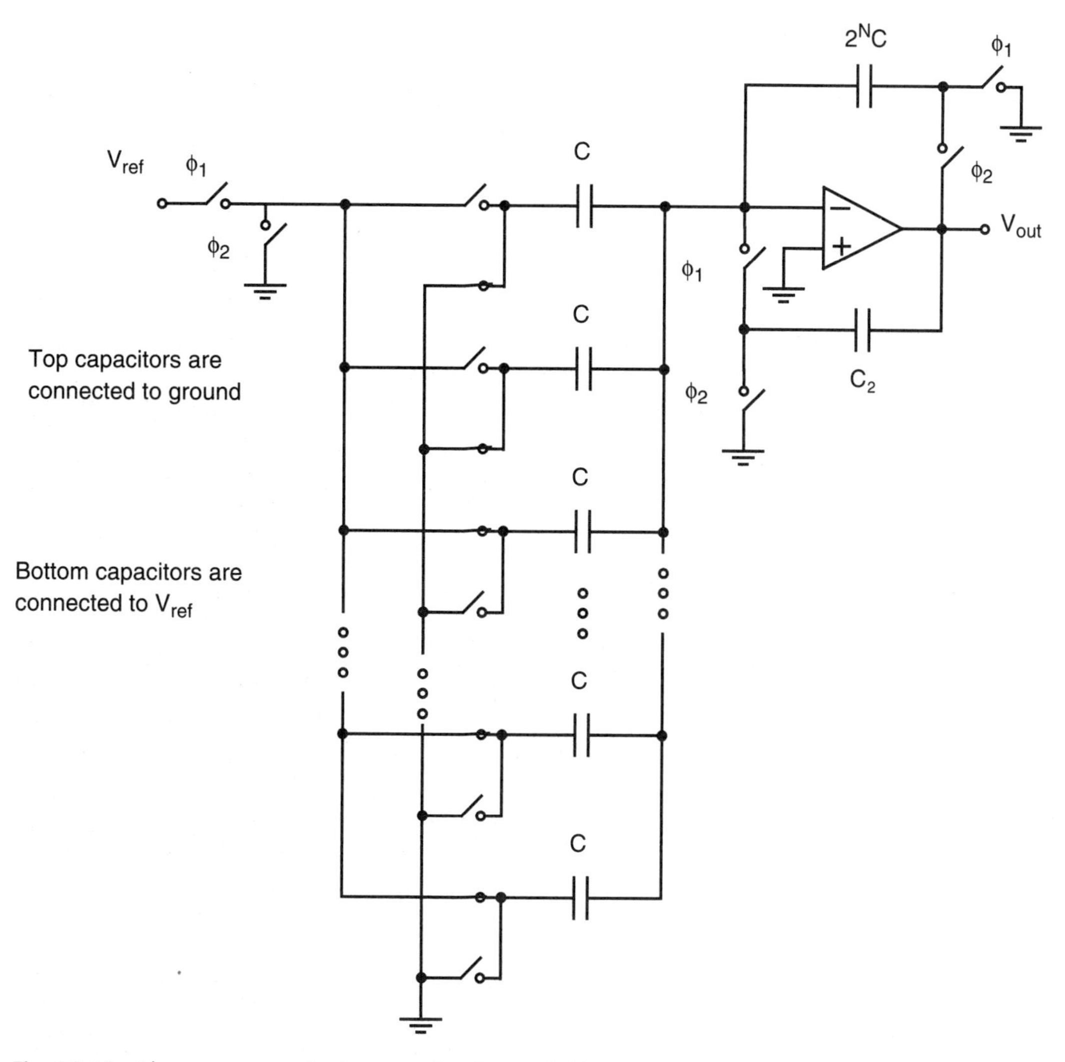

Fig. 12.16 Thermometer-code charge-redistribution D/A converter.

Thermometer-Code Current-Mode D/A Converters

A thermometer-code-based current-mode D/A converter, shown in Fig. 12.17, has been the basis for a variety of designs (see, for example, [Miki, 1986; Chi, 1986; and Letham, 1987]). Here, thermometer-code decoders are used for both the row and column decoders, resulting in inherent monotonicity and good DNL errors. Current is switched to the output when both row and column lines for a cell are high. Also, in high-speed applications, the output feeds directly into an off-chip 50-Ω or 75-Ω resistor, rather than an output opamp. Note that cascode current sources are used here to reduce current-source variation due to voltage changes in the output signal, V_{out}.

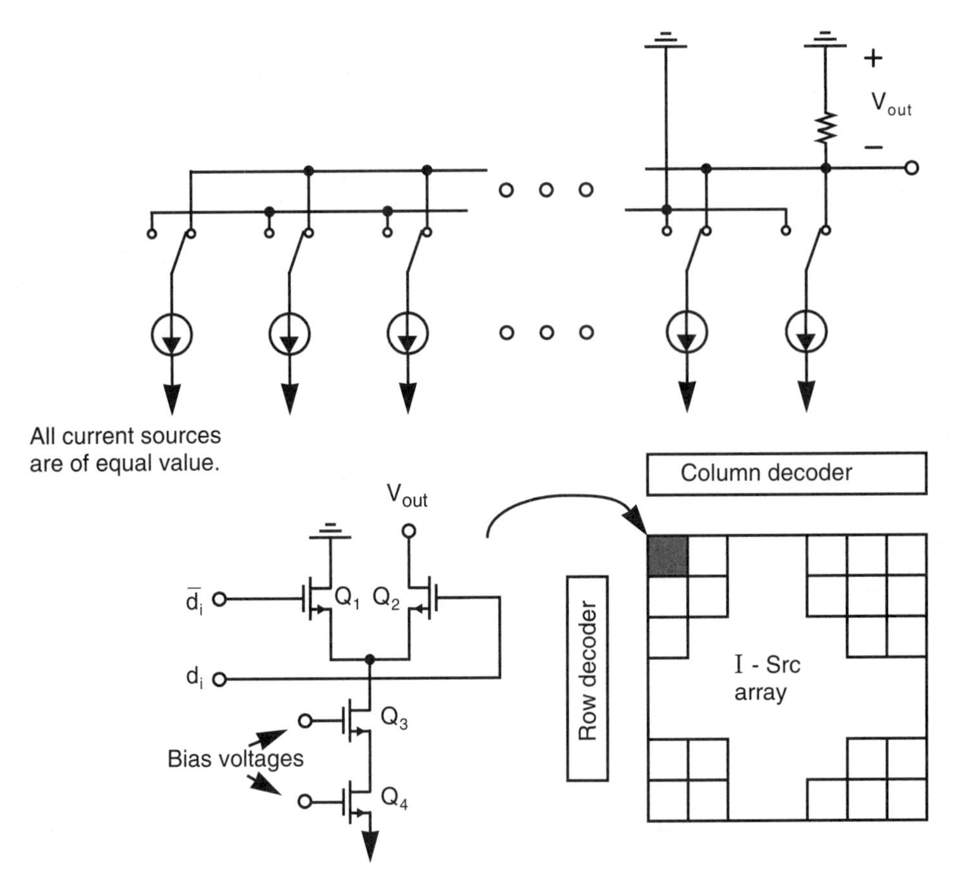

Fig. 12.17 Thermometer-code current-mode D/A converter. Q_1 and Q_2 form the current switch, whereas Q_3 and Q_4 implement a cascode current source.

Note the need here for precisely timed edges of d_i and $\overline{d_i}$. If both logic levels are high simultaneously, V_{out} is shorted to ground. If both logic levels are low at the same time, the drain of Q_3 is pulled low and the circuit takes longer to respond. To avoid the use of two logic driving levels, the gate of Q_2 should be connected to a dc bias voltage, as described in the next approach.

Single-Supply Positive-Output Converters

The architecture for a fast single-supply positive-output D/A (often used in video RAMs, which are known as RAMDACs) is shown in Fig. 12.18 [Colles, 1988]. Here, a matched feedback loop is used to set up accurate known current-source biasing. (Note that the opamp input connections appear reversed but are correct due to signal inversion by Q_4.) Also, to maintain accurate current matching that is independent of V_{out}, one side of each differential current-steering pair is connected to V_{bias}, rather than to the inversion of the bit signal. For example, when the current is steered to the output through Q_2, the drain-source voltage across the current source, Q_3, remains mostly constant if V_{out} stays near zero, such that Q_2 remains in the active region.

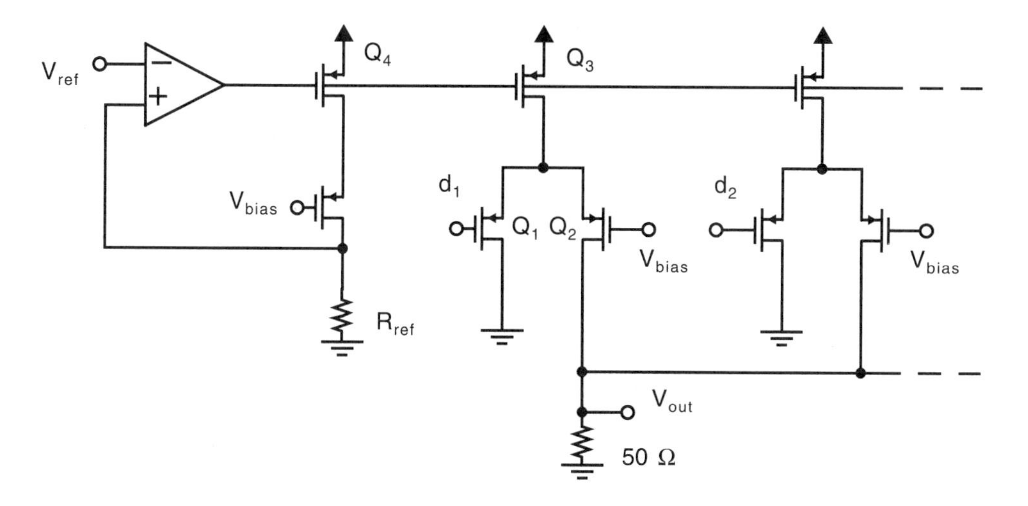

Fig. 12.18 A single-supply positive-output D/A converter.

Thus, Q_2 and Q_3 effectively form a *cascode current source* when they drive current to the output.

To maximize speed in this converter, the voltage swing at the common connections of the current switches (for example, Q_1, Q_2, and Q_3) should be small. To keep this swing small, this common connection should be at a voltage where the output transistors are just turned off when the current is steered to ground. It should also be noted that when a switch is turning on or off, the switching feedthrough from the digital input connected to the grounding transistor (for example, Q_1) actually enhances the switching. Finally, note that this design does not use two logic driving signals, d_i and $\overline{d_i}$, and can therefore be clocked at the maximum rate without the need for precisely timed edges.

Dynamically Matched Current Sources

The use of dynamic techniques with current switching is a method for realizing very well-matched current sources (up to 16-bit accuracy) for audio-frequency D/A converters [Schouwenaars, 1988].

This approach was used to design a 16-bit audio-frequency D/A converter, where the 6 MSB were realized using a thermometer code. Since the accuracy requirements are reduced for the remaining bits, a binary array was used in their implementation. The basic idea for realizing 63 accurately matched current sources for the 6 MSBs is illustrated in Fig. 12.19. Here, we want to set all the currents I_{di} to the same precise value, independent of transistor mismatches and charge injection. To accomplish this high degree of matching, each current source I_{di} is periodically calibrated with the use of a single reference current source, I_{ref}, through the use of the shift register. In other words, once calibration is accomplished on I_{d1}, the same current source, I_{ref}, is used to set the next current source, I_{d2}, to the same value as I_{d1}, and so on. Before proceeding to see how this calibration is accomplished, it is worth mentioning that the values of I_{di} need not precisely equal I_{ref} but do need to accurately match each other.

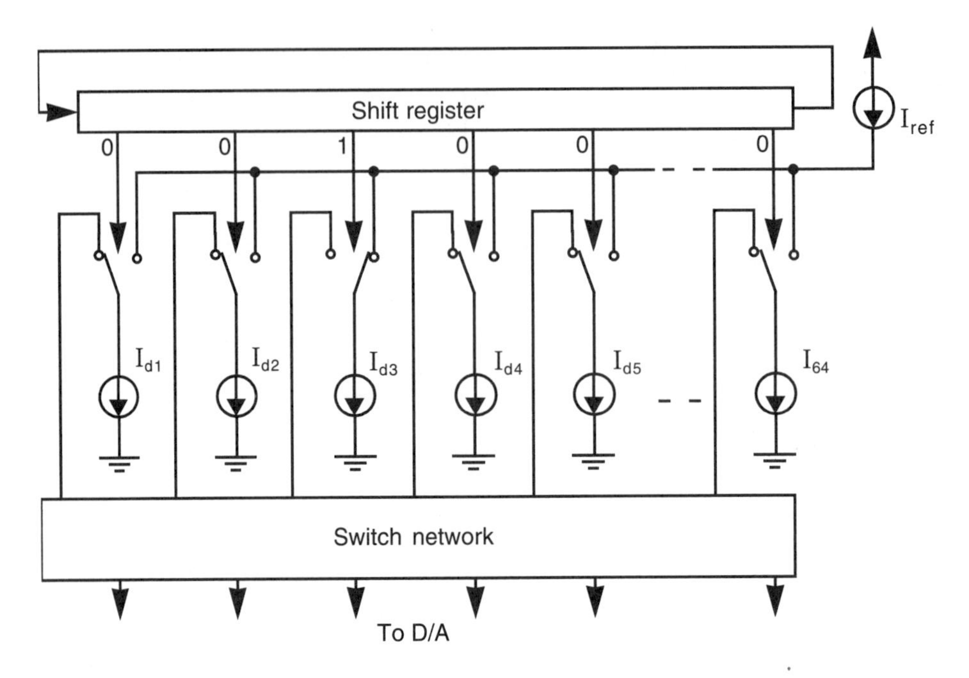

Fig. 12.19 Dynamically matching current sources for 6 MSB.

Therefore, any common errors in the calibration stage are not a problem. Also, note that 64 current sources are calibrated, even though only 63 are required for the 6 MSB. This extra current source is needed so that the D/A converter can continuously operate, even when one of the current sources is being calibrated.

The method for calibrating and using one of the current sources is shown in Fig. 12.20, for I_{d1}. During calibration, the current source is connected to the reference current I_{ref}, while Q_1 is configured in a diode-connected mode. This places whatever voltage is necessary across the parasitic C_{gs} so that I_{d1} equals I_{ref}. When S_1 is opened, I_{d1} remains nearly equal to I_{ref}, assuming the drain-source voltage of Q_1 doesn't change and the clock feedthrough and charge injection of S_1 are small. During regular system use, the gate voltage (and therefore current) is determined by the voltage stored on the parasitic capacitance, C_{gs}.

A major limitation in matching the 64 current sources is due to the differences in clock feedthrough and charge injection of the switches S_i. Since mismatches will always exist between different switches, the best way to keep all current sources equal is to minimize the total amount of clock feedthrough and charge injection. These non-ideal effects can be minimized by having the capacitance C_{gs} and bias voltage V_{GS} large. (A large V_{GS} voltage implies that a small voltage difference will cause a smaller current deviation.) To accomplish these requirements, the current source $0.9I_{ref}$ was added in parallel to Q_1, so that Q_1 needs only to source a current near $0.1I_{ref}$. With such an arrangement, a large, low-transconductance device can be used (a $W/L = 1/8$ might be used).

Finally, each current source must be recalibrated before the leakage currents (on the order of 10 pA/μm^2 of junction area) on C_{gs} cause the current source to deviate

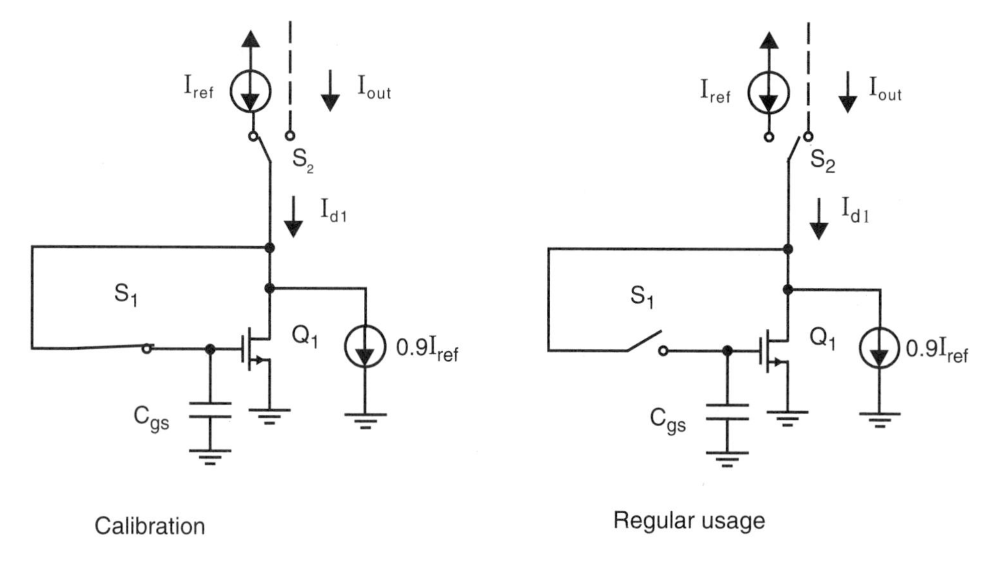

Fig. 12.20 Dynamically setting a current source, I_{d1}.

by 0.5LSB. Fortunately, having a large C_{GS} and V_{GS} has the added benefit of extending this calibration interval (every 1.7 ms in [Schouwenaars, 1988]).

Many other additional details are described in [Schouwenaars, 1988]. For example, during calibration, a p-channel-input common-gate amplifier is added to the diode-connected loop. This is typically done in dynamic switched-current circuits to control the drain-source voltage of Q_1 (i.e., to keep it constant independent of the actual current and to keep it matched to its value during regular use) and to speed up the circuit by decreasing the effect of parasitic capacitances on the large I_{ref} bus. Also, dummy switches are connected to S_1 to help minimize clock feedthrough by partially cancelling the charge injected.

The converter in [Schouwenaars, 1988] achieves 0.0025 percent distortion, 92 dB S/(N+D), 94 db S/N, and only dissipated 20 mW. The converter could also be run using only a 3-V power supply. The clocking frequency is limited to 44 kHz by the digital audio application. No mention is made on the upper limit of the clocking frequency.

12.4 HYBRID CONVERTERS

Combining the techniques discussed in Sections 12.1–12.3 for realizing different portions of a D/A converter results in hybrid designs. Hybrid designs are an extremely popular approach for designing converters because they combine the advantages of different approaches. For example, it is quite common to use a thermometer-code approach for the top few MSBs while using a binary-scaled technique for the lower LSBs. In this way, glitching is significantly reduced and accuracy is high for the MSB where it is needed most. However, in the LSBs where glitching and accuracy requirements are much reduced, valuable circuit area is saved with a binary-scaled approach. This section discusses some useful hybrid designs.

Resistor-Capacitor Hybrid Converters

It is possible to combine tapped resistor strings with switched-capacitor techniques in a number of different ways. In one approach, an SC binary-weighted D/A converter has its capacitors connected to adjacent nodes of a resistor-string D/A converter, as shown in Fig. 12.21 [Yang, 1989]. Here, the top 7 bits determine which pair of voltages across a single resistor is passed on to the 8-bit capacitor array. For example, if

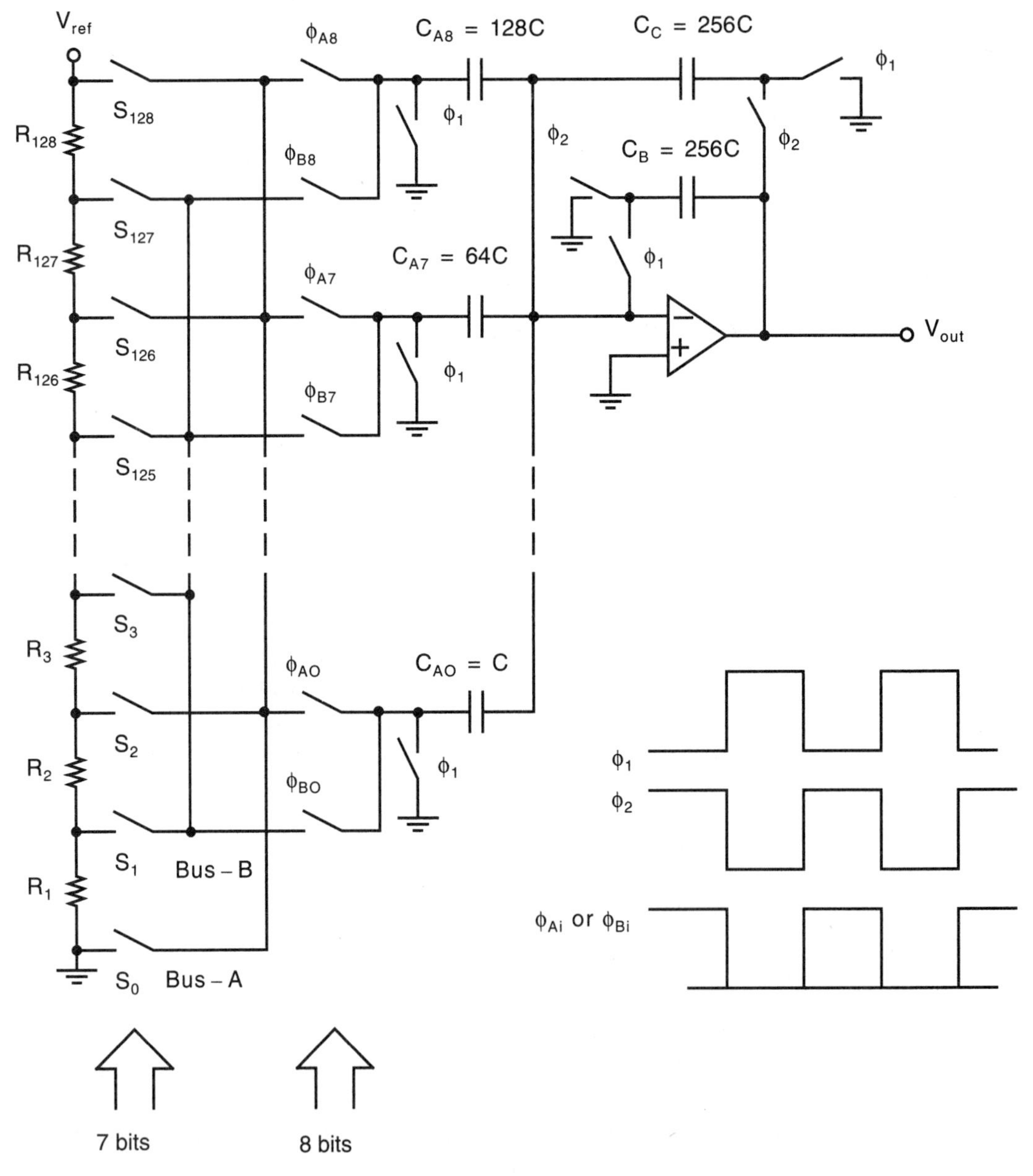

Fig. 12.21 One example of a 15-bit resistor-capacitor hybrid D/A converter.

the top 7 bits are 0000001, then switches S_1 and S_2 would be closed, while the other S_i switches would remain open. The capacitor array then performs an 8-bit interpolation between the pair of voltages by connecting the capacitors associated with a 1 to the higher voltage and the capacitors associated with a 0 to the lower voltage. This approach gives guaranteed monotonicity, assuming the capacitor array is accurate to only 8 bits. The converter in [Yang, 1989] has 15-bit monotonicity, without trimming, and 100-kHz sampling frequencies for a very-low-power 10-mW realization.

Segmented Converters

Segmented converters have probably become the most popular design approach for D/A converters [Schoeff, 1979; Grebene, 1984; Schouwenaars, 1988]. A 6-bit segmented D/A converter is shown in Fig. 12.22. In this approach, the two MSB currents are obtained in one segment from three equal current sources using thermometer coding. High bits are switched to the output, whereas low bits are switched to ground. As discussed in Section 12.3, the use of a thermometer code for the MSB currents greatly minimizes glitches. For the four LSBs, one additional current source from the MSB

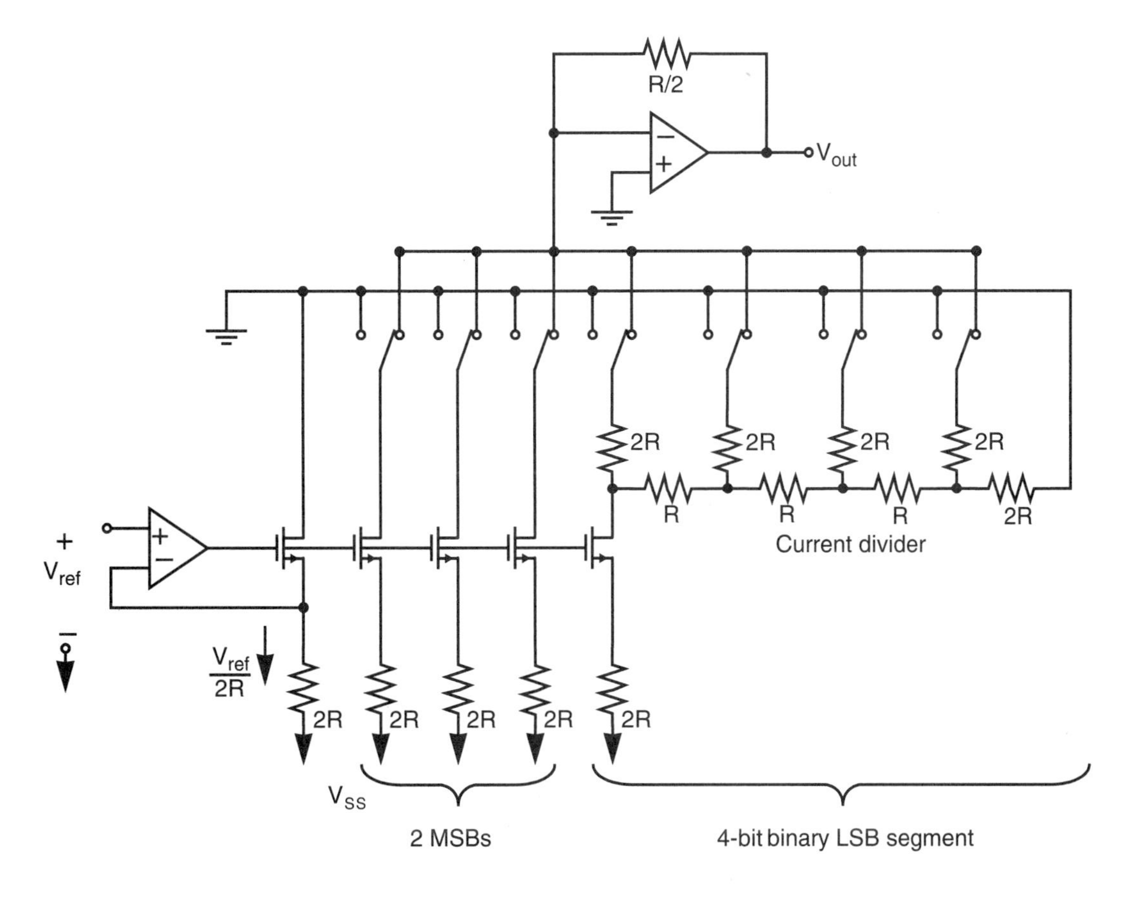

Fig. 12.22 A 6-bit segmented D/A converter.

segment is diverted where it is divided into binary-weighted currents, which are also switched to either ground or the output. Although this four-LSB segment is not guaranteed to be monotonic, its accuracy requirements are very relaxed, since it is used only for the LSBs.

12.5 REFERENCES

A. Abrial et al. "A 27-MHz Digital-to-Analog Video Processor," *IEEE J. of Solid-State Circuits,* Vol. 23, pp. 1358–1369, December 1988.

K. K. Chi et al. "A CMOS Triple 100-Mbit/s Video D/A Converter with Shift Register and Color Map," *IEEE J. of Solid-State Circuits,* Vol. 21, pp. 989–996, December 1986.

J. Colles. "TTL to CMOS Voltage Level Translator," patent no. 4,794,282, granted December 27, 1988.

A. B. Grebene. *Bipolar and MOS Analog Integrated Circuit Design.* John Wiley & Sons, New York, 1984.

D. W. J. Groeneveld, H. J. Schouwenaars, H. A. H. Termeer, and C. A. A. Bastiaansen. "A Self-Calibration Technique for Monolithic High-Resolution D/A Converters," *IEEE J. of Solid-State Circuits,* Vol. 24, pp. 1517–1522, December 1989.

A. R. Hamadé. "A Single-Chip All-MOS 8-bit A/D Converter," *IEEE J. of Solid-State Circuits,* Vol. 13, pp. 785–791, December 1978.

P. Holloway. "A Trimless 16-bit Digital Potentiometer," *IEEE Intl. Solid-State Circuits Conf.,* pp. 66–67, February 1984.

L. Letham et al. "A High-Performance CMOS 70-MHz Palette/DAC," *IEEE J. of Solid-State Circuits,* Vol. 22, pp. 1041–1047, December 1987.

K. W. Martin, L. Ozcolak, Y. S. Lee, and G. C. Temes. "A Differential Switched-Capacitor Amplifier," *IEEE J. of Solid-State Circuits,* Vol. 22, pp. 104–106, February 1987.

T. Miki et al. "An 80-MHz 8-bit CMOS D/A Converter," *IEEE J. of Solid-State Circuits,* Vol. 21, pp. 983–988, December 1986.

P. H. Saul and J. S. Urquhart. "Techniques and Technology for High-Speed D/A Conversion," *IEEE J. of Solid-State Circuits,* Vol. 19, pp. 62–68, December 1984.

A. S. Sedra and K. C. Smith. *Microelectronic Circuits*, 3rd ed. Saunders College Publishing/HRW, New York, 1991.

J. A. Schoeff. "An Inherently Monotonic 12-bit DAC," *IEEE J. of Solid-State Circuits,* Vol. 14, pp. 904–911, December 1979.

H. J. Schouwenaars, D. W. J. Groeneveld, and H. A. H. Termeer. "A Low-Power Stereo 16-bit CMOS D/A Converter for Digital Audio," *IEEE J. of Solid-State Circuits,* Vol. 23, pp. 1290–1297, December 1988.

J. W. Yang and K. W. Martin. "High-Resolution Low-Power CMOS D/A Converter," *IEEE J. of Solid-State Circuits,* Vol. 24, pp. 1458–1461, October 1989.

12.6 PROBLEMS

12.1 Derive an expression for the number of switches in a general N-bit resistor-string D/A converter similar to that shown in Fig. 12.1.

12.2 Assume that an 8-bit resistor-string D/A converter with digital decoding (see Fig. 12.3) has a total resistor-string resistance of 400 Ω, that its pass transistors have an on resistance of 400 Ω, and that the drain-source capacitances to ground of its pass transistors are 0.1 pF. Ignoring all other effects and using the open-circuit time-constant approach, estimate the worst-case settling time to 0.1 percent.

12.3 Derive an expression for the number of switches in an N-bit folded resistor-string D/A converter similar to that shown in Fig. 12.4 (Assume N is even).

12.4 Assume that the first resistor string of a 10-bit multiple-R-string D/A converter must match to 0.1 percent whereas the second string must match to 1.6 percent since the converters realize the top 4 bits and lower 6 bits, respectively. For $V_{ref} = 5$ V, how much offsets in the opamps can be tolerated?

12.5 For a binary-weighted 10-bit resistor D/A converter (see Fig. 12.7), assume that R_F is chosen such that the output goes from 0 to $V_{ref} - V_{LSB}$. What is the resistor ratio between the largest and smallest resistors? What is the ratio between the currents through the switches for b_1 and b_{10}?

12.6 It is desired to realize a binary-weighted 4-bit resistor D/A converter, as shown in Fig. 12.7, that must be linear to 10 bits. Ignoring all other nonidealities except for resistance mismatch error, what percentage matching accuracy is required for each of the $b_2 - b_4$ resistors relative to the b_1 resistor?

12.7 Consider an 8-bit D/A converter built using binary-weighted capacitors, where capacitor tolerances are ±0.5 percent. What would be the worst-case differential nonlinearity in units of LSBs, and at what transition does it occur?

12.8 Consider the reduced-resistance-ratio approach shown in Fig. 12.8, where a single series resistor is applied to an N-bit converter, and where N is an even number. What is the improvement in the resistance ratio compared to using all binary-scaled resistances?

12.9 Draw the circuit for a reduced-resistance-ratio 8-bit D/A converter (see Fig. 12.8), where a resistor is inserted between b_4 and b_5. Ignoring R_F, what is the resistance spread? Now draw a similar circuit, but add resistors between b_2 and b_3 as well as between b_6 and b_7. What is this circuit's resistance spread?

12.10 Assuming all the switches in Fig. P12.10 are MOSFETS and are scaled so that each has 100 mV across the drain source when on (including S_4, which always remains on), show that no accuracy is lost.

12.11 For the 4-bit R-2R-ladder D/A converter shown in Fig. P12.10, what is the output error (in LSBs) when $R_A = 2.01 R_B$? What is the output error when $R_C = 2.01 R$?

12.12 Show that the D/A converter circuit shown in Fig. 12.11 operates correctly. Estimate the speed of this circuit if the opamp has infinite bandwidth, $R = 10\ \text{k}\Omega$, and all nodes have a capacitance of 0.5 pF to ground.

12.13 Consider the D/A converter shown in Fig. 12.11, where $I = 1\ \text{mA}$, $R_f = 2\ \text{k}\Omega$ and $R = 10\ \text{k}\Omega$. If this converter is perfectly linear but is found to have an offset error of 0.15LSB and a gain error of 0.2LSB, find the output levels for the inputs $b_1b_2b_3b_4$ equal to 0000, 1000, and 1111.

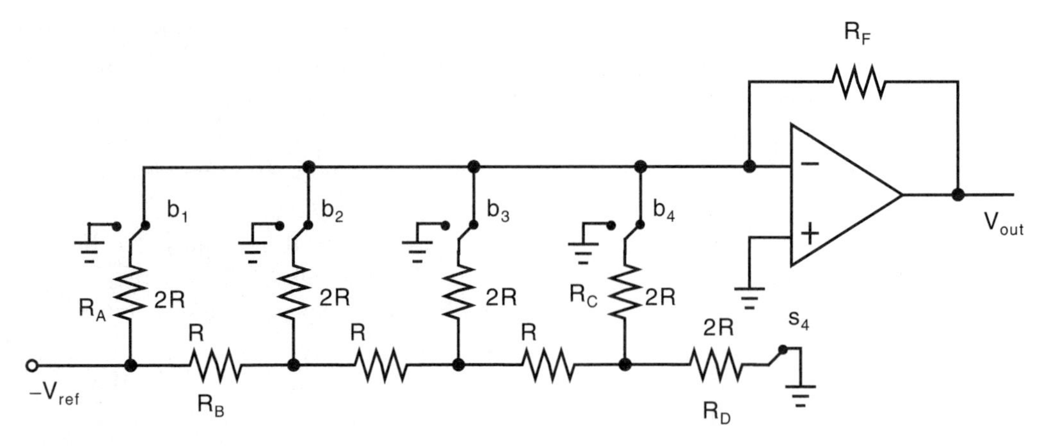

Fig. P12.10

12.14 A 12-bit binary-weighted D/A converter produces a glitch voltage of 0.5 V when the MSB is changed. What is the glitch voltage reduced to if a thermometer-code approach is used for the top 4 bits, whereas binary-weighting is used for the remaining 8 bits?

12.15 For the circuit in Fig. 12.17, sketch a typical waveform that would occur on the node connecting Q_1, Q_2, and Q_3 when the code signals, d_i and $\overline{d_i}$, are nonoverlapping. Repeat this sketch if the code signals are overlapping. Assume the code signals have logic swings from 0 to –5 V and that the threshold voltage for the transistors equals 1 V.

12.16 For the circuit in Fig. 12.18, sketch a typical waveform that would occur on the node connecting Q_1, Q_2, and Q_3 when the code signal, d_i, goes low-to-high and high-to-low. Assume here that the code signal has a logic swing of 0 to 5 V, that the threshold voltage for the transistors equals –1 V, and that V_{bias} is 1 V.

12.17 A D/A converter is realized using dynamically matched current sources, as shown in Fig. 12.20. Assuming all the transistors are ideal, find W/L for the Q_1 needed to set $V_{GS} = 3$ V when $I_{ref} = 50\ \mu A$, $V_t = 1$ V, and $\mu_n C_{ox} = 92\ \mu A/V^2$. If switch S_1 causes a random charge injection voltage of 1 mV, what is the expected percentage of random variation of the current being held on Q_1?

12.18 Repeat Problem 12.17 if the design does not incorporate the $0.9I_{ref}$ extra current source (i.e., if Q_1 must be the source for all of I_{ref}).

CHAPTER

13 Nyquist-Rate A/D Converters

Architectures for realizing analog-to-digital converters (ADC) can be roughly divided into three categories (Table 13.1)—low-to-medium speed, medium speed, and high speed. In this chapter, design details are discussed for these different approaches except for oversampling converters. Oversampling converters are best described using many signal processing concepts and are therefore discussed separately in Chapter 14.

Before proceeding, it should be noted that when discussing the design of A/D converters, we usually ignore the 0.5 LSB offset present in the A/D transfer characteristic. Such a simplification is made so as not to complicate the concepts presented.

13.1 INTEGRATING CONVERTERS

Integrating A/D converters is a popular approach for realizing high-accuracy data conversion on very slow-moving signals. These types of converters have very low offset and gain errors in addition to being highly linear. A further advantage of integrating converters is the small amount of circuitry required in their implementation. One application that has traditionally made use of integrating converters is measurement instruments such as voltage or current meters.

A simplified diagram for a *dual-slope* integrating converter is shown in Fig. 13.1. Dual-slope refers to this converter performing its conversion in two phases, (I) and (II), in the following manner (Fig. 13.2):

Phase (I) Phase (I) is a fixed time interval of length T_1 determined by running the counter for 2^N clock cycles. Thus, we have

$$T_1 = 2^N T_{clk} \tag{13.1}$$

Table 13.1 Different A/D converter architectures

Low-to-Medium Speed, High Accuracy	Medium Speed, Medium Accuracy	High Speed, Low-to-Medium Accuracy
Integrating	Successive approximation	Flash
Oversampling	Algorithmic	Two-step
		Interpolating
		Folding
		Pipelined
		Time-interleaved

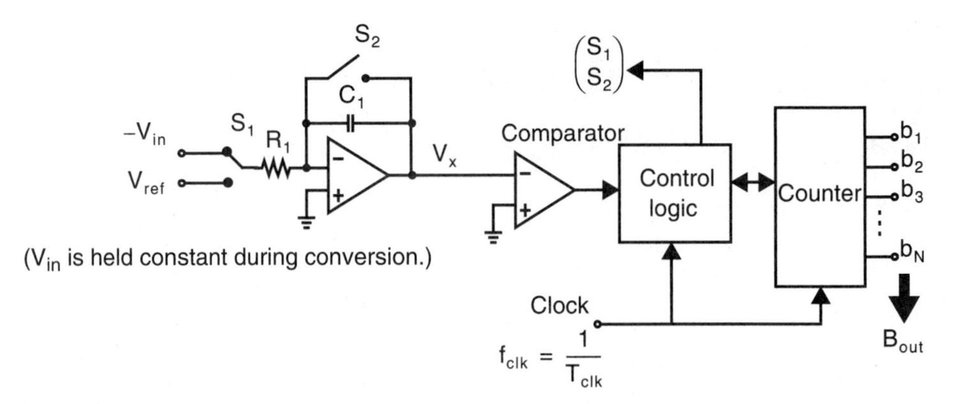

Fig. 13.1 Integrating (dual slope) A/D converter.

where T_{clk} is the period for one clock cycle. During this interval, switch S_1 is connected to $-V_{in}$ such that V_x ramps up proportional to the magnitude of V_{in}. Assuming V_x is initially equal to zero (due to a pulse on S_2), we have the following relationship for V_x:

$$V_x(t) = -\int_0^t \frac{(-V_{in})}{R_1C_1} d\tau = \frac{V_{in}}{R_1C_1} t \tag{13.2}$$

Thus, at the end of phase (I), the value of V_x is equal to $V_{in}T_1/R_1C_1$.

Phase (II) Phase (II) occurs for a variable amount of time, T_2, as shown in Fig. 13.2 for three different input voltages. At the beginning of this phase, the counter is reset and switch S_1 is connected to V_{ref}, resulting in a constant slope for the decaying voltage at V_x. To obtain the digital output value, the counter simply counts until V_x is less than zero, at which point that count value equals the digitized value of the input

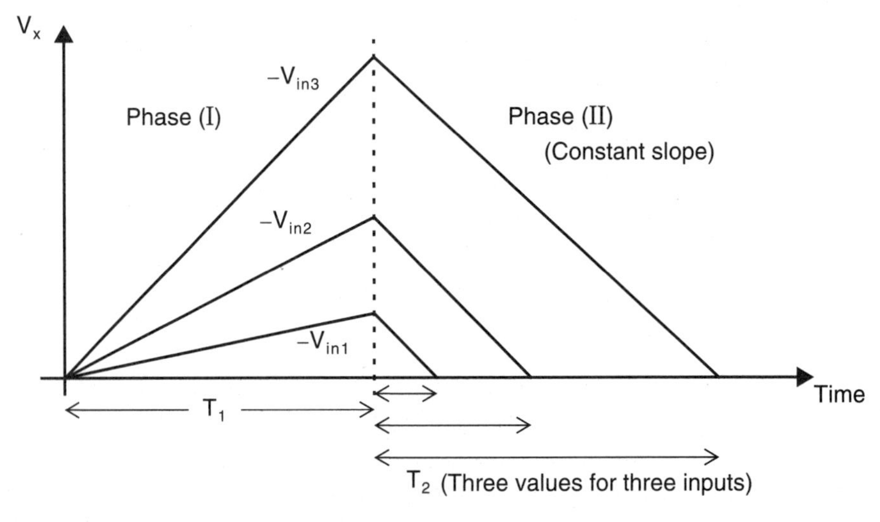

Fig. 13.2 Operation of the integrating converter for three different input voltages.

signal, V_{in}. Thus, assuming the digital output count is normalized so that the largest count is unity, the counter output, B_{out}, can be defined to be

$$B_{out} = b_1 2^{-1} + b_2 2^{-2} + \cdots + b_{N-1} 2^{-(N-1)} + b_N 2^{-N} \tag{13.3}$$

and we have

$$T_2 = 2^N B_{out} T_{clk} = (b_1 2^{N-1} + b_2 2^{N-2} + \cdots + b_{N-1} 2 + b_N) T_{clk} \tag{13.4}$$

To see why this count gives the correct value, we find the equation for V_x during phase (II) to be given by

$$\begin{aligned} V_x(t) &= -\int_{T_1}^{t} \frac{V_{ref}}{R_1 C_1} d\tau + V_x(T_1) \\ &= \frac{-V_{ref}}{R_1 C_1}(t - T_1) + \frac{V_{in} T_1}{R_1 C_1} \end{aligned} \tag{13.5}$$

Since V_x equals zero when $t = T_1 + T_2$, we can write

$$0 = \frac{-V_{ref} T_2}{R_1 C_1} + \frac{V_{in} T_1}{R_1 C_1} \tag{13.6}$$

and thus T_2 is related to T_1 by the following relationship:

$$T_2 = T_1 \left(\frac{V_{in}}{V_{ref}} \right) \tag{13.7}$$

Combining (13.7) with (13.1) and (13.4), we find

$$B_{out} = b_1 2^{-1} + b_2 2^{-2} + \cdots + b_{N-1} 2^{-(N-1)} + b_N 2^{-N} = \frac{V_{in}}{V_{ref}} \tag{13.8}$$

as expected.

From (13.8), we see that by going to a dual-slope conversion (i.e., two phases), the digital output does not depend on the time constant, $R_1 C_1$. In fact, the value of this time constant need only be stable during a single conversion for proper operation. However, R_1 and C_1 should be chosen such that a reasonable large peak value of V_x is obtained without clipping to reduce noise effects. If instead of dual-slope, a single-slope conversion was used, then a gain error would most likely occur, and this error would be a function of the time-constant value, $R_1 C_1$.

Although a dual-slope converter does not suffer from gain error, it can have an offset error due to opamp offset and other factors. Such an offset error can be calibrated out by going to a quad-slope conversion. In a quad-slope conversion, a dual-slope conversion is performed twice, once with the input connected to ground and then with the input connected to the signal to be converted. A subtraction of the two output words causes the offset error to be reduced to zero.

The conversion speed for these types of converters is quite slow. For example, in a dual-slope converter, the worst-case conversion speed occurs when V_{in} equals V_{ref}. In this case, the number of clock cycles to perform a conversion is 2^{N+1}. Thus for a 16-bit converter with a clock frequency equal to 1 MHz, the worst-case conversion rate is around 7.6 Hz.

Finally, it should be mentioned that by a careful choice for T_1, certain frequency components superimposed on the input signal can be significantly attenuated. For example, if the input signal, V_{in}, is a dc level with power line noise of 60 Hz superimposed on it, choosing T_1 equal to an integer multiple of 16.67 ms filters out the power line noise. To see why this filtering effect occurs, we write V_{in} as

$$V_{in} = V_{in(ideal)} + V_{in(60\ Hz)} \tag{13.9}$$

where $V_{in(ideal)}$ is the desired dc level and $V_{in(60\ Hz)}$ is the interfering 60 Hz noise; or mathematically,

$$V_{in(60\ Hz)} = A\sin(120\pi t + \phi) \tag{13.10}$$

where A and ϕ are arbitrary magnitude and phase values. Now substituting this relationship for V_{in} into (13.2), we have

$$V_x(T_1) = -\int_0^{T_1} \frac{(-V_{in})}{R_1C_1}d\tau = -\int_0^{T_1} \frac{(-V_{in(ideal)})}{R_1C_1}d\tau - \int_0^{T_1} \frac{(-V_{in(60\ Hz)})}{R_1C_1}d\tau \tag{13.11}$$

However, the last term in (13.11) can be shown to equal zero when T_1 is an integer multiple of $1/(60\ Hz)$ (i.e., 16.67 ms). In this way, the peak value, $V_x(T_1)$, remains correct so that the conversion is performed without error. Note that for this same value of T_1, the harmonics of 60 Hz are also suppressed (i.e., 120 Hz, 180 Hz, 240 Hz, etc.). In fact, other frequencies are also attenuated but not fully suppressed as are the harmonics of 60 Hz. To quantify the filtering effect, we note that this converter effectively "integrates and dumps" the input signal. Such an integrating-then-reset behavior results in the impulse response of this converter being a square pulse of length T_1. Since the Fourier transform of a square pulse is a "sin(x)/x" type response, we have an effective input filter, as shown in Fig. 13.3. Here we have assumed that $T_1 = 1/(60\ Hz)$, resulting in a filter transfer function, $H(f)$, being given by

$$|H(f)| = \left|\frac{\sin(\pi f T_1)}{(\pi f T_1)}\right| \tag{13.12}$$

Note that higher frequencies are attenuated more and that full suppression is achieved at harmonics of 60 Hz, as expected.

EXAMPLE 13.1

It is desired to build a 16-bit two-slope integrating A/D such that a maximum input signal of $V_{in} = 3\ V$ results in the peak voltage of V_x being 4 V. In addition, input noise signals at 50 Hz and harmonics should be significantly attenuated. Find the required RC time constant and clock rate. Also, find the attenuation of a noise signal around 1 kHz superimposed on the input signal.

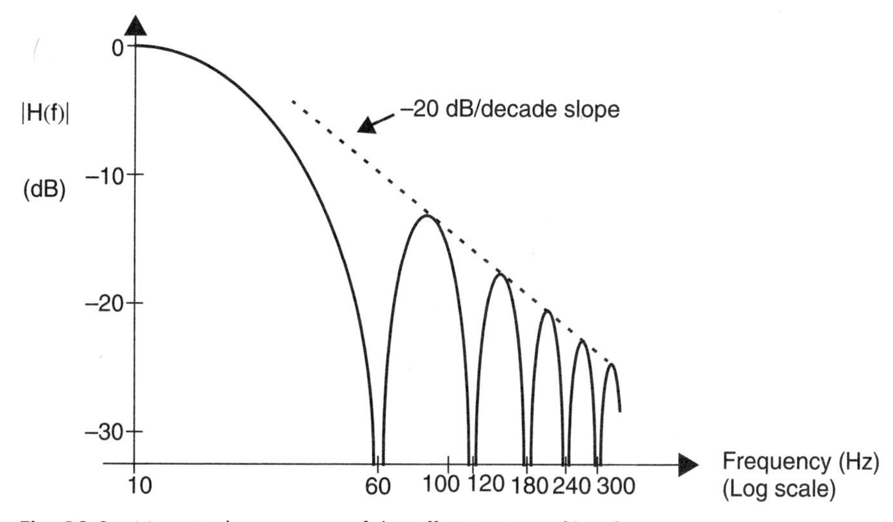

Fig. 13.3 Magnitude response of the effective input filter for an integrating-type converter with $T_1 = 1/(60\ \text{Hz})$.

Solution

Since 50 Hz and harmonics are to be rejected, we choose

$$T_1 = \frac{1}{50} = 20\ \text{ms} \tag{13.13}$$

Thus, for a 16-bit converter, we require a clock frequency of

$$f_{clk} = \frac{1}{T_{clk}} = \frac{2^{16}}{T_1} = 3.28\ \text{MHz} \tag{13.14}$$

To find the RC time constant needed, we note that at the end of phase (I), V_x is given by

$$V_x = \frac{V_{in}T_1}{R_1C_1} \tag{13.15}$$

and using the values of $V_x = 4\ \text{V}$, $V_{in} = 3\ \text{V}$, and $T_1 = 20\ \text{ms}$ results in

$$R_1C_1 = 15\ \text{ms} \tag{13.16}$$

Finally, the attenuation of a 1-kHz signal is infinite since it is a harmonic of 50 kHz. However, as seen in (13.12), attenuation is reduced halfway between harmonics, so we find the gain for an input signal at 975 Hz to be

$$|H(f)| = \left|\frac{\sin(\pi \times 975\ \text{Hz} \times 20\ \text{ms})}{\pi \times 975\ \text{Hz} \times 20\ \text{ms}}\right| = 16.3 \times 10^{-3} \tag{13.17}$$

which implies the attenuation is 36 dB.

13.2 SUCCESSIVE-APPROXIMATION CONVERTERS

Successive-approximation A/D converters are one of the most popular approaches for realizing A/D converters due to their reasonably quick conversion time, yet moderate circuit complexity. To understand the basic operation of successive-approximation converters, knowledge of the search algorithm referred to as a "binary search" is helpful. As an example of a binary search, consider the game of guessing a random number from 1 to 128 where one can ask only questions that have a "yes/no" response. The first question might be, "Is the number greater than 64?" If the answer is yes, then the second question asks whether the number is greater than 96. However, if the first answer is no, then the second question asks whether the number is greater than 32.

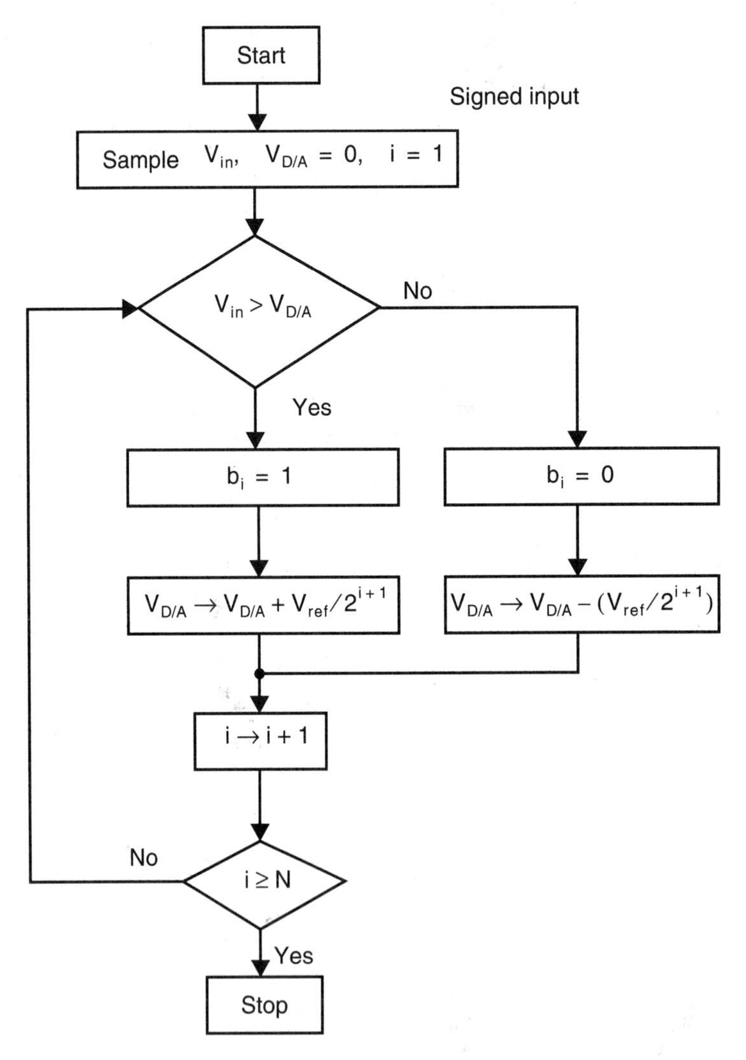

Fig. 13.4 Flow graph for the successive-approximation approach.

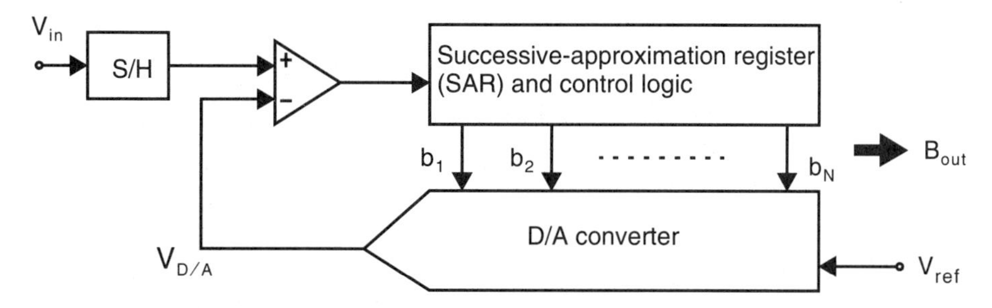

Fig. 13.5 D/A converter-based successive-approximation converter.

The third question divides the search space in two once again and the process is repeated until the random number is determined. In general, a binary search divides the search space in two each time, and the desired data can be found in N steps for a set of organized data of size 2^N.

Successive-approximation converters apply a binary search algorithm to determine the closest digital word to match an input signal. Specifically, in the first period, after possibly the reset period, the MSB, b_1, is determined. In the second period, the next bit, b_2, is determined, followed by b_3 and so on until all N bits are determined. Thus, in its most straightforward implementation, a successive-approximation converter requires N clock cycles to complete an N-bit conversion. A flow graph for a *signed* conversion using a successive-approximation approach is shown in Fig. 13.4. Here, the signed output is in offset-binary coding and the input signal is assumed to be within $\pm 0.5\ V_{ref}$. The flow graph for a *unipolar* conversion is only slightly different and is left as an exercise for the reader.

DAC-Based Successive Approximation

The block diagram for a unipolar successive-approximation A/D converter that uses a DAC is shown in Fig. 13.5. The successive-approximation register (SAR) and control logic are entirely digital and perform the necessary binary search. At the end of the conversion, the digital value in the SAR results in the voltage $V_{D/A}$ being within $0.5\ V_{LSB}$ of the input signal. With this type of architecture, the D/A converter typically determines the accuracy and speed of the A/D converter. Note that a sample and hold is required at the input so that the value to be converted does not change during the conversion time.

EXAMPLE 13.2

Consider the case where $V_{ref} = 8\ V$, $V_{in} = 2.831\ V$, and a 3-bit conversion is performed. Find intermediate D/A values and the final output.

Solution

In this case, $V_{LSB} = 1.0\ V$.

In cycle 1: $B_{out} = 100$ so that $V_{D/A} = 4.0\ V$. Since $V_{in} < V_{D/A}$, b_1 is set to 0.

In cycle 2: $B_{out} = 010$ so that $V_{D/A} = 2.0\ V$. Since $V_{in} > V_{D/A}$, b_2 is set to 1.

In cycle 3: $B_{out} = 011$ so that $V_{D/A} = 3.0\ V$. Since $V_{in} < V_{D/A}$, b_3 is set to 0.

Therefore, the resulting output is the last value of B_{out}, which is 010. Although the final quantization error is 0.831 V_{LSB}, it is greater than $\pm 0.5\ V_{LSB}$ because we have not accounted for the 0.5 LSB offset as discussed at the beginning of this chapter.

Unipolar Charge-Redistribution A/D

The straightforward approach of using a separate D/A converter and setting it equal to the input voltage (within one LSB) can be modified to the flow graph shown in Fig. 13.6. Here, the error signal V equals the difference between the input signal, V_{in}, and the D/A output, $V_{D/A}$. As a result, V is always compared to ground, as seen at the top of the flow graph, and the goal is to set this error difference with one LSB of zero.

One of the first switched-capacitor analog systems using this approach is a charge-redistribution MOS A/D converter [McCreary, 1975]. With such a converter, the sample and hold, D/A converter, and the difference portion of the comparator are all combined into a single circuit. The unipolar case is shown in Fig. 13.7 and operates as follows:

1. *Sample mode:* In the first step, all the capacitors are charged to V_{in} while the comparator is being reset to its threshold voltage through s_2, In this step, note that the capacitor array is performing the sample-and-hold operation.
2. *Hold mode:* Next, the comparator is taken out of reset by opening s_2, and then all the capacitors are switched to ground. This causes V_x, which was originally zero, to change to $-V_{in}$, thereby holding the input signal, V_{in}, on the capacitor array. (This step is sometimes merged with the first bit time during bit cycling). Finally, s_1 is switched so that V_{ref} can be applied to the capacitor array during bit cycling.
3. *Bit cycling:* Next, the largest capacitor (i.e., the $16C$ capacitor in this example) is switched to V_{ref}. V_x now goes to $(-V_{in} + V_{ref}/2)$. If V_x is negative, then V_{in} is greater than $V_{ref}/2$, and the MSB capacitor is left connected to V_{ref}. Also b_1 is considered to be a 1. Otherwise, the MSB capacitor is reconnected to ground and b_1 is taken to be 0. This process is repeated N times, with a smaller capacitor being switched each time, until the conversion is finished.

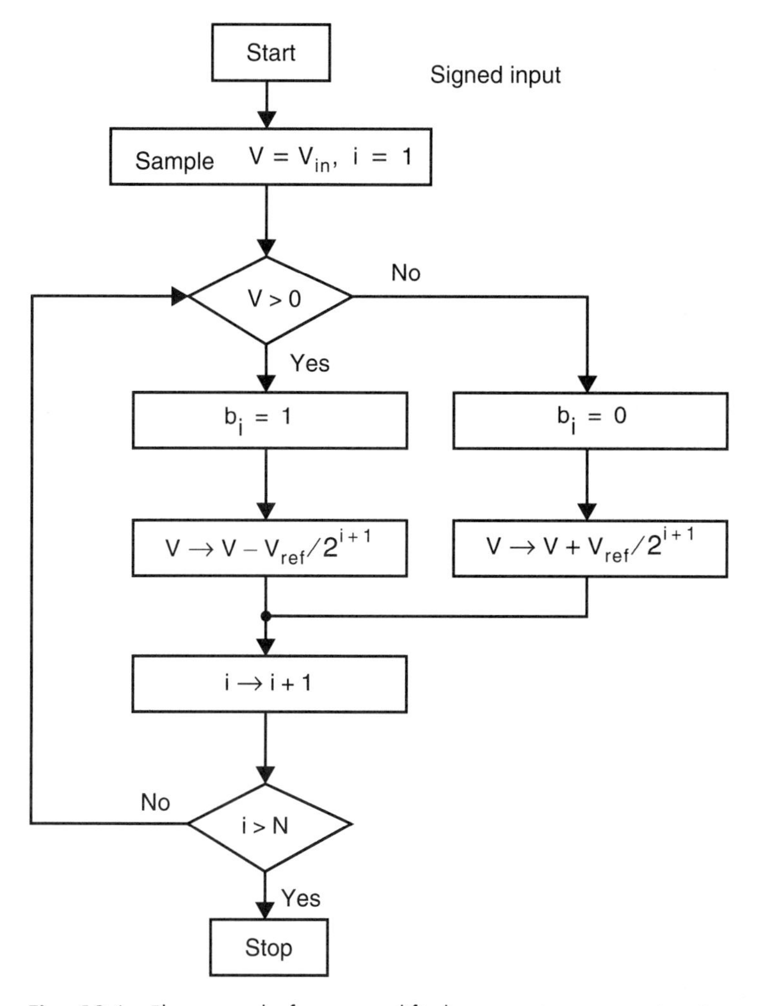

Fig. 13.6 Flow graph for a modified successive approximation (divided remainder).

To get an exact division by two, note that an additional unit capacitor of size C has been added so that the total capacitance is $2^N C$ rather than $(2^N - 1)C$. Also, the capacitor *bottom plates* should be connected to the V_{ref} side, not to the comparator side, to minimize the parasitic capacitance at node V_x. Although parasitic capacitance at V_x does not cause any conversion errors with an ideal comparator, it does attenuate the voltage V_x.

A signed A/D conversion can be realized by adding a $-V_{ref}$ input. If V_x is less than zero at the first step, then proceed as in the unipolar case using V_{ref}. Otherwise, if V_x is greater than zero, use $-V_{ref}$ and test for V_x greater than zero when deciding whether to leave the capacitors connected to $-V_{ref}$ or not at each bit cycling.

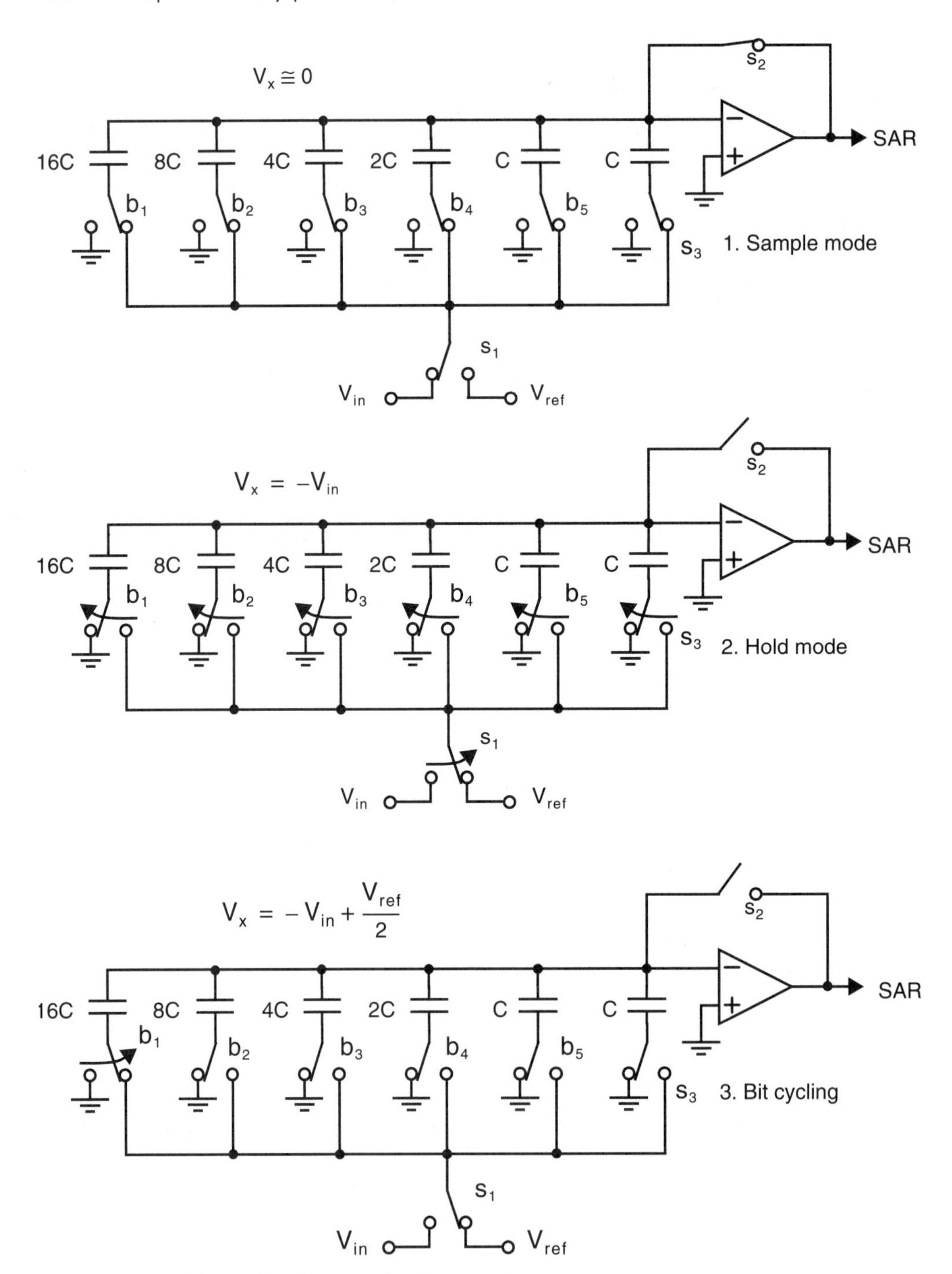

Fig. 13.7 A 5-bit unipolar charge-redistribution A/D converter.

EXAMPLE 13.3

Find intermediate node voltages at V_x during the operation of the 5-bit charge-redistribution converter shown in Fig. 13.7 when $V_{in} = 1.23$ V and $V_{ref} = 5$ V. Assume a parasitic capacitance of $8C$ exists on the node at V_x.

Solution

First, $V_x = 0$ during sample mode. Next, during hold mode, all capacitors are switched and the charge on V_x is shared between the $32C$ total converter capacitance and the parasitic $8C$ capacitance, resulting in

$$V_x = \frac{32}{32+8} \times -V_{in} = -0.984 \text{ V} \tag{13.18}$$

During the first bit cycling, b_1 is switched, resulting in

$$V_x = -0.984 + \frac{16}{40} \times 5 \text{ V} = 1.016 \text{ V} \tag{13.19}$$

Since this result is greater than zero, switch b_1 is reversed back to ground and V_x returns to –0.984 V. Thus, $b_1 = 0$.

Next, b_2 is switched, and we have

$$V_x = -0.984 + \frac{8}{40} \times 5 \text{ V} = 0.016 \text{ V} \tag{13.20}$$

It is also greater than zero, so $b_2 = 0$, and V_x is set back to –0.984 V by switching b_2 back to ground.

When b_3 is next switched, we have

$$V_x = -0.984 + \frac{4}{40} \times 5 = -0.484 \text{ V} \tag{13.21}$$

Since this result is less than zero, $b_3 = 1$, and switch b_3 is left connected to V_{ref}.

Next, b_4 is switched, resulting in

$$V_x = -0.484 + \frac{2}{40} \times 5 = -0.234 \text{ V} \tag{13.22}$$

which is also less than zero, so $b_4 = 1$, and this switch is left connected to V_{ref}.

Finally, b_5 is switched, resulting in

$$V_x = -0.234 + \frac{1}{40} \times 5 = -0.109 \text{ V} \tag{13.23}$$

which is also less than zero, so $b_5 = 1$.

Therefore, the output is given by $B_{out} = 00111$ and voltage V_x is within a V_{LSB} of ground ($V_{LSB} = 5/32$ V).

Signed Charge-Redistribution A/D with a Single Reference Voltage

The same structure as the unipolar case can be used to realize a signed A/D conversion while using only a single V_{ref} if a slightly modified switching arrangement is used. Referring to Fig. 13.8, and assuming V_{in} is between $\pm V_{ref}/2$, the conversion proceeds as follows:

1. *Sample mode:* In the first step, all the capacitors, except for the largest capacitor, are charged to V_{in} while the comparator is being reset to its threshold voltage. For the signed case, the largest capacitor is now connected to $V_{ref}/2$.
2. *Hold mode:* Next, the comparator is first taken out of reset, and then all the capacitors, except the largest one, are switched to ground. This causes V_x, which was originally zero, to change to $-V_{in}/2$. At the end of this step, the sign of the input signal is determined by looking at the comparator output.
3. *Bit cycling:* Next, the largest capacitor (i.e., the $16C$ capacitor in this example) is switched to ground *if, and only if,* V_x is larger than zero (i.e., when V_{in} is less than zero). Specifically, if V_x is less than zero, then V_{in} is positive and b_1 is set to 1, and conversion proceeds as in the unipolar case, starting with b_2, until conversion is completed. However, if V_x is larger than zero, b_1 is set to 0, the largest capacitor is switched to ground, causing V_x to become $-V_{in}/2 - V_{ref}/4$ (which is a negative value), and conversion proceeds as in the unipolar case, starting with b_2. Once conversion is completed, some digital recoding may be required to obtain the desired output code.

This approach for realizing signed A/Ds has the disadvantage that V_{in} has been attenuated by a factor of two, which makes noise more of a problem for high-resolution A/Ds. Also, any error in the MSB capacitor now causes both an offset and a sign-dependent gain error. The latter causes integral nonlinearity errors.

Resistor-Capacitor Hybrid

A combination of the resistor-string and capacitor-array approaches in a hybrid A/D converter, similar to the hybrid D/A converter [Fotouhi, 1979], is shown in Fig. 13.9.

The first step is to charge all the capacitors to V_{in} while the comparator is being reset. Next, a successive-approximation conversion is performed to find the two adjacent resistor nodes that have voltages larger and smaller than V_{in}. One bus will be connected to one node while the other is connected to the other node. All of the capacitors are connected to the bus having the lower voltage. A successive approximation using the capacitor-array network is then done. Starting with the largest capacitor, a capacitor is switched to the adjacent resistor-string node having a larger voltage. If the comparator output is a 1, it is switched back and b_i is a 0. Otherwise, the switch is left as is and b_i is a 1. Since the resistor string is monotonic, this type of converter is guaranteed monotonic if the capacitor array is monotonic.

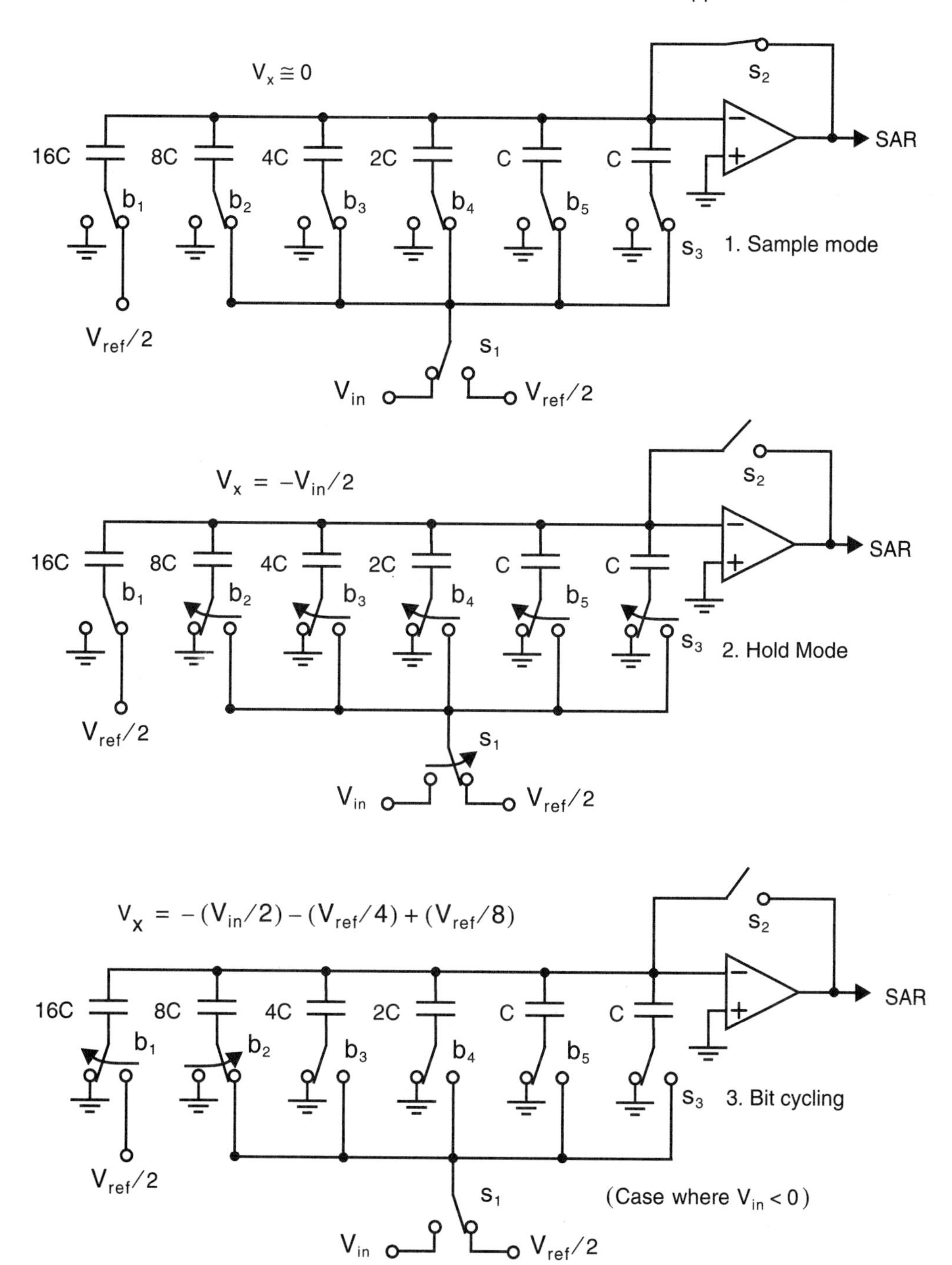

Fig. 13.8 A 5-bit signed charge-redistribution A/D converter.

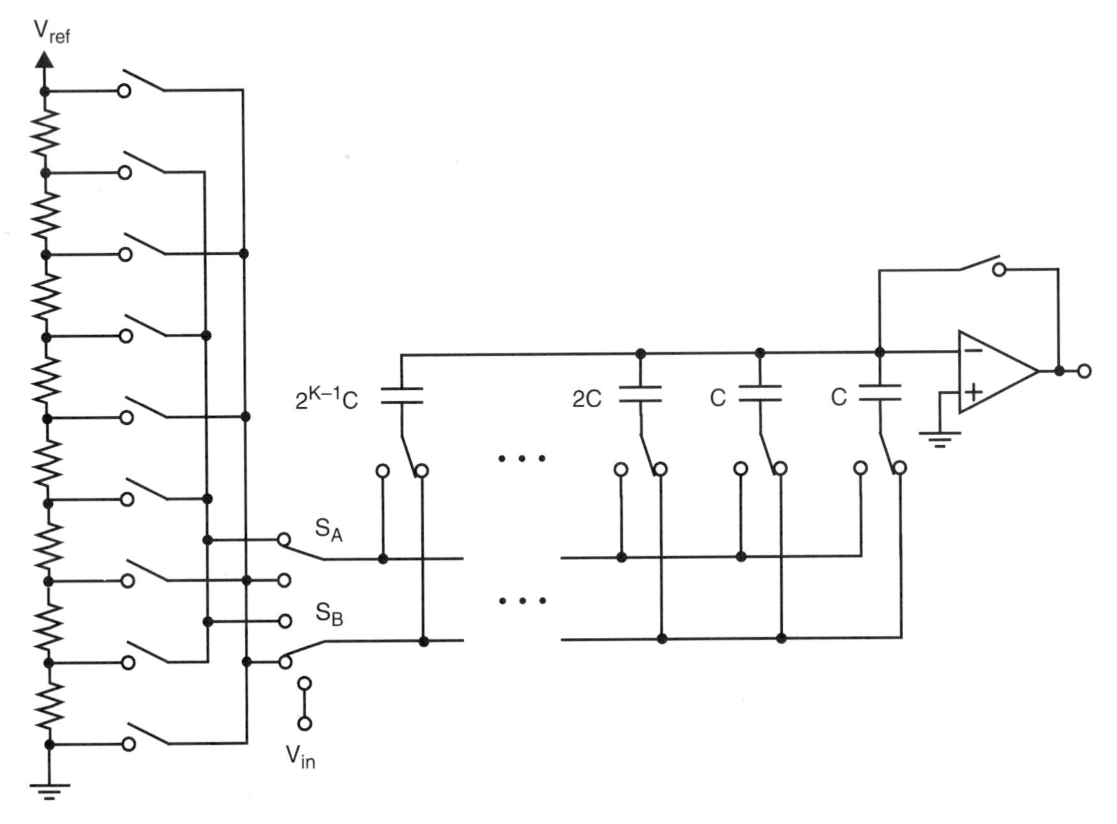

Fig. 13.9 Resistor-capacitor hybrid A/D converter.

Charge-Redistribution with Error Correction

With the best matching accuracy of on-chip elements being about 0.1 percent, one is limited to successive-approximation converters having 10-bit accuracy specifications without some sort of calibration. One error-correction technique that has been used to obtain 16-bit linear converters is shown in Fig. 13.10 [Lee, 1984]. In this approach, the MSB array is realized using binary-weighted capacitors that determine, for example, the first 10 bits. For a 16-bit converter, the final 6 bits are determined using an additional capacitor and a resistor string referred to as a *sub-dac*. Although this combination of an MSB capacitor array and an LSB resistor string is not inherently monotonic, it can be easily autocalibrated at start-up by adding a second resistor string referred to as a *cal-dac*.

Calibration is done by measuring the errors of each capacitor, starting with the largest capacitor, calculating the correction terms required, and then storing the correction terms in a data register as DV_{ei}. During a regular successive approximation operation, whenever a particular capacitor is used, its error is cancelled by adding the value stored in the data register to that stored in an accumulator register, which con-

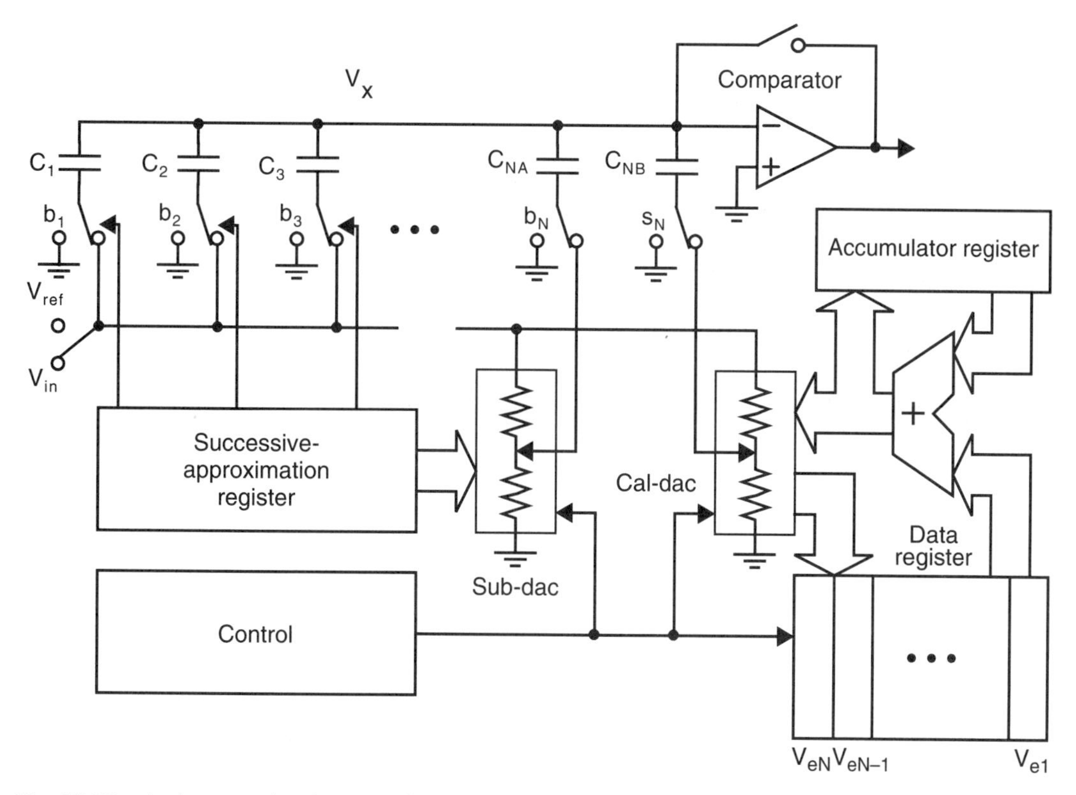

Fig. 13.10 A charge-redistribution A/D converter with error correction.

tains the sum of the correction terms for all of the other capacitors currently connected to V_{ref}. No correction terms are measured for the resistor sub-dac; its accuracy is not critical since it only determines the remaining LSBs.

The error terms are found starting with the MSB capacitor, C_1. It is connected to ground and all of the other capacitors are connected to V_{ref} while the comparator is reset. Next, the comparator is taken out of reset mode, all of the other capacitors are switched to ground, and C_1 is switched to V_{ref}. Defining C_{total} to be the sum of all the capacitors in the array, the ideal value of C_1 is $C_{total}/2$. However, in practice C_1 varies from its ideal value, and thus, C_1 can be written as

$$C_1 \equiv (C_{total}/2) + \Delta C_1 \tag{13.24}$$

where ΔC_1 is the capacitance error (either positive or negative). As a result, the simplified model shown in Fig. 13.11(*a*) occurs after the switches are reversed. Here, the remaining capacitance equals $(C_{total}/2) - \Delta C_1$ to make the total capacitance correct. With the above switch operation, V_x would remain zero if ΔC_1 equals zero. However, when ΔC_1 is not zero, the resulting voltage, V_x, is twice the error voltage, defined as V_{e1}, that would be introduced during a normal successive approximation when C_1 alone is switched to V_{ref}. A digital representation of V_{e1}, defined as DV_{e1}, is obtained by doing a successive approximation using the cal-dac shown in

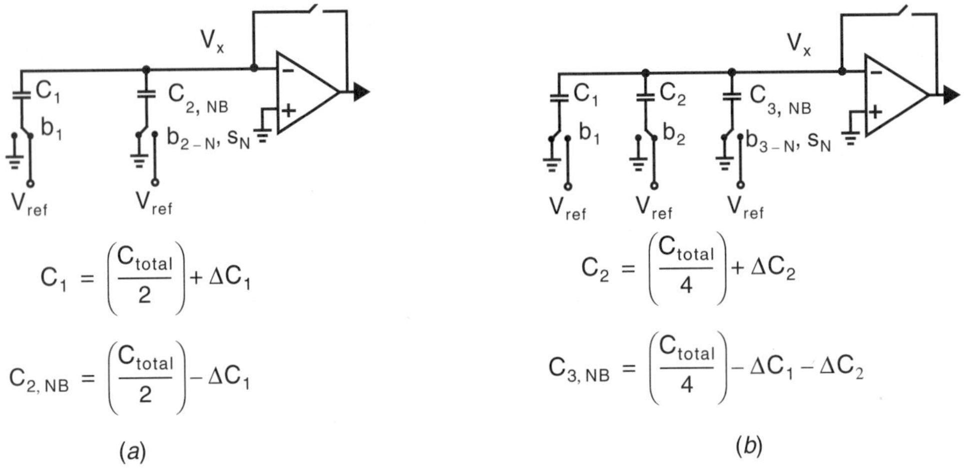

V_x
C_1
b_1
C_2, NB
b_2 – N, S_N
V_ref
V_ref
C_1 = (C_total / 2) + ΔC_1
C_2, NB = (C_total / 2) – ΔC_1
(a)
V_x
C_1
b_1
C_2
b_2
C_3, NB
b_3 – N, S_N
V_ref
V_ref
V_ref
C_2 = (C_total / 4) + ΔC_2
C_3, NB = (C_total / 4) – ΔC_1 – ΔC_2
(b)

Fig. 13.11 Equivalent models for determining capacitance errors. (*a*) During calibration for C_1. (*b*) During calibration for C_2.

Fig. 13.10 and then dividing the resulting digital value by 2 to obtain DV_{e1}. This digital correction term, DV_{e1}, is stored in the data register for use during regular conversion.

To obtain the similar correction term for C_2, defined as DV_{e2}, the model in Fig. 13.11(*b*) is used. Here, the switch procedure is to always leave the b_1 switch grounded while the comparator is reset, the b_2 switch is grounded, and the remaining switches b_{3-N}, s_N are connected to V_{ref}. Next, the b_2 and b_{3-N}, s_N switches are all reversed (i.e., b_2 set to V_{ref} and the others grounded). Note, however, that even if ΔC_2 equals zero, the measured value of V_x will be equivalent to $-DV_{e1}$, since $-\Delta C_1$ is assumed to be part of $C_{3,NB}$. Therefore, the error voltage that would be due only to C_2 can be calculated by digitizing V_x, as before, but then subtracting off DV_{e1} using digital circuitry. Mathematically,

$$DV_{e2} = \frac{1}{2}(DV_{x2} - DV_{e1}) \tag{13.25}$$

A similar procedure is used for the other capacitors. Specifically, DV_{xi} is found by connecting all capacitors smaller than C_i to V_{ref}, and connecting C_i and all larger capacitors to ground while the comparator is reset. Next, the comparator is taken out of reset mode, all capacitors smaller than C_i are switched to ground, C_i is switched to V_{ref}, and all capacitors larger than C_i are left connected to ground. DV_{xi} is then found using the cal-dac and successive approximation. DV_{ei} is finally calculated using the formula

$$DV_{ei} = \frac{1}{2}\left(DV_{xi} - \sum_{j=1}^{i-1} DV_{ej}\right) \tag{13.26}$$

and stored in the ith word of the data register.

During a regular conversion, normal successive approximation is performed with the MSB capacitor array; however, appropriate correction voltages are either added or subtracted using the cal-dac and C_{NB} capacitor. The proper correction voltage is determined through the use of a digital accumulator that stores the sum of all digital errors of those MSB capacitors deemed to be a 1 (i.e., connected to V_{ref}). In other words, when the ith bit is being tested, DV_{ei} is added to the digital accumulator that is driving the cal-dac. If the ith bit is determined to be a 0, then the digital accumulator returns to its previous value; otherwise, it maintains its new accumulated value. Finally, normal successive approximation is performed using the sub-dac to determine the final LSBs. With this approach, a digital addition must be performed during each bit cycle, and a small amount of digital RAM is required (10 bytes in the case of a 10-bit MSB array).

Finally, it should be mentioned that similar error correction techniques have also been described to account for capacitor inaccuracies (for example, [Tan, 1990]).

Speed Estimate for Charge-Redistribution Converters

The major limitation on speed with charge redistribution converters is often due to the RC time constants of the capacitor array and switches. To estimate this time, consider the simplified model of a capacitor array being reset, as shown in Fig. 13.12. Here, R, R_{s1}, and R_{s2} represent the switch-on resistances of the bit line, $s1$ and $s2$ switches, respectively. Although this circuit is easily simulated using Spice to find its settling time, it is useful to have a rough estimate of the charging time to speed up the design process. As in Section 12.1, the open-circuit time constant approach [Sedra, 1991] can be used to obtain an estimate of the high-frequency time constant by summing individual time constants due to each capacitor. For example, the individual time constant due to the capacitance $2C$ equals $(R_{s1} + R + R_{s2})2C$. Following such

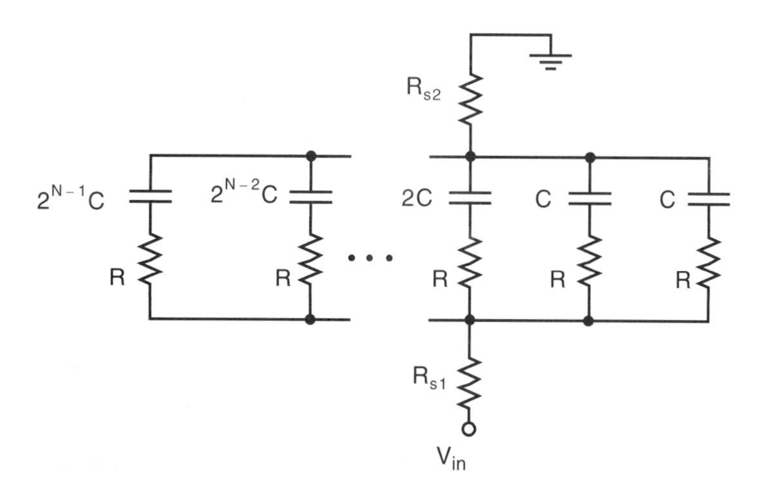

Fig. 13.12 Simplified model of a capacitor array during the sampling time.

an approach, the open-circuit time constant for the circuit shown in Fig. 13.12 is equal to

$$\tau_{eq} = (R_{s1} + R + R_{s2})2^N C \tag{13.27}$$

For better than 0.5-LSB accuracy, we need

$$e^{-T/\tau_{eq}} < \frac{1}{2^{N+1}} \tag{13.28}$$

where T is the charging time. This equation can be simplified to

$$T > \tau_{eq}(N+1)\ \ln(2) = 0.69(N+1)\tau_{eq} \tag{13.29}$$

It has been observed that (13.29) gives results about 30 percent higher than those obtained by simulating the actual RC network shown in Fig. 13.12. So while (13.29) can be used to roughly determine the maximum sampling rate, the final design should be simulated using accurate transistor models. Finally, although this result is not the same for all charge-redistribution A/D converters (for example, the size of the switches going to the larger capacitors may be increased to reduce their on resistance), the basic approach can be modified for most charge-redistribution A/D converters.

13.3 ALGORITHMIC (OR CYCLIC) A/D CONVERTER

An algorithmic converter operates in much the same way as a successive-approximation converter. However, whereas a successive-approximation converter halves the reference voltage in each cycle, an algorithmic converter doubles the error voltage while leaving the reference voltage unchanged. The flow graph for a signed algorithmic conversion is shown in Fig. 13.13.

Ratio-Independent Algorithmic Converter

The block diagram for an algorithmic converter is shown in Fig. 13.14 [McCharles, 1977; Li, 1984]. This converter requires a small amount of analog circuitry because it repeatedly uses the same circuitry to perform its conversion cyclically in time.

One of the difficulties in realizing a high-precision algorithmic converter is building an accurate multiply-by-two gain amp. Fortunately, it is possible to realize the gain amp so that it does not rely on any capacitor matching if four clock cycles are taken for the multiply-by-two operation. The operation of the multiply-by-two gain amp is shown in Fig. 13.15. Although this gain-amp circuitry is shown using single-ended circuits for simplicity, fully differential circuits are normally used.

The basic idea of this gain amp is to sample the input signal twice using the same capacitor. During the second sampling, the charge from the first capacitor is stored on a second capacitor whose size is unimportant. After the second sampling, both charges are recombined into the first capacitor which is then connected between the opamp input and output.

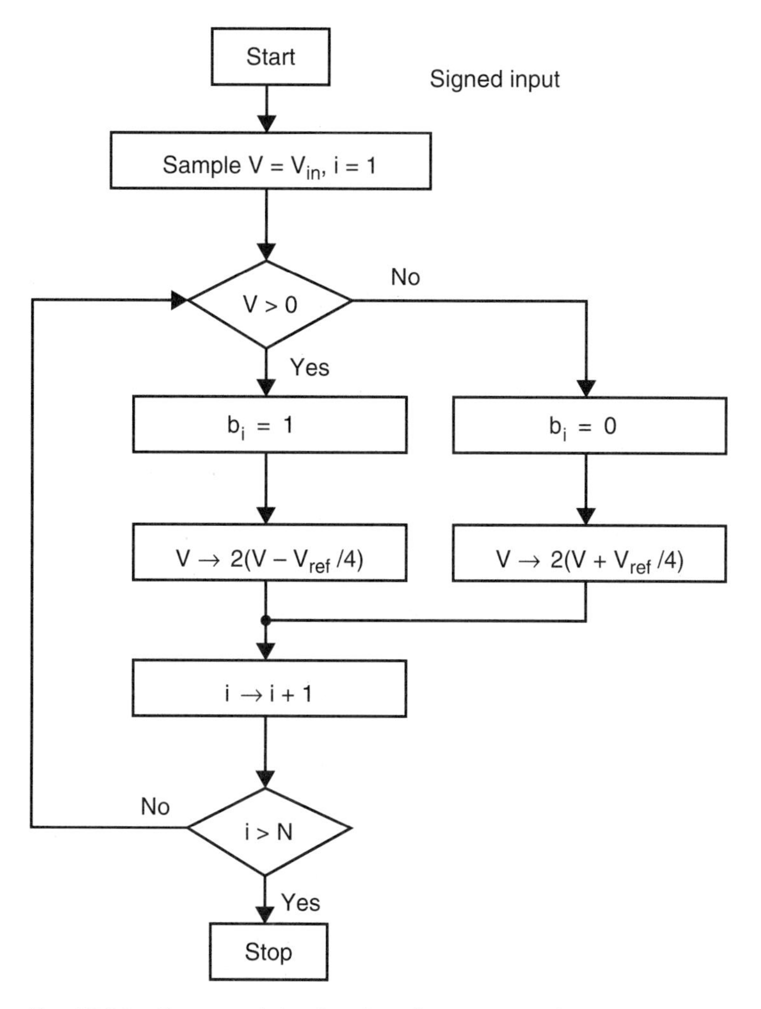

Fig. 13.13 Flow graph for the algorithmic approach.

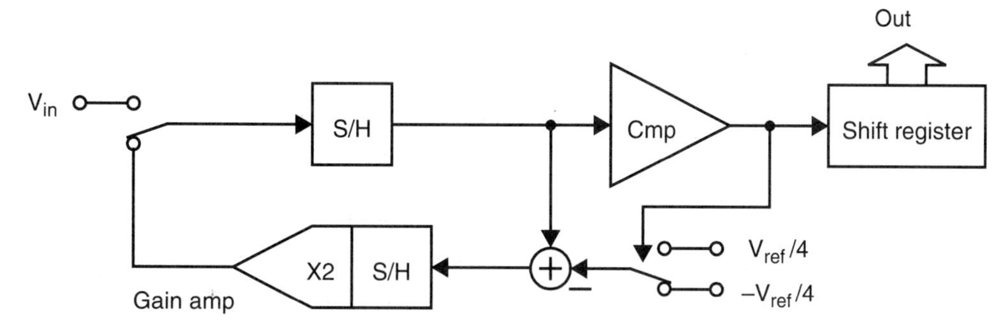

Fig. 13.14 Algorithmic converter.

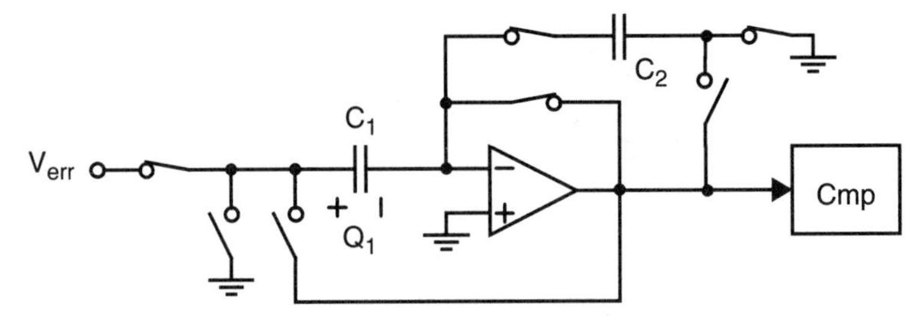

1. Sample remainder and cancel input-offset voltage.

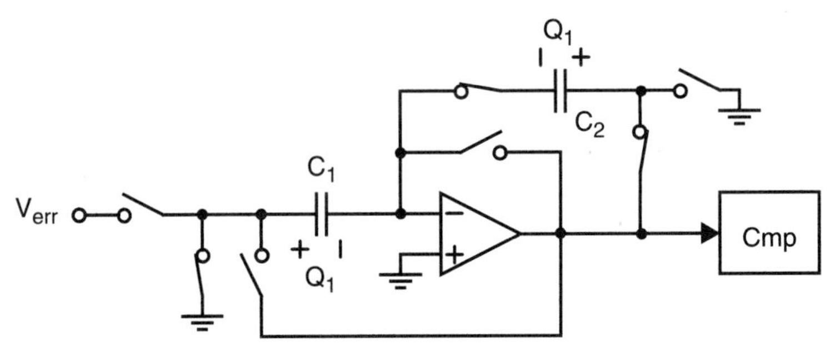

2. Transfer charge Q_1 from C_1 to C_2.

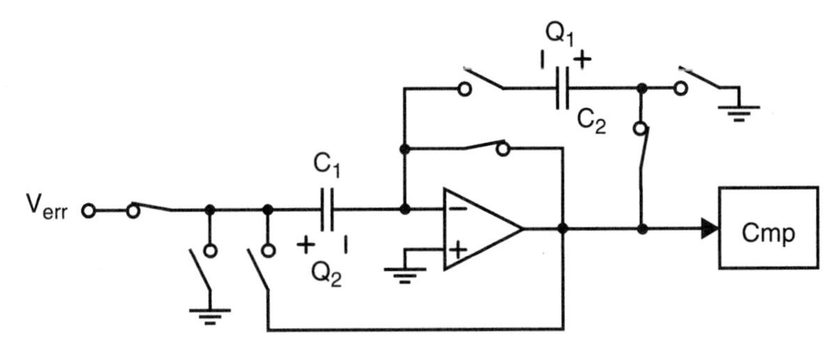

3. Sample input signal with C_1 again, after storing charge Q_1 on C_2.

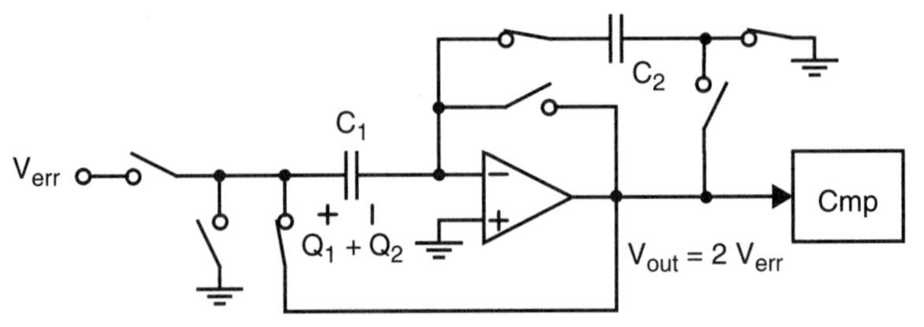

4. Combine Q_1 and Q_2 on C_1, and connect C_1 to output.

Fig. 13.15 Multiply-by-two gain circuitry for an algorithmic converter that does not depend on capacitor matching.

EXAMPLE 13.4

Consider the multiply-by-two gain circuitry shown in Fig. 13.15. Assuming the opamp has an input offset designated as V_{off}, find the values of V_{C1}, V_{C2}, and V_{out} at the end of each of the phases shown.

Solution

During phase one, the opamp output, V_{out}, is connected to the negative opamp input, resulting in

$$V_{out} = V_{off} \tag{13.30}$$

and the voltages across the two capacitors are

$$V_{C1} = V_{err} - V_{off} \tag{13.31}$$

$$V_{C2} = 0 - V_{off} = -V_{off} \tag{13.32}$$

At the end of phase two, we have

$$V_{C1} = 0 - V_{off} = -V_{off} \tag{13.33}$$

implying that its charge change was

$$\Delta Q_{C1} = C_1(V_{err} - V_{off} - (-V_{off})) = C_1 V_{err} \tag{13.34}$$

All this charge is placed on C_2, resulting in

$$V_{C2} = -V_{off} + \left(\frac{C_1}{C_2}\right)V_{err} \tag{13.35}$$

and $V_{out} = (C_1/C_2)V_{err}$.

At the end of phase three, V_{C2} remains unchanged since one side of it has been opened. Also, $V_{C1} = V_{err} - V_{off}$ and $V_{out} = V_{off}$ as in phase one.

Finally, at the end of phase four, C_2 is discharged to the same value it had in phase 1, resulting in $V_{C2} = -V_{off}$ and its change in charge being

$$\Delta Q_{C2} = C_2\left(\frac{C_1}{C_2}\right)V_{err} = C_1 V_{err} \tag{13.36}$$

All of this charge is placed back on C_1, resulting in

$$V_{C1} = 2\,V_{err} - V_{off} \tag{13.37}$$

and the output voltage being the desired result of $V_{out} = 2\,V_{err}$. Note that this final result is independent of the sizes of C_1, C_2, and the offset value, V_{off}.

13.4 FLASH (OR PARALLEL) CONVERTERS

Flash converters are the standard approach for realizing very-high-speed converters, as seen in some recent publications [Peetz, 1986; Yoshii, 1987; Hotta, 1987; and

Gendai, 1991]. The input signal in a flash converter is fed to 2^N comparators in parallel, as shown in Fig. 13.16. Each comparator is also connected to a different node of a resistor string. Any comparator connected to a resistor string node where V_{ri} is larger than V_{in} will have a 1 output while those connected to nodes with V_{ri} less than V_{in} will have 0 outputs. Such an output code word is commonly referred to as a thermometer code since it looks quite similar to the mercury bar in a thermometer. Note that the top and bottom resistors in the resistor string have been chosen to create the 0.5 LSB offset in an A/D converter.

The NAND gate that has a 0 input connected to its inverting input and a 1 input connected to its noninverting input detects the transition of the comparator outputs from 1s to 0s, and will have a 0 output. All other NAND-gate outputs will be 1, resulting in simpler encoding. It also allows for error detection by checking for more than one 0 output, which occurs during a bubble error (see the next subsection) and, perhaps, error correction.

Flash A/Ds are fast but they require a large number of comparators, which typically take up a large area and are very power hungry—especially when they are

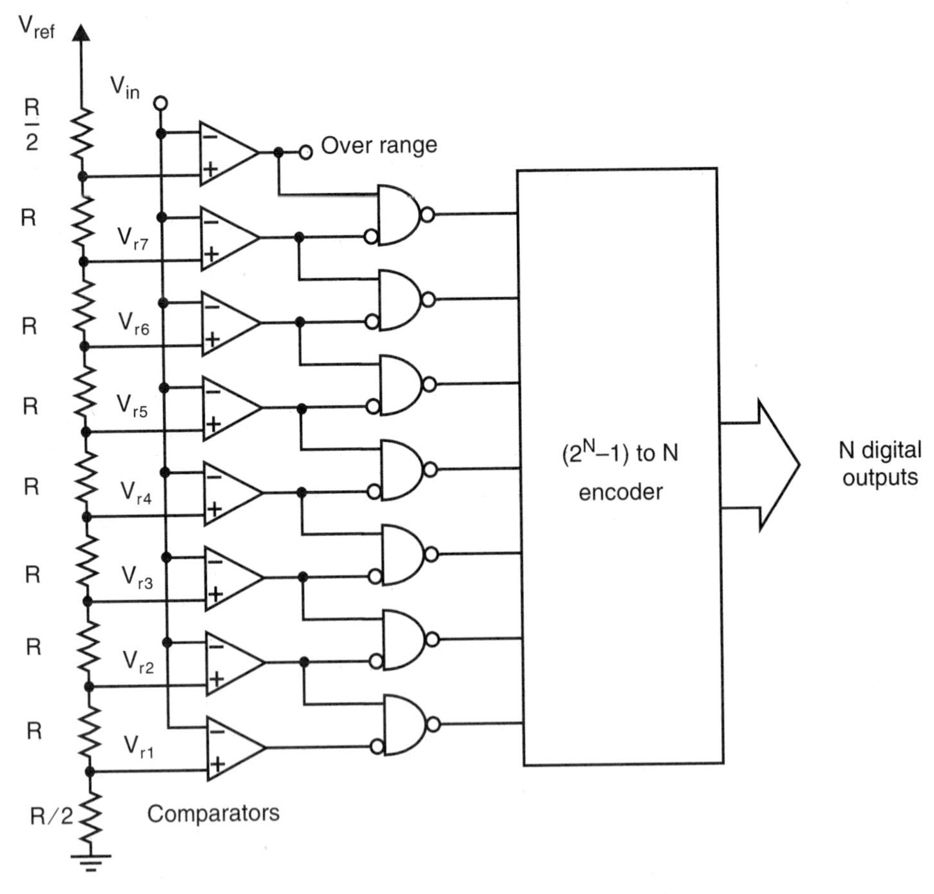

Fig. 13.16 A 3-bit flash A/D converter.

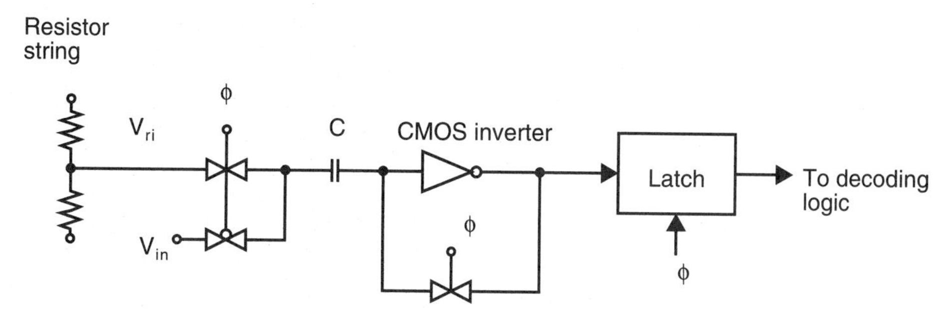

Fig. 13.17 A clocked CMOS comparator.

clocked fast. One way to realize a small clocked CMOS comparator is by using a CMOS inverter, as shown in Fig. 13.17 [Dingwall, 1979].

When ϕ is high, the inverter is set to its bistable operating point, where its input voltage equals its output voltage (i.e., its threshold voltage). Normally with an odd number of inverters, a ring oscillator is formed; however, in the case of a single CMOS inverter, the inverter operates as a single stage opamp with only one pole (no nondominant poles), so stability is guaranteed. With this inverter set to its threshold voltage, the other side of C is charged to V_{ri}. When ϕ goes low, the inverter is free to fall either high or low depending on its input voltage. At the same time, the other side of C is pulled to the input voltage, V_{in}. Since the inverter side of the capacitor is floating, C must keep its original charge, and therefore the inverter's input will change by the voltage difference between V_{ri} and V_{in}. Since the inverter's input was at its bistable point, the difference between V_{ri} and V_{in} will determine which direction the inverter's output will fall. However, it should be mentioned that this simple comparator suffers from poor power supply rejection, which is often a critical design specification in fast converters. Using fully differential inverters helps alleviate this shortcoming.

Issues in Designing Flash A/D Converters

We discuss here some important design issues that should be addressed when building high-speed flash A/D converters.

Input Capacitive Loading The large number of comparators connected to V_{in} results in a large parasitic load at the node V_{in}. Such a large capacitive load often limits the speed of the flash converter and usually requires a strong and power-hungry buffer to drive V_{in}. We shall see that this large capacitive loading can be reduced by going to an interpolating architecture.

Resistor-String Bowing The input currents of bipolar comparators cause errors in the voltages of the nodes of the resistor string. These errors usually necessitate the bias current in the resistor string being two orders of magnitude greater than the input currents of the comparators. However, these errors are greatest at the center node of the resistor string and thus considerable improvement can be obtained by using additional circuitry to force the center tap voltage to be correct.

Comparator Latch-to-Track Delay Another consideration that is often overlooked is the time it takes a comparator latch to come from latch mode to track mode when a small input signal of the opposite polarity from the previous period is present. This time can be minimized by keeping the time constants of the internal nodes of the latch as small as possible. This is sometimes achieved by keeping the gain of the latches small, perhaps only two to four. In some cases, the differential internal nodes might be shorted together temporarily just after latch time.

Signal and/or Clock Delay Even very small differences in the arrival of clock or input signals at the different comparators can cause errors. To see this, consider a 250-MHz, 1-V peak-input sinusoid. This signal has a maximum slope of 1570 V/μs at the zero crossing. If this signal is being encoded by an 8-bit A/D converter with $V_{ref} = 2$ V (i.e., the sinusoid covers the whole range), then it would only take 5 ps to change through 1 LSB. This time is roughly about the same time it takes a signal to propagate 500 μm in metal interconnect. If there is clock skew between comparators greater than this, the converter will have more than 1 LSB error. One means of easing this problem is to precede the converter by a sample-and-hold circuit. However, high-speed sample-and-hold circuits can be more difficult to realize than the flash converter itself. In addition, the clock and V_{in} should be routed together with the delays matched [Gendai, 1991]. It should also be noted that the delay differences may not be caused just by routing differences, but could also be caused by different capacitive loads, or by phase differences between the comparator preamplifiers at high frequencies.

Substrate and Power-Supply Noise For $V_{ref} = 2$ V and an 8-bit converter, only 7.8 mV of noise injection would cause a 1 LSB error. Typically, on an integrated circuit having a clock signal in the tens of MHz, it is difficult to keep power-supply noise below a few tenths of a volt. This power-supply noise can easily couple through the circuitry or substrate, resulting in errors. To minimize this problem, the clocks must be shielded from the substrate and from analog circuitry. Also, running differential clocks closely together will help prevent the signals being coupled into the substrate or through the air. Also, analog power supplies should be separated from digital power supplies including having analog power to the comparator preamps while using digital power to the latch stages. On-chip power-supply bypassing is a necessity. It is also necessary to make sure the power-supply bypassing circuitry doesn't form a resonant circuit with the bonding wires—include small resistors in series with the bypass capacitors.

Bubble Error Removal The outputs of the comparators should be a thermometer code with a single transition. However, sometimes a lone 1 will occur within the string of 0s (or a 0 within the string of 1s) due to comparator metastability, noise, cross talk, limited bandwidth, etc. These bubbles usually occur near the transition point of the thermometer code. Fortunately, these bubbles can usually be removed with little extra complexity by replacing the two-input NAND gates shown in Fig. 13.16 with three-input NAND gates, as shown as shown in Fig. 13.18 [Steyaert, 1993]. With this modification, there must now be two 1s immediately above a 0 in determining the transition point in the thermometer code. However, this circuit will not eliminate the problem of a stray 0 being two places away from the transition point, which may cause a large decoding error. Another digital approach for reducing the effect of bubble errors is to allow bubble errors in the lower 2 LSBs but have the remaining MSBs determined by looking for transitions between every *fourth* comparator [Gendai, 1991]. With this approach, bubble errors that occur within four places of the transition point do not cause any large errors. An alternate approach to reduce the effect of distant bubble errors is to create two encoders (one AND type and one OR type) rather than a single encoder [Ito, 1994]. When an unexpected output pattern occurs at the NAND outputs, the errors in two different encoders tend to be equal in magnitude but opposite in sign. Thus, the final output is taken as the average of the two encoder outputs, which is performed by adding the two outputs and dropping the LSB (to divide by two).

An alternate method to remove bubble errors that does not increase the power dissipation is shown in Fig. 13.19. Here, extra transistors have been added to the inputs of the slave latches, which are driven by the comparator master latches [van Valburg, 1992]. These extra transistors make the value stored in a slave latch not just a function of its master latch, but also a function of the two adjacent master latches. If a bubble occurs, the outputs from the two adjacent master latches are the same, but different from the center master latch. In this case, the values from the adjacent master latches overrule the center master latch. Note that with this approach, the power dissipation is not increased because the added transistors make use of existing current in the slave latch. Alternatively, this same voting scheme can be implemented entirely in digital form.

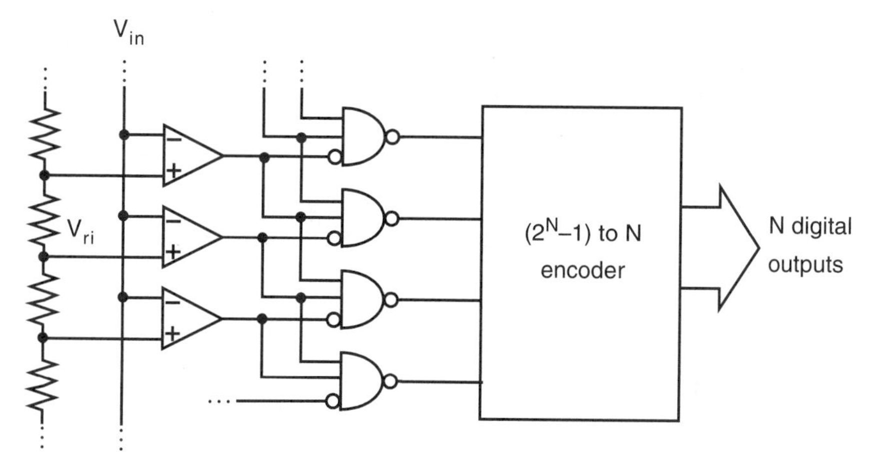

Fig. 13.18 Using three-input NAND gates to remove single bubble errors.

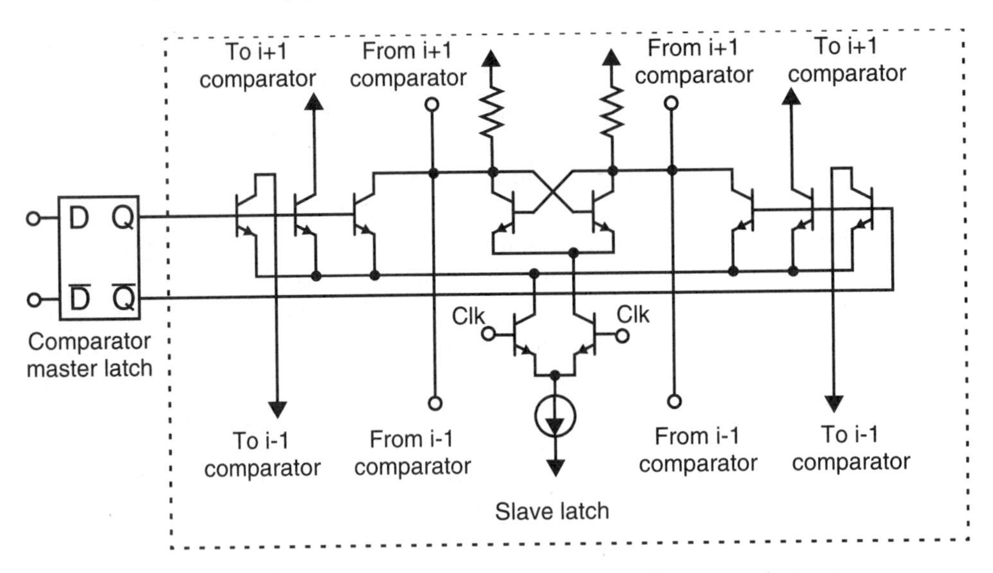

Fig. 13.19 Bubble-error voting circuit that does not increase the power dissipation.

Flashback An additional source of error is flashback. Flashback is caused by latched comparators, which are almost always used. When clocked comparators are switched from track to latch mode, or vice-versa, there is major charge glitch at the inputs to the latch. If there is no preamplifier, this will cause major errors due to the unmatched impedances at the comparator inputs (one input goes to the resistor

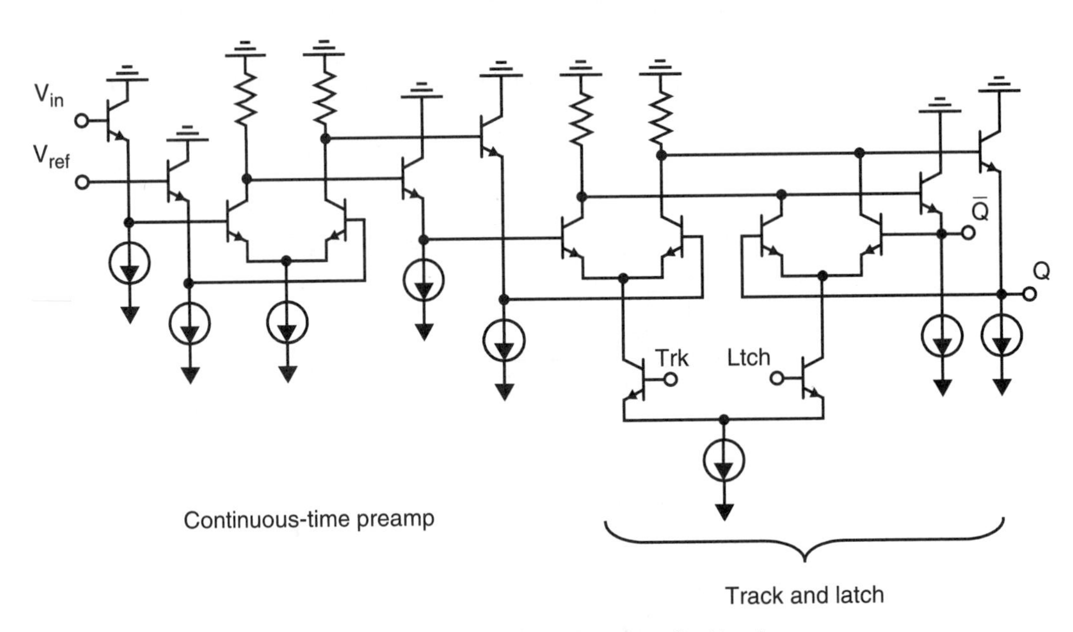

Fig. 13.20 Clocked comparator with a preamplifier to reduce flashback.

string—the other to the input signal). To minimize this effect, most modern comparators have one or two stages of continuous-time buffering and/or preamplification. For example, a commonly used comparator is shown in Fig. 13.20. Notice that it has a buffer, a low-gain preamp, and another buffer before the track-and-latch circuitry. Also notice that in the positive feedback latch the feedback is taken from the emitter-follower outputs, which minimizes the capacitances of the internal nodes of the latch.

Another technique sometimes used to minimize the effects of flashback is to match the input impedances as much as is possible. For example, it is possible to implement a second matched resistor string, with the nodes of the comparators that were originally connected to V_{in} now being connected to it, and the end nodes of the string connected together to V_{in}. This approach matches impedances and also minimizes the resistor-string bowing due to the comparator input currents. Unfortunately, it does result in different delays for V_{in} reaching the various comparators, and unless these are matched to the routing of the clock signals, these different delays may not be tolerable.

13.5 TWO-STEP A/D CONVERTERS

Two-step (or subranging) converters are currently the most popular approach for high-speed medium-accuracy A/D converters. This popularity is due to several advantages they have over their flash counterparts. Specifically, two-step converters require less silicon area, dissipate less power, have less capacitive loading, and the voltages the comparators need to resolve are less stringent than for flash equivalents. However, two-step converters do have a larger latency delay, although their throughput approaches that of flash converters.

The block diagram for a two-step converter is shown in Fig. 13.21. The operation of this two-step converter is as follows. The 4-bit MSB A/D determines the first four MSBs. To determine the remaining LSBs, the quantization error is found by reconverting the 4-bit digital signal to an analog value using the 4-bit D/A and subtracting that

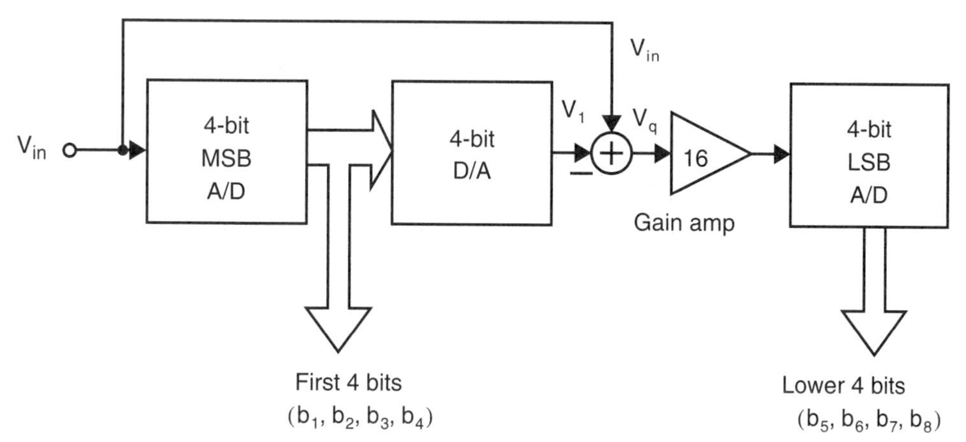

Fig. 13.21 An 8-bit two-step A/D converter.

value from the input signal. To ease the requirements in the circuitry for finding the remaining LSBs, the quantization error is first multiplied by 16 using the gain amplifier, and the LSBs are determined using the 4-bit LSB A/D. With this approach, rather than requiring 256 comparators as in an 8-bit flash converter, only 32 comparators are required for a two-step A/D converter. However, this straightforward approach would require all components to be at least 8-bit accurate. To significantly ease the accuracy requirements of the 4-bit MSB A/D converter, digital error correction is commonly used and is discussed next.

Digital Error Correction

The block diagram for a two-step converter with digital error correction is shown in Fig. 13.22. Although the second sample and hold (S/H_2) is not necessary, its purpose is to allow the first S/H_1 to sample a new input signal before the gain amplifier has finished settling. However, the first S/H_1 is critical and its performance often limits the overall linearity. The reason for digital error correction is to significantly ease the requirements placed on the 4-bit MSB A/D converter. Without error correction, this first A/D converter needs to be at least 8-bit accurate. However, with error correction, the requirements on this MSB A/D converter are that it need only be 4-bit accurate. To see how this correction works and why a second-stage 5-bit converter is needed (rather than 4-bit), consider the quantization error that occurs in an ideal converter. Defining $V_{LSB} = V_{ref}/2^8$ (i.e., always relative to 8-bit accuracy), we have for an

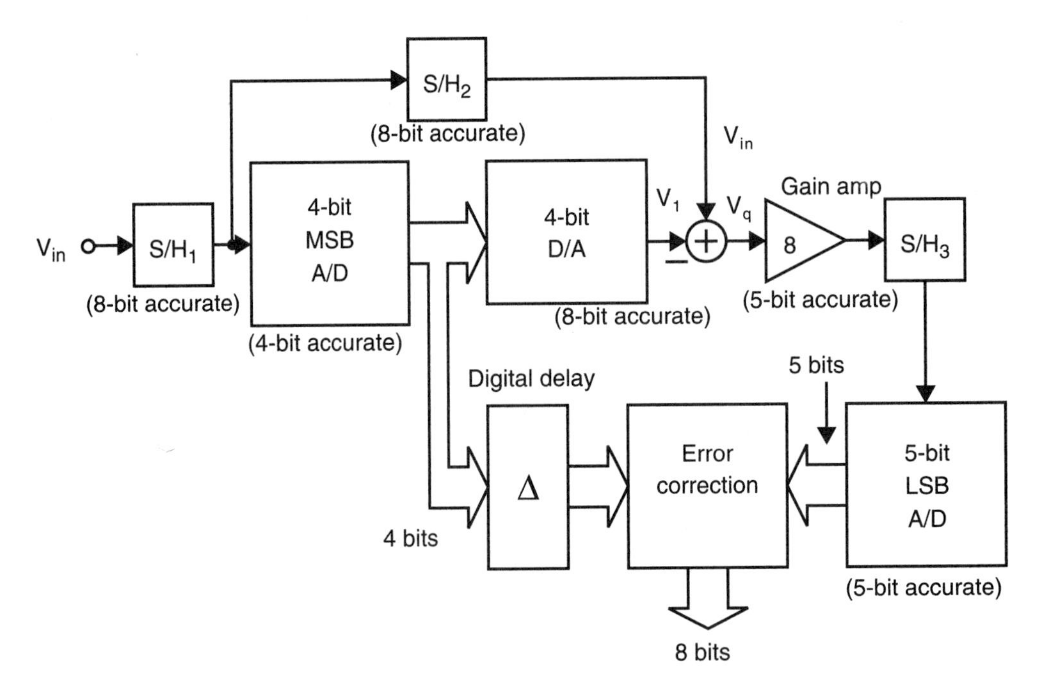

Fig. 13.22 An 8-bit two-step A/D converter with digital error correction.

ideal 8-bit converter,

$$V_{ref}B_{out} = V_{in} + V_q \quad \text{where} \quad -\frac{1}{2}V_{LSB} < V_q < \frac{1}{2}V_{LSB} \tag{13.38}$$

However, for a nonideal 8-bit converter with an absolute accuracy of 0.5 LSB, we have

$$V_{ref}B_{out} = V_{in} + V_q \quad \text{where} \quad -V_{LSB} < V_q < V_{LSB} \tag{13.39}$$

In other words, the maximum quantization signal is now twice that of the ideal case.

Similarly, for an ideal 4-bit A/D converter, we have (keeping the 8-bit definition of V_{LSB}),

$$V_{ref}B_{out} = V_{in} + V_q \quad \text{where} \quad -8V_{LSB} < V_q < 8V_{LSB} \tag{13.40}$$

Thus in the ideal case, the value of V_q can be determined (to 8-bit accuracy) using a 4-bit A/D converter since V_q must be within $16V_{LSB}$. However, for the nonideal case where the 4-bit MSB flash converter has an absolute accuracy of $8V_{LSB}$, the quantization error, V_q, is now bounded within $32V_{LSB}$. Thus, in the case of a non-ideal 4-bit MSB converter, a 5-bit LSB converter must be used; otherwise V_q may go out of range. Note that the gain amplifier of 8 is used to amplify the quantization error back to maximum signal levels to ease the requirements of the 5-bit LSB converter. Finally, to determine V_{in}, we see that the digital value of V_q has been found to within $0.5V_{LSB}$ and the digital value of V_1 is known to the same accuracy since we assumed the D/A converter to be 8-bit accurate (and the digital word applied to the D/A converter is known). Therefore, we find V_{in} from the relation

$$V_{in} - V_1 = V_q \tag{13.41}$$

Specifically, V_{in} is found by properly combining the digital equivalents of V_1 and V_q.

In summary, the MSB A/D need only be accurate to $1/2^4 = 1/16$. The only components that need 0.5 LSB accuracy at the 8-bit level are S/Hs 1 and 2, the D/A, and the subtraction circuit. Besides the difficulty in realizing the S/H circuits, another major limitation is the difficulty of designing a high-speed, accurate (to 0.5 LSB at the 5-bit level) gain amplifier. In fact, due to difficulties in realizing high-speed circuits with gain, often fewer bits are determined in the first stage of a two-step converter to reduce the amplification required. For a more detailed treatment of a two-step A/D converter, the reader is referred to a 10-bit 75 MHz implementation (with integrated S/H), described in [Petschacher, 1990].

EXAMPLE 13.5

For the two-step 8-bit A/D converter shown in Fig. 13.22, what is the maximum voltage range at V_q when the converter's full-scale input is ± 2.5 V in the case where the 4-bit MSB A/D converter is (a) 8-bit accurate and (b) 4-bit accurate? Assume all other components are ideal.

Solution

With a full-scale peak-to-peak input voltage of 5 V applied to an 8-bit A/D converter, we have

$$V_{LSB} = \frac{5}{2^8} = 19.5\text{ mV} \tag{13.42}$$

For an ideal 4-bit A/D converter, we have from (13.40) the maximum voltage range of V_q is $16V_{LSB}$, or equivalently, 312 mV.

(a) If we go to the trouble to make the 4-bit A/D converter have an absolute accuracy of $0.5V_{LSB}$ (i.e., 8-bit accurate), then V_q becomes bounded between $\pm 8.5V_{LSB}$. In other words, the maximum range of V_q would now be $17V_{LSB}$, or equivalently, 332 mV. Note that the input range of the LSB converter is $8 \times 332\text{ mV} = 2.7\text{ V}$ (a little more than half of the input range of the overall converter), so more gain could be used in this case.

(b) In the case of a 4-bit accurate MSB converter, V_q is bounded between $\pm 16V_{LSB}$, implying that the maximum range of V_q is $32\ V_{LSB}$, or equivalently, 618 mV. After the gain of 8, the input range of the LSB converter becomes $8 \times 618\text{ mV} = 4.9\text{ V}$ (the same as the input range of the overall converter).

13.6 INTERPOLATING A/D CONVERTERS

Interpolating converters make use of input amplifiers, as shown in Fig. 13.23. These input amplifiers behave as linear amplifiers near their threshold voltages but are allowed to saturate once their differential inputs become moderately large. As a result, noncritical latches need only determine the sign of the amplifier outputs since the differences between the input signal and threshold voltages have been amplified. Also, the number of input amplifiers attached to V_{in} is significantly reduced by interpolating between adjacent outputs of these amplifiers. While this approach is often combined with a "folding" architecture [van de Grift, 1987; van Valburg, 1992], the interpolating architecture has also been used quite successfully by itself [Goodenough, 1989; Steyaert, 1993].

To further understand this interpolation approach, some possible signals for the input-amplifier outputs, V_1 and V_2, as well as their interpolated values are shown in Fig. 13.24. As can be seen in the figure, the logic levels are assumed to be zero and five volts, with the input comparators having a maximum gain of about –10. Also, here the latch threshold is near the midpoint of the two logic levels (or about 2.5 volts). As V_{in} increases, the latch for V_1 is first triggered, followed by V_{2a} and so on until V_2. As a result, more reference levels have been created between V_1 and V_2. It should be noted here that for good linearity, the interpolated signals need only cross the latch threshold at the correct points, while the rest of the interpolated signals responses are of secondary importance. One way to create such correct crossing

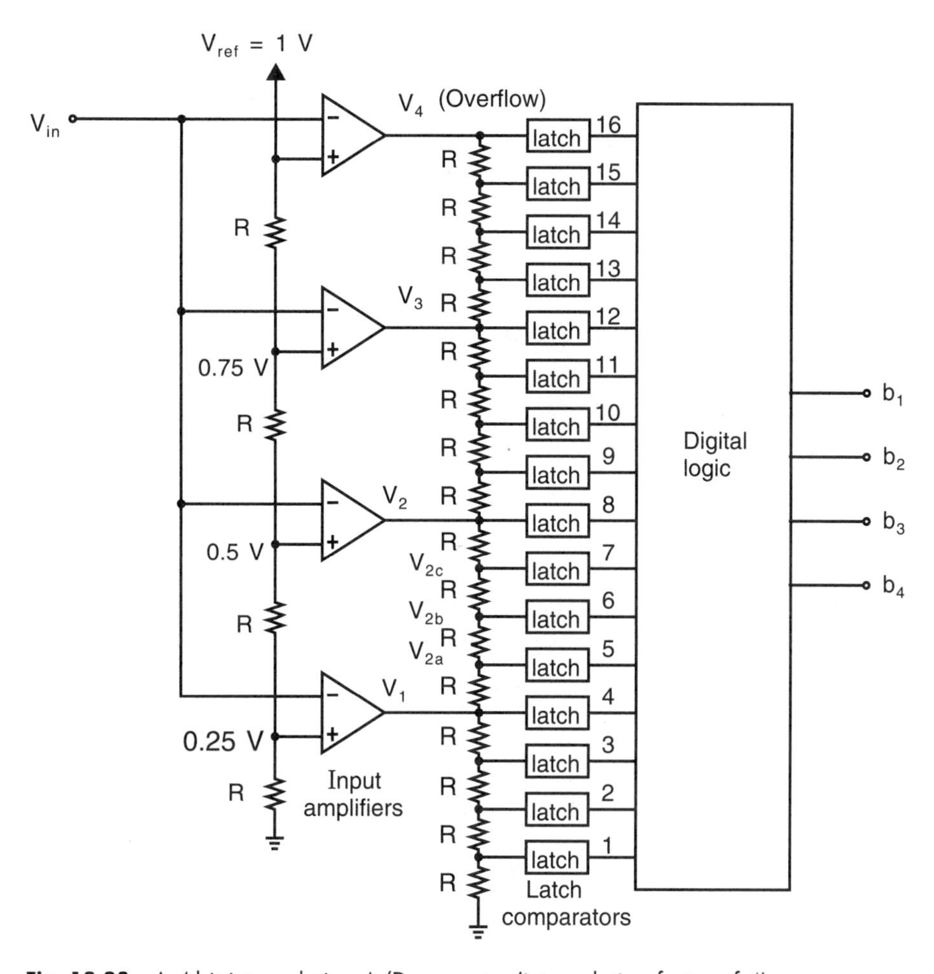

Fig. 13.23 A 4-bit interpolating A/D converter (interpolating factor of 4).

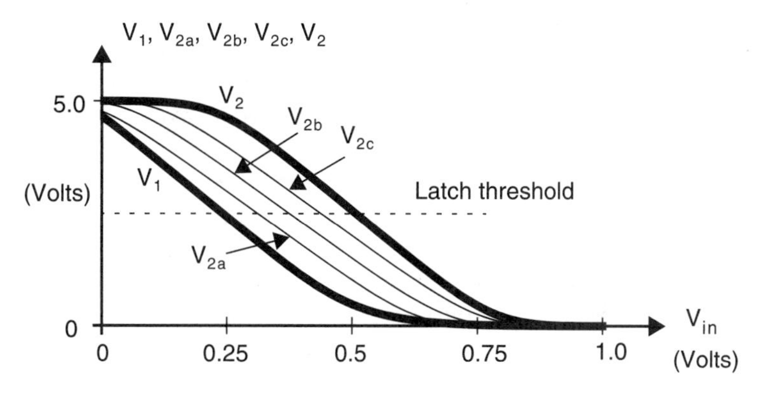

Fig. 13.24 Possible transfer responses for the input-comparator output signals, V_1 and V_2, and their interpolated signals.

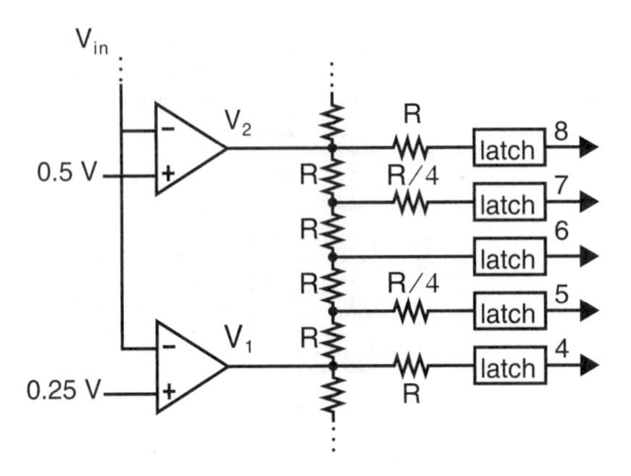

Fig. 13.25 Adding series resistors to equalize delay times to the latch comparators.

points is to ensure that V_1 and V_2 are linear between their own thresholds. In Fig. 13.24, this linear region corresponds to $0.25 < V_{in} < 0.5$.

For fast operation, it is important that the delays to each of the latches are made to equal each other as much as possible. Since the latch comparators have similar input capacitances associated with them, the delays can be made nearly equal by adding extra series resistors, as shown in Fig. 13.25. These series resistors equalize the impedances seen by each latch comparator looking back into the resistive string, assuming the input-amplifier outputs are low impedance [van de Plassche, 1988].

As mentioned earlier, the main benefit of an interpolating architecture is the reduction in the number of differential pairs attached to the input signal, V_{in}. Such a reduction results in a lower input capacitance (which is quite high for a flash converter), a slightly reduced power dissipation, and a lower number of accurate reference voltages that need to be created. Finally, it should be mentioned that circuit techniques other than resistive strings can be used to realize this interpolative approach. In [Steyaert, 1993], current mirrors were used to interpolate eight times between comparators, resulting in a 100-MHz 8-bit A/D converter realized with a 1.5-μm CMOS process. In another implementation, two stages of interpolation using capacitors to interpolate resulted in a 10-bit 20-MHz A/D low-power converter [Kusumoto, 1993].

EXAMPLE 13.6

Using current mirrors, show how one can interpolate two current outputs, I_1 and I_2, by three. What reduction in input capacitance of the converter would be expected over a traditional flash architecture?

Solution

If interpolating by three, it is desired to create two new currents, I_{2a}, I_{2b} such that

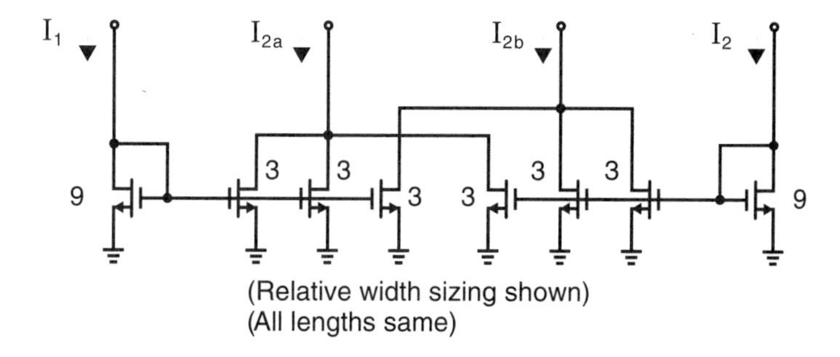

Fig. 13.26 Interpolating by three between two current outputs.

$$I_{2a} = \frac{2}{3}I_1 + \frac{1}{3}I_2 \tag{13.43}$$

$$I_{2b} = \frac{1}{3}I_1 + \frac{2}{3}I_2 \tag{13.44}$$

These output currents can be realized as shown in Fig. 13.26. These four currents can be converted back to voltages to send to the latches. Alternatively, the currents can be directly sent to latches which make use of current inputs.

Since we are interpolating by three here, this converter would require one-third the number of input amplifiers in relation to a traditional flash converter. Thus, the input capacitance for this interpolated approach would be one-third of that for a flash.

13.7 FOLDING A/D CONVERTERS

We just saw that the number of input amplifiers can be reduced through the use of an interpolating architecture. However, the number of latch comparators remains at 2^N for an N-bit converter. This large number of latch comparators can be significantly reduced through the use of a *folding architecture*. A folding A/D converter is similar in operation to a two-step (or subranging) converter in that a group of LSBs are found separately from a group of MSBs. However, whereas a two-step converter requires an accurate D/A converter, a folding converter determines the LSB set more directly through the use of analog preprocessing while the MSB set is determined at the same time.

As an example, consider the 4-bit folding converter shown in Fig. 13.27. Defining the *folding rate* to be the number of output transitions for a single folding block as V_{in} is swept over its input range, we see that the folding rate here is four. This folding rate determines how many bits are required in the MSB converter. The operation of this converter is as follows. The MSB converter determines whether the input signal, V_{in}, is in one of four voltage regions (i.e., between 0 and 1/4, 1/4 and 1/2, 1/2 and 3/4, or 3/4 and 1). Although the MSB converter is shown separately, these bits are usually

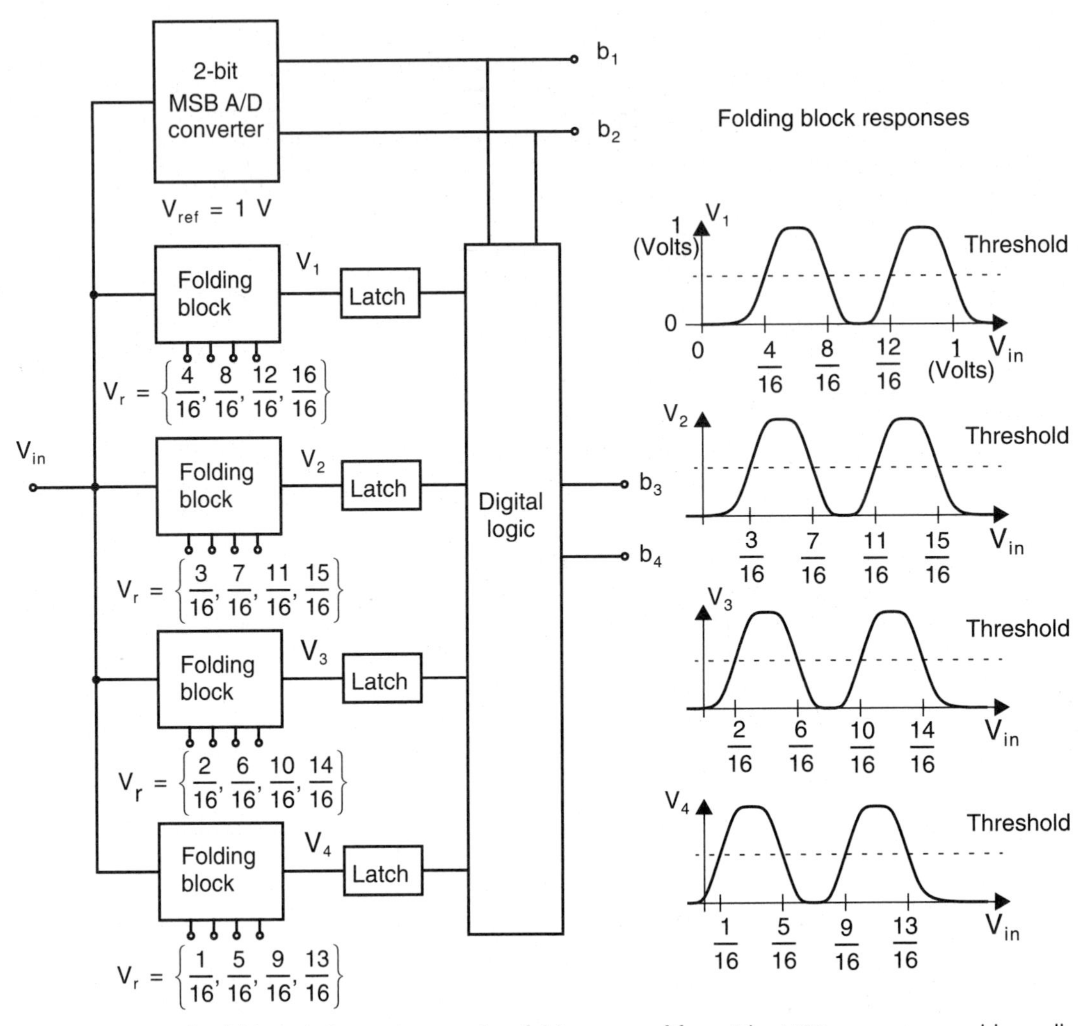

Fig. 13.27 A 4-bit folding A/D converter with a folding rate of four. (The MSB converter would usually be realized by combining some folding block signals.)

determined by combining appropriate signals within the folding blocks. To determine the 2 LSBs, V_1 to V_4 produce a thermometer code for each of the four MSB regions. Note, however, that the four LSB latches are also used for different MSB regions and the thermometer code is inverted when V_{in} is between either 1/4 and 1/2 or 3/4 and 1. For example, as V_{in} increases from 0 to 1/4, the thermometer code changes as 0000, 0001, 0011, 0111, 1111. However, as V_{in} continues to increase to 1/2, the code changes as 1110, 1100, 1000, 0000. Also, note that latch comparators can be used for the LSB set since the transitions are amplified by the folding blocks. In summary, folding reduces the number of latch comparators needed as compared to a flash converter. For example, in a flash 4-bit converter, 16 latches would be required whereas only eight are needed in the 4-bit folding example of Fig. 13.27. Specifically, four

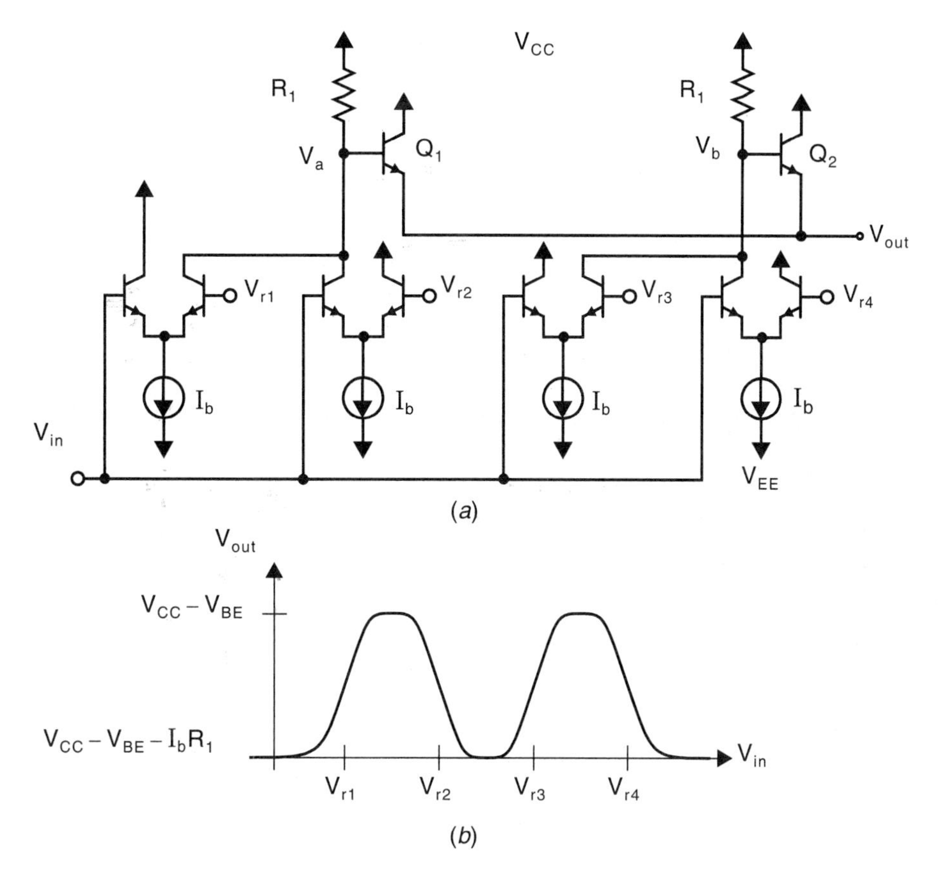

Fig. 13.28 A folding block with a folding-rate of four. (*a*) A possible single-ended circuit realization; (*b*) input-output response.

latches are used for the MSB converter and the other four are shown explicitly. In general, the savings can be greater (see Problem 13.21).

The folding blocks can be realized using cross-coupled differential pairs, as seen in the simplified bipolar circuit shown in Fig. 13.28. Here, four sets of differential-pair transistors are connected in such a way as to realize the input-output response shown in Fig. 13.28(*b*). The output signal V_{out} is related to the voltages V_a and V_b in an "or" type fashion. In other words, V_{out} is low only if both V_a and V_b are low; otherwise, V_{out} is high. With regard to the behavior of V_a and V_b, note that V_b remains low whenever V_{in} is less than V_{r3} or greater than V_{r4}. However, V_a remains low whenever V_{in} is greater than V_{r2} or less than V_{r1}. Also, the cross coupling of adjacent differential pairs causes V_a to go high when V_{in} is between V_{r1} and V_{r2}, while V_b goes high when V_{in} is between V_{r3} and V_{r4}. Such behaviors for V_a and V_b give rise to the folding output for V_{out}, as shown.

Some points worth mentioning here are that for full-scale input signals, the output signal from a folding block is at a much higher frequency than the input signal. In fact, the frequency of the folding block's output signal is equal to the multiplication of the frequency of the input signal times the folding rate. This multiplying effect limits the practical folding rate used in high-speed converters. Also, it should be mentioned that the circuit shown is a single-ended version, and differential circuits are almost always used in practical implementations.

Another point to note here is that while the folding approach reduces the number of latch comparators, a large input capacitance similar to that for a flash converter is also present with the folding circuit shown. In fact, flash converters have similar input stages of differential pairs of transistors for each comparator, but they are, of course, not cross coupled. Since the number of differential pairs in each folding block equals the folding rate, and the input signal goes to one side of each differential pair, it can be shown that the number of transistors driven by the input signal equals 2^N—the same number as for a flash converter. To reduce this large input capacitance, folding converters also make use of an interpolating architecture. With an interpolate-by-two

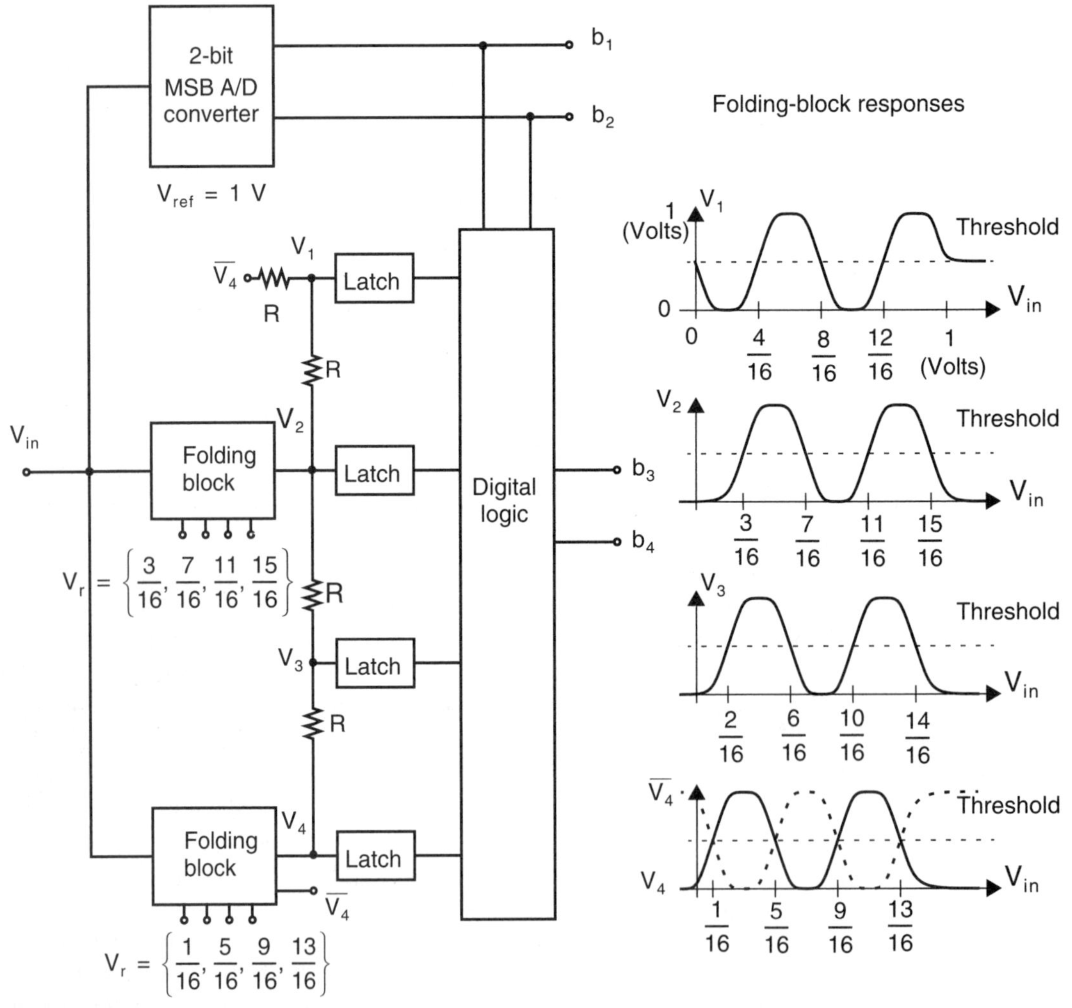

Fig. 13.29 A 4-bit folding A/D converter with a folding rate of four and an interpolate-by-two. (The MSB converter would usually be realized by combining some folding-block signals.)

technique applied to the 4-bit example, the resulting architecture would be that shown in Fig. 13.29. Note that a new inverted signal $\overline{V_4}$ is required to connect the top folding block to the bottom one. Although the creation of this inverted signal is needed in a single-ended version, no extra circuitry is required in a differential version since it can be accomplished by simply cross-coupling the differential output wires.

In [van Valburg, 1992], in order to realize a very-high-speed 8-bit converter, four folding blocks each with a folding rate of eight were used. Assuming $V_{ref} = 1$ V, then the difference between reference voltages of adjacent inputs of a folding block was 1/8 V. Each adjacent folding block was offset by 1/32 V, giving zero crossings at each 1/32 V. By interpolating between each adjacent folding block using four 8-tap resistor strings, zero crossings every 1/256 V were obtained (i.e., an 8-bit converter with 32 latch comparators connected to the interpolating resistor strings). The MSBs were realized by taking appropriate outputs from selected differential pairs and summing them separately to realize additional folding amplifiers with reduced folding rates. The final A/D converter could be clocked at 650 MHz and could resolve a 150-MHz sinusoid with 7.8 bits effective resolution, while dissipating only 0.8 W and requiring only $1.8 \times 2\ \text{mm}^2$ active area. The converter also included circuitry to prevent bubble errors, as shown in Fig. 13.19. Other examples of folding converters are given in [van de Grift, 1987, and Colleran, 1993]. In addition, folding converters have been used in the internal operation of a two-step converter [Vorenkamp, 1992].

EXAMPLE 13.7

In Fig. 13.27, show how the two MSBs can be derived using internal signals from the folding blocks.

Solution

The MSB, b_1 is determined by the input signal being above or below $V_{ref}/2$. Looking at Fig. 13.27, we see that the input is compared to the appropriate signal, $(8/16)V_{ref}$ in the top folding block, producing V_1. As a result, the MSB is easily obtained by using the collector current of the transistor connected to V_{r2} in the top folding block. Similarly, the top folding block uses references $4/16$ and $12/16$, which are needed to determine the second bit, b_2. In fact, the signal V_1 can be used directly as the second bit, b_2, as seen in Fig. 13.27. Thus, the top two bits, b_1 and b_2, can be determined using the top folding block, as shown in Fig. 13.30.

13.8 PIPELINED A/D CONVERTERS

The two-stage architecture described in Section 13.5 can be generalized to multiple stages, where each stage finds a single bit. Specifically, the first stage finds the most-significant bit, b_1, the second stage finds the next bit, b_2, and so on. Unfortunately, a straightforward implementation of this approach would be too slow, since the final bit

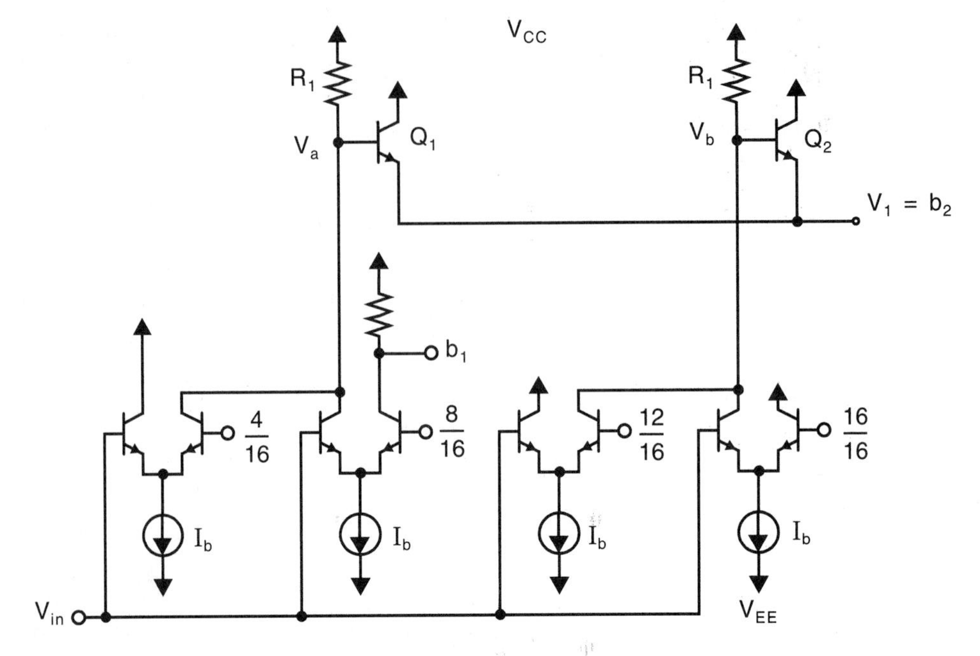

Fig. 13.30 Using the V_1 folding block to also determine the top two MSBs.

would not be available until residual errors ripple through the entire converter. A better approach is to also incorporate *pipelining* such that once the first stage completes its work, it does not sit idle while the remaining lower bits are found, but immediately starts work on the next input sample. A block diagram of a pipelined A/D converter is shown in Fig. 13.31 [Martin, 1981].

Each digital approximator (DAPRX) performs the basic operation required in the algorithmic algorithm. Specifically, for a signed conversion, the input voltage is compared to 0 V. If $V_{in} > 0$, then $V_{out} = 2V_{in} - V_{ref}/2$, and $B_{out} = 1$. Otherwise, $V_{out} = 2V_{in} + V_{ref}/2$, and $B_{out} = 0$.

Also, each DAPRX contains an S/H to store the input signal. This S/H allows the preceding DAPRX to be immediately used to process its next input signal before the succeeding DAPRX has finished, as long as the preceding DAPRX's digital output is also stored.

Although it takes N clock cycles to process each input signal (i.e., the latency is N), a new sample can be entered in the pipeline each clock cycle. Thus, the processing rate is one sample/cycle, but the complexity is only proportional to N, which is less than other architectures also processing one sample/cycle. This makes pipelined A/D converters a good choice where small area is important. The block diagram of a DAPRX is shown in Fig. 13.32. It should be mentioned here that the S/H can be incorporated into the gain of two amplifiers, if desired.

In some pipelined implementations, more than one bit is converted per stage. In this case, the DAPRX can be realized as shown in Fig. 13.33. For a multi-bit per stage pipelined converter, digital error correction can be added similar to that for a two-stage A/D converter. The major limitation on the accuracy in pipelined convert-

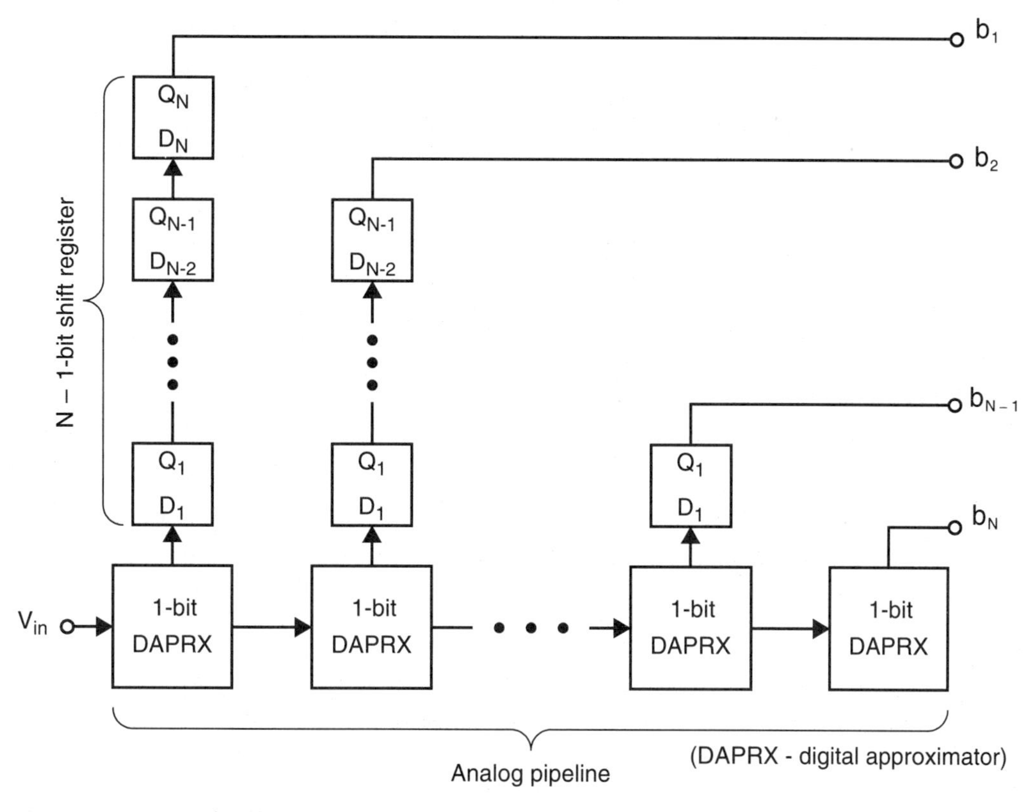

Fig. 13.31 A pipelined A/D converter.

ers is the gain amplifier, especially in the first few stages, where accuracy requirements are most stringent. For this reason, the gain is often taken smaller for the first stages (perhaps even equal to unity), which makes the realization of a high-speed gain amplifier considerably easier. The S/H is also a critical component, especially for multi-bit implementations. Also, presently most pipelined A/D converters are switched-capacitor implementations.

Current state of the art is 12 to 15 bits for pipelined converters with error correction at 1 to 2 MHz. The speed is expected to go up substantially in the near future. Some examples of pipelined A/D converters are given in [Martin, 1981; Sutarja, 1988; Song, 1988; Lin, 1991; and Karanicolas, 1993].

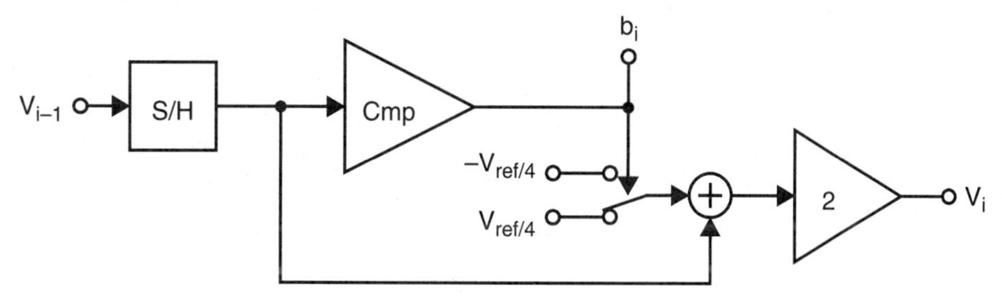

Fig. 13.32 A 1-bit digital approximator (DAPRX).

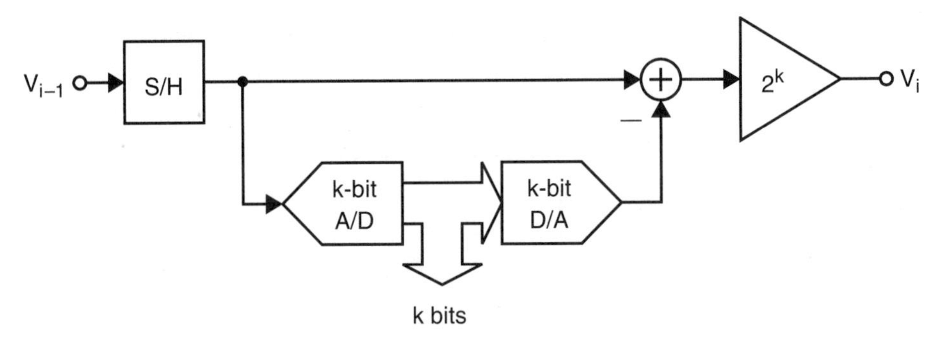

Fig. 13.33 A multi-bit digital approximator (DAPRX).

13.9 TIME-INTERLEAVED A/D CONVERTERS

Very-high-speed A/D conversions can be realized by operating many A/Ds in parallel [Black, 1980]. The system architecture for a four-channel A/D is shown in Fig. 13.34. Here, ϕ_0 is a clock at four times the rate of ϕ_1 to ϕ_4. Additionally, ϕ_1 to ϕ_4 are delayed with respect to each other by the period of ϕ_0, such that each converter will

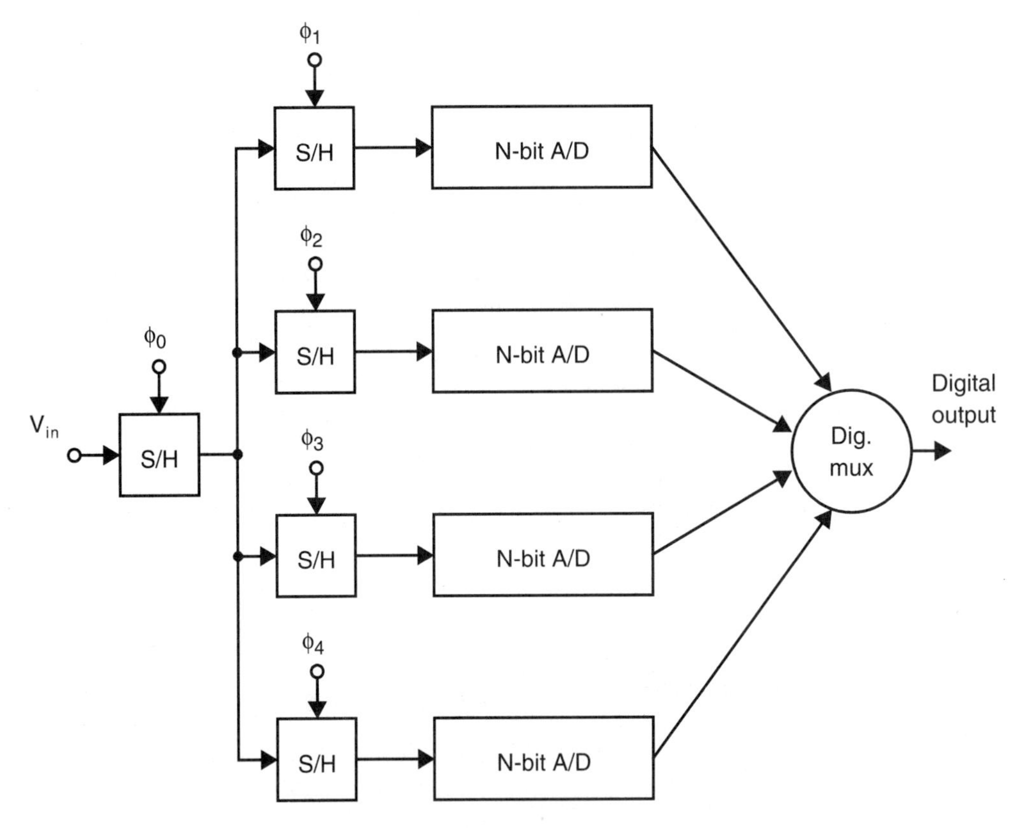

Fig. 13.34 A four-channel time-interleaved A/D converter.

get successive samples of the input signal, V_{in}, sampled at the rate of ϕ_0. In this way, the four A/D converters operate at one-quarter the rate of the input sampling frequency.

With this approach, the input S/H making use of ϕ_0 is critical, while the remaining four S/H converters can have considerable jitter since the signal is already sampled at that point. Thus, sometimes the input S/H is realized in a different technology, such as GaAs, while the remaining S/H circuits could be realized in silicon. An example of a 1 GHz 6-bit A/D converter using time interleaving and GaAs S/H circuits is described in [Poulton, 1987], where four bipolar converters operating at 250 MHz were used.

It is also essential that the channels are extremely well matched, as mismatches will produce tones at f_s/m when there are m channels. For example, consider a dc input signal in a four-channel time-interleaved converter where one converter has a dc offset of, say, 100 mV. Such a system will produce every fourth digital word different from the other three and hence a tone at $f_s/4$. Such nonideal behavior can be disastrous for many applications since the tone may reside well within the frequency of interest.

13.10 REFERENCES

Analog Devices. *Analog-Digital Conversion Handbook.* Prentice Hall, Englewood Cliffs, New Jersey, 1986.

W. C. Black, Jr., and D. A. Hodges. "Time-Interleaved Converter Arrays," *IEEE J. of Solid-State Circuits*, Vol. 15, pp. 1022–1029, December 1980.

W. T. Colleran, T. H. Phan, and A. A. Abidi. "A 10-b, 100-Ms/s Pipelined A/D Converter," *IEEE Int. Solid-State Circuits Conf.,* pp. 68–69. San Francisco, February 1993.

A. G. F. Dingwall. "Monolithic Expandable 6-bit 20-MHz CMOS/SOS A/D Converter," *IEEE J. of Solid-State Circuits*, Vol. 14, pp. 926–931, December 1979.

B. Fotouhi and D. A. Hodges. "High-Resolution A/D Conversion in MOS/LSI," *IEEE J. of Solid-State Circuits*, Vol. 14, pp. 920–925, December 1979.

Y. Gendai, Y. Komatsu, S. Hirase, and M. Kawata. "An 8-b 500-MHz ADC," *IEEE Int. Solid-State Circuits Conf.,* pp. 172–173, San Francisco, February 1991.

F. Goodenough. "Interpolators Put 10-bit 75-MHz A/D Converters on 8-bit Digital Process," *Electronic Design*, pp. 29–30, December 14, 1989.

M. Hotta et al. "A 12-mW 6-b Video Frequency ADC," *IEEE Int. Solid-State Circuits Conf.*, pp. 100–101, New York, February 1987.

M. Ito et al. "A 10b 20Ms/s 3V-Supply CMOS A/D Converter for Integration Into System VLSIs," *IEEE Int. Solid-State Circuits Conf.*, pp. 48–49, San Francisco, February 1994.

A. N. Karanicolas, H-S. Lee, and K. L. Bacrania. "A 15b 1Ms/s Digitally Self-Calibrated Pipeline ADC," *IEEE Int. Solid-State Circuits Conf.,* pp. 60–61, San Francisco, February 1993.

K. Kusumoto et al. "A 10b 20MHz 30mW Pipelined Interpolating CMOS ADC," *IEEE Int. Solid-State Circuits Conf.,* pp. 62–63, San Francisco, February 1993.

H. Lee, D. A. Hodges, and P. R. Gray. "A Self-Calibrating 15-bit CMOS A/D Converter," *IEEE J. of Solid-State Circuits*, Vol. 19, pp. 813–819, December 1984.

P. W. Li, M. J. Chin, P. R. Gray, and R. Castello. "A Ratio-Independent Algorithmic Analog-To-Digital Conversion Technique," *IEEE J. of Solid-State Circuits,* Vol. 19, pp. 828–836, December 1984.

Y-M. Lin, B. Kim, and P. R. Gray. "A 13-b 2.5-MHz Self-Calibrated Pipelined A/D Converter in 3 μm CMOS," *IEEE J. of Solid-State Circuits*, Vol. 26, pp. 628–636, December 1991.

K. W. Martin. "A High-Speed, High-Accuracy Pipelined A/D Converter," *IEEE 15th Asilomar Conf. on Circuits, Systems, and Computers*, November 1981.

R. H. McCharles, V. A. Saletore, W. C. Black, Jr., and D. A. Hodges. "An Algorithmic Analog-To-Digital Converter," *IEEE Int. Solid-State Circuits Conf.*, Philadelphia, February 1977.

J. L. McCreary et al. "All-MOS Charge Redistribution A/D Conversion Technique — Part I," *IEEE J. of Solid-State Circuits*, Vol. 10, pp. 371–379, December 1975.

B. Peetz, B. D. Hamilton, and J. Kang. "An 8-bit 250 Megasample Per Second Analog-To-Digital Converter: Operation without a Sample and Hold," *IEEE J. of Solid-State Circuits*, Vol. 21, pp. 997–1002, December 1986.

R. Petschacher et al. "A 10-b 75-MSPS Subranging A/D Converter with Integrated Sample and Hold," *IEEE J. of Solid-State Circuits*, Vol. 25, pp. 1339–1346, December 1990.

K. Poulton, J. J. Corcoran, and T. Hornak. "A 1-GHz 6-bit ADC System," *IEEE J. of Solid-State Circuits,* Vol. 22, pp. 962–970, December 1987.

A. S. Sedra and K. C. Smith, *Microelectronic Circuits*, 3rd ed. Saunders College Publishing/HRW, New Jersey, 1991.

B-S. Song, M. F. Tompsett, and K. R. Lakshmikumar. "A 12-bit 1-Msample/s Capacitor Error-Averaging Pipelined A/D Converter," *IEEE J. of Solid-State Circuits*, Vol. 23, pp. 1324–1333, December 1988.

M. Steyaert, R. Roovers, and J. Craninckx. "A 100-MHz 8-bit CMOS Interpolating A/D Converter," *IEEE Custom Integrated Circuits Conf.*, pp. 28.1.1–28.1.4, San Diego, May 1993.

S. Sutarja and P. R. Gray. "A Pipelined 13-bit, 250-ks/s, 5-V Analog-to-Digital Converter," *IEEE J. of Solid-State Circuits*, Vol. 23, pp. 1316–1323, December 1988.

K-S. Tan et al. "Error Correction Techniques for High-Performance Differential A/D Converters," *IEEE J. of Solid-State Circuits,* Vol. 25, pp. 1318–1326, December 1990.

R. E. J. van de Grift, I. W. J. M. Rutten, and M. van der Veen. "An 8-bit Video ADC Incorporating Folding and Interpolation Techniques," *IEEE J. of Solid-State Circuits*, Vol. 22, pp. 944–953, December 1987.

R. J. van de Plassche and P. Baltus. "An 8-bit 100-MHz Full-Nyquist Analog-to-Digital Converter," *IEEE J. of Solid-State Circuits*, Vol. 23, pp. 1334–1344, December 1988.

J. van Valburg and R. J. van de Plassche. "An 8-b 650-MHz Folding ADC," *IEEE J. of Solid-State Circuits*, Vol. 27, pp. 1662–1666, December 1992.

P. Vorenkamp, and J. P. M. Verdaasdonk. "A 10b 50MS/s Pipelined ADC," *IEEE Int. Solid-State Circuits Conf.*, pp. 32–33, San Francisco, February 1992.

Y. Yoshii et al. "An 8b 350MHz Flash ADC," *IEEE Int. Solid-State Circuits Conf.*, pp. 96–97, New York, February 1987.

13.11 PROBLEMS

13.1 What is the worst-case conversion time for an 18-bit dual-slope integrating A/D converter when the clock rate is 5 MHz?

13.2 Consider an 18-bit integrating A/D converter, as shown in Fig. 13.1, where V_{ref} equals 10 volts, $C_1 = 100$ pF, and a clock frequency of 1 MHz is used. What value of R_1 should be chosen such that the opamp output never exceeds 10 volts when $0\ V < V_{in} < 10\ V$?

13.3 Derive the equivalent of (13.8) for Fig. 13.1 when the opamp has an input-offset voltage of V_{off1} and the comparator has an input-offset voltage of V_{off2}.

13.4 What is the offset error (in LSBs) of the converter described in Problem 13.2 when the integrating opamp has an input-offset voltage of 20 mV?

13.5 What input-signal frequencies are completely attenuated by a dual-slope 16-bit integrating A/D having a clock frequency of 1 MHz ? For this same converter, what is the attenuation of an input signal at 60 Hz ?

13.6 Repeat Problem 13.5 with the converter's clock frequency being equal to 100 kHz.

13.7 Consider a 4-bit unipolar DAC-based successive-approximation A/D converter, as shown in Fig. 13.5. Find the sequence of the D/A converter's output levels for an input of 3.333 volts when $V_{ref} = 8$ V. What is the final digital output?

13.8 Consider a 4-bit unipolar charge-redistribution A/D converter, as shown in Fig. 13.7, where $V_{ref} = 8$ V. Find the sequence of the voltage level for V_x if the input signal is 3.333 volts.

13.9 Repeat Problem 13.8, assuming that a parasitic capacitance of 8 C is connected between the node at V_x and ground. Is the final digital result affected?

13.10 Show a method for modifying the unipolar charge-redistribution A/D converter shown in Fig. 13.7 such that the threshold offsets of $0.5V_{LSB}$ are correctly realized.

13.11 How is the digital code obtained from the signed charge redistribution A/D converter shown in Fig. 13.8 converted to a 2's complement code?

13.12 What value of error voltage V_{x1} would be measured in Fig. 13.11(*a*) if the total array capacitance equals 64 pF, the MSB (i.e., b_1) capacitor equals 31.5 pF, and there is a parasitic capacitance to ground of 10 pF on node V_x? For the same circuit, show that the error voltage, V_{e1}, that occurs when using the MSB capacitor during a normal conversion equals $0.5V_{x1}$.

13.13 For the same circuit as described in Problem 13.12, what value of error voltage V_{x2} would be measured in Fig. 13.11(*b*) if the MSB-1 (i.e., b_2) capacitor equals 16.4 pF? Show that the error voltage, V_{e2}, that occurs when using the MSB-1 capacitor during a normal conversion equals $0.5(V_{x2} - V_{e1})$.

13.14 Assuming switch resistances are all about 1 kΩ, estimate the settling time needed for a 12-bit charge-redistribution A/D converter where the total array capacitance equals 128 pF.

13.15 Draw a block diagram similar to that for Fig. 13.22 for a 10-bit two-step A/D converter where the first stage determines 4 bits. Indicate the accuracy needed in all the blocks.

13.16 Show that the circuit shown in Fig. 13.25 results in all time constants being equal when each of the latches has the same input capacitance and the amplifiers have zero output impedance.

13.17 Consider the clocked comparator shown in Fig. 13.18, where

$$\mu_n C_{ox} = 2\,\mu_p C_{ox} = 100\ \mu A/V^2$$

$$2\left(\frac{W}{L}\right)_n = \left(\frac{W}{L}\right)_p = 2$$

$$|V_{tn}| = |V_{tp}| = 1\ V$$

$$V_{DD} = 5\ V$$

$$\text{and } V_{SS} = 0\ V$$

If the MOSFETS in the inverter are ideal except for each having a 100-kΩ output impedance when active, find the minimum differential input that will cause a 1-volt change (use a linear analysis). Assuming this minimum differential input corresponds to $(1/2)$ LSB, what is the maximum number of bits that this comparator could determine in a flash A/D converter?

13.18 In Section 13.6, it was stated that the interpolated signals will have the correct threshold crossings if V_1 and V_2 in Fig. 13.24 are linear in the region $0.25 < V_{in} < 0.5$. In fact, this region is larger than what is actually necessary. What is the minimum region of V_{iu} that V_1 and V_2 should be linear over?

13.19 Many very-high-speed A/D converters do not use a sample and hold since it would limit their speed, but instead allow the input signal to be applied to the array of comparators, which are all clocked "simultaneously." Qualitatively, explain why a clocked flash or interpolating A/D converter might operate faster than a sample-and-hold circuit.

13.20 In an N-bit folding A/D converter (with no interpolation), what is the product of the folding rate times the number of folding blocks?

13.21 Find an expression for the number of latches in an N-bit folding A/D converter where the folding rate is $FR = 2^F$.

13.22 Assuming the input capacitance of a differential stage is the same in either a flash or folding/interpolating A/D converter, what reduction of input capacitance over a flash converter would be achieved with an 8-bit folding/interpolating A/D converter having four folding blocks, each with a folding rate of eight? If a straight interpolating A/D converter is to have the same reduction in input capacitance, how many resistors are required between "input comparators"?

13.23 Draw the clock waveforms of the sample-and-hold circuits for the time-interleaved A/D converter shown in Fig. 13.34.

CHAPTER

14 Oversampling Converters

Recently, oversampling A/D and D/A converters have become popular for high-resolution medium-to-low-speed applications such as high-quality digital audio. The major reasons for their popularity include the following: First, oversampling converters relax the requirements placed on the analog circuitry at the expense of more complicated digital circuitry. This trade-off becomes more desirable for modern submicron technologies with 3.3-V power supplies where complicated high-speed digital circuitry is more easily realized in less area, but the realization of high-resolution analog circuitry is complicated by low power-supply voltages and poor transistor output impedance (caused by short-channel effects). With oversampling data converters, the analog components have reduced requirements on matching tolerances and amplifier gains. A second advantage of oversampling converters is that they simplify the requirements placed on the analog anti-aliasing filters for A/D converters and smoothing filters for D/A converters. For example, usually only a first-order anti-aliasing filter is required for A/D converters, which can often be realized on chip or at worst case very inexpensively off chip. Furthermore, a sample-and-hold is usually not required at the input of an oversampling A/D converter.

In this chapter, the basics of oversampling converters are discussed first. We shall see that extra bits of resolution can be extracted from converters that sample much faster than the Nyquist rate. Furthermore, this extra resolution can be obtained with lower oversampling rates by spectrally shaping the quantization noise through the use of feedback. The use of shaped quantization noise applied to oversampling signals is commonly referred to as delta-sigma ($\Delta\Sigma$) modulation.[1] Simple first- and second-order $\Delta\Sigma$ modulators are discussed, followed by a discussion of typical system architectures for $\Delta\Sigma$ data converters. Next, two popular approaches for realizing decimation filters are described. Descriptions of some modern approaches are then described along with some practical considerations. The chapter concludes with an example design of a third-order $\Delta\Sigma$ A/D converter.

14.1 OVERSAMPLING WITHOUT NOISE SHAPING

In this section, the advantage of sampling at higher than the Nyquist rate is discussed. Here, we shall see that extra dynamic range can be obtained by spreading the

1. Delta-sigma modulation is also sometimes referred to as sigma-delta modulation.

quantization noise power over a larger frequency range. However, we shall see that the increase in dynamic range is only 3 dB for every doubling of the sample rate. To obtain much higher dynamic-range improvements as the sampling rate is increased, noise shaping through the use of feedback can be used and is discussed in the next section.

Quantization Noise Modelling

We begin by modelling a quantizer as adding quantization error $e(n)$, as shown in Fig. 14.1. The output signal, $y(n)$, is equal to the closest quantized value of $x(n)$. The *quantization error* is the difference between the input and output values. This model is exact if one recognizes that the quantization error is not an independent signal but may be strongly related to the input signal, $x(n)$. This linear model becomes approximate when assumptions are made about the statistical properties of $e(n)$, such as $e(n)$ being an independent white-noise signal. However, even though approximate, it has been found that this model leads to a much simpler understanding of $\Delta\Sigma$ and with some exceptions is usually reasonably accurate.

White Noise Assumption

If $x(n)$ is very active, $e(n)$ can be approximated as an independent random number uniformly distributed between $\pm\Delta/2$, where Δ equals the difference between two adjacent quantization levels. Thus, the quantization noise power equals $\Delta^2/12$ (from Section 11.3) and *is independent of the sampling frequency*, f_s. Also, the spectral density of $e(n)$, $S_e(f)$, is white (i.e., a constant over frequency) and all its power is within $\pm f_s/2$ (a two-sided definition of power).

Assuming white quantization noise, the spectral density of the quantization noise, $S_e(f)$ appears as shown in Fig. 14.2.

The spectral density height is calculated by noting that the total noise power is $\Delta^2/12$ and, with a two-sided definition of power, equals the area under $S_e(f)$ within

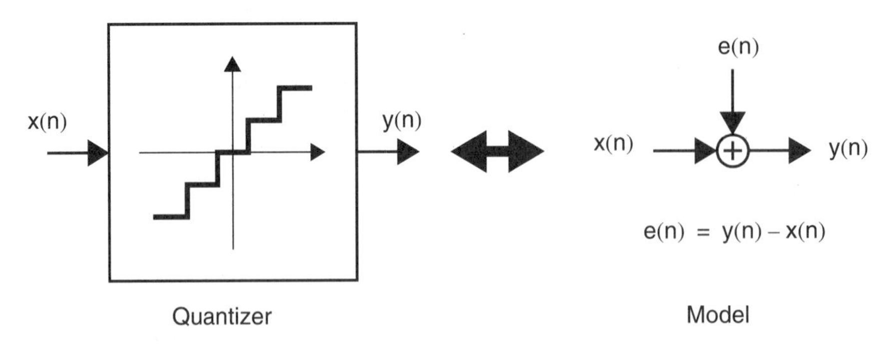

Fig. 14.1 Quantizer and its linear model.

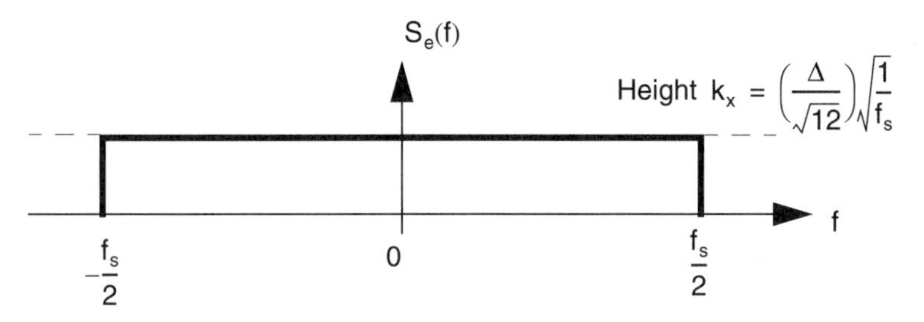

Fig. 14.2 Assumed spectral density of quantization noise.

$\pm f_s/2$, or mathematically,

$$\int_{-f_s/2}^{f_s/2} S_e^2(f)\,df = \int_{-f_s/2}^{f_s/2} k_x^2\,df = k_x^2 f_s = \frac{\Delta^2}{12} \tag{14.1}$$

Solving this relation gives

$$k_x = \left(\frac{\Delta}{\sqrt{12}}\right)\sqrt{\frac{1}{f_s}} \tag{14.2}$$

EXAMPLE 14.1

Find the output and quantization errors for two different quantizers, as shown in Fig. 14.3, when the input values are

$$x(n) = \{0.01, 0.31, -0.11, 0.80, 0.52, -0.70\} \tag{14.3}$$

Also find the expected power and the power density height, $S_e^2(f)$, of the quantization noise when the sampling frequency is normalized to 2π rad/sample.

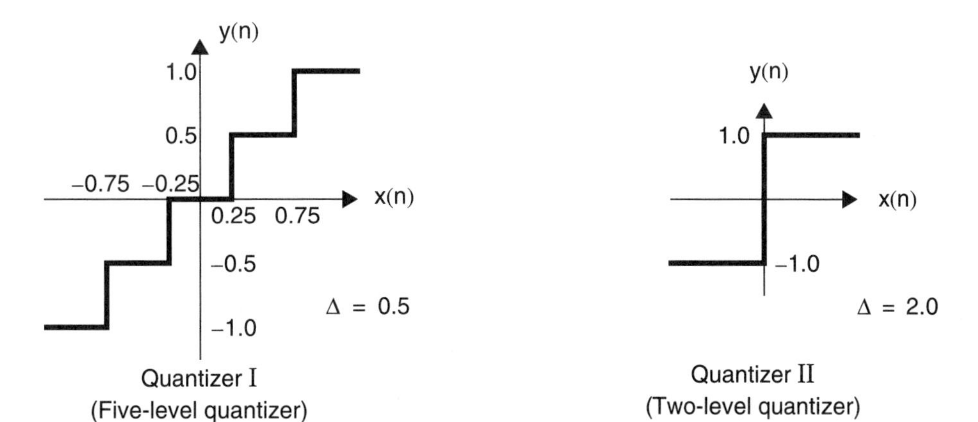

Fig. 14.3 Two example quantizers.

Table 14.1 Example signal values and quantization noise for two quantizers.

	Quantizer I (Five Levels)		Quantizer II (Two Levels)	
x(n)	y(n)	e(n)	y(n)	e(n)
0.01	0.0	–0.01	1	0.99
0.31	0.5	0.19	1	0.69
–0.11	0.0	0.11	–1	–0.89
0.80	1.0	0.2	1	0.20
0.52	0.5	–0.02	1	0.48
–0.70	–0.5	0.2	–1	–0.30

Solution

The output and quantization noise values for this example are given in Table 14.1. Note that although the output signals, $y(n)$, are well-defined values, the quantization error values can be approximated as uniformly distributed random numbers since the input signal is quite active.

Recalling that the quantization noise power is given by

$$P_e = \frac{\Delta^2}{12} \tag{14.4}$$

then the expected noise powers of the two quantizers are

$$P_I = \frac{0.5^2}{12} = 0.0208\ \mathrm{W} \tag{14.5}$$

$$P_{II} = \frac{2^2}{12} = 0.333\ \mathrm{W} \tag{14.6}$$

where we note that these values are not affected by normalizing the sampling frequency. Also note that the two power estimates for P_I and P_{II} correspond to rms power levels of 0.144 and 0.577, respectively.

For the power density, we note that the power is spread evenly between $\pm f_s/2$, resulting in a density of

$$S_e^2(f) = \frac{P_e}{f_s} \tag{14.7}$$

which results in

$$S_{eI}^2(f) = \left(\frac{0.0208}{2\pi}\right) = 0.00331\ \frac{\mathrm{W}}{\mathrm{rad/sample}} \tag{14.8}$$

$$S_{eII}^2(f) = \left(\frac{0.333}{2\pi}\right) = 0.053\ \frac{\mathrm{W}}{\mathrm{rad/sample}} \tag{14.9}$$

Oversampling Advantage

Oversampling occurs when the signals of interest are bandlimited to f_0 yet the sample rate is at f_s, where $f_s > 2f_0$ ($2f_0$ being the Nyquist rate or, equivalently, the minimum sampling rate for signals bandlimited to f_0). We define the oversampling ratio, OSR, as

$$OSR \equiv \frac{f_s}{2f_0} \tag{14.10}$$

After quantization, since the signals of interest are all below f_0, $y_1(n)$ is filtered by $H(f)$ to create the signal $y_2(n)$, as shown in Fig. 14.4. This filter eliminates quantization noise (together with any other signals) greater than f_0.

Assuming the input signal is a sinusoidal wave, its maximum peak value without clipping is $2^N(\Delta/2)$. For this maximum sinusoidal wave, the signal power, P_s, has a power equal to

$$P_s = \left(\frac{\Delta 2^N}{2\sqrt{2}}\right)^2 = \frac{\Delta^2 2^{2N}}{8} \tag{14.11}$$

The power of the input signal within $y_2(n)$ remains the same as before since we assumed the signal's frequency content is below f_0. However, the quantization noise power is reduced to

$$P_e = \int_{-f_s/2}^{f_s/2} S_e^2(f)|H(f)|^2\, df = \int_{-f_0}^{f_0} k_x^2\, df = \frac{2f_0\Delta^2}{f_s\ 12} = \frac{\Delta^2}{12}\left(\frac{1}{OSR}\right) \tag{14.12}$$

Therefore, doubling OSR (i.e., sampling at twice the rate) decreases the quantization noise power by one-half or, equivalently, 3 dB (or, equivalently, 0.5 bits).

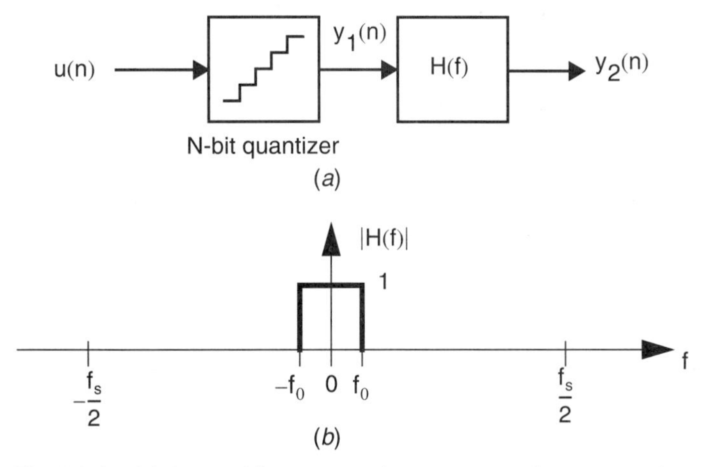

Fig. 14.4 (*a*) A possible oversampling system without noise shaping. (*b*) The brick-wall response of the filter to remove much of the quantization noise.

We can also calculate the maximum SNR (in dB) to be the ratio of the maximum sinusoidal power to the quantization noise in the signal $y_2(n)$. Mathematically, we have through the use of (14.11) and (14.12)

$$SNR_{max} = 10\log\left(\frac{P_s}{P_e}\right) = 10\log\left(\frac{3}{2}2^{2N}\right) + 10\log(OSR) \tag{14.13}$$

which is also equal to

$$SNR_{max} = 6.02N + 1.76 + 10\log(OSR) \tag{14.14}$$

The first term is the SNR due to the N-bit quantizer while the OSR term is the SNR enhancement obtained from oversampling. Here we see that straight oversampling gives a SNR improvement of 3 dB/octave or, equivalently, 0.5 bits/octave. The reason for this SNR improvement through the use of oversampling is that when quantized samples are averaged together, the signal portion adds linearly, whereas the noise portion adds as the square root of the sum of the squares.

EXAMPLE 14.2

Consider a sampled dc signal, V_s, of value 1 V where the measured voltage, V_{meas}, is V_s plus a noise signal, V_{noise}. Assume V_{noise} is a random signal uniformly distributed between $\pm\sqrt{3}$. What is the SNR for V_{meas} when looking at individual values? If eight samples of V_{meas} are averaged together, roughly what is the new SNR? To illustrate the results, use eight typical samples for V_{meas} of {0.94, –0.52, –0.73, 2.15, 1.91, 1.33, –0.31, 2.33}.

Solution

Referencing signals to 1 Ω, we calculate the power of V_s and V_{noise} to both be 1 watt. Thus the SNR for V_{meas} is 0 dB when looking at individual V_{meas} values. Note that it is difficult to see the signal value of 1 V in the example V_{meas} samples since the SNR is so poor.

If eight samples are averaged, we are realizing a modest low-pass filter, resulting in the oversampling ratio being approximately equal to 8 (this is a rough estimate since a brick-wall filter is not being used). Since each octave of oversampling results in a 3-dB SNR improvement, the averaged value should have a SNR of around 9 dB. Note that averaging the eight given V_{meas} samples results in 0.8875, which more closely represents the signal value of 1 V.

The reason oversampling improves the SNR here is that by summing eight measured values, the eight signal values add linearly to 8 V_{rms} (or 64 watts) while the eight noise values add to $\sqrt{8}$ V_{rms} (or 8 watts), since the noise values are assumed to be independent.

EXAMPLE 14.3

Given that a 1-bit A/D converter has a 6-dB SNR, what sample rate is required using oversampling (no noise shaping) to obtain a 96-dB SNR (i.e., 16 bits) if $f_0 = 25$ kHz ? (Note that the input into the A/D converter has to be very active for the white-noise quantization model to be valid—a difficult arrangement when using a 1-bit quantizer with oversampling without noise shaping).

Solution

Oversampling (without noise shaping) gives 3 dB/octave where 1 octave implies doubling the sampling rate. We require 90 dB divided by 3 dB/octave, or 30 octaves. Thus, the required sampling rate, f_s, is

$$f_s = 2^{30} \times 2f_0 \cong 54{,}000 \text{ GHz !}$$

This example shows why noise shaping is needed to improve the SNR faster than 3 dB/octave, since 54,000 GHz is highly impractical.

The Advantage of 1-bit D/A Converters

While oversampling improves the signal-to-noise ratio, it does not improve linearity. For example, if a 16-bit linear converter is desired while using a 12-bit converter with oversampling, the 12-bit converter must have an integral nonlinearity error less than $1/2^4$ LSB (here, LSB refers to that for a 12-bit converter). In other words, the component accuracy would have to match better than 16-bit accuracy (i.e., $100 \times (1/2^{16}) = 0.0015$ percent accuracy). Thus, some sort of auto calibration or laser trimming must be used to obtain the required linearity. However, as we saw in Example 14.3, with a high enough sampling rate, the output from a 1-bit converter can be filtered to obtain the equivalent of a 16-bit converter. *The advantage of a 1-bit D/A is that it is inherently linear.*[2] This linearity is a result of a 1-bit D/A converter having only two output values and, since two points define a straight line, no trimming or calibration is required. This inherent linearity is one of the major motivations for making use of oversampling techniques with 1-bit converters. In fact, the reader may be aware that many audio converters presently use 1-bit converters for realizing 16- to 18-bit linear converters (with noise shaping). In addition, 20-bit linearity has been reported without the need for any trimming [Leopold, 1991]. Finally, it should be mentioned that there are other advantages when using oversampling techniques, such as a reduced requirement for analog anti-aliasing and smoothing filters.

2. This assumes that second-order errors due to imperfections such as signal-dependent power supply, or reference voltages, or memory in the D/A converter are not present.

14.2 OVERSAMPLING WITH NOISE SHAPING

In this section, the advantage of noise shaping the quantization noise through the use of feedback is discussed. Here, we shall see a much more dramatic improvement in dynamic range when the input signal is oversampled as compared to oversampling the input signal with no noise shaping. Although this section illustrates the basic principles with reference to A/D converters, much of it applies directly to $\Delta\Sigma$ D/A converters as well.

The system architecture of a $\Delta\Sigma$ oversampling A/D converter is shown in Fig. 14.5. The first stage is a continuous-time anti-aliasing filter and is required to band-limit the input signal to frequencies less than one-half the oversampling frequency, f_S. When the oversampling ratio is large, the anti-aliasing filter can often be quite simple, such as a simple RC low-pass filter. Following the anti-aliasing filter, the continuous-time signal, $x_c(t)$, is sampled by a sample-and-hold. This signal is then processed by the $\Delta\Sigma$ modulator, which converts the analog signal into a noise-shaped low-resolution digital signal. The third block in the system is a decimator. It converts the oversampled low-resolution digital signal into a high-resolution digital signal at a lower sampling rate usually equal to twice the frequency of the desired bandwidth of the input signal. The decimation filter can be conceptually thought of as a low-pass filter followed by a down sampler, although in many systems the decimation is performed in a number of stages. It should be mentioned that in many realizations where the $\Delta\Sigma$ modulator is realized using switched-capacitor circuitry, a separate sample-and-hold is not required, as the continuous-time signal is inherently sampled by the switches and input capacitors of the SC $\Delta\Sigma$. In the next few sections the operation of the various building blocks will be described in greater detail.

Noise-Shaped Delta-Sigma Modulator

A general noise-shaped delta-sigma ($\Delta\Sigma$) modulator and its linear model are shown in Fig. 14.6. This arrangement is known as an interpolative structure and is analogous to an amplifier realized using an opamp and feedback. In this analogy, the feedback reduces the effect of the noise of the output stage of the opamp in the closed-loop amplifier's output signal at low frequencies when the opamp gain is high. At high frequencies, when the opamp's gain is low, the noise is not reduced. Note that the quan-

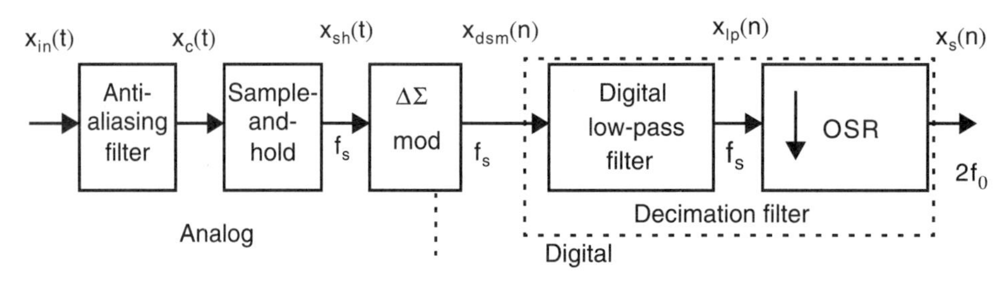

Fig. 14.5 Block diagram of an oversampling A/D converter.

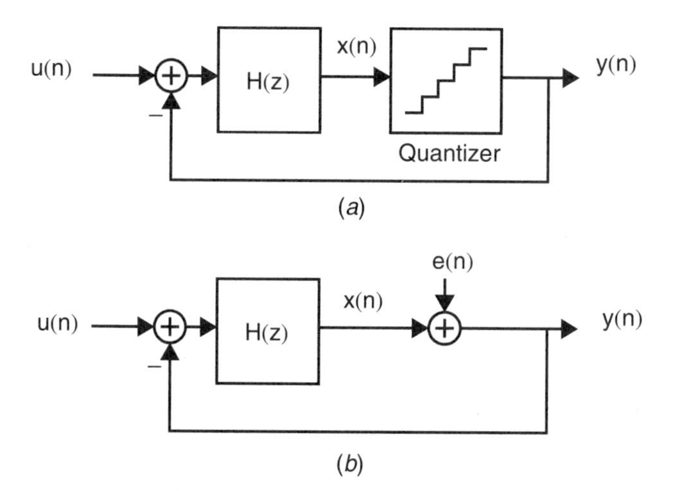

Fig. 14.6 A modulator and its linear model: (*a*) a general $\Delta\Sigma$ modulator (interpolator structure); (*b*) linear model of the modulator showing injected quantization noise.

tizer here is shown for the general case where many output levels occur. While most present oversampling converters make use of 1-bit quantizers (i.e., only two output levels) due to reasons already discussed, there is certainly no reason to restrict ourselves to such implementations. In fact, multibit oversampling converters, discussed in Section 14.8, are gaining in popularity.

Treating the linear model shown in Fig. 14.6(*b*) as having two *independent* inputs (which is an approximation), we can derive a signal transfer function, $S_{TF}(z)$, and a noise transfer function, $N_{TF}(z)$.

$$S_{TF}(z) \equiv \frac{Y(z)}{U(z)} = \frac{H(z)}{1+H(z)} \tag{14.15}$$

$$N_{TF}(z) \equiv \frac{Y(z)}{E(z)} = \frac{1}{1+H(z)} \tag{14.16}$$

Note that the zeros of the noise transfer function, $N_{TF}(z)$, will be equal to the poles of $H(z)$. In other words, when $H(z)$ goes to infinity, we see by (14.16) that $N_{TF}(z)$ will go to zero. We can also write the output signal as the combination of the input signal and the noise signal, with each being filtered by the corresponding transfer function. In the frequency domain, we have

$$Y(z) = S_{TF}(z)U(z) + N_{TF}(z)E(z) \tag{14.17}$$

To noise-shape the quantization noise in a useful manner, we choose $H(z)$ such that its magnitude is large from 0 to f_0 (i.e., over the frequency band of interest). With such a choice, the signal transfer function, $S_{TF}(z)$, will approximate unity over the frequency band of interest very similarly to an opamp in a unity-gain feedback configuration. Furthermore, the noise transfer function, $N_{TF}(z)$, will approximate zero over the same band. Thus, the quantization noise is reduced over the frequency band of interest while the signal itself is largely unaffected. The high-frequency noise is not reduced by the feedback as there is little loop gain at high frequencies. However, additional post filtering can remove the out-of-band quantization noise with little effect on the desired signal.

Before choosing specific functions for $H(z)$, note that the maximum level of the in-band input signal, $u(n)$, must remain within the maximum levels of the feedback signal, $y(n)$; otherwise the large gain in $H(z)$ will cause the signal $x(n)$ to saturate. For example, if a 1-bit quantizer having output levels of ± 1 is used, the input signal must also remain within ± 1 for frequencies where the gain of $H(z)$ is large. In fact, for many modulators the input signal needs to be significantly smaller than the bounds of the quantizer output levels to keep the modulator *stable*.[3] However, the maximum level of the input signal, $u(n)$, for frequencies where the gain of $H(z)$ is small will not necessarily cause the signal $x(n)$ to saturate. In other words, the maximum level of the out-of-band input signal can be quite a bit larger than the feedback levels (see Problem 14.4).

First-Order Noise Shaping

To realize first-order noise shaping, the noise transfer function, $N_{TF}(z)$, should have a zero at dc (i.e., $z = 1$) so that the quantization noise is high-pass filtered. Since the zeros of $N_{TF}(z)$ are equal to the poles of $H(z)$, we can obtain first-order noise shaping by letting $H(z)$ be a discrete-time integrator (i.e., have a pole at $z = 1$).[4] Specifically,

$$H(z) = \frac{1}{z-1} \tag{14.18}$$

A block diagram for such a choice is shown in Fig. 14.7.

Time Domain View From a time domain point of view, if the feedback is operating correctly and the system is stable, then the signal $x(n)$ is bounded (i.e., $\neq \infty$). Since the integrator has infinite dc gain, the *average value of the discrete-time integrator's input must exactly equal zero* (i.e., average value of $u(n) - y(n)$ equals zero). This

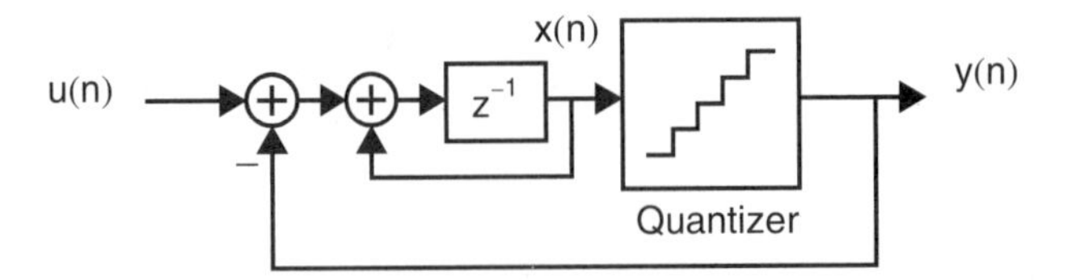

Fig. 14.7 A first-order noise-shaped interpolative modulator.

3. A modulator is defined to be stable if the input to the quantizer does not become so large as to cause the quantizer error to become greater than $\pm\Delta/2$ (which is referred to as overloading the quantizer). See Section 14.7.

4. A continuous-time integrator can also be used, but discrete-time integrators are more popular in integrated realizations as they are less sensitive to sampling jitter and have better distortion characteristics.

result implies that the average value (i.e., dc value) of $u(n)$ must equal the average value (i.e., dc value) of $y(n)$.

Again, the similarity of this configuration and an opamp having unity-gain feedback is emphasized. The open-loop transfer function of an opamp is closely approximated by a first-order integrator having very large gain at low frequencies.

EXAMPLE 14.4

Find the output sequence and state values for a dc input, $u(n)$, of $1/3$ when a two-level quantizer of ± 1.0 is used (threshold at zero) and the initial state for $x(n)$ is 0.1.

Solution

The output sequence and state values are given in Table 14.2.

Table 14.2 First-order modulator example

n	x(n)	x(n + 1)	y(n)	e(n)
0	0.1	–0.5667	1.0	0.9
1	–0.5667	0.7667	–1.0	–0.4333
2	0.7667	0.1	1.0	0.2333
3	0.1	–0.5667	1.0	0.9
4	–0.5667	0.7667	–1.0	–0.4333
5	…	…	…	…

Note that the average of $y(n)$ exactly equals $1/3$ as expected. However, also note that the output pattern is periodic, which implies that the quantization noise is not random in this example. (However, the result is much more satisfying than applying $1/3$ directly into a 1-bit quantizer using straight oversampling, which would give all 1s as its output.)

Frequency Domain View From a frequency domain view, the signal transfer function, $S_{TF}(z)$, is given by

$$S_{TF}(z) = \frac{Y(z)}{U(z)} = \frac{1/(z-1)}{1+1/(z-1)} = z^{-1} \tag{14.19}$$

and the noise transfer function, $N_{TF}(z)$, is given by

$$N_{TF}(z) = \frac{Y(z)}{E(z)} = \frac{1}{1+1/(z-1)} = (1-z^{-1}) \tag{14.20}$$

We see here that the signal transfer function is simply a delay, while the noise transfer function is a discrete-time differentiator (i.e., a high-pass filter).

To find the magnitude of the noise transfer function, $|N_{TF}(f)|$, we let $z = e^{j\omega T} = e^{j2\pi f/f_s}$ and write the following:

$$N_{TF}(f) = 1 - e^{-j2\pi f/f_s} = \frac{e^{j\pi f/f_s} - e^{-j\pi f/f_s}}{2j} \times 2j \times e^{-j\pi f/f_s}$$
$$= \sin\left(\frac{\pi f}{f_s}\right) \times 2j \times e^{-j\pi f/f_s} \tag{14.21}$$

Taking the magnitude of both sides, we have the high-pass function

$$|N_{TF}(f)| = 2\sin\left(\frac{\pi f}{f_s}\right) \tag{14.22}$$

Now the quantization noise power over the frequency band from 0 to f_0 is given by

$$P_e = \int_{-f_0}^{f_0} S_e^2(f)|N_{TF}(f)|^2 df = \int_{-f_0}^{f_0} \left(\frac{\Delta^2}{12}\right)\frac{1}{f_s}\left[2\sin\left(\frac{\pi f}{f_s}\right)\right]^2 df \tag{14.23}$$

and making the approximation that $f_0 << f_s$ (i.e., $OSR >> 1$), so that we can approximate $\sin((\pi f)/f_s)$ to be $(\pi f)/f_s$, we have

$$P_e \cong \left(\frac{\Delta^2}{12}\right)\left(\frac{\pi^2}{3}\right)\left(\frac{2f_0}{f_s}\right)^3 = \frac{\Delta^2\pi^2}{36}\left(\frac{1}{OSR}\right)^3 \tag{14.24}$$

Assuming the maximum signal power is the same as that obtained before in (14.11), the maximum SNR for this case is given by

$$SNR_{max} = 10\log\left(\frac{P_s}{P_e}\right) = 10\log\left(\frac{3}{2}2^{2N}\right) + 10\log\left[\frac{3}{\pi^2}(OSR)^3\right] \tag{14.25}$$

or, equivalently,

$$SNR_{max} = 6.02N + 1.76 - 5.17 + 30\log(OSR) \tag{14.26}$$

We see here that doubling the OSR gives an SNR improvement for a first-order modulator of 9 dB or, equivalently, a gain of 1.5 bits/octave. This result should be compared to the 0.5 bits/octave when oversampling with no noise shaping.

Switched-Capacitor Realization of a First-Order A/D Converter

It is possible to realize a first-order modulator using switched-capacitor (SC) techniques. An example of a first-order modulator where a 1-bit quantizer is used in the feedback loop is shown in Fig. 14.8. Here, the $\Delta\Sigma$ modulator consists of both analog and digital circuitry. It should be mentioned that the two input capacitances to the discrete-time integrator in Fig. 14.8 can be combined to one capacitance, as shown in Fig. 14.9 [Boser, 1988]. However, such an approach does not easily allow scaling of the feedback signal relative to the input signal.

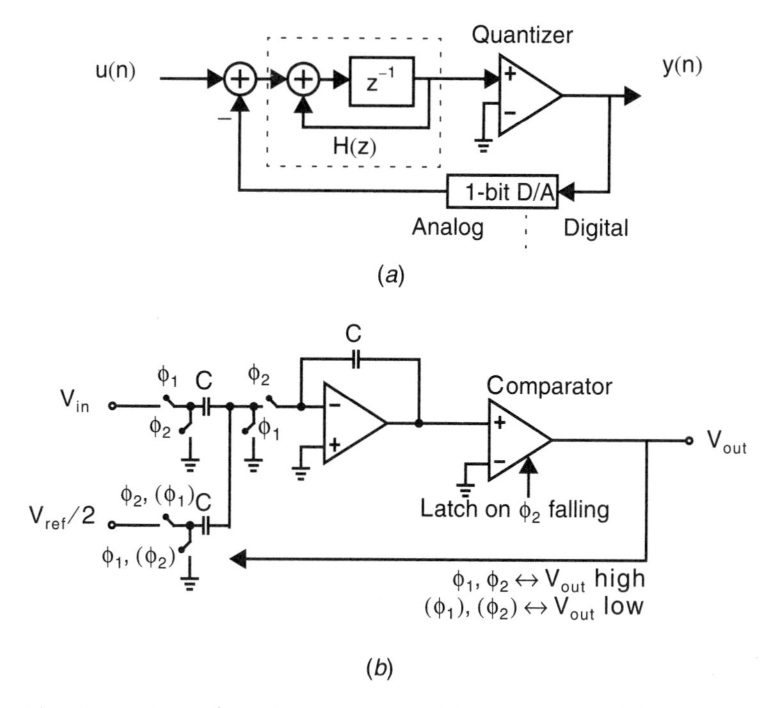

Fig. 14.8 First-order A/D modulator: (*a*) block diagram; (*b*) switched-capacitor implementation.

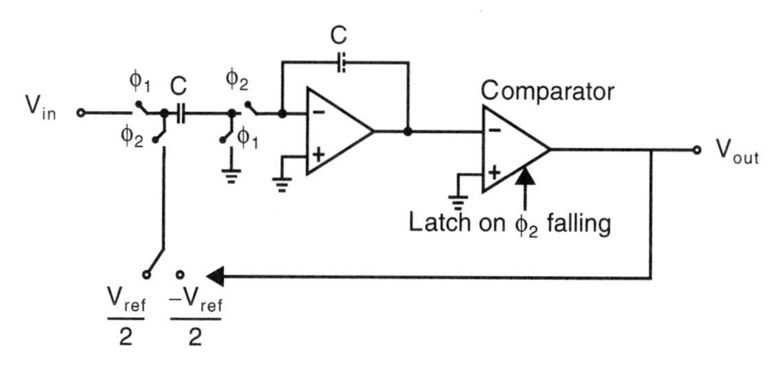

Fig. 14.9 First-order A/D modulator using only one input capacitance to the discrete-time integrator.

Second-Order Noise Shaping

The modulator shown in Fig. 14.10 realizes second-order noise shaping (i.e., the noise transfer function, $N_{TF}(z)$, is a second-order high-pass function). For this modulator, the signal transfer function is given by

$$S_{TF}(f) = z^{-1} \tag{14.27}$$

and the noise transfer function is given by

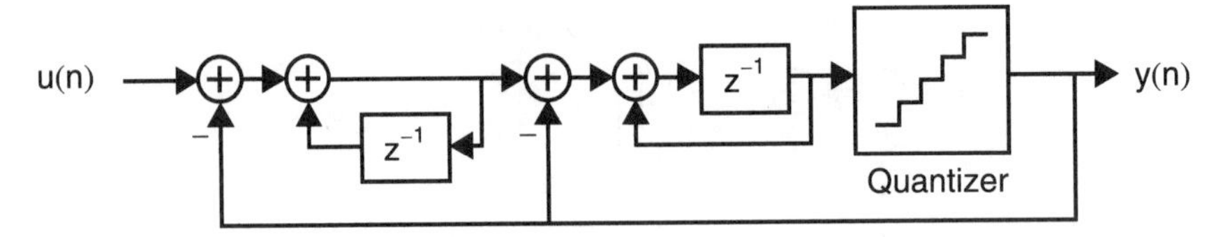

Fig. 14.10 Second-order ΔΣ modulator.

$$N_{TF}(f) = (1 - z^{-1})^2 \tag{14.28}$$

Additionally, the magnitude of the noise transfer function can be shown to be given by

$$|N_{TF}(f)| = \left[2 \sin\left(\frac{\pi f}{f_s}\right)\right]^2 \tag{14.29}$$

resulting in the quantization noise power over the frequency band of interest being given by

$$P_e \cong \frac{\Delta^2 \pi^4}{60}\left(\frac{1}{OSR}\right)^5 \tag{14.30}$$

Again, assuming the maximum signal power is that obtained in (14.11), the maximum SNR for this case is given by

$$SNR_{max} = 10 \log\left(\frac{P_s}{P_e}\right) = 10 \log\left(\frac{3}{2} 2^{2N}\right) + 10 \log\left[\frac{5}{\pi^4}(OSR)^5\right] \tag{14.31}$$

or, equivalently,

$$SNR_{max} = 6.02N + 1.76 - 12.9 + 50 \log(OSR) \tag{14.32}$$

We see here that doubling the OSR improves the SNR for a second-order modulator by 15 dB or, equivalently, a gain of 2.5 bits/octave.

The realization of the second-order modulator using switched-capacitor techniques is straightforward and is left as an exercise for the interested reader.

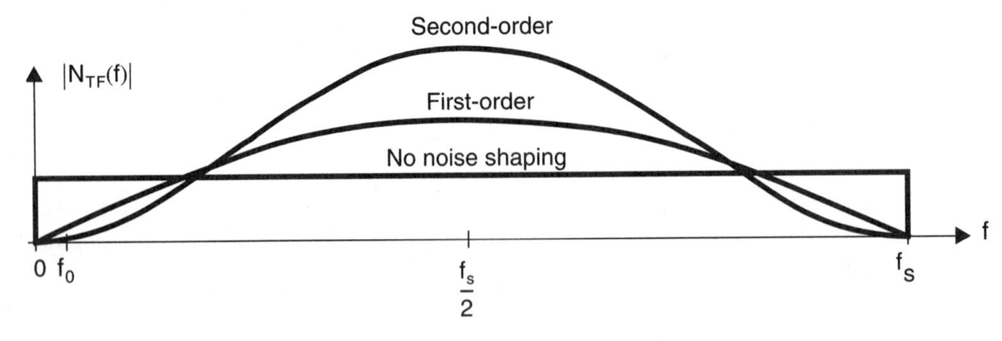

Fig. 14.11 Some different noise-shaping transfer functions.

Noise Transfer-Function Curves

The general shape of zero-, first-, and second-order noise-shaping curves are shown in Fig. 14.11. Note that over the band of interest (i.e., from 0 to f_0), the noise power decreases as the noise-shaping order increases. However, the out-of-band noise increases for the higher-order modulators.

EXAMPLE 14.5

Given that a 1-bit A/D converter has a 6-dB SNR, what sample rate is required to obtain a 96-dB SNR (or 16 bits) if $f_0 = 25$ kHZ for straight oversampling as well as first- and second-order noise shaping?

Solution

Oversampling with No Noise Shaping From before, straight oversampling requires a sampling rate of 54,000 GHz.

First-Order Noise Shaping First-order noise shaping gives 9 dB/octave where 1 octave is doubling the OSR. Since we lose 5 dB, we require 95 dB divided by 9 dB/octave, or 10.56 octaves. Thus, the required sampling rate, f_s, is

$$f_s = 2^{10.56} \times 2f_0 \cong 75 \text{ MHz}$$

This compares very favorably with straight oversampling, though it is still quite high.

Second-Order Noise Shaping Second-order noise shaping gives 15 dB/octave, but loses 13 dB. Thus we required 103 dB divided by 15 dB/octave, resulting in a required sampling rate of only 5.8 MHz. However, this simple calculation does not take into account the reduced input range for a second-order modulator needed for stability.

Quantization Noise Power of 1-bit Modulators

Assuming the output of a 1-bit modulator is ± 1, then one can immediately determine the total power of the output signal, $y(n)$, to be a normalized power of 1 watt. Since $y(n)$ consists of both signal and quantization noise, it is clear that the signal power can never be greater than 1 watt. In fact, as alluded to earlier, the signal level is often limited to well below the ± 1 level in higher-order modulators to maintain stability. For example, assuming that the maximum peak signal level is only ± 0.25, then the maximum signal power is 62.5 mW. And since the signal power plus quantization noise power equals 1 W, the maximum signal power is about 12 dB below the quantization noise power. Fortunately, as we saw above, the quantization noise power is mostly in a different frequency region than the signal power and can therefore be filtered out. Note, however, that the filter must have a dynamic range capable of accommodating the full power of $y(n)$ at its input. For a $\Delta\Sigma$ A/D converter, the filtering would be done by digital filters following the quantizer.

Error-Feedback Structure

Before leaving this section, it is of interest to look at another structure for realizing a delta-sigma modulator—an error-feedback structure, as shown in Fig. 14.12 [Anastassiou, 1989]. Using a linear model for the quantizer, it is not difficult to show that this error-feedback structure has a signal transfer function, $S_{TF}(z)$, equal to unity while the noise transfer function, $N_{TF}(z)$, equals $G(z)$. Thus for a first-order modulator, $G(z) = 1 - z^{-1}$, or in other words, the block $(G(z) - 1)$ is simply $-z^{-1}$.

Unfortunately, a slight coefficient error can cause significant noise-shaping degradation with this error-feedback structure. For example, in a first-order case, if the delayed signal becomes $-0.99z^{-1}$ (rather than $-z^{-1}$), then $G(z) = 1 - 0.99z^{-1}$, and the zero is moved off dc. Such a shift of the zero will result in the quantization noise not being fully nulled at dc and therefore would not be suitable for high oversampling ratios. Thus, this structure is not well suited to analog implementations where coefficient mismatch occurs. In contrast, an interpolative structure has the advantage that the zeros of the noise transfer function remain at dc as long as $H(z)$ has infinite gain at dc. This high gain at dc can usually be obtained in analog implementations by using opamps with large open-loop gains and does not rely on component matching. The error feedback structure is discussed here because it is useful for analysis purposes and can work well for fully digital implementations where no coefficient mismatches occur.

A second-order modulator based on the error-feedback structure is shown in Fig. 14.13. For this case we have

$$G(z) - 1 = z^{-1}(z^{-1} - 2) \tag{14.33}$$

implying that

$$\begin{aligned} G(z) &= 1 - 2z^{-1} + z^{-2} \\ &= (1 - z^{-1})^2 \end{aligned} \tag{14.34}$$

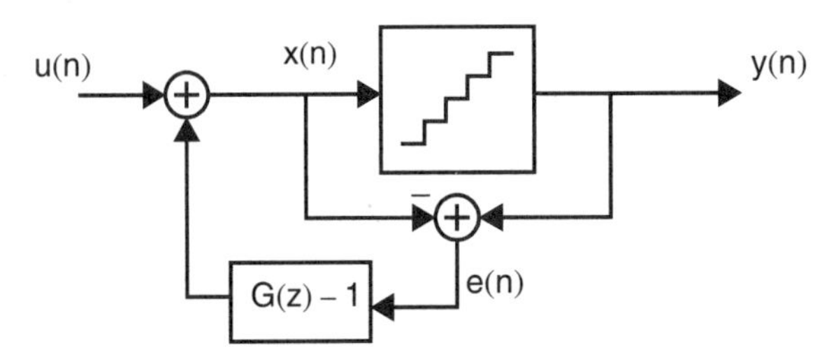

Fig. 14.12 The error-feedback structure of a general $\Delta\Sigma$ modulator.

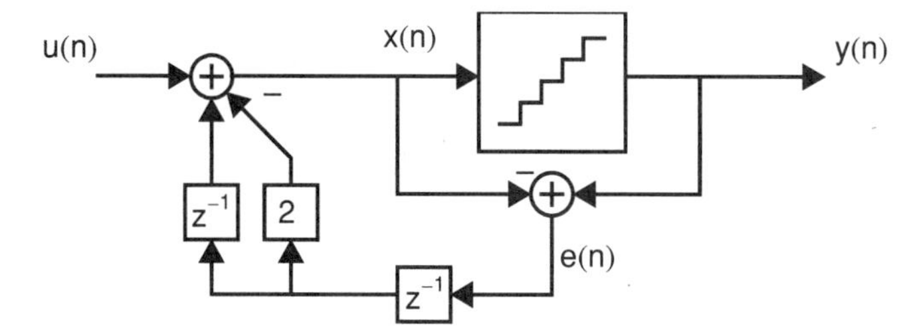

Fig. 14.13 The error-feedback structure of a second-order $\Delta\Sigma$ modulator.

which is identical to (14.28), and we see that second-order noise shaping is obtained. In practical realizations, one might have to take into account additional details such as preventing overflow and reducing complexity. For these more advanced topics, the reader is referred to [Temes, 1996].

14.3 SYSTEM ARCHITECTURES

In this section, we look at typical system architectures for oversampled A/D and D/A converters.

System Architecture of Delta-Sigma A/D Converters

The system architecture for a typical $\Delta\Sigma$ oversampling A/D converter is shown in Fig. 14.14, and some example signal spectra are shown in Fig. 14.15. In the case of digital audio, various sampling frequencies might be $f_s = 5.6448$ MHz, $2f_0 = 44.1$ kHz, which represent an oversampling ratio of 128. Here, the input signal, $x_c(t)$, is sampled and held,[5] resulting in the signal $x_{sh}(t)$. This sampled-and-held signal is then applied to an A/D $\Delta\Sigma$ modulator, which has as its output a 1-bit digital signal $x_{dsm}(n)$. This 1-bit digital signal is assumed to be linearly related to the input signal

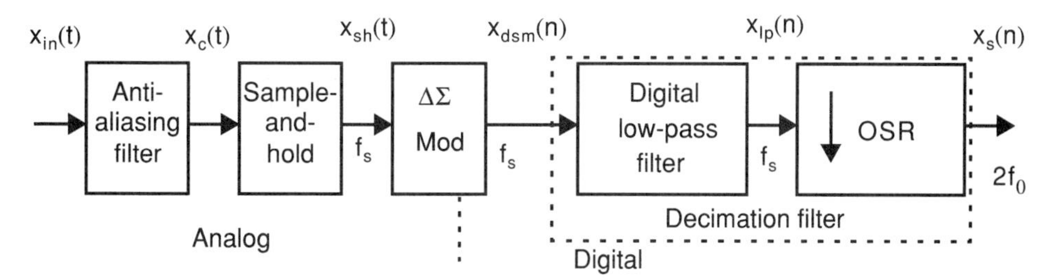

Fig. 14.14 Block diagram of an oversampling A/D converter.

5. This sample-and-hold block is often inherent in the switched-capacitor modulator. Thus, the signal $x_{sh}(t)$ may never physically exist in many realizations.

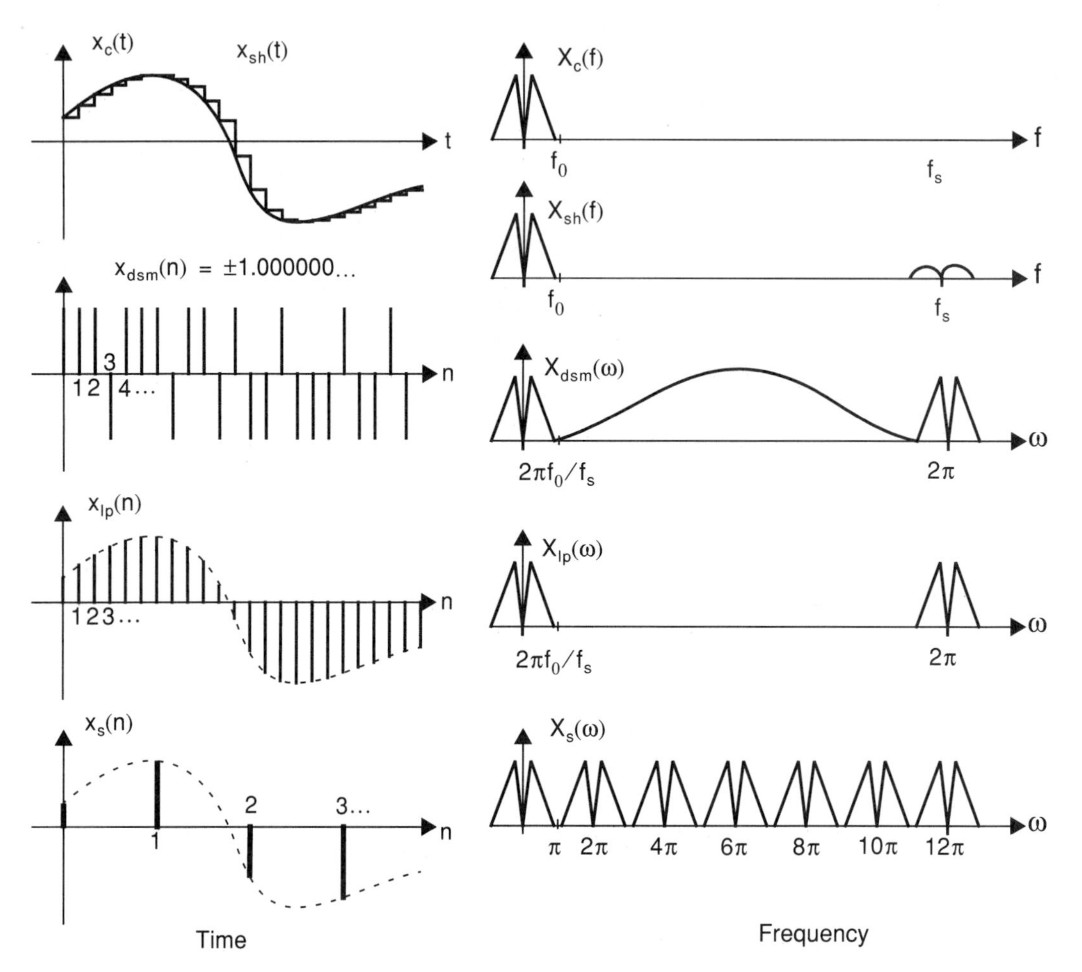

Fig. 14.15 Signals and spectra in an oversampling A/D converter.

$x_c(t)$ (accurate to many bits of resolution), although it includes a large amount of out-of-band quantization noise. To remove this out-of-band quantization noise, a digital decimation filter is used as shown. Conceptually, one can think of the decimation process as first reducing the quantization noise through the use of a digital low-pass filter, resulting in the multi-bit signal $x_{lp}(n)$. Note that this low-pass filter will also remove any higher-frequency signal content that was originally on the input signal, $x_c(t)$, and thus also acts as an anti-aliasing filter to limit signals to one-half the final output sampling rate, $2f_0$, as opposed to the anti-aliasing filter at the input, which needed to only limit signals to frequencies less than $f_s/2$. Next, $x_{lp}(n)$ is resampled at $2f_0$ to obtain $x_s(n)$ by simply keeping samples at a submultiple of the oversampling rate and throwing away the rest. In Fig. 14.15 an oversampling rate of only 6 is shown for clarity as opposed to the more typical values of 64 or 128 that are used in many commercial applications. This decimation process does not result in any loss of information, since the bandwidth of the original signal was assumed to be f_0. In other words, the signal

$x_{lp}(n)$ has redundant information since it is an oversampled signal where all of its spectral information lies well below π, and by throwing away samples, the spectral information is spread over 0 to π. Finally, it should be noted that there is no need to actually create the signal $x_{lp}(n)$, and much digital circuit complexity can be saved by combining the digital low-pass filter with the resampling block to directly produce the downsampled signal $x_s(n)$, as discussed in Section 14.4. The final signal, $x_s(n)$, would typically have 16-bit resolution in digital audio applications.

It is of interest to look at what element most strongly affects the linearity of this oversampling A/D system. Returning to the $\Delta\Sigma$ modulator, note that an internal 1-bit D/A converter is used whose output signal is combined with the input signal such that, over the frequency band of interest, these two signals are nearly the same. As a result, the overall linearity of this $\Delta\Sigma$ modulator converter depends strongly on the linearity of its internal D/A converter. For example, with a nonlinear internal D/A converter and a slow linear ramp input signal, the low-frequency content of the D/A converter's output would essentially equal that ramp. However, the low-frequency content of the digital input to the D/A converter would be a nonlinear ramp to account for the D/A converter's nonlinearity. Since the remaining digital circuitry is linear, the overall linearity of this oversampling A/D converter is most strongly dependent on realizing a linear D/A converter inside the $\Delta\Sigma$ modulator. In fact, nonlinearities in the internal A/D converter (if it was multi-bit) have only a small effect on the linearity of the overall converter, since the high gain in the feedback loop compensates for that nonlinearity.

System Architecture of Delta-Sigma D/A Converters

Most of the discussion so far has focused on $\Delta\Sigma$ A/D converters, but much of it also applies to $\Delta\Sigma$ D/A converters, with a few qualifications. A high-resolution oversampling D/A converter using a 1-bit converter can be realized as shown in the block diagram in Fig. 14.16. Some illustrative signals that might occur in this system are shown in Fig. 14.17. The digital input signal, $x_s(n)$, is a multi-bit signal and has an equivalent sample rate of $2f_0$, where f_0 is slightly greater than the highest input signal frequency. For example, in a compact disc audio application, a 16-bit input signal is used with a frequency band of interest from 0 to 20 kHz while the sample rate, $2f_0$,

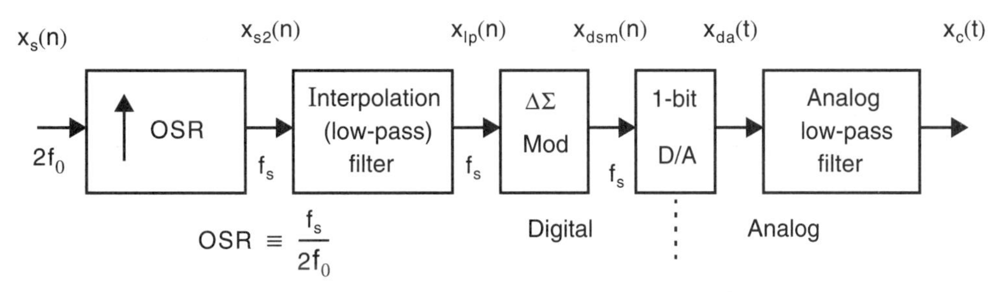

Fig. 14.16 Block diagram of a 1-bit oversampling D/A converter.

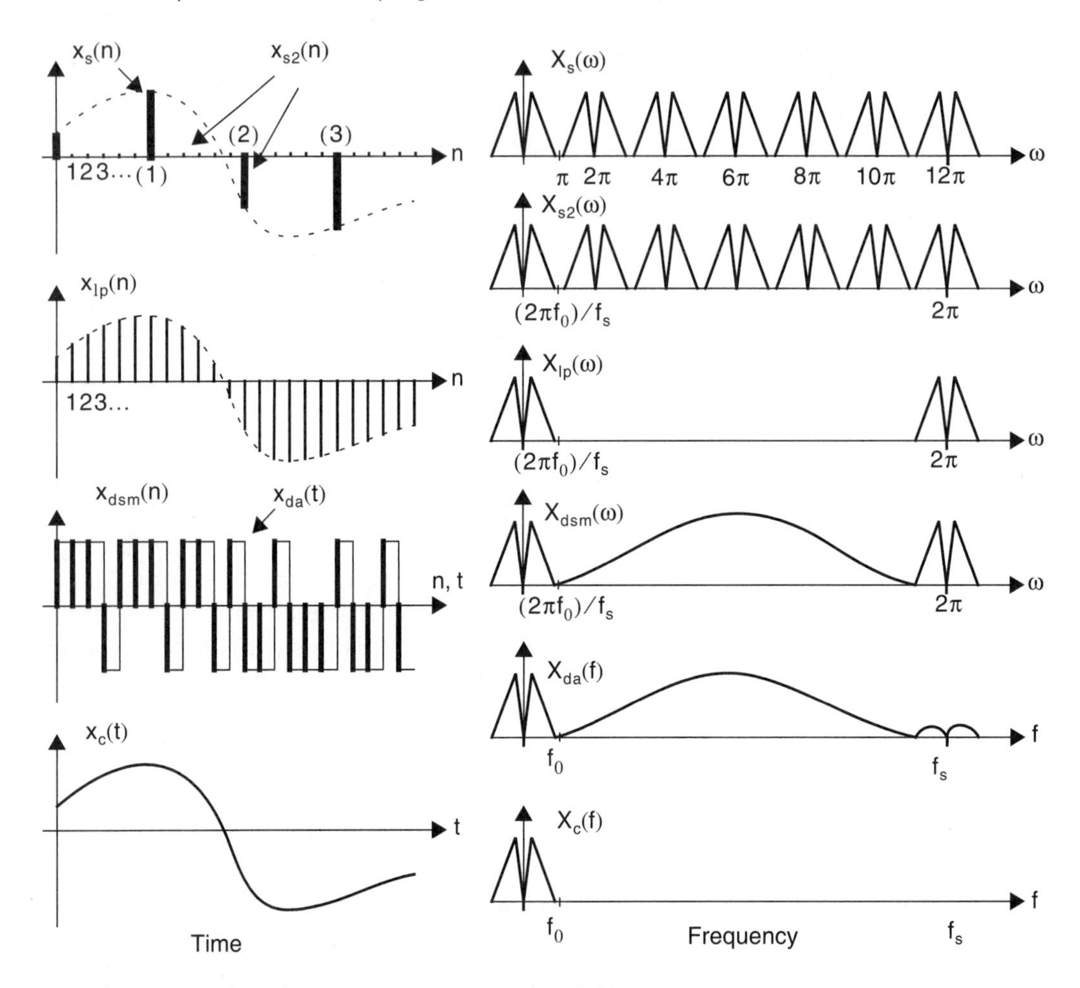

Fig. 14.17 Signals and spectra in an oversampling D/A converter.

is 44.1 kHz. Note, however, that $x_s(n)$ is actually just a series of numbers and thus its shown frequency spectrum has normalized the sample rate to 2π. Next, the input signal is upsampled to an equivalent higher sampling rate, f_s, resulting in the upsampled signal $x_{s2}(n)$. In the example shown, the oversampling rate is only six (i.e., $f_s = 6 \times 2f_0$), while in typical audio applications, f_s is often near a 5-MHz rate (i.e., oversampling rate of 128). However, $x_{s2}(n)$ has large images left in its signal so an interpolation filter is used to create the multi-bit digital signal, $x_{lp}(n)$, by digitally filtering out the images. This interpolation filter is effectively a digital brick-wall-type filter that passes 0 to $(2\pi f_0)/f_s$ and rejects all other signals. The resulting signal, $x_{lp}(n)$, is then applied to a fully digital $\Delta\Sigma$ modulator that produces a 1-bit output signal, $x_{dsm}(n)$, which has a large amount of shaped quantization noise. As discussed earlier, the main reason for going to a 1-bit digital signal is so that a 1-bit D/A converter can now be used to create $x_{da}(t)$, which has excellent linearity properties but still a large amount of out-of-band quantization noise. Finally, the desired output signal, $x_c(t)$, can be obtained by using an analog filter to filter out this out-of-band quantization noise. The analog filter may be a combination of switched-capacitor and continuous-time filtering.

Some points to note here are that while oversampling allows the use of a 1-bit D/A converter, which can have excellent linearity, the use of oversampling also relaxes some of the analog-smoothing filter specifications. For example, if a 16-bit converter was used at the Nyquist rate, $2f_0$, the analog-smoothing filter would have to remove the images in the signal instead of a precise digital interpolation filter removing these images. This specification can be very demanding, as in a digital-audio application, where a near 96-dB attenuation is required at 24 kHz, while up to 20 kHz should pass with unity gain. With the use of an oversampling approach, the digital interpolation filter is faced with this strict specification rather than an analog smoothing filter. In fact, oversampling is often used in audio applications with multi-bit D/A converters just to reduce this analog-smoothing filter's complexity.

Another point to note is that the order of the analog low-pass filter should be at least one order higher than that of the $\Delta\Sigma$ modulator. The reason for this choice is that if the analog filter's order is equal to that of the modulator, the slope of the rising quantization noise will match the filter's falling attenuation, and thus the resulting quantization noise will be approximately a constant spectral density up to one-half the sampling rate (i.e., $f_s/2$). By having a higher-order analog filter, the spectral density of the output signal's quantization noise will have a bandwidth similar to the analog filter's bandwidth, which is around f_0.

Finally, note that the analog filter must be able to strongly attenuate high-frequency noise, as much of the quantization noise is centered around $f_s/2$, and this analog filter should be linear so it does not modulate the noise back to the frequency band of interest. In many applications, the realization of these filters, especially if they are integrated, is nontrivial.

14.4 DIGITAL DECIMATION FILTERS

There are many techniques for realizing digital decimation filters for oversampling A/D converters. In this section, we discuss two popular approaches—multi-stage and single-stage.

Multi-Stage

One method for realizing decimation filters is to use a multi-stage approach, as shown in Fig. 14.18. Here, the first-stage FIR filter, $T_{sinc}(z)$, removes much of the quantization noise such that its output can be downsampled to four times the Nyquist rate (i.e., $8f_0$). This lower-rate output is applied to the second-stage filter, which may be either an IIR filter, as shown in Fig. 14.18(*a*), or a cascade of FIR filters, as shown in Fig. 14.18(*b*).

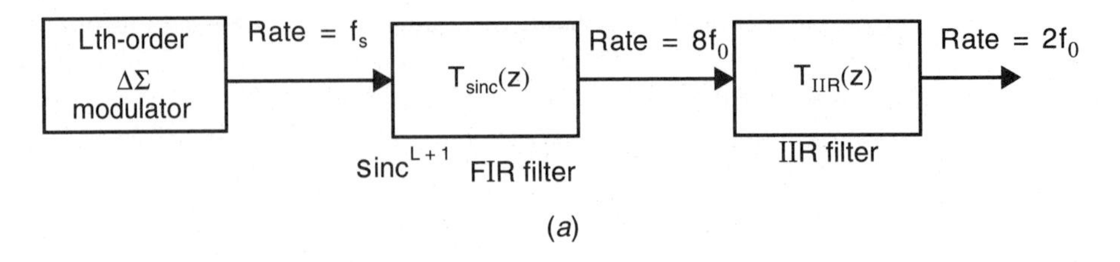

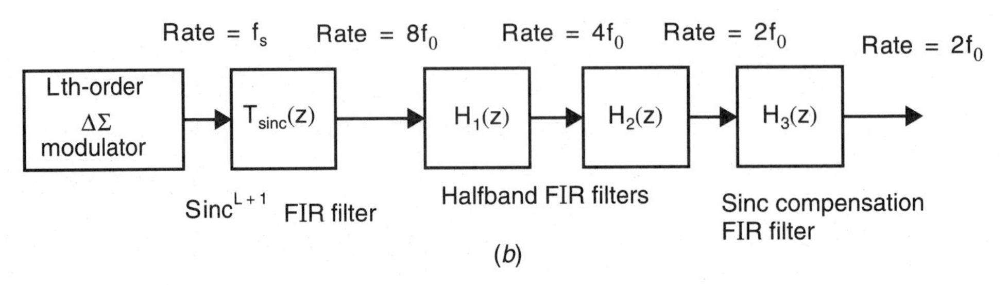

Fig. 14.18 Multi-stage decimation filters: (*a*) sinc followed by an IIR filter; (*b*) sinc followed by halfband filters.

The sinc^{L+1} FIR filter is a cascade of $L+1$ averaging filters where the transfer function of a single averaging filter, $T_{avg}(z)$, is given by

$$T_{avg}(z) = \frac{Y(z)}{U(z)} = \frac{1}{M}\sum_{i=0}^{M-1} z^{-i} \tag{14.35}$$

and M is the integer ratio of f_s to $8f_0$ (i.e., $8f_0 = f_s/M$). Note that the impulse response of this filter is finite, implying it is an FIR-type filter. In addition, all of its impulse response coefficients are symmetric (in fact, they are all equal), and thus it is also a linear-phase filter.[6] Finally, note that the $1/M$ multiplication term is easily realized by changing the location of the fractional bits when M is chosen to be a power of two.

As an illustration of a series of averaging filters reducing quantization noise, consider an average-of-four filter (i.e., $M = 4$) operating on the output of the 1-bit signal in Example 14.4. Since the output 1-bit sequence is $\{1, 1, -1, 1, 1, -1, 1, 1, -1, \ldots\}$, the first averaged output, $x_{lp1}(n)$, would be given by

$$x_{lp1}(n) = \{0.5, 0.5, 0.0, 0.5, 0.5, 0.0, \ldots\}$$

To obtain more attenuation of the quantization noise, the signal $x_{lp1}(n)$ can also be applied to a running-average-of-four filter, giving $x_{lp2}(n)$ as

$$x_{lp2}(n) = \{0.375, 0.375, 0.25, 0.375, 0.375, 0.25, \ldots\}$$

and repeating this process to get $x_{lp3}(n)$ would give

$$x_{lp3}(n) = \{0.344, 0.344, 0.313, 0.344, 0.344, 0.313, \ldots\}$$

6. A linear-phase filter results in all frequency components being delayed by the same amount and is therefore desirable in applications where phase distortion is unwanted (such as hi-fi audio).

Note the convergence of these sequences to a series of all samples equalling $1/3$ as expected.

To show that the frequency response of an averaging filter, $T_{avg}(z)$, has a sinc-type behavior, it is useful to rewrite (14.35) as

$$MY(z) = \left(\sum_{i=0}^{M-1} z^{-i}\right)U(z) = (1 + z^{-1} + z^{-2} + \cdots + z^{-(M-1)})U(z) \tag{14.36}$$

which can also be rewritten as

$$\begin{aligned} MY(z) &= (z^{-1} + z^{-2} + \cdots + z^{-M})U(z) + (1 - z^{-M})U(z) \\ &= Mz^{-1}Y(z) + (1 - z^{-M})U(z) \end{aligned} \tag{14.37}$$

Finally, we group together $Y(z)$ terms and find the transfer function of this averaging filter can also be written in the recursive form as

$$T_{avg}(z) = \frac{Y(z)}{U(z)} = \frac{1}{M}\left(\frac{1 - z^{-M}}{1 - z^{-1}}\right) \tag{14.38}$$

The frequency response for this filter is found by substituting $z = e^{j\omega}$, which results in

$$T_{avg}(e^{j\omega}) = \frac{\mathrm{sinc}\left(\frac{\omega M}{2}\right)}{\mathrm{sinc}\left(\frac{\omega}{2}\right)} \tag{14.39}$$

where $\mathrm{sinc}(x) \equiv \sin(x)/x$.

A cascade of $L + 1$ averaging filters has the response $T_{sinc}(z)$ given by

$$T_{sinc}(z) = \frac{1}{M^{L+1}}\left(\frac{1 - z^{-M}}{1 - z^{-1}}\right)^{L+1} \tag{14.40}$$

The reason for choosing to use $L + 1$ of these averaging filters in cascade is similar to the argument that the order of the analog low-pass filter in an oversampling D/A converter should be higher than the order of the $\Delta\Sigma$ modulator. Specifically, the slope of the attenuation for this low-pass filter should be greater than the rising quantization noise, so that the resulting noise falls off at a relatively low frequency. Otherwise, the noise would be integrated over a very large bandwidth, usually causing excessive total noise.

An efficient way to realize this cascade-of-averaging filters is to write (14.40) as

$$T_{sinc}(z) = \left(\frac{1}{1 - z^{-1}}\right)^{L+1}(1 - z^{M})^{L+1}\frac{1}{M^{L+1}} \tag{14.41}$$

and thus realize it as shown in Fig. 14.19 [Candy, 1992]. A point to note here is that at first glance it appears as though this circuit will not operate properly due to a dc input

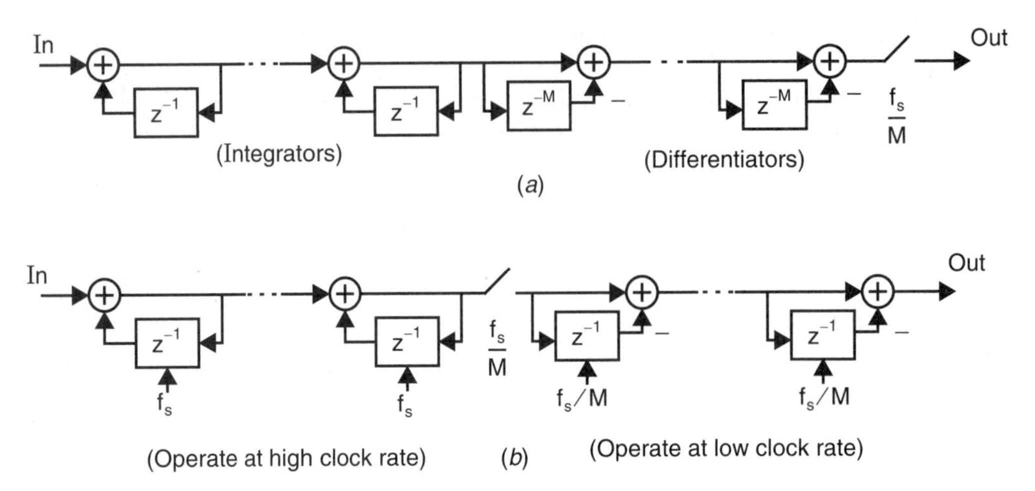

Fig. 14.19 Realizing $T_{sinc}(z)$ as a cascade of integrators and differentiators: (*a*) downsampling performed after all the filtering; (*b*) a more efficient method where downsampling is done before the differentiators.

causing saturation of the discrete-time integrators. Fortunately, such a result does not occur when 2's-complement arithmetic is used due to its wrap-around characteristic. Specifically, although the dc levels at the integrator outputs may be incorrect, the differentiators compare the present sample with that of M samples past while rejecting the dc component. Thus, as long as this difference is properly maintained (as it would be using 2's-complement arithmetic) the correct calculation is performed.

Referring once again to Fig. 14.18, the purpose of the filters following the $T_{sinc}(z)$ filter is twofold. One reason for their use is to remove any higher-frequency input signals (in effect, a sharp anti-aliasing filter) before the signal is downsampled to the final Nyquist rate (i.e., $2f_0$). In other words, while the $T_{sinc}(z)$ filter is good at filtering out the quantization noise, it is not sharp enough to act as a reasonable anti-aliasing filter for input signals slightly higher than f_0. A second reason is to compensate for the frequency drop in the passband caused by the $T_{sinc}(z)$ filter. This anti-aliasing and sinc-compensation filter might be realized using a single IIR filter, as shown in Fig. 14.18(*a*). Alternatively, a few halfband FIR filters might be used together with a separate sinc-compensation FIR filter, as shown in Fig. 14.18(*b*). A halfband FIR filter has a passband from 0 to $\pi/2$, while its stopband is from $\pi/2$ to π with every second coefficient being zero [Vaidyanathan, 1993]. Thus, with a high enough filter order, its output can be downsampled by a factor of two. It has also been shown that in some applications, these halfband and sinc-compensation filters can be realized using no general multi-bit multipliers [Saramaki, 1990].

Single Stage

Another approach for realizing decimation filters is to use a relatively high-order FIR filter. For example, in [Dattorro, 1989], a 2048 tap FIR filter was used to decimate

1-bit stereo outputs from two $\Delta\Sigma$ modulators, each having an oversampling ratio of 64. While this FIR order seems high, note that no multi-bit multiplications are needed, since the input signal is simply 1-bit, so all multiplications are trivial. In addition, the output need only be calculated at the Nyquist rate (intermediate samples would be discarded anyway) such that 2048 additions are required during one clock cycle at the Nyquist rate. However, if only one accumulator is used to perform all 2048 additions, the clock rate of that accumulator would be 2048 times the Nyquist rate. For example, if the Nyquist rate is 48 kHz, the single accumulator would have to be clocked at 98.3 MHz. To overcome this high clock rate, 32 separate FIR filters are realized (with shared coefficients) in a time-interleaved fashion, with each FIR having 2048 coefficients and each producing an output at a clock rate of 1.5 kHz. In this way, each of the 32 FIR filters uses a single accumulator operating at 3 MHz (i.e., 2048 times 1.5 kHz). The coefficient ROM values were shared between the FIR filters (as well as in the two stereo channels), and the ROM size can also be halved if coefficients are duplicated, as in the case of linear-phase FIR filtering.

Finally, it is also possible to reduce the number of additions by grouping together input bits. For example, if 4 input bits are grouped together, then a 16-word ROM lookup table can be used rather than using 3 additions. With such a grouping of input bits, each of the 32 FIR filters would require only 512 additions.

14.5 HIGHER-ORDER MODULATORS

In general, it can be shown that an Lth-order noise-shaping modulator improves the SNR by $6L + 3$ dB/octave, or equivalently, $L + 0.5$ bits/octave. In this section, we look at two approaches for realizing higher-order modulators—interpolative and MASH. The first approach is typically a single high-order structure with feedback from the quantized signal. The second approach consists of a cascade of lower-order modulators, where the latter modulators are used to cancel the noise errors introduced by the earlier modulators.

Interpolative Architecture

As discussed earlier, when compared to the error-feedback structure, the interpolative structure of Fig. 14.6(*a*) is much better suited to analog implementations of modulators due to its reduced sensitivity. One of the first approaches for realizing higher-order interpolative modulators was presented in [Chao, 1990]. It used a filtering structure very similar to a direct-form filter structure; however, a direct-form-type structure made it sensitive to component variations, which can cause the zeros of the noise transfer function to move off the unit circle. To improve component sensitivity, resonators can be used together with a modified interpolative structure, as shown in Fig. 14.20 [Ferguson, 1991]. Note here that a single 1-bit D/A signal is still used for feedback, and therefore its linearity will be excellent. The resonators in this structure are due to the feedback signals associated with f_1 and f_2, resulting in the placement of zeros in the noise transfer function spread over the frequency-of-interest band.

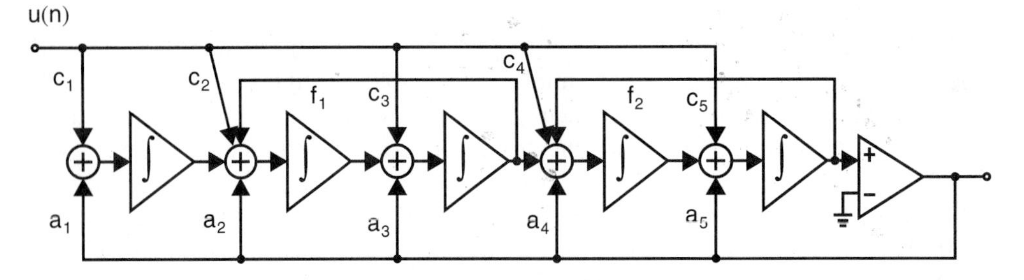

Fig. 14.20 A block diagram of a fifth-order modulator.

Such an arrangement gives better dynamic range performance than placing all the zeros at dc as we have been assuming thus far.

Unfortunately, it is possible for modulators of order two or more to go unstable, especially when large input signals are present. When they go unstable, they may never return to stability even when the large input signals go away. Guaranteeing stability for an interpolative modulator is nontrivial and is discussed further in Section 14.7.

Multi-Stage Noise Shaping (MASH) Architecture

Another approach for realizing modulators is to use a cascade-type structure where the overall higher-order modulator is constructed using lower-order ones. The advantage of this approach is that since the lower-order modulators are more stable, the overall system should remain stable. Such an arrangement has been called MASH (Multi-stAge noise SHaping) [Matsuya, 1987].

The arrangement for realizing a second-order modulator using two first-order modulators is shown in Fig. 14.21. The basic approach here is to pass along the first section's quantization error, $e_1(n)$, to another modulator and combine the outputs of both modulators in such a way that the first section's quantization noise is removed completely. The output is then left with only the second section's quantization noise, which has been filtered twice—once by the second modulator and once by the post digital circuitry. Assuming linear models for the quantizers, straightforward linear analysis gives

$$Y(z) = z^{-2}U(z) - (1 - z^{-1})^2 E_2(z) \tag{14.42}$$

Thus, a MASH approach has the advantage that higher-order noise filtering can be achieved using lower-order modulators. The lower-order modulators are much less susceptible to instability as compared to an interpolator structure having a high order with a single feedback.

Unfortunately, MASH or cascade approaches consisting of a cascade of first-order stages are sensitive to finite opamp gain and mismatches between the analog and digital circuitry. Such mismatches in the above example cause first-order noise to leak through from the first modulator and hence reduce dynamic range performance.

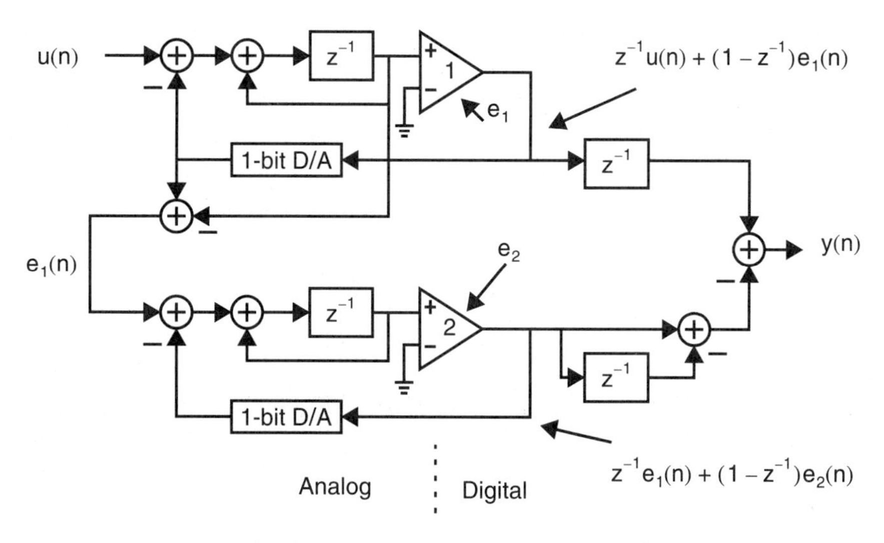

Fig. 14.21 A second-order MASH modulator using two first-order modulators. Note that the output, $y(n)$, is a four-level signal.

To alleviate this mismatch problem, often the first stage is chosen to be a higher-order modulator such that any leak-through of its noise does not have as serious an effect as it would if this first modulator was first-order. For example, a third-order modulator would be realized using a second-order modulator for the first stage and a first-order modulator for the second stage. Also, it is very important to minimize errors due to input-offset voltages that might occur because of clock feedthrough or opamp input-offset voltages. Typically, in practical realizations additional circuit design techniques will be employed to minimize these effects.

Finally, note that the use of this MASH approach results in the digital output signal, $y(n)$, being a four-level signal (rather than two-level) due to the combination of the original two-level signals. Such a four-level signal would require a linear four-level D/A converter in a D/A application. For A/D applications, it makes the FIR decimation filter slightly more complex.

14.6 BANDPASS OVERSAMPLING CONVERTERS

As we have seen, oversampling converters have the advantage of high linearity through the use of 1-bit conversion and reduced anti-aliasing and smoothing-filter requirements. However, at first glance some signals do not appear to easily lend themselves to oversampling approaches such as modulated radio signals. Such signals have information in only a small amount of bandwidth, say 10 kHz, but have been modulated by some higher-frequency carrier signal, say 1 MHz. For such applications, one can make use of bandpass oversampling converters.

In low-pass oversampling converters, the transfer function $H(z)$ in Fig. 14.6 is chosen such that it has a high gain near dc, and thus quantization noise is small around dc. In a similar way, bandpass oversampling converters are realized by choosing $H(z)$ such that it has a high gain near some frequency value, f_c, [Schreier, 1989]. With such an approach, the quantization noise is small around f_c, and thus most of the quantization noise can be removed through the use of a narrow bandpass filter of total width f_Δ following the modulator. Thus, in the case of a bandpass A/D converter intended for digital radio, post narrowband digital filtering would remove quantization noise as well as adjacent radio channels. In addition, further digital circuitry would also perform the necessary demodulation of the signal.

An important point here is that the oversampling ratio for a bandpass converter is equal to the ratio of the sampling rate, f_s, to two times the width of the narrowband filter, $2f_\Delta$, and it does not depend on the value of f_c. For example, consider a bandpass oversampling converter having a sampling rate $f_s = 4$ MHz, which is intended to convert a 10 kHz signal bandwidth centered around 1 MHz (or $f_c = f_s/4$). In this case, $f_\Delta = 10$ kHz, resulting in the oversampling ratio being equal to

$$\text{OSR} = \frac{f_s}{2f_\Delta} = 200 \tag{14.43}$$

To obtain a high gain in $H(z)$ near f_c, a similar approach to the low-pass case is taken. In a first-order low-pass oversampling converter, $H(z)$ is chosen such that it has a pole at dc (i.e., $z = 1$) and such that the noise transfer function, $N_{TF}(z)$, has a zero at dc. In a second-order bandpass oversampling converter with $f_c = f_s/4$, $H(z)$ is chosen such that it has poles at $e^{\pm j\pi/2} = \pm j$. In other words, $H(z)$ is a resonator that has an infinite gain at the frequency $f_s/4$. The zeros of the noise transfer function for this type of second-order oversampling converter are shown in Fig. 14.22, together with a similar low-pass case.

A block diagram for the example bandpass modulator is shown in Fig. 14.23, where one can derive $H(z)$ to be given by

$$H(z) = \frac{z}{z^2 + 1} \tag{14.44}$$

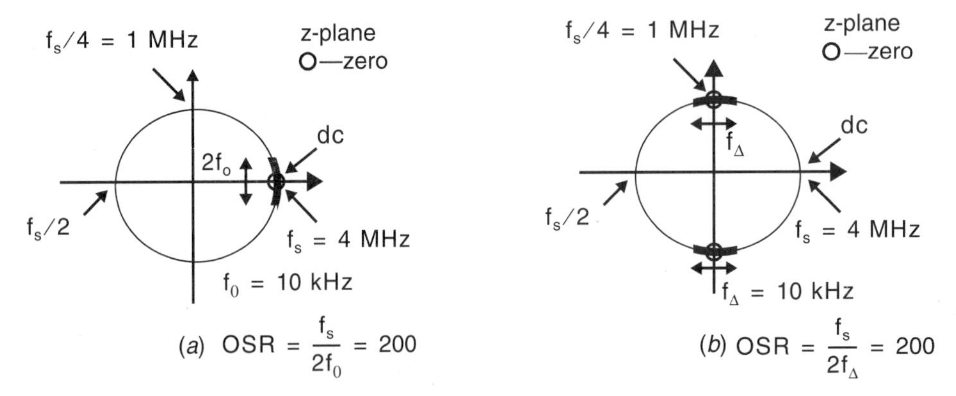

Fig. 14.22 Noise-transfer-function zeros for oversampling converters: (*a*) first-order low-pass; (*b*) second-order bandpass.

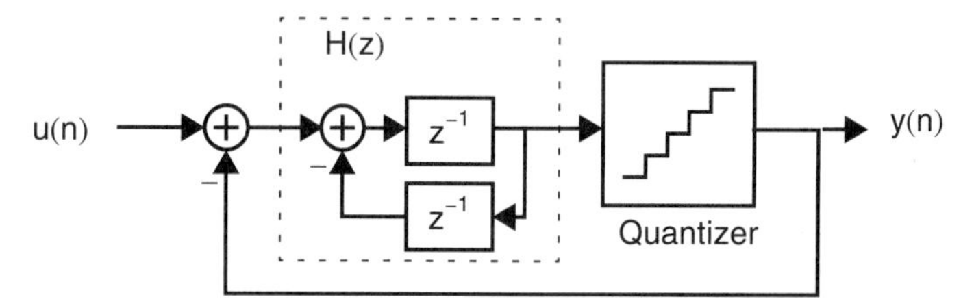

Fig. 14.23 A second-order bandpass oversampling modulator that shapes the quantization noise away from $f_s/4$.

Note that this second-order example has poles at $\pm j$, and therefore the noise transfer function, $N_{TF}(z)$, has only one zero at j and another zero at $-j$. For this reason, the dynamic range increase of a second-order bandpass converter equals that of a first-order low-pass converter that also has only one zero at dc. Specifically, the dynamic range increase of a second-order bandpass converter is 1.5 bits/octave or, equivalently, 9 dB/octave. To obtain the equivalent dynamic range increase of a second-order low-pass converter (i.e., 15 dB/octave), a fourth-order bandpass converter would have to be used. The first design of an integrated bandpass oversampling A/D converter was presented in [Jantzi, 1993].

14.7 PRACTICAL CONSIDERATIONS

Stability

As with any feedback-type system, delta-sigma modulators have the potential to go unstable. A stable modulator is defined here as one in which the input (or inputs) to the quantizer (or quantizers) remains bounded such that the quantization does not become *overloaded.* An overloaded quantizer is one in which the input signal goes beyond the quantizer's normal range, resulting in the quantization error becoming greater than $\pm\Delta/2$.

Unfortunately, the stability of higher-order 1-bit modulators is not well understood as they include a highly nonlinear element (a 1-bit quantizer), resulting in the system being stable for one input but unstable for another. As a general rule of thumb, keeping the peak frequency response gain of the noise-transfer function, $N_{TF}(z)$, less than 1.5 often results in a stable modulator [Chao, 1990].[7] In mathematical terms,

$$\left|N_{TF}(e^{j\omega})\right| \le 1.5 \qquad \text{for } 0 \le \omega \le \pi \tag{14.45}$$

should be satisfied for a 1-bit quantizer. It should be noted that this stability criterion has little rigorous justification and that there does not presently exist any necessary and sufficient stability criterion for $\Delta\Sigma$ modulators having arbitrary inputs. There is,

7. A figure of 2 was used in the reference but 1.5 appears to be a more practical choice.

however, a rigorous 1-norm stability test (which is sufficient but not necessary), but it is often too conservative, as it eliminates many "stable" modulators [Anastassiou, 1989]. For a summary of stability tests for $\Delta\Sigma$ modulators, the reader is referred to [Schreier, 1993].

It should be mentioned here that the stability of a modulator is also related to the maximum input signal level with respect to the 1-bit feedback as well as the poles of the noise transfer function. Specifically, a modulator can be made more stable by placing the poles of the system closer to the noise-transfer-function zeros. Also, a more stable modulator allows a larger maximum input signal level. However, this stability comes at a dynamic-range penalty since less of the noise power will then occur at out-of-band frequencies, but instead the noise will be greater over the in-band frequencies.

Since the stability issues of higher-order modulators are not well understood (particularly for arbitrary inputs), additional circuitry for detecting instability is often included, such as looking for long strings of 1s or 0s occurring. Alternatively, the signal amplitude at the input of the comparator might be monitored. If predetermined amplitude thresholds are exceeded for a specified number of consecutive clock cycles, then it is assumed that the converter has gone (or is going) unstable. In this case, the circuit is changed to make it more stable. One possibility is to turn on switches across the integrating capacitors of all integrators for a short time. This resets all integrator outputs to zero. A second alternative is to change the modulator into a second- or even first-order modulator temporarily by using only the first one or two integrators and resetting other integrators. Another possibility is to eliminate the comparator temporarily and feed back the input signal to the comparator. This changes the modulator into a stable filter.

Finally, it should be mentioned that modulators having multi-bit DACs appear to exhibit improved stability over their 1-bit counterparts but often require high-linearity DACs. There have been a number of architectures proposed to lessen the linearity requirements on the DACs, as described in Section 14.8.

Linearity of Two-Level Converters

It was stated earlier that 1-bit D/A converters are inherently linear since they need only produce two output values and two points define a straight line. However, as usual, practical issues limit the linearity of D/A converters having only two output levels. It should be pointed out here that most of the linearity issues discussed below are also applicable to multi-level converters, but we restrict our attention to two-level converters for clarity.

One limitation is when the two output levels somehow become functions of the low-frequency signal they are being used to create. For example, if the two voltage levels are related to power supply voltages, then these power-supply voltages might slightly rise and fall as the circuit draws more or less power. Since typically the amount of power drawn is related to the low-frequency signal being created, distortion will occur. A mechanism whereby this occurs is if a different amount of charge is being taken from the voltage references when a 1 is being output from the 1-bit D/A converter, as opposed to when a 0 is being output from the D/A converter. A somewhat more subtle but similar mechanism can occur due to the clock feedthrough of the input switches. This feedthrough is dependent on the gate voltages driving the input switches and therefore on the power-supply voltages connected to the drivers of the input switches. It is possible that these voltages can change when more 1s than 0s are being output by the D/A converter; having well-regulated power-supply voltages on the drivers for the input switches is very important. A similar argument can be made about any clock jitter in the converter. Thus, for high linearity it is important that the two output levels and associated clock jitter not be a function of the low-frequency input signal.

A more severe linearity limitation occurs if there is memory between output levels. For example, consider the ideal and typical output signals for a two-level D/A converter, shown in Fig. 14.24. The area for each symbol is defined to be the integral of the waveform over that symbol's time period. Notice that for the typical output signal, the area is dependent on the past symbol and that difference is depicted using δ_i. As an illustrative example, consider the average voltage for three periodic patterns corresponding to average voltages of 0, $1/3$, and $-1/3$ when V_1 and V_2 are ± 1 volt.

In the ideal case, both δ_1 and δ_2 are zero as the waveform is memoryless. Thus, the three periodic patterns 0, $1/3$, and $-1/3$ result in the following averages:

$$0:\quad \{1, -1, 1, -1, 1, -1, 1, -1, \ldots\} \rightarrow \overline{V_a(t)} = \frac{A_1 + A_0}{2}$$

$$1/3:\quad \{1, 1, -1, 1, 1, -1, 1, 1, -1, \ldots\} \rightarrow \overline{V_b(t)} = \frac{2A_1 + A_0}{3}$$

$$-1/3:\quad \{-1, -1, 1, -1, -1, 1, -1, -1, 1, \ldots\} \rightarrow \overline{V_c(t)} = \frac{A_1 + 2A}{3}$$

By calculating the differences, $\overline{V_b(t)} - \overline{V_a(t)}$ and $\overline{V_a(t)} - \overline{V_c(t)}$, and noting that they are identical, we conclude that $\overline{V_a(t)}$, $\overline{V_b(t)}$, and $\overline{V_c(t)}$ lie along a straight line and therefore this memoryless D/A converter is perfectly linear.

In the typical waveform case, the three periodic patterns result in the following averages:

$$0:\quad \{1, -1, 1, -1, \ldots\} \rightarrow \overline{V_d(t)} = \frac{A_1 + A_0 + \delta_1 + \delta_2}{2} = \overline{V_a(t)} + \frac{\delta_1 + \delta_2}{2}$$

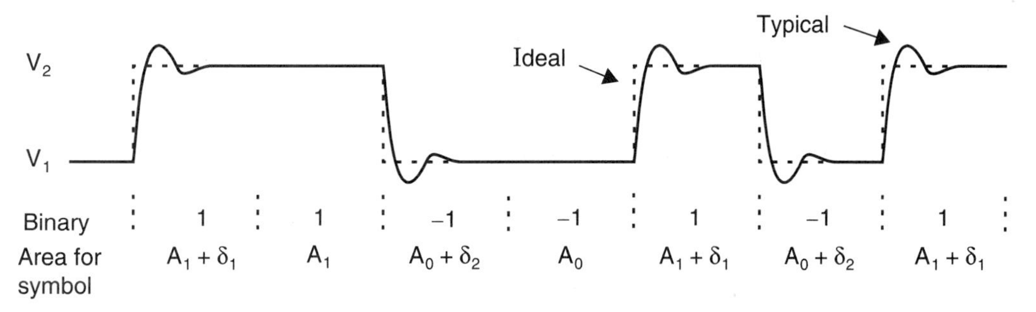

Fig. 14.24 A nonreturn-to-zero (NRZ) 1-bit D/A converter typical output.

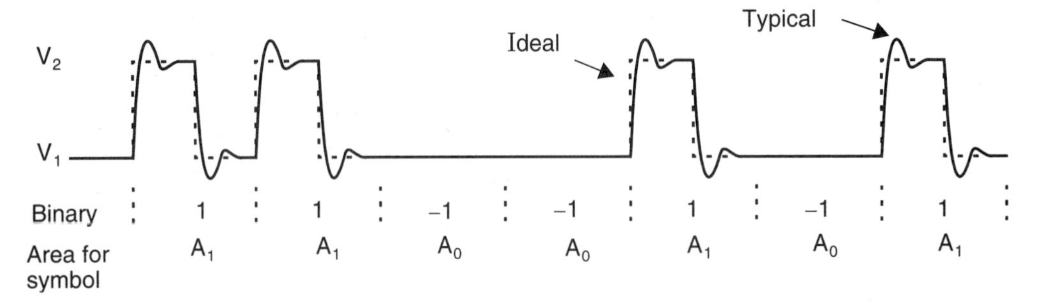

Fig. 14.25 A return-to-zero (RTZ) coding scheme, which is one example of a memoryless coding scheme.

$$1/3: \quad \{1, 1, -1, \dots\} \rightarrow \overline{V_e(t)} = \frac{2A_1 + A_0 + \delta_1 + \delta_2}{3} = \overline{V_b(t)} + \frac{\delta_1 + \delta_2}{3}$$

$$-1/3: \quad \{-1, -1, 1, \dots\} \rightarrow \overline{V_f(t)} = \frac{A_1 + 2A_0 + \delta_1 + \delta_2}{3} = \overline{V_c(t)} + \frac{\delta_1 + \delta_2}{3}$$

Now, these three averages do not lie on a straight line except when $\delta_1 = -\delta_2$. Thus, one way to obtain high linearity is to match falling and rising signals—a very difficult task since they are typically realized with different types of devices.

A better way to obtain linearity without matching rising and falling signals of the D/A converter is to use some sort of *memoryless coding scheme*. For example, consider the return-to-zero coding scheme shown in Fig. 14.25. It is not difficult to see here that since the area of one output bit does not depend on any previous bits, the coding scheme is memoryless and the output signal will remain linear. Naturally, other memoryless schemes can also be used, such as using two levels of opposite signs but ensuring that the signal settles to ground between samples.

It is of interest to see how this memoryless coding is applied in oversampling A/D converters. Presently, most oversampling A/D converters are realized using a switched-capacitor technology for linearity reasons. Switched capacitor circuits naturally realize memoryless levels as long as enough time is left for settling on each clock phase. In other words, switched-capacitor circuits implement memoryless coding since capacitors are charged on one clock phase and discharged on the next. For the same reason, the first stage of filtering in oversampling D/A converters is often realized using switched-capacitor filters, which is followed by continuous-time filtering. However, some type of memoryless coding should be used when high linearity is desired from a 1-bit D/A output either fed directly to a continuous-time filter in an oversampling D/A or used in the internal operation of a continuous-time oversampling A/D converter.

Idle Tones

Consider the case of applying a dc level of 1/3 to a first-order $\Delta\Sigma$ modulator having a 1-bit quantizer with output levels of ± 1 (as in Example 14.4). With such an input, the output of the modulator will be a periodic sequence of two 1s and one –1. In other words,

$$y(n) = \{1, 1, -1, 1, 1, -1, 1, 1, \dots\} \quad (14.46)$$

This output periodic pattern is three cycles long and its power is concentrated at dc and $f_s/3$. Although the output pattern is periodic, the post low-pass filter will eliminate the high-frequency content such that only the dc level of 1/3 remains.

Now consider applying a dc level of $(1/3 + 1/24) = 3/8$ to the same modulator. For this case, the output sequence becomes

$$y(n) = \{1, 1, -1, 1, 1, -1, 1, 1, -1, 1, 1, -1, 1, 1, 1, -1, 1, 1, -1, \dots\} \quad (14.47)$$

The period of this output pattern is now 16 cycles long and has some power at $f_s/16$. With an oversampling ratio of eight (i.e., $f_0 = f_s/16$), the post low-pass filter will not attenuate the signal power at $f_s/16$ since that frequency is just within the frequency band of interest. In other words, a dc level of $3/8$ into this modulator will produce the correct dc output signal but have a superimposed $f_s/16$ signal on it even when using a brick-wall type low-pass filter.

While the above example shows a tone occurring at $f_s/16$, it is not difficult to find other cases that would produce frequency tones at much lower frequencies, say $f_s/256$. Such low-frequency tones *will not be filtered out by the next stage low-pass filter* and can lead to annoying tones in the audible range.[8] In addition, although a first-order modulator was used in the above example, it has been shown that annoying audible tones exist even in higher-order modulators [Norsworthy, 1993]. Finally, it should be mentioned that these tones might not lie at a single frequency but instead be short-term periodic patterns. In other words, a tone appearing near 1 kHz might actually be a signal varying between 900 Hz and 1.1 kHz in a random-like fashion.

Dithering

One way to reduce the amount of idle tones in modulators is through the use of dithering. The term dithering here refers to the act of introducing some random (or pseudo-random) signal into a modulator.

Assuming the random signal to be introduced has a white-noise type spectrum, then the most suitable place to add the dithering signal is just before the quantizer, as shown in Fig. 14.26. The reason for adding it at this location is so that the dithering signal becomes noise shaped in the same manner as the quantization noise, and thus a large amount of dithering can be added. Typically, the dithering signal is realized using some sort of pseudo-random number generator with only a few bits of resolution, but its total noise power is comparable to the quantization noise power. Thus, the use of dithering in a D/A converter would require a small additional digital adder, while a small D/A converter would be needed in an oversampling A/D converter.

It should be noted that the use of dither to reduce idle tones is not an attempt to add noise to mask out the tones but instead breaks up the tones so that they never

---

8. Human hearing can detect tones buried beneath white noise.

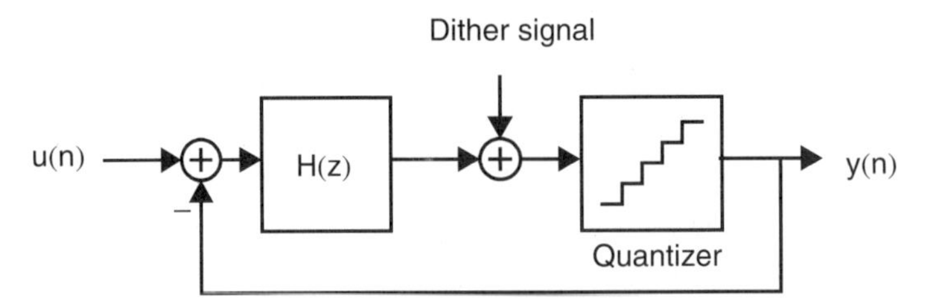

Fig. 14.26 Adding dithering to a delta-sigma modulator. Note that the dithered signal is also noise shaped.

occur. However, since the noise power of the dithering signal is similar to the quantization noise power, the use of dithering adds about 3 dB extra in-band noise and often requires rechecking the modulator's stability.

Opamp Gain

In Chapter 10, we saw that a switched-capacitor integrator with a finite opamp gain, A, results in the pole of the integrator moving to the left of $z = 1$ by an amount $1/A$ (see Problem 10.4 and assume $C_2 \approx C_1$ as is typical in oversampled converters). With this knowledge, we can determine the new NTF zeros due to finite opamp gain by substituting $z - 1$ with $z - (1 - 1/A)$. In other words, we substitute all ideal integrators with damped integrators and recalculate the transfer functions. Such an approach results in the NTF zeros, which were located at $z = 1$ for a low-pass design, to be moved to $z = (1 - 1/A)$. Thus, the shaped quantization noise does not drop to zero at dc but instead levels off near dc. Such a zero near $z = 1$ results in the 3-dB break frequency being approximately equal to $1/A$ rad/sample. Now, we note that the frequency band of interest, f_0, should be greater than $1/A$ rad/sample since the shaped quantization noise is flat below that level. In other words, if f_0 is below where the quantization noise flattens out, then we are not obtaining any further *noise-shaping* benefits. Thus, any further doubling of the oversampling ratio will only improve the SNR by 3 dB/octave. Finally, we can write the following requirement,

$$\frac{f_0}{f_s} > \frac{1/A}{2\pi} \tag{14.48}$$

and recalling that $OSR = f_s/(2f_0)$, we can rearrange the above to

$$A > \frac{OSR}{\pi} \tag{14.49}$$

Of course, some approximations have been made here, such as having the two capacitors of the integrator equal (that is $C_2 \approx C_1$) and having the 3-dB break point being sharp, as well as allowing noise to be flat from dc to f_0 (rather than shaped). In summary, designers will typically ensure that the opamp gain is at least twice the oversampling ratio, which is not usually a difficult requirement.

In addition, the above analysis only applies for modulators having a single D/A feedback and does not apply to MASH or cascade modulators, where larger opamp gains are often required to aid matching between digital and analog signal paths.

14.8 MULTI-BIT OVERSAMPLING CONVERTERS

Although 1-bit oversampling converters have the advantage that they can realize highly linear data conversion, they also have some disadvantages. For example, 1-bit oversampling modulators are prone to instability due to the high degree of nonlinearity in the feedback. Another disadvantage is the existence of idle tones, as previously discussed. Additionally, in oversampling D/A converters, the use of a 1-bit D/A converter results in a large amount of out-of-band quantization noise, which must be significantly attenuated using analog circuitry. Such a task often requires relatively high-order analog filtering (recall that the low-pass filter should be at least one order higher than the modulator order). Also, the dynamic range of the analog filter may have to be significantly higher than the converter's dynamic range since the filter's input must cope with power of both the quantization noise and signal itself. Since the quantization noise may be up to 15 dB larger in level than the signal level, such an extension in the dynamic range makes the realization of analog filters for oversampling D/A converters nontrivial. The use of a multi-bit D/A converter can significantly reduce this large amount of quantization noise, but care must be taken to ensure that the multi-bit converter remains linear. Typically, a multi-bit oversampling system would be similar to that shown in Fig. 14.16, except that an M-bit quantizer would be used in the digital modulator, and its output would drive an M-bit D/A converter. This section will briefly discuss some multi-bit oversampling converter architectures that employ methods so that the overall system remains highly linear.

Multi-bit Randomizer D/A Converter

One approach for realizing a multi-bit D/A converter intended for oversampling systems is that of dynamic element matching [Carley, 1989]. With this approach, a thermometer-code D/A converter is realized, and randomization is used to effectively spread the nonlinearity over the whole frequency range. For example, a 3-bit D/A converter using dynamic element matching is shown in Fig. 14.27. Here, the 8-line randomizer makes use of a pseudo-random code to distribute the thermometer lines equally to each of the unit capacitances, C. One point to note here is that the mismatch "noise" from this multi-bit converter is not shaped by any feedback loop and therefore appears more white than high-pass filtered. In [Carley, 1989], the resulting multi-bit oversampling D/A converter had 15-bit linearity while using peak element mismatches of 0.2 percent.

Recently, this dynamic element matching has been extended to include noise shaping the nonlinearity introduced by the DAC [Chen, 1995; Adams, 1995; Baird,

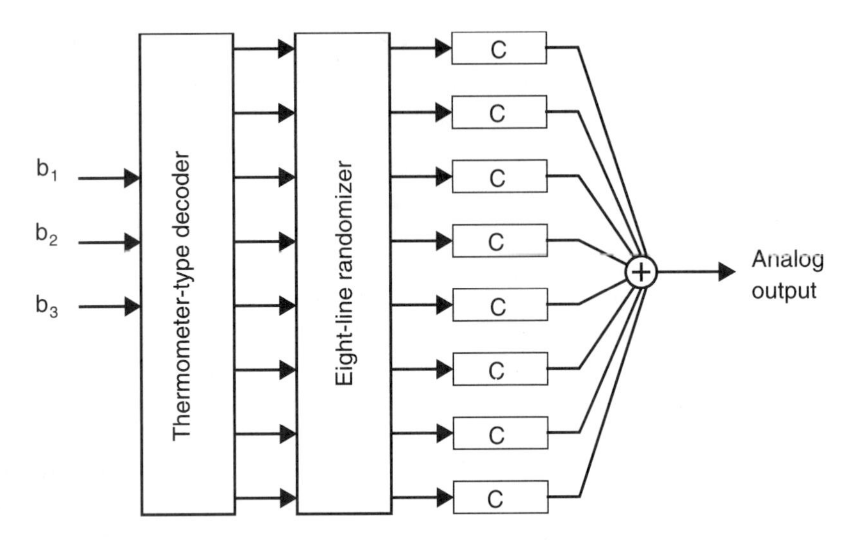

Fig. 14.27 Dynamic element matching 3-bit D/A converter.

1995]. This new approach is quite promising, since much lower oversampling ratios can be used while maintaining high linearity with a reasonably coarse DAC mismatch.

Dynamic Matched Current Sources in a D/A Converter

Another method of maintaining high linearity in an oversampling D/A system is through the use of a highly linear D/A converter operating at the fast oversampling rate. Such high linearity has been achieved using dynamically matched current sources similar to those described in Chapter 12 [Schouwenaars, 1991]. With this approach, a 5-bit D/A converter operating at 5.6 MHz was used to realize an effective 90 dB of dynamic range with a full-scale 0-dB input signal. Also, when an input signal of –30 dB (relative to full scale) was applied, the noise plus distortion was –85 dB lower than the signal power, indicating an effective dynamic range of 115 dB. The output bandwidth of the converter was dc to 20 kHz.

Digital Calibration A/D Converter

One way to make use of a multi-bit D/A converter in an oversampling A/D modulator is to have a calibration cycle where digital circuitry is used to model the static nonlinearity of the D/A converter, as shown in Fig. 14.28 [Larson, 1988]. With this approach, the multi-bit digital signal $x_2(n)$ creates 2^N output levels at $x_1(n)$, which are consistent in time but not necessarily linearly related to $x_2(n)$. However, the feedback loop of the modulator forces the in-band frequency of $x_1(n)$ to very closely equal $u(n)$, and thus $x_1(n)$ is still linearly related to $u(n)$. Through the use of the digital correction circuitry, the signal $y(n)$ also equals $x_1(n)$ if its nonlinearity equals that of the D/A con-

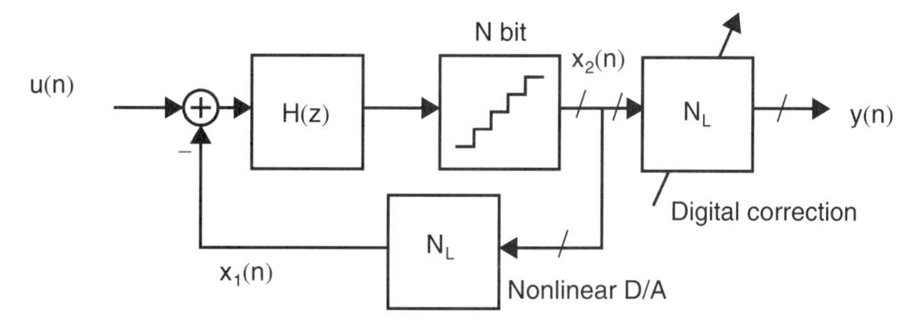

Fig. 14.28 Digitally corrected multi-bit A/D modulator.

verter. To realize this digital correction circuitry, a 2^N word look-up RAM is used where the size of each RAM word is at least equal to the desired linearity. For example, if a 4-bit D/A converter is used and 18 bits of linearity is desired, then the digital correction circuitry consists of a 16-word RAM, each of length 18 bits or greater (or equivalently, a 288-bit RAM). Furthermore, this RAM size can be reduced by digitizing only the difference between the desired linearity and the expected linearity of the D/A converter [Sarhang-Nejad, 1993]. For example, if the D/A converter is expected to have a linearity of at least 9 bits, then the RAM need only have a word length of 10 bits for 18-bit overall linearity.

The calibration of the digital correction circuitry can be accomplished by reconfiguring the ΔΣ A/D modulator to a single-bit system. Next, the input to this 1-bit converter is one of the 4-bit D/A converter levels to be calibrated. In other words, the multi-bit converter's input is held constant while its output is applied to the 1-bit A/D modulator for many clock cycles. The resulting 1-bit signal is digitally low-pass filtered and gives the desired digital equivalent value of that particular dc level for the D/A converter. This procedure is repeated 16 times to find all the 16 words of RAM.

Finally, it should be mentioned that a similar approach can be used to improve the linearity of a multi-bit oversampling D/A converter, except in this case the digital correction circuit would be in the feedback portion of the digital modulator.

A/D with Both Multi-Bit and Single-Bit Feedback

Another very interesting alternative architecture uses both multi-bit feedback and single-bit feedback [Hairapetian, 1994]. A third-order example of this A/D is shown in Fig. 14.29. If one assumes that the errors due to the quantization of the 1-bit A/D and 5-bit A/D are Q_1, and Q_5, respectively, and also that the 5-bit D/A injects errors due to its nonlinearity given by Q_d, then by applying a linear analysis to the system of Fig. 14.29 it can be shown that

$$U_s(z) = U(z)z^{-1} + Q_1(z)\frac{(1-z^{-1})^2}{1-0.5z^{-1}} - Q_5(z)\frac{z^{-1}(1-z^{-1})^{-2}}{1-0.5z^{-1}}$$
$$-Q_d(z)\frac{z^{-1}(1-z^{-1})^{-2}}{1-0.5z^{-1}} \qquad (14.50)$$

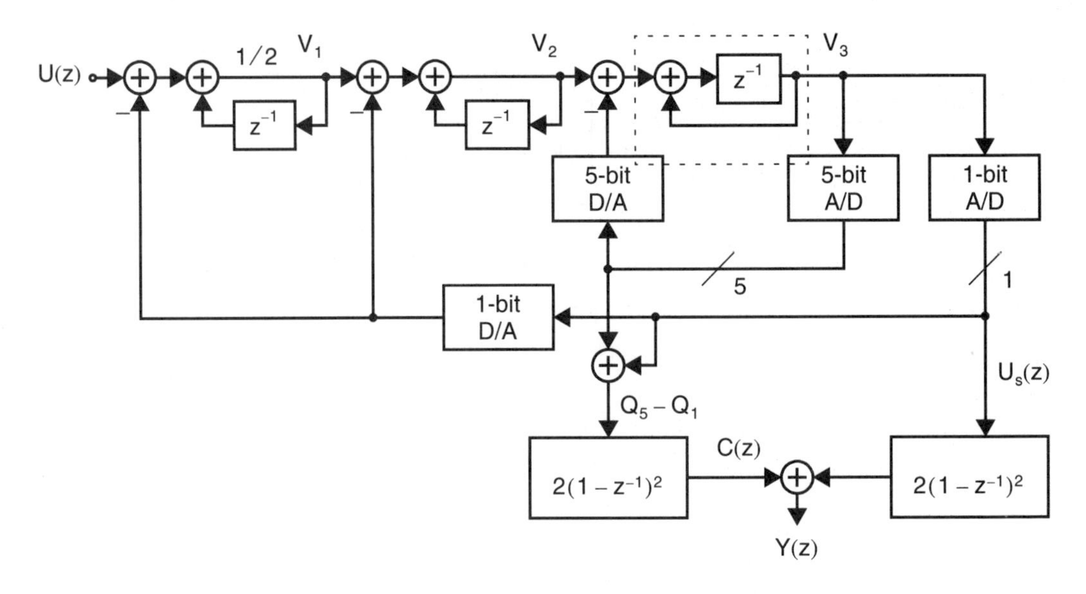

Fig. 14.29 An A/D architecture that uses both multi-bit and single-bit feedback.

After digital signal processing, the digital output signal, $Y(z)$, is given by

$$Y(z) = U(z)z^{-1} + 2Q_5(z)(1-z^{-1})^3 - 2Q_d(z)z^{-1}(1-z^{-1})^2 \tag{14.51}$$

It is seen that, based on the assumption of perfect integrators, the quantization noise due to the 1-bit A/D is cancelled exactly, the quantization noise due to the 5-bit A/D undergoes third-order noise shaping, and the noise due to the nonlinearity of the 5-bit D/A converter undergoes second-order shaping. Of course, if the integrators are not perfect, then the cancellation of the quantization noise of the 1-bit A/D will not be perfect, but any errors in the cancellation still undergo second-order noise shaping. It can be shown that the stability characteristics of this system are very similar to those of a second-order $\Delta\Sigma$ A/D converter. It should also be noted that the digital signal processing required does not require any multiplications, but that multi-bit signals must be processed by the following decimator, which complicates its realization.

14.9 THIRD-ORDER A/D DESIGN EXAMPLE

In this section, a design example is presented for a third-order oversampled switched-capacitor A/D converter. In this example, all the zeros of the noise transfer function (NTF) are placed at $z = 1$ (i.e., dc) so that the converter could be used for various oversampling ratios. In other words, the zeros are not spread over the frequency band of interest, as that would restrict the converter's usefulness to a particular oversampling ratio.

Since all the zeros are assumed to be at $z = 1$, the NTF has the following form:

$$NTF(z) = \frac{(z-1)^3}{D(z)} \tag{14.52}$$

To find the poles of the system (i.e., the roots of $D(z)$), we need to ensure modulator stability while shaping as much quantization noise away from dc as possible. To ensure modulator stability, recall that a heuristic approach is to restrict the peak NTF to be less than 1.5. However, to shape the quantization noise away from dc, we wish to make the NTF as large as possible outside the frequency band of interest. Thus, the best NTF choice would be one which has a flat gain of 1.4 at high frequencies. Such a high-pass filter is obtained (i.e., no peaking in the NTF) when the poles are placed in a Butterworth configuration, as shown in Fig. 14.30. Thus, the denominator is found using digital filter software to design a third-order high-pass digital filter where the passband is adjusted until the peak NTF is less than 1.4 (i.e., $|NTF(e^{j\omega})| < 1.4$). Specifically, with a passband edge at $f_s/20$, a third-order Butterworth high-pass filter has a peak gain equal to 1.37, and thus the $NTF(z)$ is given by

$$NTF(z) = \frac{(z-1)^3}{z^3 - 2.3741z^2 + 1.9294z - 0.5321} \tag{14.53}$$

Now, by rearranging (14.16), we can find $H(z)$ in term of $NTF(z)$, resulting in

$$H(z) = \frac{1 - NTF(z)}{NTF(z)} \tag{14.54}$$

which, for the function given in (14.53), results in

$$H(z) = \frac{0.6259z^2 - 1.0706z + 0.4679}{(z-1)^3} \tag{14.55}$$

Next, a suitable implementation structure is chosen. In this case, a cascade-of-integrators structure was used, as shown in Fig. 14.31. The α_i coefficients are included for dynamic-range scaling and are initially set to $\alpha_2 = \alpha_3 = 1$, while the last term, α_1, is initially set equal to β_1. By initially setting $\alpha_1 = \beta_1$, we are allowing the input signal to have a power level similar to that of the feedback signal, $y(n)$. In other words, if α_1 were initially set equal to one and β_1 were quite small, then the circuit would initially be stable for only small input-signal levels.

Coefficient values for β_i are then found by deriving the transfer function from the 1-bit D/A output to V_3 and equating that function to $-H(z)$ in (14.55). Specifically, assuming $\alpha_2 = \alpha_3 = 1$, we find the following $H(z)$:

$$H(z) = \frac{z^2(\beta_1 + \beta_2 + \beta_3) - z(\beta_2 + 2\beta_3) + \beta_3}{(z-1)^3} \tag{14.56}$$

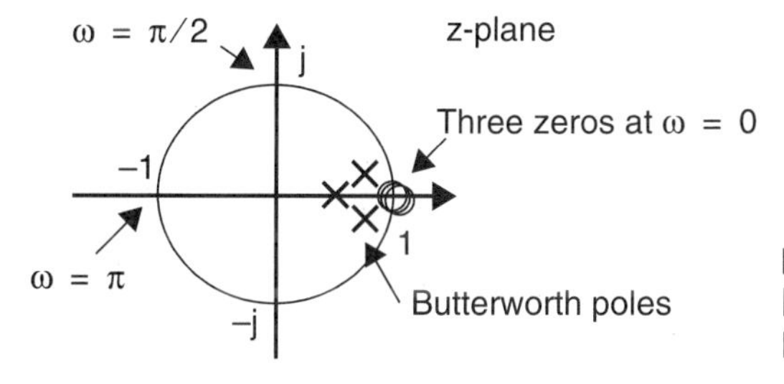

Fig. 14.30 Pole and zero locations for the third-order NTF.

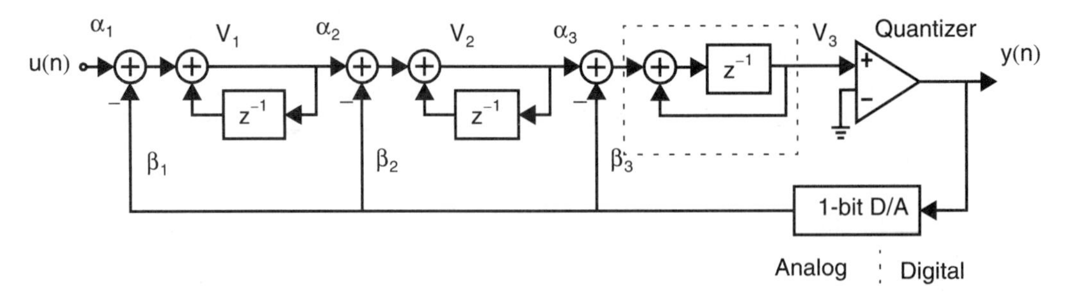

Fig. 14.31 The cascade-of-integrators structure used to realize the third-order modulator. Note that α_i coefficients are used for dynamic-range scaling and are initially set to $\alpha_1 = \beta_1$ and $\alpha_2 = \alpha_3 = 1$.

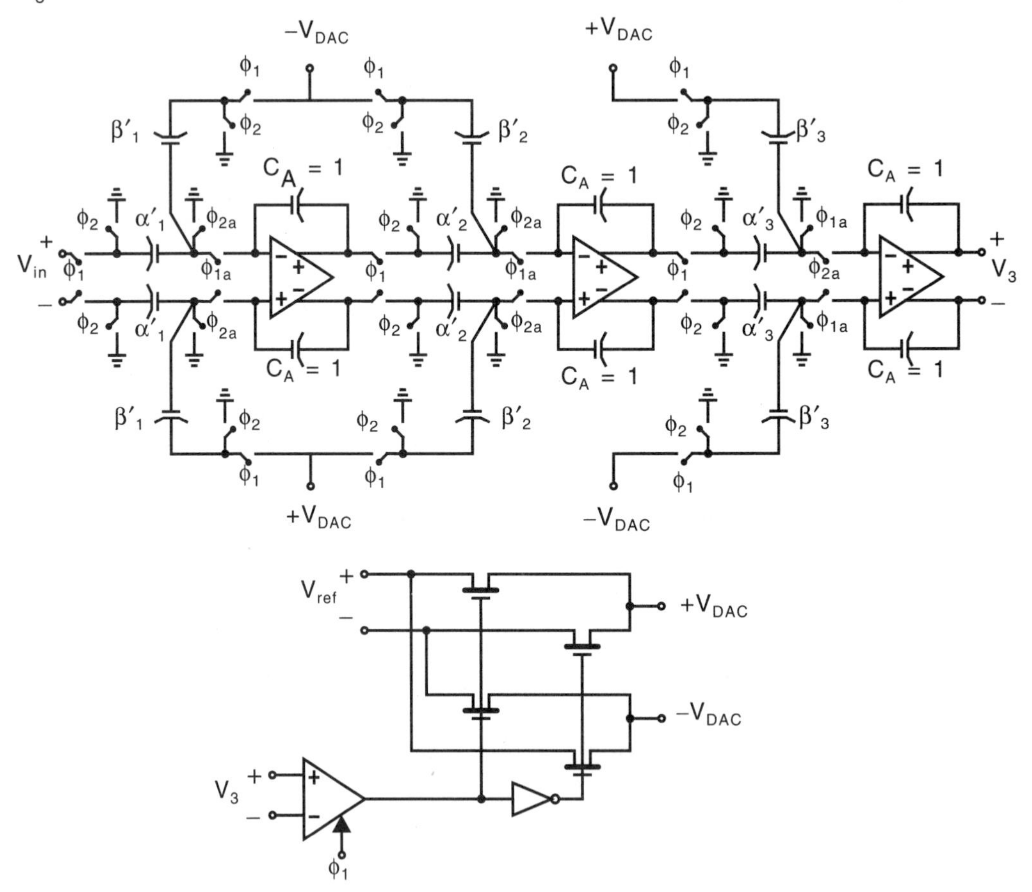

Fig. 14.32 A switched-capacitor realization of the third-order modulator example.

Equating (14.56) with (14.55), we find an initial set of coefficients to be

$$\alpha_1 = 0.0232, \quad \alpha_2 = 1.0, \quad \alpha_3 = 1.0$$
$$\beta_1 = 0.0232, \quad \beta_2 = 0.1348, \quad \beta_3 = 0.4679 \qquad (14.57)$$

It is then necessary to perform dynamic-range scaling. Dynamic scaling is necessary to ensure that all nodes have approximately the same power level, so that all nodes will clip near the same level, and there will be no unnecessarily large noise gains from nodes with small signal levels. Here, dynamic-range scaling was accomplished by simulating the system in Fig. 14.31 with the coefficients just given and an input sinusoidal wave with a peak value of 0.7 at a frequency of $\pi/256$ rad/sample. Following the simulation, the maximum values at nodes V_1, V_2, V_3 were found to be 0.1256, 0.5108, and 1.004, respectively. To increase the output level of a node by a factor k, the input branches should be multiplied by k, while the output branches should be divided by k. For example, the maximum value of node V_1 is increased to unity by multiplying both α_1 and β_1 by $1/0.1256$ and dividing α_2 by $1/0.1256$. Similarly, node V_2 is scaled by dividing $0.1256\alpha_2$ and β_2 by 0.5108 and multiplying α_3 by 0.5108. Node V_3 was left unchanged, as its maximum value was already near unity. Such a procedure results in the following final values for the coefficients.

$$\alpha'_1 = 0.1847, \quad \alpha'_2 = 0.2459, \quad \alpha'_3 = 0.5108$$
$$\beta'_1 = 0.1847, \quad \beta'_2 = 0.2639, \quad \beta'_3 = 0.4679 \qquad (14.58)$$

Finally, the block diagram of Fig. 14.31 can be realized by the switched-capacitor circuit shown in Fig. 14.32 to obtain the final oversampled A/D converter.

14.10 REFERENCES

R. W. Adams and T. W. Kwan. "Data-Directed Scrambler for Multi-Bit Noise Shaping D/A Converters," U.S. patent no. 5,404,142, April 1995.

D. Anastassiou. "Error Diffusion Coding for A/D Conversion," *IEEE Trans. on Circuits and Systems,* Vol. 36, pp. 1175–1186, September 1989.

R. T. Baird and T. S. Fiez. "Improved Delta-Sigma DAC Linearity Using Data-Weighted Averaging," *Proc. IEEE Symp. on Circuits and Systems,* pp. 13–16, Seattle, 1995.

B. E. Boser and B. A. Wooley. "The Design of Sigma-Delta Modulation Analog-To-Digital Converters," *IEEE J. of Solid-State Circuits,* Vol. 23, pp. 1298–1308, December 1988.

J. C. Candy. "Decimation for Sigma-Delta Modulation," *IEEE Trans. Commun.,* Vol. 34, pp. 72–76, January 1986.

J. C. Candy and G. C. Temes. "Oversampling Methods for A/D and D/A Conversion," *Oversampling Delta-Sigma Data Converters,* ed. J. C. Candy and G. C. Temes, IEEE Press, New York, 1992.

L. R. Carley. "A Noise-Shaping Coder Topology for 15+ Bit Converters," *IEEE J. of Solid-State Circuits,* Vol. 24, pp. 267–273, April 1989.

K. C.-H. Chao, S. Nadeem, W. L. Lee, and C. G. Sodini. "A Higher-Order Topology for Interpolative Modulators for Oversampling A/D Converters," *IEEE Trans. on Circuits and Systems,* Vol. 37, pp. 309–318, March 1990.

F. Chen and B. H. Leung. "A High-Resolution Multibit Sigma-Delta Modulator with Individual Level Averaging," *IEEE J. of Solid-State Circuits,* Vol. 30, pp. 453–460, April 1995.

J. Dattorro et al. "The Implementation of One-Stage Multirate 64:1 FIR Decimator for Use in One-Bit Sigma-Delta A/D Applications," *AES 7th Int. Conf.,* May 1989.

P. Ferguson, Jr., et al. "An 18b 20kHz Dual Sigma-Delta A/D Converter," *Int. Solid-State Circuits Conf.*, February 1991.

A. Hairapetian and G. C. Temes, "A Dual-Quantization Multi-Bit Sigma-Delta Analog/Digital Converter," *IEEE Int. Symp. on Circuits and Systems,* pp. 437–441, May 1994.

A. Hairapetian, G. C. Temes, and Z. X. Zhang. "Multibit Sigma-Delta Modulator with Reduced Sensitivity to DAC Nonlinearity," *Electron. Lett.,* Vol. 27, pp. 990–991, May 1991.

S. A. Jantzi, W. M. Snelgrove, and P. F. Ferguson Jr. "A Fourth-Order Bandpass Sigma-Delta Modulator," *IEEE J. of Solid-State Circuits,* Vol. 28, pp. 282–291, March 1993.

L. E. Larson, T. Cataltepe, and G. C. Temes. "Multibit Oversampled $\Sigma - \Delta$ A/D Converter with Digital Error Correction," *Electron. Lett.,* Vol. 24, pp. 1051–1052, August 1988.

H. A. Leopold et al. "A Monolithic CMOS 20-b Analog-to-Digital Converter," *IEEE J. of Solid-State Circuits,* Vol. 26, pp. 910–916, July 1991.

T. C. Leslie and B. Singh. "An Improved Sigma-Delta Modulator Architecture," *IEEE Int. Symp. on Circuits and Systems,* pp. 372–375, May 1990.

Y. Matsuya, et al. "A 16-bit Oversampling A-to-D Conversion Technology Using Triple-Integration Noise Shaping," *IEEE J. of Solid-State Circuits,* Vol. 22, pp. 921–929, December 1987.

S. Norsworthy. "Optimal Nonrecursive Noise-Shaping Filters for Oversampling Data Converters–Part I and II," *IEEE Int. Symp. on Circuits and Systems,* pp. 1353–1360, Chicago, May 1993.

T. Saramaki et al. "Multiplier-Free Decimator Algorithms for Super-Resolution Oversampled Converters," *IEEE Int. Symp. on Circuits and Systems,* pp. 3275–3278, New Orleans, May 1990.

M. Sarhang-Nejad and G. C. Temes, "A High-Resolution Multibit $\Sigma\Delta$ ADC with Digital Correction and Relaxed Amplifier Requirements," *IEEE J. of Solid-State Circuits,* Vol. 28, pp. 648–660, June 1993.

H. J. Schouwenaars, D. W. J. Groeneveld, C. A. A. Bastiaansen, and H. A. H. Termeer. "An Oversampled Multibit CMOS D/A Converter for Digital Audio with 115-dB Dynamic Range," *IEEE J. of Solid-State Circuits,* Vol. 26, pp. 1775–1780, December 1991.

R. Schreier. "An Empirical Study of High-Order Single-Bit Delta-Sigma Modulators," *IEEE Trans. on Circuits and Systems—II: Analog and Digital Signal Processing,* Vol. 40, pp. 461–466, August 1993.

R. Schreier and W. M. Snelgrove. "Bandpass Sigma-Delta Modulation," *Electron. Lett.,* Vol. 25, no. 23, pp. 1560–1561, November 9, 1989.

G. C. Temes, S. Shu, and R. Schreier. "Architectures for $\Delta\Sigma$ DACs," in *Oversampling $\Delta\Sigma$ Data Converters,* ed. G. C. Temes, IEEE Press, 1996.

P. P. Vaidyanathan. *Multirate Systems and Filter Banks.* Prentice Hall, Englewood Cliffs, New Jersey, 1993.

14.11 PROBLEMS

14.1 Assuming oversampling with no noise shaping and using (14.14), find the approximate sampling rate required to obtain a maximum SNR of 80 dB on a signal with a 1-kHz bandwidth using a 1-bit quantizer.

14.2 Repeat Example 14.4, except use a dc input level of $1.2/3$. At what frequency would a tone appear (relative to the sampling rate, f_s)?

14.3 Repeat Example 14.4, except use a dc input level of 1.1 to see if the internal state of the modulator saturates.

14.4 Repeat Example 14.4, except use an input sequence of $\{10, -10, 10, -10, 10, -10, \ldots\}$ to see if the internal state of the modulator saturates. This problem demonstrates that a large level input signal can be applied to a modulator if its signal power resides at a frequency where the gain of $H(z)$ is low.

14.5 Given that an 8-bit A/D converter has a SNR of 50 dB but is linear to 12 bits, what is the sampling rate required to achieve 12 bits of accuracy using straight oversampling on a signal bandwidth of 1 MHz?

14.6 Repeat Problem 14.5, assuming that the 8-bit A/D converter is placed inside a first-order delta-sigma modulator. What is the sampling rate if a second-order modulator is used?

14.7 Repeat Problem 14.1, assuming a first-order modulator is used. What is the sampling rate if a second-order modulator is used?

14.8 Assuming a 1-bit output signal of ± 1 and a sinusoidal signal of ± 0.5 peak-to-peak, what is the ratio of the signal power to quantization noise power before any filtering is applied?

14.9 What does the block $G(z) - 1$ need to be in Fig. 14.13 in order that a second-order noise transfer function is realized equal to $(1 - z^{-1})^2$?

14.10 Show a switched-capacitor modulator so that the second-order modulator shown in Fig. 14.10 is realized.

14.11 Show that the sequence of $\{1, 1, -1, 1, 1, -1, 1, 1, -1, \ldots\}$ into a cascade of three integrators and differentiators shown in Fig. 14.19 gives the correct running average. Assume that the adders have infinite width and that $M = 4$.

14.12 At first glance, the impulse response of a running average filter of length 4 is the same as that for a hold system that holds the input signal for length 4. By producing a block diagram for each system, explain the difference between the two systems.

14.13 Verify (14.42).

14.14 Show the block diagram for a MASH architecture that realizes third-order noise shaping using a second-order modulator (as shown in Fig. 14.10) and a first-order modulator.

14.15 Sketch the noise transfer function from 0 to 4 MHz for the second-order band-pass modulator shown in Fig. 14.23, assuming the sampling rate, f_s, equals 4 MHz.

14.16 Show how the block diagram in Fig. 14.23 can be modified so that the band-pass modulator places zeros at $\pm f_s/8$ rather than $\pm f_s/4$.

14.17 Using the "typical" waveform pattern shown in Fig. 14.24, find the practical average outputs when 0, $1/2$, and $-1/2$ have been applied to the first-order modulator shown in Fig. 14.7.

CHAPTER

15 Continuous-Time Filters

There are two main techniques for realizing integrated analog filters. One technique is the use of switched-capacitor circuits, as discussed in Chapter 10. A switched-capacitor circuit, although its signal remains continuous in voltage (i.e., it is never quantized), is, in fact, a discrete-time filter since it requires sampling in the time domain. Because of this time-domain sampling, the clock rate must always be at least twice that of the highest frequency being processed to eliminate aliasing. In fact, typically the clock rate is much greater than twice the signal bandwidth, to reduce the requirements of an anti-aliasing filter. As a result, switched-capacitor filters are limited in their ability to process high-frequency signals. Presently, the second most popular technique for realizing integrated analog filters is *continuous-time filtering.* As their name suggests, continuous-time filters have signals that remain continuous in time and that have analog signal levels. Since no sampling is required, continuous-time filters have a significant speed advantage over their switched-capacitor counterparts.

Although continuous-time filters can operate successfully on high-speed signals, they do have some disadvantages that have thus far restricted most of their industrial use to high-speed but otherwise lower-performance applications. One disadvantage is their need for some sort of tuning circuitry. Tuning is required because their filter coefficients are determined by the product of two dissimilar elements, such as capacitance and resistor (or transconductor) values. Thus, although switched-capacitor filters have coefficient accuracies of 0.1 percent, integrated continuous-time coefficients are initially set to only about 30-percent accuracy! Fortunately, tuning circuitry can set the filter coefficient accuracy to around 1 percent at the cost of more (and sometimes complex) circuitry, as we will see in Section 15.7. Another practical disadvantage exhibited by most continuous-time filters is their relatively poor linearity and noise performance. Although low-frequency switched-capacitor filters have been realized with nonlinear distortion and noise performance better than 90 dB, high-frequency continuous-time filters typically have distortion plus noise performance not much better than 60 dB (often it is much worse). Fortunately, there are numerous high-speed applications in which distortion and noise performances are not too demanding, such as many data communication and video circuits.

15.1 INTRODUCTION TO G_m-C FILTERS

This section gives a brief introduction to G_m-C filters. We will see here that the main circuit building blocks of G_m-C filters are transconductors. The basic functions of transconductors are described in this section, and circuit techniques for realizing transconductors are presented in the following sections.

Integrators

An integrator is the main building block for most continuous-time filters. To realize an integrator in G_m-C technology, a transconductor and a capacitor can be used as shown in Fig. 15.1. A transconductor is essentially a transconductance cell (an input voltage creates an output current) with the additional requirement that the output current be *linearly related to the input voltage.* Here, the output of the transconductor is the current i_o, and, in the ideal case, both the input and output impedance of this block are infinite. The output current is linearly related to the differential input voltage signal by

$$i_o = G_m v_i \tag{15.1}$$

where G_m is the transconductance of the cell. It should be re-emphasized that this relationship is different from that of an operational-transconductance-amplifier (OTA) in that here, the output current should be *linearly* related to the input voltage. In addition, a transconductor should have a well-known (or at least controllable) transconductance value. In the case of an OTA, the transconductance does not necessarily need to be a linear function of the input voltage or a well-controlled value (though usually the transconductance value should be as large as possible).

This output current is applied to the integrating capacitor, C_1, resulting in the voltage across C_1 being given by

$$V_o = \frac{I_o}{sC_1} = \frac{G_m V_i}{sC_1} \tag{15.2}$$

Defining ω_{ti} to be the unity-gain frequency of the integrator, we can write

$$V_o \equiv \left(\frac{\omega_{ti}}{s}\right)V_i = \left(\frac{G_m}{sC_1}\right)V_i \tag{15.3}$$

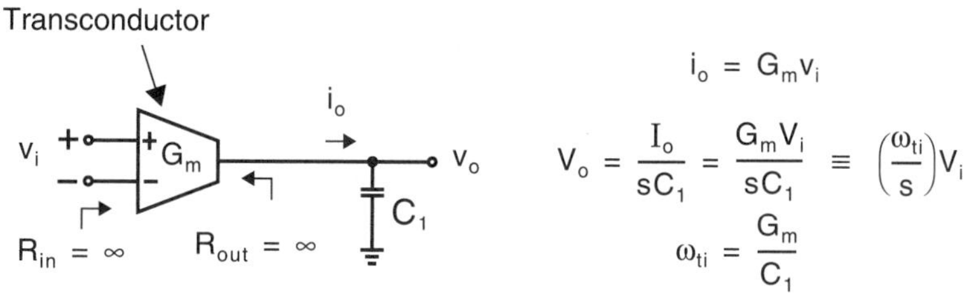

Transconductor

v_i + − G_m i_o V_o C_1

$R_{in} = \infty$ $R_{out} = \infty$

$$i_o = G_m v_i$$

$$V_o = \frac{I_o}{sC_1} = \frac{G_m V_i}{sC_1} \equiv \left(\frac{\omega_{ti}}{s}\right)V_i$$

$$\omega_{ti} = \frac{G_m}{C_1}$$

Fig. 15.1 A single-ended G_m-C integrator.

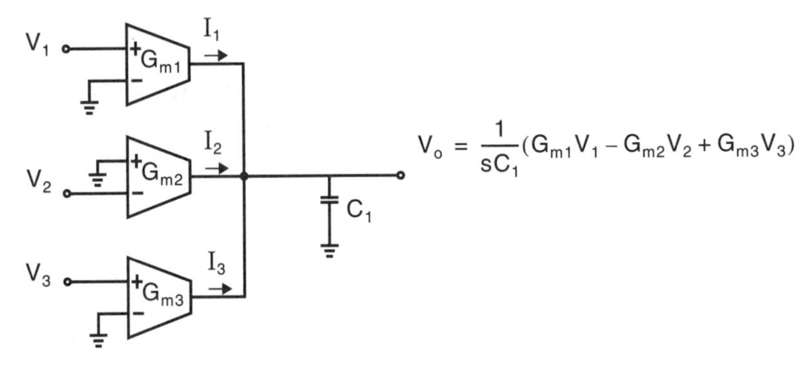

Fig. 15.2 A three-input, single-ended integrator/summer circuit.

Thus, we see that the output voltage is equal to the integration of the differential input voltage multiplied by the integrator unity-gain frequency, which is given by

$$\omega_{ti} = \frac{G_m}{C_1} \tag{15.4}$$

To realize a more general integrator/summer circuit, one need only connect multiple transconductors in parallel since the output currents all sum together. An example of a three-input integrator/summer circuit is shown in Fig. 15.2.

EXAMPLE 15.1

What transconductance is needed for an integrator that has a unity-gain frequency of 20 MHz, assuming a 2-pF integrating capacitor?

Solution

Rearranging (15.4), we have

$$\begin{aligned} G_m &= 2\pi \times 20\text{ MHz} \times 2\text{ pF} \\ &= 0.251\text{ mA/V} \end{aligned} \tag{15.5}$$

Note that this transconductance value corresponds to

$$G_m = 1/3.98\text{ k}\Omega \tag{15.6}$$

which, not surprisingly, is related to the unity-gain frequency by

$$2\pi \times 20\text{ MHz} = \frac{1}{3.98\text{ k}\Omega \times 2\text{ pF}} \tag{15.7}$$

In other words, the required transconductance value is the inverse of the resistance value needed to form an RC time constant that corresponds to the desired unity-gain frequency.

Fully Differential Integrators

Although the preceding blocks have all been shown with single-ended signals,[1] in most integrated applications it is desirable to keep the signals fully differential. As discussed in Chapter 10, fully differential circuits have better noise immunity and distortion properties. Thus, mostly fully differential circuits are discussed in the remainder of this chapter.

A fully differential transconductor has two outputs—one positive output (current flowing out for a positive input voltage) and one negative output (current flowing in for a positive input voltage). With this new degree of freedom (having two outputs instead of just one), a fully differential integrator can be realized in one of two ways, as shown in Fig. 15.3.

Note that the integrator shown in Fig. 15.3(*a*) requires one-fourth the capacitance needed for the integrator of Fig. 15.3(*b*) to achieve the same integrator coefficient. Although at first glance, this increase in size appears to be a disadvantage of the circuit shown in Fig. 15.3(*b*), it should be mentioned that fully differential circuits require some method of common-mode feedback. This requirement is due to the fact that, although the differential signal remains stable due to connections of the filter feedback network, the common-mode signal is free to drift if it is not also controlled using some feedback circuit. As a result, one must also be concerned with the stability of the common-mode feedback circuit. Since, typically, dominant pole compensation using capacitors on the output node is often used to stabilize the common-mode network, the integrator shown in Fig. 15.3(*b*) has the advantage that the integrating capacitors, $2C_1$, can also be used for this purpose. Also, recall that

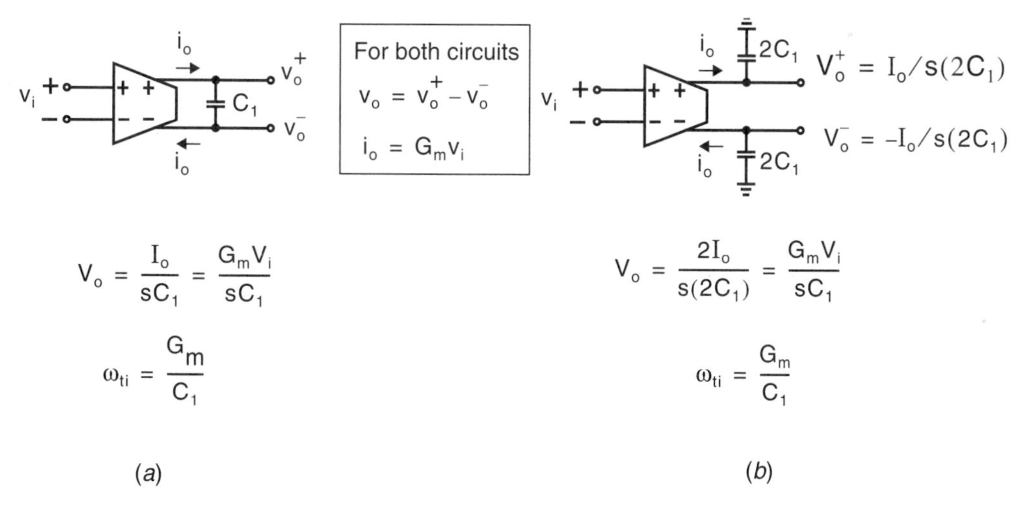

Fig. 15.3 Fully differential G_m-C integrators: (*a*) single capacitor, (*b*) two capacitors.

1. The transconductor input may be differential, but its output is single ended, and single-ended signals are typically applied to the transconductor inputs.

the single capacitor of the integrator shown in Fig. 15.3(a) must be realized using integrated capacitors, and thus top- and bottom-plate parasitic capacitances exist. Since the bottom-plate parasitic capacitance can be quite large (up to 20 percent of the capacitance value), to maintain symmetry, C_1 is often realized using two parallel capacitors, as shown in Fig. 15.4. We take the parallel combination of the two capacitances, which equals C_1, and include the effect of the parasitic capacitances, C_p, to obtain the output voltage, given by

$$V_o = \frac{G_m V_i}{s(C_1 + C_p/2)} \tag{15.8}$$

Thus, we see here that the parasitic capacitances affect the integration coefficient and must be taken into account.

Parasitic capacitances can also cause linearity problems since these capacitances are partially nonlinear (as most parasitic capacitances are). To reduce the effects of these parasitic capacitances, a common approach is to realize the integrating function using a Miller integrator, as shown in Fig. 15.5. As the figure shows, the integrating

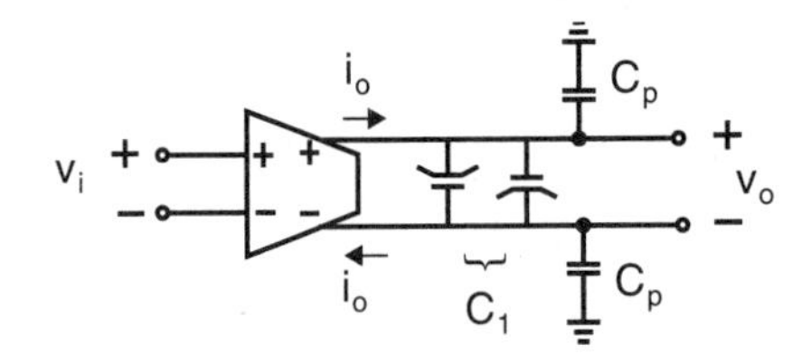

$$V_o = \frac{G_m V_i}{s(C_1 + C_p/2)} \qquad \omega_{ti} = \frac{G_m}{(C_1 + C_p/2)}$$

Fig. 15.4 A fully differential integrator maintaining symmetry and showing parasitic capacitances.

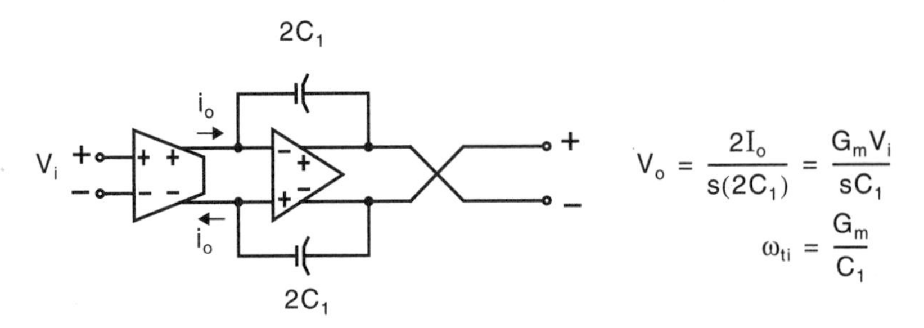

Fig. 15.5 A G_m-C opamp integrator (also called a Miller integrator). Note the cross coupling of the output wires to maintain a positive integration coefficient. Do not cross-couple the output if a negative coefficient is desired. This circuit requires that the common-mode voltage of the input nodes (as well as that of the output nodes) of the fully differential opamp be set to a specified value.

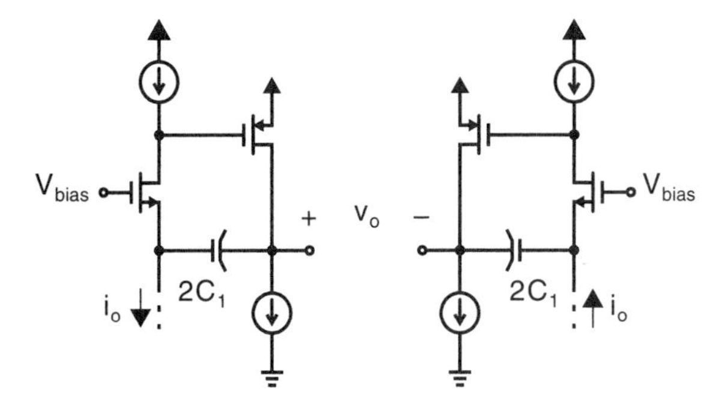

Fig. 15.6 A simple high-speed differential Miller integrator.

coefficient for this G_m-C opamp integrator is the same as that for the integrator in Fig. 15.3(*b*). The advantage of using the extra amplifier is that the effect of parasitic capacitances is reduced by the gain of the opamp. An additional advantage is that the output of the transconductance cell drives an opamp virtual ground, and so lower-output-impedance transconductance cells can be used. Another benefit is that the transconductance output nodes (i.e., the opamp input nodes) are maintained at a specific operating voltage set by a common-mode feedback circuit. As a result, the transconductance output stage can use small voltage swing circuitry, which is easier to design. Finally, the lower impedances at the transconductor output nodes make those nodes less sensitive to capacitive coupling of noise.

In summary, the addition of this extra opamp can simplify the design of the transconductance cell's output stage since the transconductor does not have to swing a large voltage range; it can have larger (and nonlinear) parasitic capacitances, it need not have a large output impedance, and it can be less sensitive to noise. The penalty paid here is a lower operating speed and a potential increase in circuit area and power. These two penalties can be reduced by using a simple high-speed design for the amplifier in the Miller integrator. For example, a current-input common-gate common-source amplifier, as shown in Fig. 15.6, is simple and moderately fast [Willingham, 1993; Padmanabhan, 1994]. The input impedance of this amplifier is small, on the order of $2/(g_m^2 r_{ds})$, due not only to the low impedance of a common-gate first stage, but also due to feedback. Finally, it should be noted that either a low- or high-output-impedance opamp can be used for the Miller integrator since it needs to drive only capacitive loads. Typically, high-output-impedance amplifiers are used since they tend to have higher unity-gain frequencies, be simpler, and have less power dissipation.

First-Order Filter

A block diagram for a general first-order continuous-time filter is shown in Fig. 15.7. It can be shown that the transfer function for this block diagram is given by

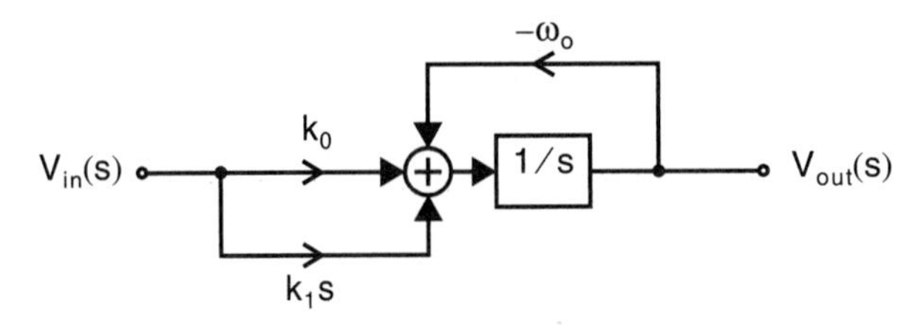

Fig. 15.7 A block diagram of a general first-order continuous-time filter.

$$H(s) \equiv \frac{V_{out}(s)}{V_{in}(s)} = \frac{k_1 s + k_o}{s + \omega_o} \tag{15.9}$$

A single-ended G_m-C filter that realizes a first-order filter is shown in Fig. 15.8. The transfer function for this first-order G_m-C filter is found by writing a current equation for the output node, $V_{out}(s)$:

$$G_{m1}V_{in}(s) + sC_X[V_{in}(s) - V_{out}(s)] - sC_A V_{out}(s) - G_{m2}V_{out}(s) = 0 \tag{15.10}$$

Rearranging to find $V_{out}(s)/V_{in}(s)$, we have

$$\frac{V_{out}(s)}{V_{in}(s)} = \frac{sC_X + G_{m1}}{s(C_A + C_X) + G_{m2}} = \frac{\left[s\left(\frac{C_X}{C_A + C_X}\right) + \left(\frac{G_{m1}}{C_A + C_X}\right)\right]}{\left[s + \left(\frac{G_{m2}}{C_A + C_X}\right)\right]} \tag{15.11}$$

Equating (15.11) with (15.9) results in the following relationships:

$$C_X = \left(\frac{k_1}{1 - k_1}\right)C_A \tag{15.12}$$

$$G_{m1} = k_0(C_A + C_X) \tag{15.13}$$

$$G_{m2} = \omega_o(C_A + C_X) \tag{15.14}$$

It should be noted from (15.12) that, for C_X to remain a positive value, the term k_1 must satisfy $0 \le k_1 < 1$. What this restriction implies is that the shown circuit cannot realize a filter that has a high-frequency gain that is either negative or greater

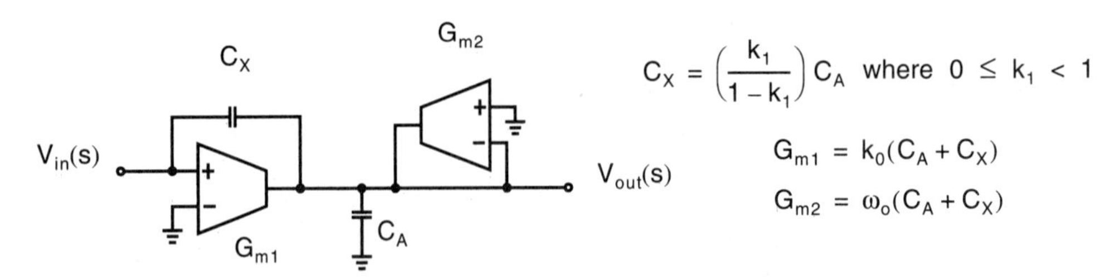

Fig. 15.8 A single-ended general first-order G_m-C filter.

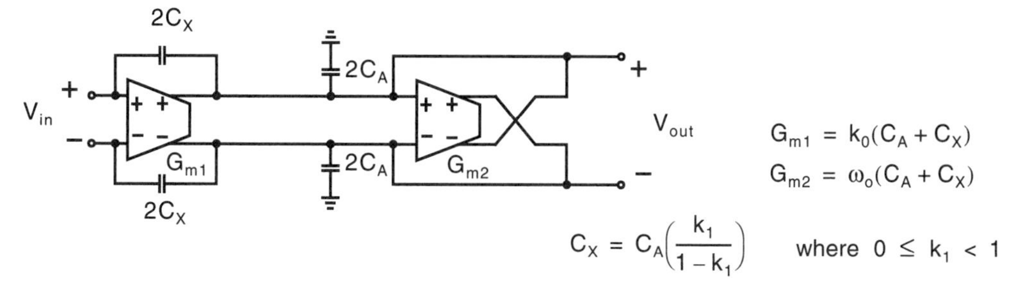

Fig. 15.9 A fully differential general first-order G_m-C filter.

than 1. Such a result should come as no surprise since the high-frequency gain of this circuit is simply a capacitor-divider circuit, and hence the restriction on k_1. Such a restriction on the maximum magnitude of the high-frequency gain would not occur if a G_m-C opamp filter were used since the feed-in capacitor could go into the virtual ground of an opamp and high-frequency gains greater than one would be possible.

Finally, the equivalent fully differential first-order G_m-C filter is shown in Fig. 15.9. Note that capacitor sizes are doubled here, in relation to the single-ended case, so that the same design equations occur. Also note that the same restriction holds on k_1, although a negative value of k_1 can be realized by cross-coupling the feed-in capacitor pair, C_X.

EXAMPLE 15.2

Based on Fig. 15.9, find component values for a first-order filter with a dc gain of 0.5, a pole at 20 MHz, and a zero at 40 MHz. Assume the integrating capacitors are sized $C_A = 2\text{ pF}$.

Solution

A first-order filter with a zero at 40 MHz and a pole at 20 MHz is given by

$$H(s) = \frac{k(s + 2\pi \times 40\text{ MHz})}{(s + 2\pi \times 20\text{ MHz})} \tag{15.15}$$

The leading coefficient is determined by setting the dc gain to 0.5, resulting in $k = 0.25$. Thus, we have

$$H(s) = \frac{0.25s + 2\pi \times 10\text{ MHz}}{s + 2\pi \times 20\text{ MHz}} \tag{15.16}$$

and equating this transfer function to (15.9) results in

$$k_1 = 0.25 \tag{15.17}$$

$$k_0 = 2\pi \times 10^7 \tag{15.18}$$

$$\omega_o = 4\pi \times 10^7 \tag{15.19}$$

Using the design equations shown in Fig. 15.9, we find

$$C_X = 2\ \text{pF} \times \frac{0.25}{1-0.25} = 0.667\ \text{pF} \quad (15.20)$$

$$G_{m1} = 2\pi \times 10^7 \times 2.667\ \text{pF} = 0.168\ \text{mA/V} \quad (15.21)$$

$$G_{m2} = 4\pi \times 10^7 \times 2.667\ \text{pF} = 0.335\ \text{mA/V} \quad (15.22)$$

where $C_A = 2\ \text{pF}$, as given.

Second-Order Filter

To realize a second-order, or *biquad,* filter in G_m-C technology, we use the block diagram derived in Chapter 10 for a general continuous-time biquad filter. The block diagram of Fig. 10.19, slightly modified to make use of positive integrators, is shown in Fig. 15.10. The transfer function for this second-order filter is given by

$$H(s) \equiv \frac{V_{out}(s)}{V_{in}(s)} = \frac{k_2 s^2 + k_1 s + k_o}{s^2 + \left(\frac{\omega_o}{Q}\right)s + \omega_o^2} \quad (15.23)$$

where it should be noted that the leading negative coefficient in (10.45) is no longer present due to the use of positive integrators. Of course, the inversion of the output signal could easily be obtained by simply cross-coupling the output wires. The fully differential version of this filter is shown in Fig. 15.11. The transfer function of this biquad filter is found to be given by

$$H(s) \equiv \frac{V_{out}(s)}{V_{in}(s)} = \frac{s^2\left(\frac{C_X}{C_X + C_B}\right) + s\left(\frac{G_{m5}}{C_X + C_B}\right) + \left(\frac{G_{m2}G_{m4}}{C_A(C_X + C_B)}\right)}{s^2 + s\left(\frac{G_{m3}}{C_X + C_b}\right) + \left(\frac{G_{m1}G_{m2}}{C_A(C_X + C_B)}\right)} \quad (15.24)$$

Equating the coefficients of (15.23) and (15.24) results in

$$k_2 = \frac{C_X}{C_X + C_B} \quad (15.25)$$

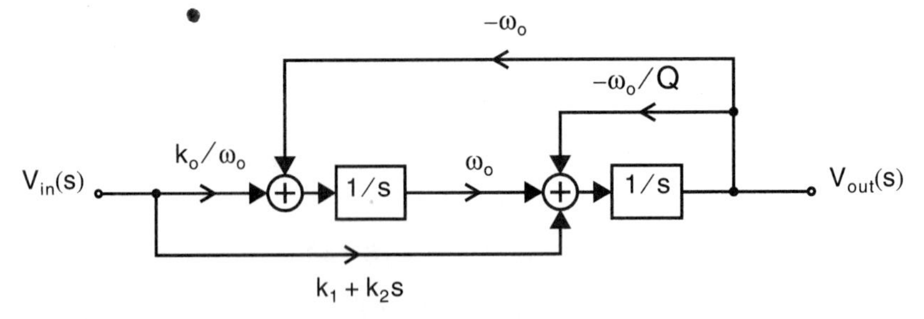

Fig. 15.10 A block diagram of a general second-order continuous-time filter.

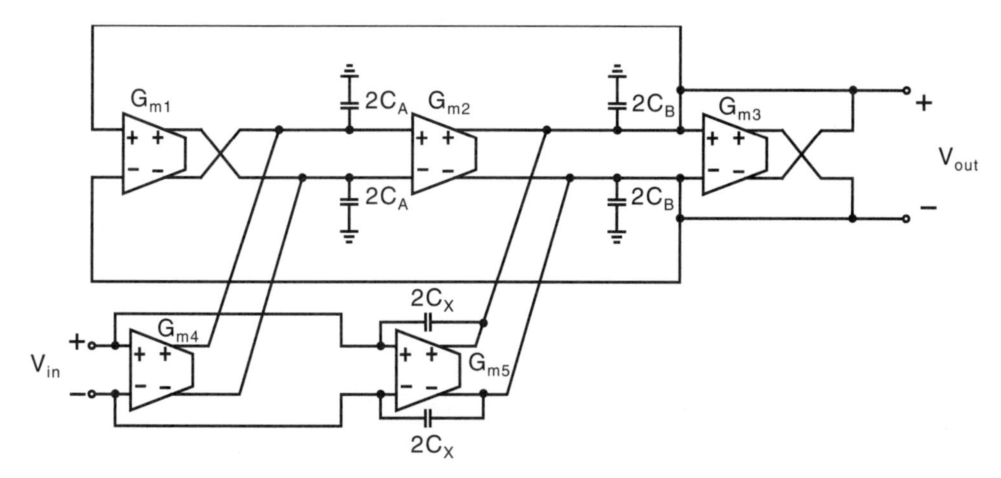

Fig. 15.11 A general second-order filter using fully differential G_m-C technology.

$$k_1 = \frac{G_{m5}}{C_X + C_B} \tag{15.26}$$

$$k_0 = \frac{G_{m2}G_{m4}}{C_A(C_X + C_B)} \tag{15.27}$$

and

$$\omega_o^2 = \frac{G_{m1}G_{m2}}{C_A(C_X + C_B)} \tag{15.28}$$

For Q, we note that

$$\frac{\omega_o}{Q} = \frac{G_{m3}}{C_X + C_b} \tag{15.29}$$

and so, by using (15.28), we can solve for Q, resulting in

$$Q = \sqrt{\left(\frac{G_{m1}G_{m2}}{G_{m3}^2}\right)\left(\frac{C_X + C_B}{C_A}\right)} \tag{15.30}$$

Although the preceding equations are for analyzing a given circuit, they can also be used for design, resulting in the following design equations:

$$C_X = C_B\left(\frac{k_2}{1 - k_2}\right) \quad \text{where } 0 \le k_2 < 1$$

$$G_{m1} = \omega_o C_A \quad G_{m2} = \omega_o(C_B + C_X) \quad G_{m3} = \frac{\omega_o(C_B + C_X)}{Q}$$

$$G_{m4} = (k_o C_A)/\omega_o \quad G_{m5} = k_1(C_B + C_X)$$

Note that for this design, there is a restriction on the high-frequency gain coefficient, k_2, similar to that occurring in the first-order case.

EXAMPLE 15.3

Find the transconductance and capacitor values for a second-order filter that has a bandpass response with a center frequency of 20 MHz, a Q value of 5, and a center frequency gain of 1. Assume that $C_A = C_B = 2$ pF.

Solution

A bandpass transfer function has the form

$$H(s) \equiv \frac{V_{out}(s)}{V_{in}(s)} = \frac{Gs\frac{\omega_o}{Q}}{s^2 + s\frac{\omega_o}{Q} + \omega_o^2} \tag{15.31}$$

where $G = 1$ is the gain at the center frequency. Using the facts that $\omega_o = 2\pi \times 20$ MHz and $Q = 5$, we find

$$k_1 = G\frac{\omega_o}{Q} = 2.513 \times 10^7 \text{rad/s} \tag{15.32}$$

Since k_0 and k_2 are zero, we have $C_X = G_{m4} = 0$ and

$$G_{m1} = \omega_o C_A = 0.2513 \text{ mA/V} \tag{15.33}$$

$$G_{m2} = \omega_o (C_B + C_X) = 0.2513 \text{ mA/V} \tag{15.34}$$

and, in this case,

$$G_{m3} = G_{m5} = k_1 C_B = 50.27\ \mu\text{A/V} \tag{15.35}$$

15.2 BIPOLAR TRANSCONDUCTORS

There are two main approaches for realizing bipolar transconductors. One approach is to use a fixed transconductor cascaded with a gain cell. The fixed transconductor is typically a differential pair linearized through resistor degeneration. The gain cell allows scaling of the output current by varying the ratio of two current sources. The other approach is to use a differential input stage with multiple inputs, where the transistors are scaled such that improved linearity occurs. In this case, the transconductance value can be tuned by varying the bias current through the input stage.

Fixed Transconductors Using Resistors

Two similar approaches for using resistors to linearize a differential pair of transistors are shown in Fig. 15.12. To understand the basic operation of the two circuits, we first make the simplifying assumption that the V_{be} voltages of the transistors are constant. As a result, we see that the differential input voltage, v_i, appears across the two $R_E/2$ resistors in Fig. 15.12(*a*) and across R_E in Fig. 15.12(*b*). Thus, the emitter current I_{E1} is given by

$$I_{E1} = I_1 + \frac{v_i}{R_E} \tag{15.36}$$

and the emitter current I_{E2} is given by

$$I_{E2} = I_1 - \frac{v_i}{R_E} \tag{15.37}$$

Ignoring base currents, these two emitter currents equal their respective collector currents. Then defining i_{o1} to be the difference in the output collector currents from I_1 (as shown), we have

$$i_{o1} = \frac{v_i}{R_E} \tag{15.38}$$

Thus, we see that the transconductance of these two differential input circuits is a fixed value of $1/R_E$. One important difference between the two fixed transconductors of Fig. 15.12 is that, in the case of $v_i = 0$, a bias current of I_1 flows through the two $R_E/2$ resistors in Fig. 15.12(*a*), whereas no bias current is flowing through R_E in Fig. 15.12(*b*). As a result, the common-mode voltage of Fig. 15.12(*b*) can be closer to ground. For example, if a bias current of $I_1 = 100\ \mu A$ is used together

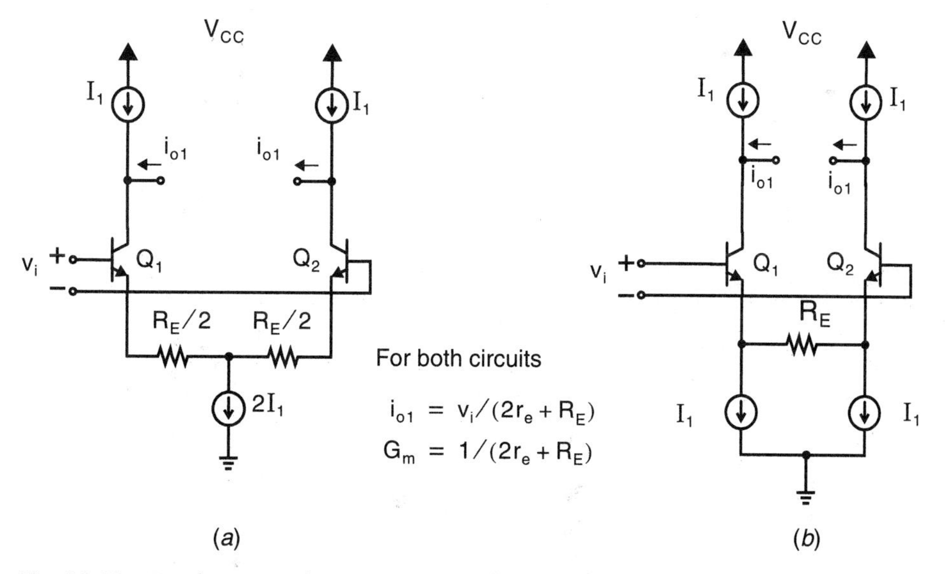

Fig. 15.12 Fixed transconductors using resistors. The lower common-mode range of the circuit in (*a*) is not as low as that of the circuit in (*b*) due to bias currents flowing through the $R_E/2$ resistors.

with $R_E = 20\ k\Omega$, then the voltage drop across each of the two $R_E/2$ resistors would result in 1 V less of common-mode range. It should be noted here that the common-mode input voltage must be far enough above ground to account for the largest expected differential input voltage, v_i. Specifically, one must ensure that the circuit remains functional while the base of one side of the differential pair is lower than the common-mode voltage by $v_i/2$.

A simple estimate of the maximum differential input range for both these circuits can be found by determining how much voltage exists across the linearizing resistors at the point where one transistor just turns off. For example, in Fig. 15.12(*a*), when Q_2 turns off due to a large positive input voltage, all of $2I_1$ flows through $R_E/2$ on the left and nothing flows through $R_E/2$ on the right. Thus, the voltage at the center of the two resistors equals that of the emitter of Q_2 (since no current flows right), and the maximum input voltage equals that voltage across the left $R_E/2$ resistor, giving

$$V_{i,max} = I_1 R_E \tag{15.39}$$

A similar argument gives the same maximum input voltage for the circuit of Fig. 15.12(*b*). It should be noted here that this simple estimate does not take into account that significant distortion occurs when this maximum input voltage is applied. Specifically, at $V_{i,max} = I_1 R_E$, one of the two transistors will be off, implying that V_{be} has not remained a relatively constant value, as we assumed earlier in this section. Typically, the maximum input range that applies to the preceding fixed transconductors might be one-half that of (15.39), or, equivalently, $(I_1/2)R_E$.

So far, we have assumed that the transistor's V_{be} voltages remain constant. Such an assumption is reasonably valid due to the exponential current-voltage relationship of bipolar transistors. Specifically, this assumption is reasonable when one restricts the minimum emitter current, $I_{E,min}$, in either transistor to be that which keeps the small-signal emitter resistance, r_e, much smaller than R_E. Recalling that

$$r_e \equiv \frac{V_T}{I_E} \tag{15.40}$$

we then have the requirement,

$$r_{e,\,max} = \frac{V_T}{I_{E,min}} << R_E \tag{15.41}$$

or equivalently,

$$I_{E,min} >> \frac{V_T}{R_E} \tag{15.42}$$

For example, if $R_E = 20\ k\Omega$ and $V_T = 25$ mV, the minimum emitter current in either transistor should be much greater than 1.25 μA. A good question is, How much is "much greater than"? The answer to this question depends on how much distortion can be tolerated since, when r_e becomes a comparable size to R_E, the varying V_{be} voltage will result in the output current, i_o, being a significant nonlinear function of the input voltage, v_i. For example, if the maximum value of r_e goes to 10 percent of that of R_E when the maximum applied input voltage is $(I_1/2)R_E$, then a linearity deviation of about 3 percent occurs. Thus, to attain a high linearity with the preceding circuits, the bias currents should be large (resulting in a small value of r_e) and the input signal should be small (to maintain the small values of r_e). Of course, the use of small input signals may result in noise becoming a problem.

In applications in which a moderate amount of nonlinearity can be tolerated, R_E might be chosen similar in size to the nominal bias value of r_e. In this case, the overall transconductance of this transconductor is most easily found using the hybrid T model for the input transistor pair. With such an approach, the output current is more accurately described by

$$i_{o1} = \frac{v_i}{r_{e1} + R_E + r_{e2}} = \frac{v_i}{2r_e + R_E} \tag{15.43}$$

where r_e is the small-signal emitter resistance (equal to α/g_m), given by

$$r_e = \frac{V_T}{I_E} = \frac{V_T}{I_1} \tag{15.44}$$

Note that (15.43) reduces to (15.38) when R_E is chosen much larger than $2r_e$. In summary, a more accurate formula for the transconductance of the two circuits of Fig. 15.2 is

$$G_m = \frac{i_{o1}}{v_i} = \frac{1}{2r_e + R_E} \tag{15.45}$$

EXAMPLE 15.4

Consider the circuit of Fig. 15.2(*b*), where $R_E = 3\ k\Omega$ (such that $G_m \approx 0.333\ mA/V$) and the maximum differential input voltage is $\pm 500\ mV$. Choose an appropriate bias current, I_1, such that $r_{e,max} < 0.1R_E$, and find a more precise value for G_m. Find the new value of I_1 and G_m if the constraint $r_{e,max} < 0.01R_E$ is used instead.

Solution

In the first case, we have $r_{e,max} < 0.1R_E = 300\ \Omega$. Thus, in the case of $v_i = 500\ mV$, we require that $r_{e2} \approx 300\ \Omega$, which is attained for an emitter current of

$$I_{E2,min} = \frac{V_T}{r_{e,max}} = \frac{26\ mV}{300\ \Omega} = 86.7\ \mu A \tag{15.46}$$

For this same input voltage, the resistor current is approximately

$$I_{RE} \cong \frac{500\ mV}{3\ k\Omega} = 167\ \mu A \tag{15.47}$$

and since the current through R_E plus I_{E2} must equal I_1 (according to the node currents at the emitter of Q_2), we have

$$I_1 = 167 + 86.7 = 253\ \mu A \tag{15.48}$$

In other words, a minimum bias current of 253 μA should be chosen. For this value of bias current, the nominal value of r_e is given by

$$r_e = \frac{V_T}{I_1} = 103\ \Omega \tag{15.49}$$

resulting in

$$G_m = \frac{1}{(2r_e + R_E)} = 0.312\ mA/V \tag{15.50}$$

For the case $r_{e,max} < 0.01R_E$, one finds $I_1 = 1.03$ mA and $G_m = 0.328$ mA/V. Note here that the nominal value of r_e is now only 26 Ω, which is significantly less than R_E. Although this second design will have less distortion than the first (about 10 times better), it uses much more power. Next, we look at using feedback to improve distortion without resorting to large bias currents.

One way to improve the distortion properties of the above fixed transconductors without using large bias currents and small input signal levels is to use an extra pair of amplifiers, as shown in Fig. 15.13. Here, the virtual ground in each opamp forces the emitter voltages of Q_1 and Q_2 to equal those of v_i^+ and v_i^-. As a result, the input

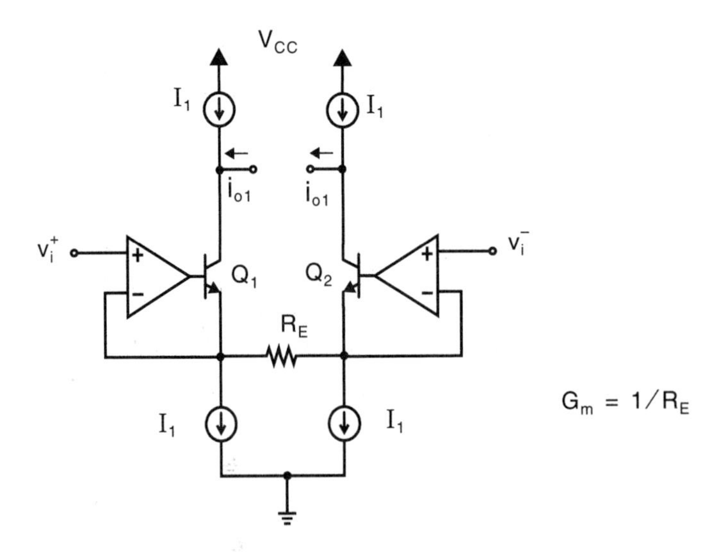

Fig. 15.13 Improving the linearity of a fixed transconductor through the use of opamps.

voltage appears directly across R_E and does not depend on the V_{be} voltages of Q_1 and Q_2. Thus, the transconductance of this circuit is given by

$$G_m = \frac{i_{o1}}{v_i^+ - v_i^-} = \frac{1}{R_E} \tag{15.51}$$

It should be mentioned here that the opamps can be quite simple (typically, a single stage) because they merely improve an input stage that is already reasonably linear.

Another method for improving the linearity of a fixed transconductor is to force constant currents through Q_1 and Q_2 such that their V_{be} voltages are fixed, as shown in Fig. 15.14 [Koyama, 1993]. Here, I_1 flows through each of Q_1 and Q_2, while Q_3 and Q_4 have varying collector currents corresponding to the current flow through R_E. The output currents can then be obtained at the collectors of Q_7 and Q_8, which mirror the current through Q_3 and Q_4. Thus, assuming the current mirror transistors consisting of Q_7 and Q_8 are equal in size to those of Q_3 and Q_4, the transconductance of this cell is given by

$$G_m = \frac{i_{o1}}{v_i^+ - v_i^-} = \frac{1}{R_E} \tag{15.52}$$

If the current mirror transistors are scaled in relation to Q_3 and Q_4, then the transconductance for this cell is also scaled by the same amount.

It should be noted that simple diode-connected transistors are not used at Q_3 and Q_4 since a feedback circuit is required to set the voltage level at the collectors of Q_1

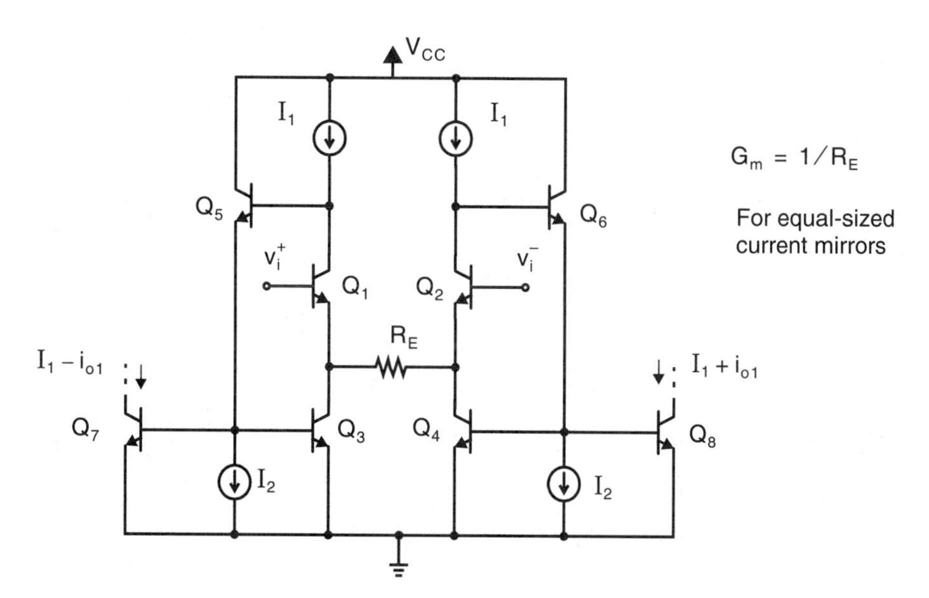

Fig. 15.14 Improving the linearity of a fixed transconductor by maintaining a constant V_{be} voltage for Q_1 and Q_2. Level shifts should be added at the emitters of Q_5, Q_6 if a larger input range is desired.

and Q_2 without drawing much current. The feedback loop occurs as follows: If the voltage at the collector of Q_1 increases because I_{C1} is less than I_1, then V_{be3} also increases, forcing more current through Q_3 and Q_1 until I_{C1} once again equals I_1. Although the circuit shown results in each of Q_1 and Q_2 collector voltages being set to two diode drops above ground (i.e., $V_{be3} + V_{be5}$), a larger voltage can be obtained by adding diode-connected transistors that operate as voltage level-shifters. It is possible to combine parts of this fixed-transconductor circuit with a gain-cell stage, as was done in [Koyama, 1993]. Finally, it should be mentioned here that the same linearization technique has been used in a CMOS variable transconductor, where the resistance R_E was realized using a MOS transistor operating in the triode region [Welland, 1994].

Gain-Cell Transconductors

The transconductors just described have a fixed transconductance value equal to approximately $1/R_E$. However, to accommodate tuning, variable transconductance values are required. One common approach for creating tunable transconductors that have a fixed transconductance input stage is to use a gain cell as described in Section 8.7. Recall that in a gain cell, the output current is equal to a scaled version of the input current, where the scaling factor is determined by the ratio of two dc bias currents. In addition, the circuit is highly linear.

A tunable transconductor that uses a gain cell is shown in Fig. 15.15. Here, the voltage sources, V_{LS}, represent voltage level shifters and could be realized using the circuit shown in Fig. 15.16. By combining the relationships given for the fixed

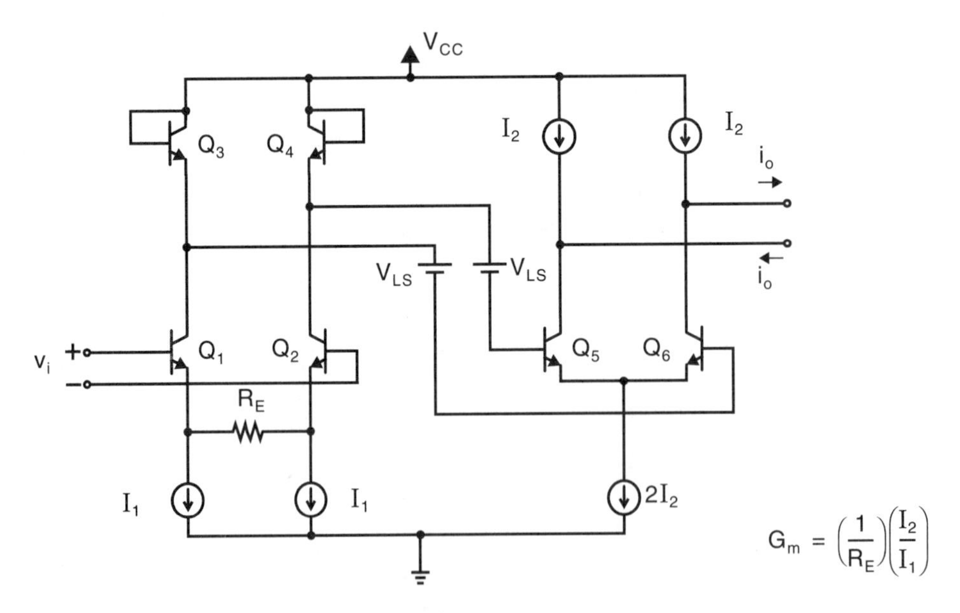

Fig. 15.15 A typical gain-cell transconductor.

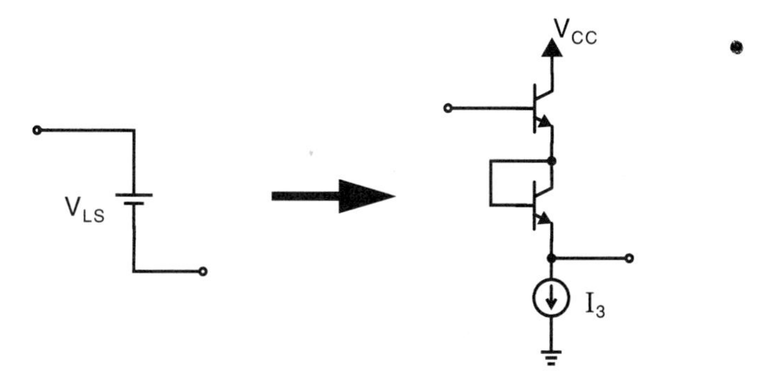

Fig. 15.16 A possible realization of the level shifters.

transconductor and the gain cell, specifically, (15.38) and (8.61), then the output current is given by

$$i_o = \frac{v_i I_2}{R_E I_1} \tag{15.53}$$

resulting in a transconductance value equal to

$$G_m = \left(\frac{1}{R_E}\right)\left(\frac{I_2}{I_1}\right) \tag{15.54}$$

Thus, the transconductance value is proportional to the ratio of I_2/I_1.

It is also possible to place the gain cell below the input differential pair, as shown in Fig. 15.17 [Wyszynski, 1993; Moree, 1993]. Note that while the previous designs

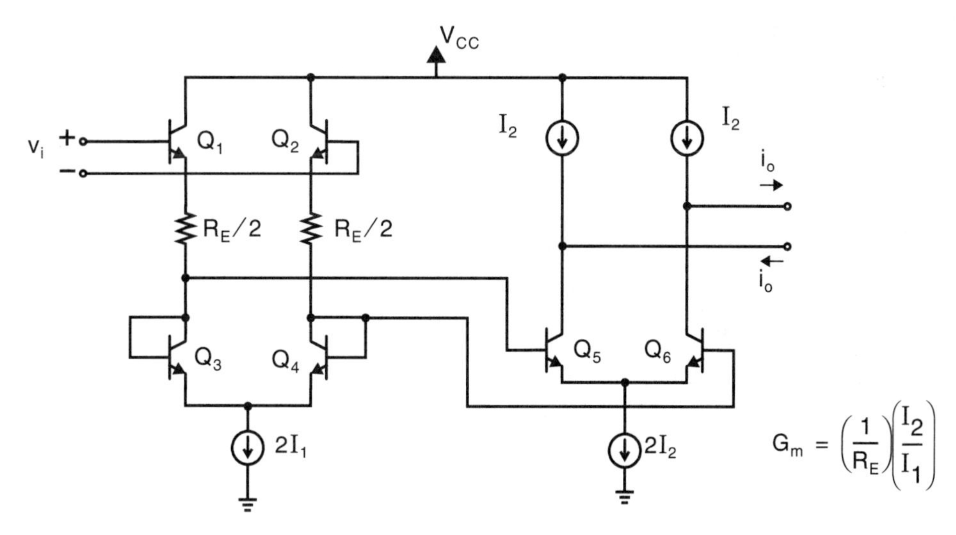

Fig. 15.17 A gain-cell transconductor with the gain cell below the input differential pair.

have $2r_e$ in series with the degeneration resistor, R_E, this circuit has $4r_e$ in series with R_E (that of $Q_1 - Q_4$), which might affect the distortion performance. However, the gain cell shown has less distortion due to finite β effects and has therefore been referred to as a *beta-immune type-A cell* [Gilbert, 1990].

Transconductors Using Multiple Differential Pairs

Quite a different approach for realizing a bipolar transconductor is by partially linearizing the input stage through the use of multiple differential pairs. To see how this linearization is achieved, consider a simple differential pair, as shown in Fig. 15.18(*a*). Ignoring base currents, we saw in Section 3.10 that the collector current i_{C2} is given by

$$i_{C2} = \frac{I_1}{1 + e^{v_i/V_T}} \tag{15.55}$$

This current is plotted in Fig. 15.18(*b*), where we note that the curve is relatively linear near $v_i = 0$. It is well known that the slope at this point equals the transconductance of this differential input stage and is given by

$$G_m = \frac{1}{2r_e} = \frac{I_1}{4V_T} \tag{15.56}$$

Thus, we see that the transconductance is proportional to the bias current, I_1. Unfortunately, this two-transistor input stage has a very limited input range when used as a linear transconductor. In fact, when v_i becomes greater than around 32 mV_{pp} (i.e., $v_i > \pm 16$ mV), the total harmonic distortion of the output current becomes greater than 1 percent (see Example 15.5). We now discuss a method to use parallel differential pairs to increase the linear input range [Calder, 1984].

Consider the input stage shown in Fig. 15.19(*a*), where two differential pairs are connected in parallel but with a dc offset voltage applied to each. The currents of the

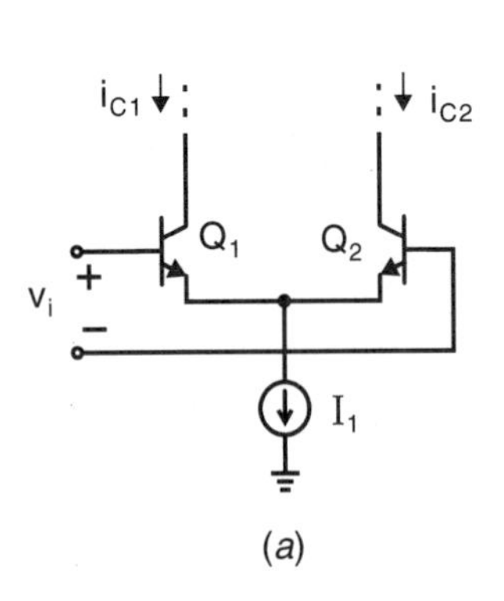

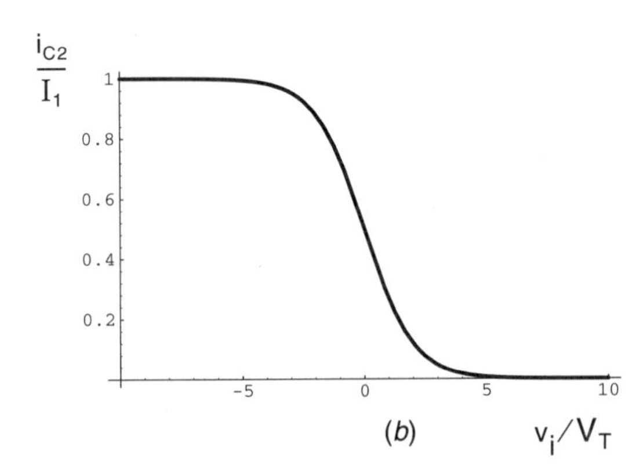

Fig. 15.18 (*a*) Differential pair. (*b*) Current output versus input voltage.

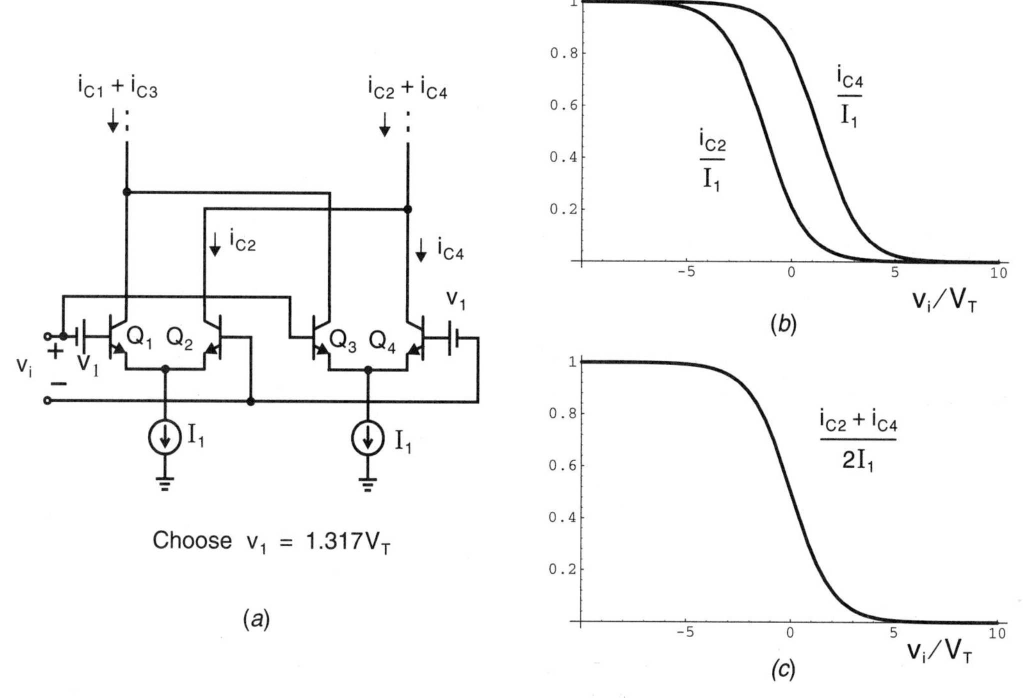

Fig. 15.19 *(a)* A linearized differential input; *(b)* individual right-side currents; *(c)* total linearized right-side current. The left-side currents are similar.

right side transistors in each transistor pair are shown in Fig. 15.19(*b*). If the dc offset voltages, V_1, are chosen carefully, then the two current curves will partially linearize each other when they are added together, as shown in Fig. 15.19(*c*). In this case, V_1 is chosen to be equal to $1.317V_T$ to maximize the linear input range [Tanimoto, 1991]. As a result of this linearization technique, the input range can now be approximately three times that of the original differential pair or, equivalently, 96 mV_{pp}, and achieve the same distortion performance.

To reduce circuitry, the two dc offset voltage sources can be eliminated by sizing the differential pair of transistors appropriately. Specifically, assuming that $v_i = 0$, then when $V_1 = 1.317V_T$, the collector currents of Q_2 and Q_4 are given by

$$i_{C2} = \frac{I_1}{1 + e^{1.317}} = 0.2113 I_1 \tag{15.57}$$

and

$$i_{C4} = i_{C1} = \frac{I_1}{1 + e^{-1.317}} = 0.7887 I_1 \tag{15.58}$$

If a two-transistor differential pair has unequal-sized transistors such that the same two currents occur for $v_i = 0$, then this unequal-sized differential pair will behave the same as that of an equal-sized differential pair with a dc offset applied. To determine the sizing of the unequal-sized differential pair, define k to be the ratio of the base-emitter area of Q_1 with respect to Q_2. Thus, for this unequal-sized case, when $v_i = 0$, we have

$$i_{C1} = kI_s e^{(v_{be}/V_T)} = 0.7887I_1 \tag{15.59}$$

and

$$i_{C2} = I_s e^{(v_{be}/V_T)} = 0.2113I_1 \tag{15.60}$$

Dividing (15.59) by (15.60), we obtain

$$k = 3.73 \tag{15.61}$$

Thus, we should make Q_1 3.73 times larger than Q_2, and similarily Q_4 should be 3.73 times larger than Q_3. Such transistor sizing achieves the same result as the dc offset voltages. This result should make intuitive sense since we require that the ratio of the two currents be $0.7887/0.2113$ when their base-emitter voltages are equal; this requirement results in the area ratio. Typically, this area ratio is set to 4 for practical reasons, and the final linearized transconductor is realized as shown in Fig. 15.20. Thus, for this circuit with $v_i = 0$, the currents through Q_1 and Q_2 are $0.8I_1$ and $0.2I_1$, respectively.

To determine the transconductance of the circuit of Fig. 15.20, we note that G_m equals the sum of the transconductance of the two differential pairs, resulting in

$$G_m = \frac{1}{r_{e1} + r_{e2}} + \frac{1}{r_{e3} + r_{e4}} \tag{15.62}$$

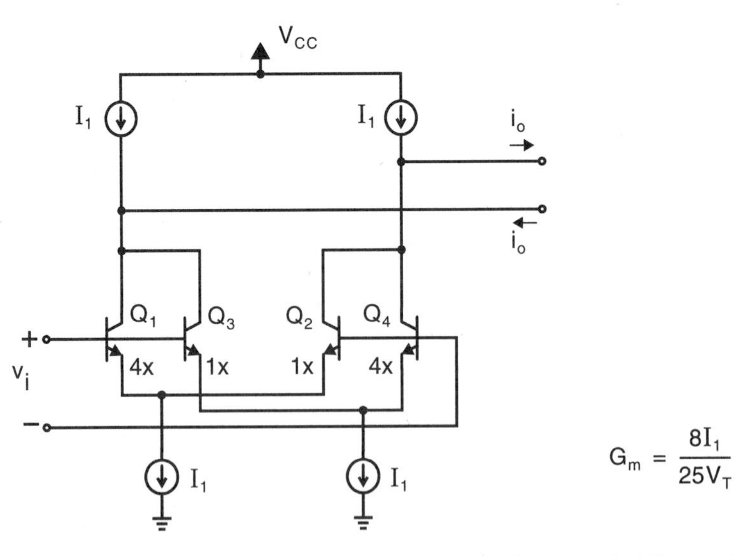

Fig. 15.20 Using transistor sizing to create a linearized differential transconductor.

Also, at $v_i = 0$, we have

$$r_{e1} = r_{e4} = \frac{V_T}{0.8I_1} \tag{15.63}$$

and

$$r_{e2} = r_{e3} = \frac{V_T}{0.2I_1} \tag{15.64}$$

Combining (15.62) to (15.64), we have

$$G_m = \frac{8I_1}{25V_T} \tag{15.65}$$

which is 28 percent greater than the transconductance value for the transistor pair shown in Fig. 15.18(*a*). However, this comparison is not fair since two differential pairs and twice the current are used in this linearized approach. If two differential pairs of Fig. 15.18(*a*) are used together, then their total transconductance would be $G_m = I_1/(2V_T)$, implying that this linearized approach has a 36 percent smaller transconductance. Also note that, as in the two-transistor pair case, this transconductance value is also proportional to the bias current, I_1, and thus no gain cell is required.

Recent circuits based on this approach have resulted in a 2- to 10-MHz programmable filter intended for disk-drive applications [De Veirman, 1992]. Finally, it should be mentioned here that this same technique can be extended to a larger number of input stages to further improve the linear input range [Tanimoto, 1991]. Using four differential input pairs in parallel, the linear input range was reported to be 16 times better than that for a single differential pair.

EXAMPLE 15.5

Consider the single differential pair shown in Fig. 15.18(*a*), where $I_1 = 2$ mA. Find the percentage error between an ideal linear transconductor and the true output current when $v_i = 16$ mV.

Solution

Using (15.56), we have

$$G_m = \frac{I_1}{4V_T} = 19.2 \text{ mA/V} \tag{15.66}$$

Thus, for an ideal linear transconductor, the output current for a 16-mV input voltage would be

$$i_o = G_m \times 16 \text{ mV} = 0.307 \text{ mA} \tag{15.67}$$

The true output current can be found from (15.55) as

$$I_{C2} = \frac{2 \text{ mA}}{1 + e^{16/26}} = 0.702 \text{ mA} \tag{15.68}$$

$$I_{C1} = \frac{2\text{ mA}}{1 + e^{-16/26}} = 1.298\text{ mA} \tag{15.69}$$

implying that the change in current from the bias current of 1 mA is 0.298 mA. Thus, the percentage error between the true output current and the ideal linear transconductor output is

$$\frac{0.298 - 0.307}{0.307} \times 100 = -3\% \tag{15.70}$$

In this case, the deviation is three-percent worst case, which results in about a one-percent total-harmonic-distortion (THD) measurement (see Section 15.8).

EXAMPLE 15.6

Consider the linearized differential pairs shown in Fig. 15.19(*a*), where $I_1 = 2$ mA. Find the percentage error between an ideal linear transconductor and the true output current when $v_i = 16$ mV.

Solution

First, we need to derive a formula for G_m in a manner similar to that for (15.65), where

$$r_{e1} = r_{e4} = \frac{V_T}{0.7887 I_1} \tag{15.71}$$

$$r_{e2} = r_{e3} = \frac{V_T}{0.2113 I_1} \tag{15.72}$$

Thus, the transconductance is given by

$$G_m = \frac{1}{r_{e1} + r_{e2}} + \frac{1}{r_{e3} + r_{e4}} = \frac{0.333 I_1}{V_T} = 25.6\text{ mA/V} \tag{15.73}$$

For an ideal linear transconductor, the output current for a 16-mV input voltage would result in

$$i_o = G_m \times 16\text{ mV} = 0.4102\text{ mA} \tag{15.74}$$

To calculate the true output current, we first note that

$$v_1 = 1.317 V_T = 34.24\text{ mV} \tag{15.75}$$

and since $v_i = 16$ mV, we use $v_i + v_1$ to find I_{C1}, resulting in

$$I_{C1} = \frac{2\text{ mA}}{1 + e^{-50.24/26}} = 1.747\text{ mA} \tag{15.76}$$

$$I_{C2} = 2\text{ mA} - I_{C1} = 0.253\text{ mA} \tag{15.77}$$

For I_{C4}, we use $v_1 - v_i$ to obtain

$$I_{C4} = \frac{2\text{ mA}}{1 + e^{-18.24/26}} = 1.337\text{ mA} \tag{15.78}$$

$$I_{C3} = 2\text{ mA} - I_{C4} = 0.663\text{ mA} \tag{15.79}$$

The true output is then equal to

$$I_{C1} + I_{C3} = 2.41\text{ mA} \tag{15.80}$$

$$I_{C2} + I_{C4} = 1.59\text{ mA} \tag{15.81}$$

implying that the change in current from the bias current of 2 mA is 0.4099 mA. Thus, the percentage error between the true output current and the ideal linear transconductor output is

$$\frac{0.4099 - 0.4102}{0.4102} \times 100 = -0.1\% \tag{15.82}$$

Note that this error is over ten times better than with a simple differential pair.

15.3 CMOS TRANSCONDUCTORS USING TRIODE TRANSISTORS

In this section, we discuss transconductor circuit techniques that rely on transistors operating in the triode region. It should be stated here that in the following circuits, not all transistors are in the triode region. As we will see, most of the transistors are biased in the active region, but the transconductance of the circuit relies on one or two key transistors biased in the triode region.

We begin this section by recalling the classical model equation for an n-channel transistor operating in the triode region.

$$I_D = \mu_n C_{ox}\left(\frac{W}{L}\right)\left[(V_{GS} - V_{tn})V_{DS} - \frac{V_{DS}^2}{2}\right] \tag{15.83}$$

Also recall that a transistor remains in the triode region as long as the drain-source voltage is lower than the effective gate-source voltage. In other words, the following condition must be satisfied:

$$V_{DS} < V_{eff} \qquad \text{where } V_{eff} = V_{GS} - V_{tn} \tag{15.84}$$

We will see next that if the preceding equations were exact, then perfectly linear circuits could be realized. However, these equations are only mildly accurate for a modern short-channel process, and therefore some distortion occurs because higher-order terms are not modelled in the preceding equations. Hence, almost all practical continuous-time circuits use fully differential architectures (see Section 10.3) to reduce even-order distortion products, leaving the third-order term to dominate.

Transconductors Using a Fixed-Bias Triode Transistor

We see from (15.83) that if a small drain-source voltage, V_{DS}, is used, the V_{DS}^2 term goes to zero quickly and the drain current is approximately linear with respect to an applied drain-source voltage. Thus, we can model a transistor in triode as a resistor given by

$$r_{DS} \equiv \left(\frac{\partial i_D}{\partial v_{DS}}\right)^{-1}\Bigg|_{V_{DS}=0} \tag{15.85}$$

which results in a small-signal resistance of

$$r_{DS} = \frac{1}{\mu_n C_{ox}\left(\frac{W}{L}\right)(V_{GS} - V_{tn})} \tag{15.86}$$

One approach for using this equivalent resistance to realize a transconductor with moderate linearity is shown in Fig. 15.21 [Welland, 1994]. This circuit has the same structure as that of the bipolar circuit of Fig. 15.14. However, here the bipolar transistors are replaced with their MOSFET counterparts (biased in the active region), and the resistor is replaced with a transistor operating in the triode region. Note that the circuit of Fig. 15.21 forces constant currents through Q_1 and Q_2 (as in the bipolar case) such that their individual gate-source voltages remain constant. As a result, the

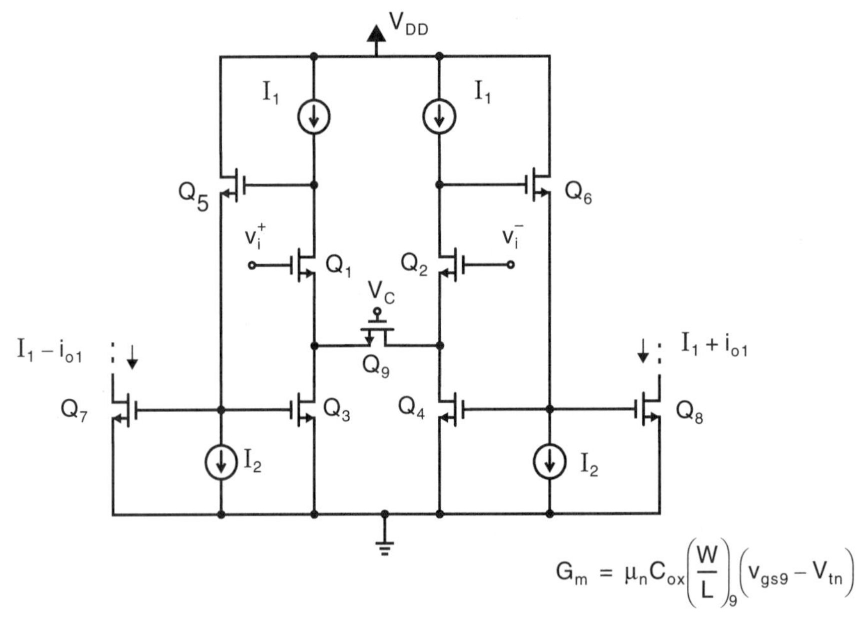

Fig. 15.21 A CMOS transconductor using Q_9 in the triode region.

transconductance of this cell is equal to the inverse of the small-signal resistance, r_{ds}, of Q_9. Using the small-signal resistance result of (15.86), the overall transconductance of this cell is given by

$$G_m = \frac{i_{o1}}{v_i^+ - v_i^-} = \mu_n C_{ox}\left(\frac{W}{L}\right)_9 (V_{gs9} - V_{tn}) \tag{15.87}$$

Thus, this transconductor has a variable transconductance value that can be adjusted by changing the value of V_{gs9}. As in the bipolar case, this transconductance value can be scaled by scaling the current mirrors Q_7 and Q_8 with respect to Q_3 and Q_4.

A similar approach but with reduced complexity and fewer p-channel inputs is shown in Fig. 15.22 [Kwan, 1991]. Here, a constant current of I_2 is forced through each of Q_1 and Q_2, whereas the dc bias current for each of Q_3 and Q_4 equals the difference between I_1 and I_2. The signal current, i_{o1}, also flows through Q_3 and Q_4, which is mirrored to the output transistors, Q_5 and Q_6. Note that this configuration has one less transistor stage around the input transistor loops. Specifically, the Q_5 and Q_6 source followers shown in Fig. 15.21 are not required in the circuit of Fig. 15.22, which usually results in a speed advantage for the latter.

An advantage of the transconductors in Figs. 15.21 and 15.22 (and other similar ones) is that they are easily modified to have multiple outputs by simply including output transistors. For example, the transconductor in Fig. 15.22 with two outputs is shown in Fig. 15.23. These additional transistors can be scaled for any desired ratio of output currents. These multiple outputs often allow filters to be realized using fewer transconductors. For example, a biquad similar to that in Fig. 15.11, but using multiple-output transconductors, is shown in Fig. 15.24. The transconductors are now associated with the outputs rather than the inputs. There is one transconductor per filter order plus an additional one for the input conversion from voltage to current. This approach has resulted in a savings of two transconductors. Not only does this approach save die area, but, often more importantly, it saves power.

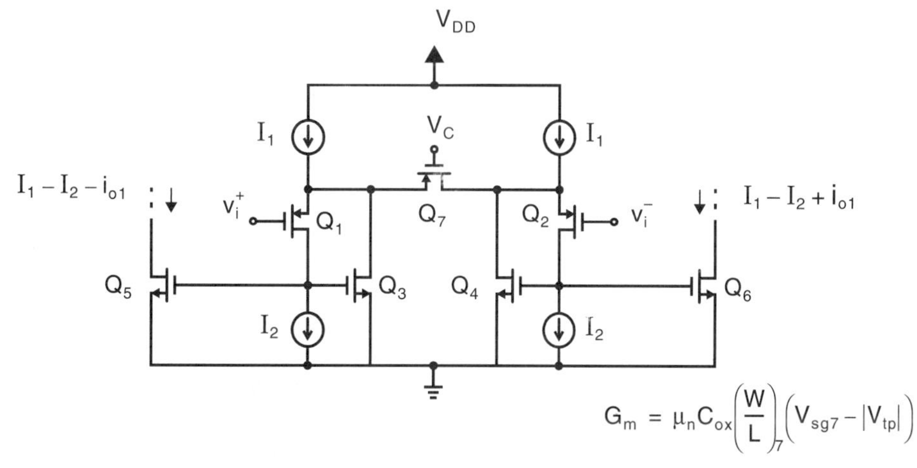

Fig. 15.22 A CMOS transconductor with p-channel inputs.

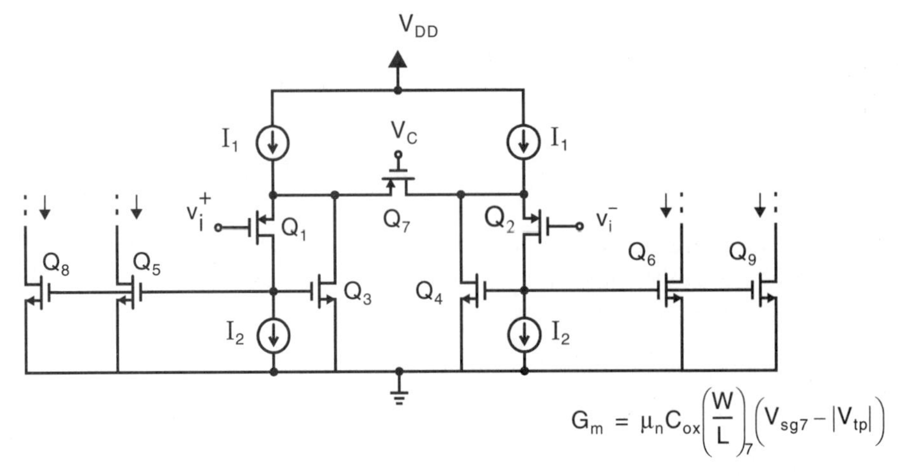

$$G_m = \mu_n C_{ox}\left(\frac{W}{L}\right)_7 (V_{sg7} - |V_{tp}|)$$

Fig. 15.23 A multiple-output CMOS transconductor.

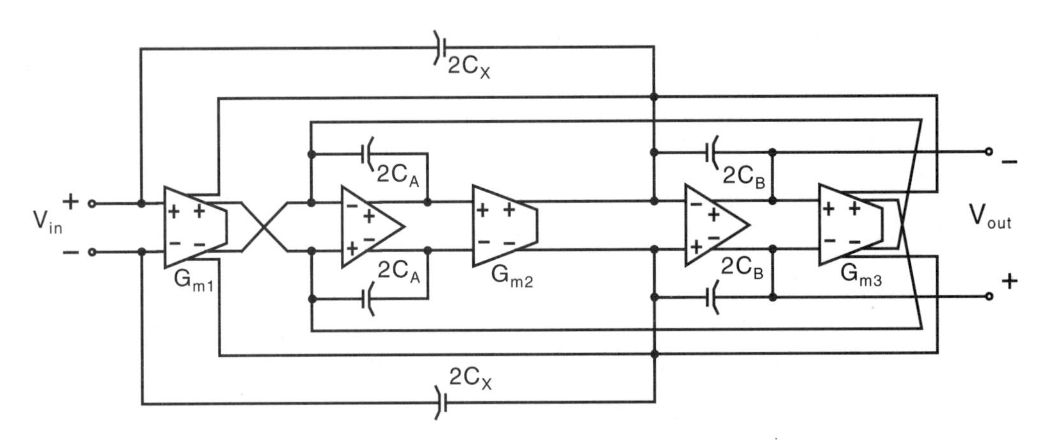

Fig. 15.24 A general biquad realized using multiple-output transconductors.

The biquad in Fig. 15.24 illustrates the use of Miller integrators. Because Miller integrators are used, the interconnects are all low-impedance nodes with current signals rather than voltage signals. This helps make these nodes insensitive to capacitively coupled noise and interconnect resistance. This current-mode style of design is increasing in popularity.

Transconductors Using Varying Bias-Triode Transistors

Another way to linearize a MOSFET differential stage, using transistors primarily in the triode region, is shown in Fig. 15.25 [Krummenacher, 1988]. Note that here the gates of transistors Q_3, Q_4 are connected to the differential input voltage rather than to a bias voltage. As a result, triode transistors Q_3 and Q_4 undergo varying bias con-

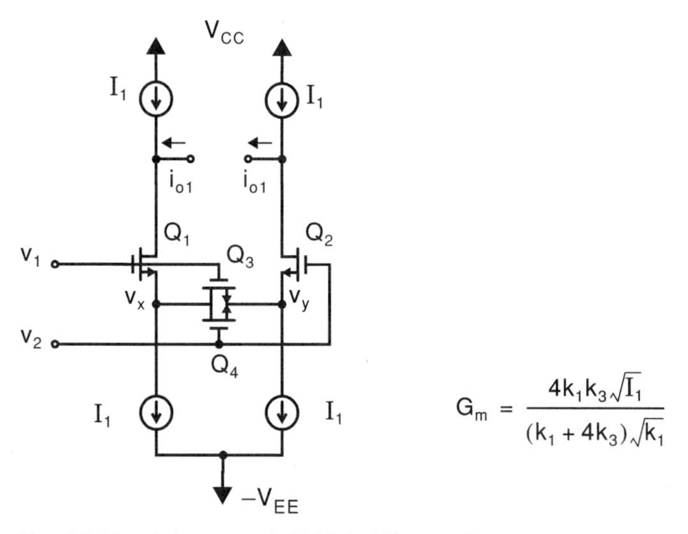

$$G_m = \frac{4k_1k_3\sqrt{I_1}}{(k_1 + 4k_3)\sqrt{k_1}}$$

Fig. 15.25 A linearized CMOS differential pair using transistors Q_3, Q_4 primarily in the triode region. The top pair of current sources can be set equal to the bottom pair through the use of a common-mode feedback.

ditions to improve the linearity of this circuit (as discussed shortly). To see that Q_3 and Q_4 are generally in the triode region, we look at the case of equal input signals (i.e., $v_1 = v_2$), resulting in

$$v_x = v_y = v_1 - v_{GS1} = v_1 - (V_{eff1} + V_{tn}) \tag{15.88}$$

Thus, the drain-source voltages of Q_3, Q_4 are both zero. However, their gate-source voltages equal those of Q_1, Q_2, implying that they are indeed in the triode region.

To determine the small-signal transconductance of this stage, we note from (15.86) that the small-signal drain-source resistance of Q_3, Q_4 is given by

$$r_{ds3} = r_{ds4} = \frac{1}{2k_3(V_{GS1} - V_{tn})} \tag{15.89}$$

where we have assumed that Q_3 and Q_4 are matched devices and have defined

$$k_3 = \frac{\mu_n C_{ox}}{2}\left(\frac{W}{L}\right)_3 \tag{15.90}$$

Note that in this circuit we cannot ignore the effect of the varying gate-source voltages of Q_1 and Q_2 since their drain currents are not fixed to a constant value (as in Fig. 15.21). The small-signal source resistance values of Q_1 and Q_2 are given by

$$r_{s1} = r_{s2} = \frac{1}{g_{m1}} = \frac{1}{2k_1(V_{GS1} - V_{tn})} \tag{15.91}$$

Using the small-signal T model, we find the small-signal output current of i_{o1} to be equal to

$$i_{o1} = \frac{v_1 - v_2}{r_{s1} + r_{s2} + (r_{ds3} \| r_{ds4})} \tag{15.92}$$

Thus, defining $G_m \equiv i_{o1}/(v_1 - v_2)$, we have

$$G_m = \frac{1}{r_{s1} + r_{s2} + (r_{ds3} \| r_{ds4})} \tag{15.93}$$

which can be shown to be

$$G_m = \frac{4k_1k_3(V_{GS1} - V_{tn})}{k_1 + 4k_3} \tag{15.94}$$

Finally, to modify this result such that it is in terms of current, we note that

$$I_1 = k_1(V_{GS1} - V_{tn})^2 \tag{15.95}$$

or, equivalently,

$$(V_{GS1} - V_{tn}) = \sqrt{\frac{I_1}{k_1}} \tag{15.96}$$

which leads to

$$G_m = \frac{4k_1k_3\sqrt{I_1}}{(k_1 + 4k_3)\sqrt{k_1}} \tag{15.97}$$

Thus, we see that the transconductance of this transconductor can be tuned by changing the bias current, I_1. Note also that the transconductance is proportional to the square root of the bias current as opposed to a linear relation in the case of a bipolar transconductor.

To see how varying bias conditions for the triode transistors improves the linearity of this circuit, first consider the case of small differences in v_1 and v_2 (say, 100 mV). In this small-input-signal case, Q_3 and Q_4 essentially act as two source degeneration resistors. Thus, linearity improves (assuming $r_{ds3} >> r_{s1}$) over that of a simple differential pair, since transistors in triode are more linear than the source resistances of transistors in the active region. In other words, for a small-input-signal value, the linearity of this circuit is similar to that for a circuit in which the gates of Q_3, Q_4 are connected to fixed-bias voltages. However, if the gates of Q_3, Q_4 were connected to fixed-bias voltages, (15.83) shows that, as the drain-source voltage is increased, the drain current is reduced below that which would occur in a linear relationship due to the squared v_{DS} term. In addition, the changing values of V_{GS1} and V_{GS2} (due to current changes) also reduce the value of V_{DS}, and hence the transconductance of the cell for large input signals. To alleviate this reduction in transconductance for large input signals, the gates of Q_3 or Q_4 are connected to the input signals as shown. Now, as the input signal is increased, the small-signal resistance of one of the two triode transistors in parallel, Q_3 or Q_4, is reduced. This reduced resistance attempts to increase the transconductance, resulting in a partial cancelling of the decreasing transconductance value. Thus, if a proper ratio of k_1/k_3 is chosen, a more stable transconductance (and hence better linearity) is achieved. For a 4-MHz filter presented in [Krummenacher, 1988], a ratio of $k_1/k_3 = 6.7$ is used to obtain distortion levels around 50 dB below the maximum output signal level of 350 mVrms. Also, an extension of this approach is presented in [Silva-Martinez, 1991], where the input signal is split over two input stages to further improve the distortion performance.

EXAMPLE 15.7

Consider the circuit of Fig. 15.25 with the following values: $I_1 = 100\ \mu A$, $\mu_n C_{ox} = 96\ \mu A/V^2$, $(W/L)_1 = (W/L)_2 = 20$, $(W/L)_3 = (W/L)_4 = 3$, and $V_2 = 0\ V$.

1. Assuming a perfectly linear transconductor, use (15.97) to find i_{o1} when V_1 equals 2.5 mV and 250 mV.
2. Assume the gates of Q_3, Q_4 are connected to ground and use classical models for both the triode and active regions. Find the true value of i_{o1} when V_1 equals 2.5 mV and 250 mV. Compare your results with those in part 1.
3. Repeat part 2, assuming the gates of Q_3, Q_4 are connected to the input signals as shown in Fig. 15.25.

Solution

1. For the given values, we have

$$k_1 = \frac{96}{2} \times 20 = 960\ \mu A/V^2 \tag{15.98}$$

$$k_3 = \frac{96}{2} \times 3 = 144\ \mu A/V^2 \tag{15.99}$$

and substituting the above values in (15.97), we have

$$G_m = \frac{4k_1k_3\sqrt{100}}{(k_1 + 4k_3)\sqrt{k_1}} = 116.2\ \mu A/V \tag{15.100}$$

Now, to find i_{o1}, we use the definition of G_m to find

$$i_{o1} = G_m(V_1 - V_2) = G_m V_1 \tag{15.101}$$

Thus, for $V_1 = 2.5$ mV, we have $i_{o1} = 0.2905\ \mu A$. However, if $V_1 = 250$ mV, then $i_{o1} = 29.05\ \mu A$. In other words, for a 250-mV input voltage, the total drain current of Q_1 would be 129.05 μA, whereas that for Q_2 would be 70.95 μA.

2. To find i_{o1} in this case, we first note that the sum of the drain currents in Q_1 and Q_2 must equal the total bias current, 200 μA. In other words,

$$i_{D1} + i_{D2} = 2I_1 = 200\ \mu A \tag{15.102}$$

And since (using $V_{effi} = V_{GS} - V_{tn}$) the drain currents in Q_1 and Q_2 are given by

$$i_{D1} = \frac{\mu_n C_{ox}}{2}\left(\frac{W}{L}\right)_1 (V_{eff1}^2) \quad \text{and} \quad i_{D2} = \frac{\mu_n C_{ox}}{2}\left(\frac{W}{L}\right)_2 (V_{eff2}^2) \tag{15.103}$$

we can write (15.102) as

$$V_{eff1}^2 + V_{eff2}^2 = \frac{5}{24}\,V^2 \tag{15.104}$$

The current from node V_x to V_y through the two triode transistors is given by

$$i_{xy} = 2 \times i_{D3} = 2\mu_n C_{ox}\left(\frac{W}{L}\right)_3 \left(V_{eff2}V_{xy} - \frac{V_{xy}^2}{2}\right) \tag{15.105}$$

where we have noted that $V_{eff3} = V_{eff2}$. The sum of i_{xy} and i_{D2} must equal I_1, resulting in

$$960V_{eff2}^2 + 576\left(V_{eff2}V_{xy} - \frac{V_{xy}^2}{2}\right) = 100 \tag{15.106}$$

Finally, we note that the input signal is equal to the following sum (the threshold voltage terms cancel):

$$-V_{eff2} + V_{xy} + V_{eff1} = V_{in} \tag{15.107}$$

Equations (15.104), (15.106), and (15.107) can be solved using an equation solver and then choosing the solution that has all positive values.

For a value of $V_{in} = 2.5\text{ mV}$, the following results are obtained.

$$V_{eff1} = 0.32322\text{ V}$$
$$V_{eff2} = 0.32228\text{ V}$$
$$V_{xy} = 0.0015648\text{ V} \tag{15.108}$$

These values of effective gate voltages imply that the drain currents are

$$i_{D1} = 100.2898\ \mu\text{A} \qquad i_{D2} = 99.7104\ \mu\text{A} \tag{15.109}$$

which, in turn, imply that

$$i_{o1} = i_{D1} - 100\ \mu\text{A} = 0.2898\ \mu\text{A} \tag{15.110}$$

Such a result agrees closely with that of part 1.

However, for the case of $V_{in} = 250\text{ mV}$, we have

$$V_{eff1} = 0.35452\text{ V}$$
$$V_{eff2} = 0.28749\text{ V}$$
$$V_{xy} = 0.18297\text{ V} \tag{15.111}$$

resulting in $i_{o1} = 20.66\ \mu A$. This result is about 30 percent lower than that expected by a linear transconductor value of $29.05\ \mu A$, as found in part 1.

3. Finally, if the gates are connected to the input signal, the preceding equations in part 2 remain the same, except for i_{xy}, which is now the sum of i_{D3} and i_{D4}, resulting in,

$$i_{xy} = \mu_n C_{ox}\left(\frac{W}{L}\right)_3\left[(V_{eff2} + V_{in})V_{xy} - \frac{V_{xy}^2}{2}\right] + \mu_n C_{ox}\left(\frac{W}{L}\right)_4\left(V_{eff2}V_{xy} - \frac{V_{xy}^2}{2}\right) \tag{15.112}$$

and so (15.106) is replaced with

$$960V_{eff2}^2 + 576[(2V_{eff2} + V_{in})V_{xy} - V_{xy}^2] = 100 \tag{15.113}$$

Once again, using an equation solver, we find that in the case of $V_{in} = 2.5\ mV$, the results are essentially the same as in parts 1 and 2. However, in the case of $V_{in} = 250\ mV$, we find that

$$V_{eff1} = 0.36620\ V$$
$$V_{eff2} = 0.27245\ V$$
$$V_{xy} = 0.15625\ V \tag{15.114}$$

These values lead to $i_{o1} = 28.74\ \mu A$, which is only 1 percent below that expected by a linear transconductor value of $29.05\ \mu A$.

Transconductors Using Constant Drain-Source Voltages

Recall, once again, the classical model for a transistor biased in the triode region,

$$i_D = \mu_n C_{ox}\left(\frac{W}{L}\right)\left[(V_{GS} - V_{tn})V_{DS} - \frac{V_{DS}^2}{2}\right] \tag{15.115}$$

We see here that if the drain-source voltage, V_{DS}, is kept constant, the drain current is linear with respect to an applied gate-source voltage (ignoring the dc offset current due to the squared V_{DS} term) [Pennock, 1985]. Unfortunately, this simple model does not account for important second-order effects (such as velocity saturation and mobility degradation) that severely limit the accuracy of the preceding equation model. However, if fully differential structures are used to cancel even-order error terms, a reasonably linear transconductor (around 50 dB) can be built using this constant-V_{DS} approach.

One possible implementation based on the linearization scheme just described is shown in Fig. 15.26. Here, Q_1 and Q_2 are placed in the triode region and their drain-source voltages are set equal to V_C through the use of Q_3, Q_4 and two extra amplifiers.

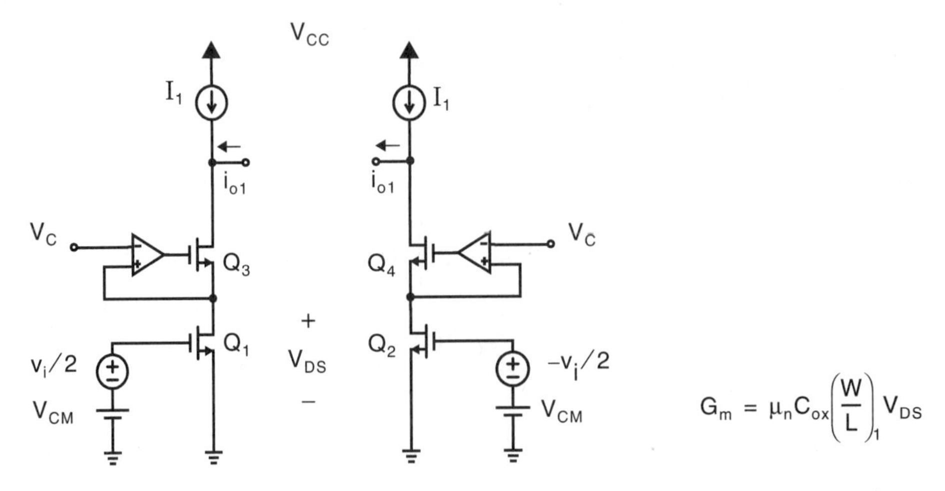

Fig. 15.26 A differential transconductor using triode transistors that have constant drain-source voltages. The control signal, V_C, adjusts the transconductance value, whereas V_{CM} is a fixed common-mode voltage. The top current pair, I_1, is set equal to the bias current through Q_1, Q_2, using a common-mode feedback circuit (not shown).

The differential input signal is applied to the gates of Q_1, Q_2 and is superimposed on a constant common-mode voltage signal, V_{CM}. An alternative design here is to use an extra opamp to create a G_m-C opamp integrator [Wong, 1989].

For the circuit of Fig. 15.26, we have

$$i_{D1} = \mu_n C_{ox}\left(\frac{W}{L}\right)_1\left[\left(\frac{v_i}{2} + V_{CM} - V_{tn}\right)V_{DS} - V_{DS}^2/2\right] \tag{15.116}$$

Noting that i_{o1} is the difference between this current and I_1, which is set via a common-mode feedback signal equal to

$$I_1 = \mu_n C_{ox}\left(\frac{W}{L}\right)_1\left[(V_{CM} - V_{tn})V_{DS} - V_{DS}^2/2\right] \tag{15.117}$$

we then have

$$i_{o1} = i_{D1} - I_1 = \mu_n C_{ox}\left(\frac{W}{L}\right)_1\left(\frac{v_i}{2}\right)V_{DS} \tag{15.118}$$

Then, defining $G_m \equiv i_{o1}/v_i$ results in

$$G_m = \mu_n C_{ox}\left(\frac{W}{L}\right)_1 V_{DS} \tag{15.119}$$

Note that here the transconductance value is proportional to the drain-source voltage, and from (15.117), we see that the bias current is also approximately proportional to the drain-source voltage (if V_{DS} is small).

EXAMPLE 15.8

Consider the circuit shown in Fig. 15.26, where $\mu_n C_{ox} = 96\ \mu A/V^2$, $(W/L)_1 = (W/L)_2 = 10$, $V_{tn} = 0.8\ V$, and $V_{CM} = 2\ V$. Find the transconductance value when $V_C = 0.2\ V$ and when $V_C = 0.4\ V$. For these two control-signal values, also find the bias currents through Q_1 and Q_2 (i.e., when $v_i = 0$) and the maximum input signal swing before either transistor exits the triode region (assume the classical transistor model).

Solution

Using (15.119), for $V_C = 0.2\ V$, we have

$$G_m = 96\ \mu A/V^2 \times 10 \times 0.2\ V = 0.192\ mA/V$$

Also, the bias current through Q_1 or Q_2 for this V_C value can be found using (15.117):

$$I_1 = 96\ \mu A/V^2 \times 10\left[(2-0.8)0.2 - \frac{0.2^2}{2}\right] = 0.211\ mA$$

Finally, to remain in the triode region, the *effective* gate-source voltages of Q_1 and Q_2 must remain above $V_{DS} = 0.2\ V$. In other words, their gate-source voltages must remain above $V_{DS} + V_t = 1.0\ V$. Since the common-mode voltage, V_{CM}, equals 2 V, the negative portion, $v_i/2$, must not go below a level of -1 V, otherwise the transistor will enter the active region. Thus, a maximum differential input voltage, v_i, of 2 V can be applied in this case.

Similarly, for $V_C = 0.4\ V$, we have twice the transconductance value of $G_m = 0.384\ mA/V$. Here, the bias current is found to be $I_1 = 0.384\ mA$ and the maximum input signal swing is a differential voltage of $v_i = 1.6\ V$. Note that doubling the drain-source voltage doubles the transconductance (as expected) but requires almost a doubling of the bias current and a loss of signal swing.

15.4 CMOS TRANSCONDUCTORS USING ACTIVE TRANSISTORS

In general, CMOS transconductors based on the use of active transistors[2] have worse linearity performance than their triode-based counterparts. However, active-based designs often have improved speed performance (assuming equal bias currents in the two cases). The linearity of active-transistor transconductors is only moderate (around 30 to 50 dB) because they rely on the square-law model of MOS transistors, which is not very accurate (particularly in short-channel processes). In addition, only the difference

2. Active transistors are those transistors that operate in the active region. The active region is also known as the pinch-off, or saturation, region. This active region for an n-channel transistor occurs when V_{GS} is greater than V_{tn} and $V_{DS} \geq V_{GS} - V_{tn} = V_{eff}$.

in output currents is ideally linear, whereas individual currents have significant second-order harmonics. Thus, second-order harmonic distortion also occurs if the difference is not taken precisely.

We should mention here that not all linearization techniques using active transistors are discussed. For example, one approach not discussed is a technique in which the tail current of a differential pair is adaptively biased [Nedungadi, 1984]. This technique is limited in its frequency response, and thus a designer would usually choose a triode-based design instead. In summary, in this section we present CMOS linearization techniques intended for high-speed operation and only moderate linearity.

CMOS Pair

Before proceeding to look at transconductor circuits that have active transistors, it is worthwhile to introduce a two-transistor circuit known as the *CMOS pair*, as shown in Fig. 15.27 [Seevinck, 1987]. As we will show, this two-transistor pair acts as a single transistor, assuming that both transistors remain in the active region. Such an observation allows one to more easily derive some of the CMOS transconductors described later in this section.

To show that this two-transistor pair acts as a single transistor, we first assume that both transistors remain in the active region. Thus, we can write their drain currents as

$$I_D = K_n(V_{GSn} - V_{tn})^2 \qquad \text{where } V_{tn} > 0 \tag{15.120}$$

and

$$I_D = K_p(V_{SGp} + V_{tp})^2 \qquad \text{where } V_{tp} < 0 \tag{15.121}$$

Rearranging these two equations in terms of their gate-source voltages gives

$$V_{GSn} = V_{tn} + \frac{1}{\sqrt{K_n}}\sqrt{I_D} \tag{15.122}$$

and

$$V_{SGp} = -V_{tp} + \frac{1}{\sqrt{K_p}}\sqrt{I_D} \tag{15.123}$$

Defining the equivalent gate-source voltage, $V_{GS\text{-}eq}$, to be the difference between the two gate voltages, we have

$$V_{GS\text{-}eq} \equiv V_{GSn} + V_{SGp} = (V_{tn} - V_{tp}) + \left(\frac{1}{\sqrt{K_n}} + \frac{1}{\sqrt{K_p}}\right)\sqrt{I_D} \tag{15.124}$$

Since this equation has the same form as those for a single transistor (equations (15.122) and (15.123)), we define the equivalent threshold voltage, $V_{t\text{-}eq}$, as

$$V_{t\text{-}eq} \equiv V_{tn} - V_{tp} \qquad \text{where } V_{tn} > 0 \text{ and } V_{tp} < 0 \tag{15.125}$$

and we define the equivalent device parameter, K_{eq}, as

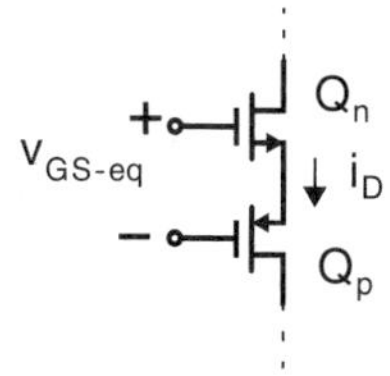

$V_{GS\text{-}eq}$ + − Q_n i_D Q_p

$$i_D = K_{eq}\left(V_{GS\text{-}eq} - V_{t\text{-}eq}\right)^2$$

$$K_{eq} = \frac{K_n K_p}{\left(\sqrt{K_n} + \sqrt{K_p}\right)^2}$$

$$V_{t\text{-}eq} = V_{tn} - V_{tp}$$

(where $V_{tn} > 0$ and $V_{tp} < 0$)

Fig. 15.27 A CMOS double pair, where both transistors are in the active region. The transistor pair is equivalent to a single transistor with an equivalent threshold voltage, $V_{t\text{-}eq}$, and device parameter, K_{eq}.

$$\frac{1}{\sqrt{K_{eq}}} \equiv \frac{1}{\sqrt{K_n}} + \frac{1}{\sqrt{K_p}} \tag{15.126}$$

which can be rewritten as

$$K_{eq} = \frac{K_n K_p}{\left(\sqrt{K_n} + \sqrt{K_p}\right)^2} \tag{15.127}$$

Finally, combining (15.124), (15.125), and (15.127) and rearranging, we see that when both transistors are in their active regions, the drain current can be written as

$$i_D = K_{eq}(v_{GS\text{-}eq} - V_{t\text{-}eq})^2 \tag{15.128}$$

where K_{eq}, $V_{GS\text{-}eq}$, and $V_{t\text{-}eq}$ are defined as just described. Clearly, this equation is in the same form as that for a single transistor, and thus this two-transistor pair can replace single transistors in many circuits. As we will see next, this replacement can be quite useful in designing linear transconductor circuits.

Constant Sum of Gate-Source Voltages

One popular approach for creating an active transistor transconductor is by using a constant sum of gate-source voltages, as shown in Fig. 15.28. Assuming the two transistors are operating in the active region (i.e., the drain voltages are high and each of v_{GS1} and v_{GS2} remain greater than V_{tn}), the output differential current is given by

$$(i_{D1} - i_{D2}) = K\left(v_{GS1} + v_{GS2} - 2V_{tn}\right)(v_{GS1} - v_{GS2}) \tag{15.129}$$

Here, the *output differential current is linear if the sum of the two gate-source voltages remains constant.* As we will see next, there are a variety of ways to make the sum of the gate-source voltages remain constant when applying an input signal. One important point to note here is that, although the differential current is linear, the individual drain currents are not linear. Thus, if the subtraction between currents has some

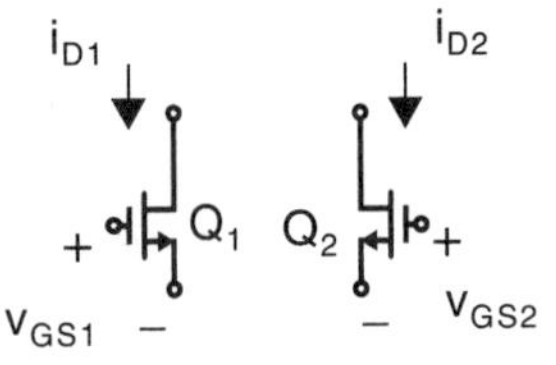

i_{D1} i_{D2}

$+$ Q_1 Q_2 $+$

v_{GS1} $-$ $-$ v_{GS2}

(Matched devices)

$$K = (\mu_n/2)C_{ox}\left(\frac{W}{L}\right)$$

$$i_{D1} = K\left(v_{GS1} - V_{tn}\right)^2 \quad \text{and} \quad i_{D2} = K\left(v_{GS2} - V_{tn}\right)^2$$

$$(i_{D1} - i_{D2}) = K\left(v_{GS1} + v_{GS2} - 2V_{tn}\right)\left(v_{GS1} - v_{GS2}\right)$$

Fig. 15.28 Two-transistor circuit in which the difference in the drain currents is linear with respect to $(v_{GS1} - v_{GS2})$ if the sum of the two gate-source voltages remains constant. Transistors are matched and in the active region.

error (as it certainly will), some distortion products will occur even if perfect square-law devices are obtainable.

Source-Connected Differential Pair

One approach to maintain a constant sum of gate-source voltages is to apply a balanced differential input voltage, as shown in Fig. 15.29. Here, the differential pair is biased using the dc common-mode voltage, V_{CM}, whereas the input signal varies *symmetrically* around this common-mode voltage. Note that, from the preceding analysis, this input signal symmetry is important in maintaining linearity. However, in practice, a slightly nonbalanced input signal is not the dominant source of nonlinearity. Specifically, the linearity achievable with this approach is quite limited because MOS transistors are not well modelled by square-law behavior. In addition, it is difficult to obtain a precise difference between the two drain currents, which also causes even-order harmonics. As a result, this technique is typically limited to less than 50 dB of linearity.

To tune the transconductance of this circuit, the common-mode voltage, V_{CM}, can be adjusted. However, it should be noted that, in a short-channel process and when moderate gate-source voltages are used, carrier velocity saturation occurs and limits the transconductance variation. In addition, one should be aware that as the input

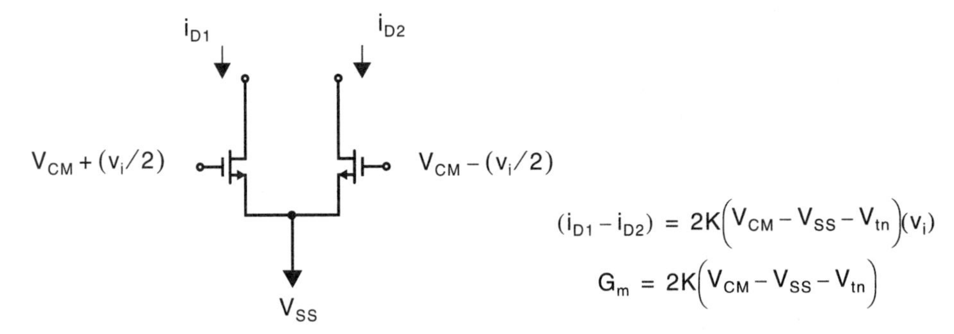

Fig. 15.29 A source-connected differential-pair transconductor. The differential input signal must be balanced around V_{CM} for linearization of the transistor square-law relation to occur.

common-mode voltage is adjusted, linearity performance and maximum signal swing levels change as well. One example of this approach is presented in [Snelgrove, 1992] where speeds up to 450 MHz are demonstrated in a 0.9-μm CMOS process with around a 30- to 40-dB dynamic range performance.

Inverter-Based

A transconductor approach similar to the source-connected differential pair is the inverter-based transconductor, shown in Fig. 15.30 [Park, 1986]. Here, two CMOS pairs, Q_{1a}, Q_{1b} and Q_{2a}, Q_{2b}, have replaced the two individual transistors in a traditional CMOS inverter, and the difference in their output currents is the overall output current, i_o. Assuming that all four transistors are in the active region and that the two CMOS pairs are matched, we can use the previous results on CMOS pairs to write

$$i_1 = K_{eq}(V_{GS\text{-}eq1} - V_{t\text{-}eq})^2 \tag{15.130}$$

and

$$i_2 = K_{eq}(V_{GS\text{-}eq2} - V_{t\text{-}eq})^2 \tag{15.131}$$

Note here that these equations are in the same form as those shown in Fig. 15.28. Thus, as in (15.129), the output current is given by

$$i_o \equiv i_1 - i_2 = K_{eq}(V_{GS\text{-}eq1} + V_{GS\text{-}eq2} - 2V_{t\text{-}eq})(V_{GS\text{-}eq1} - V_{GS\text{-}eq2}) \tag{15.132}$$

Thus, the output current is linearly related to the difference in equivalent gate-source voltages of the CMOS pairs if the sum of their equivalent gate-source voltages is a constant value. To see that a constant equivalent gate-source voltage sum occurs, we have

$$V_{GS\text{-}eq1} = V_{C1} - V_{in} \tag{15.133}$$

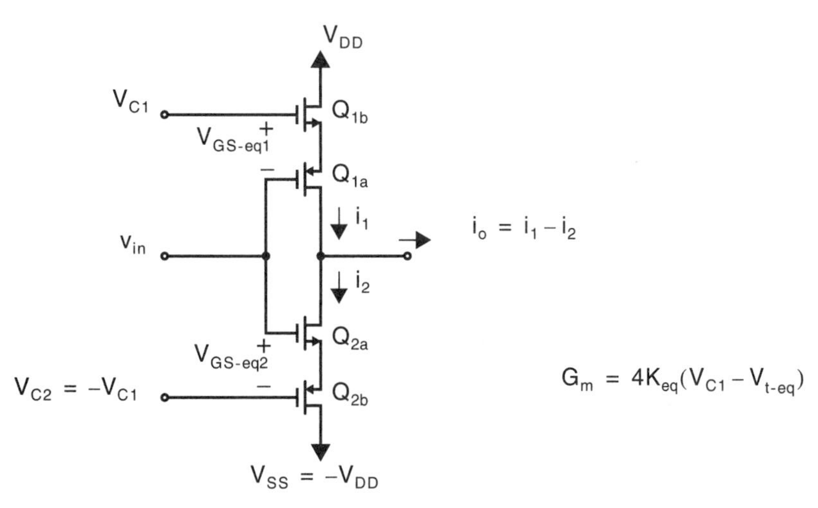

Fig. 15.30 Complementary differential-pair transconductor using CMOS pairs.

and

$$V_{GS\text{-}eq2} = v_{in} - V_{C2} \tag{15.134}$$

and therefore

$$V_{GS\text{-}eq1} + V_{GS\text{-}eq2} = V_{C1} - V_{C2} \tag{15.135}$$

Also, the difference in equivalent gate-source voltages is given by

$$V_{GS\text{-}eq1} - V_{GS\text{-}eq2} = V_{C1} + V_{C2} - 2v_{in} \tag{15.136}$$

Finally, in the case where we choose $V_{C2} = -V_{C1}$, we have

$$i_o = 4K_{eq}(V_{C1} - V_{t\text{-}eq})(v_{in}) \tag{15.137}$$

and the transconductance value can be varied by changing the control voltages, V_{C1} and V_{C2} (which, incidentally, are high-input impedance nodes and thus are easily adjustable). Note that, with this transconductor, matching need occur only for transistors of the same type. Specifically, the n-channel transistors should match and the p-channel transistors should match, but there is no need to match n- and p-channel transistors to each other.

Differential-Pair with Floating Voltage Sources

Another approach to maintain a constant sum of gate-source voltages is by using two floating dc voltage sources, as shown in Fig. 15.31 [Nedungadi, 1984]. Writing a voltage equation around the loop, we have

$$v_{GS1} - (V_x + V_{tn}) + v_{GS2} - (V_x + V_{tn}) = 0 \tag{15.138}$$

and thus,

$$v_{GS1} + v_{GS2} = 2(V_x + V_{tn}) \tag{15.139}$$

As a result, this circuit maintains a constant sum of gate-source voltages even if the applied differential signal is not balanced. Also, writing two equations that describe the relationship between v_1 and v_2, we have

$$v_1 - v_{GS1} + V_x + V_{tn} = v_2 \tag{15.140}$$

and

$$v_2 - v_{GS2} + V_x + V_{tn} = v_1 \tag{15.141}$$

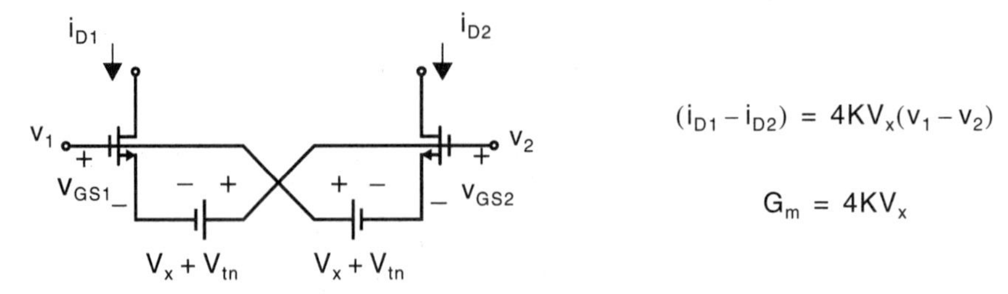

Fig. 15.31 A linearized differential pair using floating voltage sources.

Subtracting (15.140) from (15.141) and rearranging, we obtain

$$V_{GS1} - V_{GS2} = 2(v_1 - v_2) \tag{15.142}$$

Thus, the difference between the input voltages is equal to one half the difference between the gate-source voltages. Finally, the differential output current is found by combining (15.129), (15.139), and (15.142) to give

$$(i_{D1} - i_{D2}) = 4KV_x(v_1 - v_2) \tag{15.143}$$

One simple way to realize the floating voltage sources of Fig. 15.31 is to use two source followers, as shown in Fig. 15.32 [Nedungadi, 1984]. The transistors labelled nK are n times larger than the other two transistors, and thus they act as source followers when n is large (typically, when n is 5 or greater). If n is not large, the bias current through the two source followers changes as the input signal is applied and linearity is worsened. A disadvantage of this circuit is that a large dc bias current passes through the two source followers. However, if only moderate linearity is desired and the input signal is relatively small (say, 500 mV), then this approach can be quite useful.

Since it is difficult to realize a floating voltage source that drives the source of another transistor, another approach is to modify the circuit of Fig. 15.31 such that the floating voltage sources need drive only the transistor gates. Such a modification is possible by replacing the single transistors shown in Fig. 15.31 with their CMOS pairs, as shown in Figure 15.33. Note that the voltage sources must now be larger than $V_{t\text{-}eq}$ and must include a controllable V_x voltage. Fortunately, such circuits are easily accomplished by using two more CMOS pairs, as shown in Fig. 15.34, where the pairs Q_{3a}, Q_{3b} and Q_{4a}, Q_{4b} have been added together with two adjustable current sources [Seevinck, 1987]. Referring back to (15.128), we see that V_x is given by

$$V_x = \frac{1}{\sqrt{K_{eq}}}\sqrt{I_B} \tag{15.144}$$

where, for simplicity, we have assumed that all CMOS pairs are matched. As a result, the differential output current is given by

$$(i_{D1} - i_{D2}) = 4\sqrt{K_{eq}I_B}(v_1 - v_2) \tag{15.145}$$

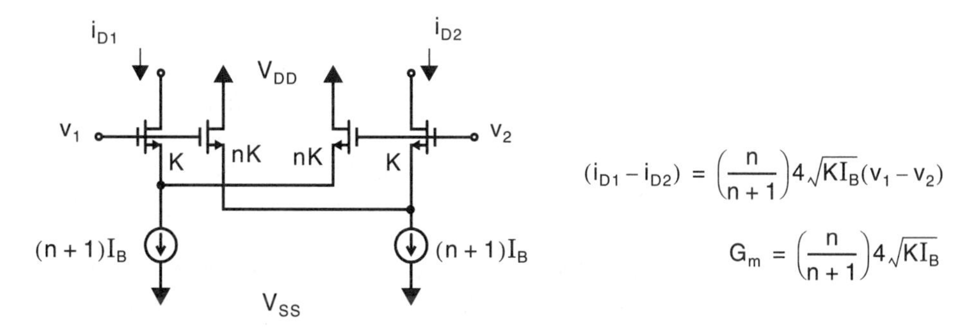

Fig. 15.32 A linear transconductor for large n.

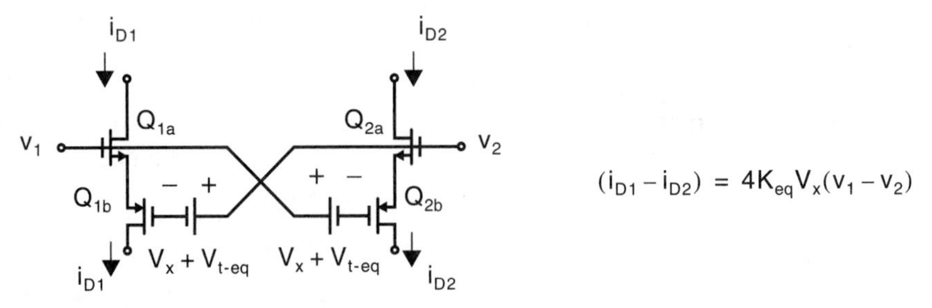

Fig. 15.33 A linearized differential pair in which the floating voltage sources need drive only the transistor gates.

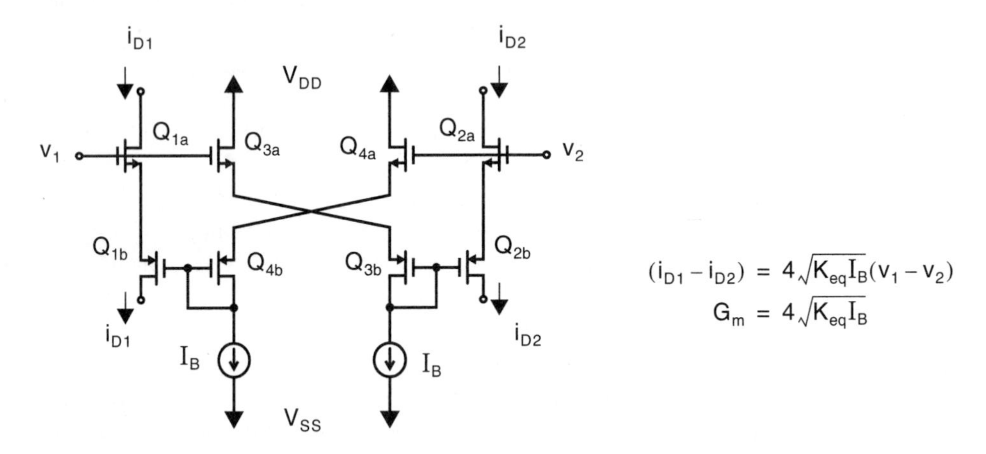

Fig. 15.34 A transconductor that uses CMOS pairs and floating voltage sources.

Thus, we see that the output current can be tuned by adjusting the dc bias current, I_B. Note that the transconductance coefficient is proportional to the square root of I_B and thus a fourfold increase in current results in only a factor of 2 coefficient increase.

Before leaving this circuit, a couple of points are worth mentioning. First, note that the output currents from this circuit are available at both the upper pair, Q_{1a}, Q_{2a}, as well as at the lower pair, Q_{1b}, Q_{2b}. Such flexibility may be useful in the remaining transconductor circuitry. Second, note that V_{SS} must be at least $V_{tn} + |V_{tp}|$ below the most negative input signal swing of either v_1 or v_2. If we assume v_1 and v_2 have ±1-V signals associated with them around ground, then, for $V_{tn} = -V_{tp} = 1$ V, V_{SS} would have to be less than −3 V. Thus, assuming we have symmetrical power supplies, a 6-V power supply would be needed for only a 2-V signal swing. Finally, note that this circuit was discussed earlier in Chapter 6 as a class-AB input stage to improve the slew rate of an opamp (see Fig. 6.16 in Section 6.5). Thus, this particular class-AB input stage has a linear transconductance, when all transistors remain in the active region.

Bias-Offset Cross-Coupled Differential Pairs

Another approach to realize a transconductor with active transistors is to use two cross-coupled differential pairs where the input into one pair is intentionally voltage offset [Wang, 1990]. One implementation of this approach is shown in Fig. 15.35, where the differential pair Q_1, Q_2 is cross coupled with the bias-offset differential pair Q_3, Q_4. Here, we assume all transistors are in the active region and are matched (i.e., we assume all transistors have equal device parameters of K and equal threshold voltages of V_{tn}).

To find the differential output current, we first find

$$i_1 = K(v_1 - V_x - V_{tn})^2 + K(v_2 - V_B - V_x - V_{tn})^2 \tag{15.146}$$

and

$$i_2 = K(v_2 - V_x - V_{tn})^2 + K(v_1 - V_B - V_x - V_{tn})^2 \tag{15.147}$$

Now finding the difference, $i_1 - i_2$, we have

$$(i_1 - i_2) = 2KV_B(v_1 - v_2) \tag{15.148}$$

Thus, the output differential current is linear with respect to the differential input voltage (as expected) and the transconductance value is proportional to V_B. Note, however, that the bias currents through Q_5, Q_6, Q_7, and Q_8 are all square-law related to V_B. Thus, the transconductance value is proportional to the square root of the changing bias current, as in the other active transistor CMOS transconductors. Also, note that the bias current, I_{SS}, does not affect the transconductance value but does determine the maximum (or minimum) output current available.

Finally, note that this offset-bias linearization technique can also be realized in a single-ended version with only a few transistors, as shown in Fig. 15.36 [Bult, 1987]. The analysis is left to the reader.

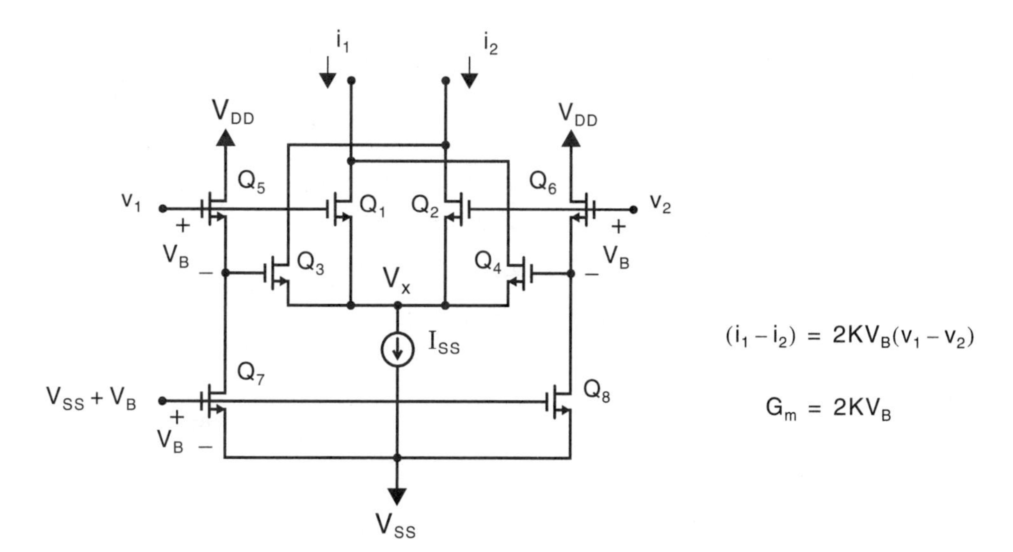

Fig. 15.35 A bias-offset cross-coupled transconductor.

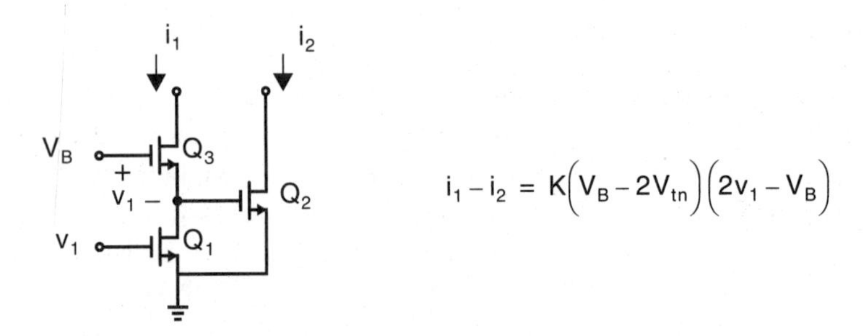

Fig. 15.36 A single-ended offset-bias transconductor.

15.5 BICMOS TRANSCONDUCTORS

In this section, we discuss some of the reported approaches for realizing BiCMOS transconductors and filters.

Tunable MOS in Triode

One approach for realizing a BiCMOS integrator is shown in Fig. 15.37 [Laber, 1993]. Here, linearization and tuning are accomplished by using transistors Q_1 through Q_4, which are all assumed to be in the triode region. As we saw in Section 15.3, the currents through these transistors are ideally linear if their drain-source voltages are kept constant. Although in practice these currents have rather large

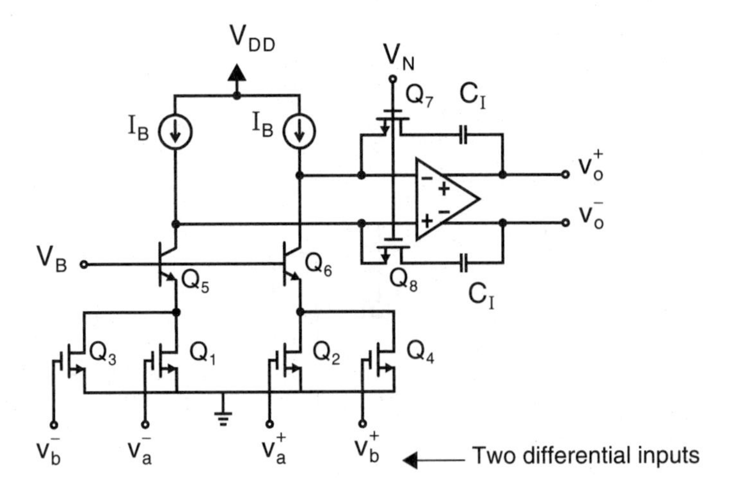

Fig. 15.37 A fully differential two-input BiCMOS integrator using MOS transistors in triode, Q_1 through Q_4. G_m-C opamp integration is used to reduce parasitic capacitance effects. Note that, in addition to a common-mode feedback circuit for the output (not shown), the common-mode voltage at the input nodes of the opamp must also be set.

second-order distortion, fortunately even-order distortion products are significantly reduced in a fully differential structure. The drain-source voltages of Q_1 through Q_4 are controlled through the use of the two emitter followers, Q_5, Q_6. This simple voltage control is one of the benefits of using a BiCMOS process since a MOS source follower would have too much voltage variation. Note that additional inputs into this integrator are added as simply two extra transistors (a two-input integrator is shown). Various coefficient ratios can be obtained by scaling the sizes of the input transistors. However, because of the shown tuning circuit connection, all coefficients in one integrator will scale together as the control voltage is varied. Note also that a G_m-C opamp integrator is used to reduce the effects of parasitic capacitance. Also, transistors Q_7, Q_8 are included to realize lead compensation. Finally, it should be mentioned that a common-mode feedback circuit is required at the output nodes of the differential opamp. Also, some method is needed for setting the common-mode voltage at the opamp inputs. The BiCMOS filter in [Laber, 1993] is built in a 1.5-μm, 4-GHz BiCMOS technology and operated around 20 MHz using a 5-V supply. The transconductor demonstrated moderate linearity of around 50 dB when processing 2-V_{pp} differential output signals, but no mention is made of any intermodulation results or the frequency of the test signals.

Fixed-Resistor Transconductor with a Translinear Multiplier

Another approach that has been successfully used to realize BiCMOS filters that have very high linearity has been to use highly linear transconductors realized using fixed resistors and then achieve tuning by following the transconductor with a translinear multiplier. As we saw in Chapter 8, the output of a translinear multiplier is linear with respect to one of its input currents, and its gain coefficient can be scaled as the ratio of two other currents in either a positive or negative fashion.

A translinear multiplier can be used for realizing a low-distortion BiCMOS integrator, as shown in Fig. 15.38 [Willingham, 1993]. Here, the output currents from fixed-value transconductors are applied to a translinear multiplier, such that tuning

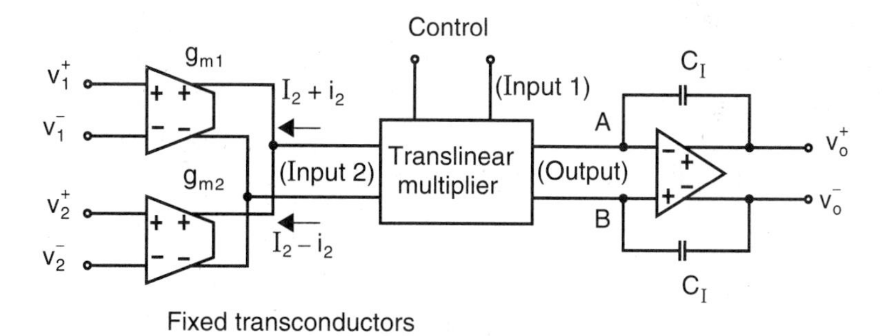

Fig. 15.38 A two-input BiCMOS integrator using fixed resistor-degenerated transconductors together with a translinear multiplier for tuning. The common-mode voltage at nodes A and B was set by using a low-input-impedance opamp.

can be accomplished while maintaining high linearity. The outputs of the translinear multiplier are applied to a Miller integrator to once again maintain high linearity. In this implementation, the input impedance of the fully differential opamp is kept low such that a fixed voltage occurs at the input nodes of the opamp, thereby setting the input common-mode voltage. Phase-lead resistors are also used in series with the integration capacitors (not shown) to improve the integrator's phase response.

The fixed transconductors are realized using resistor degeneration and local feedback, as shown in Fig. 15.39. Figure 15.39(*a*) shows that local feedback around the two shown opamps force their input nodes to be equal, and therefore the applied input voltage appears across R_x. A simplified circuit implementation of this approach is shown in Fig. 15.39(*b*), which shows that the opamps are implemented with only a PMOS transistor and current source.

This integrator was used to build a seventh-order low-pass filter in a 2-μm, 2.5-GHz BiCMOS process that featured thin-film resistors. The filter passband edge was tunable from 7.5 MHz to 10 MHz. The results of this low-distortion BiCMOS filter are quite impressive; third-order intermodulation products remained below 65 dB, with respect to the input signal, when measured using two-tone inputs near 4 MHz. Finally, it should be mentioned that the circuit also uses a center tap on each of the R_x resistors to set the common-mode voltage of the preceding output. With this simple addition, much of the traditional common-mode circuitry is eliminated since it is necessary only to compare the center tap voltage with a dc reference signal and then feed back the appropriate control signal.

Similar approaches, in which a highly linear transconductor that has local feedback is followed by a current attenuator, are also possible using ordinary CMOS and are now gaining in popularity [Chang, 1995].

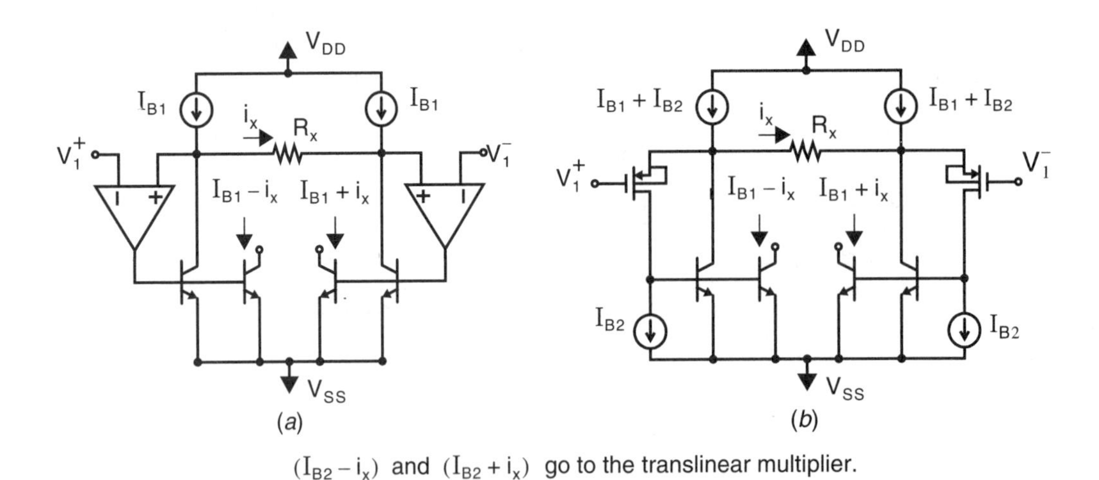

Fig. 15.39 A fixed BiCMOS transconductor: (*a*) the two local feedbacks keep the input voltage applied across R_x; (*b*) circuit implementation.

Fixed Active MOS Transconductor with a Translinear Multiplier

A BiCMOS integrator intended for applications that require high-speed operation, wide tuning range, and only moderate linearity is shown in Fig. 15.40 [Shoval, 1993]. In this circuit, the common-mode voltage, V_{CM}, is set to a constant value (it is set by a common-mode feedback circuit); the result is that the differential currents of Q_1, Q_4 and Q_2, Q_3 are linearly related to the balanced differential voltage (as discussed in Section 15.4). This linear voltage-to-current conversion remains at a fixed value (resulting in a relatively constant linearity performance and input swing level), and transconductance tuning is performed with the aid of a translinear multiplier. Note, however, that two pairs of MOS transistors are performing voltage-to-current conversion on the balanced input signal. The reason for using two pairs is to maximize the largest possible transconductance of the overall circuit for a given bias current. Specifically, if the multiplier is tuned to send all of i_{C6}, i_{C7} through to v_o^+, v_o^-, respectively, then the signal currents add together and the overall transconductance is equal to twice that which occurs for a single set of MOS transistors (without an increase in bias current). Alternatively, if the multiplier is tuned to send all of i_{C6}, i_{C7} through to v_o^-, v_o^+, respectively, then the currents subtract, causing a zero overall transconductance. Finally, if the translinear multiplier is tuned to cause a zero output (i.e., to cause two constant dc currents at the multiplier outputs), the overall transconductance is equal to that caused by transistors Q_2, Q_3, which is half the maximum value. Thus, the use of two pairs on the input signal doubles the maximum possible transconductance at the cost of being able to tune the transconductance only from zero to the maximum value. In other words, the overall transconductance value cannot change signs by changing the multiplier control signal. Of course, both positive and negative transconductance values can be obtained by cross-coupling the output signals.

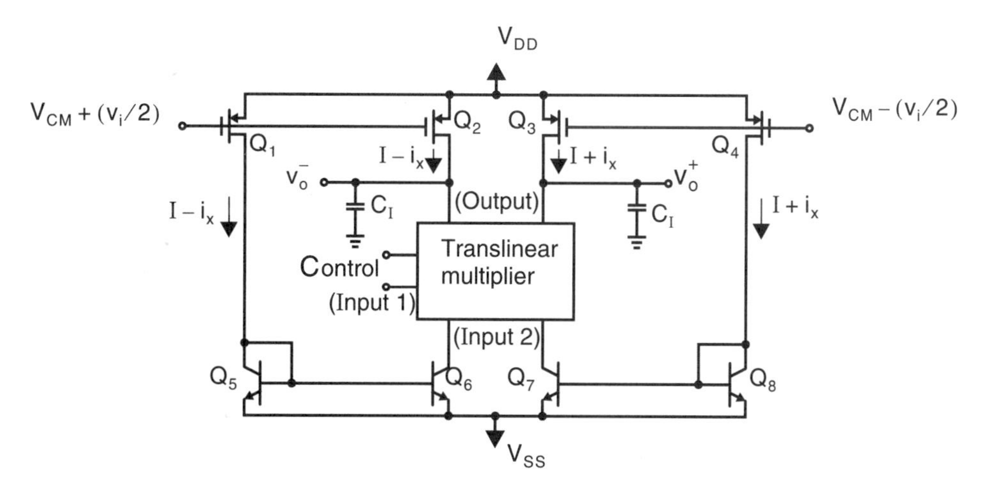

Fig. 15.40 A BiCMOS integrator using MOS transistors in the active region for linearization, and a translinear multiplier for tuning. The common-mode feedback circuit is not shown.

This BiCMOS integrator is used to realize a second-order pulse-shaping filter operating on 100-Mb/s data. Although the filter has a wide tuning range of approximately 10:1, it achieves only a moderate spurious-free dynamic range level of 35 dB.

15.6 MOSFET-C FILTERS

Although G_m-C filters are the most common type of continuous-time filters used in integrated circuit applications, another closely related technique is that of MOSFET-C filters. MOSFET-C filters are similar to fully differential active-RC filters, except resistors are replaced with equivalent MOS transistors in triode. Since active-RC and MOSFET-C filters are closely related, one benefit of MOSFET-C filters is that a designer can use the wealth of knowledge available for active-RC filters. However, MOSFET-C filters are generally slower than their G_m-C counterparts since they use Miller integration (i.e., the integrating capacitors are connected around a high-gain amplifier). Also, they are further reduced in speed in a CMOS technology because opamps capable of driving resistive loads are required as opposed to only capacitive G_m-C filters. However, they are competitive in BiCMOS technologies where high-transconductance opamps are available. Here, we discuss three variations of MOSFET-C filters—two-transistor integrators, four-transistor integrators, and R-MOSFET-C filters.

Two-Transistor Integrators

An example of a two-input fully differential active-RC integrator and its two-transistor MOSFET-C counterpart is shown in Fig. 15.41 [Banu, 1983]. Before we look at the linearization mechanism that occurrs in a MOSFET-C integrator, let us look at the equivalence of the two circuits in Fig. 15.41. In the active-RC filter, the two input

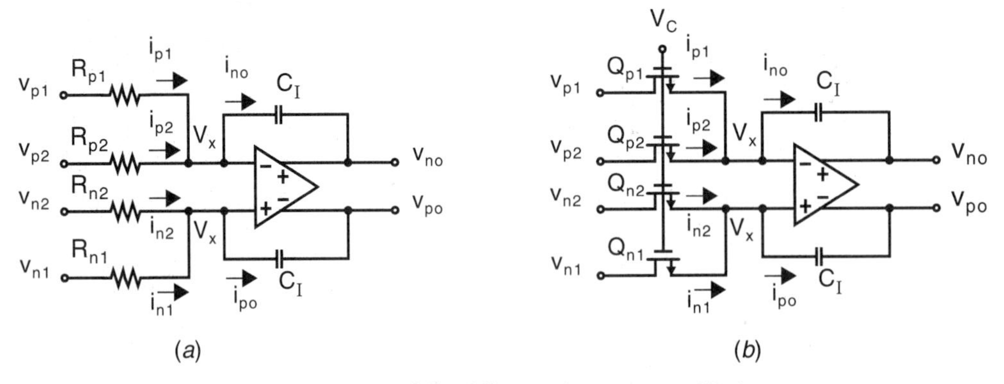

Fig. 15.41 (*a*) A two-input active-RC fully differential integrator. (*b*) A two-input two-transistor MOSFET-C integrator. The MOS transistors are in the triode region.

voltages of the opamp are equal (due to negative feedback) and, defining these two input voltages to be V_x, we can write

$$i_{no} = i_{p1} + i_{p2} = \frac{v_{p1} - V_x}{R_{p1}} + \frac{v_{p2} - V_x}{R_{p2}} \tag{15.149}$$

and

$$i_{po} = i_{n1} + i_{n2} = \frac{v_{n1} - V_x}{R_{n1}} + \frac{v_{n2} - V_x}{R_{n2}} \tag{15.150}$$

The output signals are given by

$$v_{po} = V_x - \frac{i_{po}}{sC_I} \tag{15.151}$$

and

$$v_{no} = V_x - \frac{i_{no}}{sC_I} \tag{15.152}$$

Thus, the differential output signal, v_{diff}, is

$$v_{diff} \equiv v_{po} - v_{no} = \frac{i_{no} - i_{po}}{sC_I} \tag{15.153}$$

Assuming the component values in the positive and negative half circuits are equal, or specifically, if

$$R_1 \equiv R_{p1} = R_{n1} \quad \text{and} \quad R_2 \equiv R_{p2} = R_{n2} \tag{15.154}$$

we have

$$v_{diff} = \frac{1}{sR_1C_I}(v_{p1} - v_{n1}) + \frac{1}{sR_2C_I}(v_{p2} - v_{n2}) \tag{15.155}$$

Thus, we see that the differential output signal, v_{diff}, equals positive integration of the two input differential signals. Negative integration is easily obtained by cross-coupling two input or output wires. Note that the common-mode signal, V_x, does not affect the final result.

Since the transistors in the MOSFET-C integrator, shown in Fig. 15.41(*b*), are assumed to be in the triode region, the small-signal analysis for the MOSFET-C integrator is quite similar to that for the active-RC circuit. Recall from (15.86) that the small-signal equivalent resistance of a transistor in triode is given by

$$r_{DS} = \frac{1}{\mu_n C_{ox}\left(\frac{W}{L}\right)(V_{GS} - V_{tn})} \tag{15.156}$$

Thus, assuming the transistors are matched, the differential output of the MOSFET-C integrator is given by

$$v_{diff} = \frac{1}{sr_{DS1}C_I}(v_{p1} - v_{n1}) + \frac{1}{sr_{DS2}C_I}(v_{p2} - v_{n2}) \tag{15.157}$$

where

$$r_{DSi} = \frac{1}{\mu_n C_{ox}\left(\frac{W}{L}\right)_i (V_C - V_x - V_{tn})} \tag{15.158}$$

Here, note that V_x is important because it helps to determine the small-signal resistance of the transistors. Tuning can be accomplished by varying the control voltage, V_C.

So far, we have analyzed the small-signal operation of a MOSFET-C integrator. In this MOSFET-C case, the differential input signals, (v_{p1}, v_{n1}) and (v_{p2}, v_{n2}), are assumed to be balanced around a common-mode voltage, V_{CM}, set by the output common-mode feedback from the previous integrators. The common-mode feedback of the integrator causes the integrator's two outputs to be balanced also. As a result, it can be shown that V_x is related to the square of the input signal. This nonlinear relationship at V_x together with the nonlinear current through triode transistors results in quite linear single-ended signals at V_{no} and V_{po}. In addition, since the circuit is fully differential, even-order distortion products cancel, as depicted in Fig. 10.17. Thus, two-transistor MOSFET-C integrators can be used to realize filters with around a 50-dB linearity.

Based on the block diagram shown in Fig. 15.10, an example of a general second-order MOSFET-C filter is shown in Fig. 15.42(*a*). Here, the terms G_i represent the small-signal drain-source conductance of the transistors. The transfer function for this filter can be found by drawing its equivalent active-RC half circuit, as shown in Fig. 15.42(*b*). The output of the first integrator is given by

$$V_1(s) = -\frac{G_1}{sC_A}V_i(s) - \frac{G_4}{sC_A}V_o(s) \tag{15.159}$$

whereas the second integrator's output is given by

$$sC_B V_o(s) = G_3 V_1(s) - G_2 V_i(s) - sC_1 V_i(s) - G_5 V_o(s) \tag{15.160}$$

Combining the preceding two equations and rearranging, we have

$$\frac{V_o(s)}{V_i(s)} = -\frac{\left[\left(\frac{C_1}{C_B}\right)s^2 + \left(\frac{G_2}{C_B}\right)s + \left(\frac{G_1 G_3}{C_A C_B}\right)\right]}{\left[s^2 + \left(\frac{G_5}{C_B}\right)s + \left(\frac{G_3 G_4}{C_A C_B}\right)\right]} \tag{15.161}$$

Thus, we see that the transfer function is general, and that negative coefficients can be realized by cross-coupling appropriate signal wires.

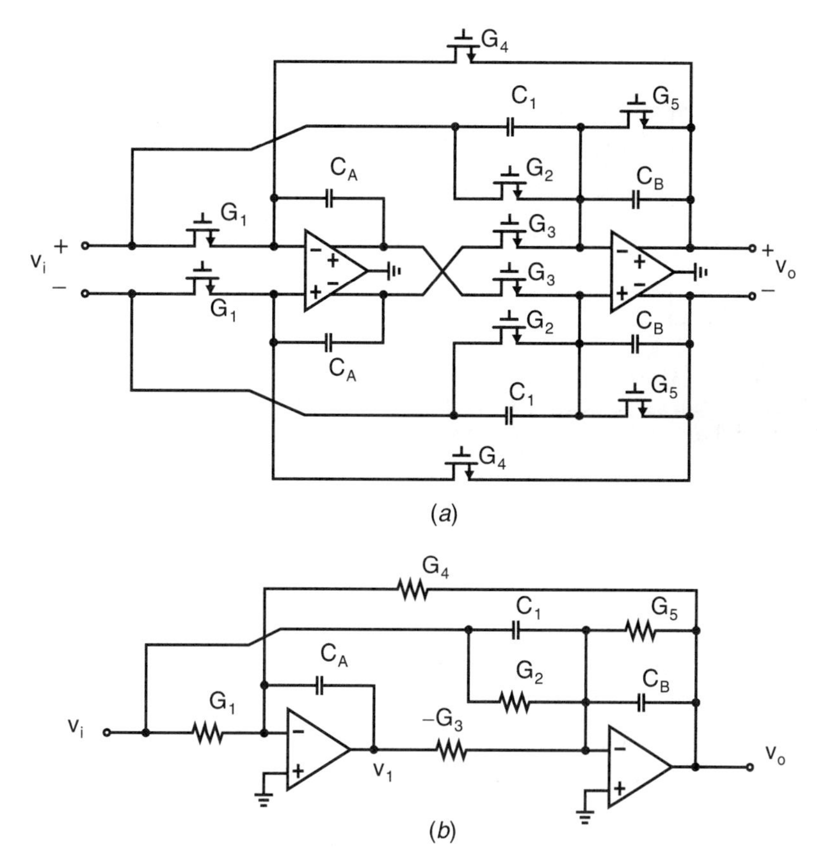

Fig. 15.42 *(a)* A general second-order MOSFET-C filter. *(b)* The equivalent active-RC half circuit.

Four-Transistor Integrators

One way to improve the linearity of a MOSFET-C filter is to use four-transistor MOSFET-C integrators, as shown in Fig. 15.43 [Czarnul, 1986]. The small-signal analysis of this four-transistor integrator can be found by treating this one-input integrator as effectively a two-input integrator in which the two input signals are $(v_{pi} - v_{ni})$ and the inverted signal is $(v_{ni} - v_{pi})$. With such an observation, and applying (15.157), the output signal is given by

$$v_{diff} = \frac{1}{sr_{DS1}C_I}(v_{pi} - v_{ni}) + \frac{1}{sr_{DS2}C_I}(v_{ni} - v_{pi}) \tag{15.162}$$

where

$$r_{DS1} = \frac{1}{\mu_n C_{ox}\left(\frac{W}{L}\right)_1 \left(V_{C1} - V_x - V_t\right)} \tag{15.163}$$

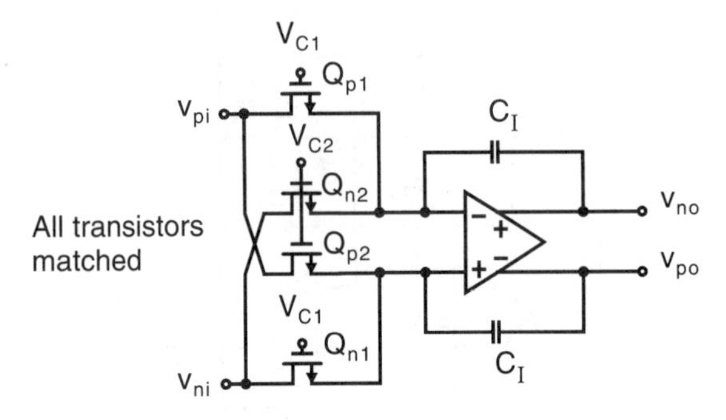

Fig. 15.43 A single-input four-transistor MOSFET-C integrator. It can cancel both even and odd distortion products if the nonlinear terms are not dependent on the control signals, V_{C1} and V_{C2}.

and

$$r_{DS2} = \frac{1}{\mu_n C_{ox}\left(\frac{W}{L}\right)_2 (V_{C2} - V_x - V_t)} \tag{15.164}$$

In the case where all four transistors are matched, we can combine the preceding equations to obtain

$$V_{diff} = V_{po} - V_{no} = \frac{1}{sr_{DS}C_I}(V_{pi} - V_{ni}) \tag{15.165}$$

where r_{DS} is given by

$$r_{DS} = \frac{1}{\mu_n C_{ox}\left(\frac{W}{L}\right)(V_{C1} - V_{C2})} \tag{15.166}$$

Thus, we see that the effective resistance is determined by the difference in the control voltages.

With regard to the distortion properties of this integrator, it has been shown for long channel-length transistors that the drain-source current in a triode transistor is reasonably well modelled by an equation in which the nonlinear distortion terms do not depend on the controlling gate voltage. In other words, although the linear term of each drain current is set by the controlling gate voltage, all distortion terms of the drain currents remain equal in the pair Q_{p1}, Q_{p2} as well as in the pair, Q_{n1}, Q_{n2} since, within each pair, the drain-source voltages are equal. As a result of the cross-coupling connection, *both the even and odd distortion terms cancel*. Unfortunately, this model does not hold as well for short-channel devices, and transistor mismatch also limits the achievable distortion performance. Hence, in practice, one might expect around a 10-dB linearity improvement using this technique over the two-transistor MOSFET-C integrator.

R-MOSFET-C Filters

Another technique intended to improve the linearity of MOSFET-C filters is the use of additional linear resistors [Moon, 1993]. An R-MOSFET-C first-order filter and its MOSFET-C counterpart are shown in Fig. 15.44. Using the equivalent active-RC half circuit, shown in Fig. 15.44(*c*), one finds the transfer function of this first-order R-MOSFET-C filter to be given by

$$\frac{V_o(s)}{V_i(s)} = -\frac{R_2/R_1}{sC_I\left[R_2\left(1+\dfrac{R_{Q1}}{R_1 \parallel R_2 \parallel R_{Q2}}\right)\right]+1} \tag{15.167}$$

where R_{Q1} and R_{Q2} represent the small-signal drain-source resistances of Q_1 and Q_2, respectively. Note that the dc gain of this first-order filter is not adjustable. However, it can be set precisely since it is determined by the ratio of two resistors. In addition, the time constant can vary from a minimum value of R_2C_I to an arbitrarily large value by changing the values of R_{Q1} and R_{Q2} via the control voltages, V_{C1} and V_{C2}. What this transfer function equation indicates is that at low-frequencies, the

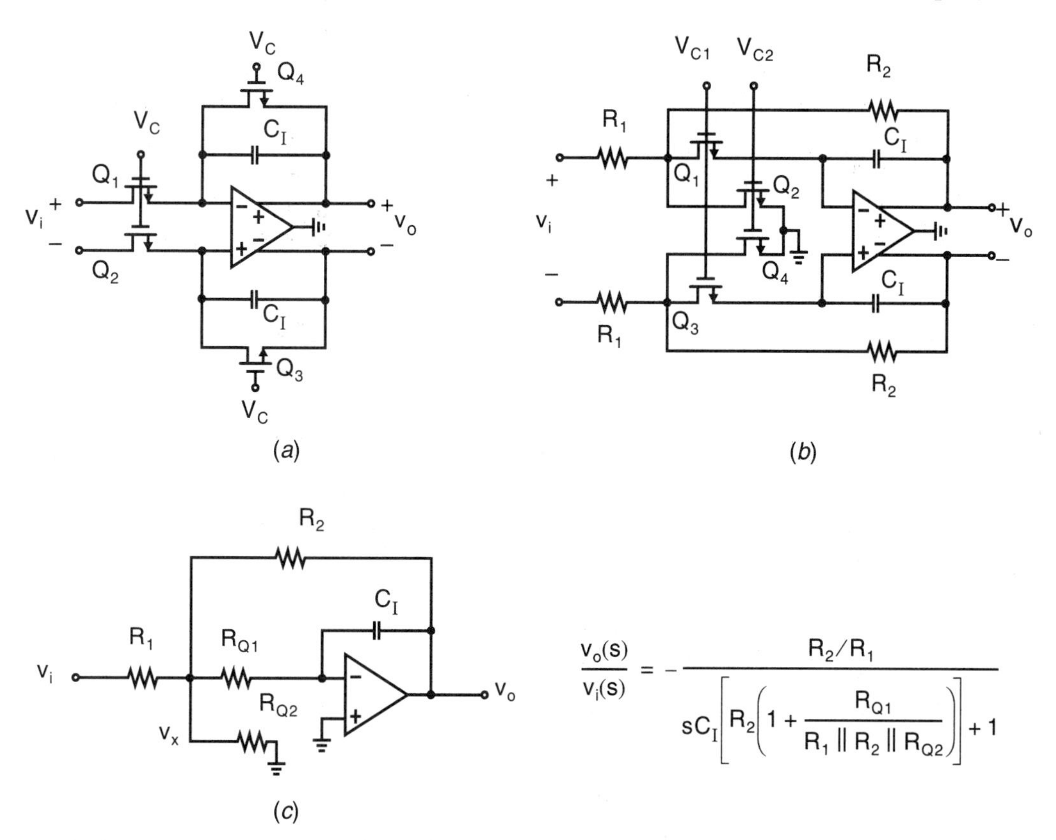

Fig. 15.44 (*a*) A first-order MOSFET-C filter. (*b*) A first-order R-MOSFET-C filter. (*c*) Equivalent active-RC half-circuit for R-MOSFET-C filter.

feedback around R_2 forces node V_x to act as a virtual ground, and therefore excellent linearity is obtained since the circuit is essentially a resistor-only circuit. In [Moon, 1993], a linearity of approximately –90 dB is observed with input signals in the range of a few kilohertz. At higher frequencies, the linearity steadily decreases, and measured results of –70 dB are reported at the passband edge of a fifth-order Bessel filter (the passband is around 20 kHz). However, it should be noted that the measured values are total-harmonic-distortion values, and no intermodulation results are reported.

One other approach that has been used to realize high-linearity continuous-time filters is simply to realize differential active-RC filters, but to realize the capacitors using digitally programmable capacitor arrays. For example, [Durham, 1993] describes a filter intended for digital-audio interfaces that achieves distortion products at a –94-dB and a 95-dB signal-to-noise ratio.

15.7 TUNING CIRCUITRY

As mentioned at the beginning of this chapter, one disadvantage of continuous-time integrated filters is the requirement of additional tuning circuitry. This requirement is a result of large time-constant fluctuations due mainly to process variations. For example, integrated capacitors might have 10 percent variations, whereas resistors or transconductance values might have around 20 percent variations. Since these elements are constructed quite differently, the resulting RC or G_m/C time-constant products are accurate to only around 30 percent with these process variations. In addition, temperature variations further aggravate the situation because resistance and transconductance values change substantially over temperature. Thus, we need tuning circuitry to modify transconductance values such that the resulting overall time constants are set to known values. It should be noted, however, that although absolute component tolerances are high, relative accuracies can be quite good. Specifically, the ratio of two capacitors (or two resistors, or two transistors) can be matched to under 1 percent. This section discusses some techniques to realize tuning circuitry for integrated continuous-time filters. Note that, although the tuning approaches are discussed while we refer to G_m-C (transconductance-C) filters, the concepts are easily extended to other integrated continuous-time filters, such as MOSFET-C filters.

Tuning Overview

Consider once again the G_m-C continuous-time biquad filter shown in Fig. 15.11. Its transfer function, repeated here for convenience, is given by

$$H(s) = \frac{s^2\left[\frac{C_X}{C_X + C_B}\right] + s\left[\frac{G_{m5}}{C_X + C_B}\right] + \left(\frac{G_{m2}G_{m4}}{[C_A(C_X + C_B)]}\right)}{s^2 + s\left[\frac{G_{m3}}{C_X + C_B}\right] + \left(\frac{G_{m1}G_{m2}}{[C_A(C_X + C_B)]}\right)} \tag{15.168}$$

First, note that the numerator coefficient for s^2 is set by the capacitor ratio of $C_X/(C_X + C_B)$ and can therefore be quite accurate (i.e., the absolute values of the capacitors are unimportant—only their relative values matter). Second, although the transconductance values, G_{mi}, have large variations in their absolute values, relative transconductance values can also be set reasonably accurately. For example, if the transconductors are realized as bipolar differential pairs with an emitter degeneration resistor (as shown in Fig. 15.11), then the relative transconductance values of two transconductors is set by the ratio of their respective emitter degeneration resistor (assuming the gain-cell currents are equal in the two transconductors). Similarly, if other transconductors of the same type are used, their relative G_m values can be accurately set by choosing appropriate transistor sizes, bias currents, or resistor values (or some combination of the three). As a result, we can write all the transconductance values, G_{mi}, as an accurate constant multiplied by some basic transconductance value. Specifically, let us write all the transconductance values, G_{mi}, as a function of G_{m1}, resulting in

$$G_{mi} = k_{mi} G_{m1} \tag{15.169}$$

Here, the constants, k_{mi}, are set reasonably accurately by appropriate transistor, resistor, or bias-current scaling. Furthermore, let us also write

$$C_X + C_B = k_{XB} C_A \tag{15.170}$$

where k_{XB} is a constant that is accurately set as the ratio of capacitance values. With such an approach, (15.168) becomes

$$H(s) = \frac{s^2\left[\dfrac{C_X}{(C_X + C_B)}\right] + s\left(\dfrac{k_{m5}}{k_{XB}}\right)\left(\dfrac{G_{m1}}{C_A}\right) + \left(\dfrac{k_{m2}k_{m4}}{k_{XB}}\right)\left(\dfrac{G_{m1}}{C_A}\right)^2}{s^2 + s\left(\dfrac{k_{m3}}{k_{XB}}\right)\left(\dfrac{G_{m1}}{C_A}\right) + \left(\dfrac{k_{m1}k_{m2}}{k_{XB}}\right)\left(\dfrac{G_{m1}}{C_A}\right)^2} \tag{15.171}$$

We see from (15.17) that four of the coefficients for this second-order filter can be accurately set if only one ratio is tuned—that of G_{m1}/C_A. The other two coefficients (those corresponding to s^2) should not need adjustment since they do not rely on any RC-type time constant. Since it is difficult to tune C_A, the ratio G_{m1}/C_A is tuned by changing G_{m1}. Of course, as G_{m1} is varied, all the other transconductance values, G_{mi}, are also changed since their values track that of G_{m1} by their respective constants, k_{mi}. It should be mentioned here that this same single transconductance tuning is also valid for higher-order filter structures.

In summary, the most common way to tune a continuous-time integrated filter is to build an extra transconductor that is tuned (tuning details are discussed next) and to use the resulting tuning signal to control the filter transconductors, as shown in Fig. 15.45. Such an approach is commonly referred to as *indirect frequency tuning*. This tuning is indirect since one relies on matching between the filter transconductors (and parasitic capacitances) and the extra transconductor. In other words, the filter is not directly tuned by looking at its output signal. Frequency tuning refers to the fact that the ratio G_{m1}/C_A is equal to the unity-gain frequency of a G_m-C integrator

(G_{m1}/C_A is also proportional to the pole frequency, ω_o, in a first- or second-order filter).

Unfortunately, this frequency-tuning approach may not be sufficient in some applications due to nonideal integrators or mismatch effects. In these applications, some extra tuning circuitry known as Q-factor tuning (discussed later in this section) can be used. It is also possible to implement *direct tuning* where the filter is tuned by directly looking at the filter's output. Two such direct approaches are also discussed in this section.

Constant Transconductance

As just discussed, if no tuning is performed, one might expect around a 30-percent tolerance on the absolute value for the G_m/C ratio. However, integrated capacitance tolerance typically accounts for only a 10-percent variation in the overall 30-percent tolerance. Thus, in applications where a 10-percent variation can be permitted (such as an anti-aliasing filter with a sample rate moderately higher than the Nyquist rate), it is sufficient to set only the G_m value with the use of a fixed external resistor. As we will see below, such tuning of a G_m value is not difficult and does not require any clock signals.

In the case where 10-percent accuracy is not enough, clearly one must also account for capacitor tolerance. One way to do so (while still not needing a clock signal) is to have a variety of external resistors available and choose the most appropriate one to also account for capacitor tolerance. With such an approach, one might expect that the overall G_m/C ratio would be accurate to under 1 percent. Although a variety of methods exist for choosing this external resistance, one way is to use a known value of external resistance and measure the step response of the filter during system production. This step-response result can then be used to determine the C value, and hence the proper resistance value can be calculated. Finally, since integrated capacitors typically have small temperature coefficients, overall temperature effects are small if the external resistor is also chosen to have a low temperature coefficient.

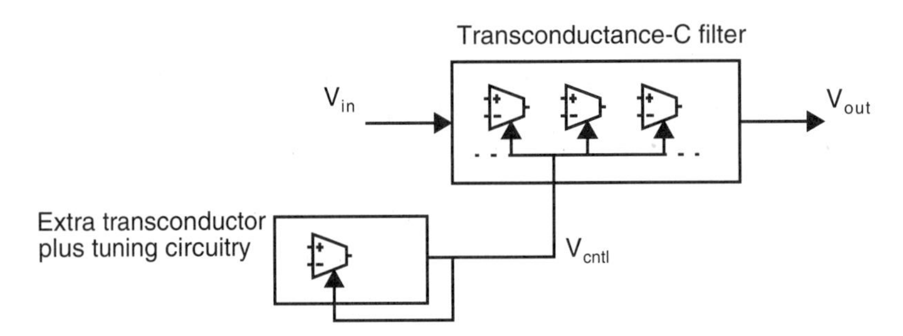

Fig. 15.45 Indirect frequency tuning. Note that, although the tuning signal is shown here as a voltage signal, it may also be realized using a set of current controlling signals.

To set a transconductance value equal to the inverse of an external resistance, (i.e., $G_m = 1/R_{ext}$), a variety of feedback circuits can be used. Two example circuits are shown in Fig. 15.46, where it should be noted that, in both cases, it is assumed the transconductor's transconductance increases as the level of the control signal is increased. In Fig. 15.46(*a*), if G_m is too small, the current through R_{ext} is larger than the current supplied by the transconductor, and the difference between these two currents is integrated with the opamp and capacitor. As a result, the control voltage, V_{cntl}, is increased until these two currents are equal (but opposite in direction) and, therefore, $G_m = 1/R_{ext}$. The circuit in Fig. 15.46(*b*) shows two voltage-to-current converters—one is the transconductor being tuned, and the other is a fixed voltage-to-current converter that simply needs a large transconductance and is not necessarily linear. (It might be implemented as a simple differential pair.) This circuit operates as follows: If G_m is too small, then the voltage at the top of R_{ext} will be less than V_B and the fixed V/I converter will increase I_{cntl}. At steady state, the differential voltage into the fixed V/I converter will be zero, resulting in $G_m = 1/R_{ext}$.

Frequency Tuning

A precise tuning of a G_m/C_A ratio can be achieved if an accurate time period is available. For example, assuming one has available an accurate clock frequency, say, f_{clk}, then one method of frequency tuning is by using a switch-capacitor circuit as shown in Fig. 15.47 [Viswanathan, 1982]. Here, the tuning circuit is quite similar to the constant-transconductance approaches of Fig. 15.46, except that the external resistance is replaced with a switched-capacitor equivalent resistance. From Chapter 10, the equivalence resistance of the switched-capacitor circuit is given by $R_{eq} = 1/(f_{clk}C_m)$, and therefore the transconductance value, G_m, is set to $f_{clk}C_m$. In this way, the G_m/C_A ratio is accurately set to $(f_{clk}C_m)/C_A$, and hence precise frequency tuning can be

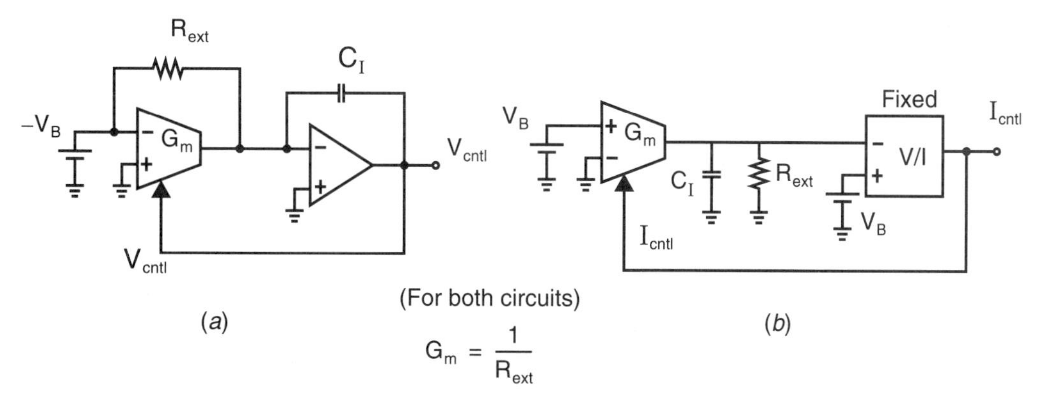

Fig. 15.46 Constant transconductance tuning. Possible tuning circuits if the transconductors are (*a*) voltage controlled or (*b*) current controlled. In both circuits, it is assumed the transconductor's transconductance increases as the level of the control signal increases. V_B is an arbitrary voltage level, whereas C_I is an integrating capacitor used to maintain loop stability.

achieved without any external components. Note that an additional low-pass filter, R_1C_1, has been added here to remove the high-frequency ripple voltage due to using a switched-capacitor circuit. It should also be noted that, since this method requires a clock signal, there is a strong possibility that some of the clock signal will leak into the continuous-time filter, either mainly through the transconductor controlling signal or through the IC substrate.

One difficulty with this switched-capacitor tuning approach is that it requires large transconductance ratios (and hence poorer matching) if high-frequency filters are desired. For example, consider the case of a 100-MHz filter, where the G_m/C_A ratio should be set to around $2\pi \times 100$ MHz. If a capacitance value, C_A, is assumed to be 1 pF, then $G_m = 2\pi \times 10^{-4}$ V/A. Thus, in the tuning circuit of Fig. 15.47, assuming that $C_m = 1$ pF, we would require the switched-capacitor circuit to be clocked at the impractical frequency of

$$f_{clk} = \frac{G_m}{C_m} = \frac{2\pi \times 10^{-4}}{1 \times 10^{-12}} = 628 \text{ MHz} \tag{15.172}$$

At first glance, one might consider reducing this clock frequency by increasing the value of C_m, but, although f_{clk} is reduced, settling-time requirements remain difficult since the capacitance, C_m, is larger. Another way to reduce f_{clk} is to let the tuning circuit set a smaller transconductance value, say, $0.1G_m$. In this way, f_{clk} is reduced by ten but the filter's transconductors must be 10 times greater than the tuning circuitry's transconductor. Thus, poorer matching occurs between the filter and the tuning circuitry, resulting in a less accurate frequency setting for the filter.

Another approach to lower the clock frequency of this switched-capacitor tuning circuitry is by using two scaled current sources, as shown in Fig. 15.48 [Silva-Martinez, 1992]. Using an equivalent resistance of $-1/f_{clk}C_m$ for the switched-capacitance circuit, and noting that the diode-connected transconductor is equivalent to a resistor of value $1/G_m$, one can show that when the average current into the integrator is zero, the transconductance value is given by

$$G_m = Nf_{clk}C_m \tag{15.173}$$

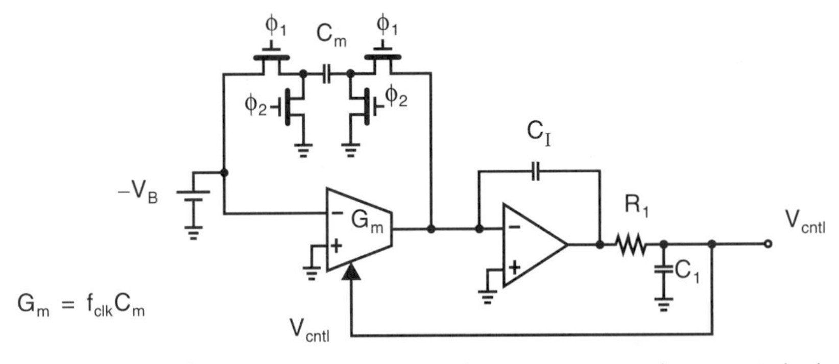

Fig. 15.47 A frequency-tuning circuit where G_m is set by a switched-capacitor circuit. The switched-capacitor branch has a clock frequency of f_{clk}. The low-pass filter, R_1C_1, is set to a frequency low enough to reduce the high-frequency ripple voltage.

Thus, the clock frequency of this circuit is N times lower than that for the circuit shown in 15.47. This circuit was used to tune a 10.7-MHz bandpass filter, where a value of $N = 148$ was used, resulting in a clock frequency of only 450 kHz. The low-pass filter was also realized by using a switched-capacitor circuit and an off-chip capacitor.

Another approach to achieving frequency tuning is to use a phase-locked loop (PLL) (described in Chapter 16) as shown in Fig. 15.49 [Tan, 1978]. Here, two continuous-time integrators are placed in a loop to realize a voltage-controlled oscillator that is placed into a phase-locked loop. After the circuit is powered up, the negative-feedback of the PLL causes the VCO frequency and phase to lock to the external reference clock. Once the VCO output is locked to an external reference signal, the G_m/C ratio of the VCO is set to a desired value, and the control voltage, V_{cntl}, can be used to tune the integrated filter. It should be noted that choosing the external reference clock is a trade-off because it affects both the tuning accuracy as well as the tuning signal leak into the main filter. Specifically, for best matching between the tuning circuitry and the main filter, it is best to choose the reference frequency that is equal to the filter's upper passband edge. However, noise gains for the main filter are typically the largest at the filter's upper passband edge, and therefore the reference-signal leak into the main filter's output might be too severe. As one moves away from the upper passband edge, the matching will be poorer, but an improved immunity to the reference signal results. Another problem with this approach is that, unless some kind of

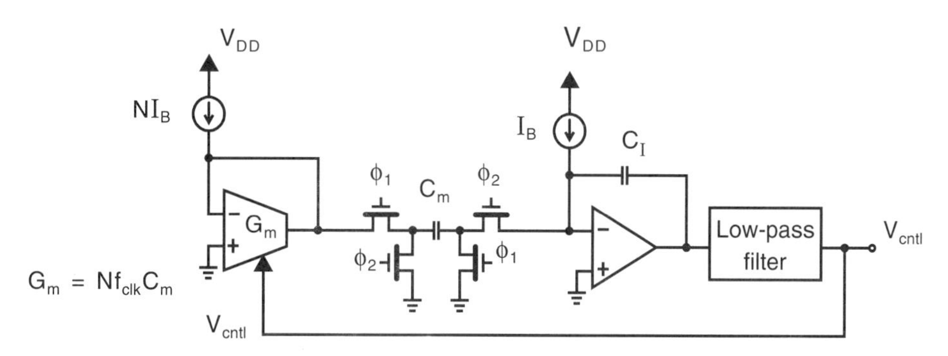

Fig. 15.48 A switched-capacitor frequency-tuning circuit that operates at a lower clock frequency.

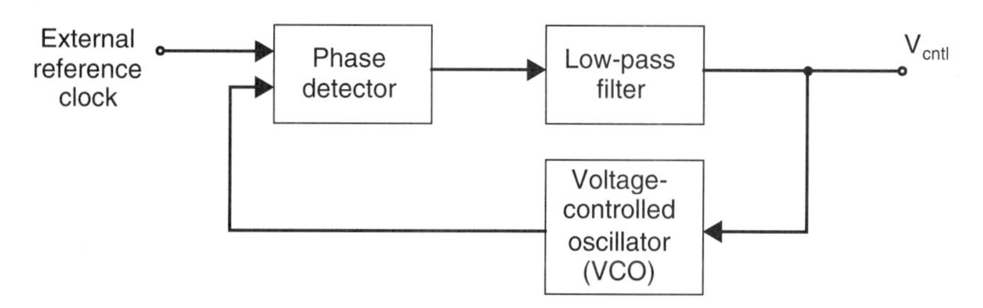

Fig. 15.49 Frequency tuning using a phase-locked loop. The VCO is realized by using transconductor-based integrators that are tuned to adjust the VCO's frequency.

power-supply-insensitive voltage control is added to the VCO, any power-supply noise will inject jitter into the control signal, V_{cntl}. The addition of amplitude control is quite complicated. For a good discussion of this phase-locked-loop approach, see [Krummenacher, 1988; Khoury, 1991]. For a better understanding of the operation of a phase-locked loop, see Chapter 16.

Another, somewhat similar, approach that has been used is based on having a tracking filter that locks onto a fixed-frequency input [Khorramabadi, 1984]. This approach is reported to be less sensitive to power-supply noise. The control voltage for the tracking filter is used to adjust the transconductors that are used in the actual system, assuming the transconductors in the tracking filter match the transconductors in the actual system. An example calibration system is shown in Fig. 15.50. When the low-pass filter has the same resonant frequency as the external reference clock, it has exactly a 90-degree phase shift. Therefore, the correlation between its input and output is zero. If the frequencies are not the same, then the correlation, as obtained by the multiplier and integrator, is nonzero. The negative feedback from the output of the integrator to the frequency-control voltage of the filter drives the resonant frequency in the proper direction to equalize the two frequencies. After lock is obtained, the same control voltage can be used for other filters. The other filters will be tuned to known time constants, assuming they match the filter in the control loop.

It is important that the integrator in the loop have a very low-frequency time constant in order to filter out the component at twice the external reference frequency that is present at the output of the multiplier. Otherwise, this component can cause clock leakage into the other filters being tuned. A variant on this scheme drives the correlation between a notch output and a low-pass filter output (obtained from the same biquad) to zero when one is obtaining the control voltage [Kwan, 1991]. This latter method has much fewer signal components at twice the frequency of the external reference clock and does not require as large a time constant for the integrator in the loop. This makes the latter method more readily integrable without external components. Also, the loop is less sensitive to phase errors; it will always converge to the notch frequency, even when the difference in phase between the low-pass and notch outputs is not exactly 90°.

Q-Factor Tuning

In some applications, such as high-speed or highly-selective filtering, nonideal integrator effects and parasitic components force the need to also tune the Q factors of the

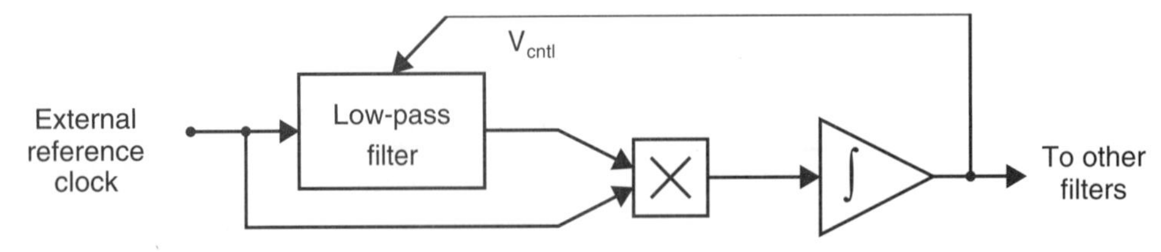

Fig. 15.50 Using a tracking filter to derive a control signal for tuning continuous-time filters.

poles of the integrated filter. However, it should be noted that most commercial integrated filters do not presently have Q-factor tuning due to increased circuit complexity and design difficulty. Also, the addition of a Q-factor tuning circuit increases the possibility that some of the reference tuning signal leaks into the main filter. However, it is likely that more applications in the near future will require Q-factor tuning, and thus we present a brief discussion of it here.

One way to perform Q-factor tuning is to tune the phase of the filter's integrators to ensure that they all have a 90-degree phase lag near the filter's passband edge. An integrator's phase response can be adjusted by introducing a tunable resistance (say, realized as a transistor in triode) in series with the integrating capacitor. The control voltage for this tunable resistance, V_Q, is generated by a Q-factor tuning circuit, shown in Fig. 15.51, and passed to all the filter integrators.

Alternatively, in the case where the integrated filter is second order, the pole Q can be adjusted by changing the transconductance of the damping transconductor in the filter. Specifically, referring back to (15.171) and equating the denominator to a second-order denominator written in the form of ω_o and Q, we have,

$$s^2 + s\left(\frac{k_{m3}}{k_{XB}}\right)\left(\frac{G_{m1}}{C_A}\right) + \left(\frac{k_{m1}k_{m2}}{k_{XB}}\right)\left(\frac{G_{m1}}{C_A}\right)^2 = s^2 + s\left(\frac{\omega_o}{Q}\right) + \omega_o^2 \tag{15.174}$$

Equating the two sides of (15.174), we find the following relationships:

$$\omega_o = \left(\frac{G_{m1}}{C_A}\right)\sqrt{\frac{k_{m1}k_{m2}}{k_{XB}}} \tag{15.175}$$

$$Q = \left(\frac{k_{XB}}{k_{m3}}\right)\sqrt{\frac{k_{m1}k_{m2}}{k_{XB}}} \tag{15.176}$$

From this Q relationship, we see that the Q factor of the filter's poles is inversely proportional to k_{m3}, and hence one need only tune G_{m3} to adjust the Q factor of this second-order filter.

The control signal, V_Q, for either of the preceding Q-factor tuning approaches can be realized using the magnitude-locked loop system shown in Fig. 15.51 [Schaumann, 1990]. The circuit's operation is based on the knowledge that a Q-factor error results in a magnitude error in a filter's frequency response. To measure the magnitude error, a sinusoidal signal of an appropriate reference frequency is applied to a Q-reference circuit, and the magnitude of its output is compared to a scaled version of the input signal's magnitude. This sinusoidal frequency might be obtained as the output of the VCO in the frequency-tuning circuitry.

Tuning Methods Based on Adaptive Filtering

Adaptive filters are commonly used in digital-signal-processing applications such as model matching, channel equalization, and noise (or echo) cancellation. It is also

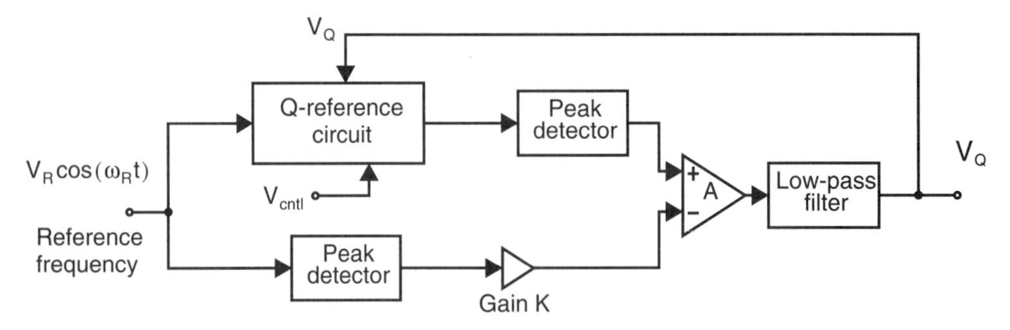

Fig. 15.51 A magnitude-locked loop for Q-factor tuning. The Q-reference circuit is typically a second-order filter that is frequency tuned through the use of V_{cntl}.

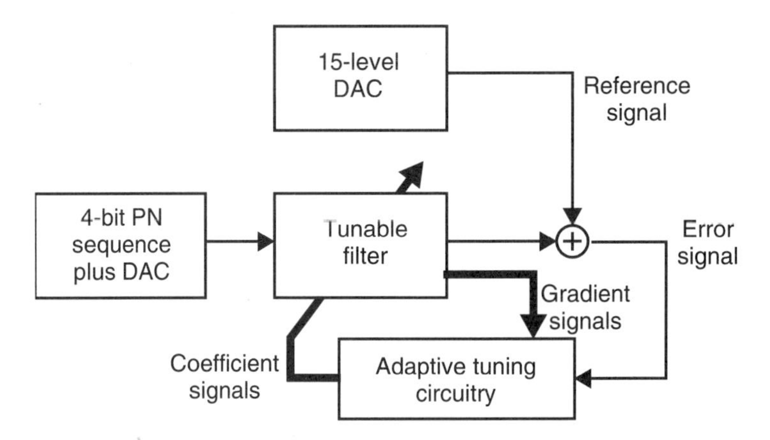

Fig. 15.52 A tuning approach in which all the poles and zeros of the tunable filter can be adjusted. The 15-level DAC outputs are determined by what the ideal filter response would generate, while the adaptive tuning circuitry minimizes the error signal.

possible to use adaptive filter techniques to tune a continuous-time integrated filter to a given specification. One example of such an approach is shown in Fig. 15.52 [Kozma, 1991], where the adaptive tuning circuitry is used to minimize the error signal. During tuning, the input to the tunable filter is a predetermined output from a pseudorandom (PN) 4-bit sequence and a D/A converter. Since the input is known and repetitive, the ideal output from a desired transfer-function can be precalculated (or simulated) and used as a reference signal for the adaptive system. After adaptation, the transfer function of the tunable filter should match that of the desired filter, and, by freezing the coefficients (assuming they are digitally controlled), the tunable filter can be disconnected from the tuning circuit and used in the desired operation. With this approach, it is possible to tune all the poles and zeros of the tunable filter to better match the desired transfer function. Although the adaptive tuning circuitry simply consists of comparators, analog multiplexors, and integrators, note that this adaptive tuning scheme has not yet been fully integrated or experimentally verified in a high-frequency application.

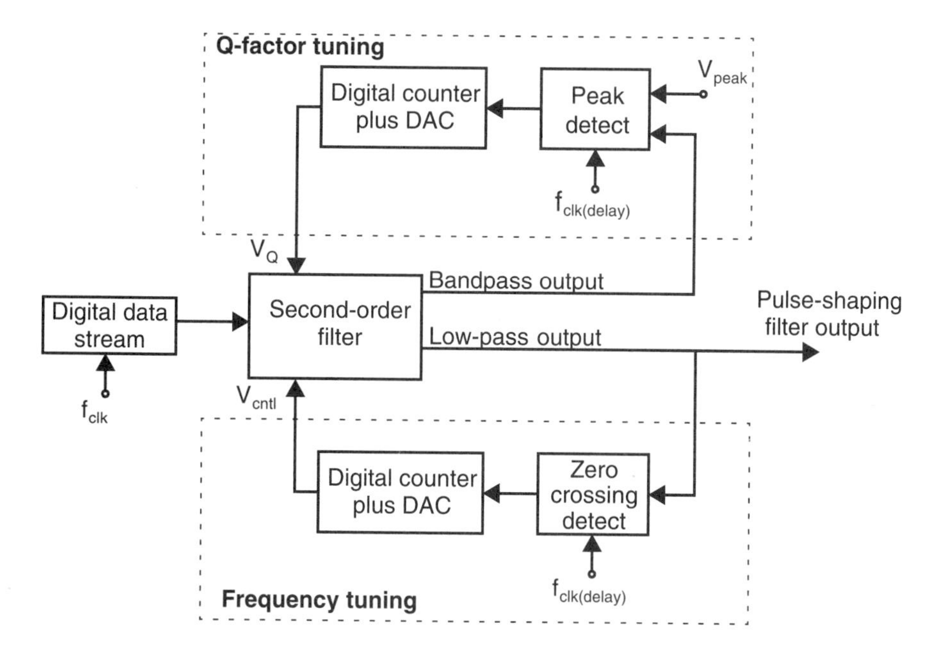

Fig. 15.53 A tuning method to implement a pulse-shaping filter needed for data communication systems.

Another adaptive-filter tuning method intended for high-speed data transmission over a twisted-wire pair is shown in Fig. 15.53 [Shoval, 1995]. In this application, the filter input is a data stream of ones and zeros that are represented by the input signal being either high or low. Before transmission, a pulse-shaping filter is required to ensure that not too much high-frequency power is radiated from the twisted-wire channel. The two tuning loops (frequency and Q-factor loops) rely on the fact that the input to the filter is a series of steps (say, ± 1 V), and thus, in fact, tune the step response of the filter. Specifically, frequency tuning is accomplished by ensuring that the zero crossing of the filter's low-pass output occurs at the correct time period after a data transition. Also, Q-factor tuning is performed by comparing the peak in the filter's bandpass output with a known voltage. Fortunately, the peak in the bandpass output occurs at approximately the same time as the zero crossing, so both detectors can be realized through the use of two clocked comparators, which are triggered at a set time after a data transition. This approach was used to tune a second-order filter intended as a pulse-shaping filter for a 100-Mb/s data transmission. In summary, clever tuning techniques can be implemented when a restricted set of input signals are applied to a tunable filter such as in data communication circuits.

15.8 DYNAMIC RANGE PERFORMANCE

Analog circuits generally have many different performance criteria that must be met so that overall system performance meets specifications. Some typical analog filter performance criteria are transfer-function accuracy, power consumption, linearity, and noise performance. In this section, we discuss the last two items—linearity and noise performance. Linearity limits the value of the largest useful signals that can occur in the filter, whereas noise limits the value of the smallest useful signals. Thus, we find that linearity and noise together determine the useful dynamic range of a filter. It should be noted here that the dynamic range of integrated continuous-time filters is often a crucial measure since its value is often low and can seriously impair system performance.

Total Harmonic Distortion (THD)

If a sinusoidal waveform is applied to a linear time-invariant system, it is well known that the output will also be a sinusoidal waveform at the same frequency, but possibly with different magnitude and phase values. However, if the same input is applied to a nonlinear system, the output signal will have frequency components at harmonics of the input waveform, including the fundamental (or first) harmonic. For example, if a 1-MHz sinusoidal input signal is used, the output signal will have power at the fundamental, 1 MHz, as well as at the harmonic frequencies, 2 MHz, 3 MHz, 4 MHz, and so on. The *total harmonic distortion* (THD) of a signal is defined to be the ratio of the total power of the second and higher harmonic components to the power of the fundamental for that signal. In units of dB, THD is found using the following relation:

$$\mathrm{THD} = 10\log\left(\frac{V_{h2}^2 + V_{h3}^2 + V_{h4}^2 + \cdots}{V_f^2}\right) \tag{15.177}$$

where V_f is the amplitude of the fundamental and V_{hi} is the amplitude of the ith harmonic component. Sometimes THD is presented as a percentage value. In this case,

$$\mathrm{THD} = \frac{\sqrt{V_{h2}^2 + V_{h3}^2 + V_{h4}^2 + \cdots}}{V_f} \times 100 \tag{15.178}$$

For example, a 0.1-percent THD value implies that the amplitude of the fundamental is 1,000 times larger than the amplitude of the harmonic components. This 0.1-percent THD value is equivalent to a value of –60 dB THD. It should also be noted that the THD value is almost always a function of the amplitude of the input (or output) signal level, and thus the corresponding signal amplitude must also be reported. Also, for practical reasons, typically the power of only the first few harmonics (say, the first 5) are included since the distortion components usually fall off quickly for higher harmonics.

One difficulty with the use of THD in reporting filter performance is that often the harmonic components fall in the stopband of the filter, and thus the THD value is falsely improved. For example, if a 21-MHz low-pass filter is being tested, then a 5-MHz input signal results in the 10-MHz, 15-MHz, and 20-MHz components falling within the passband, whereas higher distortion components will be attenuated by the filter. However, filter linearity is often worse when higher input signal frequencies are applied due to nonlinear capacitances or, worse, nonlinear signal cancellation. Thus, it is useful to measure filter linearity with input signals near the upper edge of the passband. Unfortunately, if a 20-MHz input signal is applied to the example low-pass filter, the second and higher harmonic components lie in the stopband and the THD value will indicate much better linearity than would occur for a practical application. Moreover, the testing of a narrowband filter always has this THD measurement difficulty since the use of an input frequency in the passband will be attenuated by the filter's stopband. In summary, a THD measurement is straightforward to perform but does not work well in the important test of high-frequency signals near the upper passband of the filter. Fortunately, as we will see next, one can use intermodulation tests to measure filter linearity near the upper passband edge.

EXAMPLE 15.9

Calculate the THD value of a signal in which the fundamental component is at 1 MHz and has a power level of –10 dBm, and the only significant harmonics are the first three above the fundamental. These harmonics, at 2 MHz, 3 MHz, and 4 MHz, are measured to have power levels of –53 dBm, –50 dBm, and –56 dBm, respectively.

Solution

Recalling that 0 dBm refers to a 1-mW signal, then the power of the fundamental component is 100 μW, whereas the power of the first three harmonics are 5 nW, 10 nW, and 2.5 nW, respectively. Thus, the total power of the harmonics is equal to

$$P_{harm} = 5 + 10 + 2.5 = 17.5 \text{ nW} \tag{15.179}$$

As a result, the THD is calculated to be

$$THD = 10\log\left(\frac{100\ \mu W}{17.5\ nW}\right) = 37.6 \text{ dB} \tag{15.180}$$

Third-Order Intercept Point (IP3)

Here, we introduce the concept of the *third-order intercept point* (IP3) as a measure for the third-order distortion component [Carson, 1990]. We focus here on third-order distortion since, for fully differential circuits, the even-order distortion components are ideally zero and, thus, third-order distortion typically dominates. However, in cases where second-order distortion dominates, a similar concept is also possible and is referred to as the *second-order intercept point* (IP2).

Consider a nonlinear system[3] with an input signal, $v_{in}(t)$, and an output signal, $v_o(t)$. The output signal can be written as a Taylor series expansion of the input signal:

$$v_o(t) = a_1 v_{in}(t) + a_2 v_{in}^2(t) + a_3 v_{in}^3(t) + a_4 v_{in}^4(t) + \cdots \tag{15.181}$$

Here, the linear term is a_1, whereas a_2, a_3, and a_4 characterize the second-, third-, and fourth-order distortion terms, respectively. As mentioned previously, in fully differential circuits, all even terms (i.e., a_2 and a_4) are small, so typically a_3 dominates and we approximate $v_o(t)$ as

$$v_o(t) \cong a_1 v_{in}(t) + a_3 v_{in}^3(t) \tag{15.182}$$

If $v_{in}(t)$ is a sinusoidal signal given by

$$v_{in}(t) = A\cos(\omega t) \tag{15.183}$$

the output signal can be shown to be approximated by

$$v_o(t) \cong a_1 A\cos(\omega t) + \frac{a_3}{4}A^3[3\cos(\omega t) + \cos(3\omega t)] \tag{15.184}$$

where we see a fundamental term and a third-harmonic term. Defining H_{D1} and H_{D3} to be the amplitudes of the fundamental and third-harmonic terms, respectively, we can write

$$v_o(t) \equiv H_{D1}\cos(\omega t) + H_{D3}\cos(3\omega t) \tag{15.185}$$

Since, typically, $(3/4)a_3A^3 << a_1A$, one usually approximates the linear component of the output signal as

$$H_{D1} = a_1 A \tag{15.186}$$

and the third-harmonic term as

$$H_{D3} = \frac{a_3}{4}A^3 \tag{15.187}$$

Finally, we see that the third-order distortion term results in power at the third harmonic frequency and the ratio of H_{D3}/H_{D1} is defined as the third-order harmonic distortion ratio, given by

$$HD_3 \equiv \frac{H_{D3}}{H_{D1}} = \left(\frac{a_3}{a_1}\right)\left(\frac{A^2}{4}\right) \tag{15.188}$$

Unfortunately, as just noted, this distortion term lies at $3\omega t$ for a single sinusoidal input, and thus we resort to an intermodulation test to move the distortion term back near the frequency of the input signals.

Consider now the case of an intermodulation test, where the input signal consists of two equally sized sinusoidal signals and is written as

$$v_{in}(t) = A\cos(\omega_1 t) + A\cos(\omega_2 t) \tag{15.189}$$

---

3. We assume here that the nonlinear system is memoryless and time invariant. Unfortunately, filters are not memoryless, and a Volterra series should be used; however, this assumption simplifies the analysis and usually results in good approximations for distortion figures.

In this case, the output signal can be shown to be approximated by

$$
\begin{aligned}
v_o(t) \cong & \left(a_1 A + \frac{9a_3}{4}A^3\right)[\cos(\omega_1 t) + \cos(\omega_2 t)] \\
& + \frac{a_3}{4}A^3[\cos(3\omega_1 t) + \cos(3\omega_2 t)] \\
& + \frac{3a_3}{4}A^3[\cos(2\omega_1 t + \omega_2 t) + \cos(2\omega_2 t + \omega_1 t)] \\
& + \frac{3a_3}{4}A^3[\cos(\omega_1 t - \Delta\omega t) + \cos(\omega_2 t + \Delta\omega t)]
\end{aligned}
\tag{15.190}
$$

where $\Delta\omega$ is defined to be the difference between the input frequencies (i.e., $\Delta\omega \equiv \omega_2 - \omega$) which we assume to be small. Here, we see that the first line of (15.90) is the fundamental components, the second line shows the levels at three times the fundamentals, the third line also describes distortion at nearly three times the fundamentals, and the fourth line describes the distortion levels at two new frequencies that are close to the input frequencies (slightly below ω_1 and slightly above ω_2). As a result, for a narrowband or low-pass filter, these two new distortion components (due to third-order distortion) fall in the passband and can be used to predict the third-order distortion term.

Using the same notation as in the harmonic distortion case, we have the intermodulation distortion levels given by

$$
I_{D1} = a_1 A \tag{15.191}
$$

and

$$
I_{D3} = \frac{3a_3}{4}A^3 \tag{15.192}
$$

The ratio of these two is the third-order intermodulation value, given by

$$
ID_3 = \frac{I_{D3}}{I_{D1}} = \left(\frac{a_3}{a_1}\right)\left(\frac{3A^2}{4}\right) \tag{15.193}
$$

From (15.191) and (15.192), note that, as the amplitude of the input signal, A, is increased, the level of the fundamental rises linearly, whereas the I_{D3} rises in a cubic fashion. For example, if A is increased by 1 dB, then I_{D1} also increases by 1 dB while I_{D3} increases by 3 dB. Thus, when the fundamental and intermodulation levels are plotted in a dB scale against the signal amplitude, A, the two curves are linear, but with different slopes, as shown in Fig. 15.54. The third-order intercept point is defined to be the intersection of these two lines. Note, however, that as the signal amplitude rises, the linear relationships of I_{D1} and I_{D3} are no longer obeyed due to the large amount of nonlinearities violating some of our original assumptions. For example, in (15.190), we ignored the cubic A term in estimating the fundamental level, I_{D1}. Also, other distortion terms that were ignored now become important. As a result, it is impossible to directly measure the third-order intercept point, and thus it must be extrapolated from the measured data. The third-order intercept point results in two

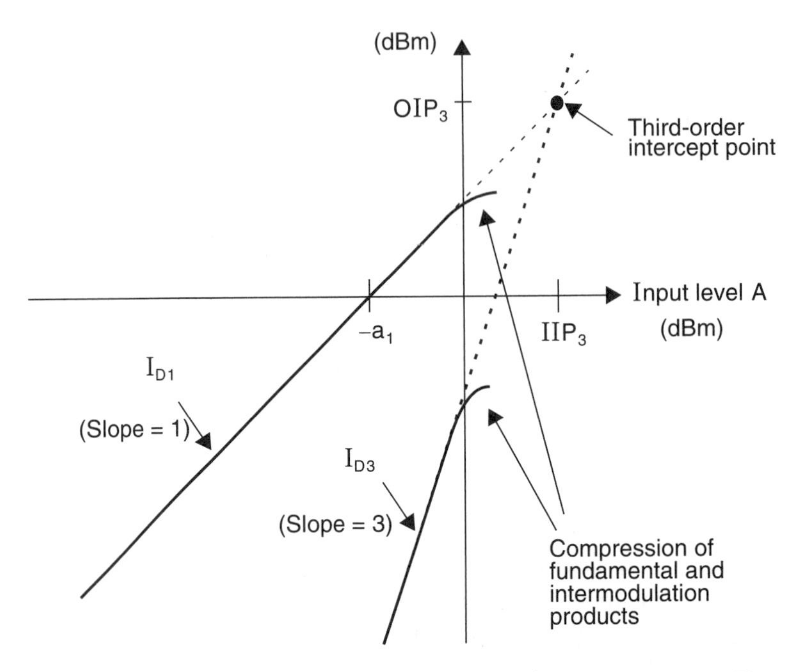

Fig. 15.54 Graphical illustration of the third-order intercept-point. IIP_3 and OIP_3 are the input and output third-order intercept points, respectively. They cannot be measured directly due to compression of the fundamental and intermodulation products at high-power levels.

values—IIP_3 and OIP_3, which are related to the input and output levels, respectively. If the linear gain term, a_1, is unity (or, equivalently, 0 dB), then $IIP_3 = OIP_3$. However, if a_1 is not unity, one must be careful to state which of the two intercept points is being reported.

Knowledge of the third-order intercept point is quite useful in determining what signal level should be chosen to achieve a desired intermodulation ratio, ID_3. Specifically, we see from (15.193) that ID_3 improves by 2 dB for every 1 dB of signal level decrease (since it is related to A^2) and that OIP_3 is defined to be the I_{D1} point where $ID_3 = 0$ dB. Thus, we have the simple relationship (all in decibels)

$$OIP_3 = I_{D1} - \frac{ID_3}{2} \tag{15.194}$$

EXAMPLE 15.10

If $OIP_3 = 20$ dBm, what output-signal level should be used such that the third-order intermodulation products are 60 dB below the fundamental?

Solution

Using (15.194) with $ID_3 = -60$ dB, we have

$$I_{D1} = OIP_3 + \frac{ID_3}{2} = -10 \text{ dBm} \tag{15.195}$$

Thus, an output level of -10 dBm should be used.

Spurious-Free Dynamic Range (SFDR)

Spurious-free dynamic range (SFDR) is defined to be the signal-to-noise ratio when the power of the third-order intermodulation products equals the noise power. In Fig. 15.55, the filter's output noise power is shown along the vertical axis as N_o. If a low enough signal level is used, I_{D3} will be well below the noise floor. However, since I_{D3} rises 3 dB for every 1 dB of signal-level increase, there will soon be a point where I_{D3} is equal to the noise power. As the figure shows, SFDR is defined to be the output SNR ratio when I_{D3} is equal to N_o. Alternatively, one can measure SFDR using the input-signal levels as the difference between the level that results in $I_{D3} = N_o$ and the level A_{No} that results in a fundamental output level equal to N_o.

To find a formula relating SFDR, OIP_3, and N_o (all in dB units), we first note that

$$SFDR = I_{D1}^* - N_o = I_{D1}^* - I_{D3}^* \tag{15.196}$$

where I_{D1}^* and I_{D3}^* refer to the output and distortion levels when $I_{D3} = N_o$, as shown in Fig. 15.55. Since the units are assumed to be in dB, we also have, from (15.193),

$$ID_3 = I_{D3} - I_{D1} \tag{15.197}$$

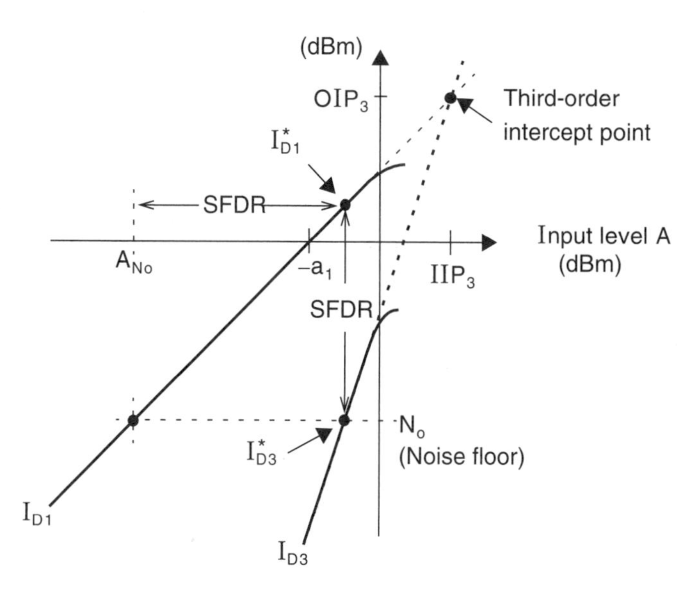

Fig. 15.55 Graphical illustration of spurious-free dynamic range (SFDR).

Using (15.194), we have

$$OIP_3 = I_{D1}^* - \frac{(N_o - I_{D1}^*)}{2} \tag{15.198}$$

and substituting in $I_{D1}^* = SFDR + N_o$ from (15.196) and rearranging, we have

$$SFDR = \frac{2}{3}(OIP_3 - N_o) \tag{15.199}$$

EXAMPLE 15.11

At an input-signal level of 0 dBm, an intermodulation ratio of –40 dB was measured in an filter with a gain of 2 dB. Calculate the value of the input and output third-order intercept points. What input-signal level should be applied if one wants an intermodulation ratio of –45 dB? If the noise power at the output is measured to be –50 dBm, what is the expected SFDR, and what output-signal level does it correspond to?

Solution

With an input level of 0 dBm and a gain of 2 dB, the output level is $I_{D1} = 2\text{ dBm}$ with a measured value of $ID_3 = -40\text{ dB}$. Using (15.194), we have

$$OIP_3 = I_{D1} - \frac{ID_3}{2} = 22\text{ dBm} \tag{15.200}$$

whereas the IIP_3 is 2 dB lower (i.e., $IIP_3 = 20\text{ dBm}$).

For signal levels corresponding to an intermodulation of $ID_3 = 45\text{ dB}$, we have

$$I_{D1} = OIP_3 + \frac{ID_3}{2} = 22 - \frac{45}{2} = -0.5\text{ dBm} \tag{15.201}$$

However, this value is the level of the output signal, so the input signal should be 2 dB lower or, equivalently, the input-signal level should be at -2.5 dBm. Note that this result could have been obtained more directly by noting that a 5-dB improvement in distortion was desired, implying that the signal level should be decreased by $(5\text{ dB})/2$.

Finally, we use (15.199) to find

$$SFDR = \frac{2}{3}(22 + 50) = 48\text{ dB} \tag{15.202}$$

from which the output level is given by

$$I_{D1}^* = SFDR + N_o = -2\text{ dBm} \tag{15.203}$$

Thus, when the output level is at –2 dBm, the third-order intermodulation power equals the noise power. If the output level is increased, the distortion products will increase and limit the dynamic-range performance. However, if the output level is decreased, the distortion products will be buried below the noise floor, and the noise will limit dynamic-range performance. Therefore, optimum dynamic-range performance is obtained at an output level of –2 dBm. However, note that the dynamic-range performance is actually 3 dB below the SFDR value since the dynamic range is based on the ratio of the signal power to the distortion plus the noise power (which are both equal at this point).

15.9 REFERENCES

M. Banu and Y. Tsividis. "Fully Integrated Active RC Filters in MOS Technology," *IEEE J. of Solid-State Circuits,* Vol. 18, pp. 644–651, December 1983.

K. Bult and H. Wallinga. "A Class of Analog CMOS Circuits Based on the Square-Law Characteristics of a MOS Transistor in Saturation," *IEEE J. of Solid-State Circuits,* Vol. 22, pp. 357–365, June 1987.

D. W. H. Calder. "Audio Frequency Gyrator Filters for an Integrated Radio Paging Receiver," in *Proc. 1984 IEE Conf. Mobile Radio Syst. Tech.,* pp. 21–26.

R. S. Carson. *Radio Communications Concepts: Analog.* John Wiley & Sons, New York, 1990.

Z. Chang, D. Macq, D. Haspeslagh, P. Spruyt, and B. Goffart. "A CMOS Analog Front-End Circuit for an FDM-Based ADSL System," *IEEE Int. Solid-State Circuits Conf.,* pp. 330–331, February 1995.

Z. Czarnul. "Modification of the Banu-Tsividis Continuous-Time Integrator Structure," *IEEE Trans. on Circuits and Systems,* Vol. 33, pp. 714–716, July 1986.

G. A. De Veirman and R. G. Yamasaki. "Design of a Bipolar 10-MHz Programmable Continuous-Time 0.05° Equiripple Linear Phase Filter," *IEEE J. of Solid-State Circuits,* Vol. 27, pp. 324–331, March 1992.

S. T. Dupuie and M. Ismail. "High-Frequency CMOS Transconductors," in *Analog IC Design: The Current-Mode Approach,* ed. C. Toumazou, F. J. Lidgey, and D. G. Haigh. Peter Peregrinus, London, UK, 1990.

A. Durham and W. Redman-White. "Integrated Continuous-Time Balanced Filters for 16-b DSP Interfaces," *IEEE J. of Solid-State Circuits,* Vol. 28, no. 7, pp. 835–839, July 1993.

B. Gilbert. "Current-Mode Circuits from a Translinear Viewpoint: A Tutorial," in *Analog IC Design: The Current-Mode Approach,* ed. by C. Toumazou, F. J. Lidgey, and D. G. Haigh. Peter Peregrinus, London, UK, 1990.

H. Khorramabadi and P. R. Gray. "High-Frequency CMOS Continuous-Time Filters," *IEEE J. of Solid-State Circuits,* Vol. SC-19, no. 6, pp. 939–948, December 1984.

J. M. Khoury. "Design of a 15-MHz CMOS Continuous-Time Filter with On-Chip Tuning," *IEEE J. of Solid-State Circuits,* Vol. 26, pp. 1988–1997, December 1991.

M. Koyama, T. Arai, H. Tanimoto, and Y. Yoshida. "A 2.5-V Active Low-Pass Filter Using All npn Gilbert Cells with a 1-$V_{p\text{-}p}$ Linear Input Range," *IEEE J. of Solid-State Circuits,* Vol. 28, pp. 1246–1253, December 1993.

K. Kozma, D. A. Johns, and A. S. Sedra. "Automatic Tuning of Continuous-Time Integrated Filters Using an Adaptive Filter Technique," *IEEE Trans. on Circuits and Systems,* Vol. 38, pp. 1241–1248, November 1991.

F. Krummenacher and N. Joehl. "A 4-MHz CMOS Continuous-Time Filter with On-Chip Automatic Tuning," *IEEE J. of Solid-State Circuits,* Vol. 23, pp. 750–758, June 1988.

T. Kwan and K. Martin. "An Adaptive Analog Continuous-Time CMOS Biquadratic Filter," *IEEE J. of Solid-State Circuits,* Vol. 26, pp. 859–867, June 1991.

C. A. Laber and P. R. Gray. "A 20-MHz Sixth-Order BiCMOS Parasitic-Insensitive Continuous-Time Filter and Second-Order Equalizer Optimized for Disk-Drive Read Channels," *IEEE J. of Solid-State Circuits,* Vol. 28, pp. 462–470, April 1993.

K. R. Laker and W. M. C. Sansen. *Design of Analog Integrated Circuits and Systems.* McGraw-Hill, New York, 1994.

U. Moon and B.-S. Song. "Design of a Low-Distortion 22-kHz Fifth-Order Bessel Filter," *IEEE J. of Solid-State Circuits,* Vol. 28, pp. 1254–1264, December 1993.

J. P. Moree, G. Groenewold, and L. A. D. van den Broeke. "A Bipolar Integrated Continuous-Time Filter with Optimized Dynamic Range," *IEEE J. of Solid-State Circuits,* Vol. 28, pp. 954–961, September 1993.

A. Nedungadi and T. R. Viswanathan. "Design of Linear CMOS Transconductance Elements," *IEEE Trans. on Circuits and Systems,* Vol. 31, pp. 891–894, October 1984.

M. Padmanabhan and K. Martin. "A CMOS Analog Multi-Sinusoidal Phase-Locked-Loop," *IEEE J. of Solid-State Circuits,* Vol. 29, no. 9, pp. 1046–1057, September 1994.

C. S. Park and R. Schaumann. "A High-Frequency CMOS Linear Transconductance Element," *IEEE Trans. on Circuits and Systems,* Vol. 33, pp. 1132–1138, November 1986.

J. L. Pennock. "CMOS Triode Transconductor for Continuous-Time Active Integrated Filters," *Electron. Lett.,* Vol. 21, no. 18, pp. 817–818, August 1985.

R. Schaumann. "Continuous-Time Integrated Filters — A Tutorial," in *Integrated Continuous-Time Filters,* IEEE Press, 1992.

R. Schaumann, M. S. Ghausi, and K. R. Laker. *Design of Analog Filters,* Prentice Hall, Englewood Cliffs, New Jersey, 1990.

E. Seevinck and R. F. Wassenaar. "A Versatile CMOS Linear Transconductor/Square-Law Function Circuit," *IEEE J. of Solid-State Circuits,* Vol. 22, pp. 366–377, June 1987.

A. Shoval, D. A. Johns, and W. M. Snelgrove. "A Wide-Range Tunable BiCMOS Transconductor," *Microelectronics Journal,* Vol. 24, pp. 555–564, Elsevier Science Publishers, UK, August 1993.

A. Shoval, W. M. Snelgrove, and D. A. Johns. "A 100-Mb/s BiCMOS Adaptive Pulse-Shaping Filter," accepted for publication in *IEEE J. on Selected Areas in Communications: Special Issue on Copper Wire Access Technologies for High-Performance Networks,* 1995.

J. Silva-Martinez, M. S. J. Steyaert, and W. C. Sansen. "A Large-Signal Very Low-Distortion Transconductor for High-Frequency Continuous-Time Filters," *IEEE J. of Solid-State Circuits,* vol. 26, pp. 946-955, July 1991.

J. Silva-Martinez, M. S. J. Steyaert, and W. C. Sansen. "A 10.7-MHz 68-dB SNR CMOS Continuous-Time Filter with On-Chip Automatic Tuning," *IEEE J. of Solid-State Circuits,* Vol. 27, pp. 1843–1853, December 1992.

W. M. Snelgrove and A. Shoval. "A Balanced 0.9-μm CMOS Transconductance-C Filter Tunable over the VHF Range," *IEEE J. of Solid-State Circuits,* Vol. 27, pp. 314–323, March 1992.

K. S. Tan and P. R. Gray. "Fully-Integrated Analog Filters Using Bipolar-JFET Technology," *IEEE J. of Solid-State Circuits,* Vol. SC-13, pp. 814–821, December 1978.

H. Tanimoto, M. Koyama, and Y. Yoshida. "Realization of a 1-V Active Filter Using a Linearization Technique Employing Plurality of Emitter-Coupled Pairs," *IEEE J. of Solid-State Circuits,* Vol. 26, pp. 937–945, July 1991.

Y. P. Tsividis. "Integrated Continuous-Time Filter Design—An Overview," *IEEE J. of Solid-State Circuits,* Vol. 29, pp. 166–176, March 1994.

Y. P. Tsividis and J. O. Voorman. *Integrated Continuous-Time Filter,* IEEE Press, New York, 1992.

T. R. Viswanathan, et al. "Switched-Capacitor Frequency Control Loop," *IEEE J. of Solid-State Circuits,* Vol. 17, pp. 775–778, August 1982.

Z. Wang and W. Guggenbühl. "A Voltage-Controllable Linear MOS Transconductor Using Bias Offset Technique," *IEEE J. of Solid-State Circuits,* Vol. 25, pp. 315–317, February 1990.

D. R. Welland, et al. "A Digital Read/Write Channel with EEPR4 Detection," *IEEE Int. Solid-State Circuits Conf.*, pp. 276–277, San Francisco, 1994.

S. D. Willingham, K. W. Martin, and A. Ganesan. "A BiCMOS Low-Distortion 8-MHz Low-Pass Filter," *IEEE J. of Solid-State Circuits,* Vol. 28, pp. 1234–1245, December 1993.

S. L. Wong. "Novel Drain-Based Transconductance Building Blocks for Continuous-Time Filter Applications," *Electron. Lett.*, Vol. 25, no. 2, pp. 100–101, January 1989.

A. Wyszynski, R. Schaumann, S. Szczepanski, and P. Van Halen. "Design of a 2.7-GHz Linear OTA and a 250-MHz Elliptic Filter in Bipolar Transistor-Array Technology," *IEEE Trans. on Circuits and Systems—II: Analog and Digital Signal Processing,* Vol. 40, pp. 19–31, January 1993.

15.10 PROBLEMS

Unless otherwise stated, assume the following:

- npn bipolar transistors:
  $\beta = 100$
  $V_A = 80\ \text{V}$
  $\tau_b = 13\ \text{ps}$
  $\tau_s = 4\ \text{ns}$
  $r_b = 330\ \Omega$
- n-channel MOS transistors:
  $\mu_n C_{ox} = 92\ \mu\text{A/V}^2$
  $V_{tn} = 0.8\ \text{V}$
  $\gamma = 0.5\ \text{V}^{1/2}$
  $r_{ds}(\Omega) = 8{,}000\text{L}\ (\mu\text{m})/I_D\ (\text{mA})$ in active region
  $C_j = 2.4 \times 10^{-4}\ \text{pF}/(\mu\text{m})^2$
  $C_{j\text{-}sw} = 2.0 \times 10^{-4}\ \text{pF}/(\mu\text{m})$
  $C_{ox} = 1.9 \times 10^{-3}\ \text{pF}/(\mu\text{m})^2$
  $C_{gs(overlap)} = C_{gd(overlap)} = 2.0 \times 10^{-4}\ \text{pF}/\mu\text{m}$
- p-channel MOS transistors:
  $\mu_p C_{ox} = 30\ \mu\text{A/V}^2$
  $V_{tp} = -0.9\ \text{V}$
  $\gamma = 0.8\ \text{V}^{1/2}$
  $r_{ds}(\Omega) = 12{,}000\text{L}(\mu\text{m})/I_D(\text{mA})$ in active region
  $C_j = 4.5 \times 10^{-4}\ \text{pF}/\mu\text{m}^2$
  $C_{j\text{-}sw} = 2.5 \times 10^{-4}\ \text{pF}/\mu\text{m}$
  $C_{ox} = 1.9 \times 10^{-3}\ \text{pF}/(\mu\text{m})^2$
  $C_{gs(overlap)} = C_{gd(overlap)} = 2.0 \times 10^{-4}\ \text{pF}/\mu\text{m}$

15.1 Show that the input impedance (at nodes i_o) of the Miller integrator shown in Fig. 15.6 is on the order of $2/(g_{m2}^2 r_{ds})$ at frequencies where capacitors $2C_1$ can be considered short circuits.

15.2 Based on the block diagram of Fig. 15.9, find transconductance values for a first-order filter with a dc gain of 10 and a pole at 15 MHz (no finite zero). Assume the integrating capacitors are sized $C_A = 5$ pF.

15.3 Find the transconductance and capacitance values of the second-order filter of Fig. 15.11 for a low-pass filter with a pole frequency of 10 MHz, $Q = 1$, and a dc gain of 5.

15.4 Derive the design equations for the biquad circuit shown in Fig. 15.24.

15.5 Consider the transconductor shown in Fig. 15.13. Find the transconductance of this circuit for finite opamp gains A.

15.6 In the transconductor of Fig. 15.12(*b*), find the minimum collector voltages of Q_1, Q_2 such that they remain in the active region for $v_i = \pm 1$ V around a common-mode voltage of 2 V. Repeat for the transconductor of Fig. 15.13.

15.7 In the transconductor of Fig. 15.14, find the new transconductance value if the base-emitter areas of Q_7, Q_8 are four times those of Q_3, Q_4.

15.8 Consider the transconductor of Fig. 15.14, where we want $G_m = 1$ mA/V. What are reasonable values for I_1 and I_2 such that all transistors do not fall below 20 percent of their nominal bias current when the peak value of $v_i = v_i^+ - v_i^-$ is 500 mV?

15.9 Consider the linearized two-differential pairs shown in Fig. 15.20, where $I_1 = 2$ mA. Find the percentage of error between an ideal linear transconductor and the true output current when $v_i = 48$ mV.

15.10 Consider the circuit Fig. 15.21, with $I_1 = 100\ \mu$A, $I_2 = 10\ \mu$A, $(W/L)_9 = 2$, $V_C = 5$ V, and $v_i^- = 2.5$ V.

a. Find the transconductance, G_m.

b. Assuming a perfectly linear transconductor, find i_{o1} when v_i^+ equals 2.6 V and 3 V.

c. Find the true value of i_{o1} when v_i^+ equals 2.6 V and 3 V by taking into account the triode equation for Q_9 (assume everything else is ideal). Compare your results with those in part b.

15.11 Choose reasonable transistor sizes for the circuit in Fig. 15.25 to realize a transconductor that has $G_m = 0.3$ mA/V, $k_1/k_3 = 6.7$, and a peak differential-input voltage of 1 V.

15.12 Consider the circuit in Fig. 15.30, where all transistors have $W/L = 10\ \mu\text{m}/2\ \mu\text{m}$, $V_{DD} = -V_{SS} = 2.5$ V, and $V_{C1} = -V_{C2} = 2.0$ V. Find the transconductance of the circuit.

15.13 Show that the transconductance of Fig. 15.32 is given by

$$G_m = \left(\frac{n}{n+1}\right)4\sqrt{kI_B}$$

15.14 Consider the transconductor shown in Fig. 15.34, where all transistors have size $W/L = 10\ \mu\text{m}/2\ \mu\text{m}$, $I_B = 50\ \mu$A, and $V_{DD} = -V_{SS} = 3$ V. What is the maximum differential input, $v_1 - v_2$ (centered around 0 V), that can be used while maintaining proper operation?

15.15 Show that the circuit of Fig. 15.36 results in

$$i_1 - i_2 = K\left(V_B - 2V_{tn}\right)\left(2v_1 - V_B\right)$$

15.16 Using two-transistor integrators, design a MOSFET-C second-order low-pass filter with a pole frequency of 1 MHz, $Q = 1$, and a dc gain of 2. Let the integrating capacitors be 10 pF, $V_C = 3$ V, and $V_x = 0$ V.

15.17 Calculate the percentage THD value of a signal in which the fundamental component is at 1 MHz and has a voltage level of 1 V_{rms}, and the only significant harmonics are the first three above the fundamental. These harmonics, at 2 MHz, 3 MHz, and 4 MHz, are measured to have voltage levels of 1 mV_{rms}, 0.5 mV_{rms}, and 0.3 mV_{rms}, respectively.

15.18 If a circuit is measured to have $IIP_3 = 10$ dBm and has a gain of 6 dB, what output-signal level should be used such that the third-order intermodulation products are 60 dB below the fundamental?

15.19 At an input-signal level of –4 dBm, an intermodulation ratio of –40 dB was measured in a filter with a gain of 6 dB. Calculate the value of the input and output third-order intercept points. What input-signal level should be applied if we want an intermodulation ratio of –50 dB? If the noise power at the output is measured to be –60 dBm, what is the expected SFDR and what output-signal level does it correspond to?

CHAPTER

16 Phase-Locked Loops

Phase-locked loops (PLL) are analog building blocks used extensively in many analog and digital systems. An application where PLLs are used is clock recovery in many communication and digital systems. A second application area is frequency synthesizers used in televisions or wireless communication systems to select different channels. A third area is in the demodulation of frequency-modulated signals. These are only a few of the application areas, but it is safe to say that phase-locked loops have become an important building block in many electronic systems.

Although PLLs are highly nonlinear systems, it has been found that when they are in lock, their transient behavior can be reasonably well approximated by linear differential equations. Indeed, analysis methods similar to those first developed to describe PLLs have now been applied to other important nonlinear systems such as adaptive filters and switching power supplies. Thus, understanding the small-signal modelling of PLLs can be very useful in understanding many important signal-processing systems.

16.1 BASIC LOOP ARCHITECTURE

The basic architecture of a PLL is shown in Fig. 16.1 The PLL usually consists of a phase detector, a low-pass filter, a gain stage, and a *voltage-controlled oscillator* (VCO) configured in a loop. The phase detector is a circuit that normally has an output voltage with an average value proportional to the phase difference between the input signal and the output of the VCO. The low-pass filter is used to extract the average value from the output of the phase detector. This average value is then amplified and used to drive the VCO. The negative feedback of the loop results in the output of the VCO being synchronized with the input signal. Depending on the type of the phase-locked loop, the output of the oscillator may or may not have a phase difference compared with the phase of the input signal.

In most PLLs, the low-pass filter is first or second order. The simplest possibility is an RC low-pass filter, which historically has been the most commonly used filter. However, as will be seen shortly, a first-order filter having a low-frequency pole and a higher-frequency zero is recommended. There are many possibilities for the phase detector, including a simple exclusive-OR gate, a sample and hold, an analog multiplier, or a combination of digital circuits such as D flip-flops. The voltage-controlled

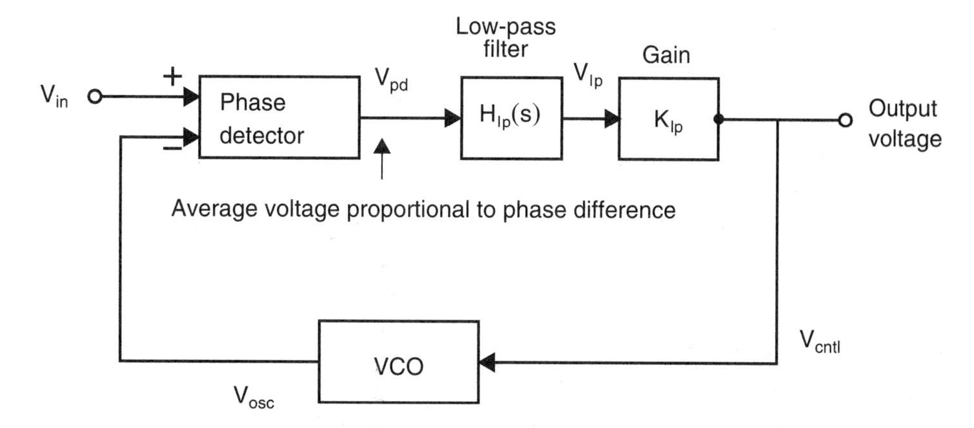

Fig. 16.1 The basic architecture of a phase-locked loop.

oscillator typically has an output that is either a square wave or a sinusoidal signal. A square-wave output might be used in applications where the input is a digital signal, whereas a sinusoidal output might be used in a wireless frequency synthesizer, for example. Normally, the amplitude of the input signal will be known, since its value can affect loop dynamics depending on the choice of phase detector. When the input signal's amplitude is not known, the input might be converted into a digital signal by first sending it through a limiter circuit, or perhaps the input signal may be amplified to a predetermined amplitude by applying it to an automatic-gain-control (AGC) circuit.

To understand phase-locked loops, a simple example will be considered. In this example, the input signal, V_{in}, is assumed to be a sinusoid with a known amplitude. The phase detector is realized as an analog multiplier having a relationship given by

$$V_{pd} = K_M V_{in} V_{osc} \tag{16.1}$$

where V_{osc} is the VCO output, V_{pd} is the phase-detector output, and K_M is a multiplication constant. It will also be assumed that the low-pass filter has a transfer function given by

$$H_{lp}(s) = \frac{1 + s\tau_z}{1 + s\tau_p} \tag{16.2}$$

where $\tau_z \ll \tau_p$. This first-order transfer function is known as a *lead-lag filter* and represents a fairly general form for PLL applications.

Let the input signal be described by a sinusoid having

$$V_{in} = E_{in} \sin(\omega t) \tag{16.3}$$

and the output of the voltage-controlled oscillator is given by a sinusoid where

$$V_{osc} = E_{osc} \sin(\omega t - \phi_d + 90°) = E_{osc} \cos(\omega t - \phi_d) \tag{16.4}$$

The argument ϕ_d represents the phase difference between the input signal and the output of the oscillator. The reason for the 90° offset is so that when the phase difference is zero the average output of the phase detector (which is an analog multiplier for our example) is zero. The combination of an analog multiplier followed by a low-pass filter performs what is commonly known as the correlation function on analog signals. When the input signal and the VCO output have a 90° phase difference (i.e., $\phi_d = 0$), they are uncorrelated and the output of the low-pass filter will be zero. From this 90° phase offset, if the phase difference changes so that the two waveforms are more in phase (i.e., $\phi_d > 0$), then the correlation and, therefore, the output of the low-pass filter will be positive. For example, from (16.4), if $\phi_d = 90°$, then the output of the oscillator and the input signal would be exactly in phase, and the average output of the phase detector is found to be the amplitudes of the two sinusoids multiplied together and divided by two, resulting in

$$V_{cntl} = K_{lp}K_M\frac{E_{in}E_{osc}}{2} \qquad (16.5)$$

This situation is approximately shown in Figure 16.2, where ϕ_d is taken slightly smaller than 90° to distinguish the two waveforms. Also shown is the output of the multiplier, V_{pd}, for the case of $E_{in} = E_{osc} = 1$ V, and $K_M = 1\ V^{-1}$. Here, it is seen that the output is at twice the frequency of the input signal and that it has an offset voltage of approximately 1/2 V. This corresponds closely to the maximum possible output of the low-pass filter. On the other hand, if the phase changes so as to make the two waveforms more out of phase, then the correlation between them and the output of the low-pass filter will be negative. Shown in Fig. 16.3 are some examples of the different waveforms. It is readily seen that $\phi_d > 0$ corresponds to the waveforms that are more in phase.

These statements can be quantified analytically. The output of the phase detector is given by

$$V_{pd} = K_M V_{in} V_{osc} = K_M E_{in} E_{osc} \sin(\omega t)\cos(\omega t - \phi_d) \qquad (16.6)$$

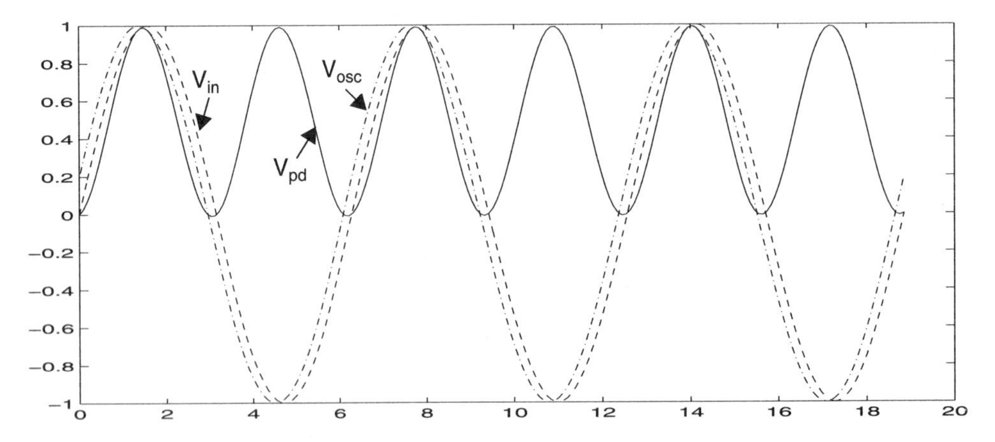

Fig. 16.2 The output of the phase detector when the input signal and the oscillator are nearly in phase.

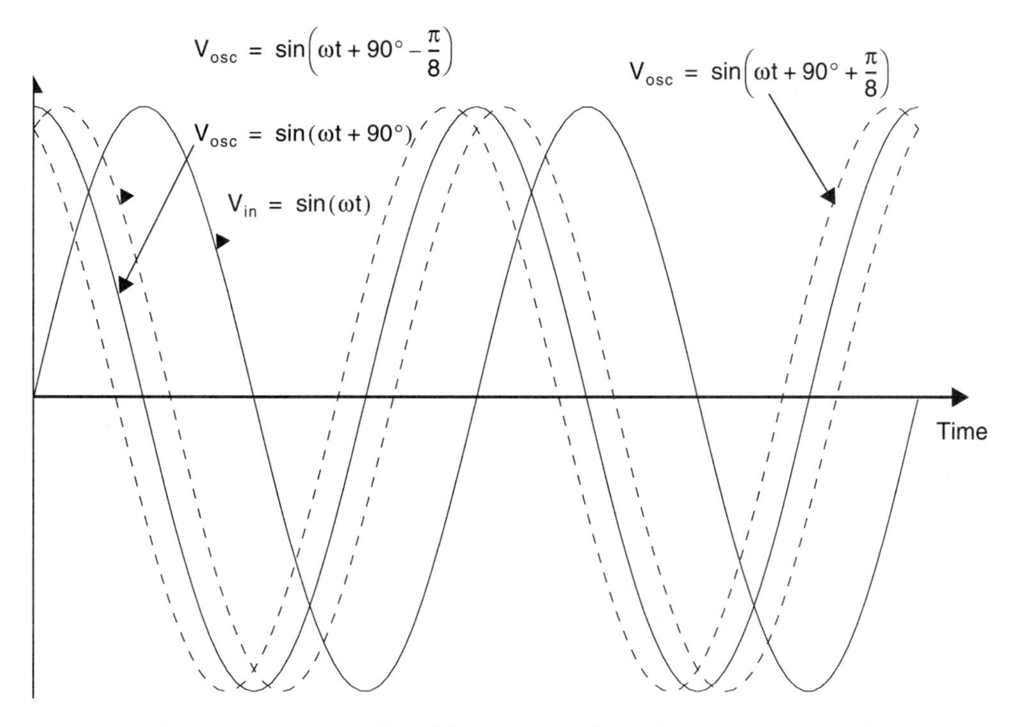

Fig. 16.3 The relative phase angles of the input signal and the output of the oscillator.

Using the trigonometric identity

$$\sin(A)\ \cos(B) = \frac{1}{2}[\sin(A+B) + \sin(A-B)] \tag{16.7}$$

we have

$$V_{pd} = K_M \frac{E_{in}E_{osc}}{2}[\sin(\phi_d) + \sin(2\omega t - \phi_d)] \tag{16.8}$$

The low-pass filter's function is to remove the second term at twice the frequency of the input signal.[1] The signal V_{cntl} is therefore given by

$$V_{cntl} = K_{lp}K_M \frac{E_{in}E_{osc}}{2}\sin(\phi_d) \tag{16.9}$$

Note that V_{cntl} is either a dc value or a slowly varying signal, assuming that the frequency of the input signal is constant or slowly varying. For small ϕ_d, we can approximate (16.9) as

$$V_{cntl} \cong K_{lp}K_M \frac{E_{in}E_{osc}}{2}\phi_d = K_{lp}K_{pd}\phi_d \tag{16.10}$$

1. The extent to which this is not true is a major source of distortion when sinusoidal VCOs are used.

Thus, the output of the low-pass filter is approximately proportional to the phase difference between the output of the oscillator and the input signal, assuming the 90° offset bias is ignored. This approximation is used in analyzing the PLL to obtain a linear model. The constant of proportionality is called K_{pd} and is given by

$$K_{pd} = K_M \frac{E_{in}E_{osc}}{2} \tag{16.11}$$

Note that with this analog-multiplier phase detector, K_{pd} is a function of the input amplitude and oscillator output amplitude. As we shall soon see, K_{pd} affects loop dynamics, and thus we see the need for having known values for E_{in} and E_{osc}.

To understand the operation of the PLL, assume the VCO has a free-running frequency ω_{fr} when its input is zero. Also, assume that the input signal is initially equal to the free-running frequency of the VCO, and the system is in lock with $\phi_d = 0$ and the output of the low-pass filter is also equal to zero. Next, assume the input frequency slowly increases. This will initially cause it to lead the VCO output (again ignoring the 90° offset), which corresponds to $\phi_d > 0$, since ϕ_d is defined as the difference between the input signal's phase minus the VCO's output signal's phase. Since for $\phi_d = 0$, the input and VCO output were 90° out of phase, now with $\phi_d > 0$, the two waveforms will become more in phase (see Fig. 16.3) and after a short time (roughly the time constant of the low-pass filter) the output of the low-pass filter will go positive. Since the two waveforms are now assumed to be at slightly different frequencies, the phase difference, and therefore the output of the low-pass filter, will slowly continue to increase. However, the frequency of the VCO is proportional to its control voltage, so the increase in the low-pass filter's output will cause the VCO's frequency to increase until it is the same as that of the input signal again, which will keep the two signals in synchronism (i.e., locked). Of course, the opposite would occur if the input signal's frequency decreased. Specifically, the average output of the phase detector would become negative, which, after being averaged by the low-pass filter, would drive the oscillator to a lower frequency until it is again identical to the frequency of the input signal and the two waveforms are again synchronized. It is readily seen here that the phase-locked loop stays in lock because of the negative feedback of the loop in which the phase of the oscillator's output is subtracted from the phase of the input signal.

Note, however, that at this new input frequency, which does not equal the free-running frequency of the VCO, their phase difference is no longer 90°. We can find the new phase difference for these two locked signals by noting that the frequency of the oscillator's output signal is given by

$$\omega_{osc} = K_{osc}V_{cntl} + \omega_{fr} \tag{16.12}$$

Here, ω_{fr} is the free-running frequency of the VCO when its control voltage is zero and K_{osc} is a constant relating the change in frequency to control voltage ratio. The output voltage of the amplified low-pass filter is now given by

$$V_{cntl} = \frac{\omega_{in} - \omega_{fr}}{K_{osc}} \tag{16.13}$$

where ω_{in} is the frequency of the input signal, which is equal to the frequency of the oscillator's output. From (16.10), we also have

$$\phi_d = \frac{V_{cntl}}{K_{lp}K_{pd}} = \frac{\omega_{in} - \omega_{fr}}{K_{lp}K_{pd}K_{osc}} \tag{16.14}$$

EXAMPLE 16.1

Consider a PLL where the amplitudes of the input signal and the oscillator output are both at a 0.75-V peak. An analog multiplier is being used for the phase detector, which would have a 2-V output when both inputs are dc values of 2 V. The VCO has a free-running frequency of 10 MHz, which would decrease to zero for $V_{cntl} = -1$ V, and the gain block is unity (i.e., $K_{lp} = 1$). When in lock, what is the phase difference between the input and oscillator output when the input frequency is 11 MHz ? What is the phase difference when the input is 9 MHz? How does the phase difference change if $K_{lp} = 2$?

Solution

Since $V_{pd} = K_M V_{in} V_{osc}$, we have $K_M = 0.5\ V^{-1}$. Using (16.11), we also have

$$K_{pd} = K_M \frac{E_{in}E_{osc}}{2} = 0.1406 \text{ V/rad} \tag{16.15}$$

In addition, we find

$$K_{osc} = \frac{\Delta\omega_{osc}}{\Delta V_{cntl}} = \frac{2\pi \times 10^7 \text{ rad/s}}{1\text{V}} = 6.28 \times 10^7 \text{ rad/Vs} \tag{16.16}$$

Making use of (16.14), we have for the case $\omega_{in} = 11$ MHz,

$$\phi_d = \frac{2\pi \times (11 \text{ MHz} - 10 \text{ MHz})}{K_{lp}K_{pd}K_{osc}} = 0.7112 \text{ rad} = 40.8° \tag{16.17}$$

resulting in the phase difference between the input and oscillator output being $90° - 40.8° = 49.2°$ (i.e., becoming closer to an in-phase difference of 0°).

For the case where $\omega_{in} = 9$ MHz, we have $\phi_d = -40.8°$, resulting in a phase difference of 130.8° (i.e., becoming closer to an out-of-phase difference of 180°).

If $K_{lp} = 2$, then half the phase difference results in the same VCO control voltage, V_{cntl}. Therefore, for $\omega_{in} = 11$ MHz, $\phi_d = 20.4°$, and for $\omega_{in} =$ 9 MHz, $\phi_d = -20.4°$.

Linearized Small-Signal Analysis

The phase-locked loop is a highly nonlinear system. Fortunately, once a PLL is in lock, its dynamic response to input-signal phase and frequency changes can be well approximated by a linear model, as long as these changes are slow and small about their operating or bias point.

Before the linear model is presented, first consider a sinusoidal signal where the frequency is slowly varying with time. This signal can be represented by the equation

$$v(t) = E\ \sin[\omega(t)t + \phi_d(t)] \tag{16.18}$$

where $\omega(t)$ and $\phi_d(t)$ are both deterministic signals that slowly change with time. The total phase at any instance in time is given by

$$\phi(t) = \omega(t)t + \phi_d(t) \tag{16.19}$$

For the time being, assume $\phi_d(t)$ is a constant. The instantaneous frequency is defined to be

$$\omega_{inst}(t) = \frac{d\phi(t)}{dt} \tag{16.20}$$

Note that in the case where $\omega(t)$ is also a constant, we see from combining (16.19) and (16.20) that we have the satisfying intuitive result that the instantaneous frequency, $\omega_{inst}(t)$, is equal to $\omega(t)$.

If we integrate both sides of (16.20), we have

$$\phi(t) = \phi(0) + \int_0^t \omega_{inst}(\tau)\, d\tau \tag{16.21}$$

and we see that the *instantaneous phase is the integral of the instantaneous frequency*. From here on, we will assume that the instantaneous frequency and $\omega(t)$ are, for practical purposes, the same and drop the subscript from $\omega_{inst}(t)$ to make our notation simpler (although this simplification is not necessary). If we take the Laplace transform of both sides of (16.20), we have

$$\phi(s) = \frac{\omega(s)}{s} \tag{16.22}$$

where $\phi(s)$ and $\omega(s)$ are the Laplace transforms of $\phi(t)$ and $\omega(t)$, respectively. This relationship will be used in modelling the relationship between the VCO's input-control voltage and the phase of its output signal.

We can now give an approximately equivalent, small-signal model for a phase-locked loop that relates the phase of the input signal and oscillator output to the low-pass filter output voltage. It should be noted that this model is a small-signal model in that it relates changes about an operating point. In other words, the phase variable represents *changes in phase* from reference signals that were identical to the true signals at the beginning of the analysis. Similarly, any frequency variables, which will be introduced shortly, also represent *changes from the original frequencies* at the beginning of the analysis. A signal-flow graph for the linearized small-signal model is shown in Fig. 16.4. The constant, K_{pd}, is given by (16.11). Recalling (16.12),

$$\omega_{osc} = K_{osc}V_{cntl} + \omega_{fr} \tag{16.23}$$

note that since we are only modelling changes about an operating point, the term ω_{fr} is absent from the signal-flow graph (SFG) of Fig. 16.4, as it is a nonchanging bias term.

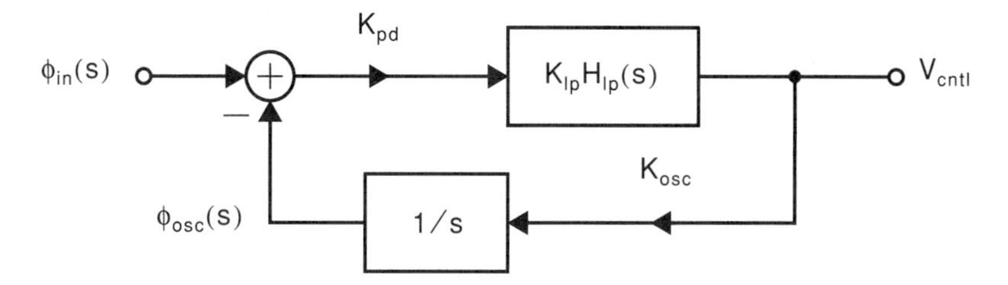

Fig. 16.4 A signal-flow graph for the linearized small-signal model of a PLL when in lock.

It is now possible to analyze the SFG of Fig. 16.4 to determine the transfer function that relates the Laplace transform of $V_{cntl}(s)$ to the Laplace transform of $\phi_{in}(s)$. We have

$$V_{cntl}(s) = K_{pd}K_{lp}H_{lp}(s)[\phi_{in}(s) - \phi_{osc}(s)] \tag{16.24}$$

and

$$\phi_{osc}(s) = \frac{K_{osc}V_{cntl}(s)}{s} \tag{16.25}$$

Combining equations (16.24) and (16.25) and rearranging gives

$$\frac{V_{cntl}(s)}{\phi_{in}(s)} = \frac{sK_{pd}K_{lp}H_{lp}(s)}{s + K_{pd}K_{lp}K_{osc}H_{lp}(s)} \tag{16.26}$$

Equation (16.26) is general in that it applies to almost every phase-locked loop. Differences between one PLL and another are determined only by what is used for the low-pass filter (which determines $H_{lp}(s)$), the phase detector (which determines K_{pd}), or the oscillator (which determines K_{osc}). Normally, most low-pass filters have a function similar to that given in (16.2). Substituting the first-order low-pass relation given in (16.2) into (16.26) and rearranging results in

$$H(s) = \frac{V_{cntl}(s)}{\phi_{in}(s)} = \frac{\dfrac{1}{K_{osc}}s(1 + s\tau_z)}{1 + s\left(\dfrac{1}{K_{pd}K_{lp}K_{osc}} + \tau_z\right) + \dfrac{s^2\tau_p}{K_{pd}K_{lp}K_{osc}}} \tag{16.27}$$

Thus, we see that the relationship is a second-order transfer function. The fact that the gain is zero at dc signifies that for phase changes (not frequency changes), the output of the low-pass filter will eventually settle to its original bias value.

To find the relationship between V_{cntl} and the input frequency, we can use (16.22) to obtain

$$\frac{V_{cntl}(s)}{\omega_{in}(s)} = \frac{\dfrac{1}{K_{osc}}(1 + s\tau_z)}{1 + s\left(\dfrac{1}{K_{pd}K_{lp}K_{osc}} + \tau_z\right) + \dfrac{s^2\tau_p}{K_{pd}K_{lp}K_{osc}}} \tag{16.28}$$

This transfer function has a dc gain of $1/K_{osc}$. This result was expected because the input voltage to the VCO must change by exactly this amount to keep the loop in lock when the input frequency changes.

Note that (16.28) has frequency appearing in two fashions. Specifically, we see that the input is frequency and the relationship between the input frequency and the low-pass filter's output voltage is also a function of the frequency variable, s. What this implies is that the low-pass filter's output voltage is a function of how fast the input frequency changes, which we have constrained to be slow. The variable s, when replaced by $j\omega$, represents the periodic rate at which the input frequency changes, assuming this change is sinusoidal.

In results of both (16.27) and (16.28), the transfer functions are second-order—a fortunate occurrence, as these transfer functions are well understood. Second-order transfer functions are often written in the form

$$H(s) = \frac{N(s)}{D(s)} \tag{16.29}$$

where $D(s)$, the denominator, is expressed as

$$D(s) = 1 + \frac{s}{\omega_0 Q} + \frac{s^2}{\omega_0^2} \tag{16.30}$$

The constant ω_0 is called the resonant frequency of the transfer function. For small Q (around 1/2 or slightly larger), the transient time constant of the complete loop for small phase or frequency changes is roughly given by

$$\tau_{pll} \cong \frac{1}{\omega_0} \tag{16.31}$$

The constant Q is called the Q factor[2] of the transfer function. Normally, for good settling behavior when step changes of the input signal occur, one would choose a Q factor equal to 1/2. Such a choice results in real poles for the transfer function. Alternatively, one might want maximally flat group delay, in which case one would choose $Q = 1/\sqrt{3} = 0.577$. A third alternative would be to have maximally flat amplitude response, in which case one would choose $Q = 1/\sqrt{2} = 0.707$. For most PLL applications, the first choice of setting $Q = 1/2$ is recommended.

Setting the term multiplying each power of s in (16.30) equal to the corresponding term in the denominator of (16.27) or (16.28), then solving for ω_0 and Q, results in

$$\omega_0 = \frac{K_{pll}}{\sqrt{\tau_p}} \tag{16.32}$$

and

$$Q = \frac{\sqrt{\tau_p}}{\dfrac{1}{K_{pll}} + \tau_z K_{pll}} \tag{16.33}$$

---

2. For those more familiar with damping factor, σ, of a second-order transfer function, note that $\sigma = 1/2Q$.

where

$$K_{pll} = \sqrt{K_{pd}K_{lp}K_{osc}} \tag{16.34}$$

For most cases, when ω_0 is much less than the oscillator's free-running frequency, we have $\tau_z >> (1/K_{pll}^2) = 1/(K_{pd}K_{osc})$, and therefore

$$Q \cong \frac{1}{\omega_0\tau_z} = \frac{\sqrt{\tau_p}}{\tau_z K_{pll}} \tag{16.35}$$

Note that taking $Q = 1/2$ and using (16.32) and (16.35) results in

$$\tau_z = \frac{2\sqrt{\tau_p}}{K_{pll}} = \frac{2}{\omega_0} \tag{16.36}$$

Thus, the frequency of the zero in the numerator of (16.28) (that is, $1/\tau_z$) is at one-half the resonant frequency of the loop. Remembering this fact makes for very simple component selection in PLLs. Note also that this zero is responsible for some peaking in the magnitude of the transfer function given in (16.28). This peaking might be a problem in some PLL applications (such as in some communication systems) but can be eliminated if necessary [Lee, 1992].

In a normal design process, the designer will either have little choice over K_{pd} and K_{osc} or will choose them based on practical considerations such as power dissipation and desired signal amplitudes. One would next choose τ_p to achieve the desired loop settling time, using (16.31) and (16.32). Finally, one would choose τ_z to obtain the desired Q factor of the loop.

Note that in the special case where $\tau_z = 0$ and the low-pass filter is a simple first-order low-pass filter with no zero (in which case it could be easily implemented using a resistor and a capacitor only), we have from (16.33)

$$Q = \sqrt{\tau_p}K_{pll} \tag{16.37}$$

Substituting (16.37) into (16.32) results in

$$\omega_0 = \frac{K_{pll}^2}{Q} = \frac{K_{pd}K_{osc}}{Q} \tag{16.38}$$

Thus, for a given K_{pll} and Q factor, the resonant frequency of the loop is completely determined. If the loop is designed to have ω_0 substantially less than $K_{pd}K_{osc}$, then Q will be large and its transient response will be extremely poor. This limitation is the major reason why most well-designed PLLs have a zero in the loop filter (i.e., $\tau_z \neq 0$).

EXAMPLE 16.2

Assuming the same conditions as in Example 16.1, design a low-pass filter for the loop so that the loop time constant is approximately 10 periods at 10 MHz and $Q = 1/2$.

Solution

Recall that $K_{pd} = 0.1406$ V/rad and $K_{osc} = 6.28 \times 10^7$ rad/Vs. Assuming $K_{lp} = 1$, then we have, using (16.34),

$$K_{pll} = \sqrt{K_{pd}K_{lp}K_{osc}} = 2{,}972\ \text{s}^{-1/2} \tag{16.39}$$

Since it is desired that the loop have a time constant of 100 periods, we have

$$\omega_0 = \frac{1}{\tau_{pll}} = \frac{10^7}{100} = 10^5\ \text{s}^{-1} \tag{16.40}$$

Using (16.32) gives

$$\tau_p = \left(\frac{K_{pll}}{\omega_0}\right)^2 = \frac{K_{pd}K_{lp}K_{osc}}{\omega_0^2} = 0.883\ \text{ms} \tag{16.41}$$

Thus, the pole of the low-pass filter is at a frequency given by

$$f_p = \frac{\omega_p}{2\pi} = \frac{1}{2\pi\tau_p} = 180\ \text{Hz} \tag{16.42}$$

Using (16.33) gives

$$\tau_z = \frac{\sqrt{\tau_p}}{QK_{pll}} - \frac{1}{K_{pll}^2} = 19.89\ \mu\text{s} \tag{16.43}$$

which is almost twice as large as the time constant of the loop. Therefore, the zero of the low-pass filter is at 8.0 kHz. Note that if the slightly less accurate but simpler equation (16.36) had been used, we would have had $\tau_z = 20\ \mu\text{s}$, an insignificant difference. The low-pass filter can be realized by the circuit shown in Fig. 16.5. This circuit has a transfer function given by

$$H_{lp}(s) = \frac{1 + sR_2C_1}{1 + s(R_1 + R_2)C_1} \tag{16.44}$$

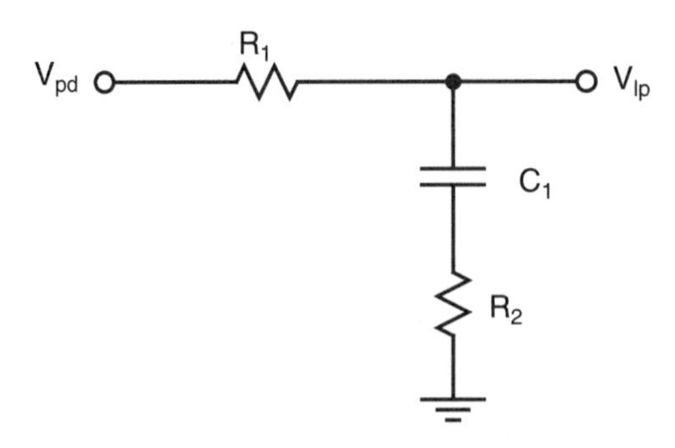

Fig. 16.5 The low-pass filter of Example 16.2.

If we somewhat arbitrarily choose $C_1 = 10\ \text{nF}$, we find

$$R_2 = \frac{\tau_z}{C_1} = 1.99\ \text{k}\Omega \tag{16.45}$$

and

$$R_1 = \frac{\tau_p}{C_1} - R_2 = 86.4\ \text{k}\Omega \tag{16.46}$$

As an aside, if there were no zero in the low-pass filter's transfer function, using (16.37) we would have

$$Q = \sqrt{\tau_p K_{pll}} = 88.4 \tag{16.47}$$

This is an extremely high Q factor, which would cause the loop to be very slow when settling after a transient in either the input signal's phase or frequency. After the transient, the phase settling would be oscillatory, with an envelope having a time constant given by [Desoer, 1969]

$$\tau_{env} = \frac{2Q}{\omega_0} = 0.156\ \text{s} \tag{16.48}$$

which is 177 times greater than desired. This example points out the importance of having a zero in the PLL's low-pass filter.

Capture Range

The preceding linearized analysis describes the loop's dynamics accurately when it is in lock. However, the process of attaining lock is highly nonlinear, and little useful quantitative analysis has been developed to describe this process. Nonetheless, a qualitative description of what happens as the PLL is attaining lock is possible.

First, assume there is no input signal. The VCO would then oscillate at its free-running frequency, since the phase detector's output will, on average, be zero. Next, assume that an input signal appears close to the free-running frequency. If the frequency of the signal is close enough to the free-running frequency of the oscillator, then eventually the loop will lock onto the input signal. The maximum difference between the input signal's frequency and the oscillator's free-running frequency where lock can eventually be attained is defined as the *capture range*. This frequency is on the order of the pole frequency of the low-pass filter.

Next, assume the input signal suddenly appears at a slightly different frequency than the oscillator's frequency. Then the input signal will slowly wander in and out of phase with respect to the oscillator's output. Let us assume the input frequency is greater than the frequency of the oscillator's output. When the signals are in phase, the correlation between the two signals, and therefore the output of the low-pass filter, will be positive and the oscillator will be driven toward lock. When the signals are out of phase, the correlation is negative and the oscillator's frequency will be driven away from lock. This phenomenon is called beating. The analog multiplier can be considered a demodulator. When the input signal's frequency is different from the oscillator's frequency, the input signal is demodulated to a frequency different from, but close to, dc. If this frequency is in the transition region of the low-pass filter, then when the oscillator is being driven toward lock, the output of the low-pass filter will increase in amplitude; whereas when the oscillator is being driven away from lock, the output of the low-pass filter decreases in amplitude. Therefore, with each beat, when the PLL is being driven toward lock, (which occurs when the two signals are more in phase than out of phase), the larger amplitude of the low-pass filter's output drives the loop closer toward lock than the smaller amplitude (which occurs when the signals are out of phase) drives the loop away from lock. Thus, during each beat, the VCO increases in frequency more than it decreases in frequency, and the loop gets closer to lock until eventually, after a number of beats, lock is attained.

Another reason for the VCO being driven toward lock, on the average, is that the demodulated beat signal is distorted, because when it is being driven away from lock it is at a higher instantaneous frequency than when it is being driven toward lock [Gardner, 1979]. This causes the time during which it is being driven toward lock to be larger than the amount of time it is being driven away from lock. The net effect is to cause the loop to become closer to lock with each beat.

However, if the initial frequency difference is quite large, the demodulated signal will not be in the transition region of the low-pass filter (assuming a zero is present). As a result, the difference in the VCO positive- and negative-going excursions will be zero or insignificant, and lock will not be attained, or only attained after an impractically long time.[3]

Analyses of the time required to attain lock are given in [Viterbi, 1966; Blanchard, 1976; and Gardner, 1979]. For the case where the initial difference between the input signal's frequency and the oscillator's frequency is moderately large, Blanchard showed that the acquisition time, t_{acq}, is approximately given by

$$t_{acq} = \frac{Q(\omega_{in} - \omega_{osc})^2}{\omega_0^3} \qquad (16.49)$$

Obviously, when a PLL is designed to have a narrow loop bandwidth, ω_0, the acquisition time can be quite large. In many applications this process, whereby lock is attained, is too slow, especially when multipliers are used for the phase detectors. It is therefore common to add circuitry that will detect when the frequencies of the input signal and the VCO are different and drive the loop toward lock much more quickly than would otherwise occur. Once lock is attained, this additional circuitry will then detect that fact and no longer have an effect on the loop. This additional circuitry is

3. In [Blanchard, 1976; Gardner, 1979], it is stated that the capture range is theoretically infinite when the low-pass filter has its pole at zero (as is the case for the charge pump PLL to be described shortly). However, even when there is no pole at zero, if the initial frequency difference is large, the acquisition time can be so large as to make the PLL unusable. Also, it is possible that, due to circuit nonidealities such as offset voltages, lock may never be attained.

usually called a frequency detector. Alternatively, some PLLs have a low-pass filter with a programmable pole frequency. During initial acquisition, the pole frequency is relatively high in order to broaden the capture range and speed up the acquisition process. Once lock is achieved, the pole is changed to a much lower frequency to increase noise rejection. A third possible alternative is to sweep the VCO's frequency through its entire range during acquisition, with the PLL disabled, until it is detected that the VCO's frequency is close to that of the input signal, at which time the sweeping circuitry is disabled and the PLL is enabled.

Lock Range

Once lock is attained, as long as the input signal's frequency changes only slowly it will remain in lock over a range that is much larger than the capture range. This range is calculated as follows. From (16.5), the maximum output of the amplified low-pass signal is given by

$$V_{cntl\text{-}max} = K_{lp}K_M\frac{E_{in}E_{osc}}{2} = K_{lp}K_{pd} \tag{16.50}$$

Therefore, the maximum excursion of the VCO's frequency away from its free-running frequency is given by

$$\omega_{lck} = \pm K_{osc}K_{lp}K_{pd} \tag{16.51}$$

This result is valid when using a multiplier for a phase detector and assumes that the lock range is limited by the phase-detector limit rather than some internal voltage limits.

The PLL will track the input signal as long as the frequency of the input signal does not exceed this range. However, one should be cautious because, if the loop is locked to a frequency very different from the free-running frequency and a sudden phase change occurs, it is possible that the loop could skip a beat and fall out of lock. Once it has gotten out of lock, lock cannot be attained again until the input signal's frequency is changed to be within the capture range of the free-running frequency, which is much smaller than the lock range. Also, when the PLL is near the edge of the lock range, the gain of the phase detector has decreased substantially from its value around the free-running frequency of the VCO and substantially changes the loop dynamics. Specifically, from (16.32), ω_0 will be decreased, and therefore the loop will be slower in responding. Also, from (16.35), Q will be larger, which may cause the loop to be underdamped and therefore exhibit ringing when the input has transients. For these reasons, one often designs the loop so that the VCO frequency changes only over half the lock range.

EXAMPLE 16.3

What is the lock range for the PLL of Example 16.1?

Solution

Recall that $K_{pd} = 0.1406\ \text{V/rad}$ and $K_{osc} = 6.28 \times 10^7\ \text{rad/Vs}$. Assuming $K_{lp} = 1$, then we have, using (16.51),

$$\omega_{lck} = \pm 2\pi \times 1.41\ \text{MHz} \tag{16.52}$$

In other words, when the input is 1.41 MHz away from the free-running frequency, the output of the phase detector is at a maximum (or minimum). Any further deviation from the free-running frequency causes the phase detector's average output to decrease (or increase), which results in the PLL's losing lock.

False Lock

Another phenomenon that may possibly occur when a multiplier is used for the phase detector is that the loop locks to harmonics of the input signal. For example, the PLL could lock to twice the frequency of the input signal, three times the frequency of the input signal, or any other multiple. Alternatively, it could lock to a submultiple of the frequency of the input signal (i.e., one-half, one-third, or other submultiple). This false lock will occur whenever the free-running frequency of the VCO is closer to a multiple or submultiple of the input signal's frequency than to its actual frequency. False lock can only be alleviated by somehow guaranteeing that the free-running frequency of the oscillator is closer to the actual input frequency than it is to some multiple or submultiple thereof. This requires a priori knowledge of the input signal and is a limitation on how much the input signal's frequency can vary. In the next section, a PLL phase comparator that does not exhibit this false-lock phenomenon will be described.

Exclusive-OR Phase Comparators

A PLL with an exclusive-OR phase comparator operates very similarly to a PLL with an analog multiplier for a phase comparator. An exclusive-OR logic gate has an output given by $V_{pd} = V_{in} \oplus V_{osc}$, where the amplitudes of the signals are determined by the logic levels. Some typical waveforms are shown in Fig. 16.6. It is seen that

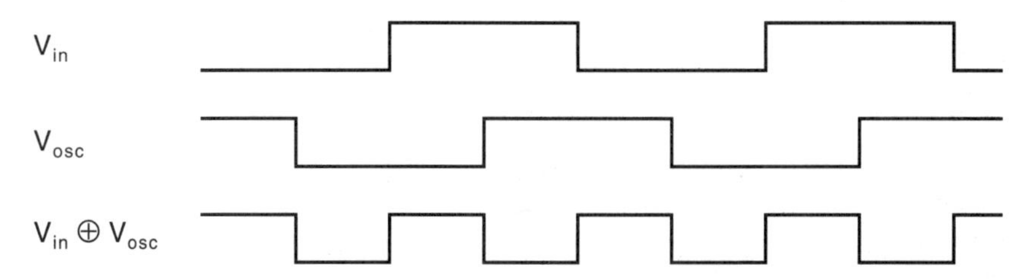

Fig. 16.6 Typical waveforms when an exclusive-OR gate is used as a phase comparator.

when the waveforms are 90° out of phase, the output is a waveform at twice the frequency of the input signal and has a 50 percent duty cycle. If all the waveforms are symmetric about 0 volts, then the output's average value as extracted by the low-pass filter would be zero. When the waveforms become more out of phase, the average value of the output signal is positive; whereas when they become more in phase, the average value of the output signal is negative. If the frequency of the input signal increases, the input signal and the oscillator become more out of phase, which causes the output of the low-pass filter to increase, thereby keeping the signals in lock. This is very similar to how the analog multiplier functions. Similar to PLLs using analog multipliers as phase comparators, PLLs having exclusive-OR gates as phase comparators also exhibit false lock.

16.2 PLLS WITH CHARGE-PUMP PHASE COMPARATORS

One phase-comparator that is popular for integrated applications is called the *charge-pump phase-comparator*. PLLs based on this phase comparator exhibit a number of desirable features. First, they do not exhibit false lock. Second, the input-signal and oscillator-output waveforms are exactly in phase when the system is in lock, even when the input frequency is not the same as the free-running frequency of the oscillator. Finally, the PLL attains lock quickly, even when the input frequency is quite different from the free-running frequency of the oscillator (assuming that a specific implementation of the sequential phase comparator is used). A charge-pump phase comparator is shown in Fig. 16.7. This phase comparator either injects, subtracts, or leaves alone the charge stored across a capacitor in the low-pass filter, depending on the output of a sequential phase-detector circuit. When S_1 is closed, I_{ch} flows into

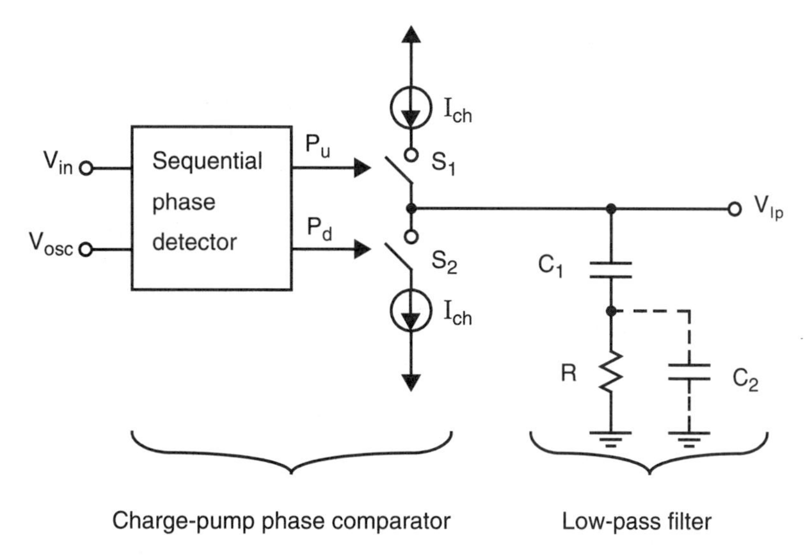

Fig. 16.7 A charge-pump phase comparator.

the low-pass filter, increasing the control voltage into the VCO; when S_2 is closed, I_{ch} flows out of the low-pass filter, decreasing the control voltage of the VCO. When both switches are open, the top plate of C_1 is open circuited and the output voltage remains constant in the steady state. The resistor, R, has been included to realize a zero in the low-pass filter's transfer function, similar to what was recommended when an analog multiplier is used for a phase comparator. This allows the designer to control the damping of the loop's transfer function separately from the time constant of the loop. The capacitor C_2 is included to suppress glitches; it has little effect on loop dynamics and will initially be ignored.

The sequential phase detector is a logic circuit that generates two outputs, P_u and P_d, depending on its inputs, V_{in} and V_{osc}. These are normally digital signals from the PLL input and the VCO, respectively.[4] Although there are many possibilities for the sequential phase detector, the one we will consider is controlled by the leading edges of V_{in} and V_{osc}. If V_{in} goes to 1 before V_{osc}, then P_u will be 1 during the time that the signals are different. This will inject some charge into the loop filter to speed up the VCO. Similarly, if V_{osc} goes to 1 before V_{in}, then P_d will be 1 during the time the inputs are different, S_2 will be closed, and charge will be subtracted from the loop filter, which decreases the VCO frequency. When the leading edges are coincident, ideally both P_u and P_d are always at 0, both S_1 and S_2 are always open, and the VCO input voltage is a constant dc value. For practical sequential phase detectors, both P_u and P_d may go high for a few gate delays when the input waveforms are coincident, but this time is usually short enough that it has very little effect on the dc output of the loop filter, and its effect is largely eliminated by the deglitching capacitor in the low-pass filter, C_2. The falling edges of both V_{in} and V_{osc} have no effect on P_u and P_d, which are both zero at this time. The waveforms when V_{in} leads V_{osc} are shown in Fig. 16.8. Notice that P_d never goes high in this case. Notice also that P_u only goes

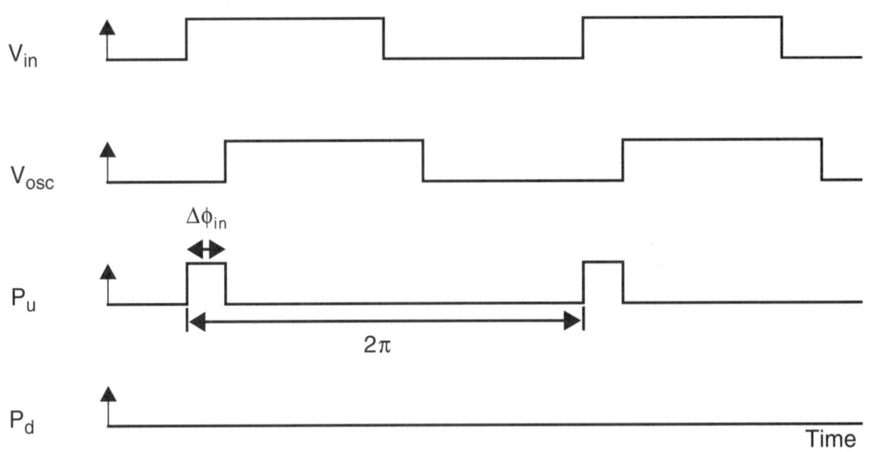

Fig. 16.8 Some typical waveforms of a charge-pump phase comparator.

4. This PLL is only used when the VCO output and the input are both digital signals or have been converted to digital signals by a limiter.

high at the leading edges but is insensitive to the falling edges. This makes the PLL insensitive to the duty cycles of the waveform.

As we shall see shortly, the resistor R in the low-pass filter is required to realize reasonable damping for the loop so that settling during transients is smooth, without oscillations. Unfortunately, it does cause a voltage glitch in the low-pass filter of magnitude $I_{ch}R$ when either S_1 or S_2 first turn on. Glitches occur because when the switches first turn on, the capacitors initially function as small-signal short circuits during the transients, and the change in the loop filter's output voltage is simply given by $I_{ch}R$. This glitch is greatly minimized by the inclusion of capacitor C_2. If C_2 is taken around $C_1/10$, then it usually has little effect on the loop dynamics, other than to decrease the damping slightly (i.e., increase the Q factor). To compensate for this, the loop is usually designed ignoring C_2, but the damping factor is taken greater than unity (or equivalently the Q factor is taken less than $1/2$) by a small factor.

To understand the operation of a charge-pump PLL, consider the case when the VCO's frequency is the same as the frequency of the input signal but their phases are different. Also, assume the input signal is leading the output of the VCO, as shown in Fig. 16.8. At each leading edge of V_{in}, S_1 turns on, which causes a current I_{ch} to flow into the low-pass filter. This current flows until the leading edge of V_{osc} goes high. The additional charge that is stored on C_1 is equal to the time that S_1 is on multiplied by I_{ch}, which causes an increase in the voltage across C_1 equal to

$$\Delta V_{C1} = \frac{\Delta Q_{C1}}{C_1} = \frac{I_{ch}\Delta t}{C_1} \tag{16.53}$$

This change increases the low-pass filter's output voltage, which in turn increases the frequency of the VCO. The VCO frequency increase causes the VCO's output to become more in phase with the input signal. Eventually they become locked in phase. At this time, ideally neither S_1 or S_2 will turn on anymore, and the output voltage of the low-pass filter, and therefore the control voltage of the PLL, will remain constant from then on, assuming the characteristics of the input signal remain constant.

Small-Signal Analysis of a Charge-Pump PLL

To do a small-signal analysis of a charge-pump PLL, it is necessary to make the assumption that the phase of the input and VCO signals change little in any given period. Assume that the frequencies of both signals are the same but the input signal leads the VCO output by a phase difference equal to $\Delta\phi_{in}$, as shown in Fig. 16.8. During this time, I_{ch} flows into the low-pass filter, whereas during the rest of the period, no charge will flow. Precise analysis of this system is difficult because it is time variant. Therefore, the analysis will be done on a different time-invariant system that has the same average charge flowing into the low-pass filter. As long as the dynamics of both systems change only slightly in a given period, the two systems are approximately equivalent. It might be mentioned that this type of approximation is also the basis for analyzing many other systems, such as switching power-supply regulators.

The average charge flowing into the low-pass filter for the charge-pump PLL is given by

$$I_{avg} = \frac{\Delta\phi_{in}}{2\pi} I_{ch} \tag{16.54}$$

So, we will analyze an approximately equivalent system that has a current I_{avg} flowing into an impedance $R + 1/(sC_1)$ (assuming C_2 is zero). We can then define a phase comparator for the approximately equivalent system, where

$$I_{avg} = K_{pd}(\phi_{in} - \phi_{osc}) = K_{pd}\Delta\phi_{in} \tag{16.55}$$

and

$$K_{pd} = \frac{I_{ch}}{2\pi} \tag{16.56}$$

If we further define $H_{lp}(s)$ to be given by

$$H_{lp}(s) = \frac{V_{lp}(s)}{I_{avg}(s)} = R + \frac{1}{sC_1} = \frac{1 + sRC_1}{sC_1} \tag{16.57}$$

we can substitute (16.55) and (16.57) into the general transfer function of a PLL, namely (16.26), to find the transfer function of a charge-pump PLL. Repeating (16.26) for clarity, we have

$$\frac{V_{lp}(s)}{\phi_{in}(s)} = \frac{sK_{pd}H_{lp}(s)}{s + K_{pd}K_{osc}H_{lp}(s)} \tag{16.58}$$

Note that K_{lp} has been absorbed into $H_{lp}(s)$ here for simplicity. Substituting (16.57) into (16.58) and rearranging gives

$$\frac{V_{lp}(s)}{\phi_{in}(s)} = \frac{1}{K_{osc}} \frac{s(1 + sRC_1)}{1 + sRC_1 + \frac{s^2C_1}{K_{pd}K_{osc}}} \tag{16.59}$$

Setting the coefficient of the s^2 term in the denominator of (16.59) equal to $1/\omega_0^2$ and solving for ω_0 results in

$$\omega_0 = \sqrt{\frac{K_{pd}K_{osc}}{C_1}} \tag{16.60}$$

Setting the coefficient of the s term in the denominator of (16.59) equal to $(1/\omega_0 Q)$ and solving for Q gives

$$Q = \frac{1}{RC_1\omega_0} = \frac{1}{R\sqrt{C_1K_{pd}K_{osc}}} = \frac{1}{R}\sqrt{\frac{2\pi}{C_1I_{ch}K_{osc}}} \tag{16.61}$$

When designing a charge-pump PLL one normally chooses I_{ch} based on practical considerations such as power dissipation and speed. Also, ω_0 is chosen so that the loop has the desired transient settling-time constant according to

$$\omega_0 = \frac{1}{\tau_{pll}} \tag{16.62}$$

Next, C_1 is chosen using (16.60) and R is chosen using (16.61). At this point, it is a good idea to use a value for Q slightly less than what is eventually desired. This reduction accounts for the fact that when C_2 is eventually added to minimize glitches, the Q factor will be enhanced somewhat. A rule of thumb might be to take Q, say, 20 percent smaller than eventually desired. The final step is to add C_2. Another rule of thumb here is to take C_2 around one-eighth to one-tenth the size of C_1. The effect this will have on the transfer function is to make it third order by introducing a real pole at frequencies quite a bit higher than the resonant frequency of the loop. This could be taken into account during the design phase by substituting

$$H_{lp}(s) = \frac{R}{1 + sRC_2} + \frac{1}{sC_1} \tag{16.63}$$

into (16.58) and then solving for the poles. Since it is known that the additional pole will be real and at a frequency substantially higher than the original resonant frequency without C_2, the third-order system can be approximated by a second-order system. However, now the Q factor will be enhanced slightly by C_2,which is easily taken into account by taking R slightly larger than would be the case if C_2 were not present. Normally, this extra complication is not merited, but the complete system with C_2 should be simulated to ensure that the dynamics are not seriously degraded by the inclusion of C_2. The simulation of phase-locked loops is nontrivial and is the topic discussed in Section 16.4

EXAMPLE 16.4

Let $K_{osc} = 2\pi \times 50$ Mrad/V and $I_{ch} = 10\ \mu A$. Assume that the free-running frequency of the oscillator is 50 MHz and that it is desired that the loop have a time constant of 100 cycles, or 2 μs. Design the components of the low-pass filter shown in Fig. 16.7.

Solution

A time constant of 2 μs corresponds to

$$\omega_0 = \frac{1}{2\ \mu s} = 500\ \text{krad/s} \tag{16.64}$$

Now, using (16.56) and (16.60), we have

$$C_1 = \frac{1}{\omega_0^2} \frac{I_{ch}}{2\pi} K_{osc} = 2\ \text{nF} \tag{16.65}$$

Let $C_2 = C_1/10 = 200$ pF and $Q = 0.4$. From (16.61), we have

$$R = \frac{1}{Q}\sqrt{\frac{2\pi}{C_1 I_{ch} K_{osc}}} = 2.5\ \text{k}\Omega \tag{16.66}$$

Phase / Frequency Detector

There are many ways to realize a sequential phase detector. Perhaps the most common sequential phase detector is the *phase/frequency detector* shown in Fig. 16.9. It is an asynchronous sequential logic circuit, and its operation is moderately complicated to follow. Although a detailed description of its operation will not be attempted here, a few comments are in order. The cross-coupled NOR gates basically form RS flip-flops. One of the top flip-flops will be set if either P_u or P_d is 1. One of the bottom flip-flops will be set whenever P_u or P_d must be disabled. When the PLL is in lock, this disabling occurs after the opposite signal has gone high. This disabling can also occur when the PLL is not in lock, by detecting a frequency difference between the input signal and the oscillator. The disabling flip-flops can only be reset when the appropriate input signal has changed to 0. The four-input NOR gate is primarily used to reset the top two flip-flops. This reset occurs when V_{in} and V_{osc} both become 1 when the loop is in lock. At this time, reset goes high, which resets the top two flip-flops, causing both P_u and P_d to go low.

To better understand the operation of the phase comparator, assume the PLL is in lock with V_{in} leading V_{osc}. Assume also that P_u, P_d, $P_{u\text{-}dsbl}$, $P_{d\text{-}dsbl}$, and reset are all low. In other words, all of the flip-flops are reset low. Furthermore, assume V_{in} and

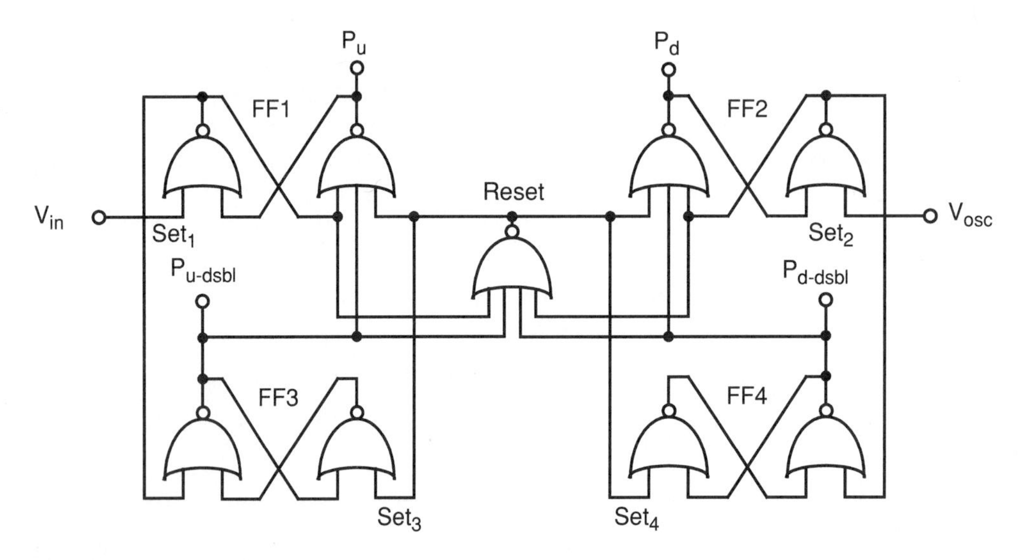

Fig. 16.9 A phase/frequency sequential phase detector based on NOR gates.

V_{osc} are both low, and then V_{in} goes high. The first thing that happens is FF1 gets set, causing P_u to go high. This causes the charge-pump to turn on and charge the low-pass filter, ultimately increasing the frequency of the VCO. Next, when V_{osc} goes to 1, FF2 is also set temporarily. Just after V_{osc} goes high, reset goes high. This causes P_u and P_d to go low one gate delay later. Also, reset going high causes FF3 and FF4 to be set. Thus, $P_{u\text{-}dsbl}$ and $P_{d\text{-}dsbl}$ go high two delays after reset goes high. These signals cause reset to go low one delay later. More importantly, they guarantee that FF1 and FF2 are kept in the reset mode, with both P_u and P_d low. It is only when V_{in} goes back to 0 that FF3 is reset and $P_{u\text{-}dsbl}$ is turned off. Similarly, when V_{osc} goes low, FF4 is reset and $P_{d\text{-}dsbl}$ goes low. We are now back in the original state, with all FFs in the reset state. The operation is very similar if V_{osc} leads V_{in}.

Now consider what happens when the frequencies are very different. Assume V_{in} is at a much higher frequency than V_{osc}. Whenever a positive-going edge of V_{in} occurs (which is quite often), P_u goes high, which causes the VCO frequency to increase. However, when V_{in} goes back to 0, P_u is still high if the much lower-frequency V_{osc} has not changed. P_u will stay high until both V_{in} and V_{osc} go to 1. At this time, reset goes high and $P_{u\text{-}dsbl}$ and $P_{d\text{-}dsbl}$ both get set, causing both P_u and P_d to go low. The next time V_{in} goes to 0, FF1 is reset, which in turn resets FF3, turning $P_{u\text{-}dsbl}$ off. However, if V_{osc} is still at a much lower frequency, V_{osc} is still at 1 and FF4 has not been reset. This guarantees that reset cannot go high. Thus, when V_{in} goes back to 1, FF1 is again set and remains set until V_{osc} eventually goes low (resetting FF2 and FF4) and then goes back high. Thus, it can be seen that most of the time P_u is high, with P_d seldom high, causing V_{osc} to quickly increase in frequency until lock is achieved.

The various waveforms when V_{in} is at a much higher frequency than V_{osc} are shown in Fig. 16.10. Note that most of the time P_u is at 1. It can only shut off if both $P_{u\text{-}dsbl}$ and $P_{d\text{-}dsbl}$ are in their reset states. For this to be the case, however, both V_{in}

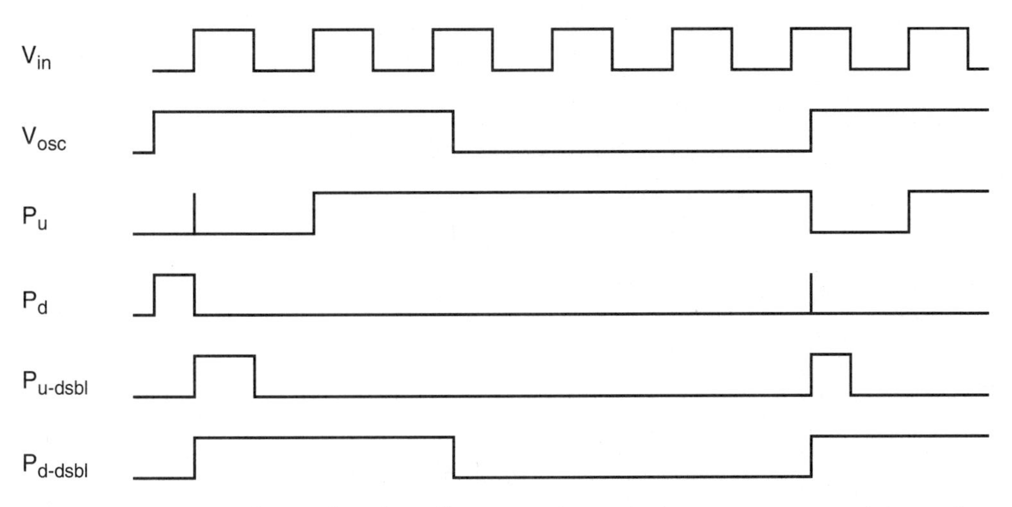

Fig. 16.10 The waveforms of a phase/frequency detector when V_{in} is at a much larger frequency than V_{osc}.

(for $P_{u\text{-dsbl}}$) and V_{osc} (for $P_{d\text{-dsbl}}$) must have been zero to reset FFs 3 and 4. Only after this happens, and then only after V_{in} and V_{osc} both go to 1, can reset go high, turning off P_u. Thus, this phase detector is very fast in acquiring lock. Also, it does not exhibit false lock as do multiplier or exclusive-OR phase detectors. Its only disadvantages are that it isn't suitable for nondigital or sinusoidal inputs and it can be "fooled" by noisy digital inputs having false edges or glitches because it is sensitive to positive-going edges of the input signal. If false edges are a possibility, one might put the input signal into a buffer to go through hysteresis before going into the PLL. Although this helps substantially, there will still be more jitter introduced than would be the case with a PLL having an analog multiplier, which averages the continuous waveforms rather than being sensitive to just the edges.

There are many other phase comparators that have been proposed for phase-locked loops. For example, sample and holds and D flip-flops. The interested reader is referred to [Buchwald, 1993; Meyr, 1990].

16.3 VOLTAGE-CONTROLLED OSCILLATORS

The first step in building a VCO, or *voltage-controlled oscillator*, is to realize an oscillator and then add a means whereby its frequency of oscillation can be controlled by an input voltage (although, in some cases the control signal could be a current). Oscillators come in a variety of different types, but perhaps the two major classifications are those that directly create sinusoidal outputs as opposed to those with square (or triangular) wave outputs. *Sinusoidal-output oscillators* are usually realized using some kind of frequency-selective or tuned circuit in a feedback configuration, whereas square-wave-output oscillators are usually realized using a nonlinear feedback circuit such as a relaxation oscillator or a ring counter. Sinusoidal oscillators realized using tuned circuits can be further classified into RC circuits, switched-capacitor circuits, LC circuits, and crystal oscillators. A table showing the classification of oscillators is shown in Fig. 16.11.

The design of tuned oscillators usually involves placing a tuned circuit into a feedback loop in which, at the oscillation frequency, the loop gain becomes positive and exactly equal to unity. Most integrated oscillators based on tuned circuits have

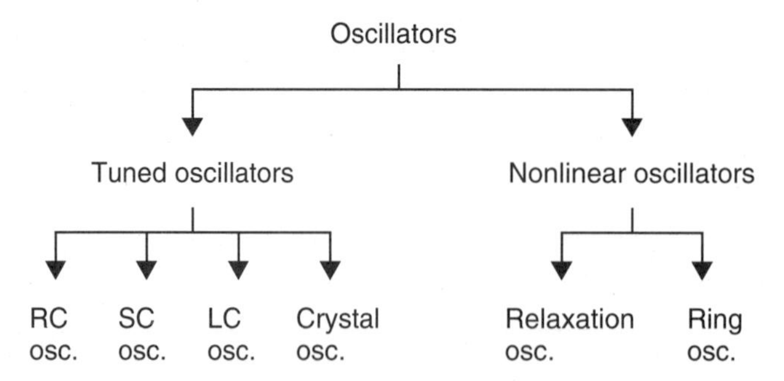

Fig. 16.11 Classification of oscillators.

been realized using either switched-capacitor circuits or integrated amplifiers with off-chip crystals or LC circuits. A switched-capacitor VCO has already been described in Chapter 10, and a number of other SC oscillators are described in [Gregorian, 1986]. The reference [Sedra, 1991] contains a good discussion on designing a variety of tuned oscillators that use RC circuits, LC circuits, and crystals. Most realizations of the first two types have been discrete circuits and were not integrated, although recently, integrated LC filters at frequencies of 1 GHz and up have been realized [Nguyen, 1992]. Another excellent reference on integrated oscillators, although somewhat dated, is [Grebene, 1984].

There are a variety of different methods used in maintaining the loop gain exactly at unity in tuned oscillators. Most of these techniques also determine the amplitude of the oscillator's output signal. The easiest method is to make use of some nonlinear element in the feedback loop. For small amplitude signals, the loop gain is greater than unity. The amplitude of the oscillator increases until the nonlinearity causes the effective loop gain to decrease to exactly unity. The simplest possible nonlinearity might be the limiting of the amplifier in the oscillator when its output gets too large, although for low distortion and a well-controlled amplitude, a more well-controlled nonlinearity is required. The former method is usually used for crystal oscillators. Other techniques might involve measuring the amplitude of the output signal, perhaps using a peak detector, and adjusting the Q factor of the poles of the circuit to be infinite when the amplitude is correct. When the amplitude is too large, the Q factor is changed to be large but positive, whereas when the amplitude is too small, the Q factor is adjusted to be large but negative. The first case corresponds to poles right on the $j\omega$ axis, the second case corresponds to poles close to the axis but in the left half-plane, whereas the last case corresponds to right-half-plane poles. In these cases, the oscillator's output amplitude remains constant, decreases, or grows, respectively. The amplitude-control loop can often be almost as complicated as a complete PLL. After this very cursory introduction to tuned amplifiers, they will be covered no further, as they are deemed to be out of the range of the subject matter of this book because very few of them have been integrated. Rather, a few examples of nonlinear digital-output VCOs will be described.

Bipolar Emitter-Coupled Multivibrator

One of the most popular configurations for a bipolar VCO has been the emitter-coupled multivibrator [Grebene, 1984]. This emitter-coupled oscillator is a form of a relaxation oscillator and is shown in Fig. 16.12. In this oscillator, Q_1 and Q_2 alternately turn on and off. When one of these transistors is off, the node at its emitter slowly changes to a more negative voltage due to the current flowing from the current source, I_1, connected to the emitter out of the capacitor, C_{osc}, which is also connected to the emitter. The node voltages for the case when Q_1 is cut off are shown in Fig. 16.13. During this time, the current through Q_2 is $2I_1$. The voltage at the collector of Q_2 will be one diode drop below the positive power-supply voltage, which is assumed to be 5 V. Here, we have assumed

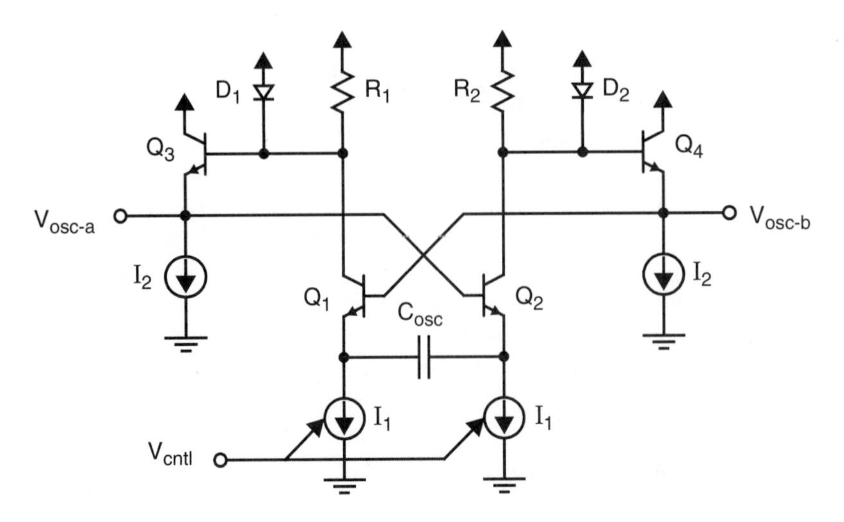

Fig. 16.12 An emitter-coupled relaxation oscillator.

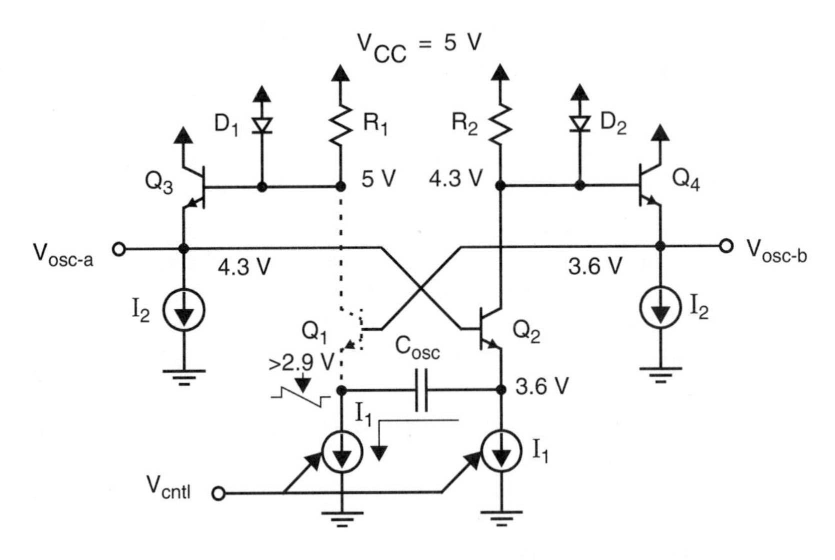

Fig. 16.13 The various voltages and currents of the oscillator in Fig. 16.12 when $V_{osc\text{-}a}$ is high.

$$2I_1R_2 > V_D \cong 0.7\text{ V} \tag{16.67}$$

where V_D is the voltage drop of a conducting p-n junction. The voltage at the collector of Q_1 is approximately 5 V because Q_1 is off. Transistors Q_3 and Q_4 both operate as emitter-follower buffers and, thus, both have approximately 0.7 V base-emitter voltage drops. As a result, the right side of the capacitor C_{osc} is at 3.6 V and the left-side is at some voltage between 2.9 V and 4.3 V. This left-side voltage is changing in the negative-going direction with a rate of change given by

$$\frac{dV_{E1}}{dt} = \frac{I_1}{C_{osc}} \tag{16.68}$$

Its total voltage change will be from 3.6 V to 2.9 V or $2V_D \cong 1.4$ V. The time it takes for this voltage drop to occur is equal to one-half the period of oscillation and is given by

$$\frac{T}{2} = \frac{2V_D C_{osc}}{I_1} \tag{16.69}$$

After the emitter voltage reaches 2.9 V, Q_1 begins to conduct. The conduction of Q_1 causes its collector voltage to decrease from approximately V_{CC}, which in turn causes Q_2 to conduct less. A reduction in the conduction of Q_2 causes the collector voltage of Q_2 to increase, thereby turning Q_1 on more quickly. It is therefore seen that during the transition the circuit is in a positive-feedback loop, which causes it to change state very quickly. Thus, during the transition, Q_1 goes quickly from off to on, and Q_2 goes from on to off. When Q_2 turns off, its collector voltage goes to $V_{CC} = 5$ V, which, in turn, causes the left side of C_{osc} to quickly change from 2.9 V (its voltage when the transition began) to 3.6 V (its voltage after the change has taken place), via Q_4 and Q_1. The right side of C_{osc} must have the same voltage change, and it therefore changes from its 3.6-V initial voltage to 4.3 V. Since Q_2 is now turned off, the right side of C_{osc} is now discharged with a current equal to I_1, whereas the left side of C_{osc} is now held fixed at 3.6 V via Q_4 and Q_1. This process repeats itself indefinitely, resulting in the waveforms shown in Fig. 16.14. Thus, making use of (16.68), we have

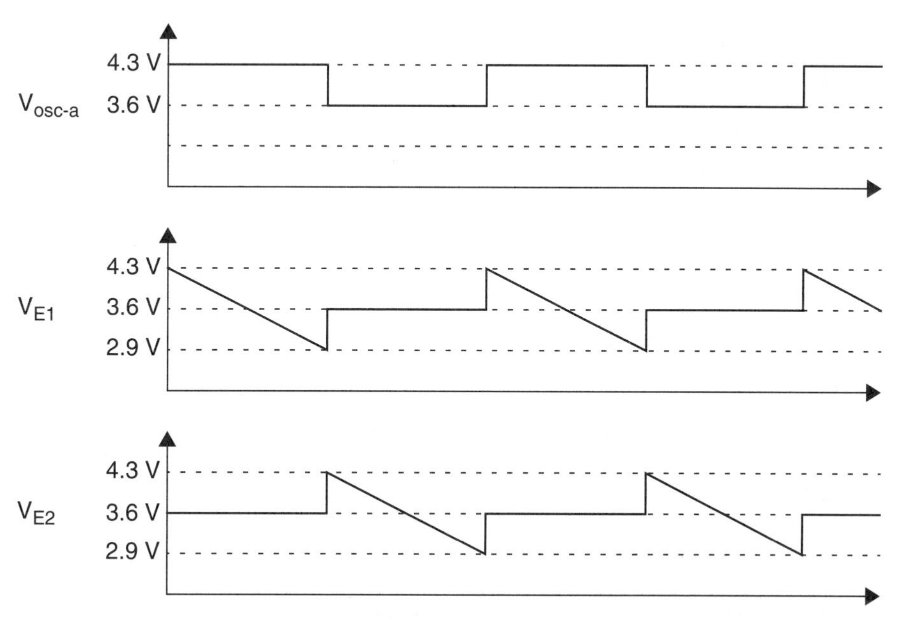

Fig. 16.14 The waveforms of the oscillator in Fig. 16.12.

$$f_{osc} = \frac{I_1}{4V_D C_{osc}} \tag{16.70}$$

To control the oscillator frequency, the current sources used to discharge C_{osc} (i.e., those labelled as I_1) are made proportional to a control voltage. Although methods for implementing linear voltage-controlled current sources were described extensively in Chapter 15 with respect to continuous-time filters, a simple example for making some reasonably linear, voltage-controlled current sources is shown in Fig. 16.15. This circuit uses emitter degeneration to obtain output currents at the collectors of Q_3 and Q_4, where

$$I_{C3} = I_{C4} \cong \frac{V_{cntl} - V_{ref}}{R} + I_1 \tag{16.71}$$

Combining equations (16.70) and (16.71), we have

$$f_{osc} = \frac{V_{cntl} - V_{ref}}{4V_D R C_{osc}} + f_{fr} \tag{16.72}$$

where

$$f_{fr} = \frac{I_1}{4V_D C_{osc}} \tag{16.73}$$

Therefore,

$$K_{osc} = \frac{1}{4V_D R C_{osc}} \tag{16.74}$$

These equations are all based on the assumption that the impedance looking into the emitters of the transistors is much less than R. When this is not the case, the linearity can be greatly improved by using techniques similar to those described in Section 15.2 on bipolar transconductors.

It should be noted that this oscillator can be strongly temperature dependent due to the temperature dependence of V_D—the voltage drop across a forward-biased

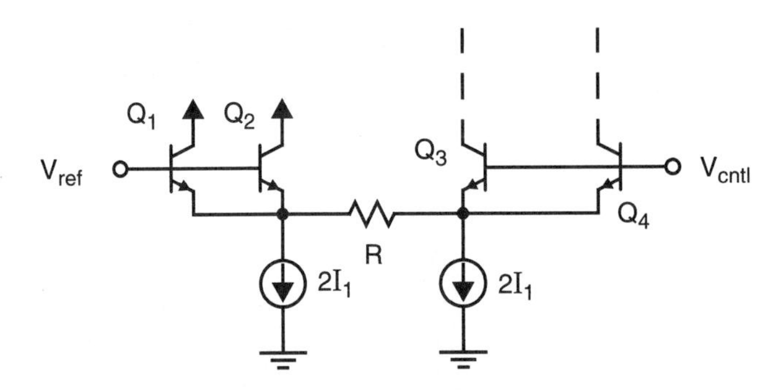

Fig. 16.15 A simple circuit for realizing approximately linear, voltage-controlled current sources.

diode. This temperature dependence can be greatly minimized by making all currents proportional to absolute temperature. The reader is referred to [Grebene, 1984] for details. Another source of error for this VCO is that as the currents I_1 of Fig. 16.12 become smaller, the base-emitter voltage drops of Q_1 and Q_2 also decrease. These V_{be} decreases cause the voltage changes across the capacitor C_{osc} to also become smaller. This source of nonlinearity can also be improved by circuit-design changes, but this topic is beyond the scope of this text.

Ring Oscillators

One of the most popular ways of realizing digital-output MOS VCOs is to use a *ring oscillator* and add voltage control to it. A ring oscillator is realized by placing an odd number of open-loop inverting amplifiers in a feedback loop. The simplest type of amplifier that can be used is a simple digital inverter, as shown in Fig. 16.16. This circuit is a form of negative feedback, but since each inverter has approximately 90° phase shift at its unity-gain frequency (assuming that all the loads on the inverters are matched and that there are at least three inverters in the loop) it is guaranteed that the loop gain will still be greater than unity when the phase shift around the loop becomes greater than 180°. As a result, the circuit is unstable and oscillations occur. Each half-period, the signal will propagate around the loop with an inversion. For example, assume the output of the first inverter changes to 1. This change will propagate through all five inverters in a time of $T/2$, at which time the output of the first inverter will change to 0; after an additional time of $T/2$, the first inverter's output will change back to 1, and so on. Assuming each inverter has a delay of τ_{inv} and that there are n inverters, we then have

$$\frac{T}{2} = n\tau_{inv} \tag{16.75}$$

Thus,

$$f_{osc} = \frac{1}{T} = \frac{1}{2n\tau_{inv}} \tag{16.76}$$

By making the delay of the inverters voltage controlled, the ring oscillator's frequency can be made voltage controlled.

It is very important that in CMOS technology, ordinary CMOS inverters not be used. This avoidance is because ordinary CMOS inverters have a gate-threshold voltage proportional to the power-supply voltage. Thus, when the power supply increases

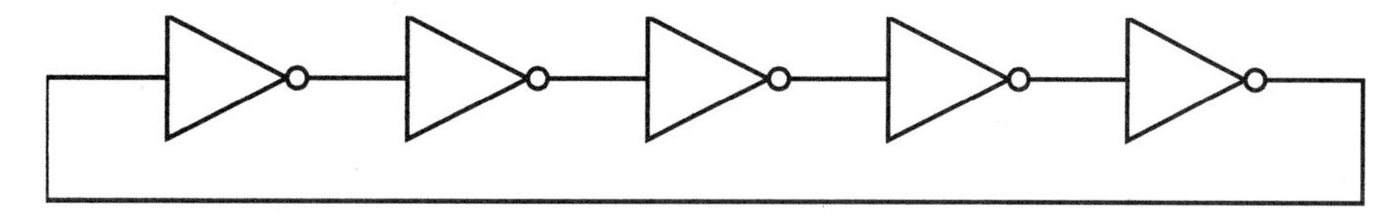

Fig. 16.16 A ring oscillator realized using five digital inverters.

in voltage, the voltage excursions increase in voltage proportionally, the currents of the inverters increase in proportion to the square of the power-supply voltages, and the frequency of oscillation increases proportionally to the power-supply voltage. Thus, any noise on the power-supply voltage will cause the oscillation frequency to have jitter. This poor power-supply rejection is perhaps the major source of jitter in integrated oscillators. In integrated oscillators, it is extremely important that the frequency of oscillation be independent of the power-supply voltage if low jitter is required.

In most integrated ring oscillators, fully differential inverters are used to obtain better power-supply insensitivity. When this is done, an even number of inverters can be used and the inversion required around the loop can be achieved by simply interchanging the outputs of the last inverter before feeding them back to the input, as shown in Fig. 16.17. This realization has a very important advantage in that the outputs of the middle inverters will have a quadrature phase relationship as compared with the last inverters (assuming all inverters and their loads are carefully matched) [Buchwald, 1991]. These quadrature outputs are very useful in many communication applications such as quadrature modulators and some clock-extraction circuits.

An example of realizing a fully differential inverter with a programmable delay is shown in Fig. 16.18, where the current-source loads are made voltage programmable. Cascode transistors, Q_3 and Q_4, have been added to increase the output impedance of the programmable current sources to make them more insensitive to power-supply

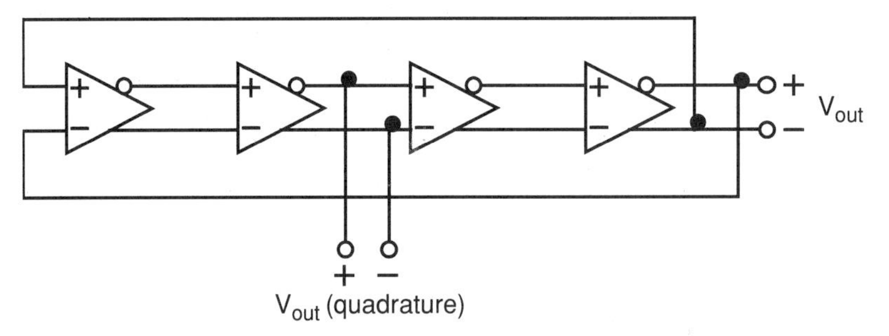

Fig. 16.17 Realizing a ring oscillator using fully differential inverters for improved power-supply rejection.

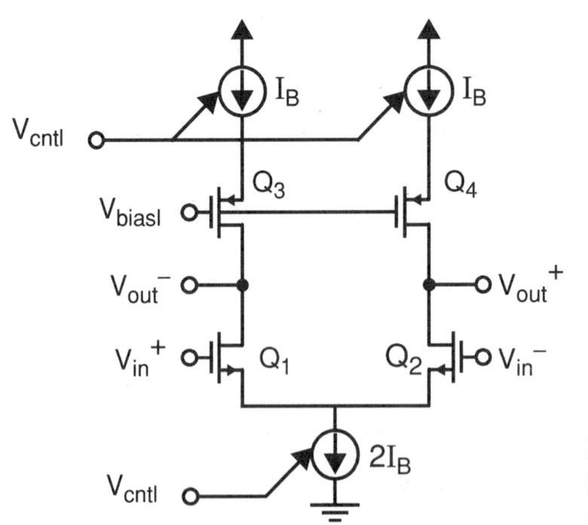

Fig. 16.18 A fully differential inverter with a programmable delay (CMFB circuitry not shown).

noise. Assuming the current-source loads are proportional to V_{cntl} with a constant of proportionality of K_{bias}, we have

$$I_B = K_{bias} V_{cntl} \tag{16.77}$$

The delay of each inverter is proportional to the unity-gain frequency of that inverter. In other words,

$$\tau_{inv} \propto \frac{C_L}{g_m} \tag{16.78}$$

where C_L is the load capacitance of the inverter and g_m is the transconductance of the drive transistors of the inverter. Since $g_m \propto \sqrt{I_B}$, we know that the delay is proportional to $1/\sqrt{I_B} \propto 1/\sqrt{V_{cntl}}$ and therefore, $f_{osc} \propto \sqrt{V_{cntl}}$. Thus, the relationship between the oscillation frequency and the control voltage is not very linear, but in many VCOs, this nonlinearity is not a problem. The major reason for nonlinearity is that at higher frequencies, the voltage changes at the outputs of each stage increase in amplitude because I_B increases. If the amplitudes of the voltage changes could be stabilized, the linearity would be greatly improved, as will be shown shortly.

The circuit shown in Fig. 16.19 can be used to realize the voltage-controlled current sources. The high gain of the opamp guarantees that its differential input voltage is very small, and, therefore, the voltage across the resistor R is very closely equal to V_{cntl}. Thus, the current through R is $I_R = V_{cntl}/R$. This current is mirrored by the wide-swing current mirrors consisting of Q_8 and Q_9 on the input side, and Q_3 and Q_5 for one output current source and Q_4 and Q_6 for the second output current source. The operation of the wide-swing current mirrors was described in Chapter 6, and will not be repeated here. In noncritical applications, it may be possible to use simpler current mirrors.

It is possible to realize a ring-oscillator VCO with good voltage-to-frequency linearity if one ensures that the voltage changes of the delay elements are independent of

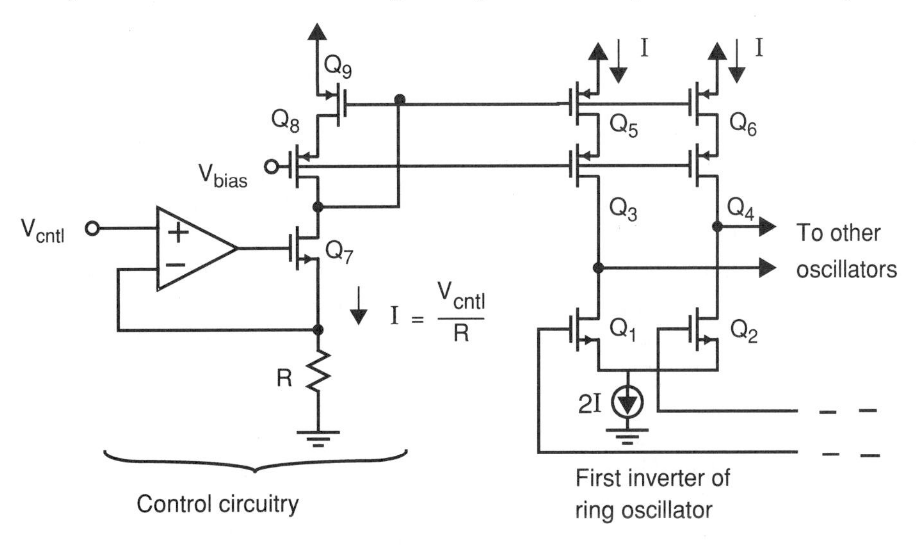

Fig. 16.19 A circuit that can be used to realize the voltage-controlled current sources required in Fig. 16.18 (CMFB circuitry not shown).

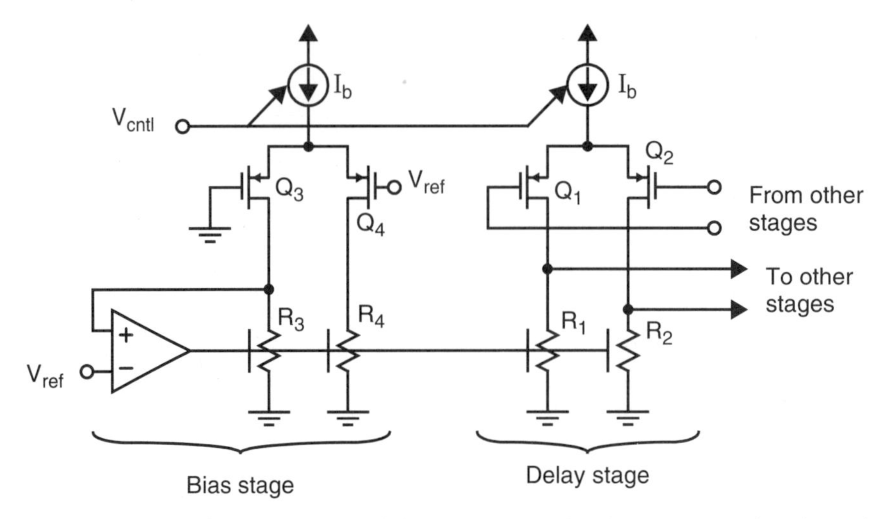

Fig. 16.20 A differential delay cell having resistive loads consisting of n-channel transistors.

the oscillation frequency [Kim, 1990; Young, 1992; Reynolds, 1994]. An alternative design for a differential delay stage where this linearization has been realized is shown in Fig. 16.20 [Young, 1992], along with the biasing circuitry that is used to control the voltage swings. Each stage consists of a differential amplifier with p-channel input resistors and resistive loads. The resistive loads are realized using n-channel transistors biased in the triode region. Their resistance is adjusted by the bias circuit so that when all of I_b is flowing through them, they will have V_{ref} across them, where V_{ref} is obtained from a temperature-independent voltage reference. This is achieved by using a replica of the delay cell in the bias circuit. The replica consists of Q_3, Q_4, R_3, and R_4. In the replica, all of I_b flows through Q_3 and R_3, so Q_4 and R_4 could have been eliminated from the circuit—they have been included in the figure for clarity only. Recalling that R_3 is actually realized by an n-channel transistor, it is seen that the negative feedback loop including the opamp will cause the voltage across R_3 to be equal to V_{ref} when I_b is going through it. Since the output of the opamp is also used to control the resistance of all the triode-region resistances used for the delay cells of the ring oscillator, they will also have V_{ref} across them when I_b flows through them. Next, if the VCO frequency changes because of a change in V_{cntl}, and therefore in I_b, the bias loop will change the resistance of the delay-stage loads so the maximum voltages across them will still equal V_{ref}. Assuming the capacitive loads of each stage are constant (a good assumption), the delay of each stage will be inversely proportional to I_b. This makes the frequency of oscillation proportional to I_b.

In an interesting modification of the ring oscillator of Fig. 16.17, two multipliers have been added in Fig. 16.21. This modification achieves a doubling in the output frequency while still realizing two quadrature outputs [Buchwald, 92]. Indeed, if quadrature outputs are not required, it is possible to connect the two multiplier outputs to an additional third multiplier. This achieves another doubling of frequency, but there is now only a single output. Another interesting variation of a ring oscillator is described

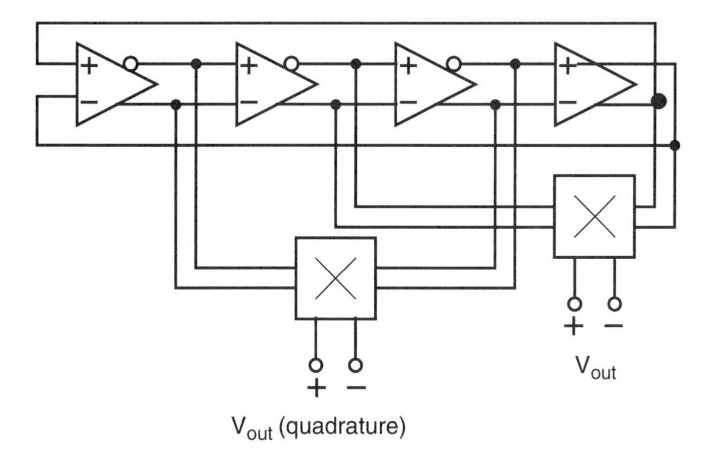

Fig. 16.21 Modifying the ring oscillator of Fig. 16.17 to achieve a doubling of frequency.

in [Razavi, 1994], where the outputs of a three-stage oscillator are combined in a novel manner to obtain a single output with a period equal to only two inverter delays.

CMOS Relaxation Oscillators

Another popular method for realizing digital-output VCOs in CMOS technology is a CMOS relaxation oscillator [Wakayama, 1989; Sun, 1987]. A typical example is shown in Fig. 16.22. In this oscillator, alternatively, C_1 and then C_2 are discharged by the voltage-controlled wide-swing cascode current-mirror consisting of Q_5 to Q_8, R, and the opamp. Due to the negative feedback around the opamp, the current through Q_5 and Q_7 is equal to V_{cntl}/R. Assume the S–R latch is in its reset state as shown in Fig. 16.22 (the S–R latch is simply two cross-coupled NOR gates). This state will cause Q_1 and Q_4 to be on and Q_2 and Q_3 to be off. Since Q_4 is on, the voltage across C_2 is clamped at V_{DD}. However, the voltage across C_1 will change in a negative-going direction because the current from the voltage-to-current converter is flowing through Q_1. This voltage, which started at V_{DD}, reaches the trip point of the comparator (i.e., V_{ref}) after one-half of a clock period. At this time, the output of the comparator goes positive, which causes the S–R latch to become set. This set state causes Q_2 and Q_3 to turn on and Q_1 and Q_4 to turn off. When Q_3 turns on, the voltage across C_1 is very quickly charged back up to V_{DD}. This increased voltage causes the output of the corresponding comparator to go to low (i.e., logic 0), so that both inputs to the S–R latch are again zero. At the same time, since Q_2 turns on, C_2 begins to discharge until its voltage reaches V_{ref}. This completes one period of oscillation. After the voltage across C_2 reaches V_{ref}, the S–R latch is reset, C_2 is quickly charged up to V_{DD}, and C_1 begins to discharge again. The reason for including the comparators, as opposed to connecting C_1 and C_2 directly to the S–R latch, is to make the voltage changes across the capacitors less sensitive to power-supply noise. This is also the major reason for making the current source consisting of Q_5 and Q_7 a cascode current source.

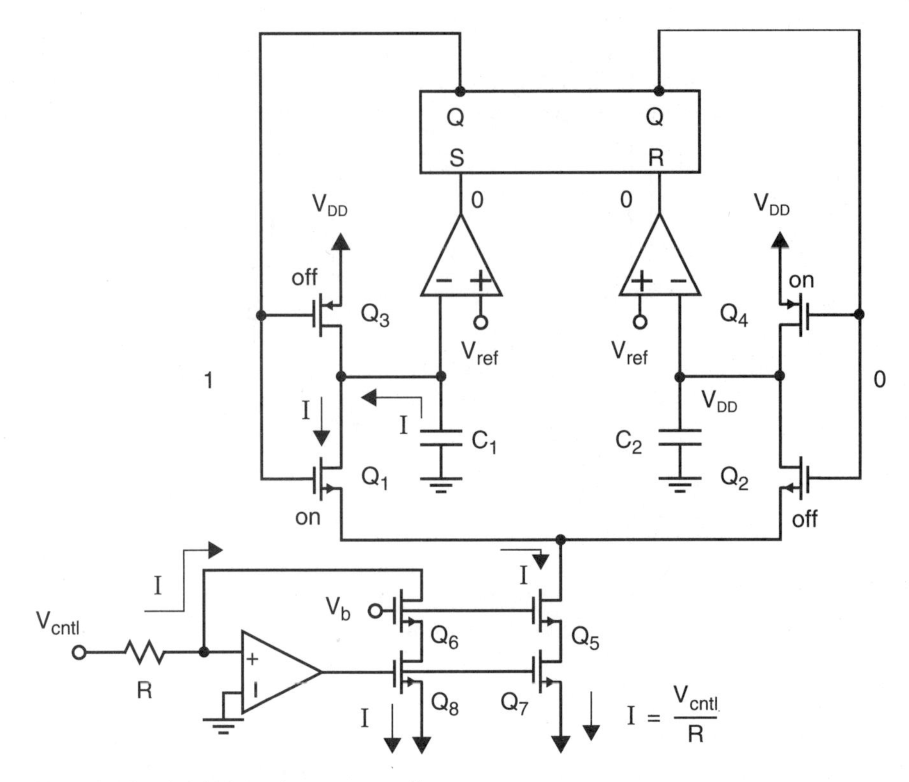

Fig. 16.22 A CMOS relaxation oscillator.

16.4 COMPUTER SIMULATION OF PLLS

Simulating the operation of PLLs is nontrivial because of the very wide range of time constants present in PLLs. Specifically, the voltage-controlled oscillator typically has nodes with important time constants much smaller than the period of its output waveform, whereas the time constants of the low-pass filter are orders of magnitude greater than the VCO's period. If SPICE is used for the simulation, the time steps must be less than the smallest time constant of the VCO, whereas the total simulation time must be greater than the largest time constant of the low-pass filter. This would take an exorbitant amount of time for simulation and is usually impractical. What normally is done is to have SPICE simulate the individual components over a few periods of the VCO's output waveform. The complete system is simulated based on simplified models in which the continuous-time components are replaced by approximately equivalent difference-equation models. This approach can be done by using difference-equation simulators, such as the Simulink program in Matlab [Simulink, 1992] or by writing custom difference-equation simulations using a computer language such as C. The former approach has the advantage that it doesn't require much expertise. The latter approach has the advantages that it runs much faster and can more easily be modified by the designer to include additional second-order effects for greater accuracy. It turns out that writing a difference-equation simulation is not too difficult and that the total amount of code that needs to be written is small.

The approach taken here is generally applicable to the simulation of many other analog systems as well. For example, it is the same approach typically used to simulate oversampling A/D converters. Thus, this section is useful for general analog IC design. However, for designers willing to use a commercial simulator such as Simulink, this section can be safely ignored.

The difference-equation simulation of PLLs will be explained by using a charge-pump PLL as an example. Other types of PLLs can be simulated by making fairly trivial changes to the example, most of them required to describe the operation of different phase detectors.

Describing Continuous-Time Components Using Difference Equations

When doing a simulation, a fixed time step, T, is chosen that might be between one-tenth and one-hundredth of the period of the VCO's output waveform. Ideally, during this fixed time step, all voltage and current changes are small. In the simulation, the voltage–current relationships of all continuous-time elements such as capacitors and resistors are described by approximately equivalent difference equations. In fact, it is not the voltage–current relationships that are usually approximated; rather, it is the voltage–charge relationships. As long as the currents don't change too much during the simulation time step, the current is simply given by the change in charge divided by the time step. The reason for approximating the voltage–charge relationships rather than the voltage–current relationships is that it is often easier to calculate total charge changes during a simulation time step than the average currents. For example, discrete-time analog circuits, such as switched-capacitor circuits, are much easier to analyze using voltage–charge relationships than by using voltage–current relationships. Thus, a single formulation can be used for both continuous-time and discrete-time circuits, and even for simulating systems containing both types of circuits.

To see how to transform the continuous-time voltage–current relationships of a component to the discrete-time simulation domain, consider an arbitrary component having a voltage–current relationship described in the frequency domain by impedance $Z(s)$. Since charge is the integral of current, the relationship between the impedance's charge and current in the frequency domain is given by

$$Q(s) = \frac{I(s)}{s} \tag{16.79}$$

Thus, we can define the voltage–charge transfer function, $M(s)$, as

$$M(s) = \frac{V(s)}{Q(s)} = sZ(s) \tag{16.80}$$

To approximate this relationship in the discrete-time domain, one must replace s with a polynomial of z (the discrete-time variable) in such a way that all relationships are approximately equivalent when $\omega T << 1$ and, therefore, $z = e^{j\omega T} \cong 1 + j\omega T$. One possible means of maintaining this approximation is to use the bilinear

z-transform[5], where every occurrence of s is replaced by the function

$$s \leftarrow \frac{2}{T}\frac{1-z^{-1}}{1+z^{-1}} \tag{16.81}$$

For $\omega T << 1$, we have

$$\frac{2}{T}\frac{1-z^{-1}}{1+z^{-1}} \cong \frac{2}{T}\frac{1-(1-j\omega T)}{1+(1-j\omega T)} \cong \frac{2}{T}\frac{j\omega T}{2-j\omega T} \cong j\omega = s \tag{16.82}$$

as desired. Furthermore, the bilinear transform has the desirable features that stable functions remain stable and that the discrete-time transfer functions are simply frequency-warped versions of the continuous-time transfer functions in which higher frequencies are compressed. The discretization process therefore consists of taking $M(s) = sz(s)$ and replacing each occurrence of s according to (16.82) to obtain $M(z)$, where

$$M(z) = \frac{V(z)}{Q(z)} = sZ(s)\Big|_{s=\frac{2}{T}\frac{1-z^{-1}}{1+z^{-1}}} \tag{16.83}$$

During each simulation time step, one normally calculates the charge going into or out of the component. Thus, we are actually interested in the relationship between changes in charge rather than the total charge stored in a component. Noting that the change in charge is equal to the previous charge value subtracted from the present charge, we can write

$$\Delta Q(z) = (1-z^{-1})Q(z) \tag{16.84}$$

Now, defining the voltage-to-charge-difference transfer function to be $P(z)$, we have

$$P(z) = \frac{V(z)}{\Delta Q(z)} = \frac{M(z)}{1-z^{-1}} \tag{16.85}$$

The equivalent time-domain difference equation can be found by taking the inverse z-transform of (16.85). This difference equation allows the voltage across the component at time nT to be calculated from previous voltages across the component (i.e., at times $nT-T$, $nT-2T$, $nT-3T$, etc.) and the change in charge into the component at time nT (and, possibly, previous times).

When it is easier to calculate the currents than charge changes, the charge going into the component during one period is easily approximated by averaging the current at the beginning of the period with the current at the end of the period and then multiplying by T. Specifically, we can use the time-domain relationship

$$\Delta q(nT) = T\left[\frac{i(nT)+i(nT-T)}{2}\right] \tag{16.86}$$

5. The bilinear z-transform and the discrete z-transform, which is commonly abbreviated as the z-transform, are different.

This approach is general enough to handle mixed continuous-time systems and discrete-time systems such as those composed of switched-capacitor filters and continuous-time components. To make equations simpler, from here on, the variable T will be dropped from the time index. That is, $i(nT)$ will be called $i(n)$ and the difference between sampling times of T will not be given explicitly.

We are now ready to apply this simulation procedure to a charge-pump PLL. The system being simulated is shown in simplified form in Fig. 16.23. We will first derive a difference-equation model for a VCO and then derive the difference-equation model for the loop filter.

Difference-Equation Modelling of a VCO

The difference-equation modelling of the VCO is based on the circuit shown in Fig. 16.24. This model is a simplified model of a relaxation oscillator. It can approximately model most VCOs where the output is a digital signal.

The model operates as follows. First assume $v_{cntl} = V_{th1}$, where V_{th1} is the input voltage to the VCO when its output frequency is equal to its free-running frequency. Up to this point, we have assumed that $V_{th1} = 0$, but we will now be more general to allow for the simulation of PLLs that have only a single power-supply voltage. Since $v_{cntl} = V_{th1}$, the current source $g_{osc}v_1 = g_{osc}(v_{cntl} - V_{th1})$ is zero. Next, assume the capacitor C_{osc} is initially discharged. It will be slowly charged by current source I_{osc}, with the voltage v_2 increasing by an amount given by $\Delta v_2 = \Delta q_{Cosc}/C_{osc} = (I_{osc}T)/C_{osc}$ each analysis time step. Once the voltage across C_{osc} exceeds V_{th2}, then the output of the comparator, which was initially 0 will go high. This causes the output of the one-shot to go high for one analysis period, which has two effects. It first of all causes the output of the toggle flip-flop (which is the output

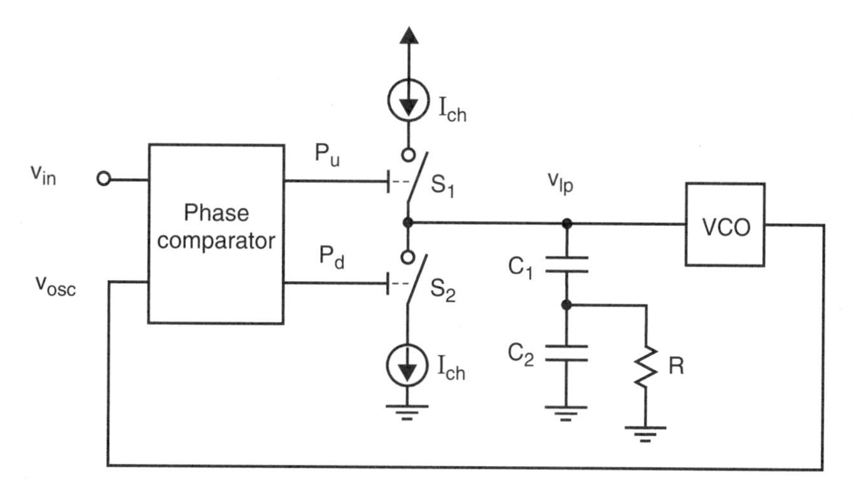

Fig. 16.23 The charge-pump PLL used as an example to illustrate how to simulate PLLs.

of the VCO, that is, v_{osc}) to change its state. It also causes the switch S_{vco} to come on. This quickly discharges C_{osc} to zero, which in turn causes the output of the comparator to go back to 0. This process repeats itself every half-period, $T_{osc}/2$. The number of analysis time steps, n, required for C_{osc} to charge from 0 to V_{th2} is given by

$$n = \frac{V_{th2}C_{osc}}{I_{osc}T} \tag{16.87}$$

Since

$$\frac{T_{osc}}{2} = nT \tag{16.88}$$

we have, after substituting (16.87) into (16.88),

$$T_{osc} = \frac{2V_{th2}C_{osc}}{I_{osc}} \tag{16.89}$$

and, therefore, the free-running frequency of the VCO, f_{fr}, is given by

$$f_{fr} = \frac{1}{T_{osc}} = \frac{I_{osc}}{2V_{th2}C_{osc}} \tag{16.90}$$

Next, consider a case when the control voltage to the VCO is not equal to the bias voltage V_{th1}. When v_{cntl} is a nonvarying value, the oscillation frequency is given by

$$f_{osc} = \frac{I_{osc} + g_{osc}(v_{cntl} - V_{th1})}{2V_{th2}C_{osc}} \tag{16.91}$$

By taking first-order derivatives of both sides of (16.91), we have

$$K_{osc} = 2\pi\frac{df_{osc}}{dv_{cntl}} = \frac{2\pi g_{osc}}{2V_{th2}C_{osc}} \tag{16.92}$$

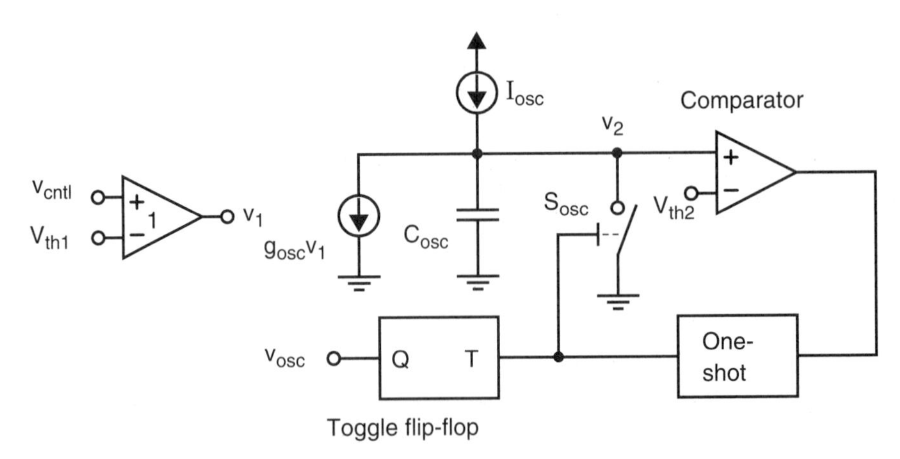

Fig. 16.24 The model used for the VCO.

The difference equations describing the system will be derived for the case where v_{cntl} can change with time using the difference-equation modelling presented above. Although this VCO is such a simple system that this formalism is unnecessary, the formal approach will still be used so that a simple example can be presented.

The capacitor, C_{osc}, has an impedance given by

$$z(s) = \frac{1}{sC_{osc}} \tag{16.93}$$

Using (16.80), we have

$$M(s) = sz(s) = \frac{1}{C_{osc}} \tag{16.94}$$

Since this is not a function of s but a constant, transforming it to the discrete-time domain is trivial. We have

$$M(z) = \frac{1}{C_{osc}} \tag{16.95}$$

Next, using (16.85), we have

$$P(z) = \frac{V_2(z)}{\Delta Q_{Cosc}(z)} = \frac{1/C_{osc}}{1-z^{-1}} \tag{16.96}$$

which implies

$$V_2(z)(1-z^{-1}) = \frac{\Delta Q_{Cosc}(z)}{C_{osc}} \tag{16.97}$$

Taking the inverse z-transform, we have

$$v_2(n) = v_2(n-1) + \frac{\Delta q_{Cosc}(n)}{C_{osc}} \tag{16.98}$$

Now, since

$$\Delta q_{Cosc}(n) = TI_{Cosc\text{-}avg} \tag{16.99}$$

and since

$$I_{Cosc\text{-}avg} = I_{osc} + g_{osc}\left[\frac{v_{cntl}(n) + v_{cntl}(n-1)}{2} - V_{th1}\right] \tag{16.100}$$

we have, using (16.98),

$$v_2(n) = v_2(n-1) + \frac{T}{C_{osc}}\left(I_{osc} + g_{osc}\left[\frac{v_1(n) + v_1(n-1)}{2} - V_{th1}\right]\right) \tag{16.101}$$

This is the basic difference equation that will be calculated once each analysis time step. After each updating of $v_2(n)$, its value is then compared with V_{th2}, and if it is larger than 0, then $v_2(n)$ is reset to zero and the logic value of v_{osc} is changed. The system can be initialized at time zero by setting $v_{cntl}(0) = V_{th1}$ and $v_2(0) = 0$.

Any nonlinearities between the VCO's control voltage and its frequency can be added to the model by making g_{osc} a function of V_1. This nonlinear modelling is most easily done by using a table lookup, whose entries are obtained from doing a number of simulations of the complete VCO at the transistor level using SPICE.

Difference-Equation Simulation of a Loop Filter

The loop-filter simulation is somewhat more complicated. Part of this complication comes from a need to maintain high numerical accuracy. The impedance of the loop filter is given by

$$Z_{lp}(s) = \frac{1}{sC_1} + \frac{1}{sC_2 + G} \tag{16.102}$$

where $G = 1/R$. Using (16.80), we have

$$\frac{V_{lp}(s)}{Q_{lp}(s)} = \frac{G + s(C_1 + C_2)}{GC_1 + sC_1C_2} \tag{16.103}$$

Next, each term of s is replaced using the bilinear z-transform according to (16.81). After this transform, we have

$$M(z) = \frac{V_{lp}(z)}{Q_{lp}(z)} = \frac{2(1 - z^{-1})(C_1 + C_2) + GT(1 + z^{-1})}{2C_1C_2(1 - z^{-1}) + C_1GT(1 + z^{-1})} \tag{16.104}$$

Using (16.85) and rearranging gives

$$P(z) = \frac{M(z)}{1 - z^{-1}} = \frac{GT + 2(C_1 + C_2) + z^{-1}[GT - 2(C_1 + C_2)]}{2C_1C_2 + GC_1T - z^{-1}4C_1C_2 + z^{-2}(2C_1C_2 - GC_1T)} \tag{16.105}$$

which can be rewritten as

$$\begin{aligned} P(z) = \frac{V_{lp}(z)}{\Delta Q_{lp}(z)} &= \frac{m_1 + m_2 z^{-1}}{1 + z^{-1}(k - 2) + z^{-2}(1 - k)} \\ &= \left(\frac{1}{1 - z^{-1}}\right)\left(\frac{m_1 + m_2 z^{-1}}{1 - z^{-1} + kz^{-1}}\right) \end{aligned} \tag{16.106}$$

where

$$k = \frac{2GC_1T}{D} \tag{16.107}$$

$$m_1 = \frac{2(C_1 + C_2) + GT}{D} \tag{16.108}$$

$$m_2 = \frac{-2(C_1 + C_2) + GT}{D} \tag{16.109}$$

and

$$D = 2C_1C_2 + GC_1T \tag{16.110}$$

This transfer function can be simulated by a discrete-time system that has the signal-flow graph shown in Fig. 16.25. This network is a "lossless-digital-integrator" (LDI) [Kingsbury, 1973] realization of the transfer function (16.106). This type of realization has very good numerical properties when the poles of the low-pass filter being simulated are at frequencies much less than the inverse of the analysis time step (which is certainly the case when simulating PLLs). A direct-form implementation would have quite poor numerical properties in this situation.

It is now possible to give the difference-equation updates for the low-pass filter, which are performed once each analysis time step based on the signal-flow graph of Fig. 16.25. Each simulation time step, do the following:

1. If P_u is 1, then take $\Delta q_{lp}(n) = TI_{ch}$. Or if P_d is 1, then take $\Delta q_{lp}(n) = -TI_{ch}$. Otherwise, take $\Delta q_{lp}(n) = 0$.
2. Update the difference equations for simulating the low-pass filter according to the following:

$$x_1(n) = \Delta q_{lp}(n) + x_1(n-1) \tag{16.111}$$

$$x_2(n) = x_1(n) + x_2(n-1) - kx_2(n-1) \tag{16.112}$$

$$V_{lp}(n) = m_1x_2(n) + m_2x_2(n-1) \tag{16.113}$$

Notice that (16.113) is a function of $x_2(n-1)$. This means that if a single variable is used to store x_2, its value must be temporarily stored, since x_2 is updated in (16.112) before (16.113).

There is one other complication. If it is desired at initialization that the low-pass filter have an initial output other than zero, as would be needed in simulations of PLLs realized using a single power-supply voltage, then the internal state variables of the difference-equation simulation should be initialized according to the following equations:

$$x_2(0) = \frac{V_{lp}(0)}{m_1 + m_2} \tag{16.114}$$

$$x_1(0) = kx_2(0) \tag{16.115}$$

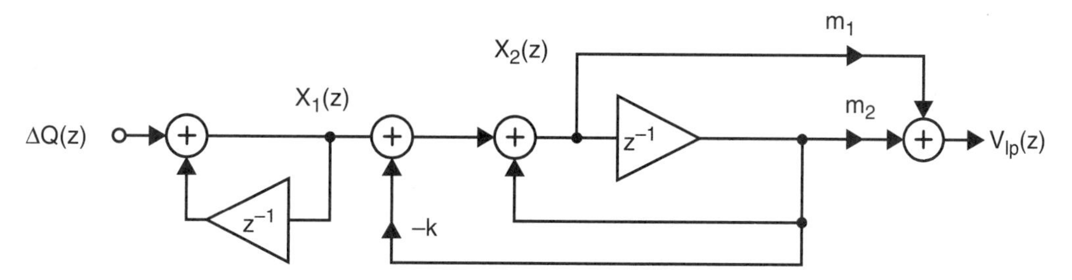

Fig. 16.25 The signal-flow graph of the difference equations used to simulate the PLL's low-pass filter.

Phase/Frequency Comparator Simulation

The phase/frequency comparator was described previously. Based on this description, the following formal definition specifies how its state variables, P_u, P_d, $P_{u\text{-dsbl}}$, and $P_{d\text{-dsbl}}$ should be updated each analysis time-step.

1. If V_{in} is 1 and $P_{u\text{-dsbl}}$ is 0, then set P_u to 1.
2. If V_{osc} is 1 and $P_{d\text{-dsbl}}$ is 0, then set P_d to 1.
3. If V_{in} and V_{osc} are both 1 and $P_{u\text{-dsbl}}$ and $P_{d\text{-dsbl}}$ are both 0, then set reset to 1; otherwise set it to 0.
4. If reset is 1, then set $P_{u\text{-dsbl}}$ and $P_{d\text{-dsbl}}$ to 1 and set P_u, P_d, and reset to 0.
5. If V_{in} and P_u are both 0, then set $P_{u\text{-dsbl}}$ to 0.
6. If V_{osc} and P_d are both 0, then set $P_{d\text{-dsbl}}$ to 0.

Complete Simulation

The simulation of the complete PLL is now accomplished by updating the difference equations for each block once each analysis time step. A C-program realization of this simulation is given in the appendix of this chapter. It should be relatively easy to modify this simulation so that it can also be used to simulate other PLLs as well.

EXAMPLE 16.5

Let $I_{ch} = 10\ \mu A$, $I_{osc} = 0.5\ mA$, $V_{th1} = V_{th2} = 2.5\ V$, and $f_{fr} = 50\ MHz$. Also, assume that $K_{osc} = 2\pi \cdot 25\ MHz \cdot V^{-1} = 1.57\times10^8\ rad \cdot V^{-1}$.

Solution

First, using (16.90), we have

$$C_{osc} = \frac{I_{osc}}{2f_{fr}V_{th2}} = 2\ pF \tag{16.116}$$

Next, using (16.92), we have

$$g_{osc} = \frac{2V_{th2}C_{osc}K_{osc}}{2\pi} = 0.25\ ms \tag{16.117}$$

The values of the components of the low-pass filter were already found in Example 16.5. They are $C_1 = 25.3\ pF$, $C_2 = 2.5\ pF$, and $R = 31.4\ k\Omega$. These are the values used in the C-code simulation of this PLL given in the appendix of this chapter. The simulation was started with the VCO having a frequency of zero. It took 28.6 μs to obtain lock. Next, after 5 μs, the phase of the input signal was given a step change of 180°. It took the PLL 0.8 μs to settle to this new phase. In both cases, the transient responses were well behaved.

16.5 APPENDIX

The source code for a C-program simulation of a charge-pump PLL having a phase/frequency converter follows. Readers who wish to obtain this source code using ftp via internet should send e-mail to martin@eecg.toronto.edu for instructions.

```
/*  This is a difference-equation simulation of a charge pump PLL having a
    phase/frequency comparator. v1n is the output of the low-pass filter, which is
the input to the VCO.
    v2n is the internal voltage of the VCO across its capacitor. x1n and x2n are
the internal state variables of the low-pass filter simulation. pu, pd, pu_d, and
pd_d are the internal state variables of the charge pump. delt_q is the charge
change of the low-pass filter. The simulation is for 500,000 analysis time steps,
and then has a step change in phase of the input voltage. Currently the input fre-
quency is 50 MHz.
*/

#define IN_FREQ = 50e6 // The input frequency is 50 MHz
#define PI 3.141592654
#define VTH1 2.5 // the VCO input voltage to obtain the free-running frequency
#define VTH2 2.5 // the trip-point for the VCO relaxation oscillator
#define ICH 1e-5 // the charge-pump current is 10 µA
#define IOSC 5e-4 // the dc current in the VCO is 0.5 mA
#define GOSC 2.5e-4 // the transconductance of the VCO
#define C1 25e-12 // the large capacitor in the loop filter
#define C2 2.5e-12 // the deglitching capacitor in the loop filter
#define C3 2e-12 // the capacitor used in the VCO simulation
#define G 3.18e-5 // the transconductance of the loop filter, R=31.4 k
#define T 1e-11 // An analysis step is 10 ps
#define NSTEPS 1e7 // Analyze for 1 msec (i.e., 10 million steps)

#include <stdio.h>
#include <math.h>
main()
    {
    double D,k,m1,m2,v1n,v1n_1,v2n,v2n_1,x1n,x1n_1,x2n,x2n_1,delt_-
q,i_avg;
    double in_freq,in_per,half_per,time,per_tim,osc_freq;
    int i,j,inp_val,osc_val,pu,pu_d,pd,pd_d,rst;
```

```
//  Initialize all variables

    D = 2*C1*C2 + C1*G*T;
    k = 2*C1*G*T/D;
    m1 = (2*(C1+C2) + G*T)/D;
    m2 = (-2*(C1+C2) + G*T)/D;
    v1n = v1n_1 = VTH1 - 2.4; // The VCO is initialized less than 50 MHz
    v2n = v2n_1 = 0.0;
    x2n = x2n_1 = v1n/(m1+m2);
    x1n = x1n_1 = k*x2n;

     inp_val = osc_val = 0;
    pu = pu_d = pd = pd_d = rst = 0;
    in_freq = 50e6;
    in_per = 1/in_freq;
    half_per = in_per/2.0;
    j = 0;

//  Start the simulation

    for (i=0; i < NSTEPS; i++)
     {
     time = T * (double) (i);

//  The first set of statements is to produce the input square wave

     per_tim = time - in_per*(double)j;
     if (per_tim >= half_per) // change value of input signal
    {
    inp_val = (inp_val == 0)? 1:0;
    j++;
    if (j%4 == 0)
     {
     osc_freq = i_avg/(2.0*C3*VTH2);

//  Print out the oscillator frequency once every four input periods
```

```
        printf("time: %.1lf, period: %.2lf, v1n: %.4lf, in_freq: %.2lf, osc_freq:
%.2lf\n",time * 1e9, (time/in_per), v1n, in_freq * 1e-6, osc_freq * 1e-6);
        }
    }

//  Have a 180-degree phase shift at analysis step 500000, or after 5 µs,

        if(i == 500000)
    {
    inp_val = (inp_val == 0)? 1:0;
    }

//  The next few statements simulate the phase/frequency comparator

        if (inp_val==1 && pu_d==0 && rst==0) pu=1;
        if (osc_val==1 && pd_d==0 && rst==0) pd=1;
        rst = (inp_val==1 && osc_val==1 && pu_d==0 && pd_d==0)? 1:0;
        if (rst==1) {pu=0; pd=0; pu_d=1; pd_d=1; rst=0;}
        if (inp_val==0 && pu==0) pu_d=0;
        if (osc_val==0 && pd==0) pd_d=0;

//  Now we simulate the charge pump and low-pass filter

        if (pu==1) delt_q = ICH*T;
        else if (pd==1) delt_q = -ICH*T;
        else delt_q = 0.0;

        x1n = delt_q + x1n_1;
        x2n = x1n + x2n_1 - k*x2n_1;
        v1n = m1*x2n + m2*x2n_1;

//  Now we simulate the VCO

        i_avg = IOSC + GOSC*((v1n + v1n_1)/2.0 - VTH1);
        v2n = v2n_1 + i_avg*T/C3;
        if (v2n >= VTH2)
    {
    v2n = v2n_1 = 0.0;
```

```
        osc_val = (osc_val == 0)? 1:0;
        }

    //  Update the state variables for the next iteration

            x1n_1 = x1n;
            x2n_1 = x2n;
            v1n_1 = v1n;
            v2n_1 = v2n;

            i++;
          }
        }
```

16.6 REFERENCES

A. Blanchard. *Phase Locked Loops.* John Wiley & Sons, New York, 1976.

A. Buchwald. "Design of Integrated Fiber-Optic Receivers Using Hetrojunction Bipolar Transistors," Ph.D. thesis, University of California, Los Angeles, 1993.

A. W. Buchwald and K. W. Martin. "High-Speed Voltage-Controlled Oscillator with Quadrature Outputs," *Electron. Lett.,* Vol. 27, no. 4, pp. 309–310, February 1991.

A. Buchwald, K. Martin, A. Oki, and K. Kobayashi. "A 6-GHz Integrated Phase-Locked Loop Using AlGaAs/GaAs Hetrojunction Bipolar Transistors," *IEEE J. Solid-State Circuits,* Vol. 27, no. 12, pp. 1752–1761, December 1992.

C. Desoer and E. Kuh. *Basic Circuit Theory.* McGraw Hill, New York, 1969.

F. Gardner. *Phaselock Techniques.* John Wiley & Sons, New York, 1979.

A. Grebene. *Bipolar and MOS Analog Integrated Circuit Design.* Wiley Interscience: John Wiley & Sons, New York, 1984.

R. Gregorian and G. Temes. *Analog MOS Integrated Circuits.* Wiley Interscience: John Wiley & Sons, New York, 1986.

B. Kim, D. Helman, and Paul Gray. "A 30-MHz Hybrid Analog-Digital Clock Recovery Circuit Is 2-µm CMOS, *IEEE J. of Solid-State Circuits,* Vol. 25, no. 6, pp. 1385–1394, December 1990.

N. G. Kingsbury. "Second-Order Recursive Filter Elements for Poles Near the Unit Circle and the Real Axis," *Electron. Lett.* Vol. 8, no. 6, pp. 271–273, June 1973.

T. Lee and J. Bulzacchelli. "A 155-MHz Clock Recovery Delay- and Phase-Locked Loop," *IEEE J. of Solid-State Circuits,* Vol. 27, no. 12, pp. 1736–1746, December 1992.

The MathWorks, Inc., *Simulink User Guide.* Natick, Massachusetts, 1992.

H. Meyr and G. Ascheid. *Synchronization in Digital Communications, Vol.1—Phase-Frequency-Locked Loops and Amplitude Control.* Wiley Interscience: John Wiley & Sons, New York, 1990.

N. M. Nguyen and R. G. Meyer. "A 1.8-GHz Monolithic LC Voltage-Controlled Oscillator," *IEEE J. Solid-State Circuits,* Vol. 27, no. 3, pp. 444–450, March 1992.

B. Razavi and J. Sung. "A 6-GHz 60-mW BiCMOS Phase-Locked Loop, *IEEE J. Solid-State Circuits,* Vol. 29, no. 12, pp. 1560–1571, December 1994.

D. Reynolds. "A 320-MHz CMOS Triple 8-bit DAC with On-Chip PLL and Hardware Cursor," *IEEE J. of Solid-State Circuits,* Vol. 29, no. 12, pp. 1545–1551, December 1994.

A. Sedra and K. C. Smith. *Microelectronic Circuits.* 3rd ed. Holt, Rinehart, & Winston/Saunders, New York, 1991.

S. Y. Sun. "An Analog PLL-Based Clock and Data-Recovery Circuit with High Input Jitter Tolerance," *IEEE J. of Solid-State Circuits,* Vol. 22, no. 6, pp. 1074–1081, December 1987.

A. Viterbi. *Principles of Coherent Communication.* McGraw Hill, New York, 1966.

M. Wakayama and A. Abidi. "A 30-MHz Low-Jitter High-Linearity CMOS Voltage-Controlled Oscillator," *IEEE J. of Solid-State Circuits,* Vol. 24, no. 2, pp. 325–330, April 1989.

I. Young, J. Greason, and K. Wong. "A PLL Clock Generator with 5 to 110 MHz of Lock Range for Microprocessors," *IEEE J. of Solid-State Circuits,* Vol. 27, no. 11, pp. 1599–1607, November 1992.

16.7 PROBLEMS

16.1 A CMOS exclusive-OR gate is being used as phase-detector in a PLL. The power-supply voltages are 0 V and 5 V. The low-pass filter is composed of discrete components. The VCO has a 10 MHz free-running frequency for when $V_{cntl} = 2.5$ V, and it has $K_{osc} = 5$ MHz/V. Assume we want the time-constant of the loop to be 50 μs. Design the low-pass filter for $Q = 0.5$. Accurately sketch the voltage waveforms of all nodes in the loop for when the loop is in lock and the input frequency is 7 MHz, both with and without the deglitching capacitor, C_2, of the low-pass filter. Label the maximum and minimum voltages of each node voltage in your sketch. What is the phase difference between the input signal and the oscillator's output signal? What is the lock range of the loop, assuming the VCO is linear for $1\text{V} < V_{cntl} < 4$ V?

16.2 A PLL has an amplifier with a gain of A following its low-pass filter, as shown in Fig. P16.2. The conversion factor of the phase detector is 3 V/rad and the conversion factor of the VCO is 50 KHz/V. The phase detector is a multiplier. The design for the low-pass filter and the amplifier gain are to be chosen. The PLL is intended to be used to demodulate a 5-KHz signal that

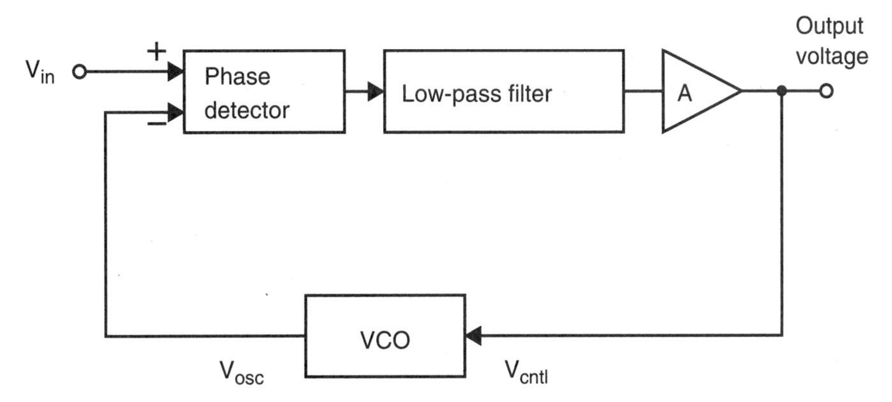

Fig. P16.2 A PLL in which an amplifier has been added after the low-pass filter.

has been frequency modulated on a 5-MHz carrier having a ±50-KHz deviation. Design the filter so that the magnitude of the transfer function of the small-signal model of the loop is maximally flat. The –3-dB frequency should be the same as the band width of the signal that has been modulated. The lock range should be chosen so that the VCO's frequency never exceeds one-half the maximum possible range. Explain what happens to the lock range if the amplifier gain is increased?

16.3 A PLL with a multiplier phase detector has been designed so its linearized small-signal transfer function has a Q factor of 1/2 and a resonant frequency of one-hundredth the free-running frequency of the VCO, when the VCO is at its free-running frequency. What are the resonant frequency and Q factor when the VCO's frequency is different from its free-running frequency by one-half the lock range?

16.4 The low-pass filter designed in Example 16.4 has V_{in} leading V_{osc} by $\pi/8$. Accurately sketch the voltage waveforms of P_u, P_d, and V_{lp} both with and without the deglitching capacitor, C_2. Label the values of the voltages at each major transition point. (*Hint:* The general equation that describes the transient relaxation of a first-order circuit may be useful).

16.5 Let $K_{osc} = 2\pi \cdot 25\ \text{MHz} \cdot V^{-1}$ and $I_{ch} = 10\ \mu A$ in a charge-pump PLL. Assume the time free-running frequency of the oscillator is 100 MHz and that we want the loop to have a time constant of 50 cycles, or $0.5\ \mu s$. Design the components of the low-pass filter. The filter should include the second deglitching capacitor, C_2.

16.6 Sketch the waveforms of V_{in}, V_{osc}, P_u, P_d, $P_{u\text{-}dsbl}$, and $P_{d\text{-}dsbl}$ of the phase/frequency comparator for the case when V_{in} has a much smaller frequency than V_{osc}.

16.7 Find reasonable values for all current sources, the resistors, and C_{osc} of the emitter-coupled multivibrator in Fig. 16.12 so that the multivibrator has a free-running frequency of 10 MHz. Assuming that the on voltage of a diode changes by –2 mV for every °C increase in temperature, estimate the change in frequency if the temperature increases by 20 °C.

16.8 Assume a ring oscillator is realized by five CMOS inverters that have a threshold voltage equal to $V_{DD}/2$. What would the change in frequency be if the power-supply voltage changes from 5 V to 5.5 V? If this happens right at the beginning of a period, what would the corresponding time jitter of output waveform be at the end of the period?

16.9 Derive an equation for the frequency of operation of a ring oscillator that has delay stages of the type shown in Fig. 16.20. Assume the capacitive load of each output is constant and equal to C_L. Also, assume that, when an inverter is switching, its bias current, I_b, changes very quickly from one side to the other.

16.10 Shown in Fig. P16.10 is the bias circuitry and one stage of a four-stage ring oscillator. Assume the loading at each output of the differential inverter is

0.1 pF. Assume transistors are all in the active region. Given that $V_{cntl} = 1$ V, $V_{ref} = 1$ V, and $R = 100$ kΩ, derive and give a value for the frequency of oscillation and for the oscillator's gain constant (in radians/volt).

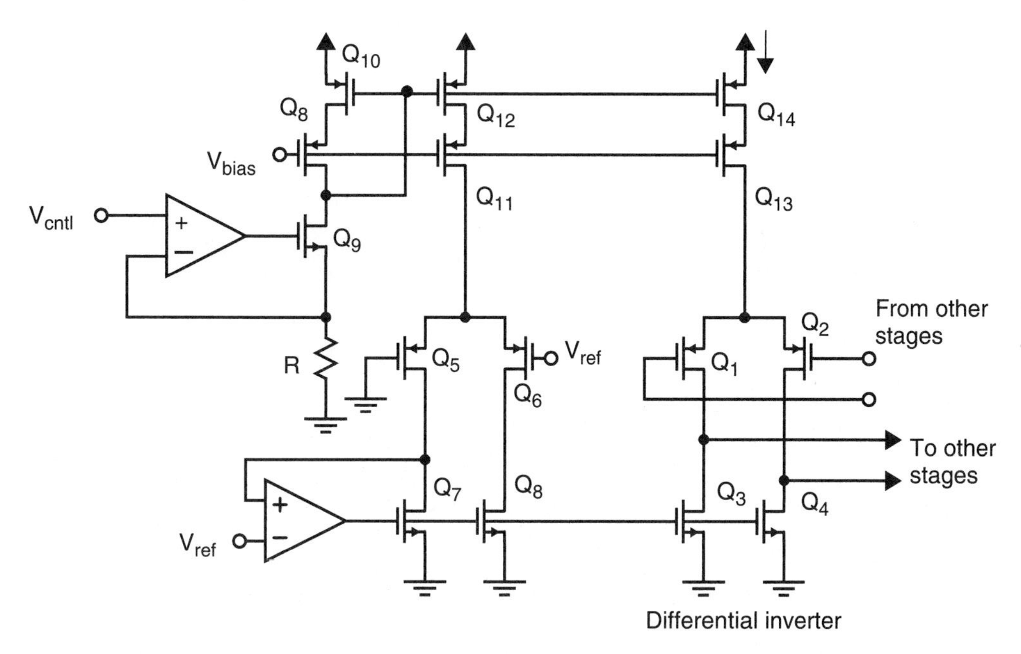

Fig. P16.10

16.11 Derive an equation for the frequency of operation of the CMOS relaxation oscillator shown in Fig. 16.22.

Index

A
Abrial, A., 466
Acceptors, 2
Accumulated channel, 20
Adams, R. W., 565
ADC, *see* Converters, data, A/D
Admittance-divider rule, 134
AGC, *see* Automatic gain control
Allstot, D., 112, 404
Amplifier(s):
  bipolar common-emitter, noise analysis example of, 207–210
  closed-loop, first-order model of, 232–234
  closed-loop, time constant of, 234–235
  common-drain, current mirrors used in, 129–132
  common-gate:
    current mirrors used in, 132–134
    frequency response of, 163
  common-source:
    current mirrors used in, 128–129
    frequency response of, 154–156
  current-mirror, unity-gain frequency of, 278–280
  differential-input, 263
  folded-cascode, unity-gain frequency of, 278–280
  noise bandwidth of, 210
  operational, *see* Opamps
  single stage, 125–180
  source-follower, current mirrors used in, 129–132
  source-follower, frequency response of, 156–163
  telescopic-cascode, 140–141
Analog-to-digital converters (ADCs), *see* Converters, data, A/D
Analysis:
  of noise, *see* Noise, analysis and modelling
  small-signal, of folded-cascode opamps, 266–273
  zero-value time-constant, 155, 164
Anastassiou, D., 546, 560
Annealing, 87, 93–94, 106
Aperture jitter (uncertainty), 335, 458
Approximator, digital (DAPRX), 524
Architecture(s):
  of delta-sigma ($\Delta\Sigma$) A/D converters, 547–549
  of delta-sigma ($\Delta\Sigma$) D/A converters, 549–551
  folding, 519–523
  interpolative, 555–556
  multi-stage noise shaping (MASH), 556–557
  for oversampling data converters, 547–551
  of phase-locked loop, 648–663
Audio, digital, 531
Automatic gain control (AGC), 649

B
Babanezhad, J. N., 256, 286
Baird, R. T., 565
Bandgap voltage reference, 334, 353–364
  bipolar, 357–359
  CMOS, 360–364
Banu, M., 289, 620
Base charge storage, 77
Base-transit time constant, 77
Beam, electron, 83
Beat test, 335
Beta-immune type-A cell, 592
BETA parameter, 62
BF parameter, 64
BiCMOS, *see* Bipolar CMOS technology
  process, 94
  technology, and sample-and-holds, 343
BIF parameter, 64
Binary coding, *see* Coding, offset binary
Binary search, 492
Bipolar CMOS (BiCMOS) technology, 43
Bipolar processing, 95–96
Bipolar transistor exponential relationship, derivation of equations of, 74–77
Bit, carry-in, 454
BJTs, *see* Transistors, bipolar-junction
Black, W. C., 526
Blanchard, A., 660
Body effect, 71
  derivation of equations of, 69–71
  in MOS transistors, 27–28
Boltzmann's constant, 197
Bosner, B. E., 542
Bound positive charge, 3
Bowers, D. F., 292
Bowing, resistor-string, in A/D converters, 509–510
Brackett, P. O., 416
Brick-wall bandpass filter, 192
Brokaw, P., 357, 359
BR parameter, 64
Brugler, J., 354
Bruton, L. T., 418
Bubble errors, in A/D converters, 510–512
Buchwald, A. W., 670, 676, 678
Buffer(s), voltage, 130
Buffer stage, of an opamp, 223

Bult, K., 263, 615
Butting contact, 102–103

C

Cal-dac, 500
Calder, D. W. H., 592
Candy, J. C., 553
Capacitance:
 bypass, 118
 depletion, 5–6, 8, 34
 diffusion, 13
 diode-diffusion, 67–69
 in double-poly capacitors, 395
 drain-bulk, 35
 fringing, 33
 gate, 21
 junction, 13–14
 large-signal, 9–12
 Miller, 34, 155, 222
 nonlinear effects of, 397
 parasitic, 99, 395
 sidewall, 100
 small-signal, 6
 source-bulk, 35
 top plate, 395
Capacitors:
 double-poly, 395
 matching, 108–112
 noise models for, 202–203
 programmable array, 473
 switched, *see* Circuits, switched-capacitor
  resistor equivalence of, 398–400
  and sample-and-holds, 346–349
 in switched-capacitor circuits, 395–396
Capture range, of a phase-locked loop, 659–661
Carley, L. R., 565
Carrier(s), hot, 40
Carrier-velocity saturation, 39
Carry-in bit, 454
Carson, R. S., 638
Cascode configuration, gain stage, 140–142
Cascode current source, 479
Castello, R., 284, 291
Caves, J. T., 394
CDS, *see* Sampling, correlated double
Cell, beta-immune type-A, 592
CGDO parameter, 63
CGSO parameter, 63
Chang, Z., 618
Channel:
 accumulated, 20
 inversion of, 70
 inverted, 20
 pinched-off, 23
 short, effect, 26
Channel charge injection, 397
Channel length, 16
Channel-length modulation, 25, 31
Channel region, 16
Chao, K. C.-H., 555, 559
Charge:
 bound positive, 3
 minority, 12
 trapping, input-transistor, 326–328
Charge injection, 21
 charge-injection errors, 308–315
 at higher frequencies, 426
 making signal independent, 311–312
 minimizing, 312–315
 in switched-capacitor circuits, 423–427
Charge-voltage relationship, 6
Charging time, 7
Chen, F., 565
Chi, K. K., 477
Circuits:
 analog, layout of, 105–118
  matching issues, 106–115
  noise considerations, 115–118
 for bandgap references, 357–364
 common-mode feedback, 287–291
 continuous-time CMFB, 287–290
 elements:
  noise analysis examples, 204–216
  noise models for, 196–204
 gain, *see* Gain circuits
 Norton equivalent, 157
 proportional-to-absolute-temperature (PTAT), 353
 sample-and-hold, *see* Sample-and-hold (circuits)
 start-up, 250, 260
 switched-capacitor, 291, 394–444, 574
  amplitude modulators, 435
  basic operation and analysis of, 398–408
  biquad filters, 415–423
  capacitors in, 395–396
  and charge injection, 423–427
  correlated double-sampling, 433–434
  first-order filters used in, 409–415
  gain circuits, 427–433
  for miscellaneous linear applications, 434–441
  nonoverlapping clocks in, 397–398
  opamps in, 395
  settling time of, 234–236
  signal-flow-graph analysis of, 407–408
  switches in, 396–397
 Thévenin-equivalent, 126
 track-and-hold, *see* Sample-and-hold (circuits)
 tuning, *see* Tuning
 wide-swing constant-transconductance bias, 259–260
CJC parameter, 64
CJE parameter, 64
CJO parameter, 62
CJ parameter, 62, 63
CJS parameter, 64
CJSW parameter, 63
Clocks:
 feedthrough, 308
 nonoverlapping, in switched-capacitor circuits, 397–398
 recovery, 648
CMFB, *see* Feedback, common-mode (CMFB)
CMOS:
 processing, 82–96
 (transistor) pair, 608–609
 variations of process, 93–94
Coban, A., 264, 286
Code, thermometer, 508
Coding:
 memoryless, 562
 offset binary, 453
 1's complement, 453
 signed, in data converters, 452–454
 2's complement, 454
Colleran, W. T., 352, 523
Colles, J., 478
Comlinear Corporation, 292
Comparators, 304–333
 bipolar, examples of, 328–330
 CMOS, 509

CMOS and BiCMOS, 321–326
latched, 317–321
multi-stage, speed of, 315–317
opamps used for, 304–308
phase:
charge-pump, 663–670
exclusive-or, *vs.* phase-locked loops, 662–663
phase/frequency simulation of, 688
Comparison phase, 305
Compensation:
dominant-pole, 240, 268
lead, 240, 242–246, 278
process and temperature sensitivity, 246–248
pole-splitting, 242
Computer simulation, *see* SPICE, simulation examples
of charge-pump PLL, 689–692
of phase-locked loops, 680–692
source code, 689–692
Constants, table of, 56
Contact, butting, 102–103
Continuous-time components, 681–683
Conversion, quad-slope, 489
Conversion time, 457
Converters, data:
absolute *vs.* relative accuracy of, 455–456
A/D:
algorithmic (cyclic), 504–507
architectures compared, 487
bowing, resistor-string, 509–510
bubble error removal, 510–512
charge-distribution with error-correction, 500–503
charge-redistribution, speed estimates for, 503–504
conversion time and sampling rate, 457
DAC-based successive-approximation, 493–494
digital calibration, 566–567
digital error correction in, 514–516
dual-slope integrating, 487–491
error-feedback realization of, 546–547
first-order, realization of, 542–543
flashback in, 512–513
flash (parallel), 507–513
folding, 519–523
ideal, 447–448
input capacitive loading in, 509
integrating, 487–491
interpolating, 516–519
latch-to-track comparator delay in, 510
missing codes, 457
multi-bit and single-bit feedback, 567–568
Nyquist-rate, 487–530
oversampling, signals and spectra of, 548
pipelined, 523–526
quantization noise in, 448–452
ratio-independent, 504–507
resistor-capacitor hybrid, 498–500
second-order noise shaping in, 543–544
signal and/or clock delay, 510
signed charge-redistribution with single reference voltage, 498
SNR in, 451
substrate and power supply noise in, 510
successive-approximation, 492–504
third-order design example, 568–571
time-interleaved, 526
two-step (subranging), 513–516
unipolar charge-distribution, 494–497
bandpass oversampling, 557–559
D/A:
binary-scaled, 469–475
binary-weighted resistor, 470
charge-redistribution switched-capacitor, 473–474
current-mode, 474
decoder-based, 463–469
dynamically matched current sources, 479–481, 566
folded resistor-string, 466–467
glitches in, 474–475
hybrid, 481–484
ideal, 445–447
monotonicity of, 457
multi-bit randomizer, 565–566
multiple resistor-string, 467–469
Nyquist-rate, 463–486
offset and gain error of, 455
1-bit, advantage of, 537
oversampling, signals and spectra of, 550
reduced-resistance-ratio ladders, 470–471
resistor-capacitor hybrid, 482–483
resistor-string, 463–466
R-2R-based, 471–473
segmented, 483–484
settling time and sampling rate, 457
signed outputs, 469
single-supply positive-output, 478–479
thermometer-code, 475–481
thermometer-code current-mode, 477–478
transfer response of, 454–455
unipolar *vs.* signed, 446
dynamic range of, 458–459
fundamentals of, 445–462
multi-bit oversampling, 565–568
nonlinearity errors, 456–457
Nyquist-rate *vs.* oversampling, 445
opamp gain in, 564–565
oversampling, 373, 531–573
without noise shaping, 531–537
performance limitations of, 454–460
resolution of, 455
sampling-time uncertainty, 458
signed coding used in, 452–454
two-level, linearity of, 560–562
Convolution, 377
Current:
drain, 40
drift, 71
leakage, 42
reversed-biased, 5
scale, 13, 45
source, cascode, 479
source, independent, 126
subthreshold, 20
mirrors, *see* Mirrors, current
CVD, *see* Deposition, chemical vapor
Czarnul, Z., 623
Czochralski method, 83

D

DAC, *see* Converters, data, D/A
Damping factor, 238*n*
DAPRX, *see* Digital approximator
Darlington pair transistors, 295
Dattorro, J., 554
dBm, 183
Decimation process, 548–549
Degeneration, source, 289
Degrauwe, M., 364
Delay, comparator, latch-to-track, in A/D converters, 510
Delay, signal and/or clock, in A/D converters, 510
Delta function, Dirac, 376
Delta-sigma ($\Delta\Sigma$) modulation, 531
Demodulation, 648
Depletion:
  capacitance, *see* Capacitance, depletion
  region, 3
  transistors, 16
Deposition, chemical vapor (CVD), 87
Design, rules concerning, 96–105
Desoer, C., 659
Detectors:
  peak, switched-capacitor circuits as, 436–437
  phase, 334, 648
  phase/frequency, 668–670
De Veirman, G. A., 595
Difference equations, 681–683
  modelling a VCO, 683–686
  simulation of a loop filter, 686–687
Differential nonlinearity error (DNL), 456–457
Differential pair, *see* Pair, differential
Diffusion capacitance, diode, derivation of equations of, 67–69
Diffusion implantation, 85–87
Digital-to-analog converters, *see* Converters, data, D/A
Digital approximator (DAPRX), 524
Dingwall, A. G. F., 509
Diode(s), 3–8
  bridge, bipolar, in sample-and-hold, 343
  clamps, 344
  equations for, 56–57
  exponential relationship, derivation of equations of, 65–67
  forward-biased, junction capacitance of, 13–14
  forward-biased, small-signal model of, 14–15
  noise models for, 198
  reverse-biased, 5–8, 213
  Schottky, 3, 15–16
  single-sided, 5
  SPICE parameters for, 61–62
  transit time of, 13, 62
  zener, 353
Diode-bridge switch, used in sample-and-hold, 349–350
Dirac delta function, 376
Discrete-time signal(s), *see* Signals, discrete-time
Discrete-time signal processing (DSP), *see* Signals, discrete-time
Distortion, total harmonic (THD), 636–638
Dithering, 563–564
DNL, *see* Differential nonlinearity error
Domino CMOS logic, 322
Donors, 2
Dopant, concentration profile of, 85–87
Doping, 2
Downsampling, 379–381
Drain-bulk capacitance, 35
Drain current, 40
Drain-induced barrier lowering (DIBL), 40
Drain voltage, 21
Drift, 12–13
  current, 71
Droop rate, 335
DSP, *see* Signals, discrete-time
DTMF, *see* Signals, dual-tone multifrequency
Duque-Carillo, J. F., 289
Durham, A., 626

E

e beam, 83
Electrons, 1
Emission coefficient, 61
Emitter followers:
  bipolar, 149–151
  bipolar-transistor, 160–163
  overshoot in, 159
  ringing in, 159
Equations, derivations of, 65–77
Equations, summary of, 56–61
Erdi, G., 349
Error:
  bubble, in A/D converters, 510–512
  charge-injection, 308–315
  correction, digital, in A/D converters, 514–516
  differential nonlinearity (DNL), 456–457
  integral nonlinearity (INL), 456
  quantization, 448, 532
Etch, plasma, reactive, 91

F

False lock, of a phase-locked loop, 662
Feedback:
  common-mode (CMFB), 280, 287–291
  current, in opamps, 291–295
  and opamp compensation, 232–251
  opamps used with, 222
Feedthrough, clock, 308
Ferguson, P., 555
Fermi level, 224
Fermi potential, 69
Fernandes, J., 353
Field implants, 34, 87–88
Field-oxide, 87–89
Field-oxide isolation, 96
Filters:
  anti-aliasing, 373
    analog, 531
  Bessel, 626
  biquad, 582–584
    using switched-capacitor circuits, 415–423
  brick-wall bandpass, 192
  continuous-time, 574–647
    adaptive, tuning methods based on, 634–636
    dynamic range performance of, 636–643
    first-order, 579–582
    G_m-C, 575–584, 578*n*
    integrators in, 575–579
    MOSFET-C, 620–626
    R-MOSFET-C, 625–626
    second-order, 582–584
    spurious-free dynamic range (SFDR), 641–643

third-order intercept point (IP3), 638–641
tuning, 574
digital decimation, 551–555
multi-stage, 551–554
single stage, 554–555
discrete-time, 382–389
stability of, 385–386
first-order, in switched-capacitor circuits, 409–415
fully differential, 414–415
finite-impulse-response (FIR), 387
G_m-C, 575–584
high-Q biquad, 420–423
infinite-impulse-response (IIR), 386–387
lead-lag, 649
loop, difference-equation simulation of, 686–687
low-Q biquad, 415–420
smoothing, 531
switched-capacitor, 373
FIR, *see* Filters, finite-impulse-response
Flashback, in A/D converters, 512–513
Fleischer, P. E., 440
Flicker noise, 189, 197
Folded-cascode stage, 140–141
Folding block, 519
Folding rate, 519
Fotouhi, B., 498
Fourier transforms, 376
Foxall, T., 427
Frequency:
clock (sampling), 394
pole, 159
resonant, 238
response:
of current mirrors, 154–168
of sample-and-hold, 389–391
unity-gain, 54
Fried, D. L., 394
Fringing capacitance, 33

G

Gain cell:
Gilbert, 364*n*
translinear, 334, 364–366
Gain circuits:
capacitive-reset, 431–433
parallel resistor-capacitor, 427–428
resettable, 428–430
switched-capacitor, 427–433
Gain control, automatic (AGC), 649
Gain stage:
bipolar, 149–154
cascode, frequency response of, 163–168
cascode, high-frequency operation of, 167–168
GAMMA parameter, 63
Gamma radiation, 94
Gardner, F., 660
Gate:
capacitance, 21
polysilicon, forming, 17, 90–91
realization of, 17
region, 16
voltage, 16
Gate-oxide, 89
Gate-source voltage, effective, 20
Gatti, U., 264, 347, 348
Gaussian profile, 85–87
Geiger, R., 41
Gendai, Y., 508, 510, 511
Gilbert, B., 364, 366, 592
Glasser, A., 113
Glitches, in D/A converters, 474–475
Goodenough, F., 516
Gray, P. R., 40, 148, 155, 164
Grebene, A. B., 113, 483, 671, 675
Gregorian, R., 423, 428, 436, 441, 671
Guard rings, 115, 120

H

Haigh, D. G., 311
Hairapethian, A., 567
Hamadé, A. R., 463
Hodges, D., 49
Hogervorst, R., 286
Hold step, 334
Holes, 1
Holes, contact, opening, 91–93
Holloway, P., 467
Hosticka, B. J., 263, 394, 400
Hotta, M., 507
Hurst, P. J., 434
Hybrid-π model, 50
Hysteresis, 319

I

Idle tones, 562–564
IIR, *see* Filters, infinite-impulse-response
Impact ionization, 261
Impedance, output, enhancement of, 265
Implant(s), field, 34
Implantation, diffusion and ion, 85–87
Inductors, noise models for, 202–203
Ingot, pulling, 83
Injection, channel charge, 397
Injection, charge, 21
INL, *see* Integral nonlinearity error
Insulator region, 16
Integral nonlinearity error (INL), 456
Integrated circuits, device model summary, 56–61
Integrated circuits, high-accuracy, 404
Integrators:
BiCMOS, 616–617
in continuous-time filters, 575–579
four-transistor, 623–625
fully differential, 577–579
G_m-C opamp, 578*f*
inverting, parasitic-insensitive, 404
lossless digital (LDI), 687
lossless discrete (LDI), 418
Miller, 578
used in transconductors, 600
parasitic-sensitive, 400–407
two-transistor, 620–623
Intercept point, second-order (IP2), 638
Intercept point, third-order (IP3), 638–641
Interpolative structure, 538
Inversion, weak, moderate, and strong, 24
Inversion of channels, 20, 70
Ion implantation, 85–87
Ionization, impact, 261
IP2, *see* Intercept point, second-order
IP3, *see* Intercept point, third-order
Ishikawa, M., 343, 344
Isolation, field-oxide, 96
IS parameter, 61, 64
ISS parameter, 64
Ito, M., 511

J

Jacobs, G. M., 404
Jantzi, S. A., 559

JFET, *see* Transistors, junction field-effect
Jitter, aperture, 335, 458
JS parameter, 61
Junctions, *see* pn junctions
  abrupt *vs.* graded, 8–9
  implanting, 91–93
  minimizing, 101

K
Karanicolas, A. N., 525
Kennedy, E. J., 196
Khorramabadi, H., 632
Khoury, J. M., 632
Ki, W. H., 433
Kickback, 318
Kim, B., 678
Kingsbury, N. G., 687
Koyama, M., 589, 590
Kozma, K., 634
KP parameter, 62
Krambeck, R. H., 322
Krummenacher, F., 600, 603, 632
Kujik, K., 360
Kusumoto, K., 518
Kwan, T., 599, 633

L
Laber, C. A., 616, 617
Laker, K. R., 423
LAMBDA parameter, 63
Laplace transforms, 374–377
  of sample-and-hold, 389
Larson, L. E., 566
Latch-mode time constant, 319
Latch-to-track comparator delay, in A/D converters, 510
Latch-up, 94, 118–120
Law, S., 270
Layout, 82, 96–120
  automatic, 96
  common centroid, 107
  resistor, 112–115
  rules concerning, 96–105
  transistor, 107–108
LDI, *see* Lossless digital integrator
LD parameter, 63
Lead compensation, *see* Compensation, lead
Leakage currents, 42
Least significant bit (LSB), 446
Lee, H., 500
Lee, T., 657
Leopold, H. A., 537
Letham, L., 477
LEVEL parameter, 62
Li, P. W., 504
Lim, P., 341, 344
Lin, Y-M, 525
Loading, input, capacitive, in A/D converters, 509
Lock range, of a phase-locked loop, 661–662
Logic, domino CMOS, 322
Loop, phase-locked, *see* Phase-locked loop
Lossless digital integrator (LDI), 687
LSB, *see* Least significant bit

M
Maloberti, F., 107, 112, 115
Martin, K. W., 264, 287, 311, 322, 342, 347, 349, 363, 398, 404, 406, 416, 419, 420, 431, 433, 437, 524, 525
Mask, 97
Masking, *see* Photolithography
Mathworks, Inc., 680
Matlab, 680
Matsumoto, H., 431
Matsuya, Y., 556
Matsuzawa, A., 349, 351
McCharles, R. H., 504
McCreary, J. L., 304, 339, 494
Meijer, G., 363
Melt, silicon, 83
Memory, static random-access (SRAM), 94
Metal, pattern, depositing, 93–94
Metastability, 321
Meyr, H., 670
Michejda, J., 362, 363
Miki, T., 477
Miller:
  approximation, 155
  capacitance, 155
  capacitor, 34
  effect, 168
  integration, 620
  integrators, *see* Integrators, Miller
Miller, G., 327
Miller's Theorem, 226
Mirrors, current:
  advanced, 256–266, 273–278
  basic, 125–180
  bipolar, 146–148
  cascode, 137–139, 260
  in common-drain amplifiers, 129–132
  in common-gate amplifiers, 132–134
  in common-source amplifiers, 128–129
  enhanced output-impedance, 260–264
  frequency response of, 154–168
  high-output impedance, 137–140
  high-output impedance, frequency response of, 163
  opamp, 273–278
  simple CMOS, 125–128
  source-degenerated, 135–136
  in source-follower amplifiers, 129–132
  SPICE example of, 169–176, 297–299
  wide-swing, 256–260
    with enhanced output impedance, 264–266
  Wilson, 139–140
MJC parameter, 64
MJE parameter, 64
MJ parameter, 62, 63
MJS parameter, 64
MJSW parameter, 63
Model:
  bipolar T, 149
  hybrid-π, 50
  MOS, large-signal, 24–27
  small-signal, of forward-biased diode, 14–15
  T, 32, 52, 55–56
Modelling:
  BJTs:
    large-signal, 46–47
    small-signal, 50–56
  MOS, advanced, 39–42
    leakage currents, 42
    subthreshold operation, 41–42
  MOS, small-signal, 28–36
  noise, *see* Noise, analysis and modelling
  semiconductors, 1–81
  small-signal, in triode and cutoff regions, 36–39
Modulation, channel-length, 25, 31
Modulation, delta-sigma ($\Delta\Sigma$), 531
Modulators, 334
  amplitude, switched-capacitor

circuits as, 435
delta-sigma ($\Delta\Sigma$), practical considerations, 559–565
delta-sigma ($\Delta\Sigma$), stability of, 559–560
higher-order, 555–557
noise-shaped delta-sigma ($\Delta\Sigma$), 538–540
1-bit, quantization noise power of, 545–546
square-wave, switched-capacitor circuits as, 435–436
Moon, U., 625, 626
Moraveji, F., 353
Moree, J. P., 591
MOS differential pair and gain stage, 142–146
MOSFETs, *see* Transistors, MOSFET
MOS triode relationship, derivation of equations of, 71–74
Multiplier:
analog, 366
Gilbert, 366
translinear, 334, 366–368
two-quadrant, 364*n*
Multivibrators, bipolar emitter-coupled, 671–675

N

Nagaraj, K., 433
Nayebi, M., 342, 343
Nedungadi, A., 608, 612, 613
Nguyen, N. M., 671
NMOS, *see* Transistors, n-channel
Noise, 181–220
in A/D converters, 448–452
analysis and modelling, 181–220
analysis examples, 204–216
frequency-domain analysis, 186–196
models for circuit elements, 196–204
time-domain analysis, 181–186
bandwidth of, 192–194
and circuit layout, 115–118
filtered, 190–192
in fully differential opamps, 280
inherent *vs.* interference, 181
Johnson or Nyquist, 197
in MOS opamps, 232
normalized power of, 182
1/f (flicker), 189, 197
tangent principle, 196
piecewise integration of, 194–196
quantization, 448–452
in 1-bit modulators, 545–546
modelling, 532
random, 182
rms value of, 182–183
sampled signal, 204
shaped quantization, 531
shaping:
in data converters, 445
first-order, 540–542
in higher-order modulators, 556–557
in oversampling data converters, 531–537, 540–544
Shot (Schottky), 197
signal-to-noise ratio, 183
spectral density of, 186–188
substrate and power supply, in A/D converters, 510
summation of, 184–186
thermal, 196–197
transfer-function curves, 545
white, 188–189, 532–534
Norsworthy, S., 322, 563
Norton equivalent circuit, 157
NSUB parameter, 63
n-well process, 18, 82
Nyquist, H., 197

O

O'Leary, P., 107, 112, 115
1's complement, *see* Coding, 1's complement
Opamps:
advanced, 266–299
biasing for stable transconductance, 248–251
CMOS:
inherent input-offset voltage, 229–231
noise analysis example of, 210–213
as comparators, 304–308
compensation of, 221, 232–251
current-feedback, 291–295
current-mirror, 273–278
design of, 221–255
dominant-pole compensated, 232
folded-cascode, 266–273
slew rate of, 270–273
small-signal analysis of, 266–273
SPICE example of, 295
fully differential, 280–287
folded-cascode, 281–282
gain of, 223–225
in converters, 564–565
dc, 395
MOS, noise in, 232
noise modelling and analysis, 202
example of, 204–207
phase margin of, 395
slew rate of, 227–229, 395
SPICE simulation examples of, 251–252
in switched-capacitor circuits, 394–395
transconductance, 267
two-stage CMOS, 221–232
compensating, 240–242
frequency response of, 225–227
n-channel or p-channel input stage, 231–232
unity-gain frequency of, 395
Open-circuit constant-time approach, 464
Operational transconductance amplifiers (OTAs), 267
vs. transconductance cells, 575
Oscillators:
CMOS relaxation, 679–680
crystal, 394
ring, 675–679
sinusoidal, switched-capacitor circuits as, 440–441
sinusoidal-output, 670
square-wave-output, 670
voltage-controlled (VCOs), 648, 670–680
bipolar, 671
difference-equation modelling of, 683–686
digital-output MOS, 675–679
switched-capacitor circuits as, 437–440
OSR, *see* Oversampling, ratio
OTAs, *see* Operational transconductance amplifiers
Overetching, 106
minimizing, 108
Overglass deposition, 93–94

Overloading, quantizer, 448
Oversampling, 204
  advantage, without noise shaping, 535–537
  with noise shaping, 538–547
  ratio (OSR), 535
Overshoot, 157
  in emitter followers, 159
  in source followers, 159
Oxide:
  field, 87–89
  gate, 89
  thermal, 89

P

Padmanabhan, M., 579
Pair, differential, 142–146
  bias-offset cross-coupled, 615–617
  bipolar, 151–154
  with floating voltage sources, 612–614
  source-connected, 610–611
  transconductors using, 592–597
Palmer, C., 363
Parasitics, integrator sensitivity to, 400–407
Park, C. S., 611
Passivation, 93–94
PB parameter, 62, 63
PCA, *see* Programmable capacitor array
Peak detectors, switched-capacitor circuits as, 436–437
Pedestal, sampling, 334
Peetz, B., 507
Pennock, J. L., 605
Permittivity, electrical, 71
Petschacher, R., 353, 515
PHA parameter, 62
Phase detector, 334, 648
Phase-locked loops (PLLs), 437, 648–695
  architecture of, 648–663
  capture range of, 659–661
  with charge-pump phase comparators, 663–670
  vs. exclusive-OR phase comparators, 662–663
  false lock of, 662
  linearized small-signal analysis of, 653–659
  lock range of, 661–662
  transient settling time, 667
Phase margin (PM), 237
PHI parameter, 62, 63
Photodetector, 213
Photolithography, 83–85
Photoresist, 84
Pinch-off:
  of a channel, 23
  region, 25
Pipelining, 524
Planarizing, 94
Plasma etch, reactive, 91
PLL, *see* Phase-locked loops
Plot, pole-zero, 384
PM, *see* Phase margin
pn junctions, 1–16
  diodes, 3–8
  forward-biased, 12–15
  graded, 8–9
  large-signal capacitance, 9–12
  Schottky diodes, 15–16
Pole frequency, 159
Pole-splitting compensation, 242
Pole-zero plot, 384
Polymerization, 84
Polysilicon gates, 17
Poschenreider, W., 394
Potential, *see* Voltage
  Fermi, 69
Poujous, R., 313, 314, 323
Poulton, K., 527
Power-supply connections, 115
Preamp, fiber-optic, noise analysis example of, 213–216
Proakis, J. G., 378
Processes:
  BiCMOS, 94
  CMOS, 93–94
  epitaxial, 117
  n-well, 82
  twin-tub, 93
Processing, 82–96
  bipolar, 95–96
Programmable capacitor array (PCA), 473
PTAAT, *see* Circuits, proportional-to-absolute-temperature

Q

Q factor, 159, 238
Quantization error, 448, 532
Quantization noise, 448–452

R

Radiation, 94
RAMDACs, *see* Converters, data, D/A, single-supply positive-output
Razavi, B., 323, 326, 679
RB parameter, 64
RC parameter, 64
RD parameter, 63
Real, P., 346, 353
Rebeschini, M., 434
Rectifiers, full-wave, switched-capacitor circuits as, 435–436
Reference, bandgap voltage, *see* Bandgap voltage reference
Region:
  active (pinch-off), 23, 25
    defining, 87
    equations for, 25, 58–60
  depletion, 3
  epitaxial, 48
  gate, insulator, and channel, 16
  saturation, 23*n*
    in BJTs, 47
  subthreshold, 41
  triode, 23–24
    equations for, 57–58
Register, successive-approximation, 493
Reinhard, D., 113
RE parameter, 64
Reset phase, 305
Resistance per square, 73
Resistors:
  equivalence of switched capacitor, 398–400
  layouts, 112–115
  noise models for, 197–198
  off-chip, 248
  triode-region transistors used to realize, 248
Reynolds, D., 678
Ring(s), guard, 115, 120
Ringing, 157
  in emitter followers, 159
  in source followers, 159
  during transients, 265
Roberge, J. K., 243
Robertson, D., 353
Root mean square voltage, 182
Root spectral density, 187
  vs. spectral density, 190
RS parameter, 61, 63
Rules for layout and design, 96–105

S

Säckinger, E., 263, 264
Sample-and-hold (circuits), 334–353
  bipolar and BiCMOS, 349–353
  CMOS, examples of, 343–349
  diode-bridge switch, based on, 349–350
  errors due to charge injection, 341–342
  frequency response of, 389–391
  Laplace transform of, 389
  MOS, 336–343
  switched-capacitor technology, based on, 346–349
  testing, 335–336
Sampling:
  correlated double, 395, 433–434
  pedestal, 334
  rate, 457
  time, ideal *vs.* true, 339
  upsampling and downsampling, 379–381
SAR, *see* Successive-approximation register
Saramaki, T., 554
Sarhang-Nejad, M., 567
Saturation:
  carrier-velocity, 39
  region, 23*n*
    in BJTs, 47
  velocity, 26
Scale current, 13, 45
Schaumann, R., 634
Schoeff, J. A., 483
Schottky, W., 197
Schottky diodes, *see* Diodes, Schottky
Schouwenaars, H. J., 479, 481, 483, 566
Schreier, R., 558, 560
Search, binary, 492
Sedra, A. S., 139, 155, 159, 226, 233, 236, 238, 464, 503, 671
Seevinck, E., 608, 613
Semiconductor, operation and modelling of, 1–81
Senderowitz, D., 291
Settling time, 265, 457
  linear, of folded-cascode and current-mirror amplifiers, 278–280
  of switched-capacitor circuits, 234–236
SFDR, *see* Spurious-free dynamic range
Shieh, J., 309, 337
Shields, 116–117
Short-channel effects, 26, 39–41
Shoval, A., 619, 634
Sigma-delta modulation, *see* Delta-sigma (ΔΣ) modulation
Signal(s):
  discrete-time, 373–393
    Laplace transforms of, 374–377
  dual-tone multi-frequency (DTMF), 440
  power, in a sinusoidal wave, 535
  spectra, 373–374
    of discrete-time signals, 376–377
Signal-to-noise ratio (SNR), 183, 451
  calculating, 536
Silicon:
  dioxide, depositing, 91–93
  doping, 2
  fabrication of wafers, 82–83
  intrinsic, 1
    Fermi level of, 224
  melt, 83
Silva-Martinez, J., 603, 631
Simulation, *see* SPICE, simulation examples
  comparator phase/frequency, 688
Simulink, 680
Sinc response, 390
Sinusoid, signal power in, 535
Slew rate:
  of folded-cascode opamps, 270–273
  limiting, 270
  of an opamp, 227–229
Snelgrove, W. M., 611
SNR, *see* Signal-to-noise ratio
Sone, K., 344, 353
Song, B-S, 321, 363, 525
Sooch, N. S., 256, 348
Source:
  degeneration, 289
  follower, 223
    vs. bipolar emitter-follow, 149
    overshoot in, 159
    ringing in, 159
  terminal, 16–17
Source-bulk capacitance, 35
Spacer, 96
Spectra, signal, *see* Signal spectra
Spectral density:
  of noise, 186–188
  vs. root spectral density, 190
SPICE:
  and lead compensation, 243–244
  modelling parameters, 61–65
  simulation examples:
    current mirror, 169–176
    current-mirror opamp, 297–299
    folded-cascode opamp, 295–297
    of opamps, 251–252
    of phase-locked loops, 680–692
Spurious-free dynamic range (SFDR), 641–643
SRAM, *see* Memory, static random-access
Stafford, K. R., 341
Start-up circuits, *see* Circuits, start-up
Steininger, J. M., 248
Steyaert, M., 511, 516, 518
Storage, base charge:
  in BJTs, 48–49
  derivation of equations of, 77
Structure, interpolative, 538
Sub-dac, 500
Substrate, well, 18
Substrate tie, 92
Subthreshold current, 20
Subthreshold region, 41
Successive-approximation register (SAR), 493
Sun, S. Y., 679
Sutarja, S., 525
Swanson, E. J., 327
Switches:
  MOSFETs as, 396
  sharing, 413
  in switched-capacitor circuits, 396–397
Symbols for MOS transistors, 17–19
Synthesizers, frequency, 648
Sze, S. M., 66

T

Tan, K-S, 503, 631
Tanimoto, H., 593, 595
Temes, G. C., 433, 547
Test, beat, 335
Tewksbury, T. L., 327
TF parameter, 64
THD, *see* Total harmonic distortion
Thermal oxide, 89
Thermometer code, 508
Thermometer-code converters,

475–481
Thévenin equivalent, 126
Threshold voltage, 20
MOS, derivation of equations of, 69–71
Tie, substrate, 92
Time:
constant, latch-mode, 319
charging, 7
transit, of diodes, 13
T model, 32, 52, 55–56
bipolar, 149
Tones, idle, 562–564
Total harmonic distortion (THD), 636–638
TOX parameter, 62
Track-and-latch stage, 317
Transconductance:
in BJT small-signal model, 50
cells, *vs.* OTAs, 575
and opamp biasing, 248–251
transistor, 30
Transconductors:
BiCMOS, 616–620
bipolar, 584–597
CMOS:
active transistors used in, 607–616
triode transistors used in, 597–607
constant drain-source voltages used in, 605–607
constant sum of gate-source voltages used in, 609–610
differential pairs used in, 592–597
fixed:
active MOS with translinear multiplier, 619–620
resistors used in, 584–590
fixed-resistor, with translinear multiplier, 617–618
gain-cell, 590–592
inverter-based, 611–612
tunable, 590
varying bias-triode transistors used in, 600–605
Transforms:
bilinear, 387–389
Fourier, 376
Laplace, 374–377
z, 377–379
Transients, ringing due to, 265
Transistors:
active, in CMOS transconductors, 607–616
basic operation of, 1–81
bipolar:
differential pair, 151–154
emitter-follower, 160–163
exponential relationship, derivation of equations of, 74–77
noise models for, 198–200
bipolar-junction (BJT), 16, 42–56
base-charge storage in, 48–49
basic operation of, 44–46
equations for, 60–61
large-signal modelling, 46–47
processing, 95–96
small-signal modelling, 50–56
speed of, 54
SPICE parameters for, 63–65
CMOS pair, 608–609
common-emitter current gain, 76
Darlington pair, 295
depletion, 16, 353
differential pair, 142–146
diode-connected, 126
enhancement, 353
enhancement n-channel, 16
junction field-effect (JFET), 18
large, layout of, 103–105
lateral, 147
layouts, 103–105, 107–108
MOS, 16–39
vs. BJT, 30
advanced modelling of, 39–42
basic operation of, 19–24
body effect in, 27–28
equations for, 57–60
large-signal modelling of, 24–27
modelling in active region, 28–36
modelling p-channel transistors, 28
small-signal modelling in triode and cutoff regions, 36–39
SPICE parameters for, 62–63
symbols for, 17–19
threshold-voltage and body-effect equations, 69–71
MOSFETs, 24
noise models for, 200–201
as switches, 396
n-channel (NMOS), 16
p-channel, 16
modelling, 28
threshold voltage, 20
transconductance, 30
triode, CMOS transconductors using, 597–607
varying bias-triode, 600–605
well, 360
Transit time, 62
of diodes, 13
Trapping, charge, by input transistors, 326–328
Triode region, 23–24
Trip point, 319
TR parameter, 64
Tsividis, Y., 33, 69, 354
TT parameter, 62
Tuning, 626–636
adaptive filtering, based on, 634–636
constant transconductance, 628–629
continuous-time filters, 574
frequency, 630–633
direct *vs.* indirect, 628
overview of circuitry, 627–628
Q-factor, 633–636
Twin-tub process, 93
2's complement, *see* Coding, 2's complement
Tzanateas, G., 364

U

UO parameter, 62
Upsampling, 379–381

V

VAF parameter, 64
Vaidyanathan, P. P., 554
Van de Grift, R. E. J., 516, 523
Van de Plassche, R. J., 329, 518
Van Valburg, J., 511, 516, 523
Vapor deposition, 87
VAR parameter, 64
VCOs, *see* Oscillators, voltage-controlled
Velocity saturation, 26
Viswanathan, T. R., 630
Viterbi, A., 660
Vittoz, E. A., 313, 323
VJ parameter, 62
Voltage, *see* Potential
bandgap, reference, *see* Bandgap voltage reference

buffers, 130
drain, 21
early, 46–47
effective gate-source, 20
gate, 16
gate-to-source, 16
  in transconductors, 609–610
inherent (systematic) input-offset, 229–231
input-offset, errors in, 307–308
native transistor threshold, 71
relationship to charge, 6
rms value of, 182
threshold:
  adjusting, 89
  MOS, derivation of equations of, 69–71
  transistor, 20
Vorenkamp, P., 353, 523
VTO parameter, 62

W

Wafer, epitaxial, 83
Wafers, fabrication of, 82–83
Wakayama, M., 352, 679
Wakimoto, T., 328
Wang, Z., 615
Well, 7
  definition of, 83–85
  substrate, 18
  transistors, 360
Welland, D. R., 590, 598
Whatley, R. A., 287
Wiener–Khinchin theorem, 188
Willingham, S. D., 579, 617
Wilson current mirrors, 139–140
Wolf, S., 39
Wong, S. L., 606
Wyszynski, A., 591

Y

Yang, J. W., 482, 483
Ye, R., 360
Yee, Y. S., 304
Yoshii, Y., 507
Young, I. A., 394, 678
Yukawa, A., 317

Z

Zener diode, 353
Zero-value time-constant analysis, 155, 164
z-transform, 377–379